FUNDAMENTALS OF DAIRY CHEMISTRY

THIRD EDITION

FUNDAMENTALS OF DAIRY CHEMISTRY

THIRD EDITION

Editor

Noble P. Wong
Agricultural Research Service
U.S. Department of Agriculture

Associate Editors

Robert Jenness
formerly Department of Biochemistry
University of Minnesota

Mark Keeney
Department of Chemistry and Biochemistry
University of Maryland

Elmer H. Marth
Department of Food Science
University of Wisconsin

VNR VAN NOSTRAND REINHOLD
_____ New York

Printed in the United States of America

Van Nostrand Reinhold
115 Fifth Avenue
New York, New York 10003

Van Nostrand Reinhold International Company Limited
11 New Fetter Lane
London EC4P 4EE, England

Van Nostrand Reinhold
480 La Trobe Street
Melbourne, Victoria 3000, Australia

Nelson Canada
1120 Birchmount Road
Scarborough, Ontario M1K 5G4, Canada

16 15 14 13 12 11 10 9 8 7 6 5 4 3 2

Library of Congress Cataloging-in-Publication Data

Fundamentals of dairy chemistry.

"An AVI book."
Includes bibliographies and index.
1. Milk—Composition.   2. Dairy products—
Composition.   I. Wong, Noble P.
SF251.F78 1988   637   87-21586
ISBN 0-442-20489-2

To
BYRON H. WEBB

for his outstanding dedicated service to the dairy industry that spans half a century and whose persistence and guidance has led to another edition of Fundamentals.

Contributors

Judith S. Acosta, formerly Dept. of Animal Science and Industry, Kansas State University, Manhattan, Kansas 66506.

Richard Bassette, formerly Dept. of Animal Science and Industry, Kansas State University; present address 811 Kiowa Dr. East, Kiowa, Texas 76240.

Rodney J. Brown, Dept. of Nutrition and Food Sciences, Utah State University, Logan, Utah 84322.

Richard M. Clark, Dept. of Nutritional Sciences, University of Connecticut, Storrs, Connecticut 06268.

Daniel P. Dylewski, Kraft Technical Center, 801 Waukegan Rd., Glenview, Illinois 60025.

C. A. Ernstrom, Dept. of Nutrition and Food Sciences, Utah State University, Logan, Utah 84322.

Harold M. Farrell, Jr., USDA, Eastern Regional Research Center, 600 E. Mermaid Lane, Philadelphia, Pennsylvania 19118.

Joseph F. Frank, Dept. of Animal and Dairy Science, University of Georgia, Athens, Georgia 30602.

Virginia H. Holsinger, USDA, Eastern Regional Research Center, 600 E. Mermaid Lane, Philadelphia, Pennsylvania 19118.

Robert Jenness, formerly Dept. of Biochemistry, University of Minnesota; present address 1837 Corte del Ranchero, Alamogordo, New Mexico 88310.

Robert G. Jensen, Dept. of Nutritional Sciences, University of Connecticut, Storrs, Connecticut 06268.

Mark E. Johnson, Dept. of Food Science, University of Wisconsin, Madison, Wisconsin 53706.

Thomas W. Keenan, Dept. of Biochemistry and Nutrition, Virginia Polytechnic Institute and State University, Blacksburg, Virginia 24061.

Mark Keeney, Dept. of Chemistry and Biochemistry, University of Maryland, College Park, Maryland 20742.

Elmer H. Marth, Dept. of Food Science, University of Wisconsin, Madison, Wisconsin 53706.

Ian H. Mather, Dept. of Animal Sciences, University of Maryland, College Park, Maryland 20742.

Lois D. McBean, National Dairy Council, 6300 N. River Road, Rosemont, Illinois 60018.

Charles V. Morr, Dept. of Food Science, Clemson University, Clemson, South Carolina 29631.

Ronald L. Richter, Dept. of Animal Science, Texas A&M University, College Station, Texas 77843.

John W. Sherbon, Dept. of Food Science, Cornell University, Ithaca, New York 14853.

Elwood W. Speckmann, National Dairy Council, 6300 N. River Road, Rosemont, Illinois 60018.

John L. Weihrauch, USDA, Human Nutrition Information Service, Federal Center Building, Hyattsville, Maryland 20782.

Robert McL. Whitney (deceased), Food Science Dept., University of Illinois, Urbana, Illinois 61801.

Noble P. Wong, USDA, Beltsville Agricultural Research Center, Beltsville, Maryland 20705.

Preface

Fundamentals of Dairy Chemistry has always been a reference text which has attempted to provide a complete treatise on the chemistry of milk and the relevant research. The third edition carries on in that format which has proved successful over four previous editions (*Fundamentals of Dairy Science* 1928, 1935 and *Fundamentals of Dairy Chemistry* 1965, 1974). Not only is the material brought up-to-date, indeed several chapters have been completely re-written, but attempts have been made to streamline this edition. In view of the plethora of research related to dairy chemistry, authors were asked to reduce the number of references by eliminating the early, less significant ones. In addition, two chapters have been replaced with subjects which we felt deserved attention: "Nutritive Value of Dairy Foods" and "Chemistry of Processing." Since our society is now more attuned to the quality of the food it consumes and the processes necessary to preserve that quality, the addition of these topics seemed justified. This does not minimize the importance of the information in the deleted chapters, "Vitamins of Milk" and "Frozen Dairy Products." Some of the material in these previous chapters has been incorporated into the new chapters; furthermore, the information in these chapters is available in the second edition, as a reprint from ADSA (Vitamins in Milk and Milk Products, November 1965) or in the many texts on ice cream manufacture.

Originally, *Fundamentals of Dairy Science* (1928) was prepared by members of the Dairy Research Laboratories, USDA. Over the years, the trend has changed. The present edition draws heavily from the expertise of the faculty and staff of universities. Ten of the 14 chapters are written by authors from state universities, three from ARS, USDA, and one from industry.

It seems fitting that this is so. The bulk of future dairy research, if it is to be done, appears destined to be accomplished at our universities. Hopefully the chapter authors have presented appropriate material and in such a way that it serves best the principal users of this book, their students. As universities move away from specific product technology and food technology becomes more sophisticated, a void has been created where formerly a dairy curriculum existed. It is hoped that this edition of *Fundamentals of Dairy Chemistry* which incorporates a good deal of technology with basic chemistry can help fill this void.

Preparation of this volume took considerably longer than anticipated. The exigencies of other commitments took its toll. Originally the literature was supposed to be covered to 1982 but many of the chapters have more recent references.

I wish to acknowledge with appreciation the contribution made by the chapter authors and the associate editors. Obviously without their assistance, publication of this edition would not have been possible. Dr. Jenness was responsible for Chapters 1, 3, 8, and 9; Dr. Keeney, Chapters 4, 5, and 10; Dr. Marth, Chapters 2, 13, and 14; and Dr. Wong, Chapters 6, 7, 11 and 12.

Contents

Preface / ix

1. Composition of Milk, Robert Jenness / 1

2. Composition of Milk Products, Richard Bassette and Judith S. Acosta / 39

3. Proteins of Milk, Robert McL. Whitney / 81

4. Lipid Composition and Properties, Robert G. Jensen and Richard M. Clark / 171

5. Lipids of Milk: Deterioration, John L. Weihrauch / 215

6. Lactose, Virginia H. Holsinger / 279

7. Nutritive Value of Dairy Foods, Lois D. McBean and Elwood W. Speckmann / 343

8. Physical Properties of Milk, John W. Sherbon / 409

9. Physical Equilibria: Proteins, Harold M. Farrell / 461

10. Physical Equilibria: Lipid Phase, Thomas W. Keenan, Ian H. Mather, and Daniel P. Dylewski / 511

11. Milk Coagulation and Protein Denaturation, Rodney J. Brown / 583

12. Milk-Clotting Enzymes and Cheese Chemistry, PART I—Milk-Clotting Enzymes, Rodney J. Brown and C. A. Ernstrom / 609 PART II—Cheese Chemistry, Mark E. Johnson / 634

13. Fermentations, Joseph F. Frank and Elmer H. Marth / 655

14. Chemistry of Processing, Charles V. Morr and Ronald L. Richter / 739

Index / 767

Composition of Milk

Robert Jenness

Milk is secreted by all species of mammals to supply nutrition and immunological protection to the young. It performs these functions with a large array of distinctive compounds. Interspecies differences in the quantitative composition of milk (Jenness and Sloan 1970) probably reflect differences in the metabolic processes of the lactating mother and in the nutritive requirements of the suckling young.

Human beings consume large amounts of milk of a few species besides their own. The principal ones are cows, water buffaloes, goats, and sheep, which furnish annually about 419, 26, 7.2, and 7.3 million metric tons of milk, respectively, for human consumption *(FAO Production Yearbook* 1979). This chapter, and indeed this entire volume, deals primarily with the milk of western cattle—*Bos taurus.* References to reviews concerning milk of other important species are: Indian cattle—*B. indicus* (Basu *et al.* 1962); water buffalo—*Bubalus bubalis* (Laxminarayan and Dastur 1968); goat—*Capra hircus* (Parkash and Jenness 1968; Jenness 1980; Ramos and Juarez 1981); sheep—*Ovis aries* (Ramos and Juarez 1981); and humans—*Homo sapiens* (Macy *et al.* 1953; Jenness 1979; Blanc 1981; Gaull *et al.* 1982; Packard 1982).

In the United States, milk is defined for commercial purposes as the lacteal secretion, practically free from colostrum, obtained by the complete milking of one or more healthy cows, which contains not less than 8.25% of milk-solids-not-fat and not less than 3.25% milk fat. Minimal standards in the various states may vary from 8.0 to 8.5% for milk-solids-not-fat and from 3.0 to 3.8% for milk fat (U.S. Dept. Agr. 1980).

CONSTITUENTS OF MILK

Milk consists of water, lipids, carbohydrates, proteins, salts, and a long list of miscellaneous constituents. It may contain as many as 10^5 different kinds of molecules. Refinement of qualitative and quantitative techniques continues to add new molecular species to the list. The constituents fall into four categories:

1. Organ and species specific—most proteins and lipids.
2. Organ but not species specific—lactose.
3. Species but not organ specific—some proteins.
4. Neither organ nor species specific—water, salts, vitamins.

The following sections summarize the constituents of milk and indicate how they are quantitated operationally. Detailed descriptions and properties of lipids, lactose, and proteins will be found in later chapters.

Water

Milks of most species contain more water than any other constituent. Certainly this is true of the milks consumed by humans. The other constituents are dissolved, colloidally dispersed, and emulsified in water. The dissolved solutes in bovine milk aggregate about 0.3 M and depress the freezing point by about 0.54°C (see Chapter 8). The activity of water in milk, a_w, which is the ratio of its vapor pressure to that of air saturated with water, is about 0.993. A small amount of the water of milk is "bound" so tightly by proteins and by the fat globule membrane that it does not function as a solvent for small molecules and ions. Water content is usually determined as loss in weight upon drying under conditions that minimize decomposition of organic constituents, e.g., 3 hr at 98–100°C (Horwitz 1980).

Lipids

The lipids of milk, often simply called "fat," consist of materials that are extractable by defined methods. Simple extraction with a nonpolar solvent like ether or chloroform is not efficient because the fat is located in globules protected by a surface membrane. A widely used gravimetric method is the Roese-Gottlieb extraction (Walstra and Mulder 1964) using NH_4OH, ethanol, diethyl ether, and petroleum ether. Volumetric methods such as that of Babcock and Gerber (Ling 1956; Horwitz 1980) use H_2SO_4 to liberate the fat, which is then measured. Rapid determination of the amount of fat in milk can be done by measurement of the absorption of infrared radiation at 3.4 or 5.7 μm (Chapter 8; Goulden 1964; Horwitz 1980).

The lipids of milk are composed of about 98% triglycerides, with much smaller amounts of free fatty acids, mono-and diglycerides, phospholipids, sterols, and hydrocarbons. Chapter 4 deals in detail with the composition of milk lipids.

The fat in milk is almost entirely in the form of globules, ranging

from 0.1 to 15 μm in diameter. Size distribution is an inherited characteristic that varies among species and among breeds of cattle. Bovine
milk contains many very small globules that comprise only a small
fraction of the total fat. The total number is about 15×10^9 globules
per milliliter of which 75% are smaller than 1 μm in diameter. The fat
globules and their protective membrane of phospholipids and proteins
are described in Chapter 10.

Carbohydrates

In bovine milk, and indeed in all milks consumed by humans, the overwhelming carbohydrate is lactose. This disaccharide, 4-0-β-D-galacto-
pyranosyl-D-glucopyranose, is a distinctive and unique product of the
mammary gland. It has been found in milks of almost all of the species
analyzed to date (Jenness *et al.* 1964) and nowhere else in nature except
in low concentration in the fruits of some of the Sapotaceae (Reithel
and Venkataraman 1956). Lactose is discussed in detail in Chapter 6.

Lactose in milk can be quantitated by oxidation of the aldehyde of
the glucose moiety (Hinton and Macara 1927; McDowell 1941; Perry
and Doan 1950; Horwitz 1980), by polarimetry of a clarified solution
(Grimbleby 1956; Horwitz 1980), by colorimetry of the product of reaction with phenolic compounds (Marier and Boulet 1959), by infrared
absorption at 9.6 μm (Goulden 1964; Horwitz 1980), by enzymatic assay with β-galactosidase and galactose dehydrogenase (Kurz and Wallenfels 1974), and by chromatography (Reineccius *et al.* 1970; Beebe
and Gilpin 1983; Brons and Olieman 1983). Only the last two of these
methods are specific for lactose, but bovine milk contains so little other
material that is oxidizable, that exhibits optical rotation, that reacts
with phenolic compounds, or that absorbs at 9.6 μm that the first four
give reasonable estimates of lactose. Older analyses were made by oxidation or polarimetry. Published values for lactose contents obtained
with these methods must be scrutinized carefully because some were
calculated on the basis of lactose monohydrate and are thus 5.26% too
high (360/342).

Carbohydrates other than lactose in milk include monosaccharides,
neutral and acid oligosaccharides, and glycosyl groups bound to proteins and lipids. Glucose and galactose are detectable by thin layer
chromatography (TLC) and gas-liquid chromatography (GLC) of bovine milk. Of course, hydrolysis of lactose is an obvious source of these
two monosaccharides, but with precautions taken to avoid hydrolysis,
concentrations of 100–150 mg/liter of each have been found by GLC
(Reineccius *et al.* 1970). Specific enzymatic methods, however, have indicated considerably lower concentrations—about 30 mg glucose and

90 mg galactose per liter (Faulkner *et al.* 1981). Free *myo*-inositol has been found in the milks of several species (in addition to that bound in phosphatidyl inositols; (see Chapter 4). Bovine milk has only 40–50 mg of *myo*-inositol per liter, but milks of some other species contain much more (Byun and Jenness 1982).

The carbohydrates L(-)fucose (Fuc), N-acetylglucosamine (2-acetami-do-2-deoxy-D-glucose), N-acetyl galactosamine (2-acetamido-2-deoxy-D-galactose), and N-acetylneuraminic acid occur in milk almost entirely in the form of oligosaccharides and glycopeptides. Only small concentrations are present in the free state, although there is one report of 112 mg of N-acetylglucosamine per liter of bovine milk (Hoff 1963). The total (free and combined) content of N-acetylneuraminic acid is 100–300 mg/liter (de Koning and Wijnand 1965).

Bovine milk contains 1–2 g of oligosaccharides per liter, human milk 10–25 g/liter. Colostrums of both species have higher concentrations. A recent review (Blanc 1981) lists more than 60 oligosaccharides which have been detected in human milk. They range from 3 to 20 monosaccharide units per molecule; not all have been characterized structurally. Five oligosaccharides have been detected and characterized in bovine milk or colostrum. All of the oligosaccharides that have been characterized in either bovine or human milk have a lactose moiety, D-Gal-β-(1-4)-D-Glc in the reducing terminal position. (In a few, the terminal residue is an N-acetylglucosamine). The simplest are the trisaccharides fucosyllactose [L-Fuc α-(1–2)-D-Gal-β-(1–4)-D-Glc] and N-acetylneuraminyllactose [NANA-(2–3)-D-Gal-β-(1–4)-D-Glc]. More complex ones have longer chains with various kinds of branching (Ebner and Schanbacher 1974; Blanc 1981).

Various sugar phosphates occurring in milk are listed later under miscellaneous constituents. The glycosyl groups of several of the milk proteins are described in Chapter 3.

Proteins

The proteins of milk fall into several classes of polypeptide chains. These have been delineated most completely in bovine milk, and a system of nonmenclature has been developed for them (Chapter 3; Eigel *et al.* 1984). One group, called "caseins," consists of four kinds of polypeptides: α_{s1}-, α_{s2}-, and β-, and κ- with some genetic variants, post translational modifications, and products of proteolysis. Almost all of the caseins are associated with calcium and phosphate in micelles 20–300 μm in diameter (see Chapter 9). The other milk proteins, called "whey proteins," are a diverse group including β-lactoglobulin, α-lactalbumin, blood serum albumin, and immunoglobulins (Chapter 3). Almost all milk proteins of nonbovine species defined to date appear to be evolutionary homologs of those of the bovine and are named accordingly.

Classically, milk protein content has been determined by Kjeldahl analysis for nitrogen (N) (Horwitz 1980). This has the advantages that N is a major constituent, comprising about one-sixth of the mass of the protein, and that the N contents of the individual milk proteins are nearly the same. Multiplication by 6.38 has been used commonly to convert the N content to protein. This is based on an old determination of 15.67% N in milk proteins, but a modern weighted average of the N contents of individual milk proteins indicates that the factor should be 6.32 (Walstra and Jenness 1984). Thus older results may be nearly 1% too high. A more serious error is that protein contents have often been calculated as 6.38 × total N. Such "crude protein" values are 4–8% too high because they include N from nonprotein nitrogenous constituents.

Various procedures are used to separate milk proteins into fractions or individual components that can quantitated separately. A classic method of fractionation is by precipitation at pH 4.6, which separates the proteins into two groups—caseins in the precipitate and whey proteins in the supernatant. All proteins are precipitated from a second aliquot with trichloroacetic acid at 12% (w/v) concentration (Rowland 1938), and concentrations of casein and whey proteins are calculated as follows:

$$Casein = 6.38 \ (TN\text{-}NCN)$$
$$Whey\ protein = 6.38 \ (NCN\text{-}NPN)$$

where TN is total nitrogen, NCN is nitrogen in the pH 4.6 filtrate, and NPN is nitrogen in the trichloroacetic acid filtrate. About 80% of the proteins of bovine milk fall into the category of caseins, but the proportions differ greatly among species (see Table 1.10).

Numerous other methods have been proposed for routine determination of protein on large numbers of samples. Several are reviewed by Booy et al. (1962). They include colorimetric determination of ammonia, colorimetric determination of peptide linkages by the biuret method, analysis for tyrosyl groups, titration of protons released from lysyl groups upon reaction with formaldehyde, binding of anionic dyes to cationic protein groups, turbidimetric procedures (Kuramoto et al. 1959), and absorption of infrared radiation of 6.46 μm (Goulden 1964; Horwitz 1980). Individual milk proteins can be assayed by specific immunological tests (Larson and Twarog 1961; Larson and Hageman 1963; Babajimopoulos and Mikolajcik 1977; Guidry and Pearson 1979; Devery-Pocius and Larson 1983), by ion-exchange chromatography (Davies and Law 1977), by gel filtration (Davies 1974), by zone electrophoresis (Swaisgood 1975; West and Towers 1976, Bell and Stone 1979), and by high performance liquid chromatography (Diosady *et al.* 1980; Bican and Blanc 1982).

Salts

For the purpose of this discussion, milk salts are considered as ionized or ionizable substances of molecular weight 300 or less. Ionizable groups of proteins are not included here, although, of course, they must be taken into account in a complete description of ionic balance and equilibria. Trace elements, some of which are ionized or partially so in milk, are considered in a later section of this chapter. Milk salts include both inorganic and organic substances; thus they are not equivalent to either minerals or ash. The principal cations are Na, K, Ca, and Mg, and the anionic constituents are phosphate, citrate, chloride, carbonate, and sulfate. Small amounts of amino cations and organic acid anions are also present.

General methods for quantitating minerals (especially metals) use absorption or emission of radiation of specific wavelengths (Wenner 1958; Murthy and Rhea 1967). The former is a measure of absorption of the energy required to raise electrons to a higher energy-excited state and the latter is a measure of the energy released when excited electrons revert to their original state. These methods are particularly suitable for Na and K, for neither of which are volumetric or gravimetric methods of sufficient sensitivity available. Calcium and magnesium can also be determined by emission or absorption but often are analyzed by specific chemical methods. Dry ashing or wet digestion with H_2SO_4-H_2O_2 or HNO_3–$HClO_4$ are often used to destroy organic material before analysis for minerals, but in some procedures diluted, unfractionated samples are injected directly into the flame photometer. Defatted and deproteinized extracts, usually acid, are used to determine the content of organic salts such as citrate; they are sometimes used for analyses of mineral constituents as well.

Classically, calcium was determined by precipitation as calcium oxalate, which was then titrated in H_2SO_4 solution with $KMnO_4$ but this has been largely replaced by titration with the chelating agent ethylenediamine tetraacetate (EDTA), using as the indicator a dye (murexide) which changes color when it binds calcium (White and Davies 1962). Another more sensitive method for Ca determination is a colorimetric procedure using glyoxal *bis* (2-hydroxyanil), whose calcium complex absorbs strongly at 524 nm (Nickerson *et al.* 1964). Phosphate interferes with both methods; it can be removed by treatment with an anion exchanger or by precipitation with potassium *meta*-stannate. Alternatively, the calcium can be precipitated as oxalate before titration with EDTA.

Magnesium, formerly determined by precipitation as magnesium ammonium phosphate and determining P in the latter, can be analyzed readily by EDTA titrations. It can be obtained either as the difference between titrations for (Ca and Mg) and Ca alone or by titrating the supernatant after Ca is precipitated as oxalate (White and Davies 1962).

Phosphate is determined almost universally by its reaction with molybdate to form phosphomolybdate. The latter can be reduced to a blue compound that absorbs at various wavelengths, of which 640 and 820 nm are often used for colorimetric quantitation (Allen 1940; Sumner 1944; Meun and Smith, 1968).

Chloride is analyzed by some form of reaction with silver to form insoluble silver chloride. Direct titration of milk with silver nitrate yields erroneously high and variable results, and pre-ashing cannot be used because chloride is lost by volatilization. Satisfactory procedures involve adding an excess of standardized $AgNO_3$ directly to milk and back titrating with potassium thiocyanate (KSCN), using a soluble ferric salt as the indicator (Sanders 1939).

Citrate may be oxidized with $KMnO_4$ and brominated and decarboxylated to form the relatively insoluble pentabromacetone; certain methods for detecting citrate in milk, including that of the Association of Official Analytical Chemists (Horwitz 1980), employ this reaction for a gravimetric analysis. It is, however, cumbersome, and pentabromacetone is somewhat more soluble and volatile than desired in a gravimetric analysis. In another method, lead citrate is precipitated from a sulfuric acid–alcohol filtrate from milk and titrated with ammonium perchlorato-cerate (Heinemann 1944). A simpler and more sensitive procedure utilizes the Furth-Herrmann reaction, in which a yellow-colored condensation product of citrate with pyridine is formed in the presence of acetic anhydride (White and Davies 1963). Citrate may also be determined enzymatically by cleavage with a bacterial citrate lyase to oxaloacetate; decarboxylation of the latter to pyruvate with oxaloacetate decarboxylase; and finally, formation of malate and lactate with specific NAD-coupled dehydrogenases (Dagley 1974). The enzymatic method is the most specific method yet employed. About 10% of the total apparent citrate of milk actually is isocitrate (Faulkner and Clapperton 1981).

Inorganic sulfate, SO_4^{2-}, is present in milk in a concentration of about 1mM; it may be determined turbidimetrically after adding barium ion to a deproteinized filtrate (Koops 1965).

The total carbonate system (mostly HCO_3^- in equilibrium with CO_2)

in milk varies with the time after milking and the extent of exposure to heat and vacuum treatments. It is about 2mM in mixed raw milk in equilibrium with air. It can be released by acidification, collected by aspiration into $Ba(OH)_2$, and determined titrimetrically (McDowall 1936), or it can be released and measured in manometric apparatus (Frayer 1940).

Table 1.1 gives the mean salt composition for 12 bulk milk samples taken from a herd at approximately monthly intervals during a year (White and Davies 1958). The means and ranges of the constituents are similar to those observed by other workers. The sum of Na, K, Ca, Mg, Cl, and total phosphate as PO_4 from these data plus 0.01 g/100 g of inorganic sulfate is 0.73 g/100 g, which is a little short of the reported ash content of 0.76 g/100 g. However, the composition of ash likely differs somewhat from that of the mixture of the components summed because of loss of chloride, conversion of some of the organic S to SO_4^{2-}, retention of a little organic C as CO_3^{2-}, and formation of metallic oxides during ashing. The extent to which these processes occur depends on the temperature of incineration. The former practice of reporting the composition of ash in terms of oxides of the metals and of P (as done in previous editions of this volume) should be regarded only

Table 1.1. Salt Composition of Milk.

| | Concentration | | | | |
Constituent	Mean (mg/100 g)	Range (mg/100 g)	SD (mg/100 g)	Mean (mM)	Percent Diffusible
Cationic					
Sodium	58	47–77	10	25.2	
Potassium	140	113–171	14	35.8	
Calcium	118	111–120	2.5	29.5	31
Magnesium	12	11–13	0.6	4.9	65
Amines				~1.5	
Anionic					
Phosphorus[a]	74	61–79	—	23.9	53
Inorganic (P_i)	63	52–70	—	20.4	53
Ester	11	8–13	1.7	3.5	
Chloride	104	90–127	11.4	29.3	
Citrate	176	166–192	9	9.2	90
Carbonate				~2.0	
Sulfate				~1.0	
Organic acids				~2.0	

SOURCE: White and Davies (1958). Twelve samples of herd bulk milk (except for amines, carbonate, sulfate, organic acids; see text).
[a] Excluding casein P.

as a mode of expression and not as an indication that these oxides are actually present in ash.

Not all of the salt constituents are found in the dissolved state in milk. Calcium, magnesium, phosphate, and citrate are partitioned between the solution phase and the colloidal casein micelles (see Chapter 9 for the composition and structure of these micelles). For analytical purposes, partition of the salt constituents can be achieved by equilibrium dialysis or by pressure ultrafiltration. In the latter technique, pressures must be limited to about 1 atmosphere to avoid the so-called sieving effect (pushing water through the filter faster than the dissolved components (Davies and White 1960).

Table 1.1 shows the proportion of the several constituents found in the dissolved, diffusible state. Actually, phosphate is present in five classes of compounds: inorganic dissolved, inorganic colloidal, water-soluble esters, ester-bound in caseins, and lipid. These can be determined by making the following analyses:

I. Total P in the dry-or wet-ashed sample.
II. Lipid P in digested Roese-Gottlieb extract.
III. Dissolved P in digested ultrafiltrate.
IV. Inorganic dissolved P in undigested ultrafiltrate.
V. Acid-soluble P in undigested 12.5% trichloroacetic acid filtrate.

Then:

$$\text{Inorganic dissolved P} = \text{IV}$$
$$\text{Inorganic colloidal P} = \text{V} - \text{IV}$$
$$\text{Water soluble ester P} = \text{III} - \text{IV}$$
$$\text{Casein P} = \text{I} - (\text{II} + \text{V})$$
$$\text{Lipid P} = \text{II}$$

The total inorganic phosphate (Pi) is, of course, V.

The salt constituents in the dissolved state (ultrafilterable or diffusible) interact with each other to form various complexes. The concentrations of each of these constituents can be calculated (with suitable computer programs) from a knowledge of their several interaction or association constants. Results of such a calculation (Holt et al. 1981) for an ultrafiltrate similar to that of Table 1.1 are given in Table 1.2. Na, K, and Cl are primarily present as free ions but Ca, Mg, phosphate, and citrate are distributed throughout many complexes; those in the highest concentration are $CaCit^-$, $Mg\,Cit^-$, $H_2PO_4^-$ HPO_4^{2-} and $CaHPO_4$. The calculation yields Ca^{2+} and Mg^{2+} concentrations of 2.0

Table 1.2. Concentrations of Ions and Complexes in Typical Milk Diffusate.

Anion	Free Ion (mmol/liter)	Complex with Cation (mmol/liter)			
		Ca^{2+}	Mg^{2+}	Na^+	K^+
H_2Cit^-	$+$ [a]	$+$	$+$	$+$	$+$
$HCit^{2-}$	0.04	0.01	$+$	$+$	$+$
Cit^{3-}	0.26	6.96	2.02	0.03	0.04
$H_2PO_4^-$	7.50	0.07	0.04	0.10	0.18
HPO_4^{2-}	2.65	0.59	0.34	0.39	0.52
PO_4^{3-}	$+$	0.01	$+$	$+$	$+$
$Glc-1-P^{2-}$	1.59	0.17	0.07	0.10	0.14
H_2CO_3	0.11	$-$ [b]	$-$	$-$	$-$
HCO_3^-	0.32	0.01	$+$	$+$	$+$
SO_4^{2-}	0.96	0.07	0.03	0.04	0.10
RCOOH	0.02	$-$	$-$	$-$	$-$
$RCOO^-$	2.98	0.03	0.02	0.02	0.04
Free ion	$-$	2.00	0.81	20.92	36.29

SOURCE: Holt et al. (1981).
[a] $+$ = <0.005.
[b] $-$ = not determined.

and 0.8 mM respectively. These are close to the values found independently by determining the amounts of Ca^{2+} and Mg^{2+} bound by a resin equilibrated against milk (Holt et al. 1981). $[Ca^{2+}]$ determined by a specific calcium-ion electrode is slightly higher and that determined by colorimetry of a dye (murexide) complexing with the calcium ions is somewhat lower than the calculated values. The calculated ionic strength of the ultrafiltrate, $1/2\ m_i z_i^2$, based on the ionic distribution in Table 1.2, is 0.08.

Trace Elements

In addition to the major salt constituents discussed up to this point, the elements listed in Table 1.3 have been detected in normal bovine milk by spectroscopic and chemical analyses. They include a large number of metals, the metalloids As, B, and Si, and the halogens F, Br, and I. The subject of trace elements in milk has been reviewed comprehensively (Archibald 1958; Murthy 1974; Underwood 1977). Their significance for human nutrition is discussed in Chapter 7.

Reported concentrations of the trace elements exhibit large ranges. For some of them (e.g., I, Mo, Zn), the concentration in the milk depends markedly on that in the diet consumed by the cow. The concentrations of some of them are increased by contamination by utensils

Table 1.3. Trace Elements in Bovine Milk.

	Conc. (μg/liter)	
Element	Range	Typical Value
Aluminum (Al)	150–1000	500
Arsenic (As)	30–60	
Barium (Ba)		Trace
Boron (B)	100–1000	300
Bromine (Br)	500–20,000	
Cadmium (Cd)	1–30	
Cesium (Cs)		Trace
Chromium (Cr)	5–80	15
Cobalt (Co)	0.4–1.0	0.5
Copper (Cu)	10–200	75
Fluorine (F)	70–220	
Iodine (I)	10–1000	
Iron (Fe)	100–1500	300
Lead (Pb)	20–80	40
Lithium (Li)		Trace
Manganese (Mn)	20–100	50
Mercury (Hg)		Trace
Molybdenum (Mo)	20–120	70
Nickel (Ni)	0–30	
Rubidium (Rb)	100–3400	
Selenium (Se)	4–1200	12
Silicon (Si)		~1400
Silver (Ag)	15–50	45
Strontium (Sr)	40–500	170
Tin (Sn)		Trace
Titanium (Ti)		Trace
Vanadium (V)		Trace
Zinc (Zn)	2000–5000	3300

SOURCE: Compilations and review of Archibald (1958), Murthy (1974), and Underwood (1977).

and equipment to which milk is exposed in handling and processing. Some high values for the concentrations of certain trace elements may have resulted from contamination during laboratory analysis.

The distribution of trace elements among the compounds and physical phases in milk has not been elucidated completely. Molybdenum appears to be found exclusively in xanthine oxidase and Co in vitamin B_{12}. Iron is an essential component of xanthine oxidase, lactoperoxidase, and catalase. About half of the total Fe and 10% of the Cu are in the fat globule membrane. Copper has been studied extensively in relation to oxidation of milk lipids. The trace metal present in highest concentration in milk is Zn; its concentration of 3.5 mg/liter is about 3% of that of Mg, the major salt constituent present in lowest concentration. About 85% of the Zn is associated with casein micelles. Alkaline phosphatase, a Zn-containing enzyme, is located primarily in the fat globule membrane but accounts for only a small fraction of the total Zn. Manganese is required for fermentation of citrate by certain lactic acid bacteria, and with some milks the bacterial formation of diacetyl in cultures is inhibited by lack of sufficient Mn. Apparently, iodine is present in milk solely in the form of iodide ion; its concentration depends markedly on the amount consumed by the cow.

Radionuclides

Potentially hazardous radioactive isotopes of certain elements may enter the food chain from radioactive fallout arising from the testing of nuclear weapons or from accidents in nuclear power plants. Hazardous radionuclides that may be transferred to the human consumer in milk are those of Sr, I, Cs, and Ba listed in Table 1.4. All are β- emitters. Nonradioactive isotopes of these elements occur regularly in milk in traces, and extremely low levels of the radioactive isotopes undoubtedly occur even in the absence of fallout. Radionuclides in milk are discussed thoroughly by Lengemann et al. (1974). The physical half-life of isotope (Table 1.4) is the time required for the radioactive emission to fall to half of its original level. The biological half-life is the time required for excretion of half of an ingested dose. Actually, since an element will be distributed among several pools in the body, the values for biological half-life in Table 1.4 may concern only a small tenaciously held fraction of the total dose. The radionuclides become physically distributed in milk in much the same way as with related elements. Thus [137] Cs, like Na and K, is largely present as ions in solution.[131] I is

Table 1.4. Radionuclides that May Contaminate Milk.

Nuclide	Physical Half-Life	Biological Half-Life	Location in Milk
[89]Sr	52 days	1–50 years	>80% in micelles
[90]Sr	28 years	1–50 years	>80% in micelles
[131]I	8 days	~100 days	In solution
[133]I	21 hours	~100 days	In solution
[137]Cs	30 years	~30 days	In solution
[140]Ba	13 days	<3 years	Partly in micelles

mostly present as iodide ion, a little being bound by proteins and fat globules. The strontium isotopes and ^{140}Ba behave like Ca in that large portions are bound in the casein micelles. Actually, the proportion of Sr in micelles is greater than that of Ca because strontium phosphates are less soluble than calcium phosphates.

The greatest concern regarding the health hazard of radionuclides in milk is posed by ^{90}Sr and ^{131}I. The long physical half-life of the former and the fact that it accumulates and persists in bone make it especially hazardous. ^{131}I accumulates to high concentrations in the thyroid gland, where it can produce intensive radiation. It is especially danger-ous for relatively short periods after heavy fallout.

The cow acts somewhat as a filter and discriminator in the transmis-sion of radionuclides from feed to milk. Of the daily quantities of I and Cs ingested, about 1% is secreted in every kilogram of milk produced. For Ba and Sr the rate of transmission is less. Furthermore the cow discriminates against some of the radionuclides in favor of related ele-ments. Thus the Sr/Ca ratio is reduced by a factor of about 10 in the passage from feed to milk.

Miscellaneous Compounds

Milk contains many components in low concentration (generally less than 100 mg/liter) which do not fall into any of the categories discussed in previous sections. Some of these materials are natural and some are contaminants. Compounds may be considered natural if they are pres-ent in freshly drawn milk and have been detected in most samples in which they have been sought. The groups of such compounds consid-ered here include gases, alcohols, carbonyl compounds, carboxylic acids, conjugated compounds, nonprotein nitrogenous compounds, phosphate esters, nucleotides, nucleic acids, sulfur-containing com-pounds, and hormones. Others that might be included, but are dis-cussed in other sections of this chapter or in other chapters, are minor lipids (Chapter 4), enzymes (Chapter 3), vitamins (Chapter 7), and mi-nor carbohydrates (see the section "Carbohydrates" in this chapter).

The gases CO_2, N_2, and O_2 are present in anaerobically drawn milk in concentrations of about 6, 1, and 0.1% by volume (120, 13, and 1.4 mg/liter), respectively (Noll and Supplee 1941). Upon exposure to air, CO_2 is lost and N_2 and O_2 are gained rapidly. Mixed raw milk contains about 4.5, 1.3, and 0.5% by volume or 90, 15 and 6 mg/liter of CO_2, N_2, and O_2, respectively (Frayer 1940; Noll and Supplee 1941).

Carbon dioxide is, of course, in equilibrium with bicarbonate ion; al-most the entire CO_2–HCO_3– system can be removed by heat or vacuum treatment (Smith 1964). The oxygen content of pasteurized bottled milk is about 6 mg/liter (Herreid and Francis 1949). Dearation treatments to reduce O_2 levels to nearly zero preserve ascorbic acid and prevent the development of oxidized flavors in pasteurized milk (Guthrie 1946). Removal of oxygen from evaporated milk with glucose oxidase has been proposed (Tamsma and Tarassuk 1957).

Ethanol and a long list of carbonyl compounds and aliphatic acids occur in fresh milk (Table 1.5). Some of them have been detected in only a few of the samples in which they were sought. Techniques for detecting such compounds include derivatization with 2,4-dinitrophenylhydrazine and various methods of volatilization, extraction, and chromatography (Harper and Huber 1956; Morr et al. 1957; Harper et al. 1961; Wong and Patton 1962; Scanlan et al. 1968; Marsili et al. 1981). The sum of the concentrations of acids listed in Table 1.5 is only 1–3 mmol/liter, compared to the citrate concentration of 10 mmol/liter. Oxalate has been reported to occur in milk (Zarembski and Hodgkinson 1962) on the basis of a certain colorimetric reaction, but positive identification has not been made.

Table 1.5. Alcohols, Carbonyls, Acids, and Esters in Milk.

Compound	Conc. (mg/liter)	Ref.	Compound	Conc. (mg/liter)	Ref.
Ethanol	3	1	Octanoic acid	12.5–38	1, 4
Formaldehyde	0–0.003	2, 5	Decanoic acid	+	1, 7
Acetaldehyde	0–0.016	2, 5	Lactic acid	34–104	3
Hexanal	+	1	β-Hydroxybutyric acid	+	6
Benzaldehyde	+	1	σ-Decalactone	+	1
Acetone	0–1.1	1, 2, 5	σ-Dodecalactone	+	1
Butanone	0.8	1, 5	Glyoxylic acid	+	7
Diacetyl	+	1	Pyruvic acid	0–25	2, 3, 7
2-Pentanone	0.007–0.030	1, 5	Acetoacetic acid	0–trace	2
2-Hexanone	0.007–0.010	5	α-Ketoglutaric acid	0.1–15	2, 8
2-Heptanone	+	1, 5	Oxalic acid	3–7	9
2-Nonanone	+	1	Oxaloacetic acid	0–trace	2
Formic acid	10–85	3, 4	Oxalosuccinic acid	0–trace	2
Acetic acid	3–50	3, 4	Citric acid	1750	Table 1.1
Propionic acid	0–3	3, 4	Benzoic acid	2.3–4.0	10
Butyric acid	0–9.5	3, 4	Ethyl acetate	+	1
Valeric acid	0–3.8	4	Methyl palmitate	+	1
Hexanoic acid	4–10	1, 4			

SOURCE: References as indicated.
1. Scanlan et al. (1968)
2. Harper and Huber (1956)
3. Morr et al. (1957)
4. Harper et al. (1961)
5. Wong and Patton (1962)

6. Knodt et al. (1942)
7. Kreula and Virtanen (1956)
8. Patton and Potter (1956)
9. Zarembski and Hodgkinson (1962)
10. Vogel and Deshusses (1965)

Many compounds conjugated with glucuronate or sulfate have been detected in milk by the technique of absorption of the conjugates on a neutral resin, elution, hydrolysis with glucuronidase and aryl sulfatase, and GLC. In one such study (Brewington *et al.* 1974), some 42 such compounds were identified. They include many phenolic compounds, the aromatic aldehyde, vanillin and some of its esters, various fatty acids and lactones of hydroxy acids, *p*-hydroxyacetophenone, benzoic acid, phenylacetic acid, hippuric acid, and indole. A few of these compounds have also been detected in the free unconjugated state in milk. The conjugates have not been separated into glucuronate-bound and sulfate-bound groups, and glucuronate itself has not been detected in milk in either the bound or the free state.

A large number of N-containing compounds of low molecular weight are not precipitated with proteins by 12% trichloroacetic acid. Some small peptides are included in this group. These nonprotein nitrogen (NPN) constituents aggregate about 1 g/liter and account for about 6% of the total N (i.e., 250–350 mg of N per liter). The principal NPN components are listed in Table 1.6 (Wolfschoon-Pombo and Klostermeyer 1981). The wide variations in concentrations that have been reported for these constituents probably arise from the fact that many of them are metabolites of amino acids and nucleic acids and from the fact that their concentrations in milk depend on the amounts of those substances consumed by the cow.

Table 1.6. Principal NPN Compounds in Milk.

| | Nitrogen (mg/liter milk) | | |
| | Recent Analysis[a] | | Range in Literature |
Compound	Mean	SD	
Total NPN	296.4	37.7	229–308
Urea-N	142.1	32.6	84–134
Creatine N	25.5	6.4	6–20
Creatinine N	12.1	6.8	2–9
Uric acid N	7.8	3.3	5–8
Orotic acid N	14.6	5.9	12–13
Hippuric acid N	4.4	1.2	4
Peptide N	32.0	14.9	—
Ammonia N	8.8	6.1	3–14
α-Amino acid N	44.3	8.2	39–51

SOURCE: Wolfschoon-Pombo and Klostermeyer (1981).
[a]273 samples, each representing a single milking.

About half of the NPN of milk is accounted for by urea. Orotic acid is a particular hallmark of the milks of ruminants; milks of other species contain little if any of it (Larson and Hegarty 1977). The free amino acids constituting the α-amino N fraction in Table 1.6 include those that are also found in proteins, as well as ornithine, citrulline, and α-amino butyric acid. Quantitative analyses of the mixture of free amino acids have been published (Deutsch and Samuelsson 1958; Armstrong and Yates 1963; Rassin et al. 1978).

Table 1.7 lists a number of nitrogenous compounds that have been detected in milk, in addition to those listed in Table 1.6.

Table 1.7.   Some Nitrogeneous Substances in Milk.

Compound	Conc. (mg/liter)	Ref.
Amines		
1-Propylamine	3–15	Cole et al. (1961)
1-Hexylamine	5–24	Cole et al. (1961)
Ethanolamine	0.5–8.5	Armstrong and Yates (1963), Rassin et al. (1978)
Choline	43–285[a]	Hartman and Dryden (1974)
Putrescine } Cadaverine }	0.003–0.021	Sanguansermsri et al. (1974)
Spermidine	0.009–0.028	Sanguansermsri et al. (1974)
Spermine	0.006–0.017	Sanguansermsri et al. (1974)
Amino acid derivatives		
N-Methylglycine	+	Schwartz and Pallansch (1962A)
Histamine	0.03–0.05	Wrenn et al. (1963)
Salicyluric acid	0.016	Booth et al. (1962)
Phenylacetyl glutamine	>0.01	Schwartz and Pallansch (1962B)
Kynurenine	0.023	Parks et al. (1967)
Indoxylsulfuric acid	0.124	Spinelli (1946)
Taurine	1–7	Armstrong and Yates (1963), Rassin et al. (1978)
Other compounds		
Carnitine	10–17	Erfle et al. (1970), Snoswell and Linzell (1975)
Acetyl carnitine	2–12	Erfle et al. (1970)
Morphine	0.0002–0.0005	Hazum et al. (1981)
N-Acetylneuraminic acid (NANA)	120–270[b]	de Koning and Wijnand (1965), Kiermeier and Freisfeld (1965), Morrissey (1973)
N-Acetylglucosamine	11	Hoff (1963)

[a]Total. About 25 mg/liter is in phospholipids.
[b]Total. About 30 mg/liter is free dialyzable NANA.

About one-tenth of the P in milk (i.e., about 100 mg/liter) is in the form of water-soluble organic esters of orthophosphoric acid. A list of such esters that have been detected in milk is presented in Table 1.8. Most of them are sugar phosphates and constituents of phospholipids. Reported concentrations of some of the compounds vary considerably, and complete quantitation of the group has not been made.

Several nucleotides have been detected in milk (see Table 1.9). The list includes the common mono-and dinucleotides, 3′, 5′ cyclic AMP, and adenosine triphosphate (ATP). The ATP is located entirely in the casein micelles (Richardson et al. 1980). Several nucleotide sugars, undoubtedly excess intermediates left over from mammary synthesis of glycoproteins, are present. Both DNA and RNA have been detected in milk (Swope and Brunner 1965; Swope et al. 1965; Langen 1967); they are probably found primarily in milk leukocytes.

The total sulfur content of milk is about 360 mg/liter, of which about 300 mg/liter is in the cysteinyl and methionyl residues of milk proteins and about 35 mg/liter is inorganic sulfate, SO_4^{2-} (Table 1.1). The remainder, amounting to 25 mg of S per liter, is in the form of several organic compounds in which S is found in various states of oxidation. Thiocyanate ion (SCN^-) frequently has been reported in milk. The concentra-

Table 1.8. Phosphate Esters in Milk.

Compound	Conc. (mg/liter)	Ref.
Phosphopyruvate	0.1	1
O-Phosphoethanolamine	83	3
Phosphoglycerol ethanolamine	46	3
Phosphoserine	0.7	3
Glucose-1-phosphate	1	1, 4
Glucose-6-phosphate	12	2
N-Acetylglucosamine-1-phosphate	89	2
Galactose-1-phosphate	45	2
Fructose-6-phosphate	4	2, 4
Fructose-1,6-diphosphate	15	4
Lactose-1-phosphate	0.1	1
Lactose-3′-phosphate[a] ⎫ Lactose-6′-phosphate[a] ⎬	15	5

SOURCE: References as indicated.
1. McGeown and Malpress (1952)
2. Hoff and Wick (1963)
3. Deutsch and Samuelsson (1959)
4. Ganguli and Iya (1963)
5. Kumar et al. (1965)
[a] 3′ and 6′ refer to positions on the galactose moiety.

Table 1.9. Nucleic Acids and Nucleotides in Milk.

Compound	Conc. (mg/liter)	Ref.
Adenosine-5'-monophosphate (AMP)	0.7–21	2, 3
Adenosine-3',5'-cyclic monophosphate (cAMP)	0.3–9	1, 2, 3
Guanosine-5'-monophosphate (GMP)	0.5	3
Cytidine-5'-monophosphate (CMP)	1–20	1
Uridine-5'-monophosphate (UMP)	4.2–60	2
Adenosine-5'-diphosphate (ADP)	3–10	2, 3
Adenosine-5'-triphosphate (ATP)	0.1	4, 5
Guanosine-5'-diphosphate mannose (GDP-Man)	5.4	2
Guanosine-5'-diphosphate fucose (GDP-Fuc)	24–40	3
Cytidine-5'-diphosphate choline (CDP-choline)	3–12	3
Uridine-5'-diphosphate glucose (UDP-Glc)	4.5–200	2, 3
Uridine-5'-diphosphate galactose (UDP-Gal)	4.5–180	2, 3
Uridine-5'-diphosphate glucosamine (UDP-GlcNAc) ⎫ Uridine-5'-diphosphate galactosamine (UDP-GalNAc) ⎬	18	2
Uridine-5'-diphosphate glucuronate	+?	2
Nicotinamide adenine dinucleotide (NAD)	+	2
Deoxyribonucleic acid (DNA)	11–39	6
Ribonucleic acid (RNA)	54–176	7

SOURCE: References as indicated
1. Kolbata *et al.* (1962)
2. Johke (1963)
3. Gil and Sanchez-Mendina (1981)
4. Richardson *et al.* (1980)
5. Zulak *et al.* (1976)
6. Langen (1967)
7. Swope *et al.* (1965)

tion is about 5 mg/liter (Lawrence 1970), although higher values and seasonal variation have been observed (Han and Boulange 1963).

Indoxylsulfate

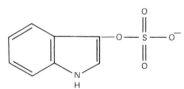

and taurine, $NH_2CH_2CH_2SO_3H$, contain N as well as S and are listed with the NPN compounds in Table 1.7. Methyl sulfide, $(CH_3)_2S$, which is primarily responsible for the "cowy" flavor of milk, has been found in concentrations of 10–40 mg/liter (Patton *et al.* 1956) and dimethyl-sulfone, $(CH_3)_2SO_2$, in concentrations of 6–8 mg/liter (Williams *et al.* 1966). Free cysteine and methionine are regularly detected among the amino acids but are found in extremely low concentrations. Lipoic acid, H_2C -CH_2 -CH_2 -$(CH_2)_4COOH$, has been identified in milk but has not

$$\underset{\displaystyle S\text{———}S}{\overset{\displaystyle |\qquad\quad |}{}}$$

been quantitated adequately (Bingham *et al.* 1967).

Hormones detected in milk include some from the peptide and steroid classes but none of the amino hormones. Prolactin, a protein of 199 amino acid residues, is normally present in a concentration of about 50 μg/liter, and the hexapeptide gonadotropin-releasing hormone of the hypothalmus at about 1.5 μg/liter. Steroid hormones from the adrenal cortex include the glucocorticoids cortisol and corticosterone, totaling 0.2–0.6 μg/liter. Those from the ovary—progesterone, estrone, and estradiol—have concentrations of 10–30, 30, and 175 μg/liter, respectively. Hormones in milk have been reviewed by Koldovsky (1980) and Pope and Swinburne (1980). The prolactin in milk is biologically active (Gala *et al.* 1980).

Many substances called "contaminants" in milk are present in concentrations of only a few micrograms per liter. They include not only substances foreign to normal milk but also extra amounts of substances that also occur normally. There are several routes of access for contaminants, including (1) passage through the cow of ingested, inhaled, or absorbed substances; (2) introduction of drugs and antibiotics into the udder; (3) entrance from air or utensils and equipment; (4) addition of chemicals required in manufacturing processes or purposeful adulterations; and (5) the action of bacteria.

The types of compounds that are of concern as contaminants are chlorinated insecticides, organophosphates, herbicides, fungicides, fasciolicides (phenolic compounds administered to cattle to control liver flukes), antibiotics and sulfonamides, detergents and disinfectants, and polychlorinated biphenyls (PCBs). Contaminants in milk have been reviewed by Kroger (1974) and Snelson (1979). In several cases, allowable levels for specific contaminants in milk have been set by the World Health Organization. Surveys have seldom revealed levels in excess of such standards.

GROSS COMPOSITION

Milk composition expressed in terms of the contents of water (or total solids = 100 − water), fat, protein, lactose, and ash is called "gross composition." Protein is often calculated as crude protein by multiplying total N by 6.38, but sometimes it is corrected to true protein 6.38 (TN -NPN); in a few studies, casein and whey protein have been calculated separately. Lactose should be expressed on an anyhydrous basis, but as pointed out previously, this has not always been done. For bovine milk the sum of fat, true protein, anhydrous lactose, and ash would be expected to fall about 0.2–0.3 percentage units short of the total solids contents because of the materials (citrate, NPN, and miscellaneous compounds) included in total solids but not in any of the four categories. In some studies, one of the components (usually lactose) is calculated by determining the difference between the total solids and the sum of the others. The result is too high for the component so calculated.

Interspecies Differences in Composition

Table 1.10 presents the gross composition of the milks of all species regularly consumed by humans, as well as those of the pig and three important laboratory species; the rat, the guinea pig, and the rabbit. These data provide a general picture of species differences, but not all of them present a true average composition for the milk of the species because of inadequate sampling. Large differences are apparent in the contents and ratios of fat, protein, and lactose and in the proportions of casein and whey proteins. Energy contents range from 40 to 200 kcal/100 g in these milks.

Intraspecies Variation in the Composition of Bovine Milk

The gross composition of bovine milk has been reviewed many times (Rook 1961; Laben 1963; Jenness 1974; Johnson 1975; Moore and Rook 1980). These reviews have delineated thoroughly the extent of the variation and factors affecting it. One of the reviews is presented in a previous edition of this volume (Johnson 1975). Consequently, the present discussion is somewhat abbreviated.

Surveys of the composition of either individual cow samples or of bulked herd milk samples within a geographic area over periods of a few months exhibit considerable variation, particularly in fat and protein contents. Lactose and ash contents vary within narrower limits. The extent of the variation can be expressed as the standard deviation σ and by the coefficient of variation (CV), which is the standard deviation divided by the mean:

$$CV = \sigma / \overline{X}$$

In early studies, Tocher (1925) analyzed 676 single milking samples from individual cows in Scotland over 14 months and Overman *et al.* (1939) analyzed 2426 3-day composite samples from 147 cows of several breeds in the University of Illinois herd over several years. Data from these two studies (Table 1.11) indicate higher CVs for fat and

Table 1.10. Gross Composition of Milks of Various Species.

Species		Water	Fat	Casein	Whey Protein	Lactose	Ash	Energy (kcal/100 g)
				Composition (g/100 g)				
Human	Homo sapiens	87.1	4.5	0.4	0.5	7.1	0.2	72
Rabbit	Oryctolagus cuniculus	67.2	15.3	9.3	4.6	2.1	1.8	202
Rat	Rattus norvegicus	79.0	10.3	6.4	2.0	2.6	1.3	137
Guinea pig	Cavia porcellus	83.6	3.9	6.6	1.5	3.0	0.8	80
Horse	Equus caballus	88.8	1.9	1.3	1.2	6.2	0.5	52
Donkey	Equus asinus	88.3	1.4	1.0	1.0	7.4	0.5	44
Pig	Sus scrofa	81.2	6.8	2.8	2.0	5.5	1.0	102
Camel	Camelus dromedarius	86.5	4.0	2.7	0.9	5.0	0.8	70
Reindeer	Rangifer tarandus	66.7	18.0	8.6	1.5	2.8	1.5	214
Cow	Bos taurus	87.3	3.9	2.6	0.6	4.6	0.7	66
Zebu	Bos indicus	86.5	4.7	2.6	0.6	4.7	0.7	74
Yak	Bos grunniens	82.7	6.5	5.8	—	4.6	0.9	100
Water buffalo	Bubalus bubalis	82.8	7.4	3.2	0.6	4.8	0.8	101
Goat	Capra hircus	86.7	4.5	2.6	0.6	4.3	0.8	70
Sheep	Ovis aries	82.0	7.2	3.9	0.7	4.8	0.9	102

SOURCE: Mostly from compilation of Jenness and Sloan (1970).

Table 1.11. Variations in the Composition of Milk.

Samples		Tocher (1925) 676 Individual Milkings	Overman et al. (1939) 2426 3-Day Composites	Herrington et al. (1972) 868 Bulk Plant
Fat (%)	Mean	3.95	4.37	3.53
	SD	0.78	0.82	0.28
	CV	0.20	0.19	0.08
Crude	Mean	3.24	3.74	3.13
protein	SD	0.40	0.52	0.14
(%)	CV	0.12	0.14	0.05
Lactose (%)	Mean	4.64	4.89[a]	4.82
	SD	0.37	0.38	0.16
	CV	0.08	0.08	0.03
Ash (%)	Mean	0.70	0.72	0.72
	SD	0.05	0.05	0.01
	CV	0.07	0.07	0.02
Total solids	Mean	—	13.73	12.02
(%)	SD	—	1.23	0.63
	CV	—	0.09	0.05

SOURCES: References indicated.
[a] Calculated by difference.

protein and lower ones for lactose and ash. Analyses of nearly 900 samples (not all analyzed for all constituents) from receiving station bulk tanks in New York in 1959–1961 (Herrington et al. 1972) are also given in Table 1.11. As expected, the CVs are less for each constituent than in the individual cow samples.

The possibility of changes in the composition of milk in a given geographic area over a period of time has been assessed by compilation and comparison of data obtained in various surveys. Such surveys for the first half of the twentieth century for the Netherlands (Janse 1950), England (Davies 1952; Griffiths and Featherstone 1957), Scotland (Waite and Patterson 1959), and United States (Armstrong 1959) revealed no great changes except a fall in solids-not-fat in England, which was probably due to an increase in the proportion of Friesian cattle. Some other surveys (Gaunt 1980) indicates changes in the fat content of Friesians of 3.35–3.7% between 1933 and 1960 in the United States and of 3.10–4.00% between 1900 and 1970 in the Netherlands.

The composition of bovine milk is influenced by a number of factors and conditions which may be classified as follows:

1. Inherited
2. Physiological
   a. Stage of lactation
   b. Pregnancy
   c. Age
   d. Nutrition
   e. Season
   f. Udder infection
3. Milking procedure
   a. Within milking
   b. Between milkings

Inherited Variation. Genetically controlled variation in milk composition is evident from both interbreed and intrabreed studies. Some comparative studies of breed differences involve sampling from individual cows; others use herd milk comparisons. Surveys with both kinds of sampling, summarized in Table 1.12, give a reasonably consistent picture of differences among breeds in the gross composition of milk.

Inherited differences among cows within a breed are comparable in magnitude to those between breeds. For example, the CVs within individual breeds in Overman's (1939) studies are comparable to the overall CVs shown in Table 1.11.

Variability in fat and protein contents among cows within a breed are shown in Table 1.13. These data represent over 23,000 lactation records for cows milked twice a day for 305 days in 22 states (Gaunt 1980). The standard deviations included genetic and environmental variances, and in this particular survey, some variance due to analytical methods as well. If, however, one selects from the compiled data only those of cows of a given breed in a single herd calving at a uniform age and date, the standard deviations are reduced by 12–18% (Touchberry 1974). From the total phenotypic variance (σ^2) remaining, a fraction may be assigned to heredity by comparing the variances among half-sibling progeny of a given bull or by comparing daughters with their dams. For the data on which Table 1.13 was based, the genetic variance, σ_p^2, was calculated to be 58 and 49% of the total phenotypic variance for fat and protein contents, respectively. Heritability = h^2 = σ_g^2/σ_p^2 = 0.58 and 0.49. Several other studies corroborate these data (Gaunt 1980), and a few give estimates of the heritability of lactose as well. Mean values for h^2 are 0.61, 0.58, and 0.55 for fat, protein, and lactose, respectively. This evidence for heritability of milk composition is, of course, statistical. There is little information on actual inheri-

Table 1.12. Gross Composition of Milk of Various Breeds (g/100 g).

	Ayrshire	Brown Swiss	Guernsey	Holstein	Jersey	Shorthorn	Red Poll
Fat							
1a	4.14 (208)[a]	4.02 (428)	5.19 (321)	3.54 (268)	5.17 (199)		
1b		3.97 (494)					
2	3.97 (70)	3.80 (23)	4.58 (23)	3.56 (75)	4.96 (72)	3.47 (18)	4.24 (20)
3	4.00 (25)	4.16 (33)	4.62 (24)	3.62 (26)	5.26 (25)	3.53	
4	3.69	—	4.49	3.46	—		
Crude protein							
1a	3.58	3.61	4.01	3.42	3.86		
1b		3.52					
2	3.59	3.51	3.83	3.33	3.97	3.32	3.70
3	3.37	3.93	3.59	3.13	4.10	3.32	
4	3.38	—	3.57	3.28	—		
Lactose							
1a	4.69	5.04	4.91	4.85	4.94		
1b		4.90					
2	4.63	4.80	4.78	4.61	4.70	4.66	4.77
3	4.53	5.00	4.52	4.79	4.84	4.51	
4	4.57	—	4.62	4.46	—		

Ash

1a	0.68	0.73	0.74	0.68	0.70		0.72
1b		0.74					
2	0.72	0.72	0.75	0.73	0.77		
3	0.75	0.78	0.80	0.74	0.83	0.74	
4	0.74	—	0.77	0.75	—	0.76	

Total solids

1a	13.11	13.41	14.87	12.50			13.28
1b		13.13			14.69		
2	12.69	12.69	13.69	11.91	14.15		
3	—	—	—	—	—	—	
4	12.51	—	13.57	12.07	—	12.27	

SOURCES: 1a. Overman et al. (1939). Three-day composites from individual cows. Lactose calculated by difference.
1b. Overman et al. (1953). Herd composites from 1 day each month; 39 herds. Lactose calculated by difference.
2. Reinart and Nesbitt (1956A–C). Herd composites from 1 day each month.
3. Cerbulis and Farrell (1975, 1976). Single milking samples, except fat from lactation average.
   Corrected from g/100 ml to g/100 g.
4. Rook (1961). Herd samples; number not given.
[a]Numbers in parentheses are number of samples.

Table 1.13. Lactation Average Fat and Protein Contents of Milks of Five Breeds.

Breed	No. Records	Fat (g/100 g)		Protein (g/100 g)	
		Mean	SD	Mean	SD
Ayrshire	3362	3.99	0.33	3.34	0.29
Brown Swiss	2621	4.16	0.35	3.53	0.26
Guernsey	6956	4.87	0.45	3.62	0.29
Holstein	9102	3.70	0.39	3.11	0.25
Jersey	6354	5.13	0.54	3.80	0.30

SOURCE: Gaunt (1980).

tance of differences in specific aspects of the synthetic and secretory mechanisms. Fat and protein contents of milk are positively correlated. For the study reported in Table 1.13, the correlations between these two parameters range from +0.20 to +0.60 in the five breeds. Since the correlation is far from perfect, there appears to be an opportunity to increase the protein content without simultaneously increasing the fat content by using breeding animals that transmit the potential for higher than average protein/fat ratios.

Changes During Lactation. The changes in milk composition during the lactation cycle of the cow have been described many times (Rook 1961). Colostrum, the initial mammary secretion after parturition, contains more mineral salts and protein and less ash than later milk. Its fat content is often, but not always, higher than that of milk. The composition of colostrum differs more among individual animals than does the composition of milk. Of the individual minerals, Ca, Na, Mg, P, and chloride are higher in colostrum but K is lower (Garrett and Overman 1940). The most remarkable difference between colostrum and milk is the high concentration of immunoglobulins (Ig's) in the former. They accumulate in the gland before parturition and serve to transfer immunity to the newborn suckling. All of the types of Ig's are found in higher concentration in colostrum than in milk. The one present in highest concentration in both is IgGI, at about 50 and 0.6 mg/ml in colostrum and milk, respectively (Butler 1974). Table 1.14 shows the extent of change in the concentration of some of the principal constituents in the first few milkings after parturition (Parrish et al. 1948, 1950). These changes continue, although at reduced rates, for approximately 5 weeks. Thereafter, the fat and protein contents rise gradually until, near the end of lactation, they increase more sharply; lactose diminishes gradually throughout lactation (Bonnier et al. 1946). Ca, P,

Table 1.14. Transition from Colostrum to Milk.

	Milking							
	1	2	3	4	5 + 6	7 + 8	15 + 16	27 + 28
	First study							
Protein (%)	16.5	10.3	5.9	4.6	4.1	4.0	3.5	3.2
Casein (%)	6.4	4.9.	3.8	3.4	3.3	3.2	2.8	2.6
Whey protein (%)	10.1	5.5	2.2	1.2	0.8	0.8	0.7	0.6
	Second study							
Fat (%)	5.3	5.4	4.4	4.5	4.5	4.8	4.8	4.6
Protein (%)	14.6	9.4	5.5	4.5	4.2	4.1	3.6	3.3
Lactose (%)	2.6	3.5	4.3	4.6	4.8	4.9	4.9	5.1
Ash (%)	1.16	1.03	0.92	0.87	0.85	0.85	0.81	0.78

SOURCE: First study: Parrish *et al.* (1948): 10 cows.
Second study: Parrish *et al.* (1950): 111 cows, not all analyzed for all constituents.

and chloride contents follow the pattern of fat and protein (Sharp and Struble 1935; Ellenberger *et al.* 1950). The increases in fat and protein contents in the latter part of the lactation period are much smaller or do not occur at all if the cow is not pregnant (Wilcox *et al.* 1959; Legates 1960; Wheelock *et al.* 1965; Parkhie *et al.* 1966).

Age. It has been commonly observed that average fat and solids-not-fat contents of milk for a lactation period decline with successive periods. The fat decreases by about 0.2% and the solids-not-fat by about 0.4% over five lactations (Legates 1960; Rook 1961). The reasons for these effects of age have not been elucidated, but Legates (1960) suggested udder deterioration with usage, increasing incidence of mastitis, and selective culling for high milk production as possible causes.

Nutrition of the Cow. Both the plane of nutrition and the physical form of the ration influence the composition of milk. Feeding at a level less than that required for maintenance and maximum production reduces the yield of milk, but since the yield of fat tends to be maintained, the percentage of fat may increase. Significant effects of the plane of nutrition on the solids-not-fat content have been found (Rook 1961; Moore and Rook 1980). Overfeeding by 25–35% above a standard may increase the solids-not-fat content by about 0.2%, whereas underfeeding by 25% decreases it by as much as 0.4–0.5%. The effect is primarily on the protein content; the greatest change is observed in the percentage of casein. Some small effects on lactose have been noted (Dawson and Rook 1972). It was demonstrated long ago that the fat content of milk can be influenced materially by the physical form of the ration. Powell (1939, 1941) was the first to demonstrate that feeding a ration low in roughage decreases the fat content but does not greatly affect the yield of milk or the solids-not-fat content. This finding has been confirmed many times (Moore and Rook 1980). The important feature is the physical state of the ration rather than the composition. Thus, either finely ground forage or concentrate will depress the fat content. The degree of depression depends somewhat on the composition of the ration, but sometimes the fat content is reduced by one-half or more. This effect is caused by changes in rumen fermentation in the animal. On low-roughage rations, the production of acetate, from which the mammary gland ordinarily synthesizes much fatty acid, is reduced and that of propionate is increased. When these fatty acids are infused directly into the rumen, acetate increases the percentage of fat and propionate decreases it (and also increases the protein content). The decrease in mammary fat synthesis on low-roughage rations is in part due to alteration of the secretion of insulin, which stimulates the synthesis of triglycerides in adipose tissue and thus deprives the mammary gland (Moore and Rook 1980). The ratio of roughage to concentrate does not affect the protein content of the milk or the ratio of individual proteins (Grant and Patel 1980).

The amount of fat in the ration has little effect on the fat content of milk, but the fatty acid composition of the dietary fat greatly influences that of milk fat (Chapter 4). Neither the total protein content of milk nor the proportions of the individual proteins is greatly influenced by the amount or kind of protein in the diet except at very greatly reduced intakes. Overfeeding with protein does, however, increase the NPN content of the milk (Thomas 1980). Milk of normal composition with the normal content and proportions of proteins can be produced on protein-free diets with urea and ammonium salts as the only sources of nitrogen (Virtanen 1966).

Seasonal Variation and the Influence of Temperature. In temperate latitudes, rather characteristic seasonal variations in milk composition are commonly observed. Both fat and solids-not-fat contents are lower in summer than in winter. In the survey by Overman (1945) of individual cows at the University of Illinois, monthly extremes for fat were 4.24 and 3.81% in January and August and for protein were 3.61 and 3.37% in January and July, respectively. Nickerson (1960) found significant seasonal differences in 18 components of bulk milks from six areas in California. Seasonal differences in fat and protein contents were similar to those observed in Illinois. Seasonal variations in milk composition could conceivably be caused by differences in temperature, nutrition, and stage of lactation and the interactions among them. Cobble and Herman (1951), who kept cows in rooms at constant temperature, found little effect on milk composition between –1° and 21°C, but above 30°C they noted increases in fat and chloride contents and decreases in milk yield, solids-not-fat, protein, and lactose. At temperatures down to −15°C, fat, solids-not-fat, and protein increased but chloride and lactose were unaffected. Maximum changes were of the order of 1% for fat, solids-not-fat, and lactose; the responses of individual cows varied widely. Wayman et al. (1962) found no significant difference in fat or solids-not-fat percentage when cows were alternated between 18 and 31°C. High environmental temperatures reduce the milk yield because the cows eat less, but it is not clear that temperature has any effect on the percentage composition of milk. It is likely that observed seasonal effects on milk composition result from the composite effect of stage of lactation (tendency for fairly synchronous spring freshening) and differences in the fibrousness of the ration.

Infection of the Udder. Infection of the mammary gland greatly influences the composition of milk. A general reference on bovine mastitis is Schalm et al. (1971). The concentrations of fat, solids-not-fat, lactose, casein, β-lactoglobulin, α-lactalbumin, and K are lowered and those of blood serum albumin, immunoglobulins, Na, and chloride are increased (Barry and Rowland 1953; Leece and Legates 1959; Bortree et al. 1962; Carroll et al. 1963). The ability to synthesize lactose and the specific milk proteins is impaired, the tight junctions between secretory cells become "leaky," and blood salts and protein pass into the milk. Mastitis severe enough to be detectable clinically results in milk of clearly abnormal composition (e.g., casein below 78% of total protein, chloride above 0.12%). Probably some of the variability reported for "normal" milks may be due to undetected subclinical mastitis (Rook 1961). Undoubtedly mastitis is largely responsible for reported differences in composition from separate quarters of the cow's udder (Rowland et al. 1959; Waite 1961).

Variations Due to the Milking Procedure. The discussion to this point has concerned factors affecting the composition of milk as secreted. Variations in composition also occur due to peculiarities of the milking procedures. Wheelock (1980) has reviewed these effects. It has long been known that the fat content increases continuously during the milking process, foremilk being very low and strippings rich in fat. The solids-not-fat content calculated as a percentage of fat-free plasma does not change during the milking process. The increase in fat content during milking apparently results from the tendency of the fat globules to cluster and be trapped in the alveoli (Whittlestone 1953). The effect was demonstrated admirably by Johansson *et al.* (1952) with special apparatus which made it possible to measure milk flow and sample continuously during the process of milking; the fat content increased continuously during milking.

From the above considerations, it follows that for an incomplete milking the fat content will be lower than normal, but for a subsequent complete milking the fat content will be higher than normal. Furthermore, when the intervals between milkings are unequal, the milk yield is greater and the fat content lower following the longer interval (Wheelock 1980). Since the usual practice is for a longer night than day interval, the fat content of the morning milk is lower than that of evening milk. Some tendency has been observed for morning milk to be lower in fat content even when the intervals are equal, but this finding does not seem to be entirely substantiated. For intervals longer than about 15 hr, the rate of milk secretion decreases and the concentrations of fat, whey proteins, Na, and chloride increase; solids-not-fat, lactose, and potassium decrease (Wheelock 1980).

REFERENCES

Allen, R. J. L. 1940. The estimation of phosphorus. *Biochem. J. 34*, 858–865.

Archibald, J. G. 1958. Trace elements in milk: A review. *Dairy Sci. Abstr. 20*, 712–725, 798–808.

Armstrong, T. V. 1959. Variations in the gross composition of milk as related to the breed of the cow: A review and critical evaluation of literature of the United States and Canada. *J. Dairy Sci. 42*, 1–19.

Armstrong, M. D. and Yates, K. N. 1963. Free amino acids in milk. *Proc. Soc. Exp. Biol. Med. 113*, 680–683.

Babajimopoulos, M. and Mikolajcik, E. M. 1977. Quantification of selected serum proteins of milk by immunological procedures. *J. Dairy Sci. 60*, 721–725.

Barry, J. M. and Rowland, S. J. 1953. Variations in the ionic and lactose concentrations of milk. *Biochem. J. 54*, 575–578.

Basu, K. P., Paul, T. M., Shroff, N. B. and Rahman, M. A. 1962. *Composition of Milk and Ghee.* Indian Counc. Agr. Res. Report Series No. 8.

Beebe, J. M. and Gilpin, R. K. 1983. Determination of α-and β-lactose in dairy products by totally aqueous liquid chromatography. *Anal. Chim. Acta 146*, 255–259.

Bell, J. W. and Stone, W. K. 1979. Rapid separation of whey proteins by cellulose acetate electrophoresis. *J. Dairy Sci. 62*, 502–504.

Bican, P. and Blanc, B. 1982. Milk protein analysis—a high-performance chromatography study. *Milchwissenschaft 37*, 592–593.

Bingham, R. J., Huber, J. D. and Aurand, L. W. 1967. Thioctic acid in milk. *J. Dairy Sci. 50*, 318–323.

Blanc, B. 1981. Biochemical aspects of human milk—comparison with bovine milk. *Wld. Rev. Nutr. Dietet. 36*, 1–89.

Bonnier, G., Hansson, A. and Jarl, F. 1946. Studies in the variations of the calory content of milk. *Acta Agr. Suecana 2*, 159–169.

Booth, A. N., Robbins. D. J. and Dunkley, W. L. 1962. Occurrence of salicyluric acid in milk. *Nature* (Lond.) *194,* 290–291.

Booy, C. J., Klijn, C. J. and Posthumus, G. 1962. The rapid estimation of the protein content of milk. *Dairy Sci. Abstr. 24,* 223–228, 275–279.

Bortree, A. L., Carroll, E. J. and Schalm, O. W. 1962. Whey protein patterns of milk from cows with experimentally produced mastitis. *J. Dairy Sci. 45,* 1465–1471.

Brewington, C. R., Parks, O. W. and Schwartz, D. P. 1974. Conjugated compounds in cow's milk. II. *J. Agr. Food. Chem. 22,* 293–294.

Brons, C. and Olieman, C. 1983. Study of the high-performance liquid chromatographic separation of reducing sugars, applied to the determination of lactose in milk. *J. Chromatog. 259,* 79–86.

Butler, J. E. 1974. Immunoglobulins of the mammary secretions. *In: Lactation: A Comprehensive Treatise,* Vol. 3. B. L. Larson and V. R. Smith (Editors). Academic Press, New York.

Byun, S. M. and Jenness, R. 1982. Estimation of free *myo*-inositol in milks of various species and its source in milk of rats *(Rattus norvegicus) J. Dairy Sci. 65,* 531–536.

Carroll, E. J., Schalm, O. W. and Lasmanis, J. 1963. Experimental coliform *(Aerobacter aerogenes)* mastitis: Distribution of whey proteins during the early acute phase. *J. Dairy Sci. 46,* 1236–1242.

Cerbulis, J. and Farrell, H. M., Jr. 1975. Composition of milks of dairy cattle. I. Protein, lactose, and fat contents and distribution of protein fraction. *J. Dairy Sci. 58,* 817–827.

Cerbulis, J. and Farrell, H. M., Jr. 1976. Composition of the milks of dairy cattle. II. Ash, calcium, magnesium and phosphorus. *J. Dairy Sci. 59,* 589–593.

Cobble, J. W. and Herman, H. A. 1951. *The Influence of Environmental Temperature on the Composition of the Milk of the Dairy Cow.* Mo. Agr. Exp. Sta. Res. Bull 485.

Cole, D. D., Harper, W. J. and Hankinson, C. L. 1961. Observations on ammonia and volatile amines in milk. *J. Dairy Sci. 44,* 171–173.

Dagley, S. 1974. Citrate—UV spectrophotometric determination. *In: Methods of Enzymatic Analysis,* 2nd ed. H. U. Bergmeyer (Editor). Academic Press, New York.

Davies, D. T. 1974. The quantitative partition of the albumin fraction of milk serum proteins by gel chromatography. *J. Dairy Res. 41,* 217–228.

Davies, D. T. and Law, A. J. R. 1977. An improved method for the quantitative fractionation of casein mixtures using ion-exchange chromatography. *J. Dairy Res. 44,* 213–221.

Davies, D. T. and White, J. C. D. 1960. The use of ultrafiltration and dialysis in isolating the aqueous phase of milk and in determining the partition of milk constituents between the aqueous and disperse phases. *J. Dairy Res. 27,* 171–190.

Davies, J. G. 1952. The chemical composition of milk between 1900 and 1950. *Analyst 77,* 494–524.

Dawson, R. R. and Rook, J. A. F. 1972. A note on the influence of stage of lactation on the response in lactose content of milk to a change of plane of energy nutrition in the cow. *J. Dairy Res. 39,* 107–111.

deKoning, P. J. and Wijand, H. P. 1965. The effect of sugars and heat treatment on the determination of *N*-acetyl neuraminic acid in milk. *Neth. Milk Dairy J. 19,* 73–81.

Deutsch, A. and Samuelsson, E. G. 1958. Amino acids and low molecular weight amino-acid derivatives in cows' milk. *Int. Dairy Cong. 1958. 3,* 1650–1652.

Devery-Pocius, J. E. and Larson. B. L. 1983. Age and previous lactations as factors in the amount of bovine colostral immunoglobulins. *J. Dairy Sci. 66,* 221–226.

Diosady, L. L., Bergen, I. and Harwalkar, V. R. 1980. High performance liquid chromatography of whey proteins. *Milchwissenschaft 35,* 671–674.

Ebner, K. E. and Schanbacher, F. L. 1974. Biochemistry of lactose and related carbohydrates. In: *Lactation: A Comprehensive Treatise.*, Vol. 2. B. L. Larson and V. R. Smith (Editors) Academic Press, New York.

Eigel, W. N., Butler, J. E., Ernstrom, C. A., Farrell, H. M., Jr. Harwalker, V. R., Jenness, R. and Whitney, R. McL. 1984. Nomenclature of the proteins of cow's milk. Fifth revision. *J. Dairy Sci. 67*, 1599-1631.

Ellenberger, H. B., Newlander, J. A. and Jones, C. H. 1950. *Variations in the Calcium and Phosphorus Contents of Cow's Milk.* Vt. Agr. Exp. Sta. Bull. 556.

Erfle, J. D., Fisher, L. J. and Sauer, F. 1970. Carnitine and acetyl-carnitine in the milk of normal and ketotic cows. *J. Dairy Sci. 53*, 486-492.

FAO Production Yearbook. 1979. Vol. 33. Food and Agriculture Organization of the United Nations.

Faulkner, A., Chaiybutr, N., Peaker, M., Carrick, D. I. and Kuhn, N. J. 1981. Metabolic significance of milk glucose. *J. Dairy Res. 48*, 51-56.

Faulkner, A. and Clapperton, J. L. 1981. Changes in the concentration of some minor constituents of milk from cows fed low or high-fat diets. *Comp. Biochem. Physiol. 68A* 281-283.

Frayer, J. M. 1940. *The Dissolved Gases in Milk and Dye Reduction.* Vt. Agr. Expt. Sta. Bull 461.

Gala, R. R., Forsyth, I. A. and Turney, A. 1980. Milk prolactin is biologically active. *Life Sci. 26,*987-993.

Ganguli, N. C. and Iya, K. K. 1963. The occurrence of sugar phosphates in milk. *Ind. J. Chem. 1*, 145-146.

Garrett, O. F. and Overman, O. R. 1940. Mineral composition of colostral milk. *J. Dairy Sci. 23*, 13-17.

Gaull, G. E., Jensen, R. G., Rassin, D. K. and Malloy, M. 1982. Human milk as food *Adv. Perinatal Med. 2*, 47-120.

Gaunt, S. N. 1980. Genetic variation in the yields and contents of milk constituents. *In: Factors Affecting the Yield and Contents of Milk Constituents of Commercial Importance.* P. C. Moore and J. A. F. Rook (Editors). Internat. Dairy Fed. Doc. 125.

Gil, A. and Sanchez-Mendina, F. 1981. Acid-soluble nucleotides of cow's, goat's and sheep's milks at different stages of lactation. *J. Dairy Res. 48*, 35-44.

Grant, D. R. and Patel, R. R. 1980. Changes of protein composition of milk by ratio of roughage to concentrate. *J. Dairy Sci. 63*, 756-761.

Griffiths, T. W. and Featherstone, J. 1957. Variations in the solids-not-fat content of milk. Investigations into the nature of the solids-not-fat problems in the West Midlands. *J. Dairy Res. 24*, 201-209.

Grimbleby, F. H. 1956. The determination of lactose in milk. *J. Dairy Res.* 229-237.

Goulden, J. D. S. 1964. Analysis of milk by infra-red absorption. *J. Dairy Res. 31*, 273-284.

Guidry, A. J. and Pearson, R. E. 1979. Improved methodology for quantitative determination of serum and milk proteins by single radial immuno-diffusion. *J. Dairy Sci. 62*, 1252-1257.

Guthrie, E. S. 1946. The results of deaeration on the oxygen, vitamin C and oxidized flavors of milk. *J. Dairy Sci. 29*, 359-369.

Han, K. and Boulange, M. 1963. Determination of thiocyanate in milk by a technique of immuno-diffusion. *Clin. Chim. Acta 8*, 779-785.

Harper, W. J., Gould, I. A. and Hankinson, C. L. 1961. Observations on the free volatile acids in milk. *J. Dairy Sci. 44*, 1764-1765.

Harper, W. J. and Huber, R. M. 1956. Some carbonyl compounds in raw milk. *J. Dairy Sci. 39*, 1609.

Hartman, A. M. and Dryden, L. P. 1974. The vitamins in milk and milk products. *In:*

Fundamentals of Dairy Chemistry, 2nd ed. B. H. Webb, A. H. Johnson and J. A. Alford (Editors). AVI Publishing Co. Westport, Conn.

Hazum, E., Sabatka, J. J., Chang, K. J., Brent, D. A., Findlay, J. W. A. and Cuatrecasas, P. 1981. Morphine in cow and human milk: Could dietary morphine constitute a ligand for specific morphine (μ) receptors? *Science 213,* 1010–1012.

Heinemann, B. 1944. The determination of citric acid in milk products by cerate oxidimetry. *J. Dairy Sci. 27,* 377–384.

Herreid, E. O. and Francis, J. 1949. The effect of handling and processing of milk on its oxygen content. *J. Dairy Sci. 32,* 202–208.

Herrington, B. L., Sherbon, J. W. Ledford, R. A. and Houghton, G. E. 1972. Composition of milk in New York State. *New York's Food Life Sci. Bull. 18,* 1–23.

Hinton, C. L. and Macara, T. 1927. The determination of aldose sugars by means of chloramine-T, with special reference to the analysis of milk products. *Analyst 52,* 668–688.

Hoff, J. E. 1963. Determination of N-acetylglucosamine–1-phosphate and *N*-acetylglucosamine in milk. *J. Dairy Sci. 46,* 573–574.

Hoff, J. E. and Wick, E. L., 1963. Acid soluble phosphates in cow milk. *J. Food Sci. 28,* 510–518.

Holt, C., Dalgleish, D. G. and Jenness, R. 1981. Calculation of the ion equilibria in milk diffusate and comparison with experiment. *Anal. Biochem. 113,* 154–163.

Horwitz, W. (Editor). 1980. *Official Methods of Analysis,* 13th ed. Association of Official Analytical Chemists, Washington, D.C.

Janse, L. C. 1950. Composition of Friesian milk. *Neth. Milk Dairy J. 4,* 1–9.

Jenness, R. 1974. The composition of milk. *In: Lactation: A Comprehensive Treatise,* Vol. 3. B.L. Larson and V.R. Smith (Editors). Academic Press, New York.

Jenness, R. 1979. The composition of human milk. *Semin. Perinatol. 3,* 225–239.

Jenness, R. 1980. Composition and characteristics of goat milk: A review, 1968–1979. *J. Dairy Sci. 63,* 1605–1630.

Jenness, R., Regehr, E. A. and Sloan, R. E. 1964. Comparative biochemical studies of milks. II. Dialyzable carbohydrates. *Comp. Biochem. Physiol. 13,* 339–352.

Jenness, R. and Sloan, R. E. 1970. The composition of milks of various species. A review. *Dairy Sci. Abstr. 32,* 599–612.

Johansson, I., Korkman, N. and Nelson, N. J. 1952. Studies on udder evacuation of dairy cows I and II. *Acta Agr. Scand. 3,* 43–81, 82–102.

Johke, T. 1963. Acid-soluble nucleotides of colostrum, milk and mammary gland. *J. Biochem. 54,* 388–397.

Johnson, A. H. 1975. The composition of milk. *In: Fundamentals of Dairy Chemistry,* 2nd ed. B. H. Webb, A. H. Johnson and J. A. Alford (Editors) AVI Publishing Co., Wesport, Conn.

Kiermeier, F. and Freisfeld, I. 1965. Neuraminic acid content of cows' milk. *Z. Lebensmitt. Untersuch. Forsch., 128,* 207–217.

Knodt, C. B., Shaw, J. C. and White. G. C. 1942. Studies on ketosis in dairy cattle. II. Blood and urinary acetone bodies of dairy cattle in relation to parturition, lactation, gestation and breed. *J. Dairy Sci. 25,* 851–860.

Kolbata, A., Ziro, S. and Kida, M. 1962. The acid-soluble nucleotides of milk. *J. Biochem. 51,* 277–287.

Koldovsky, O. 1980. Hormones in milk. *Life Sci. 26,* 1833–1836.

Koops, J. 1965. Rapid turbidimetric determination of inorganic sulphate in milk. *Neth. Milk Dairy J. 19,* 59–62.

Kreula, M. and Virtanen, A. I. 1956. α-keto acids in cow's milk. *14th Int. Dairy Cong. 1,* 802–806.

Kroger, M. 1974. General environmental contaminants occurring in milk. *In: Lactation: A Comprehensive Treatise,* Vol. 3. B. L. Larson and V. R. Smith (Editors). Academic Press, New York.

Kumar, F. A., Ferchmin, P. A. and Caputto, R. 1965. Isolation and identification of a lactose phosphate ester from cow colostrum. *Biochem. Biophys. Res. Commun. 20,* 60–62.

Kuramoto, S., Jenness, R., Coulter, S. T. and Choi, R. P. 1959. Standardization of the Harland-Ashworth test for whey protein nitrogen. *J. Dairy Sci. 42,* 28–38.

Kurz, G. and Wallenfels, K. 1974. D-Galactose. UV-assay with galactose dehydrogenase. *In: Methods of Enzymatic Analysis,* 2nd ed. H. U. Bergmeyer (Editor). Academic Press, New York.

Laben, R. C. 1963. Factors responsible for variation in milk composition. *J. Dairy Sci. 46,* 1293–1301.

Langen, H. de. 1967. Determination of DNA in milk. *Aust. J. Dairy Technol. 22,* 36–40.

Larson, B. L. and Hageman, E. C. 1963. Determination of α-lactalbumin in complex systems. *J. Dairy Sci. 46,* 14–18.

Larson, B. L. and Hegarty, H. M. 1977. Orotic acid and pyrimidine nucleotides in ruminant milks. *J. Dairy Sci. 60,* 1223–1229.

Larson, B. L. and Twarog, J. M. 1961. Determination of β-lactoglobulin in complex systems by a simple immunological procedure. *J. Dairy Sci. 44,* 1843–1856.

Lawrence, A. J. 1970. The thiocyanate content of milk. *18th Internat. Dairy Congr.* IE. 99.

Laxminarayan, H. and Dastur, N. N. 1968. Buffaloes' milk and milk products. *Dairy Sci. Abstr. 30,* 177–186, 231–241.

Leece, J. G. and Legates, J. E. 1959. Changes in the paper electrophoretic whey-protein pattern of cows with acute mastitis. *J. Dairy Sci. 42,* 698–704.

Legates, J. E. 1960. Genetic and environmental factors affecting the solids-not-fat composition of milk. *J. Dairy Sci. 43,* 1527–1532.

Lengemann, F. W., Wentworth, R. A. and Comar, C. L. 1974. Physiological and biochemical aspects of the accumulation of contaminant radionuclides in milk. *In: Lactation: A Comprehensive Treatise,* Vol 3. B. L. Larson and V. R. Smith (Editors). Academic Press, New York.

Ling, E. R. 1956. *A Textbook of Dairy Chemistry,* 3rd ed. Vol. 2, *Practical.* Chapman Hall, London.

Macy, I. G., Kelly, H. J. and Sloan, R. E. 1953. *The Composition of Milks.* Nat. Acad. Sci.-Nat. Res. Counc. Publ. 254.

Marier, J. R. and Boulet, M. 1959. Direct analysis of lactose in milk and serum. *J. Dairy Sci. 42,* 1390–1391.

Marsili, R. T., Ostapenko, H., Simmons, R. E. and Green, D. E. 1981. High performance liquid chromatographic determination of organic acids in dairy products. *J. Food Sci. 46,* 52–57.

Mc Dowall, F. H. 1936. The determination of carbon dioxide in biological fluids, more particularly milk and cream. *Analyst 61,* 472–473.

McDowell, A. K. R. 1941. The estimation of lactose in milk. *J. Dairy Res. 12,* 131–138.

McGeown, M. G. and Malpress, F. H. 1952. Studies on the synthesis of lactose by the mammary gland. 2. The sugar phosphate esters of milk. *Biochem. J. 52,* 606–611.

Meun, D. H. C. and Smith, K. C. 1968. A microphosphate method. *Anal. Biochem. 26,* 364–368.

Moore, J. H. and Rook, J. A. F. (Editors). 1980. Factors affecting the yields and contents of milk constituents of commercial importance. *Internat. Dairy Fed. Bulletin Doc. 125,* 1–167.

Morr, C. V., Harper, W. J. and Gould, I. A. 1957. Some organic acids in raw and heated skim milk. *J. Dairy Sci. 40*, 964–972.

Morrissey, P. A. 1973. The *N*-acetyl neuraminic acid content of the milk of various species. *J. Dairy Res. 40*, 421–425.

Murthy, G. K. 1974. Trace elements in milk. *CRC Crit. Rev. Environmental Control 4*, 1–37.

Murthy, G. K. and Rhea, U. 1967. Determination of major cations in milk by atomic absorption spectrophotometry. *J. Dairy Sci. 50*, 313–317.

Nickerson, T. A. 1960. Chemical composition of milk. *J. Dairy Sci. 60*, 598–606.

Nickerson, T. A., Moore, E. E. and Zimmer, A. A. 1964. Spectrophotometric determination of calcium in milk using 2,2′-(ethanediylidene-dinitrilo) diphenol (glyoxal *bis* (2-hydroxyanil). *Anal. Chem. 36*, 1676–1677.

Noll, C. I. and Supplee, G. C. 1941. Factors affecting the gas content of milk. *J. Dairy Sci. 24*, 993–1013.

Overman, O. R. 1945. Monthly variations in the composition of milk. *J. Dairy Sci. 28*, 305–309.

Overman, O. R., Garrett, O. F., Wright, K. E. and Sanmann, F. P. 1939. *Composition of the Milk of Brown Swiss Cows.* Ill. Agr. Expt. Sta. Bull. 457.

Overman, O. R., Keirs, R. J. and Craine, E. M. 1953. *Composition of Herd Milk of the Brown Swiss Breed.* Ill. Agr. Exp. Sta. Bull. 567.

Packard, V. S. 1982. *Human Milk and Infant Formula.* Academic Press, New York.

Parkash, S. and Jenness, R. 1968. The composition and characteristics of goats' milk: A review. *Dairy Sci. Abstr. 30*, 67–87.

Parkhie, M. R., Gilmore, L. O. and Fechheimer, N. S. 1966. Effect of successive lactations, gestation, and season of calving on constituents of cows' milk. *J. Dairy Sci. 49*, 1410–1415.

Parks, O. W., Schwartz, D. P., Nelson, K. and Allen, C. 1967. Evidence for kynurenine in milk. *J. Dairy Sci. 50*, 10–11.

Parrish, D. B., Wise, G. H., Hughes, E. S. and Atkeson, F. W. 1948. Properties of the colostrum of the dairy cow. II. Effect of prepartal rations upon the nitrogenous constituents. *J. Dairy Sci. 31*, 889–895.

Parrish, D. B., Wise, G. H., Hughes, E. S. and Atkeson, F. W. 1950. Properties of the colostrum of the dairy cow. V. Yield, specific gravity and concentrations of total solids and its various components of colostrum and early milk. *J. Dairy Sci. 33*, 457–465.

Patton, S., Forss, D. A. and Day, E. A. 1956. Methyl sulfide and the flavor of milk. *J. Dairy Sci. 39*, 1469–1470.

Patton, S. and Potter, F. E. 1956. The presence of α-ketoglutaric acid in milk. *J. Dairy Sci. 39*, 611–612.

Perry, N. A. and Doan, F. J. 1950. A picric acid method for the simultaneous determination of lactose and sucrose in dairy products. *J. Dairy Sci. 33*, 176–185.

Pope, G. S. and Swinburne, J. K. 1980. Hormones in milk: Their physiological significance and value as diagnostic aids. *J. Dairy Res. 47*, 427–449.

Powell, E. B. 1939. Some relations of the roughage intake to the composition of milk. *J. Dairy Sci. 22*, 453–454.

Powell, E. B. 1941. Progress report on the relation of the ration to the composition of milk. *J. Dairy Sci. 24*, 504–505.

Ramos, M. and Juarez, M. 1981. The composition of ewe's and goat's milk. *Internat. Dairy Fed. Bull. Doc. 140*, 1–19.

Rassin, D. K. Sturman, J. A. and Gaull, G. E. 1978. Taurine and other free amino acids in milk of man and other mammals. *Early Human Dev. 2*, 1–13.

Reinart, A. and Nesbitt, J. M. 1956A. The distribution of nitrogen in milk in Manitoba. *14th Int. Dairy Cong. 1,* 925–933.

Reinart, A. and Nesbitt, J. M. 1956B. The composition of milk in Manitoba. *14th Int. Dairy Cong. 1,* 946–956.

Reinart, A. and Nesbitt, J. M. 1956C. The lactose content of milk in Manitoba. *14th Int. Dairy Cong. 1,* 957–964.

Reineccius, G. A., Kavanagh, T. E. and Keeney, P. G. 1970. Identification and quantitation of free neutral carbohydrates in milk products by gas-liquid chromatography and mass spectrometry. *J. Dairy Sci. 53,* 1018–1022.

Reithel, F. J. and Venkataraman, R. 1956. Lactose in the Sapotaceace. *Science 123,* 1083–1084.

Richardson, T., McGann, T. C. A. and Kearney, R. D. 1980. Levels and location of adenosine 5'-tri-phosphate in bovine milk. *J. Dairy Res. 47,* 91–96.

Rook, J. A. F. 1961. Variations in the chemical composition of the milk of the cow. *Dairy Sci. Abstr. 23,* 251–258, 303–308.

Rowland, S. J. 1938. the determination of the nitrogen distribution in milk. *J. Dairy Res. 9,* 42–46.

Rowland, S. J., Neave, F. K., Dodd, F. H. and Oliver. J. 1959. The effect of *Staphlphococcus pyogenes* infections on milk secretion. *15th Int. Dairy Cong. 1,* 121–127.

Sanders, G. P. 1939. The determination of chloride in milk. *J. Dairy Sci. 22,* 841–851.

Sanguansermsri, J., Gyorgy, P. and Zilliken, F. 1974. Polyamines in human and cow's milk. *Am. J. Clin. Nutr. 27,* 859–865.

Scanlan, R. A., Lindsay, R. C., Libbey, L. M. and Day, E. A. 1968. Heat-induced volatile compounds in milk. *J. Dairy Sci. 51,* 1001–1007.

Schalm, D. W., Carroll, E. J. and Jain, N. C., 1971. *Bovine Mastitis.* Lea & Febiger, Philadelphia.

Schwartz, D. P. and Pallansch, M. J. 1962A. Identification of some nitrogenous constituents of cow's milk by ion exchange and paper chromatrography. *J. Ag. Food Chem. 10,* 86–89.

Schwartz, D. P. and Pallansch, M. J. 1962B. Occurrence of phenylacetyl-glutamine in cow's milk. *Nature (Lond.) 194,* 186.

Sharp, P. F. and Struble, E. B. 1935. Period of lactation and the direct titratable chloride value of milk. *J. Dairy Sci. 18,* 527–538.

Smith, A. C. 1964. The carbon dioxide content of milk during handling, processing and storage and its effect upon the freezing point. *J. Milk Food Tech. 27,* 38–41.

Snelson, J. J. (Editor). 1979. *Chemical Residues in Milk and Milk Products.* International Dairy Federation Bulletin, Document 113, 1–69.

Snoswell, A. M. and Linzell, J. L. 1975. Carnitine secretion into milk of ruminants. *J. Dairy Res. 42,* 371–380.

Spinelli, F. 1946. Indican in cow and goat milks. *Boll. Soc. Ital. Sper. 21,* 210–211.

Sumner, J. B. 1944. A method for the colorimetric determination of phosphorus. *Science 100,* 413.

Swaisgood, H. (Editor). 1975. *Methods of Gel Electrophoresis of Milk Proteins.* Am. Dairy Sci. Assn. Champaign, Ill.

Swope, F. C. and Brunner, J. R. 1965. Identification of ribonucleic acid in the fat globule membrane. *J. Dairy Sci. 48,* 1705–1707.

Swope, F. C., Brunner, J. R. and Vadhera, D. V. 1965. Riboflavin and its natural derivatives in the fat-globule membrane. *J. Dairy Sci. 48,* 1707–1708.

Tamsma, A. and Tarassuk, N. P., 1957. Removal of oxygen from evaporated milk with glucose oxidase. *J. Dairy Sci. 40,* 1181–1188.

Thomas, P. C. 1980. Influence of nutrition on the yield and content of protein in milk: Dietary protein and energy supply. *In: Factors Affecting the Yields and Content of*

Milk Constituents of Commercial Importance. J. H. Moore and J. A. F. Rook (Editors). Internat. Dairy Fed. Doc. 125.

Tocher, J. F. 1925. *Variations in the Composition of Milk.* H. M. Stationery Office, London.

Touchberry, R. W. 1974. Environmental and genetic factors in the development and maintenance of lactation. *In: Lactation: A Comprehensive Treatise,* Vol. 3. B.L. Larson and V. R. Smith (Editors). Academic Press, New York.

Underwood, E. J. 1977. *Trace Elements in Human and Animal Nutrition,* 4th ed. Academic Press, New York.

U.S. Dept. Agr. 1980. *Federal and State Standards for the Composition of Milk Products.* Agr. Handbook 51. Government Printing Office, Washington, D.C.

van der Have, A. J., Deen, J. R. and Mulder, H. 1979. The composition of cow's milk. 1. The composition of separate milkings of individual cows. *Neth. Milk Dairy J. 33,* 65–81.

Virtanen, A. I. 1966. Milk production of cows on protein-free feed. *Science 153,* 1603–1614.

Vogel, J. and Deshusses, J. 1965. Polarographic estimation of small amounts of benzoic acid: The application of the method for the estimation of benzoic acid in normal milk and yoghurt. *Mitt. Geb. Lebensmitt. Untersuch Hyg. 56,* 63–67. *(Dairy Sci. Abstr. 27,* 2944, 1965.)

Waite, R. 1961. A note on a method of overcoming the effect of udder disease or injury in experiments involving milk yield and composition. *J. Dairy Res. 28,* 75–79.

Waite, R. and Patterson, J. A. 1959. The composition of bulk milk measured on Scottish milk-recorded farms. *J. Soc. Dairy Technol. 12,* 117–122.

Walstra, P. and Jenness, R. 1984. *Dairy Chemistry and Physics.* John Wiley & Sons, New York.

Walstra, P. and Mulder, H. 1964. Gravimetric methods for the determination of the fat content of milk and milk products. V. Comparison of methods. *Neth. Milk Dairy J. 18,* 237–242.

Wayman, O., Johnson, H. D., Merilan, C. P. and Berry, I. L. 1962. Effect of ad libitum or force-feeding of two rations on lactating dairy cows subject to temperature stress. *J. Dairy Sci. 45,* 1472–1478.

Wenner, V. R. 1958. Rapid determination of milk salts and ions. I. Determination of sodium, potassium, magnesium and calcium by flame spectrophotometry. *J. Dairy Sci. 41,* 761–768.

West, D. W. and Towers, G. E. 1976. Cellulose acetate electrophoresis of casein proteins. *Anal. Biochem. 75,* 58–66.

Wheelock, J. V. 1980. Influence of physiological factors on the yields and contents of milk constituents. *In: Factors Affecting the Yields and Contents of Milk Constituents of Commercial Importance.* J. H. Moore and J. A. F. Rook (Editors). Internat. Dairy Fed. Doc. 125.

Wheelock, J. V., Rook, J. A. F. and Dodd, F. H. 1965. The effect of milking throughout the whole of pregnancy on the composition of cow's milk. *J. Dairy Res. 32,* 249–254.

White, J. C. D. and Davies, D. T. 1958. The relation between the chemical composition of milk and the stability of the caseinate complex. I. General introduction, description of samples, methods and chemical composition of samples. *J. Dairy Res. 25,* 236–255.

White, J. C. D. and Davies, D. T. 1962. The determination of calcium and magnesium in milk and milk diffusate. *J. Dairy Res. 29,* 285–296.

White, J. C. D. and Davies, D. T. 1963. The determination of citric acid in milk and milk sera. *J. Dairy Res. 30,* 171–189.

Whittlestone, W. G. 1953. Variations in the fat content of milk throughout the milking process. *J. Dairy Res. 20*, 146–153.

Wilcox, C. J., Pfau, K. O., Mather, R. E. and Bartlett, J. W. 1959. Genetic and environmental influences upon solids-not-fat content of cow's milk. *J. Dairy Sci. 42*, 1132–1146.

Williams, K. I. H., Burstein, S. H. and Layne, D. S. 1966. Dimethyl sulfone: Isolation from cows' milk. *Proc. Soc. Exp. Biol. Med. 122*, 865–866.

Wolfschoon-Pombo, A. and Klostermeyer, H. 1981. The NPN-fraction of cow milk I. Amount and composition. *Milchwissenschaft 36*, 598–600.

Wong, N. P. and Patton, S. 1962. Identification of some volatile compounds related to the flavor of milk and cream. *J. Dairy Sci. 45*, 724–728.

Wrenn, T. R., Bitman, J., Cecil, H. C. and Gilliam, D. R. 1963. Histamine concentration in blood, milk and urine of dairy cattle. *J. Dairy Sci. 46*, 1243–1245.

Zarembski, P. M. and Hodgkinson, A. 1962. The determination of oxalic acid in food. *Analyst 87*, 698–702.

Zulak, I. M., Patton, S. and Hammerstedt, R. H. 1976. Adenosine triphosphate in milk. *J. Dairy Sci. 59*, 1388–1391.

2

Composition of Milk Products

Richard Bassette and Judith S. Acosta

Raw milk is a unique agricultural commodity. It contains emulsified globular lipids and colloidally dispersed proteins that may be easily modified, concentrated, or separated in relatively pure form from lactose and various salts that are in true solution. With these physical-chemical properties, an array of milk products and dairy-derived functional food ingredients has been developed and manufactured. Some, like cheese, butter, and certain fermented dairy foods, were developed in antiquity. Other dairy foods, like nonfat dry milk, ice cream, casein, and whey derivatives, are relatively recent products of science and technology. This chapter describes and explains the composition of traditional milk products, as well as that of some of the more recently developed or modified milk products designed to be competitive in the modern food industry.

Although many newly developed dairy products have been reported from research laboratories around the world, only those currently on the market are discussed here. For additional information on recent worldwide developments in the manufacture of new and modified dairy products, see the proceedings of the Twentieth International Dairy Congress (1978).

Milk products are manufactured from fluid milk by various methods: (1) by removing an appreciable amount of water, as in condensed and evaporated milk or dry milk powder; (2) by removing one or more natural constituents and concentrating the remaining material, as in butter and nonfat dry milk; (3) by altering the natural constituents by bacterial or chemical action, as in cheese or fermented foods; in most cheeses, the casein/fat ratio remains essentially the same in the cheese as in the milk, while lactose disappears; and (4) by blending milk and milk products with sugar, flavoring agents, and stabilizers to make ice cream and ice milk.

Traditional dairy products are fairly common throughout the world, and their compositions are basically the same from one country to another. International standards proposed by the International Dairy

Federation, the Food and Agriculture Organization of the United Nations, and the World Health Organization have done much to standardize dairy products throughout the world. Some dairy foods, such as certain fermented milks and cheeses, are indigenous to particular countries or areas of the world. In the United States, the composition of most dairy products is regulated by federal and state standards (USDA 1981B). Where there are no federal standards regulating interstate commerce for a particular product, state standards prevail. Typical chemical analyses of dairy foods are given in Tables 2.2 through 2.8.

Table 2.1 lists the approximate percentages of the total milk supply used for various products in the United States and in nine major milk-producing countries. In such countries as New Zealand and Ireland, where per capita production of milk is high, most milk is used in storable manufactured products like butter, cheese, and nonfat dry milk. Where per capita production is low, as in the United States and the United Kingdom, greater amounts are used as fluid milk and creams.

FLUID MILKS AND CREAMS

Milk and milk products purchased by the consumer in liquid or semi-liquid form generally are classified as fluid milk or cream. Fluid milks include all of the plain milk products, with fat contents varying from those of whole to skim milk, as well as flavored and fermented milks. Creams include products varying in fat content from half and half to heavy whipping cream to fermented sour cream. Products from each category are described briefly, with information on their composition.

In the United States, the composition of fluid milk products and cream is regulated primarily by state and federal standards. (See Table 2.2 for these legal standards.) Sanitary quality is regulated by sanitary codes established by states and local health departments. The basis for most U.S. codes is the Grade A Pasteurized Milk Ordinances published by the U.S. Department of Health, Education and Welfare Public Health Service (1978) and recommended to states and local health agencies for legal adoption. The recommended ordinance also regulates grade A milk and milk products in interstate commerce. Approximately 40% of the milk produced in the United States is consumed as fluid milk and creams (Table 2.1).

Plain Milks

Fresh raw milk that complies with sanitary standards for consumption as fluid milk or cream is usually clarified, standardized to a certain fat test, or separated into cream and skim milk.

Table 2.1. Production and Per Capita Consumption of Fluid Milk, Cheese, Butter, and Nonfat Dry Milk in Selected Countries.

	Production of Milk and Milk Products						Per Capita Consumption [a]			
	Billions of Pounds	Per Capita	Fluid [a,b,d] Milk (%)	Cheese [c,d] (%)	Butter [c,d] (%)	Nonfat [c,d] Dry Milk (%)	Fluid Milk (lb)	Cheese (lb)	Butter (lb)	Nonfat Dry Milk (lb)
United States	128.4	581	39.7	30.7	18.8	9.9	231	17.1	4.1	3.0
Canada	17.5	716	39.4	22.5	27.3	15.6	282	16.5	9.7	4.5
Denmark	11.3	2200	10.6	41.4	47.5	8.0	233	20.2	23.6	13.3
Finland	7.2	1513	35.5	21.3	47.3	20.1	538	16.6	27.7	24.9
France	74.1	1443	16.8	26.7	31.0	20.0	243	38.6	21.1	26.3
Ireland	10.6	3207	15.4	10.1	52.9	32.1	494	6.0	26.7	6.7
Netherlands	26.0	1847	17.3	37.6	37.3	14.3	319	27.4	10.9	39.1
New Zealand	15.1	4652	9.6	15.5	81.7	27.7	447	19.1	29.3	2.7
United Kingdom	35.7	639	47.4	14.3	23.1	16.4	303	12.9	14.3	5.3
Soviet Union	199.5	751	27.5	7.6	31.6	4.1	207	5.8	11.7	3.1

SOURCE: Milk Industry Foundation, Milk Facts (1981), USDA (1981D).
[a] Fluid milk and cream based upon whole milk equivalent (fat solids basis).
[b] Percent of fluid milk calculated from per capita consumption/per capita production × 100 (USDA, 1981D).
[c] Calculated from metric tons produced and converted to pounds milk equivalent with factors 21.2 lb of milk/pound of butter, 10 lb/lb of cheese and 11 lb/lb of NFDM. Percent of production calculated by dividing total pounds of milk product by total pounds of milk produced × 100 (Milk Industry Foundation, 1981).
[d] Production figures do not total 100% because other milk products (frozen desserts, condensed, etc.) are not included, and there is an overlap of milk used for both butter and nonfat dry milk, as well as whey butter from cheese.

41

Table 2.2. U.S. Federal Standards for Fluid Milk Products.

	Milkfat		Solids TS		Vitamin A (IU/qt)	Vitamin D (IU/qt)	Stabilizer		Titratable Acidity[a] (Min. %)
	Min. (%)	Max. (%)	Min. (%)	Max. (%)			% Min.	% Max.	
Whole milk	3.25		8.25		2000[b]	400[b,c]			
Lowfat milk	0.5	2.0	8.25		2000	400[b]			
Skim milk	<0.5		8.25		2000	400[b,c]			
Acidified milk[d]	3.25		8.25		2000[b]	400[b]			0.5
Cultured milk	3.25		8.25		2000[b]	400[b]			0.5
Acidified skim milk[d]	<0.5		8.25		2000[b]	400[b]			0.5
Cultured skim milk	<0.5		8.25		2000[b]	400[b]			0.5
Yogurt	3.25		8.25						0.9
Lowfat yogurt	0.5	2.0	8.25						0.9
Nonfat yogurt	0.5		8.25						0.9
Egg nog[e]	6.0								
Light cream	18.0	30.0							
Light whipping cream	30.0	36.0							
Heavy cream	36.0	>36.0							
Sour cream[f]	18.0						0.1	0.5	
Acidified sour cream	18.0						0.1	0.5	
Half and half	10.5	18.0							
Sour half and half	10.5	18.0							0.5
(Acidified sour half and half)	10.5	18.0							0.5

SOURCE: FDA (1981A), USDA (1981B).

[a]Expressed as lactic.
[b]Optional, but when added, not less than the quantity shown.
[c]Quantity shown is amount specified when added. Federal labeling laws must be followed.
[d]Contains one or more optional acidifying ingredients: adipic acid, citric acid, fumaric acid, glucono-delta lactone, hydrochloric acid, lactic acid, malic acid, phosphoric acid, succinic acid, and tartaric acid.
[e]Not less than 1% egg yolk solids.
[f]If sweetener or flavoring is added, the weight of milkfat is not >18%; never <14.4% milk fat.

Whole Milk.   Most fluid milk is consumed in the form of pasteurized, homogenized, vitamin D–fortified whole milk. After standardization of the milk fat, which may vary from 3.0 to 3.8% (usually 3.25% in the United States), the milk is pasteurized, homogenized, packaged, and stored under refrigeration until sold. Its shelf life, as well as that of most other fluid milk products, is 10 to 14 days. The milk solids-not-fat content of 8.25% is required by most states in the United States, as well as in most other countries.

Lowfat Milks.   Per capita consumption of lowfat and skim milk has increased substantially over the past decade. In the United States it represented almost 30% of the total fluid milk consumed in 1980. Milk with the fat content reduced below that of whole milk falls into the general category of lowfat or skim milk. Most lowfat milks contain a designated amount of fat between 0.5 and 2.5%; frequently, 1 to 2% additional milk solids with vitamin D are added. The milk is pasteurized, homogenized, packaged, and refrigerated until sold. In the United States most states allow lowfat milks with fat contents of 0.5, 1.0, 1.5, or 2.0% but require that the percentage be shown on the label.

Skim Milk.   After all or most of the milk fat is removed from whole milk by continuous centrifugal separation, the resulting skim milk is fortified with 2000 International Units (IU) of vitamin A per quart and often with additional milk solids and vitamins, then pasteurized, packaged, and refrigerated until sold. The addition of vitamin D is optional but, when added, it must be not less than 400 IU/quart, and this must be shown on the label.

Low-Sodium Milk (Hargrove and Alford 1974).   Low-Sodium milk is available in some areas as a specialty product for consumers who require low-sodium foods. It is produced by passing normal milk over an ion-exchange resin which replaces the sodium of the milk with potassium. The normal sodium content of milk is reduced from 50 mg/100 ml to approximately 3 mg/100 ml; other components of the milk remain essentially the same.

Ultra-High-Temperature (UHT) Sterile Milk.   Rapid increases in the production and sales of sterile, fluid milk in Europe and its entry into new markets around the world merit its consideration in this chapter. Although the gross composition of UHT sterile, aseptically packaged milk is essentially the same as that of its pasteurized counterpart, differences in its properties and minor constituents should be mentioned.

UHT milk differs from pasteurized milk mainly in the heat treatment employed for sterilization. Usually UHT milk is heated at 130° to 150°C for 2 to 8 seconds and is then aseptically packaged. In the final heating stage, steam is injected directly into the milk, or the milk is infused into a steam chamber, followed by flash evaporation to remove added water (steam). An alternative procedure, the indirect method, involves heating milk across a stainless steel barrier, using high-pressure steam as the heating medium (Mehta 1980).

Probably the most important difference between UHT and pasteurized milk is flavor. UHT milk has an intensely cooked flavor immediately after processing that dissipates in about 1 week; a stale flavor develops 3 to 4 weeks after the milk is processed and becomes progressively worse. Not all researchers agree on the intensity and significance of the flavor of UHT milk (Anon. 1981B) and the many factors that influence it (Mehta 1980).

The other difference that has been observed is some alteration in minor chemical and biochemical components. Burton (1969) reported that, in general, vitamins are more stable under UHT processing than with pasteurization; however, UHT milk loses significant amounts of riboflavin and ascorbic acid during prolonged storage (Mehta 1980). Although free calcium is reduced, the availability of calcium does not change in UHT milk processed by the indirect method (Mehta 1980). According to Hansen and Melo (1977), cysteine and cystine (as cysteic acid) and methionine concentrations were reduced by about 34% by UHT processing (Aboshana and Hansen 1977). There is a decrease (<10%) in the chemically available lysine in whey protein and an insignificant difference in available lysine in casein protein subjected to UHT processing (Douglas *et al.* 1981). However, a study in Holland reported that 200 infants drinking UHT milk gained 7 g more per day than 200 infants drinking pasteurized milk. Also, the normal weight loss immediately after birth was regained sooner by the UHT-fed infants (Anon. 1979). For additional information on the chemical and biochemical changes associated with UHT sterile milk or the processing procedures, the reviews by Burton (1969, 1977) and Mehta (1980) are recommended.

A ruling by the U.S. Food and Drug Administration to approve the use of hydrogen peroxide and heat as sterilizing agents for aseptic packaging has encouraged some U.S. dairy companies to enter the UHT market (Anon. 1981A). Resistance to this investment among other dairy industry leaders stems from concerns about flavor stability, economics, and package size (liter-size containers) of UHT milk.

FLAVORED FLUID MILK PRODUCTS

Fluid milk and fluid milk products may be flavored with such ingredients as chocolate, vanilla, eggnog, and fruit juices. In the United States, all of the previously described fluid milk products (milk, lowfat milk, and skim milk) may be flavored. Characteristic flavoring ingredients such as fruit and fruit juices, natural and artificial food flavorings with or without coloring, nutritive sweeteners, emulsifiers, and stabilizers may be added as optional ingredients (FDA 1981A).

The most popular flavored milk or milk drink in the United States is chocolate milk or chocolate lowfat milk. Typically, chocolate milk contains about 1% cocoa, 6% sucrose, and 0.2% stabilizer such as vegetable gum, vanilla, and salt, all added to whole milk. Particular attention must be given to stabilizing the chocolate flavoring ingredients against sedimentation.

Eggnog is a flavored dairy drink with seasonally maximum sales in November and December in the United States. It must contain 6% milk fat and 1% egg yolk, with up to 0.5% stabilizer and about 7% sugar. Flavorings include nutmeg, cinnamon, vanilla, and rum concentrate.

FERMENTED AND ACIDIFIED MILKS

Fermented milks are cultured dairy products manufactured from whole, partly skimmed, skim, or slightly concentrated milk. Specific lactic acid bacteria or food-grade acids are required to develop the characteristic flavor and texture of these beverages. Fermented milks are either fluid or semifluid in consistency, with various proportions of lactic acid. Fermented products are regulated by federal standards in the United States, as stated in Table 2.2. Other fermented milks without established federal standards are regulated by state standards. Compositional standards for fermented milks have been proposed by the International Dairy Federation (Hargrove and Alford 1974). Typical analyses of various fermented milks, as well as of their condensed and dried counterparts, are given in Table 2.4.

U.S. federal standards have recently been established for several acidified fluid milk products that simulate such cultured products as acidified milk, acidified lowfat milk, and acidified skim milk (FDA 1981A).

Acidified milks are made by souring the product with one or more acidifying ingredients, with or without the addition of characterizing microorganisms. Specified acidulants are food-grade citric acid, fumaric acid, glucono-delta-lactone, hydrochloric acid, lactic acid, malic acid, phosphoric acid, succinic acid, and tartaric acid.

Buttermilk

Cultured buttermilk is manufactured by fermenting whole milk, reconstituted nonfat dry milk, partly skimmed milk, or skim milk with lactic acid bacteria. Most commercial cultured buttermilk is made from skim milk. Mixed strains of lactic streptococci are used to produce lactic acid and leuconostocs for development of the characteristic diacetyl flavor and aroma. Buttermilk is similar to skim milk in composition, except that it contains about 0.9% total acid expressed as lactic acid. The percentage of lactose normally found in skim milk is reduced in proportion to the percentage of lactic acid in the buttermilk. According to White (1978), the fat content of buttermilk usually varies from 1 to 1.8%, sometimes in the form of small flakes or granules to simulate churned buttermilk, the by-product of butter churning. Usually 0.1% salt is added.

A few U.S. states require buttermilk to be labeled as cultured wholemilk buttermilk with a minimum of 3.5% milk fat or as cultured lowfat milk with a minimum of 0.5% and a maximum of 2.0% milk fat, or as cultured skim milk with a maximum of 0.5% milkfat. When buttermilk is made with low-heat powder, higher solids (10%) usually are used to give a firmer body with less shrinkage and less whey separation during storage (White 1978).

A product similar to cultured buttermilk may be prepared by direct acidification. Food-grade acids and acid anhydrides are added to unfermented milk to obtain a product with uniform acidity and smooth body. Flavoring materials are used to improve the flavor and aroma.

Bulgarian buttermilk is similar to cultured buttermilk, except that the whole or partly skimmed milk is fermented by *Lactobacillus bulgaricus*. With a titratable acidity of 1.2 to 1.5% expressed as lactic acid, it is more acidic than cultured buttermilk.

Sour Cream

Sour cream is cream that has been soured by lactic acid bacteria or by directly adding food-grade acids. According to U.S. federal standards, both sour cream and acidified sour cream must contain not less than 18% milk fat with a titratable acidity of not less than 0.5% expressed as lactic acid (FDA 1981A). Optional ingredients are used to improve texture, prevent syneresis, and extend shelf life. Sour cream may contain not more than 0.1% sodium citrate, salt, rennet, nutritive sweeteners, flavoring, or coloring.

Sour cream is prepared by mixing milk fat and skim milk, nonfat solids, and other ingredients. The mix is then pasteurized, homogenized, and cooled to the setting temperature; starter and enzymes are then added, and the cream is allowed to ripen. After ripening, it is cooled for 24 hr before packaging.

In direct acidification, the cream is pasteurized, homogenized, and cooled to setting temperature, and the food-grade acid is added. Then cream is packaged and the characteristic body is formed as the product cools in the container (Schanback 1977).

Acidophilus Milk

Acidophilus milk is a sharp, harsh, acidic cultured milk produced by fermenting whole or skim milk with active cultures of *Lactobacillus acidophilus*. Honey, glucose, and tomato juice may be added as nutrients to stimulate bacterial growth and contribute flavor. Plain acidophilus milk has the same composition as whole milk or skim milk, except that part of its lactose is converted to 0.6 to 1% lactic acid by the culture organisms. Speck (1976), who proposed the addition of *L. acidophilus* to pasteurized milk (sweet acidophilus milk), described the beneficial effects of implanting the organisms in the human intestines.

Sweet acidophilus milk differs from conventional acidophilus milk in that a high concentration of viable *L. acidophilus* organisms is added to cold pasteurized milk and kept cold. At the low storage temperature (4.4°C) these organisms do not multiply, so the flavor and other properties of sweet acidophilus are identical to fresh fluid milk. The inoculated milk is promoted largely because it contains several million viable *L. acidophilus* cells per milliliter.

One of the acidophilus products, called "Di-gest," is a pasteurized, homogenized, lowfat milk with added *L. acidophilus* and fortified with vitamins A and D. In the United States an "acidophilus yogurt" flavored in the conventional manner is also manufactured. Denmark has a cultured product consisting of 90% normal yogurt and 10% acidophilus, and the Soviet Union produces "Biolact," a product particularly suitable for children. According to Lang and Lang (1978), it is made with selected cultures of *L. acidophilus* with high proteolytic activity and "antibiotic properties."

A thick, milky, white-to-creamy coagulum with a pleasant lactic acid odor and a refreshing, clean, aromatic taste is produced in northern

Bohemia. This product, after ripening, has organoleptic qualities similar to those of kefir. It is produced with two different cultures, and the incubation temperatures with these coagula are subsequently blended; one culture of *L. acidophilus* ferments milk to produce a thick, typically sharp, acid-tasting coagulum with an acidity of 1.9 to 2.3% lactic acid; the other culture, identified as "strains of cream cultures," yields a thick, aromatic coagulum with a lactic-acid flavor and an acidity of 0.8 to 0.9% lactic acid from a culture cream. The two coagulated products are mixed in the following proportions: one part of the thick cream culture to nine parts of the acidic *L. acidophilus* culture. The similarity to kefir probably occurs because the ratio of streptococci to lactobacilli is similar to that of kefir; however, the ripening process is simpler (Lang and Lang 1978).

Yogurt

Yogurt is a fermented milk product made by culturing whole or partly defatted milk to which either nonfat dry milk solids or a skim milk concentrate has been added. Its texture may vary from a rennet-like custard to a creamy, highly viscous liquid, depending on the milk solids and fat content. A mixed culture of *Streptococcus thermophilus* and *L. bulgaricus* growing together symbiotically produces its approximately 0.9% lactic acid and the characteristic yogurt flavor.

Keogh (1970) reported that yogurt-type products (yaaurt, jugurt, yeart, yaoert, yogurt, yahourt, and yourt) have been made for centuries, originating in countries on the eastern Mediterranean. Similar products, such as leben of Egypt, madzoon of Armenia, and dahi of India, all are fermented by *L. bulgaricus* and *S. thermophilus*. Turkish yogurt differs only in that a lactose-fermenting yeast is included in the culture. Interest in yogurt in the United States is recent. According to the Milk Industry Foundation, *Milk Facts* (1981), from 1969 to 1979 per capita sales of yogurt in the United States increased by 211%, and yogurt now represents about 1% of fluid milk sales.

A wide variety of yogurt-type products have found their way to supermarket shelves. In addition to plain or natural yogurt, the following products are marketed in the United States (Tamine and Deeth 1980):

Fruit yogurts—both Swiss-style, with fruit, flavoring, and color uniformly distributed, and sundae-style, with fruit in the bottom of the cup and yogurt on the top.
Pasteurized/UHT yogurt—heat treated after incubation for longer shelf life.

Concentrated (frozen) yogurt—resembles either soft or hard ice cream.

Dried yogurt—produced by sun-, spray-, or freeze-drying yogurt.

Low-calorie yogurt—contains 9% solids-not-fat, 0.1% fat, and 0.5–1% stabilizer.

Low-lactose yogurt—made with β-D-galactosidase, with lactose hydrolyzed for a sweeter product with no added sugar:

Kefir Cultured Milk

Kefir is a self-carbonated beverage popular in the Soviet Union, Poland, Germany, and other European countries in plain and flavored forms (Kosikowski 1978B). Made with whole, part skim, or skim milk, it contains about 1% lactic acid and 1% alcohol. Kefir exists in various forms: whole milk-, cream-, skim milk-, whey-, acidophilus-, pepsin-, grape-sugar-, and fruit-flavored kefir. Kefir buttermilk is a kefir-like product that contains less CO_2 and alcohol than normal kefir.

Basically, kefir is made with the fermenting agent called "kefir grains," which consists of casein and gelatinous colonies of microorganisms growing together symbiotically. The dominant microflora of kefir consist of *Saccharomyces kefir, Torula kefir, Lactobacillus caucasicus, Leuconostoc* species, and lactic streptococci. The microbial population is 5 to 10% yeast. Often the surface of kefir grains is covered with the white mold, *Geotrichum candidum*, which apparently does not detract much from its quality (Kosikowski, 1978A). Kefir milk differs in composition from the original milk as some of the lactose is converted to lactic acid, alcohol, and carbon dioxide. It has a definite yeasty aroma with limited proteolysis in the milk. The taste of kefir differs markedly from that of yogurt (Kosikowski. 1978B). The production of lactic acid, which is accompanied by the production of alcohol and CO_2, may be regulated by the incubation temperature.

Kumiss

Kumiss (koumiss, kymys) is an effervescent lactic acid-alcoholic fermented milk similar to kefir. It originated in the asiatic steppes and is traditionally made from mare's milk. Due to a shortage of mare's milk in the Soviet Union, large quantities of kumiss are made from cow's milk, so it differs little from kefir. According to Keogh (1970), it is enjoying a reputation comparable to that of yogurt at the time of Metchnikoff. Starter organisms used for its manufacture are *L. bulgaricus, L. acidophilus,* and *Saccharomyces lactis,* a lactose-fermenting yeast. Puhan and Gallman (1980) reported that a modified kumiss can be made with cow's milk diluted 20 to 25% with whey or water to adjust the protein concentration to approximately that of mare's milk; 2–3% glucose or sucrose is added, and the mixture is pasteurized at 90–96°C and cooled to 45°C. This modified milk is fermented at 37°C with 3–5% added culture. Four to 8 hr later, 0.5% acid is developed; the milk is then cooled and stirred at 30°C while 2.5 g of yeast per liter is added. The product is incubated for 4 to 16 hr depending on the alcohol concentration desired. Traditional kumiss from mare's milk contains about 2.5% alcohol and approximately 1% lactic acid. Any movement of the protein particles in milk produces an unstable product because CO_2 escapes during the fermentation. An unsatisfactory fermentation can be improved by hydrolyzing lactose with a β-galactosidase.

CONCENTRATED FERMENTED MILKS

Danish Ymer and Swedish Lactofil are very soft, white, fermented milk products, smooth and light, with a mild aromatic and acidic flavor. They are made from whole or skim milk and are used for desserts topped with fruits and in salads and dips. Ymer and Lactofil are made by fermenting the milk with lactic acid cultures, including *Streptococcos lactis* subspecies *diacetilactis* and *Leuconostoc* species, to form a curd. After coagulation, the curd is cooked moderately until 55% of the whey is removed. Then cream is added and the mass is homogenized to a smooth, creamy consistency before cooling. The products contain 3% fat, 7% protein, and 12% solids-not-fat (Lang and Lang 1978).

FLUID CREAM

Several types of fluid creams are manufactured and sold directly to consumers. The most significant difference in the creams is the level of milk fat they contain: from as little as 10% in half and half to 40% in whipping cream. All commercial creams are produced by centrifugally separating the less dense, higher-fat products from the residual skim milk. Most often, cream with 40% fat is separated and then standardized with skim milk to give creams with the desired fat contents. In some instances, creams with desired fat contents are collected directly from the separator; in others, as in "plastic cream" (made for manufacturing purposes), a 40% cream is pasteurized and reseparated while hot to yield a product with 80% fat.

Fluid creams for consumer markets throughout the world fall into certain classes, depending upon their fat content: 10 to 12%, 18 to 20%, 25 to 30%, 34 to 36%, and 48%, although not all countries market all types. Their names are not always the same; for example, a product with 20 to 30% fat is designated "medium cream" in the United States but "reduced cream" in Australia. The term "cream" officially varies from the 18% Food and Agriculture Organization (FAO) standard to 40% (New Zealand) in its content of milk fat.

Half and Half

The composition and properties of particular types of creams depend upon their intended use. Half and half, with 10 to 12% fat, is used as a coffee whitener and cereal cream. It may have additional milk solids and a stabilizer added. Usually half and half is homogenized and either pasteurized or ultrapasteurized for longer shelf life. In some countries it is sterilized. Most states in the United States require it to contain a minimum of 10.5% and a maximum of 18% milk fat.

Table Creams

Table or coffee creams are those of intermediate fat content. In the United States they are classified as "light" (18% fat) and "medium" (30% fat), whereas in the United Kingdom they are designated "cream" (20% fat) and "sterilized cream" (23% fat). Other than sterilized creams, the table creams are standardized to the desired fat test, pasteurized, and packaged. To extend their shelf life, sterilized creams, as well as some table creams that are ultrapasteurized, are aseptically packaged.

Whipping Cream

Whipping cream varies in fat content from 30% for light whipping cream to 36% for heavy whipping cream. It is usually processed to increase its viscosity and thickness and to enhance its whipping ability. Increasing the fat content, aging the cream, and adding nonfat dry milk solids will improve its whipping ability.

Standards for creams in the United States are presented in Table 2.2 and typical compositions of these products in Table 2.3.

CONCENTRATED MILK PRODUCTS

Whole milk, skim milk, and buttermilk are concentrated by removal of water and may be preserved by heat, addition of sugar, or refrigeration. Typical analyses of these products are given in Table 2.4.

Table 2.3. Typical Composition of Market Creams, Butter, and Frozen Desserts.

	Moisture (%)	Protein (%)	Total Fat (%)	Total Carbo-hydrate (%)	Ash (%)	Calcium (%)	Phosphorus (%)	Sodium (%)
Market creams								
Fluids: Half and half (milk and cream)	80.6	3.0	11.5	4.3	0.7	0.10	0.09	0.04
Light, coffee or table	73.7	2.7	19.3	3.7	0.6	0.10	0.08	0.04
Medium, 25% fat	68.5	2.5	25.0	3.5	0.5	0.10	0.07	0.04
Light whipping	63.5	2.2	30.9	3.0	0.5	0.07	0.06	0.03
Heavy whipping	57.7	2.0	37.0	2.8	0.4	0.06	0.06	0.04
Whipped: Cream topping, pasteurized								
Sour half and half	61.3	3.2	22.2	12.5	0.8	0.10	0.09	0.13
(cultured)	80.1	2.9	12.0	4.3	0.7	0.10	0.09	0.04
Butter and butter oil								
Regular or whipped	15.9	0.85	81.1	0.06	2.1			
Butter oil, anydrous	0.2	0.3	99.5	0.0	0.0			
Ghee	0.1	0.1	99.8	0.0	0.0	0.02	0.02	0.83
Frozen desserts								
Ice cream, vanilla hardened Regular (10% fat)	60.8	3.6	10.8	23.8	1.0	0.13	0.10	0.09
Rich (16% fat)	58.9	2.8	16.0	21.6	0.7	0.10	0.08	0.07
Ice cream, French vanilla, soft served	59.8	4.0	13.0	22.1	1.0	0.14	0.11	0.09
Ice milk, vanilla Hardened	68.6	3.9	4.3	22.1	1.0	0.13	0.10	0.08
Soft served	69.6	4.6	2.6	21.9	1.2	0.16	0.11	0.09
Sherbet, orange	66.1	1.1	2.0	30.4	0.4	0.05	0.04	0.05

SOURCE: Hargrove and Alford (1974), USDA (1981C).

Table 2.4. Percent Composition of Concentrated Milk and Dried Products.

Milk Products	Moisture (%)	Protein (%)	Total Fat (%)	Total Carbo- hydrate (%)	Ash (%)	Calcium (%)	Phosphorus (%)	Sodium (%)	Potassium (%)	Lactic Acid (%)
Concentrated										
Evaporated milk										
Whole	74.0	6.8	7.6	10.0	1.5	0.26	0.20	0.10	0.30	0
Skim	79.4	7.5	0.2	11.3	1.5	0.29	0.19	0.11	0.33	0
Sweetened condensed										
Whole	27.1	7.9	8.7	54.4	1.8	0.30	0.25	0.13	0.37	0
Plain condensed skim	73.0	10.0	0.3	14.7	2.3	0.25	0.20			0
Sweetened condensed skim	28.4	10.0	0.3	58.3	2.3	0.30	0.23			0
Condensed buttermilk (acid)	72.0	9.9	1.5	12.0	2.2					5.7
Condensed skim (acid)	72.0	10.2	0.2	9.4	2.1					6.08
Condensed whey	48.1	7.0	2.4	38.5	4.0					2.4
Sweetened condensed whey	24.0	5.0	1.7	66.5	2.8					0
Dried										
Whole	2.5	26.3	26.7	38.4	6.1	0.91	0.80	0.40	1.33	0
Nonfat										
Regular	3.2	36.2	0.8	52.0	7.9	1.26	0.97	0.53	1.79	0
Instantized	4.0	35.1	0.7	52.2	8.0	1.23	0.99	0.55	1.70	0
Buttermilk (sweet cream)	3.0	34.3	5.8	49.0	7.9	1.18	0.93	0.52	1.59	0
Buttermilk (acid)	4.8	37.6	5.7	38.8	7.4					5.7
Malted milk										
Natural flavor (powder)	2.6	13.1	8.5	72.5	3.4	0.27	0.37	0.46	0.76	0
Chocolate flavor (powder)	2.0	6.5	4.5	84.9	2.1	0.06	0.18	0.23	0.62	0
Cream	0.8	65.0	13.4	18.0	2.9					0
Whey (acid) cottage	3.5	11.7	0.5	73.4	10.8	2.05	1.35	0.97	2.29	8.6
Whey (sweet) cheddar	3.2	12.9	1.1	74.5	8.3	0.80	0.93	1.08	2.08	2.3
Casein (commercial)	7.0	88.5	0.2	0	3.8					
Casein (coprecipitate)	4.0	83.0	1.5	1.0	10.5	2.5				

SOURCE: Hargrove and Alford (1974), Posati and Orr (1976).

Evaporated Milk

This product is made by evaporation of water from whole milk under vacuum. Low percentages of sodium phosphate, sodium citrate, calcium chloride, and/or carageenan may be added to improve its stability. The concentrate is homogenized, canned, and then sterilized under pressure at 117°C for 15 min or at 126°C for 2 min. Ultra-high temperatures (130 to 150°C for a few seconds), followed by aseptic packaging, have been used with some success but have found limited commercial application.

U.S. standards of identity require that evaporated milk contain not less than 7.5% milk fat and 25% total milk solids. In addition, it must contain 25 IU of vitamin D per fluid ounce. Addition of vitamin A is optional; if added, it must be present in a concentration of 125 IU per fluid ounce (FDA 1981A). U.S. of standards identity for evaporated and condensed milks are essentially the same as those published by FAO/WHO, Codex Alimentarius (FAO 1973), and similar organizations throughout the world.

Plain Condensed Milk

Plain condensed milk or concentrated milk has the same standard of identity in the United States as evaporated milk, except that it is not given additional heat processing after concentration. This product is shipped in bulk containers and is perishable. Technology is available to produce it in a sterile or almost sterile manner, and its extended shelf life gives it a potential, but as yet undeveloped, market as a source of beverage milk. Whole milk can be successfully concentrated up to 45% total solids, and these higher concentrations have found some use in the bulk product market.

Sweetened Condensed Milk

Sweetened condensed milk is made by the addition of approximately 18% sugar to whole milk, followed by concentration under vacuum to approximately one-half its volume. The product is canned without sterilizing, for the sugar acts as a preservative.

Federal standards of identity require 8.5% fat, 28.0% total milk solids, and sufficient sugar to prevent spoilage. State standards range from 7.5 to 8.5% fat and 25.0 to 28% total milk solids (USDA 1981B).

Condensed Skim Milk

Plain condensed skim milk is usually sold in bulk in the United States for increasing milk solids in ice cream, bakery goods, and many other foods. It is usually less expensive, though more perishable, than nonfat dry milk. There are no federal standards, but states require 18 to 20% total solids-not-fat.

Sweetened condensed skim milk is prepared from skim milk in a process similar to that used for whole milk. The final product contains at least 60% sugar and 72 to 74% solids. U.S. federal standards require not less than 0.5% milk fat and 24% total milk solids. It must have sufficient sweetner to prevent spoilage (USDA 1981B).

Condensed skim milk, acid, is a product manufactured primarily for animal feed. It is made from skim milk by developing about 2% acidity with a *Lactobacillus* culture and a yeast and then concentrating the milk to about one-third of its weight.

Condensed Buttermilk

Condensed semisolid buttermilk is a creamery buttermilk (usually from sweet cream) which is allowed to ripen to an acidity of 1.6% or more and then condensed. It has found limited use in the baking industry. There are no federal standards, but a typical product contains about 28% total solids (Hargrove and Alford 1974).

DRIED MILK PRODUCTS

Typical analyses of dried milk products are given in Table 2.4.

Nonfat Dry Milk

Nonfat dry milk (NDM) is an important commodity of the dairy industry. According to the American Dry Milk Institute (1982), the 1.2 billion lb manufactured in the United States in 1980 accounts for nearly 10% of the total milk supply. In several other leading milk-producing countries, 20% or more of the milk supply is used to produce NDM (Table 2.1). It provides a convenient way for countries with fluid milk surpluses to market their milk.

Nonfat dry milk is produced from skim milk by condensing it with conventional equipment followed by spray or drum drying. The drum-dried product is relatively insoluble and is used principally for animal feeds. Over 95% of nonfat dry milk in the United States is used for human foods (American Dry Milk Institute 1982) and is produced by spray drying. Most instant NDM is made by rewetting the conventionally spray-dried product, allowing the particles to agglomerate, and then reducing the moisture content with added heat. Foam spray drying by spray drying a pressurized concentrated milk also gives a very acceptable product.

Nonfat dry milk has only its fat and water removed. Federal standards of identity in the United States and FAO/WHO allow a maximum of 5% moisture and not more than 1.5% milk fat (FDA 1981A; FAO 1973).

Dried Whole Milk

Dried whole milk is prepared by conventional spray or roller drying, with some modifications of the preheat treatment of the milk. The product is usually stored under nitrogen to delay lipid oxidation and off-flavor development. In spite of the processing changes, flavor defects and short storage life have limited the markets for dried whole milk; most of it is used in the confectionery and baking industries. Federal and FAO standards require a minimum of 20% but less than 40% milk fat and a maximum of 5% moisture.

Dry Buttermilk

Most dry buttermilk is prepared from sweet cream buttermilk, and is produced in a manner similar to that of nonfat dry milk. Dry buttermilk has a higher phospholipid content than other dry milk products and therefore is a natural emulsifier for use in the dairy and baking industries and for dry mixes and other foods. A dry, high-acid buttermilk can be produced from milk fermented by *L. bulgaricus*. It is difficult to dry, however, and has found only limited use in the baking industry. There are no United States and FAO standards for this product, although typically the moisture content is less than 5%.

Dry Cream

Dry cream may be produced by spray drying or foam drying a good-quality, standardized cream. Higher heat treatments and gas packaging to reduce the oxygen in the head space to 0.75% or less make the product more resistant to oxidation. U.S. standards require a minimum of 40% but less than 75% milk fat and a maximum of 5% moisture (FDA 1981A). FAO standards require a minimum of 65% milk fat (FAO 1973). The solids-not-fat content is usually higher than that of normal market creams. A foam spray-dried sour cream has also been manufactured. A cream tablet has been produced containing added lactose to aid tableting, but the commercial acceptance of this product is negligible.

Malted Milk Powder

Malted milk powder is made by concentrating a mixture of milk and an extract from a mash of ground barley malt and wheat flour to obtain a solid which is ground to powder. It usually contains less than 7.5% milk fat and not more than 3.5% moisture. One pound is considered equivalent to 2.65 lb of fluid milk on the basis of fat content. This difference in equivalents results from the use of milk containing approximately 2.0% fat in making malted milk (Hargrove and Alford, 1974).

BUTTER, BUTTER OIL, SPREADS

Butter

Most creamery butter is produced by churning sweet cream so that the fat globules coalesce into a soft mass. The federal standard for butter (USDA 1981B) requires not less than 80% milk fat. FAO/WHO standards specify 80% milk fat, as well as no more than 16% water and a maximum of 2.0% nonfat milk solids (FAO 1973). The required fat level is universal. A typical analysis of butter is given in Table 2.3. Whey butter has a similar composition but is derived from the milk fat recovered from cheese whey.

Butter Oil

Butter oil or anhydrous milk fat is a refined product prepared by centrifuging melted butter or by separating the milk fat from high-fat cream. There are no federal standards in the United States, but the FAO has published, in the Codex Alimentarius, standards of 99.3% fat and 0.5% moisture for butter oil and 99.8% fat and 0.1% moisture for anhydrous butter oil (FAO 1973).

Ghee

Ghee is a nearly anhydrous milk fat used in many parts of India and Egypt. It is usually made from buffalo milk, and much of the typical flavor comes from the burned nonfat solids remaining in the product. Ghee is made in the United States from butter, and recently a procedure has been developed for its production from cheese (Hargrove and Alford 1974).

Miscellaneous Spreads

Several dairy spreads and products simulating butter have emerged in the past decade. Butterine, developed in Wisconsin in 1967, is composed of at least 40% milk fat, 38 to 40% margarine, 1% milk solids, salt, and added vitamins A and D. The spreadability of this product is improved over that of butter and its flavor is improved over that of margarine. It is legal only in Wisconsin, where state standards require that it contain a minimum of 40% butterfat. South Dakota requires that "Dairy Spread" or "Dari Spread" contain not less than 38% or more than 44% milk fat and not less than 30% milk solids-not-fat, with optional ingredients of salt, artificial flavoring, coloring, and thickening agents (USDA 1981B). A lowfat spread that is allowed under Ohio state standards is required to contain at least 30% milk fat, but the word "butter" must not be used on the label.

Other low-calorie spreads containing about 50% moisture and 40% milk fat have been developed in the United States, Canada, Ireland, and Sweden.

A product was developed at the University of South Dakota that contains 44% moisture, 40% milk fat, 14 to 16% nonfat dry milk, synthetic butter flavor, high-acid starter distillate, salt, butter coloring, and a combination of gelatin and sodium carboxymethylcellulose as a stabilizer.

A spread-type product, "Bregott," in which 15% of the total fat is soybean oil, is marketed in Sweden. The oil is added to cream and then churned, with minor adjustments in temperature and time. Although it is not competitive with the best margarines in price, it is a well-accepted spread (Hargrove and Alford 1974).

CHEESE

Cheese is a concentrated dairy food produced from milk curds that are separated from whey. The curds may be partially degraded by natural milk or microbial enzymes during ripening, as in cured cheeses, or they may be consumed fresh, as in uncured cheeses like cottage cheese. Most commonly, a bacterial culture with the aid of a coagulating enzyme like rennin is responsible for producing the initial curd. The starter culture also provides important proteolytic and lipolytic enzymes to produce the characteristic texture and flavor during ripening.

Although cow's milk (whole, lowfat, skim, whey, cream, nonfat dry milk, or buttermilk) generally is used for manufacturing cheese in the United States, a small quantity of ewe's and goat's milk is also used (USDA 1978). Certain other countries use milk from camels, asses, mares, buffaloes, and reindeer, in addition to ewes and goats, to make cheese.

Casein, the major protein in milk and cheese, is coagulated by acid that is produced by selected microorganisms and/or by coagulating enzymes to form curds. Acidification by food-grade acidulants is also used for the manufacture of some types of cheese, like cottage cheese. Lactalbumin and lactoglobulin are water-soluble proteins comprising one-fifth of the total protein in milk. These two proteins do not coagulate with the acidity and temperatures used in the manufacture of most cheese. The amount of whey retained in the cheese curd will determine the amount of residual water-soluble nutrients such as water-soluble protein and lactose in the cheese.

Minerals found in milk which are insoluble remain in water in the curd and are more concentrated in the cheese than in milk. About two-thirds of the calcium and one-half of the phosphorus of milk remains in cheese. A major portion of the milk calcium is retained in the curd of cheese made with coagulating enzymes. Acid coagulation alone results in the loss of portions of both calcium and phosphorus salts in the acid whey, since these minerals are more soluble in the acidic medium. Most milk fat and fat-soluble vitamins are retained in the curd, but a considerable amount of water-soluble vitamins is lost during cheese manufacture. Retention of part of some B-complex vitamins in curd is due to their extended association with casein in the original milk.

A Cheddar-type cheese retains 48% of total solids of milk, 96% casein, 4% soluble proteins, 94% fat, 6% lactose, 6% H_2O, 62% calcium, 94% vitamin A, 15% thiamin, 26% riboflavin, and 6% vitamin C (National Dairy Council 1979). The lactose content varies in freshly prepared cheeses and decreases rapidly during ripening, completely disappearing in four to six weeks. The enzymes and ripening agents responsible for the rate and extent of fat and protein breakdown are fully discussed in Chapter 12, and vitamin variation is discussed in Chapter 7.

Classification

More than 400 cheeses are known throughout the world. They are usually named after the town or community of manufacture. Successful classification is difficult, if not impossible. There are probably 18 types or kinds of natural cheeses that differ distinctively in their method of manufacture, including setting of milk and cutting, stirring, heating, draining, and pressing of curd, which results in the characteristic qualities of each cheese. Examples are brick, Camembert, Cheddar, cottage, cream, Edam, Gouda, Hand, Limburger, Neufchatel, Parmesan, Provolone, Romano, Roquefort, Sapsago, Swiss, Trappist, and whey cheeses (Mysost and Ricotta) (USDA 1978).

The classification presented here is based upon consistency brought about by differences in moisture content (soft, semisoft, hard, very hard), the manner of ripening (bacteria, mold, yeast, surface or interior microorganisms, combinations or unripened), the method by which the curd is produced (acid or coagulating enzymes, or by acid and high heat, or combinations), and the type of milk employed (National Dairy Council 1979).

The most significant and distinguishing characteristic is used for the classification. Typical analyses of two or three representative cheeses that are classified on the basis of moisture content and manner of ripening are presented in Table 2.5. Federal standards of identity are given in Table 2.6 for some selected cheeses. For an in-depth study of cheeses, Kosikowski's (1978A) book, *Cheese and Fermented Milk Foods,* should be consulted.

Cottage Cheese

Cottage cheese is a soft, unripened, acid cheese made primarily in the United States, Canada, and England with the coagulated curd from various combinations of skim milk, partially condensed skim milk, and/or reconstituted low-heat, nonfat dry milk. In some countries like Japan, cottage cheese is made principally from reconstituted nonfat dry milk. The curd is formed by the action on milk of either a combination of lactic acid from lactic acid–producing bacteria and an enzyme coagulator like rennin or by adding edible food-grade acids and a coagulator. Finished cottage cheese consists of approximately two parts of dry curd and one part of cream dressing. The dressing usually contains salt, flavoring, and stabilizers, in addition to the cream. Commercial cottage cheese is either small or large curd, depending upon the size of the curd particles cut before cooking.

A relatively new procedure for making cottage cheese, the direct-acid-set method, currently accounts for about one-fifth of all cottage cheese made in the United States. Using food-grade acids to effect coagulation eliminates problems associated with bacterial cultures and

Table 2.5. Typical Analyses of Cheeses.

Type	Cheese	Moisture (%)	Protein (%)	Total Fat (%)	Total Carbohydrate (%)	Fat in Dry Matter (%)	Ash (%)	Calcium (%)	Phosphorus (%)	Sodium (%)	Potassium (%)
Soft unripened	Cottage (dry curd)	79.8	17.3	0.42	1.8	2.1	0.7	0.03	0.10	0.01	0.03
Lowfat	Creamed cottage	79.0	12.5	4.5	2.7	21.4	1.4	0.06	0.13	0.40	0.08
	Quarg	72.0	18.0	8.0	3.0	28.5		0.30	0.35		
	Quarg (highfat)	59.0	19.0	18.0	3.0			0.30	0.35		
Soft, unripened	Cream	53.7	7.5	34.9	2.7	75.4	1.2	0.08	0.10	0.29	0.11
Highfat	Neufchatel	62.2	10.0	23.4	2.9	62.0	1.5	0.07	0.13	0.39	0.11
Soft, ripened by surface bacteria	Limburger	48.4	20.0	27.2	0.49	52.8	3.8	0.49	0.39	0.80	0.13
	Liederkranz	52.0	16.5	28.0	0	58.3	3.5	0.30	0.25		
Soft, ripened by external molds	Camembert	51.8	19.8	24.3	0.5	50.3	3.7	0.39	0.35	0.84	0.19
	Brie	48.4	20.7	27.7	0.4	53.7	2.7	0.18	0.19	0.63	0.15
Soft, ripened by bacteria, preserved by salt	Feta	55.2	14.2	21.3	4.1	47.5	5.2	0.49	0.34	1.12	0.06
	Domiati	55.0	20.5	25.0		55.5					
Semisoft, ripened by bacteria with surface growth	Brick	41.1	23.3	29.7	2.8	50.4	3.2	0.67	0.45	0.56	0.14
	Muenster	41.8	23.4	30.0	1.1	51.6	3.7	0.72	0.47	0.63	0.13
Semisoft, ripened by internal molds	Blue	42.4	21.4	28.7	2.3	49.9	5.1	0.53	0.39	1.39	0.26
	Roquefort	39.4	21.5	30.6	2.0	50.5	6.4	0.66	0.39	1.81	0.09
	Gorganzola	36.0	26.0	32.0		50.0	5.0				
Hard, ripened by bacteria	Cheddar	36.7	24.9	33.1	1.3	52.4	3.9	0.72	0.51	0.62	0.09
	Colby	38.2	23.8	32.1	2.6	52.0	3.4	0.68	0.46	0.60	0.13

(continued)

Table 2.5. (*continued*)

Type	Cheese	Moisture (%)	Protein (%)	Total Fat (%)	Total Carbohydrate (%)	Fat in Dry Matter (%)	Ash (%)	Calcium (%)	Phosphorus (%)	Sodium (%)	Potassium (%)
Hard, ripened by eye-forming bacteria	Swiss	37.2	28.4	27.4	3.4	43.7	3.5	0.96	0.60	0.26	0.11
	Edam	41.4	25.0	27.8	1.4	47.6	4.2	0.73	0.54	0.96	0.19
	Gouda	41.5	25.0	27.4	2.2	46.9	3.9	0.70	0.55	0.82	0.12
Very hard, ripened by bacteria	Parmesan (hard)	29.2	35.7	25.8	3.2	36.5	6.0	1.18	0.69	1.60	0.09
	Romano	30.9	31.8	26.9	3.6	39.0	6.7	1.06	0.76	1.20	
	Provolone	40.9	25.6	26.6	2.1	45.1	4.7	0.76	0.50	0.88	0.14
Pasta filata (stretch cheese)	Mozzarella	54.1	19.4	21.6	2.2	47.1	2.6	0.52	0.37	0.37	0.067
Lowfat or skim milk cheese (ripened)	Euda	56.5	30.0	6.5	1.0						
	Sapsago	37.0	41.0	7.4							
Whey cheese	Ricotta	71.7	11.3	13.0	3.0	45.9	1.0	0.21	0.16	0.08	0.10
	Primost	13.8	10.9	30.2	36.6	35.0					
Processed cheese	American pasteurized processed cheese	39.2	22.1	31.2	1.6	51.4	5.8	0.62	0.74	1.43	0.16
	American cheese food, cold pack	43.1	19.7	24.5	8.3	43.0	4.4	0.50	0.40	0.97	0.36
	American pasteurized processed cheese spread	47.6	16.4	21.2	8.7	40.5	6.0	0.56	0.71	1.34	0.24
	Pimento pasteurized processed cheese	39.1	22.1	31.2	1.7	51.2	5.8	0.61	0.74	1.42	0.16
	Swiss pasteurized processed cheese	42.3	24.7	25.0	2.1	43.3	5.8	0.77	0.76	1.37	0.22
	Swiss pasteurized processed cheese food	43.7	21.9	24.1	4.5	42.8	5.8	0.72	0.53	1.55	0.28

SOURCE: Hargrove and Alford (1974), Posati and Orr (1976).

Table 2.6. Federal Standards of Identity for Cheese

Cheese Type	Moisture (Maximum) (%)		Milk Fat[a] (Minimum in Solids) (%)	Milk Fat[b] (Minimum in Cheese) (%)
Cottage curd	80		—	< .5
Lowfat cottage	82.5		—	.5–2
Creamed cottage	80		— (20)	4
Cream	55		— (73.3)	33
Limburger	50		50	— (25)
Camembert	—		50[c]	—
Feta	—		50[c]	—
Brick	44		50	— (28)
Blue	46		50	— (27)
Cheddar	39		50	— (30.5)
Swiss	41		43	— (25.4)
Parmesan	32		32	— (21.8)
Provalone	45		45	— (24.8)
Ricotta (pasteurized)	80		11	11
Process Cheddar	40		50	— (30.5)
Process Swiss	42		43	— (25.4)
Process cheese food	44		— (42.6)	23
Process cheese spread	60	(44 minimum)	— (50)	20

SOURCE: FDA (1981A).
[a] Federal standards set for milk fat in solids. Figures in parentheses calculated from standard of minimum milk fat in cheese.
[b] Federal standards set for milk fat in finished cheese. Figures in parentheses calculated from standard of minimum in solids.
[c] Federal standards for cheese class only.

reduces manufacturing time. Sharma *et al.* (1980) reported a 5% increase in yield with the direct acidification method.

The United States federal standard requires that cottage cheese contain not less than 4% butterfat and not more than 80% moisture. The standard does not specify how much of the solids and fat must come from the dressing and the curd.

Cream Cheese

Cream cheese is a soft, unripened, high-fat, lactic-type cheese prepared from a homogenized milk and cream mixture containing about 16% milk fat. A lactic acid–producing bacterial culture, with or without rennet, is added to the mixture, which is held until coagulation. The coagulated mass is drained from whey by centrifugal separators or by muslin bags. Federal standards require a minimum of 33% milk fat and a maximum of 55% moisture (FDA 1981A). Addition of 0.5% stabilizer to prevent whey leakage is allowed.

Limburger Cheese

Limburger is a semisoft, surface-ripened cheese usually made from cow's milk. It originated in the provide of Luttich, Belgium, and is named after the town of Limburg, where originally much of the cheese was marketed. According to some authorities, surface organisms are responsible for its characteristic flavor and aroma, which develop after two months of ripening. Yeast predominates at first and reduces the acidity of the cheese; this is followed by growth of *Brevibacterium linens*, with the production of a characteristic reddish-yellow pigment (USDA 1978). During ripening there is extensive protein decomposition accompanied by a strong odor and flavor.

Feta Cheese

Feta cheese is a white, soft, brine-ripened ("pickled") variety, usually made from ewe's and goat's milk. Although it originated in Greece, a Bulgarian-type feta, Egyptian domiati, and a feta cheese made in the United States with cow's milk have similar compositions and properties. Lactic acid bacteria and rennet are used to produce the feta curd. After the curd is cut, drained, matted, milled, and heavily salted, it is molded and ripened in brine for about one month before being eaten. Lloyd and Ramshaw (1979) described ripened feta as soft, short, but not crumbly, with few fermentation holes and a fresh acid and clean salty flavor. Karlikanova *et al.* (1978) studied seven salt-resistant strains of streptococci in an attempt to increase and improve the flavor of feta cheese. Denkov and Kr''stev (1970) stated that farmakhim, a dried rennin/pepsin mixture from Bulgaria, could be used in place of rennin to increase the yield by increasing the moisture 0.2%.

Camembert Cheese

Camembert cheese is a soft cheese ripened by surface molds. It was first made by Marie Fontaine of Camembert and named by Napoleon. Its interior has a distinctive, characteristically yellow, waxy, creamy, or almost fluid consistency, depending upon the degree of ripening. Its exterior is a thin, gel-like layer of gray mold and dry cheese interspersed with patches of reddish yellow (USDA 1978). Ripening results from growth of the mold *Penicillium camemberti, P. candidum*, or *P. caseiocolum*, yeast, and *Brevibacterium linens*, which also grow in association with the mold for secondary fermentation and provide the color change. Hydrolysis of casein and an increase in water-soluble proteins accompany softening of the cheese. The body texture and flavor characteristics are evident in four to five weeks.

Brick Cheese

Brick cheese is a semisoft cheese ripened with surface growth and is one of the few cheeses of American origin. It is known for its semisoft, sweet-curd, and mild but rather pungent, sweet flavor. The flavor is intermediate between those of Cheddar and Limburger, not as sharp as Cheddar or as strong as Limburger. The body is soft and firm enough to slice without crumbling. It has an open structure with several round, irregular holes. The name might have been derived from its brick shape or perhaps from the bricks used to press the curds (USDA 1978). The surface growth of yeast and *B. linens* is responsible for its flavor. The ripening process takes two to three months and involves relatively little proteolysis.

Blue-Veined Cheese

Blue-veined cheese is a semisoft, mold-ripened cheese made from cow's milk in the United States. Throughout the world it is known by various names, such as French Bleu and Roquefort, Italian Gorgonzola, American Blue, Danish Blue, and English Stilton (Kosikowski 1978A). Each differs slightly in characteristics as well as in manufacturing process, but basically all are internally mold-ripened cheeses. Ripening blue-vein cheese by the mold *Penicillium roqueforti* is a highly complex process that usually requires 16 to 18 weeks. Use of the name "Roquefort" is officially limited to the original blue-veined cheese manufactured from sheep's milk in a small area near Roquefort in southeastern France. Growth of the mold *P. roqueforti* and its subsequent metabolic activity are mainly responsible for the ripening and characteristic flavor development in blue-veined cheeses.

A water suspension of mold spores is added to the milk before setting or the spores are dusted onto the curds. The inoculated curd is incubated for four weeks; then the surface slime is scrubbed off. Surface slime organisms are proteolytic and may contribute to flavor production. Curing continues after the cheese is punctured with slender needles to allow the escape of carbon dioxide and to make air available for mold growth for about 16 to 18 weeks. During curing the lipolytic activity of *P. roqueforti* breaks down milk fat to provide free fatty acids and methyl ketones, which are largely responsible for the aroma and flavor of blue-veined cheese. The organism is also the main contributor to the proteolytic breakdown for development of a soft, smooth, full-flavored cheese (Kinsella and Hwang 1976). Coghill (1979) reported that homogenizing milk for blue-veined cheese manufacture increases the rate of flavor development, produces a lighter-colored product, accelerates fat hydrolysis, and speeds ripening.

Cheddar Cheese

Cheddar cheese originated in a little village in Cheddar, England. It was initially made as a stirred curd product without matting (Kosikowski 1978A).

Cheddar is a hard, close-textured, bacteria-ripened cheese that requires several months of curing at about 10°C to develop its characteristic flavor. Rennet and a lactic culture are used with whole milk to form curds that are warmed and pressed. Cheddaring is an important step in the manufacturing process. It involves piling and repiling of the warm curds to increase lactic acid production, which contributes to the destruction of coliform bacteria. The milk for Cheddar is often standardized to a definite fat-to-casein ratio. The starter organisms are primarily responsible for the ripening and the characteristic mild flavor. During ripening, part of the casein is converted to water-soluble proteoses, peptones, and amino acids. The firm structure becomes more integrated, softer, and smoother as the flavor develops. Good-quality Cheddar cheese is ripened at 2 to 16°C at 85% relative humidity. Most Cheddar cheese is ripened at 4°C for 4 to 12 months. In Canada, the curing time may be extended to 24 months (Kosikowski 1978A). Storage at 3°C effectively prolongs the usable mature life of good-quality Cheddar cheese after an initial high (10°C) curing temperature. Gripon *et al.* (1977) concluded that adding microbial enzymes to cheese curds improves their quality and hastens ripening. But the types and optimum amounts of enzymes to be added to produce the fine flavor are still in question.

Swiss Cheese

The manufacturing process for Swiss cheese was developed in Emmenthal, Switzerland, hence the name "Emmentaler cheese" (known as "Swiss cheese" in the United States). It is hard, pressed-curd cheese with an elastic body and a mild, nut-like, sweetish flavor. Swiss cheese is best known for the large holes or eyes that develop in the curd as the cheese ripens. *S. thermophilus* and *L. bulgaricus* or *Lactobacillus helveticus* are used for acid production, which aids in expelling whey from the curd, whereas *Propionibacterium shermanii* is largely responsible for the characteristic sweet flavor and eye formation.

To increase curd elasticity and improve eye formation, the milk used to produce Swiss cheese must be clarified. Standardization of the fat content of the milk after clarification ensures uniform composition. Rennet and lactic acid from the bacteria cause casein coagulation. Swiss cheeses made in the United States are cured for three to four months (2 months minimum). Cheeses made in Switzerland, however, are cured for up to 10 months and have a more pronounced flavor (USDA 1978) than does U.S. Swiss cheese.

Mozzarella Cheese

Mozzarella is an Italian cheese which was traditionally made from highfat milk of the water buffalo. In southern Italy the water buffalo still supplies milk for this type of cheese. In the United States, however, the cheese is produced from whole or partly skimmed milk. Small amounts of starter or organic acids followed by rennet extract are added. The curd thus formed is not cooked but simply cut, and the whey is drained. The matted curds are formed into blocks, drained, and at warm temperatures undergo mild acid ripening at pH 5.2 to 5.4. At a critical pH or acidity the curd is heated in water, stretched or molded, placed in proper forms, and slightly salted. Artificial flavor and flavor-producing enzymes normally are not added to Mozzarella cheese (Kosikowski 1978A). Mozzarella accounted for 17.3% of the total cheese production in the United States in 1980 (USDA 1981A).

Provolone Cheese

Provolone is an Italian, plastic curd cheese that originated in southern Italy. It is light in color, mellow, and smooth, with a hard, compact, flaky, thread-like texture; it slices without crumbling and has a mild, agreeable flavor. Stringy textured cheeses are made by cooking the curds at a relatively high temperature and, while hot, molding them into various shapes. Provolone represents the group of acid-bacterial-ripened cheeses that are cooked at a relatively high temperature. The curds are kneaded and stretched until they are shiny, smooth, and elastic before being molded into various shapes. The curds are then chilled, salted in brine, smoked, waxed, and ripened like Cheddar. The typical flavor stems from lipolysis of milk fat brought about by added special mammalian lipases.

Parmesan Cheese

Parmesan or "Grana," as it is known in Italy, is a group of very hard bacteria-ripened, granular-textured cheeses made from partially skimmed cow's milk. They originated in Parma, near Emilia, Italy, hence the name. Special lipolytic enzymes derived from animals are used, in addition to rennet, to produce the characteristic rancid flavor.

Starter cultures of heat-resistant lactobacilli and *S. thermophilus* are added, along with rennet, to form the curds. Manufacture and salting of the cheeses take about 20 days, with 12–15 days for brining. They are then stored in cool, ventilated rooms to ripen in one or two years. A fully cured Parmesan keeps indefinitely, is very hard and thus grates easily, and is used for seasoning. Low moisture and low fat contents contribute to its hardness. Parmesan cheese made in the United States is cured for at least ten months.

Skim Milk or Lowfat Cheeses

Sapsago is manufactured chiefly in Switzerland and made from slightly soured skim milk. It is a small, very dry and hard, cone-shaped cheese. Powdered clover leaves are added to the curds to give a sharp, pungent flavor, a pleasing aroma, and a light green or sage green color. Fully cured Sapsago dry cheese is used for grating. In contrast, Euda cheese, developed by the U.S. Department of Agriculture, is a ripened, lowfat, semisoft, skim milk cheese. It has a mild flavor and a soft body, resembling Colby cheese in appearance. Lactic acid bacteria are responsible for its ripening. Predevelopment of lipolysis in the small amount of milk fat used contributes much to the flavor of this cheese.

Ricotta Cheese

After most types of cheese are manufactured, about 50% of the milk solids (most of the lactose and lactalbumin) remain in the whey. Cheese-like products can be made from these residual solids. One of the two methods commonly used to make these whey cheeses consists of contentrating the whey through evaporation with heat to obtain a mass with a firm, sugary consistency that, when cooled, forms a cheese (Primost and Ghetost). The other method is employed in the manufacture of Ricotta cheese.

Ricotta is made from coagulable constituents (principally albumin) in the whey from cheese like Cheddar, Swiss, and Provolone; hence, it also is known as "whey cheese" or "albumin cheese." Ricotta is a soft, bland, semisweet cheese that originated in Italy. All of the fat of the milk is usually left in the whey in its manufacture. Also, about 5 to 10% whole or skim milk is added to the whey when making Ricotta in the United States. Whole milk is added to make fresh Ricotta, while skim milk is used to produce dry Ricotta, which is usually used for grating. Incorporating fat with coagulable albumin improves the body, flavor, and food value of the cheese. The milk and whey proteins are coagulated by acid (lactic or acetic) and high heat (80 to 100°C). Fresh

Ricotta has a bland flavor and a body resembling cottage cheese in consistency. No U.S. federal standards exist for Ricotta cheese; however, some states require that it be made from whole milk and have a minimum of 11% fat and a maximum of 80% moisture.

Processed Cheese, Cheese Foods, and Spreads

The first soft processed cheese was patented in 1899. In 1916, Kraft was issued a patent for heating natural Cheddar cheese and emulsifying it with alkaline salts, which was the beginning of the processed cheese industry in the United States (Kosikowski 1978A).

Pasteurized processed cheese is made by changing the physical state of one or more varieties of cheese by comminuting and blending them with the aid of heat and a suitable emulsifying agent into a homogeneous plastic mass. Heating the cheese above the pasteurization temperature stops ripening and destroys most bacteria. The high temperature and a slow cooling period aid in producing a nearly sterile product. Processed cheese which contains only Cheddar cheese is called "pasteurized processed Cheddar cheese." Of the various process cheeses available, some are fabricated from a single variety of cheese, while others may be blends of two or more. As a general rule, the milk fat content in processed cheese is the same as that of the type of cheese in one-variety cheese or an average of the milk fat contents of the cheeses used in multivariety cheese. The moisture concentration of processed cheese is usually not more than 1% above that of the "parent" cheese or 1% above the average moisture level when more than one cheese is used. The formulation of a typical cheese spread is presented by Kosikowski (1978A). Legal requirements for the various processed cheeses, cheese foods, and cheese spreads are given in the Code of Federal Regulations (FDA 1981A), and since there are a number of exceptions to the general rules for the composition of milk fat and moisture described above, those interested in legal requirements are advised to refer to this code.

Pasteurized processed cheese foods are softer and may contain optional ingredients not permitted in processed cheese, including skim milk, cream, cheese whey, lactalbumin, and albumin from cheese whey. Emulsifiers, acidifying agents, water, salt, coloring agents, fruits, vegetables, spices, and flavorings may also be added. Salts act as a taste modifier but also as an inhibitor of microbial growth in processed cheese. If whey is added to processed cheese, it is generally in dried form.

Pasteurized processed cheese spread may contain the same optional ingredients as cheese foods, but may have additional moisture and stabilizing agents such as gums, gelatin, and algin.

A cheese-like spread, which is similar to processed cheese spread, is prepared by combining hydrolyzed Swiss-or Cheddar-whey protein and cultured cream. The whey protein is precipitated by heat and acid. The granular, chalky precipitate then acquires a smooth texture by enzymatic hydrolysis with Rhozyme P-11 at 39.5 to 40.5°C for 30 min. The product is heated to 85°C for 15 min to inactivate the enzymes; then it is homogenized and blended with an equal quantity of 45° cream culture containing *Lactobacillus casei* (Webb and Whittier 1970).

FROZEN DESSERTS

Frozen desserts containing milk products include ice cream and frozen custard, ice milk, sherbet and mellorine. A brief description of each of these products is presented here. An entire chapter in the previous edition of this text is devoted to frozen desserts and their properties, composition, and technology (Keeney and Kroger 1974). Additional comments on these products appear in the last chapter of this book.

Ice Cream

The most popular of all frozen desserts in the United States is ice cream. In a survey of selected supermarket products in 1979, more than 86% of the households involved reported using ice cream or ice milk during a 30-day period. The per capita production in the United States was 14.6 quarts in 1980, as well as 5.15 quarts of ice milk, 0.8 quart of sherbet, and 0.2 quart of mellorine. Several countries, including New Zealand, Australia, and Canada, have per capita production values comparable to those in the United States (International Association of Ice Cream Manufacturers 1981).

By definition, ice cream is a frozen food product made from a mixture of dairy ingredients such as milk, cream, and nonfat milk that are blended with sugar, flavoring, fruit, and nuts. It contains a minimum of 10% milk fat and weighs not less than 4.5 lb/gal.

Table 2.7 gives the principal U.S. standards for ice cream and other frozen dessert containing milk products.

Ice Milk

Ice milk is a frozen dessert similar to ice cream, except that it contains 2 to 7% milk fat and about 20% fewer calories. There was a substantial

Table 2.7. Selected Federal Standards for Frozen Desserts.

Product	Weight (lb/gal)	Total Food Solids (lb/gal)	Total Milk Solids (%)[a]	Milk Fat (%)	Whey Solids (%)[b]	Egg Yolk Solids (%)
Ice cream	>4.5	>1.6	>20	>10	<2.5	<1.4
Bulky flavored ice cream	>4.5	>1.6	>16	>8	<2.0	<1.4
Frozen custard[c]	>4.5	>1.6	>20	>10	<2.5	<1.4
Mellorine[d]	>4.5	>1.6	—[e,g]	—[f]	—[g]	—[d]
Ice milk	>4.5	>1.3	>11	>2	<2.25	—[d]
Ice milk	>4.5	>1.3	>11	<7	<1.0	—[d]
Bulky flavored ice milk	>4.5	>1.3	>9	>2	<1.75	—[d]
Sherbet	>6.0	—[e]	2–5	1–2	0–4	—[d]

SOURCE: Tobias and Nuck (1981). Reprinted with permission from the American Dairy Science Association.

[a] Caseinates may not be used to satisfy any part of the total milk solids requirement. Increases in milk fat may be offset by corresponding decreases in nonfat milk solids, but the latter must be at least 6% in frozen custard and ice cream and 4% in ice milk. Corresponding adjustments may be made in bulky flavored products.

[b] Solids from concentrated or dried whey may not exceed 25% of the nonfat milk solids.

[c] Also designated "French ice cream" or "French custard ice cream."

[d] Permitted.

[e] No standard.

[f] Milk fat replaced by a minimum of 6% vegetable or animal fat.

[g] At least 2.7% milk-derived protein having a protein efficiency ratio (PER) not less than that of whole milk protein, 108% of that of casein.

increase in the use of ice milk in the United States between 1955 and 1970, when the per capita production increased from 2.2 to 5.6 quarts, but after 1970 there was little change and in fact a slight decrease to 5.15 quarts per capita in 1980. Only a few of the 56 countries surveyed by the International Association of Ice Cream Manufacturers (1981) reported appreciable per capita production of ice milk.

United States federal standards for plain and bulky flavored ice milk are shown in Table 2.7.

Sherbets

Frozen desserts made from sugar, water, fruit acid, color, fruit or fruit flavoring, and stabilizer, and containing a small amount of milk solids added in the form of skim milk, whole milk, condensed milk, or ice cream mix, are known as "sherbets." Federal standards for these products are included in Table 2.7.

Mellorine

This food is similar to ice milk in that the milk fat content is between 1 and 2%, the vegetable fat content is at least 6%, and the product weighs not less than 6 lb/gal. As a filled dairy product, it is illegal in those states which still have filled milk laws. Its use declined from about a quart per capita in 1970 to 0.2 quart in 1980. Table 2.7 presents federal standards for mellorine.

CASEIN

Commercial casein is usually manufactured from skim milk by precipitating the casein through acidification or rennet coagulation. Casein exists in milk as a calcium caseinate–calcium phosphate complex. When acid is added, the complex is dissociated, and at pH 4.6, the isoelectric point of casein, maximum precipitation occurs. Relatively little commercial casein is produced in the United States, but imports amounted to well over 150 million lb in 1981 (USDA 1981C). Casein is widely used in food products as a protein supplement. Industrial uses include paper coatings, glues, plastics and artificial fibers. Casein is typed according to the process used to precipitate it from milk, such as hydrochloric acid casein, sulfuric acid casein, lactic acid casein, coprecipitated casein, rennet casein, and low-viscosity casein. Differences in the composition of casein result mostly from differences in the manufacturing process and the care taken in precipitation and washing of the product.

The U.S. standards for grades of edible dry casein (acid) are presented in Title 7, Part 28, of the Code of Federal Regulation (FDA 1981B) with the following specifications: Grades are determined on the basis of flavor and odor, physical appearance, bacterial estimates [standard plate count (SPC) and coliform count], protein content, moisture content, milk fat content, extraneous material, and free acid.

Characteristic	Extra Grade	Standard Grade
Moisture (not more than)	10 %	12 %
Milk fat (not more than)	1.5%	2%
Protein (not less than)	95%	90%
Ash (not more than)	2.2%	2.2%
Free acid (not more than)	0.20 ml of 0.1 N NaOH/g	0.27 ml of 0.1 N NaOH/g
Bacterial estimates		
SPC (not more than)	30,000/g	100,000/g
Coliform	neg./0.1g	2/0.1g
Flavor and odor	Bland, natural flavor and odor and free from offensive flavors and odors	Not more than slight unnatural flavors and odors, free from offensive flavors and odors
Physical appearance	White to cream colored; if pulverized, free from lumps that do not break up under slight pressure (extra grade); moderate pressure (standard grade)	

Dry casein (acid) that fails to meet the requirements of U.S. standard grade, or contains *Salmonella* or coagulase-positive staphylococci, is considered unsuitable for human food and is not assigned a U.S. grade.

Australian standards have been established for both acid and rennet caseins. The standards for acid casein are much the same as those for U.S. casein. Rennet casein usually has between 7.0 and 8.3% ash compared with 2.2% for acid casein. The fact that rennet casein is essentially a calcium caseinate accounts for this comparatively large ash value.

Sodium Caseinate

Sodium caseinate, edible grade, is made from isoelectric casein which has been prepared to meet the sanitary standards for edible casein. Casein is solubilized with food-grade caustic soda, and the resulting soluble product (20 to 25% solids) is spray-dried. Spray-drying procedures are adjusted to obtain a product with 5% or less moisture content. Dry sodium caseinate usually contains about 90 to 94% protein, 3 to 5% moisture, 6 to 7% ash, and 0.7 to 1% fat. The best flavor in dried sodium caseinate is obtained when the product is made directly from fresh wet curd. The calcium and lactose contents and moisture in fresh curd should be as low as possible, since all three adversely affect the resulting dried product. Isoelectric casein usually has better keeping qualities than sodium caseinate. The uses for sodium caseinate are much the same as those of commercial casein. Increasing quantities of sodium caseinate are being used as a protein supplement in dietetic and bakery products, as well as in stews, soups, and imitation milk.

LACTOSE

Lactose is the characteristic carbohydrate of milk, averaging about 4.9% for fluid whole cow's milk and 4.8% for sheep and goat's milk. The commercial source of lactose today is almost exclusively sweet whey, a by-product of cheese making. Details of its production are given in Chapter 6.

Standards for anhydrous lactose are presented in Recommended International Standards for Lactose by the Food and Agriculture Organization, Codex Alimentarius, 1969.

Lactose anhydrous	99% min. (on dry basis)
Sulfated ash	0.3% max. (on dry basis)
Loss on drying (16 hr at 120°C)	6% max.
pH, 10% solution	4.5–7.0
Arsenic	1 mg/kg max.
Lead	1 mg/kg max.
Copper	2 mg/kg max.

WHEY

With the increase in the production of cheese, not only in the United States but throughout the world (USDA 1981C), and more stringent controls on disposal of waste materials, the use of surplus cheese whey is one of the most critical problems facing the dairy industry. Whey, the liquid that remains after casein and fat are separated as curds in the cheese-making process, contains most of the salts, lactose, and water-soluble proteins of the milk. It varies in composition with the type of cheese from which it comes, heat treatment, handling, and other factors. The predominant type is "sweet whey," which is derived from the manufacture of ripened cheeses (Cheddar, Swiss, Provolone, etc.), so named because its pH is only slightly less than that of fresh milk. "Acid whey," on the other hand, has a pH of approximately 4.7; it is similar in composition to sweet whey, except that up to 20% of the lactose is converted to lactic acid by lactose-fermenting bacteria in the manufacture of products like cottage cheese. The reduced pH may also be achieved by the addition of food-grade acids to replace the lactic cultures, as in directly acidified cottage cheese. Although only about 10% of the whey produced in the United States in 1980 was of the acid type (Whey Products Institute 1981), it presents a serious disposal problem. Acid whey that results from the manufacture of cottage cheese is not of sufficient volume to make its further processing (concentrating or drying) economical. On the other hand, since the fluid milk plants that manufacture cottage cheese are usually located in cities, it must be disposed of in already overloaded municipal sewers. An additional complication in disposing of acid whey comes from its high acidity; this interferes with its subsequent processing. Sweet whey from hard cheese manufacture is not hampered by these constraints.

Even though liquid whey has been successfully commercialized in the form of alcoholic and nonalcoholic beverages, these are still a rarity in most countries. Most whey is converted to whey solids as ingredients for human food or animal feeds by traditional processes such as spray drying, roller drying, concentration to semisolid feed blocks, or production of sweetened condensed whey. Jelen (1979) reported other traditionally established processes including lactose crystallization from untreated or modified whey, production of heat-denatured whey protein concentrate, or recovery of milk fat from whey cheese in "whey butter."

Nearly 60% of the whey and whey products produced in the United States in 1980 were used in human food products, over 80% as dry whey. More than 65% of the whey used for animal feed was a dried whey product. Principal users of whey products for human foods are dairies and bakeries. Lactose, which is derived primarily from whey, is used mainly in infant foods and pharmaceuticals (Whey Products Institute 1980).

Some limitations in the functional properties of dried whey for human foods, including high salt and lactose concentrations, have led to its fractionation and blending into a variety of new products. Recent

Table 2.8. Chemical Composition of Selected Commercial Whey-Based Food Ingredients.

Product	Typical Analysis			Source[a]
	% Protein	% Lactose	% Ash	
Products manufactured from whey only				
Dried sweet whey	12	74	8.5	1, 2, 3, 4, 5, 6
Partially demineralized whey	13	75	5.5	1
Demineralized whey	14	82	0.8	1
Demineralized/delactosed whey	36	56	2.4	1
Whey protein concentrate[a]	53	36	4.0	2, 3, 4, 5
Whey protein concentrate[b]	85	4	1.2	4
Traditional (heated) lactalbumin	80	5	2.5	4
Blends of whey with other materials				
Whey, skim milk (and/or buttermilk)	22	54	10.0	2, 6
Whey, caseinates	34	52	8.0	2, 6
Whey, soy (and/or corn) solids	28	60	8.0	2, 6
Whey, soy protein isolate	35	52	8.0	2, 6, 7

SOURCE: Jelen (1979). Reprinted with permission from the American Chemical Society (1982).
[a]Technical literature on which this table is based. There are other suppliers of similar products whose literature was not available: (1) Foremost Foods Co., California; (2) Dairyland Products, Minnesota; (3) Stauffer Chemical Co., Connecticut; (4) New Zealand Dairy Board, Wellington, and/or N.Z. DRI, Palmerston North; (5) Purity Cheese Co., Wisconsin; (6) Land-O-Lakes Co., Minnesota; (7) Ralston-Purina, Missouri.
[b]Total carbohydrates.

developments in molecular separation techniques such as ultrafiltration, reverse osmosis, gel filtration, electrodialysis, and ion exchange have made possible the fractionation, modification, or reconstruction and blending of a variety of whey products. Table 2.8 illustrates the type and composition of whey products currently available (Jelen 1979).

Craig (1979) has summarized the functional and nutritional properties of most of these whey-based food ingredients. A comprehensive symposium (Clark 1979A) and several excellent reviews on whey and whey utilization are recommended for further studies (Clark 1979B).

REFERENCES

Aboshana, K. and Hansen, A. P. 1977. Effect of ultra-high-temperature steam injection processing on sulfur-containing amino acids in milk. *J. Dairy Sci. 60*, 1374–1378.

American Dry Milk Inst. 1982. *Census of 1981 Dry Milk Distribution and Production Trends.* Bull. 1000. American Dry Milk Institute, Chicago.

Anon. 1979. UHT milk successful in Canada. *Am. Dairy Rev. 41* (1), 22–25.

Anon. 1981A. FDA approves hydrogen peroxide levels as sterilant for polyethylene containers. *Dairy Field 164* (2), 18.

Anon. 1981B. Aseptic—will it take dairy out of the case? *Dairy Field 164* (6), 53–55, 60.

Burton, H. 1969. Ultra-high temperature processed milk. *Dairy Sci. Abstr. 31*, 287–297.

Burton, H. 1977. An introduction to ultra-high temperature processing and plant. *J. Soc. Dairy Technol. 30*, 135–142.

Clark, W. S., Jr. 1979A. Symposium on the chemical and nutritional aspects of dairy wastes. *J. Agr. Food Chem. 27*, 653–698.

Clark, W. S., Jr. 1979B. Whey processing and utilization. *J. Dairy Sci., 62*, 96–116.

Coghill, D. 1979. The ripening of blue-vein cheese. *Aust. J. Dairy Technol. 34*, 72–75.

Craig, T. W. 1979. Dairy derived food ingredients—functional and nutritional considerations. *J. Dairy Sci. 62*, 1695–1702.

Denkov, Ts. and Kr"stev, I. 1970. Cited by G.T. Lloyd and E.H. Ramshaw. The manufacture of Bulgarian Feta cheese. *Aust. J. Dairy Technol. 34*, 180–183.

Douglas, F. W., Jr., Greenberg, R., Farrell, H. M., Jr. and Edmondson, L. F. 1981. Effects of ultra-high-temperature pasteurization on milk proteins. *J. Agr. Food Chem. 29*, 11–15.

FDA. 1981A. Code of Federal Regulations, Title 21, Parts 100–169. Food and Drugs. Food and Drug Administration. Government Service Administration, Washington, D.C.

FDA. 1981B. Code of Federal Regulations, Title 7, Chapter 28. Agriculture. Food and Drug Administration. Government Service Administration, Washington, D.C.

Food and Agriculture Organization of the United Nations—World Health Organization. 1969. Joint FAO/WHO Food Standards Programme, Codex Alimentarius Commission. Recommended International Standards for Lactose. Food and Agriculture Organization, Rome.

Food and Agriculture Organization of the United Nations, 1973. *Code of Principles Concerning Milk and Milk Products and Associated Standards,* 7th ed. Food and Agriculture Organization, Rome.

Gripon, J. C., Desmazaud, M. J., Le Bars, D. and Bergere, J. L. 1977. Role of proteolytic enzymes of *Streptococcus lactis, Penicillium roqueforti,* and *Penicillium caseicolum* during cheese ripening. *J. Dairy Sci., 60*, 1532–1538.

Hansen, A. P. and Melo, T. S. 1977. Effect of ultra-high-temperature steam injection upon constituents of skim milk. *J. Dairy Sci. 60*, 1368–1374.

Hargrove, R. E. and Alford, J. A. 1974. Composition of milk products. *In: Fundamentals of Dairy Chemistry,* 2nd ed. B.H. Webb, A. H. Johnson and J.A. Alford (Editors). AVI Publishing Co., Westport, Conn., pp. 58–86.

International Association of Ice Cream Manufacturers. 1981. *The Latest Scoop.* International Association of Ice Cream Manufacturers, Washington, D.C.

Jelen, P. 1979. Industrial whey processing technology: an overview. *J. Agr. Food Chem. 27*, 658–661.

Karlikanova, S. N., Ramazanov, I. U. and Ramazanova, D. P. 1978. Production of salt-resistant mutants of lactic acid streptococci and their use in starters for pickled cheese. Twentieth International Dairy Congress. *Brief Communication* IE, 525–526.

Keeney, P. G. and Kroger, M. 1974. Frozen dairy products. *In: Fundamentals of Dairy Chemistry*, 2nd ed. B.H. Webb, A. H. Johnson and J.A. Alford (Editors). AVI Publishing Co., Westport, Conn., pp. 873–913.

Keogh, B. P. 1970. Micro-organisms in dairy products—friend or foe? *Aust. J. Dairy Technol. 33*, 41–45.

Kinsella, J. E. and Hwang, D. H. 1976. Methyl ketone formation during germination of *Penicillium roqueforti. J. Agr. Food Chem. 24*, 443–448.

Kosikowski, F. 1978A. *Cheese and Fermented Milk Foods*, 2nd ed. F.V. Kosikowski and associates, Brooktondale, N.Y.

Kosikowski, F. 1978B. Cultured milk foods in the future. *Cultured Dairy Prod. J. 13*(3), 5–7.

Lang, F. and Lang, A. 1978. New methods of acidophilus milk manufacture and the use of bifidus bacteria in milk processing. *Aust. J. Dairy Technol. 33*, 66–68.

Lloyd, G. T. and Ramshaw, E. H. 1979. The manufacture of Bulgarian feta cheese. *Aust. J. Dairy Technol. 34*, 180–183.

Mehta, R. S. 1980. Milk processed at ultra-high-temperature—a review. *J. Food Prot. 43*, 212–225.

Milk Industry Foundation. 1981. *Milk Facts.* Milk Industry Foundation, Washington, D.C.

National Dairy Council. 1979. *Newer Knowledge of Cheese and Other Cheese Products*, 3rd ed. National Dairy Council, Rosemont, Ill.

Posati, L. and Orr, M. L. 1976. *Composition of Foods, Dairy and Egg Products, Raw, Processed, Prepared.* Agriculture Handbook No. 8-1. ARS. U.S. Department of Agriculture, Washington, D.C.

Puhan, Z. and Gallmann, P. 1980. Ultrafiltration in the manufacture of kumys and quark. *Cult. Dairy Prod. J. 15*, (1), 12–16.

Schanback, M. 1977. Manufacture of superior quality sour cream. *Cult. Dairy Prod. J. 12* (2), 19–20.

Sharma, H. S., Bassette, R., Mehta, R. S. and Dayton, A. D. 1980. Yield and curd characteristics of cottage cheese made by the culture and direct set methods. *J. Food Prot. 43*, 441–446.

Speck, M. L. 1976. Interactions among lactobacilli and man. *J. Dairy Sci. 59*, 338–343.

Tamine, A. Y. and Deeth, H. C. 1980. Yoghurt technology and biochemistry. *J. Food Prot., 43*, 939–977.

Tobias, J. and Muck, G. A. 1981. Ice cream and frozen desserts. *J. Dairy Sci. 64*, 1077–1086.

Twentieth International Dairy Congress, 1978. *Brief Communication*, Vol. E. International Dairy Congress, Paris.

USDA. 1978. *Cheese Varieties and Descriptions.* Agriculture Handbook No. 54. ARS. U.S. Department of Agriculture, Washington, D.C.

USDA. 1981A. *Dairy Products Annual Summary.* ESS, Da 2-1 (81). U.S. Department of Agriculture, Washington, D.C.

USDA. 1981B. *Federal and State Standards for the Composition of Milk Products (and Certain Non-Milkfat Products).* Agriculture Handbook No. 51. AMS. U.S. Department of Agriculture, Washington, D.C.

USDA. 1981C. *Foreign Agriculture Circular,* FAS, FD9-81. U.S. Department of Agriculture, Washington, D.C.

USDA. 1981D. *Foreign Agriculture Circular (Dairy)*. FAS, FD 1–81. U.S. Department of Agriculture, Washington, D.C.

USDHEW. 1978. *Grade A Pasteurized Milk Ordinance*. U.S. Department of Health, Education, and Welfare, Washington, D.C.

Webb, B. and Whittier, E. 1970. *Byproducts from Milk,* 2nd ed. AVI Publishing Co., Westport, Conn.

Whey Products Institute. 1980. *Whey Products—A Survey of Utilization and Production Trends 1980*. Bulletin No. 25. Whey Products Institute, Chicago.

Whey Products Institute. 1981. Personal communication. Chicago.

White, C. H. 1978. Manufacturing better buttermilks. *Cult. Dairy Prod. J. 13* (1), 16–20.

3

Proteins of Milk

Robert McL. Whitney

The proteins of milk are of great importance in human nutrition and influence the behavior and properties of the dairy products containing them. They have been studied more extensively than any other proteins except possibly those of blood. Since milk contains a number of different proteins, they must be fractionated and the proteins of interest isolated before definitive work can be done on their composition, structure, and chemical and physical properties.

This chapter considers the classification, nomenclature, primary structure, and chemical and physical properties of the individual protein as they occur in the milk from the genus *Bos*. While some studies have been done of the proteins of the milk of other mammals, bovine milk, due to its commercial importance, has been most extensively investigated. A more comprehensive treatment of the subject is in the two-volume treatise on the chemistry and molecular biology of the milk proteins edited by McKenzie (1970, 1971A). Other reviews are Whitney (1977), Brunner (1981), and Swaisgood (1982).

CLASSIFICATION, NOMENCLATURE, AND PRIMARY STRUCTURE OF MILK PROTEINS

During the nineteenth and early twentieth centuries, separation of the proteins was limited to casein and the classical lactalbumin and lactoglobulin fractions of the whey proteins. Subsequent work has resulted in the identification and characterization of numerous proteins from each of these fractions. A classification system of the known proteins in milk developed by the American Dairy Science Association's (ADSA) Committee on Milk Protein Nomenclature, Classification, and Methodology (Eigel *et al.* 1984) is summarized and enlarged to include the minor proteins and enzymes in Table 3.1.

Except for bovine serum albumin, immunoglobulins, and fat globule membrane proteins, the nomenclature employed in this classification consists of the use of a Greek letter, with or without a numerical sub-

Table 3.1. Classification and Distribution of the Milk Proteins—Genus *Bos* (30–35 G/Liter).

I. Caseins (24–28 g/liter)
  A. α_{s1}-Caseins (12–15 g/liter)
    1. α_{s1}-Casein X^a—8P (genetic variants—A, B, C, D-9P, and E)
    2. α_{s1}-Casein X^a—9P (genetic variants—A, B, C, D-10P, and E)
    3. α_{s1}-Casein fragments[c]
  B. α_{s2}-Caseins (3–4 g/liter)
    1. α_{s2}-Casein X^a—10P (genetic variants—A, B, C-9P, and D-7P)
    2. α_{s2}-Casein X^a—11P (genetic variants—A, B, C-10P, and D-8P)
    3. α_{s2}-Casein X^a—12P (genetic variants—A, B, C-11P, and D-9P)
    4. α_{s2}-Casein X^a—13P (genetic variants—A, B, C-12P, and D-10P)
  C. β-Caseins (9–11 g/liter)
    1. β-Casein X^a—5P (genetic variants—A^1, A^2, A^3, B, C-4P, D-4P, and E)
    2. β-Casein X^a—1P (f 29–209) (genetic variants—A^1, A^2, A^3, and B)
    3. β-Casein X^a—(f 106–209) (genetic variants—A^2, A^3, and B)
    4. β-Casein X^a—(f 108–209) (genetic variants—A and B)
    5. β-Casein X^a—4P (f 1–28)[b]
    6. β-Casein X^a—5P (f 1–105)[b]
    7. β-Casein X^a—5P (f 1–107)[b]
    8. β-Casein X^a—1P (f 29–105)[b]
    9. β-Casein X^a—1P (f 29–107)[b]
  D. κ-Caseins (2–4 g/liter)
    1. κ-Casein X^a—1P (genetic variants—A and B)
    2. Minor κ-caseins X^a—1, −2, −3, etc. (genetic variants—A and B)
II. Whey proteins (5–7 g/liter)
  A. β-Lactoglobulins (2–4 g/liter)
    1. β-Lactoglobulins X^a (genetic variants—A, B, C, D, Dr, E, F, and G)
  B. α-Lactalbumins (0.6–1.7 g/liter)
    1. α-Lactalbumin X^a (genetic variants—A and B)
    2. Minor α-Lactalbumins
  C. Bovine serum albumin (0.2–0.4 g/liter)
  D. Immunoglobulins (0.5–1.8 g/liter)
    1. IgG immunoglobulins
      a. IgG_1 immunoglobulins
      b. IgG_2 immunoglobulins
      c. IgG fragments
    2. IgM immunoglobulins
    3. IgA immunoglobulins
      a. IgA immunoglobulins
      b. Secretory IgA immunoglobulins
    4. IgE immunoglobulins
    5. J-chain or component
    6. Free secretory component
III. Milk fat globule membrane (MFGM) proteins
  A. Zone A (MFGM) proteins
  B. Zone B (MFGM) proteins
  C. Zone C (MFGM) proteins
  D. Zone D (MFGM) proteins
IV. Minor proteins
  A. Serum transferrin

B. Lactoferrin
C. β_2-Microglobulin
D. M_1-glycoproteins
E. M_2-glycoproteins
F. α_1-Acid glycoprotein or orosomucoid
G. Ceruloplasmin
H. Trypsin inhibitor
I. Kininogen
J. Folate-binding protein (FBP)
K. Vitamin B_{12}-binding protein
V. Enzymes
(See Table 3.2)

---

[a] X represents the genetic variant.
[b] Genetic variants of these fragments have not been specifically identified.
[c] Nomenclature has not been established for these fragments.

script preceding the class name when necessary, to identify the family of proteins. The genetic variant of the protein is indicated by an upper-case arabic letter, with or without a numerical superscript, immediately following the class name. Posttranslational modifications are added in sequence. For example, β-casein B-5P (f 1–105) indicates that the protein belongs to the β-family of caseins, is the B genetic variant, contains five posttranslational phosphorylations, and is the fragment of the entire β-casein B amino acid sequence from the N-terminal amino acid (residue 1) through residue 105.

Caseins

Originally, the caseins were defined as those phosphoproteins which precipitate from raw skim milk upon acidification to pH 4.6 at 20°C, and the individual families were identified by alkaline urea gel electrophoresis (Whitney *et al.* 1976). With the resolution of their primary structure, it became possible to classify them according to their chemical structure, rather than on the basis of an operational definition. When one does this, it is apparent that not all of the caseins contain phosphorus (Table 3.1); some are also found in the acid whey after removal of the precipitated caseins.

α_{s1}-**Caseins.** The α_{s1}-caseins, along with the α_{s2}-caseins, make up the previously designated calcium-sensitive α_s-casein fraction originally precipitated by Waugh *et al.* (1962) with 0.4 M $CaCl_2$ at pH 7 and 4°C from the old α-casein fraction of Warner. They consist of one major and one minor component previously identified as α_{s1}-casein and

α_{s0}-casein, respectively, and some fragments of α_{s1}-casein that are still not clearly defined. The primary sequence of the amino acid residues in the major component was established (Mercier *et al.* 1971, Grosclaude *et al.* 1973), and the minor component possesses the same primary sequence but has one more phosphorylated serine residue (Manson *et al.* 1977).

At the present time, five genetic variants of α_{s1}-caseins are known: A, D, B, C, and E, in the order of their decreasing mobility in gel electrophoresis in alkaline urea media (Thompson 1970, 1971; Grosclaude *et al.* 1976A). The E variant has been definitely observed only in the milk from some yaks (*B. grunniens*). However, recently, an α_{s1}-variant has been observed in Bali cattle (*B. javanicus*) which possesses a mobility less than that of the C variant in an alkaline urea-starch gel system (Bell *et al.* 1981A). Possibly it is also the E variant, but more information is needed before this can be definitely established. The polymorphs are breed specific, with the B variant predominant in *B. taurus* and the C variant predominant in *B. indicus* and *B. grunniens* (Thompson 1971; Grosclaude *et al.* 1974A).

The primary structure of α_{s1}-casein B-8P, the major component of this variant, is shown in Figure 3.1. The additional phosphorylated serine residue in the minor component, α_{s1}-casein B-9P, occurs at position 41 (Manson *et al.* 1976, 1977). The B-variant consists of 199 amino acid

10 14 20
H.Arg - Pro - Lys - His - Pro - Ile - Lys - His - Gln - Gly - Leu - Pro - Gln - Glu - Val - Leu - Asn - Glu - Asn - Leu -
Absent in variant A
-26 30 40
Leu - Arg - Phe - Phe - Val - Ala - Pro - Phe - Pro - Gln - Val - Phe - Gly - Lys - Glu - Lys - Val - Asn - Glu - Leu -
41 50 53 59 60
Ser - Lys - Asp - Ile - Gly - Ser - Glu - Ser - Thr - Glu - Asp - Gln - Ala - Met - Glu - Asp - Ile - Lys - Gln - Met -
P (α_{s1}-Cn B(9P)) P P ThrP (variant D) Lys (variant E)
70 80
Glu - Ala - Glu - Ser - Ile - Ser - Ser - Ser - Glu - Glu - Ile - Val - Pro - Asn - Ser - Val - Glu - Gln - Lys - His -
P P P P P
90 100
Ile - Gln - Lys - Glu - Asp - Val - Pro - Ser - Glu - Arg - Tyr - Leu - Gly - Tyr - Leu - Glu - Gln - Leu - Leu - Arg -
110 120
Leu - Lys - Lys - Tyr - Lys - Val - Pro - Gln - Leu - Glu - Ile - Val - Pro - Asn - Ser - Ala - Glu - Glu - Arg - Leu -
P
130 140
His - Ser - Met - Lys - Glu - Gly - Ile - His - Ala - Gln - Gln - Lys - Glu - Pro - Met - Ile - Gly - Val - Asn - Gln -
150 160
Glu - Leu - Ala - Tyr - Phe - Tyr - Pro - Glu - Leu - Phe - Arg - Gln - Phe - Tyr - Gln - Leu - Asp - Ala - Tyr - Pro -
170 180
Ser - Gly - Ala - Trp - Tyr - Tyr - Val - Pro - Leu - Gly - Thr - Gln - Tyr - Thr - Asp - Ala - Pro - Ser - Phe - Ser -
190 192 199
Asp - Ile - Pro - Asn - Pro - Ile - Gly - Ser - Glu - Asn - Ser - Glu - Lys - Thr - Thr - Met - Pro - Leu - Trp.OH
Gly (variants C & E)

Figure 3.1. Primary Structure of Bos α_{s1}-casein B-8P. (From Mercier *et al.* 1971; Grosclaude *et al.* 1973; Eigel *et al.* 1984. Reprinted with permission of the American Dairy Science Association.)

residues with a calculated molecular weight of 23,614 (Mercier *et al.* 1971; Grosclaude *et al.* 1973), and the other genetic variants differ from B as indicated in Figure 3.1. One should note that, for the D variant, the major component is named α_{s1}-casein D-9P and the minor components α_{s1}-casein D-10P due to the additional phosphorylation in this variant.

Peptides extracted from casein with N, N-dimethyl formamide have complex electrophoretic patterns identical to those of the fraction first prepared by Long and co-workers and called λ-casein (El-Negoumy 1973). These peptides are identical electrophoretically to those released by the action of plasmin, which is present in fresh raw milk, upon α_{s1}-casein (Aimutis and Eigel 1982). Two of these peptides have tryptic peptide maps and molecular weights identical to those of a pair of the peptides produced by plasmin degradation of α_{s1}-casein. These peptides appear to be fragments of α_{s1}-casein which are present in milk as the result of plasmin proteolysis. More definitive information on their primary structure is needed before nomenclature for these fragments can be established.

α_{s2}-**Caseins.** Currently there appear to be five components in the α_{s2}-casein family which have relative mobilities between those of the α_{s1}-caseins and β-casein in alkaline urea gel electrophoresis. Previously these were called α_{s2}-, α_{s3}-, α_{s4}-, α_{s5}-, and α_{s6}-casein in order of decreasing mobility (Annan and Manson 1969; Whitney *et al.* 1976), but new evidence indicates that they all have the same primary amino acid sequence with different degrees of posttranslational phosphorylation (Brignon *et al.* 1976, 1977). The exact location and perhaps even the number of phosphate groups in the various components still remain to be established. On the basis of this evidence, they have been tentatively renamed as follows: α_{s2}-casein to α_{s2}-casein X-13P; α_{s3}-casein to α_{s2}-casein X-12P; α_{s4}-casein to α_{s2}-casein X-11P; and α_{s6}-casein to α_{s2}-casein X-10P (Eigel *et al.* 1984). From its amino acid composition and the effect of mercaptoethanol upon its electrophoretic properties, α_{s}^{5}-casein has been tentatively identified as a dimer of the 11P and 12P components linked together by a disulfide bond (Hoagland *et al.* 1971).

Four genetic variants of the α_{s2}-caseins are known: A, B, C, and D. As with the α_{s1}-components, the genetic variants of all α_{s2}-caseins in a specific milk from a homozygote cow are identical. The electrophoretic mobilities of the various D-variant bands are slower than those of the corresponding bands in the A variant at pH 8.6 but are faster at pH 3.0 (Grosclaude *et al.* 1978, 1979). Variants A and D have been observed in European breeds (*B. taurus*), with D in the Vosgienne and Montbeliarde breeds (Grosclaude *et al.* 1978). In addition to variant A, variant

B is found in *B. taurus* and *B. indicus* in a high Nepalese valley, and variant C is observed specifically in yaks (*B. grunniens*) in the same region (Grosclaude *et al.* 1976B). Variant C has also been found in yaks from the Republic of Mongolia and variant B in zebu from the Republic of South Africa (Grosclaude *et al.* 1979).

The primary structure of α_{s2}-casein A-11P is shown in Figure 3.2. It consists of 207 amino acid residues with a calculated molecular weight of 25,230 (Brignon *et al.* 1977). The differences between the primary sequences of the polypeptide chains of the genetic variants still remain to be established. As a result of a comparison of the amino acid and phosphate contents of the C variant with those of α_{s2}-casein A-10P, a possible substitution of a glycine residue for a phosphoserine has been proposed (Grosclaude *et al.* 1976B). Variant D, like α_{s1}-casein A, has a peptide of nine residues deleted from the sequence (Grosclaude *et al.* 1978, 1979), but the exact peptide has not been established because three peptides are similar, as indicated in Figure 3.2. Since these peptides contain three phosphoserine residues, the D-variant corresponding to α_{s2}-casein A-11P is named α_{s2}-casein D-8P, and corresponding changes have been made in the nomenclature of the other members of the family for this variant.

```
                                          10                                              20
H . Lys – Asn – Thr – Met – Glu – His – Val – Ser – Ser – Ser – Glu – Glu – Ser – Ile – Ile – Ser – Gln – Glu – Thr – Tyr –
                                         P    P    P                                 P
                                          30                                              40
     Lys – Gln – Glu – Lys – Asn – Met – Ala – Ile – Asn – Pro – Ser – Lys – Glu – Asn – Leu – Cys – Ser – Thr – Phe – Cys –
                               50   51   52                                    58   59   60
     Lys – Glu – Val – Val – Arg – Asn – Ala – Asn – Glu – Glu – Glu – Tyr – Ser – Ile – Gly – Ser – Ser – Ser – Glu – Glu –
                                    (absent in variant D)              P    P    P
                                          70                                              80
     Ser – Ala – Glu – Val – Ala – Thr – Glu – Glu – Val – Lys – Ile – Thr – Val – Asp – Asp – Lys – His – Tyr – Gln – Lys –
     P
                                          90                                              100
     Ala – Leu – Asn – Glu – Ile – Asn – Glu – Phe – Tyr – Gln – Lys – Phe – Pro – Gln – Tyr – Leu – Gln – Tyr – Leu – Tyr –
                                          110                                             120
     Gln – Gly – Pro – Ile – Val – Leu – Asn – Pro – Trp – Asp – Gln – Val – Lys – Arg – Asn – Ala – Val – Pro – Ile – Thr –
                                          130                                             140
     Pro – Thr – Leu – Asn – Arg – Glu – Gln – Leu – Ser – Thr – Ser – Glu – Glu – Asn – Ser – Lys – Lys – Thr – Val – Asp –
                                   P         P
                                          150                                             160
     Met – Glu – Ser – Thr – Glu – Val – Phe – Thr – Lys – Lys – Thr – Lys – Leu – Thr – Glu – Glu – Glu – Lys – Asn – Arg –
                 P
                                          170                                             180
     Leu – Asn – Phe – Leu – Lys – Lys – Ile – Ser – Gln – Arg – Tyr – Gln – Lys – Phe – Ala – Leu – Pro – Gln – Tyr – Leu –
                                          190                                             200
     Lys – Thr – Val – Tyr – Gln – His – Gln – Lys – Ala – Met – Lys – Pro – Trp – Ile – Gln – Pro – Lys – Thr – Lys – Val –

     Ile – Pro – Tyr – Val – Arg – Tyr – Leu.OH
```

Figure 3.2. Primary structure of Bos α_{s2}-casein A-11P. (From Brignon *et al.* 1977; Eigel *et al.* 1984. Reprinted with permission of the American Dairy Science Association.)

β-Caseins. The β-casein family consists of one major component with at least seven genetic variants and eight minor components which are proteolytic fragments of the major component. The relative mobility of the major β-casein in alkaline urea gel electrophoresis is less than that of the α_{s2}-caseins.

In alkaline urea gel electrophoresis, variants A^1, A^2, and A^3 migrate with the same relative mobility but more rapidly than the other variants, which migrate in the following order: $B > D$, $E > C$ (Thompson 1970; Voglino 1972). The mobilities of the D and E variants have not been compared with each other in this medium. To differentiate the A variants, their mobilities must be determined by acid urea gel electrophoresis (Thompson 1970; Kiddy 1975). In this medium, the genetic variants move in the following order: $C > B = D > A^1 = E > A^2 > A^3$. Another possible variant β-casein, B_z, has been observed in Indian and African zebu cattle (*B. indicus*). It has the same electrophoretic behavior as the B variant but appears, from its chymotryptic digests, to have different peptide maps (Aschaffenburg *et al.* 1968). More recently, in a study of Choa zebu cattle, it was observed that the A^1 variant substitution was the same for *B. indicus* and *B. taurus* and that the B variant in zebu cattle differed from the A^1 variant by the same amino acid substitution that is observed in *B. taurus* (Grosclaude *et al.* 1974A). More work is needed before the existence of the B_z variant can be confirmed. Recently, another β-casein has been observed in the milk of a couple of Bali cattle (*B. javanicus*) which appears to have a slightly lower mobility in acid urea starch-gel electrophoresis than the A^3 variant (Bell *et al.* 1981A). While this variant has been designated β-casein A^4 by these authors, more information is needed before it can be established unequivocally. The A variants are the predominant polymorphs in all species and strains of *Bos* investigated (Thompson 1971).

The complete sequence of amino acid residues in β-casein A^2-5P (Figure 3.3) indicates a single polypeptide chain of 209 residues with a calculated molecular weight of 23,983 (Ribadeau-Dumas *et al.* 1972; Grosclaude *et al.* 1973). The differences between the primary structures of the other genetic variants are also indicated in Figure 3.3. The failure of the serine residue at position 35 in variant C to be phosphorylated is unusual and may be attributed to the substitution of lysine for glutamic acid at position 37. The positive charge at this position in the C variant is thought to hinder the phosphorylation of the serine, while the negative charge at position 37 in all other genetic variants may facilitate it. The C and D variants have one less phosphate group than the others, and this should be indicated in their nomenclature: β-casein C-4P and β-casein D-4P.

Based on the primary structure of the major β-casein component,

```
                                          10                           18        20
H.Arg - Glu - Leu - Glu - Glu - Leu - Asn - Val - Pro - Gly - Glu - Ile - Val - Glu - Ser - Leu - Ser -|Ser|- Ser - Glu -
                                                                                  P            P     P
                                                                                              Lys (variant D)
                                   28↓  29    30                  35   36   37                         40
Glu - Ser - Ile - Thr - Arg - Ile - Asn - Lys - Lys - Ile - Glu - Lys - Phe - Gln - Ser -|Glu|-|Glu|- Gln - Gln - Gln -
                                   (absent in variant C)              P    Lys  Lys (Variant C)
                                             50                                (variant E)            60
Thr - Glu - Asp - Glu - Leu - Gln - Asp - Lys - Ile - His - Pro - Phe - Ala - Gln - Thr - Gln - Ser - Leu - Val - Tyr -
                         67              70                                                           80
Pro - Phe - Pro - Gly - Pro - Ile -|Pro|- Asn - Ser - Leu - Pro - Gln - Asn - Ile - Pro - Pro - Leu - Thr - Gln - Thr -
     (variants C, A¹, and B) His
                                             90                                                     100
Pro - Val - Val - Val - Pro - Pro - Phe - Leu - Gln - Pro - Glu - Val - Met - Gly - Val - Ser - Lys - Val - Lys - Glu -
             105 | 106   107 | 108       110                                                        120
Ala - Met - Ala - Pro - Lys -|His|- Lys - Glu - Met - Pro - Phe - Pro - Lys - Tyr - Pro - Val - Gln - Pro - Phe - Thr -
                            Gln (variant A³)
     122                                  130                                                       140
Glu -|Ser|- Gln - Ser - Leu - Thr - Leu - Thr - Asp - Val - Glu - Asn - Leu - His - Leu - Pro - Pro - Leu - Leu - Leu -
     Arg (variant B)
                                             150                                                    160
Gln - Ser - Trp - Met - His - Gln - Pro - His - Gln - Pro - Leu - Pro - Pro - Thr - Val - Met - Phe - Pro - Pro - Gln -
                                   170                                                              180
Ser - Val - Leu - Ser - Leu - Ser - Gln - Ser - Lys - Val - Leu - Pro - Val - Pro - Glu - Lys - Ala - Val - Pro - Tyr -
                                   190                                                              200
Pro - Gln - Arg - Asp - Met - Pro - Ile - Gln - Ala - Phe - Leu - Leu - Tyr - Gln - Gln - Pro - Val - Leu - Gly - Pro -
                                   209
Val - Arg - Gly - Pro - Phe - Pro - Ile - Ile - Val.OH
```

Figure 3.3.   Primary structure of Box β-casein A²-5P. (From Grosclaude *et al.* 1973; Ribadeau-Dumas, *et al.* 1972; Eigel *et al.* 1984. Reprinted with permission of the American Dairy Science Association.)

Gordon *et al.* (1972) and Groves *et al.* (1972, 1973) established that the electrophoretically slower-moving proteins in alkaline urea gel electrophoresis which were designated γ_1-, γ_2-, and γ_3-casein in the 1976 ADSA Protein Nomenclature Committee Report (Whitney *et al.* 1976) are fragments of β-casein. These fragments are now classified in the β-casein family on the basis of their primary structure, with γ_1-casein designated as β-casein X-1P (f 29–209), γ_2-casein as β-casein X (f 106–209), and γ_3-casein as β-casein X (f 108–209) (Eigel *et al.* 1984). These fragments are formed by the action of the plasmin in milk (Kaminogawa *et al.* 1972; Eigel 1977; Eigel *et al.* 1979). The N-terminal fragments resulting from the plasmin proteolysis of β-casein have also been identified in milk. They occur in the proteose-peptone fraction and have been detected in both casein and whey. The proteose-peptone component called "8-fast," which moves in alkaline urea gel electrophoresis with a mobility approaching that of the dye front, is the fragment currently called β-casein X-4P (f 1–28) Andrews (1978B). The proteose-peptone component called "8-slow," which moves as a multizonal band somewhat more slowly than "8-fast" in alkaline urea gels, consists of two fragments: β-casein X-1P (f 29–105) and β-casein X-1P (f 29–107) (Eigel and Keenan 1979). The remaining fragments, β-casein X-5P (f 1–

105) and β-casein X-5P (f 1–107), have been identified with the pro-teose-peptone component "5," which moves electrophoretically as a doublet in front of the α_{s1}-caseins in alkaline urea gels (Andrews 1978A; Eigel 1981). While it is known that plasmin is present in the lumen, the exact point at which these fragments are first formed has not been definitely established. Under suitable conditions they will continue to be formed after milk is drawn, since plasmin can remain active unless inhibitors are present or unless it has been inactivated by heating at 80°C for 10 min.

The genetic variants of these proteins are correlated with those of the major β-casein component from which they were formed. At present, the fragments (f 29–209), (f 106–209), and (f 108–209) have been identified in milks containing the β-casein variants A^1, A^2, A^3, and B (Gordon et al. 1972; Groves et al. 1972, 1973). Fragments (f 106–209) for the A^1 and A^2 variants are identical, and the Protein Nomenclature Committee called this fragment from both variants β-casein A^2 (f 106–209). Similarly, the (f 108–209) fragments of the A^1, A^2, and A^3 variants are identical, and all are designated β-casein A (f 108–209). Examination of milks containing the C variant of the major β-casein indicates that β-casein C-1P (f 29–209) is missing (Groves et al. 1972). It has been suggested that the charge reversal in this variant at residue 37 may be related to this phenomenon (Groves et al. 1975). While similar fragments would be expected in the D variant and possibly the E variant, they have not been reported. Also, the specific genetic variants of the other β-casein fragments—(f 1–105), (f 1–107), (f 29–105), and (f 29–107)—have not yet been identified, but it is expected that their genetic nomenclature will be comparable to that of the above fragments.

κ-**Caseins.** The κ-casein family comprises that portion of the α-casein fraction of Warner that is soluble in 0.4 M $CaCl_2$ at pH 7.0 and 4°C and occurs as a mixture of polymers held together by disulfide bonds. An equilibrium is established between the polymers and monomers in a few hours (Vreeman et al. 1977).

When the κ-caseins are converted completely to monomers by reduction of the disulfide bond with mercaptoethanol or another suitable disulfide reducing agent, they possess considerable heterogeneity, consisting of a major carbohydrate-free component and at least six minor components (Vreeman et al. 1977; Doi et al. 1979A,B). This heterogeneity arises from several different sources: genetic differences, variation in carbohydrate content and/or phosphate content, and a possible variation in the para-κ-portion of the molecule. In addition, some para-κ-casein has been observed in purified κ-casein preparations. This undoubtedly is due to a chymosin-like proteolysis subsequent to translation, but more work must be done before it can be concluded that the para-κ-casein observed is a natural constituent of milk or an artifact.

Two genetic variants of the κ-caseins are known: A and B (Thompson 1970; Mackinlay and Wake 1971). In alkaline urea gel electrophoresis in the presence of mercaptoethanol, both variants show multiple bands, with the corresponding A-variant bands possessing greater mobility (Swaisgood 1975B). The A variant tends to be the predominant variant in most breeds (Aschaffenburg 1968A).

The primary structures of the genetic variants of the reduced form of the carbohydrate-free component are illustrated in Figure 3.4. (Jollès et al. 1972A, B; Mercier et al. 1973). κ-Casein B-1P consists of 169 amino acid residues with a calculated molecular weight of 19,007. There is still some question about the presence of the N-terminal pyroglutamyl residue in the native protein, since cyclization may occur during isolation (Swaisgood 1975A).

The bond that is sensitive to rennin (chymosin) hydrolysis has been definitely identified as the bond between the phenylalanine residue at position 105 and the following methionine residue (MacDonald and Thomas 1970; Polzhofer 1972). The hydrolytic products are para-κ-

10   20

PyroGlu – Glu – Gln – Asn – Gln – Glu – Gln – Pro – Ile – Arg – Cys – Glu – Lys – Asp – Glu – Arg – Phe – Phe – Ser – Asp –

30   40

Lys – Ile – Ala – Lys – Tyr – Ile – Pro – Ile – Gln – Tyr – Val – Leu – Ser – Arg – Tyr – Pro – Ser – Tyr – Gly – Leu –

50   60

Asn – Tyr – Tyr – Gln – Gln – Lys – Pro – Val – Ala – Leu – Ile – Asn – Asn – Gln – Phe – Leu – Pro – Tyr – Pro – Tyr –

70   80

Tyr – Ala – Lys – Pro – Ala – Ala – Val – Arg – Ser – Pro – Ala – Gln – Ile – Leu – Gln – Trp – Gln – Val – Leu – Ser –

90   100

Asp – Thr – Val – Pro – Ala – Lys – Ser – Cys – Gln – Ala – Gln – Pro – Thr – Thr – Met – Ala – Arg – His – Pro – His –

105↓106   110   120

Pro – His – Leu – Ser – Phe – Met – Ala – Ile – Pro – Pro – Lys – Lys – Asn – Gln – Asp – Lys – Thr – Glu – Ile – Pro –

130   136   140

Thr – Ile – Asn – Thr – Ile – Ala – Ser – Gly – Glu – Pro – Thr – Ser – Thr – Pro – Thr – Ile – Glu – Ala – Val – Glu –
              Thr (variant A)

148   150   160

Ser – Thr – Val – Ala – Thr – Leu – Glu – Ala – Ser – Pro – Glu – Val – Ile – Glu – Ser – Pro – Pro – Glu – Ile – Asn –
      (variant A) Asp P

169

Thr – Val – Gln – Val – Thr – Ser – Thr – Ala – Val.OH

Figure 3.4. Primary structure of Bos κ-casein B-1P. As indicated, the A variant has a threonine residue at position 136 and an asparic acid residue at position 148. The arrow indicates the point of attack of rennin (MacDonald and Thomas 1970; Polzhofer 1972). Reprinted with permission of the American Dairy Science Association. (From Mercier et al. 1973; Eigel et al. 1984. Reprinted with permission of the American Dairy Science Association.)

casein (residues 1–105) and a macro- or glycomacropeptide (residues 106–169).

The structures of the minor κ-casein components are still uncertain, and considerable disagreement exists between the results of various investigators. It is generally believed that they differ from the major component in that while they have the same primary amino acid sequence, they contain various amounts and types of carbohydrate moieties attached to the polypeptide chain by posttranslational glycosylation (Vreeman *et al.* 1977; Doi *et al.* 1979A, B). Other investigators have concentrated on the structure and point of attachment of the carbohydrate moieties (Tran and Baker 1970; Wheelock and Sinkerson 1970, 1973; Fiat *et al.* 1972; Jollès *et al.* 1972A, 1973, 1978; Fournet *et al.* 1975, 1979; Jollès and Fiat 1979). Wheelock and Sinkerson (1970, 1973) observed the following carbohydrates in κ-casein from which the free sugars had been removed: D-glucose, D-mannose, D-galactose, N-acetyl-D-galactosamine, and N-acetylneuraminic acid. They separated the para-κ-casein and the glycomacropeptides after rennin action and observed that while the N-acetylneuraminic acid was almost completely associated with the glycomacropeptide, the D-mannose was primarily attached to the para-κ-casein and D-galactose and N-acetyl-D-galactosamine were present in both. Fournet *et al.* (1979) isolated three oligosaccharides from κ-casein and established the following structures for two of them and their point of attachment to the κ-casein polypeptide chain:

$$
(1)\ \text{NeuNAc} \xrightarrow{\ \alpha\,2,3\ } \text{Gal} \xrightarrow{\ \beta\,1,3\ } \text{GalNAc} \xrightarrow{\ \beta\,1\ } \text{Thr(133)}
$$

$$
(1)\ \text{NeuNAc} \xrightarrow{\ \alpha\,2,3\ } \text{Gal} \xrightarrow{\ \beta\,1,3\ } \text{GalNAc} \xrightarrow{\ \beta\,1\ } \text{Thr(133)}
$$
$$
\uparrow\ \alpha\,2,6
$$
$$
\text{NeuNAc}
$$

The ADSA Protein Nomenclature Committee (Eigel *et al.* 1984) recommends that these minor κ-caseins be identified temporarily according to their genetic variant and numbered consecutively in their order of increasing relative electrophoretic mobility in alkaline urea gels in the presence of mercaptoethanol (Yaguchi *et al.* 1968) as κ-casein A-1 or κ-casein B-1, etc.

Whey Proteins

Traditionally, the term "whey proteins" has described those milk proteins remaining in the serum or whey after precipitation of the caseins at pH 4.6 and 20°C. The major families of proteins included in this class were originally the β-lactoglobulins, α-lactalbumins, serum albumins, immunoglobulins, and proteose-peptones. However, the proteose-peptone components 5, 8-slow, and 8-fast, are currently assigned to the β-casein family, since they are fragments of β-casein (Andrews 1978A, B; Eigel and Keenan 1979). The proteose-peptone component 3 is a possible breakdown product of the milk fat globule membrane protein (Kanno and Yamauchi 1979). Therefore, the term "whey proteins" should either be employed only as an operational definition or should refer to the major protein families present in whey other than the classical proteose-peptones. The classification of the β-lactoglobulins, α-lactalbumins, and serum albumin should be based on the primary sequence of the amino acids in their polypeptide chains, although gel electrophoresis can still be used to characterize and identify the individual member of each family (Swaisgood 1975B). The immunoglobulins, due to their microheterogeneity, are characterized by their antigenic determinants in accordance with the World Health Organization Nomenclature Report as revised (Nezlin 1972).

β-Lactoglobulins. Aschaffenburg and Drewry (1957A) demonstrated electrophoretically the existence of β-lactoglobulins A and B in western cattle. Since that time, the C variant has been observed in Australian Jersey cattle (Bell 1962) and the D variant in Montbeliarde cattle in France (Grosclaude *et al.* 1966). These genetic variants differ in their electrophoretic mobilities in alkaline starch or polyacrylamide gels in the following order. A > B > C > D (Thompson 1970). In a study of the milk of yaks (*B. grunniens*) in Nepal, Grosclaude *et al.* (1976A) observed an additional variant which they called β-lactoglobulin D_{yak}. While it appears to have the same electrophoretic mobility in alkaline gels as β-lactoglobulin D, it differs in primary structure; therefore, the ADSA Committee on Protein Nomenclature has recommended that it be designated β-lactoglobulin E (Eigel *et al.* 1984). The same variant appears to occur, along with two additional variants, F and G, in the milks of some Balinese cattle (*B. javanicus*) in the Northern Territory of Australia (Bell *et al.* 1981B). The proposed G variant has the same electrophoretic mobility as the E variant on starch gels at pH 8.5, while the proposed F variant moves more slowly. Bell *et al.* (1970) in Australia observed a β-lactoglobulin in the milk of some Droughtmaster cattle in combination with equal amounts of the A or B variant. Its electrophoretic mobility on starch gels at pH 8.5 is slower than that of β-lactoglobulin C. The authors named it β-lactoglobulin Dr (β-lg Dr). Originally they believed that β-lg Dr differed from β-lg A only in containing covalently bound carbohydrates. More recent evidence suggests that the β-lg Dr differs from β-lg A in primary amino acid sequence as well (Bell and McKenzie 1976; Bell *et al.* 1981B).

The primary amino acid sequence of β-lactoglobulin B (Figure 3.5) consists of 162 amino acid residues with a calculated molecular weight of 18,277 (Braunitzer and Chen 1972; Braunitzer *et al.* 1972, 1973; Grosclaude *et al.* 1976A; Préaux *et al.* 1979). It has been proposed that residue 11 is asparagine rather than aspartic acid (Grosclaude *et al.* 1976A), but this was not confirmed by others (Préaux *et al.* 1979; Bell *et al.* 1981B). The amino acid substitutions as they occur in the other genetic variants are indicated in Figure 3.5. The carbohydrate moiety in β-lactoglobulin Dr has been shown to consist of N-acetylneuraminic acid, glucosamine, galactosamine, mannose, and galactose in the ratio of 1.0:3.4:0.9:1.9:0.8 (Bell *et al.* 1970).

α-Lactalbumins. This family of proteins consists of a major component and possibly several minor components. Three genetic variants of

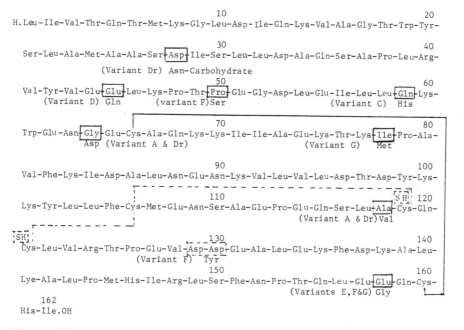

Figure 3.5. Primary Structure of Bos β-Lactoglobulin B. The SH group is postulated to be equally distributed between positions 119 and 121, with the -S-S-bridge location depending upon the position of the SH-group (McKenzie *et al.* 1972). (From Eigel *et al.* 1984. Reprinted with permission of the American Dairy Science Association.)

α-lactalbumin have been identified. The B variant, which is the slower-moving variant in alkaline zonal electrophoresis, is the only one observed in the milk of Western cattle and yaks. Both the A and B variants have been noted in African Fulani and Zebu cattle. Recently, Bell *et al.* (1981A) have observed an α-lactalbumin in the milk of Balinese cattle with an electrophoretic mobility in alkaline gels slower than that of the B variant. They have proposed that it be designated as the C variant and suggest that it possesses one more amide residue, which would explain the slower mobility.

Several minor components have been observed in recrystallized preparations of α-lactalbumin from bovine milk. Aschaffenburg and Drewry (1957B) observed a faster-moving band in paper electrophoresis at pH 8.6 and isolated the protein. It was found to have the same amino acid composition as the major component but contained one hexosamine residue per molecule (Gordon 1971). These investigators tentatively called this protein "satellite" α-lactalbumin. Other researchers have observed three minor components in their α-lactalbumin preparations on starch-gel electrophoresis at pH 7.7 (Hopper and McKenzie 1973A,B). One of these components moves faster than the major component and has the same amino acid composition, except for possibly one less amide group. The other two move more slowly than the major component and have the same amino acid composition but contain carbohydrates. The faster of these components contains N-acetylneuraminic acid in the carbohydrate moiety, while in the slower component, this acid is absent. These authors identified these proteins as α-lactalbumin (F) for the fast component and α-lactalbumin (S1) and (S2) for the slow components in order of decreasing mobility. Barman (1970) isolated a glycosylated form of α-lactalbumin which contained mannose, galactose, fucose, N-acetylglucosamine, N-acetylgalactosamine, and N-acetylneuraminic acid in the ratio of 4.0:1.4:1.0:3.1:1.1:0.64. Later he isolated another protein which he designated α-lactalbumin III (Barman 1973). It differed from α-lactalbumin B in amino acid composition and had only three disulfide bridges instead of four. Considerably more work needs to be done on these minor components, even to the extent of determining their number, before a satisfactory nomenclature can be established.

The complete primary structure of the major α-lactalbumin is shown in Figure 3.6 (Brew and Hill 1970; Brew *et al.* 1970; Vanaman *et al.* 1970). The B variant consists of 123 amino acid residues with a calculated molecular weight of 14,174, and the A variant differs from it only by having Gln instead of Arg at position 10.

α-Lactalbumin is necessary for the synthesis of lactose by its interaction with galactosyltransferase, an enzyme which catalyzes the trans-

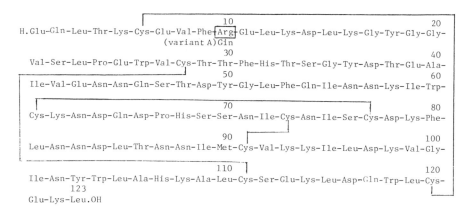

Figure 3.6. Primary structure of Bos α-Lactalbumin B. According to Schewale *et al.* (1984), the structure of α-lactalbumin should be corrected as follows: residue number 43 to Gln, 46 to Asp, 49 to Glu and 82, 83, 87, and 88 to Asp. (From Brew *et al.* 1970; Vanaman *et al.* 1970; Eigel *et al.* 1984. Reprinted with permission of the American Dairy Science Association.)

fer of galactose from uridine diphosphate galactose to N-acetylglucose either as a monomer or as the terminal residue in an oligosaccharide or glycoprotein (Ebner 1971; Ebner and Schanbacher 1974; Brew and Hill 1975; Hill and Brew 1975; Jones 1977). Without α-lactalbumin, glucose is an extremely poor substrate for galactosyltransferase.

Bovine Serum Albumin.   Since Polis *et al.* (1950) crystallized bovine serum albumin from whey and demonstrated that it was identical in all properties investigated to blood serum albumin, except in its electrophoretic behavior at pH 4.0, very little work has been done on this protein as isolated from milk. However, much work has been done on the protein isolated from bovine blood plasma. There is considerable evidence that serum albumin is heterogeneous. For example, Spencer and King (1971) have demonstrated several protein bands by electrophoretic focusing, with two major isoelectric components differing by one unit of charge. The chemical nature of this difference is not known.

In spite of this heterogeneity, a number of investigators have studied the amino acid composition and the primary sequence of serum albumin (King and Spencer 1970, 1972; Brown *et al.* 1971; Spencer 1974; Brown 1975, 1977; Peters and Feldhoff 1975; Reed *et al.* 1980). The total primary amino acid sequence is illustrated in Figure 3.7 and consists of 582 residues with a calculated molecular weight of 66,267

<pre>
                                        10   20
H.Asp-Thr-His-Lys-Ser-Glu-Ile-Ala-His-Arg-Phe-Lys-Asp-Leu-Gly-Glu-Glu-His-Phe-Lys-

                                        30   40
  Gly-Leu-Val-Leu-Ile-Ala-Phe-Ser-Gln-Tyr-Leu-Gln-Gln-Cys-Pro-Phe-Asp-Glu-His-Val-

                                        50   60
  Lys-Leu-Val-Asn-Glu-Leu-Thr-Glu-Phe-Ala-Lys-Thr-Cys-Val-Ala-Asp-Glu-Ser-His-Ala-

                                        70   80
  Gly-Cys-Glu-Lys-Ser-Leu-His-Thr-Leu-Phe-Gly-Asp-Glu-Leu-Cys-Lys-Val-Ala-Ser-Leu-

                                        90   100
  Arg-Glu-Thr-Tyr-Gly-Asp-Met-Ala-Asp-Cys-Cys-Glu-Lys-Glu-Gln-Pro-Glu-Arg-Asn-Glu-

                                        110   120
  Cys-Phe-Leu-Ser-His-Lys-Asp-Asp-Ser-Pro-Asp-Leu-Pro-Lys-Leu-Lys-Pro-Asp-Pro-Asn-

                                        130   140
  Thr-Leu-Cys-Asp-Glu-Phe-Lys-Ala-Asp-Glu-Lys-Lys-Phe-Trp-Gly-Lys-Tyr-Leu-Tyr-Glu-

                                        150   160
  Ile-Ala-Arg-Arg-His-Pro-Tyr-Phe-Tyr-Ala-Pro-Glu-Leu-Leu-Tyr-Ala-Asn-Lys-Tyr-Asn-

                                        170   180
  Gly-Val-Phe-Gln-Glu-Cys-Cys-Gln-Ala-Glu-Asp-Lys-Gly-Ala-Cys-Leu-Leu-Pro-Lys-Ile-

                                        190   200
  Glu-Thr-Met-Arg-Glu-Lys-Val-Leu-Thr-Ser-Ser-Ala-Arg-Gln-Arg-Leu-Arg-Cys-Ala-Ser-

                                        210   220
  Ile-Gln-Lys-Phe-Gly-Glu-Arg-Ala-Leu-Lys-Ala-Trp-Ser-Val-Ala-Arg-Leu-Ser-Gln-Lys-

                                        230   240
  Phe-Pro-Lys-Ala-Glu-Phe-Val-Glu-Val-Thr-Lys-Leu-Val-Thr-Asp-Leu-Thr-Lys-Val-His-

                                        250   260
  Lys-Glu-Cys-Cys-His-Gly-Asp-Leu-Leu-Glu-Cys-Ala-Asp-Asp-Arg-Ala-Asp-Leu-Ala-Lys-

                                        270   280
  Tyr-Ile-Cys-Asx-Asx-Glx-Asx-Thr-Ile-Ser-Ser-Lys-Leu-Lys-Glu-Cys-Lys-Asp-Pro-Cys-

                                        290   300
  Leu-Leu-Glu-Lys-Ser-His-Cys-Ile-Ala-Glu-Val- Glu-Lys-Asp-Ala-Ile-Pro-Glu-Asp-Leu-

                                        310   320
  Pro-Pro-Leu-Thr-Ala-Asp-Phe-Ala-Glu-Asp-Lys-Asp-Val-Cys-Lys-Asn-Tyr-Gln-Glu-Ala-

                                        330   340
  Lys-Asp-Ala-Phe-Leu-Gly-Ser-Phe-Leu-Tyr-Glu-Tyr-Ser-Arg-Arg-His-Pro-Glu-Tyr-Ala-

                                        350   360
  Val-Ser-Val-Leu-Leu-Arg-Leu-Ala-Lys-Glu-Tyr-Glu-Ala-Thr-Leu-Glu-Glu-Cys-Cys-Ala-

                                        370   380
  Lys-Asp-Asp-Pro-His-Ala-Cys-Tyr-Thr-Ser-Val-Phe-Asp-Lys-Leu-Lys-His-Leu-Val-Asp-

                                        390   400
  Glu-Pro-Gln-Asn-Leu-Ile-Lys-Gln-Asn-Cys-Asp-Gln-Phe-Glu-Lys-Leu-Gly-Glu-Tyr-Gly-

                                        410   420
  Phe-Gln-Asn-Ala-Leu-Ile-Val-Arg-Tyr-Thr-Arg-Lys-Val-Pro-Gln-Val-Ser-Thr-Pro-Thr-

                                        430   440
  Leu-Val-Glu-Val-Ser-Arg-Ser-Leu-Gly-Lys-Val-Gly-Thr-Arg-Cys-Cys-Thr-Lys-Pro-Glu-

                                        450   460
  Ser-Glu-Arg-Met-Pro-Cys-Thr-Glu-Asp-Tyr-Leu-Ser-Leu-Ile-Leu-Asn-Arg-Leu-Cys-Val-

                                        470   480
  Leu-His-Glu-Lys-Thr-Pro-Val-Ser-Glu-Lys-Val-Thr-Lys-Cys-Cys-Thr-Glu-Ser-Leu-Val-

                                        490   500
  Asn-Arg-Arg-Pro-Cys-Phe-Ser-Ala-Leu-Thr-Pro-Asp-Glu-Thr-Tyr-Val-Pro-Lys-Ala-Phe-

                                        510   520
  Asp-Glu-Lys-Leu-Phe-Thr-Phe-His-Ala-Asp-Ile-Cys-Thr-Leu-Pro-Asp-Thr-Glu-Lys-Gln-

                                        530   540
  Ile-Lys-Lys-Gln-Thr-Ala-Leu-Val-Glu-Leu-Leu-Lys-His-Lys-Pro-Lys-Ala-Thr-Glu-Glu-

                                        550   560
  Gln-Leu-Lys-Thr-Val-Met-Glu-Asn-Phe-Val-Ala-Phe-Val-Asp-Lys-Cys-Cys-Ala-Ala-Asp-

                                        570   580
  Asp-Lys-Glu-Ala-Cys-Phe-Ala-Val-Glu-Gly-Pro-Lys-Leu-Val-Val-Ser-Thr-Gln-Thr-Ala-

  Leu-Ala.OH
</pre>

96

(Brown 1975, 1977; Reed *et al.* 1980). There is still some uncertainty as to the form of the three Asx and one Glx residues. Reed *et al.* (1980) suggest that they are equally divided between the carboxyl and amide forms.

Immunoglobulins. While blood sera of the various species have been used in much of the work done on the immunoglobulins, the basic information concerning them can be translated to the immunoglobulins in milk, since there is considerable evidence that these proteins are identical in the serum and lacteal secretions of a given species. The immunoglobulins in both systems are complex mixtures of proteins with antibody activity and closely related structures. However, due to their individual heterogeneity, the usual physical-chemical parameters used to characterize and identify the other milk proteins are of little value for these proteins. Their nomenclature must be based on their immunological cross-reactivity with reference proteins primarily of human origin. The system of nomenclature proposed for the bovine immunoglobulins is based on that proposed by the World Health Organization as revised (Aaland *et al.* 1971; Butler *et al.* 1971; Nezlin 1972).

All of the immunoglobulins appear to be glycoproteins that are monomers or polymers of a four-chain molecule consisting of two light polypeptide chains ($\sim$ 20,000 MW) and two heavy chains (50,000–70,000 MW) linked together by disulfide bonds (Figure 3.8) (Gally 1973; Butler 1974; Lascelles 1977). In each molecule of a class or subclass of immunoglobulins, the two heavy chains are identical and have a constant region of 310 to 500 amino acid residues and a variable region of 107 to 115 residues. The light chains in the molecule also are identical and consist of constant and variable regions of equal length, 107 to 115 amino acid residues each. As indicated, the variable N-terminal region of both types of chains are associated with antigen binding, while complement fixation, membrane transport, and species-specific and class-specific antigenic determinants are related to the constant C-terminal region of the heavy chains (Butler 1974; Lascelles 1977). Each class of immunoglobulin has a distinctive heavy chain: γ for IgG, μ for IgM, and α for IgA. Other structural differences among the classes and subclasses reside in the constant region and include variations in covalently bound carbohydrate, heavy chain molecular weight, half-cystine content, and the binding of other polypeptides

---

Figure 3.7. Primary structure of Bovine Serum Albumin. (From Brown 1975, 1977; Reed *et al.* 1980; Eigel *et al* 1984. Reprinted with permission of the American Dairy Science Association.)

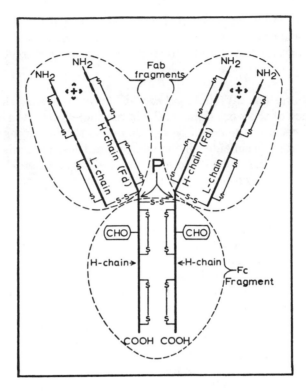

Figure 3.8.   Basic Four—polypeptide chain structural unit of an immuno-globulin. ←⊥→represents the antibody-combining sites.(From Butler 1969. Reprinted with permission of the American Dairy Science Association.)

such as the J-chain or secretory component. Differences also exist which are associated with the light chain constant regions which characterize light chain types. Two such types (κ and λ) are found in bovine immunoglobulins.

Bovine IgG_1 and IgG_2, as individually characterized by immunoelectrophoresis, are normally monomers of the four-chain unit containing 2 to 4% carbohydrate. IgG_1 is the principal immunoglobulin in lacteal secretions (Guidry *et al.* 1980). Elevated levels of IgG_2 may occur in milk during inflammation of the udder (Butler *et al.* 1972; Watson 1976; McGuire *et al.* 1979). Ion-exchange chromatographic patterns obtained during the preparation and fractionation of these immunoglobulins indicate considerable charge heterogeneity (Butler and Maxwell 1972). Characterizations of the IgG subclasses based only on such patterns, without proof of antigenic homogeneity, are subject to question. Ion-exchange patterns typically show three IgG components, which are currently identified as IgG_1, IgG_{2a}, and IgG_{2b}, but there is no definitive evidence that IgG_{2b} is actually a true subclass. Lacteal secretions contain low molecular weight fragments of bovine IgG, some of which are similar in immunoelectrophoretic behavior and molecular size to the fragments of IgG produced by pepsin and papain proteolysis (Goodger 1971; Beh 1973; Butler 1973). Currently, no definitive nomenclature has been assigned to these fragments.

Bovine IgM as identified by immunoelectrophoresis is a pentamer of four-chain units linked together by disulfide bonds between the constant C-terminal regions of the heavy chains of the monomers and has a carbohydrate content of 12.3% (Kumar and Mikalajcik 1973). Each mole of the pentamer contains one mole of covalently bound J-chain and a maximum of 1.2 moles of noncovalently bound J-chain. The J-chain has a molecular weight of 16,500 and contains 9.7 sulfhydryl groups per mole (Komar and Mukkur 1974). The IgM output in bovine milk in mid-lactation is approximately 1 g/day (Guidry et al. 1980).

Bovine IgA in lacteal secretions appears to exist primarily as a dimer of the four-chain unit linked together by disulfide bonds, although some aggregates and degradation products are also present (Porter and Noakes 1970; Duncan et al. 1972; Butler and Maxwell 1972; Butler et al. 1980). It has a carbohydrate content of 8–9%. During Sephadex G-200 chromatography of whey, it is eluted between IgM and IgG. IgA is capable of binding a glycoprotein known as "secretory component" to form a complex called "secretory IgA (SIgA)" (Mach 1970; Radl et al. 1971). This component, when first isolated from milk, was called "glycoprotein a" (Groves and Gordon 1967) but was soon shown to be the free secretory component (FSC), with a molecular weight of 79,000 (Labib et al. 1976). In SDS polyacrylamide gel electrophoresis, some heterogeneity is observed in FSC, probably due to differences in glycosylation. Like IgM, IgA contains covalently bound J-chains (Goodger 1971; Kobayaski et al. 1973).

A protein which possesses reaginic activity but does not appear to belong to the other classes of immunoglobulin is transmitted to suckled calves through colostrum (Hammer et al. 1971; Benton et al. 1976). It has been tentatively accepted as IgE, since it cross-reacts with human IgE (Nielson 1977). In molecular size it lies between IgG and IgM (Hammer et al. 1971; Wells and Eyre, 1971). This immunoglobulin needs to be further investigated and characterized.

While some of the immunoglobulins have been subjected to amino acid and carbohydrate analysis (Kumar and Mikalajcik 1973) and the amino acid sequence of certain portions of the heavy chains of IgG_1 and IgG_2 have been investigated, the heterogeneity observed within each class of immunoglobulins makes the use of this knowledge in the explanation of their physical behavior difficult (Josephson *et al.* 1972); results of these analyses have been omitted.

Milk Fat Globule Membrane (MFGM) Proteins

A thin membrane surrounds the fat globules in milk (King 1955). It contains a complex mixture of proteins and lipids. Some of its proteins are enzymes and are classified, along with the other enzymes present in milk, according to the nomenclature proposed by the International Commission of Enzymes set up by the Union of Biochemistry and adopted by the ADSA Committee on Enzyme Nomenclature (Shahani *et al.* 1973).

The ADSA Committee on Milk Protein Nomenclature (Eigel *et al.* 1984) presented a tentative nomenclature for the new enzyme membrane proteins. While the primary structures of these proteins have not been established, sufficient information exists to obtain an operational definition. The total protein complement of the membrane as observed is dependent upon the past history of the membrane from its formation to its analysis. Both the temperature and the time of storage before analysis can alter the membrane composition and physical state (Wooding 1971). In addition, plasmin has been shown to be associated with preparations of the membrane, and proteolytic products of the membrane protein have been observed in milk (Hoffman *et al.* 1979; Kanno and Yamauchi 1979). Therefore, one should use fresh warm raw milk for the study of the native MFGM protein.

Singer (1974) suggested that membrane proteins can be considered as falling into two classes, either integral or peripheral, depending upon the strength of their association with the membrane. Most, if not all, of the proteins in the aqueous phase of milk can be adsorbed on the surface of the fat globule and, according to Singer's definition, can be considered peripheral membrane proteins. Therefore, the term "milk fat globule membrane proteins" should be limited to Singer's integral membrane proteins. Most workers subject the milk fat globules to a separation and washing procedure before investigating the membrane protein components (Anderson and Cheeseman 1971; Kobylka and Carraway 1972; Mather and Keenan 1975; Kanno *et al.* 1975; Basch *et al.* 1976; Nielson and Bjerrum 1977; Snow *et al.* 1977; Mather *et al.* 1977; Freudenstein *et al.* 1979). The various washing procedures employed may alter the quantitative results obtained, but they appear to be qualitatively similar (Figure 3.9) (Eigel *et al.* 1984).

After the cream has been washed sufficiently to be essentially free of peripheral proteins and contains only integral membrane proteins, the emulsion is broken by freeze thawing or churning by mechanical

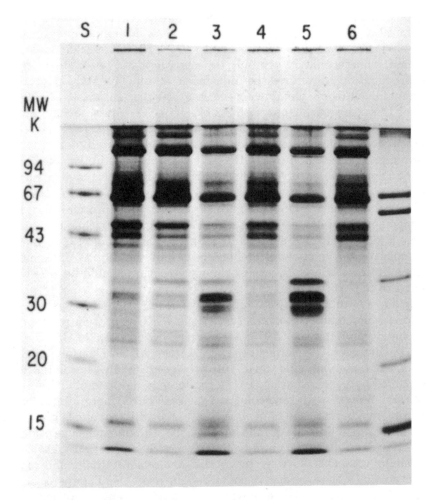

Figure 3.9. SDS gel electrophoresis of MFGM proteins prepared by different washing procedures. S reference proteins: phosphorylase B, bovine serum albumin, ovalbumin, carbonic anhydrase, trypsin inhibitor, α-lactalbumin. (Courtesy of J. Basch and H. M. Farrell, Jr.)

1. 0.1 M imidazole, pH 7, 2 mM $MgCl_2$, and 0.25 M sucrose (Kobylka and Carraway 1973).
2. 0.15 M phosphate, pH 7, 0.15 M NaCl, and 0.25 M sucrose (Anderson and Cheesman 1971).
3. 10 mM Tris, pH 7.5, 1 mM $MgCl_2$, and 0.28 M sucrose (Mather and Keenan 1975).
4. 50 mM phosphate, pH 6.8, and 0.15 M NaCl (Nielsen and Bjerrum 1977).
5. Deionized water (Herald and Brunner 1957).
6. 10 mM Tris, pH 7.4, 2 mM $MgCl_2$, and 0.15 M NaCl (Jarasch et al. 1977).

means at $10°C$ (Anderson and Cheeseman 1971; Kobylka and Carraway 1972, 1973; Mather and Keenan 1975; Mangino and Brunner 1975; Basch et al. 1976; Snow et al. 1977). The fat is then separated from the buttermilk by warming the mixture to $37°-40°C$, followed by centrifugation to remove the milk fat as an oil. The MFGM proteins are removed from the aqueous phase by centrifugation at $100,000 \times g$ for 1 hr or by salting out with 2.2 M $(NH_4)_2SO_4$ (Anderson and Cheeseman 1971; Kobylka and Carraway 1973; Kanno et al. 1975; Basch et al. 1976). Because the operational definition of the individual MFGM proteins is based upon their SDS gel electrophoresis in the presence of 2-mercaptoethanol, any proteins which survive the above extraction procedures will not be observed as MFGM proteins unless they are dispersible in 1% SDS (1.4 g SDS per gram of protein) and 10% 2-mercaptoethanol and detected by staining with either Coomassie blue and/or periodate acid-Schiff reagent (Eigel et al. 1984). Such a complex operational definition has the definite possibility of excluding some proteins which, by their primary structure, would be closely related to the proteins included in this definition.

While a number of investigators have used various SDS gel electrophoretic procedures to characterize their MFGM protein preparation (Mangino and Brunner 1975; Mather and Keenan 1975; Basch et al. 1976; Kitchen 1977; Shimizu et al. 1978), the ADSA Committee (Eigel et al. 1984) recommends the use of the procedure of Laemmli (1970) as modified by Wyckoff et al. (1977). By comparing the mobilities of the MFGM protein with molecular weight standards containing α-lactalbumin, ovalbumin, and phosphorylase b, four distinct zones can be defined (Figure 3.9). Zone A falls between the stacking gel-separating gel interface and phosphorylase b; zone B, between phosphorylase b and ovalbumin; zone C, between ovalbumin and α-lactalbumin; and zone D contains all of the lower molecular weight proteins. The resolution of zone A may require gels of lower acrylamide content. At different gel concentrations, the bands may cross over and yield anomalous results (Anderson et al. 1974). Thus the first differentiation to be made in the classification of the MFGM proteins is on the basis of zone and gel strength. For example, a band occurring in zone A in a 15% gel would be designated MFGM-A_{15}. The members of this group of proteins would be designated by their apparent molecular weight in kilodaltons as interpolated from standard gels containing molecular weight markers other than those employed to establish the zones. After electrophoretic separation on the gel, the staining ability of the protein with Coomassie blue (C) and/or periodate acid-Schiff reagent (S) should be indicated. An individual protein band would be designated as MFGM-A_{15}-127, C, S, if it had an apparent molecular weight of 127,000 and was capable of being stained by both Coomassie blue and periodate acid Schiff reagent (Eigel *et al.* 1984).

Minor Proteins

In addition to the major protein fractions indicated above, some minor proteins have been isolated or identified in milk.

A small amount of the iron-binding protein transferrin, which appears to be electrophoretically and immunologically identical to blood serum transferrin, has been demonstrated in milk (Groves 1971). Disc-gel electrophoresis of transferrin isolated from either blood serum or whey indicates the presence of multiple bands and the existence of polymorphism. The genetic basis of this polymorphism has been investigated in Danish, Banteng, African Watusi, and Ukrainian cattle (Gazia and Agergoard 1980; Steklenev and Marinchuk 1981). This protein contains a covalently bound carbohydrate moiety consisting of N-acetylglucosamine, mannose, galactose, and N-acetylneuraminic acid (Putnam 1975). There appear to be two moles of Fe^{3+} bound per mole of transferrin (Jenness 1982). Molecular weights reported by different investigators differ somewhat but probably are in the range 75,000 to 77,000 (Putnam 1975; Leger *et al.* 1977).

Another iron-binding protein in milk, lactoferrin, was first isolated as a red protein by Sorensen and Sorensen (1939). It appears to be distributed between the casein, whey, and probably the fat globule membrane fraction of the milk, and can be isolated by a number of different procedures either from the acid-precipitated casein or from the whey (Groves 1971). Like transferrin, lactoferrin shows a number of bands on gel electrophoresis which can be partially resolved on diethylaminoethyl (DEAE)-cellulose columns. The fractions show similar absorption spectra and some evidence of polymorphism but differ in their sedimentation behavior. Lactoferrin exists in both a colorless, iron-free and a red, iron-containing form. In the red form, its iron content is 0.12%, indicating a molecular weight of 93,000 based on 2 moles of iron per mole of protein. This value is in good agreement with the sedimentation-diffusion data found by some investigators (Groves 1971; Weiner and Szuchel 1975). However, other investigators report a molecular weight of 77,000 (Leger *et al.* 1977). The amino acid and carbohydrate contents of both this protein and bovine serum transferrin have been compared and found to be significantly different (Gordon *et al.* 1963). They also differ immunologically (Groves 1971). Considerable interspecies cross-reaction among lactoferrins has been observed, but insufficient data have been obtained to establish the degree of homology (Jenness 1982). The lactoferrin content of cow's milk appears to be greatest immediately after parturition, decreasing to a minimum at approximately 60 days and then gradually increasing (Senft and Klobasa 1973).

β_2-Microglobulin, which is homologous to the constant regions of the light and heavy polypeptide chains of immunoglobulin IgG (Peterson et al. 1972), was first isolated in very small amounts by DEAE-cellulose chromatography of the red protein fraction obtained from acid-precipitated casein (Groves 1971). Initially it was called "lactollin," but more recently this crystalline protein was demonstrated to be a stable tetramer of bovine β_2-microglobulin (Groves and Greenberg 1977). The same workers subsequently established the primary sequence (Figure 3.10) and compared it with those of β_2-microglobulins isolated from guinea pig, rabbit, mouse, and human (Groves and Greenberg 1982). The molecule consists of 98 amino acid residues with a calculated molecular weight of 11,636. At very low concentration it is present as a monomer, but at higher concentrations it undergoes a concentration-dependent monomer to tetramer-reversible association (Kumosinski et al. 1981). Preliminary crystallographic studies have also been presented (Becker et al. 1977).

A family of M-1 glycoproteins that are negatively charged at pH 4.5 have been isolated from milk and colostrum by Bezkorovainy (1965, 1967). Upon zonal electrophoresis they yield multiple bands and have an average molecular weight of 10,000. Further fractionation of the M-1 glycoproteins from colostrum (Bezkorovainy and Grohlich 1969) resulted in the isolation of one fraction with a molecular weight of 7200 containing 28.4% carbohydrate and another fraction with a molecular weight of 12,000 containing 39.0% carbohydrate. These proteins contain phosphate and have large amounts of glutamic acid, proline, and

<pre>
                              10   20
H.Ile-Gln-Arg-Pro-Pro-Lys-Ile-Gln-Val-Tyr-Ser--Arg-His-Pro-Pro-Glu-Asn-Gly-Lys-Pro-

                              30   40
  Asn-Tyr-Leu-Asn-Cys-Tyr-Val-Tyr-Gly-Phe-His-Pro-Pro-Gln-Ile-Glu- Ile-Asp-Leu-Leu-

                              50   60
  Lys-Asn-Gly-Glu-Lys-Ile-Lys-Ser-Glu-Gln-Ser-Asp-Leu-Ser-Phe-Ser-Lys-Asp-Trp-Ser-

                              70   80
  Phe-Tyr-Leu-Leu-Ser-His-Ala-Glu-Phe-Thr-Pro-Asp-Ser-Lys-Asp-Glu-Tyr-Ser-Cys-Arg-

                              90   98
  Val-Lys-His-Val-Thr-Leu-Glu-Gln-Pro-Arg-Ile-Val-Lys-Trp-Asp-Arg-Asp-Leu.OH
</pre>

Figure 3.10.   Primary structure of Bos β_2-Microglobulin. (From Groves and Greenberg 1982; Eigel *et al.* 1984. Reprinted with permission of the American Dairy Science Association.)

threonine, but no tryptophan or cysteine. Histidine, tyrosine, and arginine are absent from the larger protein. The carbohydrate moiety in the M-1 glycoproteins apparently contains galactose, glucosamine, galactosamine, and sialic acid. In addition to the M-1 glycoproteins, M-2 glycoproteins have been observed in bovine colostrum, as well as another acid glycoprotein called "α_1-acid glycoprotein" or "orosomucoid" (Jenness 1982). The latter protein consists of a polypeptide chain of 181 residues with five heteropolysaccharide groups linked to asparagine residues (Schmid 1975).

A copper-binding protein, ceruloplasmin, which is a blood serum protein, has been demonstrated in milk by immunodiffusion techniques (Hanson et al. 1967; Poulik and Weiss 1975). It may be the enzyme ferroxidase (EC 1.16.3.1).

Laskowski and Laskowski (1950, 1951) found a trypsin inhibitor in colostrum in relatively large amounts on the first day after parturition and in decreasing amounts thereafter. They crystallized both the inhibitor and the trypsin–inhibitor complex, which consists of a trimer containing three molecules of trypsin and three molecules of the inhibitor (Laskowski et al. 1952).

A kininogen, which when incubated with trypsin or snake venom releases a material with a kinin-like ability to cause the contraction of smooth muscle, has been found in whey and concentrated by DEAE-cellulose chromatography (Leach et al. 1967).

A folate-binding protein (FBP) has been isolated from cow's milk by affinity chromatography on sepharose to which folate has been attached (Salter et al. 1972). Cow's milk contains ~8 mg FBP per liter. The protein has a molecular weight of ~35,000. Cow's milk has also been shown to bind vitamin B_{12}, but the protein responsible has not been isolated from this source, although it has been obtained from the milks of other species (Burger and Allen 1974).

The Enzymes

To complete the picture of the protein complement of milk, one should include the numerous enzymes that have been demonstrated to be present in milk (Table 3.2). Only those enzymes normally present in milk are listed, including those that are constituents of the leukocytes and those that are transferred from the blood of the animal to its milk. Those that result from microbial contamination or other foreign sources are not listed. Some additional enzymes have been detected in milk, but insufficient work has been done to demonstrate conclusively their presence in milk as it comes from the cow. The distribution of the various enzymes in the milk system is rather specific for the particular

Table 3.2. Enzymes of Bovine Milk.

E.C. No.	Enzyme	Reaction	Location
1.1.1.27	Lactate dehydrogenase	$\text{L-CH}_3\text{-CH(OH)-C(O)-O}^- + \text{NAD}^+ \rightleftharpoons \text{CH}_3\text{-C(O)-C(O)-O}^- + \text{NADH} + \text{H}^+$	Plasma
1.1.1.37	Malate dehydrogenase	$\text{L}^-\text{-O-C(O)-CH-CH}_2\text{-C(O)-O}^- + \text{NAD}^+ \rightleftharpoons \text{O-C(O)-C-CH}_2\text{-C(O)-O}^- + \text{NADH} + \text{H}^+$	MFGM
1.2.3.2	Xanthine oxidase	$\text{Xanthine} + \text{H}_2\text{O} + 2\text{O}_2 \rightleftharpoons \text{uric acid} + 2\text{O}_2^- + 2\text{H}^+$	
1.4.3.6	Amine oxidase (Cu containing)	$\text{RCH}_2\text{NH}_2 + \text{H}_2\text{O} + \text{O}_2 \rightleftharpoons \text{RCHO} + \text{H}_2\text{O}_2 + \text{NH}_3$	
1.6.4.3	Lipoamide dehydrogenase (NAD$^+$) (diaphorase)	$\text{NADH} + \text{H}^+ + \text{lipoamide} \rightleftharpoons \text{NAD}^+ + \text{dehydrolipoamide}$	MFGM
1.6.99.3	NADH dehydrogenase (cytochrome C reductase)	$\text{NADH} + \text{H}^+ + \text{acceptor} \rightleftharpoons \text{NAD}^+ + \text{reduced acceptor}$	Serum
1.8	Sulfhydryl oxidase (not 1.8.3.2 thiol oxidase)	$2\text{RSH} + \text{O}_2 \rightleftharpoons \text{R-S-S-R} + \text{H}_2\text{O}_2$	Leuko-cytes Serum
1.11.1.6	Catalase	$2\text{H}_2\text{O}_2 \rightleftharpoons \text{O}_2 + 2\text{H}_2\text{O}$	MFGM Serum
1.11.1.7	Lactoperoxidase	$\text{Donor} + \text{H}_2\text{O}_2 \rightleftharpoons \text{oxidized donor} + 2\text{H}_2\text{O}$	
1.15.1.1	Superoxide dismutase	$2\text{O}_2^- + 2\text{H}^+ \rightleftharpoons \text{O}_2 + \text{H}_2\text{O}_2$	
2.3.2.2	γ-Glutamyl transferase	$\text{L-}\gamma\text{-Glutamyl-peptide} + \text{amino acid} \rightleftharpoons \text{peptide} + \text{L-glutamyl-amino acid}$	
2.4.1.22	Lactose synthase	$\text{UDP-galactose} + \text{D-glucose} \rightleftharpoons \text{UDP} + \text{lactose}$	
2.4.99.1	CMP-N-acetylneuraminate-galactosyl-glycoprotein sialyl transferase	$\text{CMP-}N\text{-acetylneuraminate} + \text{D-galactosyl-glycoprotein} \rightleftharpoons \text{CMP} + N\text{-acetylneuraminyl-D-galactosyl-glycoprotein}$	
2.6.1.1	Aspartate aminotransferase	$\text{L-Aspartate} + \text{2-oxoglutarate} \rightleftharpoons \text{oxaloacetate} + \text{L-glutamate}$	Plasma
2.6.1.2	Alanine aminotransferase	$\text{L-Alanine} + \text{2-oxoglutarate} \rightleftharpoons \text{pyruvate} + \text{L-glutamate}$	
2.7.1.26	Riboflavin kinase	$\text{ATP} + \text{riboflavin} \rightleftharpoons \text{ADP} + \text{FMN}$	
2.7.1.30	Glycerol kinase	$\text{ATP} + \text{glycerol} \rightleftharpoons \text{ADP} + \text{glycerol-3-phosphate}$	
2.7.7.2	FMN adenylyltransferase	$\text{ATP} + \text{FMN} \rightleftharpoons \text{FAD} + \text{P}_2\text{O}_7^{4-}$	
2.8.1.1	Thiosulfate sulfur transferase (Rhodanase)	$\text{S}_2\text{O}_3^{2-} + \text{CN}^- \rightleftharpoons \text{SO}_3^{2-} + \text{SCN}^-$	
3.1.1.	Carboxylesterase (B-Esterase)	$\text{R-C(O)-OR}' + \text{H}_2\text{O} \rightleftharpoons \text{ROH} + \text{RC-OH}$	Plasma
3.1.1.2	Arylesterase (A-Esterase)	$\text{A phenyl acetate} + \text{H}_2\text{O} \rightleftharpoons \text{a phenol} + \text{CH}_3\text{C(O)-OH}$	
3.1.1.3	Triacylglycerol lipase	$\text{Triglyceride} + \text{H}_2\text{O} \rightleftharpoons \text{diglyceride} + \text{fatty acid}$	Serum

EC number	Enzyme	Action	Source
3.1.1.7	Acetylcholine esterase	$CH_3C\!-\!O\!-\!(CH_2)_2\!-\!N^+(CH_3)_3 + H_2O \rightleftharpoons HO\!-\!(CH_2)_2\!-\!N^+(CH_3)_3 + CH_3C\!-\!OH$	MFGM
3.1.1.8	Cholinesterase	$RC\!-\!O\!-\!(CH_2)_2\!-\!N^+(CH_3)_3 + H_2O \rightleftharpoons HO(CH_2)_2\!-\!N^+(CH_3)_3 + RC\!-\!OH$	Serum
3.1.1.34	Lipoprotein lipase	Triglyceride + $H_2O \rightleftharpoons$ diglyceride + fatty acid	Casein
3.1.3.1	Alkaline phosphatase	$R\!-\!O\!-\!PO_3H_2 + H_2O \rightleftharpoons ROH + H_3PO_4$	MFGM
3.1.3.2	Acid phosphatase	$R\!-\!O\!-\!PO_3H_2 + H_2O \rightleftharpoons ROH + H_3PO_4$	MFGM
3.1.3.5	5'-Nucleotidase	A 5' ribonucleotide + $H_2O \rightleftharpoons$ a ribonucleoside + H_3PO_4	MFGM
3.1.3.9	Glucose-6-phosphatase	D-glucose-6-phosphate + $H_2O \rightleftharpoons$ D-glucose + H_3PO_4	MFGM
3.1.3.16	Phosphoprotein phosphatase	Protein phosphate + $H_2O \rightleftharpoons$ protein + H_3PO_4	Plasma
3.1.4.1	Phosphodiesterase	$R\!-\!O\!-\!PO_2H + H_2O \rightleftharpoons R\!-\!O\!-\!PO_3H_2 + R'\!-\!OH$	MFGM
3.1.27.5	Ribonuclease (pancreatic)	Endonucleolytic cleavage to 3' phosphomono- and oligonucleotides ending in Cp or Up	Serum
3.2.1.1	α-Amylase	Hydrolyzes α-1-4 glucan links in polysaccharides at random	Serum
3.2.1.2	β-Amylase	Hydrolyzes α-1-4 glucan links in polysaccarides by removing successive maltose units from the non-reducing end	
3.2.1.17	Lysozyme	Hydrolyzes the β-1-4 glycosidic bond between N-acetylglucosamine and N-acetylmuraminic acid units in mucopolysaccharides	Serum
3.2.1.24	α-D Mannosidase	Hydrolyzes α-D-mannosides by removing α-D mannose from the nonreducing end	
3.2.1.30	β-N-Acetyl-D-glucosaminidase	Hydrolyzes chitobiose and higher analogs and protein derivatives by removing N-acetyl-D-glucosamine from the nonreducing end	
3.2.1.31	β-Glucuronidase	A β-D-glucuronide + $H_2O \rightleftharpoons$ alcohol + D-glucuronic acid	Casein
3.4.21.7	Plasmin	Hydrolyzes peptide bond, preferentially at Lys > Arg	
3.4	Acid protease	Hydrolyzes peptide bond	
3.6.1.1	Inorganic pyrophosphatase	$H_4P_2O_7 + H_2O \rightleftharpoons 2H_3PO_4$	
3.6.1.3	Adenosine triphosphatase (Mg²⁺ activated)	$ATP + H_2O \rightleftharpoons ADP + H_3PO_4$	MFGM
3.6.1.9	Nucleotide pyrophosphatase	A dinucleotide + $H_2O \rightleftharpoons$ 2 mononucleotides	
4.1.2.13	Fructose-bisphosphate aldolase	D-Fructose-1,6-phosphate $\rightleftharpoons$ dihydroxyacetone-phosphate + D-glyceraldehyde-3-phosphate	
4.2.1.1	Carbonic dehydratase (carbonic anhydrase)	$H_2CO_3 \rightleftharpoons CO_2 + H_2O$	
5.3.1.9	Glucose phosphate isomerase	D-Glucose-6-phosphate $\rightleftharpoons$ D-fructose-6-phosphate	

SOURCE: Shahani et al. (1973) and Walstra and Jenness (1982). Reprinted with permission of John Wiley and Sons.

enzyme. Some enzymes are associated with the casein micelles, fat globules, or leukocytes, while others are dispersed in the serum. The report of the ADSA Committee on Enzyme Nomenclature (Shahani *et al.* 1973) and the text by Walstra and Jenness (1984) present more detailed discussions of individual milk enzymes.

STRUCTURE AND CONFORMATION OF MILK PROTEINS

The size, shape, and configuration of the protein molecule are determined not only by its primary structure and composition but also by stearic effects and secondary binding forces such as electrostatic, hydrogen, and hydrophobic bonding. These forces are influenced by the environment of the protein molecule, including such factors as the temperature, pH, and composition of the dispersing medium.

α_{s1}-Caseins

From the primary structure of the α_{s1}-caseins, it can be noted that there are a large number of proline residues (8.5%) which are relatively uniformly distributed throughout the molecule and, therefore, would minimize the formation of such secondary structure as the α-helix. Swaisgood (1982) suggests the possible presence of some β-conformation and a significant number of β-turns. In addition, there is a sufficiently large number of hydrophobic residues, such as proline, valine, leucine, isoleucine, phenylalanine, and tryptophan to yield a Bigelow parameter of hydrophobicity of 1170. These two factors result in a thermodynamically unstable number of nonpolar groups exposed at the surface which, due to hydrophobic bonding, encourage association or polymer formation (Creamer *et al.* 1982; Dosaka *et al.* 1980A; Kato and Nakai 1980; Keshavarz and Nakai 1979). This phenomenon is very complex and is dependent upon temperature, pH, ionic strength, composition of the medium, and the particular genetic variant involved.

Monomeric dispersions of α_{s1}-casein are obtained by employing strongly dissociating media such as 6 M urea at pH 7.3 (McKenzie and Wake 1959), 3 M guanidine-HCl at pH 7.0 (Noelken 1967), anhydrous formic acid, and 26% aqueous methanol with 0.05 N sodium trichloracetate (Swaisgood and Timasheff 1968). However, if sufficiently low ionic strengths or high pH values are employed so that the electrostatic forces can overcome the hydrophobic attraction, monomers can also be obtained (Swaisgood 1982), but at greater ionic strengths, more alkaline pH values are necessary (Dreizen *et al.* 1962; Schmidt *et al.* 1967).

With regard to the conformation of the monomer, attempts have been made to calculate the α-helix content of α_{s1}-casein A and B (Bloomfield and Mead 1975). Depending upon the method employed, values ranging from 8 to 19.6% were predicted for α_{s1}-casein B and 6.4 to 19.67% for α_{s1}-casein A. Optical rotatory dispersion studies of α_{s1}-casein B indicated 4 to 15% α-helix. The degree of helicity is greatly enhanced in organic solvents such as acidic methanol and 2-chloroethanol. From the hydrodynamic properties of the monomer (Table 3.3) and its primary structure, Swaisgood (1982) suggests that the native molecule is neither a globular protein nor a random coil but possesses regions which may approach random coil behavior. In contrast, the monomer in denaturing solvents and at pH 12 behaves as a random coil.

The association of α_{s1}-casein B in the neutral range, pH 6.6, appears to occur in a series of association steps at ionic strengths greater than 0.01 (Figure 3.11) (Schmidt 1970A, 1982). As the ionic strength in-

Table 3.3. Hydrodynamic Properties of the Caseins.

Casein	Experimental Values			Calculated Values	
	$S_{20w} \times 10^{13}$ sec	$[\eta]$ ml/g	Re (nm)	Globular $S_{20w} \times 10^{13}$ sec	Random coil Re (nm)
α_{s1}-Casein					
Native monomer	2.4	10	$3.4_{[\eta]}$ 2.3_s	2.53	4.3
Monomer: pH 12	1.35	19.5	$4.2_{[\eta]}$ 4.3_s	—	4.3
Monomer: 6 M Gdn · Cl	—	19.2	4.2	—	4.3
α_{s2}-Casein					
Native monomer	—	11.4	3.6	2.70	4.4
β-Casein					
Native monomer	1.5	23	$4.4_{[\eta]}$ 3.7_s	2.38	4.5
Monomer: 6 M Gdn · Cl	—	22.2	4.4	—	4.5
κ-Casein					
Native polymer	15.6	9.5	9.7_s	—	—
Monomer: pH 12	1.4	15.1	$3.6_{[\eta]}$ 3.4_s	2.11	4.1
Monomer: 5 M Gdn · Cl	1.88	—	2.5	2.11	4.1
Monomer: 67% HAc, 0.15 M NaCl	1.26	—	3.7	2.11	4.1

SOURCE: Swaisgood (1982). Reprinted with permission of Elsevier Applied Science Publishers, Ltd.

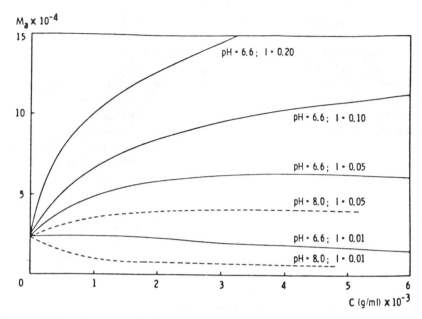

Figure 3.11. The association of α_{s1}-casein B at different values of pH and ionic strength. Molecular weights determined at 20°C using the light-scattering technique. (From Schmidt 1982. Reprinted with permission of Elsevier Applied Science Publishers, Ltd.)

creases, the ζ-potential of the molecule decreases, allowing the hydrophobic attraction to overcome the repulsive forces due to the charge on the protein. The steps in the reaction are accompanied by the formation of about 13 hydrophobic and 1–2 hydrogen bonds. The association of α_{s1}-casein C is stronger than that of the B variant. This has been ascribed to its lower charge resulting from the substitution of a glycine residue for glutamic acid at position 192 (Schmidt 1970B). However, the D variant associates only as strongly as the B variant, in spite of its greater negative charge. This has not yet been adequately explained (Schmidt 1982). Ultracentrifugal studies of genetic variants B and C at 2°, 9°, and 14°C indicate a positive linear dependence of the mean molecular weight upon the concentration over the range studied, with the slopes increasing with increasing temperature, suggesting the rapid endothermic association characteristic of hydrophobic bonding.

At pH 2.5 on the positive side of the isoelectric point, strong association is observed. At concentrations as low as 5×10^{-5} g/ml, the apparent molecular weight did not decrease below that of the dimer (Schmidt 1970C).

On the alkaline side of the isoelectric point between pH 8.0 and 9.85, the behavior is complex and is dependent upon the ionic strength and species of the ions present. In a detailed study of this phenomenon, Swaisgood and Timasheff (1968) demonstrated that at low ionic strengths (0.02–0.05) a time-dependent association occurs between the monomers and dimers and some higher polymers, while at higher ionic strengths (0.1–0.3) equilibrium is rapidly attained, with the concentration of polymers present increasing with the increasing concentration of protein. At an ionic strength of 0.2 and 2°C, the association decreases markedly between pH 7.5 and 10.5, with the sedimentation constant decreasing from ~6 to ~1 Svedberg units. This change has been attributed to the dissociation of protons from tyrosine and the ϵ-amino group of lysine, which would greatly increase the negative charge of the α_{s1}-casein molecule. Viscosity data suggest that each step in the association is accompanied by a conformational change from a compact globular dimer to a rigid rod or stiff coiled tetramer to a random coiled hexamer.

Changes in the ionic environment also influence the association phenomenon. When sodium trichloroacetate is used in place of NaCl, the observed sedimentation constant is decreased at pH 8.0 and 10.0. Replacement of the NaCl by Tris-Cl at 0.1 ionic strength also lowers the sedimentation constant, possibly due to the absence of sodium ions, which have been observed to bind to α_{s1}-casein (Swaisgood and Timasheff 1968). When calcium ions are present, the association of α_{s1}-casein increases considerably at pH 6.6, and at sufficiently high calcium levels precipitation occurs. From studies of the interaction of calcium with α_{s1}-casein (Holt et al. 1975; Parker and Dalgleish 1977A,B; Dalgleish and Parker 1979, 1980; Horne 1979), one can deduce that as the calcium ions bind to the active sites on the α_{s1}-casein, the negative charge is reduced and reactive sites are formed through which association may take place. The coagulation may be attributed to a polyfunctional condensation reaction which is dependent upon the functionality of the complexes and their charge. Dalgleish and Parker (1979) obtained a number-average functionality for the complexes of ~2. This was confirmed by electron microscopy, which demonstrated that the aggregation occurs through the formation of bent chains (Dosaka et al. 1980B). Elevated pressures were observed by Schmidt and Payens (1972) to reduce the tendency of α_{s1}-caseins to associate.

α_{s2}-Caseins

Studies of the structure and conformation of the α_{s2}-caseins have been limited. From their primary structure, one can conclude that they are the most hydrophilic of the caseins, with a hydrophobicity comparable to that of many globular proteins. The hydrodynamic properties (Table 3.3) suggest that the configuration of the native monomer is similar to that of α_{s1}-caseins.

The α_{s2}-caseins associate in a series of consecutive steps in a manner similar to that of α_{s1}-casein (Snoeren *et al.* 1980). Ultracentrifugal studies at 20°C and pH 6.7 suggest that the association increases with ionic strength to a maximum at 0.2 and then decreases at higher ionic strengths (Figure 3.12). However, the hydrodynamic volume of the molecule increases with ionic strength throughout the range investigated, due to the peculiar amino acid sequence of α_{s2}-casein, in which the C-terminal end has a cluster of 13 positive charges at this pH and the residual portion has a strong negative charge. At low ionic strengths the intramolecular attractions between these regions would be strong, yielding a more compact molecule, while the electrostatic repulsion due to the residual negative charges would discourage association. As the ionic strength increases, the repulsion due to residual charges is weakened and electrostatic interaction may take place between the positive C-terminal end of one molecule and the negative

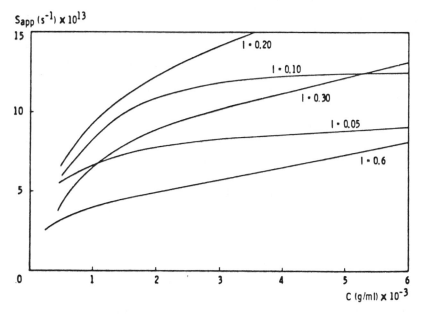

Figure 3.12. The association of α_{s2}-casein at 20°C and pH 6.7 at different ionic strengths. (From Schmidt 1982. Reprinted with permission of Elsevier Applied Science Publishers, Ltd.)

portion of an adjacent molecule, thus increasing not only the association but also the molecular hydrodynamic volume. Any further increase in ionic strength weakens the intermolecular electrostatic interaction and association decreases, while the molecular hydrodynamic volume continues to increase (Snoeren *et al.* 1980).

β-Caseins

The outstanding physical characteristic of the major component of the β-casein family is its temperature-dependent association (Swaisgood 1973). From its primary structure, it can be seen to be the most hydrophobic of the major caseins. The A^2 variant has a Bigelow hydrophobicity of 1335, and the 21 N-terminal amino acid residues at neutral pH carry a net negative charge. The balance of the molecule is essentially neutral, with a large number of hydrophobic residues. In addition, there is a relatively large number of β-turns, with the resultant exposure of a considerable number of the nonpolar groups. Therefore, the tendency for hydrophobic bonding should be great, and an increase in association with increasing temperature is to be expected.

In dissociating solvents such aa 6.6 M urea and 3 M guanidine-HCl, the major β-casein exists as a monomer independent of temperature, pH, or ionic strength. However, when dissociating solvents are absent, it is very dependent upon temperature and somewhat dependent on pH and ionic strength (Payens and van Markwijk 1963). From sedimentation and intrinsic viscosity measurements, the monomer appears to be either highly asymmetric, with an axial ratio of 12.2–15.9, or a random coil (Evans *et al.* 1971). Small-angle x-ray scattering data suggest a radius of gyration of 4.6 nm, which corresponds to that of a 16-nm rod (Andrews *et al.* 1979). This asymmetry is not consistent with the spherical shapes observed by electron microscopy (Buchheim and Schmidt 1979). Therefore, the random coil, which is consistent with the optical rotatory dispersion measurements of Herskovits (1966) and the viscosity measurements of Noelken and Reibstein (1968), seems more likely even though the predicted radius of gyration would be larger than observed, 5.1–5.7 nm (Swaisgood 1982). In dissociating solvents such as 6 M guanidine-HCl, the viscosity is essentially unchanged, suggesting little folded structure in the monomer. However, based on Chou-Fasman analyses of the primary sequence, it has been calculated that the monomer should contain 10% α-helix, 13% β-sheet, and 77% unordered structure (Andrews *et al.* 1979). The α-helix content was also calculated by Bloomfield and Mead (1975); depending upon the method used, values ranging from 8.6 to 19.6% were obtained. Optical rotatory disperson studies indicated 3 to 14% α-helix.

With heating from 5 to 45°C, thermal changes in conformation in the major β-casein are observed by spectral methods (Garnier 1966). From measurements of the optical density at 286 nm and of the specific optical rotation at 436 nm, a rapidly reversible endothermic transition (ΔH 30 kcal/mole) with a half-transition temperature of 23–24°C is observed. The optical rotatory dispersion data suggest a decrease in the poly-L-proline II structure (12 to 5%) and a slight increase in α-helix (11 to 16%) with increasing temperature. This transition probably occurs prior to association, since it is rapid, and the carboxyacyl derivative of the monomer, which does not polymerize with increasing temperature, also demonstrates the optical rotatory disperson thermal transition.

The strong temperature-dependent association of the major β-casein has been investigated by combined ultracentrifugal and light-scattering measurements (Payens *et al.* 1969; Payens and Heremans 1969; Schmidt and Payens 1972; Payens and Vreeman 1982). The rate of equilibration, in contrast to that of α_{s1}-casein, is not rapid compared to the rate of sedimentation. Also, unlike α_{s1}-casein, β-casein does not form intermediate-sized polymers; instead, a rather narrow-sized distribution of polymers results. This type of association suggests micelle formation similar to that of ionic detergents, especially since the primary structure of the monomer, with its highly charged N-terminal end and the very hydrophobic character of the remainder of the molecule, resembles an ionic detergent such as sodium dodecyl sulfate (Schmidt 1982). This is further emphasized by the observation that, below a certain critical micelle concentration, association does not occur (Figure 3.13). Similar to ionic detergents, the critical micelle concentration decreases with increasing temperature and/or ionic strength. The effect of pressure on the association of the major β-casein is complex, resulting in a decrease in polymerization up to 1500 kg/cm² followed by a rapid increase above that value (Payens and Heremans 1969).

The configuration and association of the minor β-caseins have not been investigated to any extent, but from their primary structure one can conclude that β-casein X-4P-(f 1–28), -(f 1–105), and -(f 1–107) and β-casein X-1P-(f 29–105) and -(f 29–107) are more hydrophilic than the main β-casein component and that β-casein X-1P-(f 29–209) and β-casein-(f 106–209) and -(f 108–209) will be more hydrophobic. The hydrophobicities of the last three caseins are the highest of all the caseins, with calculated Bigelow hydrophobicities of 1386 to 1511 (Swaisgood 1973). Therefore, we would expect these to demonstrate a strong temperature-dependent association; β-casein B-(f 106–209) and -(108–209) are insoluble at pH 8.0 and 25°C but soluble at 3°C, while β-casein A-(f 108–209) is soluble at both temperatures. Sedimentation studies

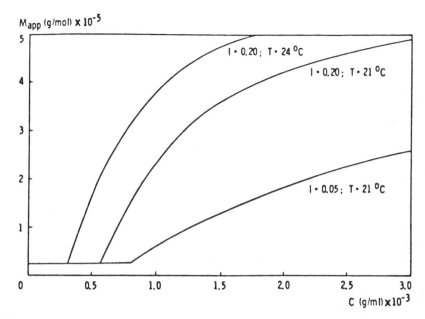

Figure 3.13. The association of β-casein at low concentrations at different values of ionic strength and temperature. From light-scattering measurements at pH 7.0. (From Schmidt and Payens 1972; Schmidt 1982. Reprinted with permission of Elsevier Applied Science Publishers, Ltd.)

of these proteins indicate that, except for β-casein B-(f 108–209), which shows nonideal behavior at pH 8.0, they exist as monomers at alkaline pH values and low ionic strengths (Groves and Townend 1970). In the presence of guanidine-HCl, β-casein A^3-(f 29–209) sediments as a monomer. The more hydrophilic minor β-caseins listed above have been observed as monomers in veronal buffer, pH 7.0 and ionic strength 0.1 (Kolar and Brunner 1970).

κ-Caseins

The heterogeneity of κ-caseins due to posttranslational glycosylation and disulfide interchange makes the interpretation of information regarding their size, shape, and conformation difficult. Evidence obtained in dissociating solvents such as 5 M guanidine-HCl, pH 5.0, 7 M, urea pH 8.5, and 33% or more acetic acid in 0.15 M NaCl indicates a heterogeneous mixture of polymers linked together by intermolecular disulfide bonds with mean molecular weights of 88,000 to 118,000.

In aqueous salt systems, the polymer sizes are larger than predicted by disulfide bonds alone and yield sedimentation constants of 13.2 to 19 Svedberg units, depending upon the preparation with molecular weights of the order of 650,000. The polymer size is not homogeneous, but the distribution of sizes is not great. The primary structure of the κ-caseins indicates that a monomer unit possesses an amphiphilic character similar to that of the other caseins (Hill and Wake 1969). The N-terminal portion (the para-κ-casein) is hydrophobic, with a Bigelow hydrophobicity of 1310, and the C-terminal portion (the macropeptide) is hydrophilic, with a hydrophobicity of 1083. Thus, in addition to the disulfide bridges, the molecules can orient themselves in the association complexes with their hydrophobic ends in the interior of the polymers and their hydrophilic ends on the surface (Hill and Wake 1969). The concentration dependency of their reduced viscosity supports this concept by indicating spherical geometry in the polymers; in the presence of dissociating solvents, similar data indicate a random chain behavior (Swaisgood 1973). Clark and Nakai (1972) obtained additional evidence of this picture of the κ-casein polymers from fluorescence studies. Increases in temperature enhance the amount of aggregation (Yoshida 1969).

At pH 12, the disulfide and noncovalent bonds are both broken, and the monomer with a sedimentation constant of 1.45 Svedberg units is released. From frictional ratios, the monomer appears to exist as a coil with a diameter of 16 Å and a length of 150 Å. Analysis of the primary structure of κ-casein (Loucheux-Lefebvre et al. 1978) suggests considerable secondary structure in the monomer. 23% α-helix, 31% β-sheets, and 24% β-turns. In contrast, other investigators, using several different approaches, obtained α-helix contents ranging from 0 to 20.8% (Bloomfield and Mead 1975). Circular dichroism spectra on the monomer indicated 14 and 31% for α-helix and β-sheet, respectively (Loucheux-Lefebvre et al. 1978). An earlier study of the optical rotatory dispersion of the κ-casein monomer yielded values for the α-helix content ranging from 2 to 16% (Herskovits 1966).

Reduction of the disulfide bonds and alkylation of the κ-caseins result in the elimination of any disulfide polymers. These reduced and alkylated κ-caseins sediment as monomers in dissociating agents. Ultracentrifugal analysis of the carbohydrate-free κ-casein B-1P in a disulfide-reducing buffer system indicated an association comparable to that of β-casein (Vreeman et al. 1977; Vreeman 1979; Payens and Vreeman 1982). A critical micelle concentration is observed which decreases with increasing ionic strength (Figure 3.14). The micelle appears to consist of 30 monomers and is independent of ionic strength. The mechanism of micelle formation is probably similar to that indicated for the polymers observed above in the absence of disulfide-

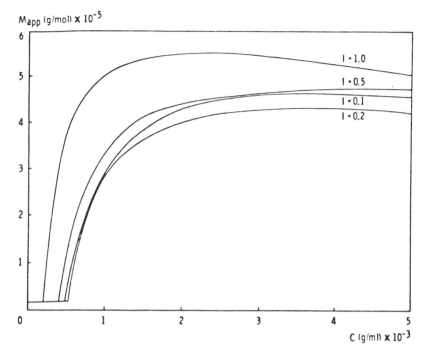

Figure 3.14. The association of carbohydrate-free sh-κ-casein at 20°C and pH 7.0 at different ionic strengths. Molecular weights determined by ultracentrifugation using the approach to equilibrim method. (From Vereman 1979; Schmidt 1982. Reprinted with permission of Cambridge University Press.)

reducing agents, except that in this case no disulfide bonds are present. The electrical forces which should be responsible for limiting micellar growth are small. Therefore, Vreeman suggests the possibility of a spatial requirement or an entropic repulsion of the hydrophilic macropeptide of the molecule.

Bovine Serum Albumin

The configuration of bovine serum albumin isolated from milk has not been investigated, but extensive investigations of this protein isolated from bovine blood serum have been made. The protein exhibits at least three different kinds of heterogeneity: (1) due to polymer formation, (2) related to the sulfur linkages in the molecule, and (3) microheterogeneity. Fractionation of bovine serum albumin on DEAE-Sephadex A-50 resulted in a monomer and two dimer fractions (Janatova *et al.* 1968), one with a relatively high sulfhydryl content which could not be converted to the monomer by thioglycolate reduction and the other free of sulfhydryl groups which could be partially reduced to the monomer. Small quantities of higher polymers were observed. The degree of polymer formation depends upon the freshness of the sample and the method of preparation (Freeman 1970). The monomer exists in at least four forms: one with a reactive sulfhydryl group called "mercaptalbumin" and three nonmercaptalbumins (Noel and Hunter 1972; Hagenmaier and Foster 1971). Two of the nonmercaptalbumins are mixed disulfides with cysteine and glutathione and are relatively *stable to* splitting by mercaptoethanol, probably due to disulfide pairing such that the cysteine and glutathione are buried in the molecule. The other is a relatively easily reduced mixed disulfide of cysteine that may be a product of the method of preparation or the age of the sample (Foster 1977; Noel and Hunter 1972). This heterogeneity has also been investigated by isoelectric focusing (Salaman and Williamson 1971; Spencer and King 1971; Ui 1971; Wallevik 1973). The microheterogeneity is apparent in the conformational transition occurring near pH 4 (Bhargava and Foster 1970; Kaplan and Foster 1971). Foster (1977) postulated that the protein consists of a continuum of molecular species which, while grossly similar, differ in their inherent stability and in the pH range in which they undergo the transition from the native form (N) to the acid form (F). This explanation was further substantiated by studies of the thermal denaturation of bovine serum albumin and its solubility behavior. The source of the microheterogeneity has been intensively investigated (Foster 1977), and possible factors such as differences in their three-dimensional conformation, the presence of bound impurities, and variations in the primary sequence have been tentatively ruled out. He proposed that it is due to posttranslational modification of amino acid side chains and/or disulfide isomerization.

Due possibly to the above mentioned heterogeneity, there is some variability with regard to the conclusions reached by various workers concerning the structure and configuration of bovine serum albumin. Brown (1977) proposed two possible models based on the primary sequence of the protein. He demonstrated that the molecule could possess a triple domain structure with three very similar domains: residues 1–190, 191–382, and 383–582. Each domain could then consist of five helical rods of about equal length arranged either in a parallel or an antiparallel manner. His second model consisted of the following: (1) a lone subdomain (1–101); (2) a pair of antiparallel subdomains, with their hydrophobic faces toward each other (113–287); (3) another pair of subdomains (314–484); and (4) a lone subdomain (512–582). These structures are supported by the observed helical content of bovine serum albumin (54–68%) (Reed *et al.* 1975) and by the location of the proline residues and reactive binding sites (Anderson *et al.* 1971; Taylor and Vatz 1973; Taylor *et al.* 1975A,B). Other investigators suggest two to nine domains (Foster 1977).

As the pH of bovine serum albumin is lowered below its isoelectric point, numerous changes in its physical and chemical properties occur. The intrinsic viscosity and molecular volumes increase markedly (Raj and Flygare 1974), the solubility in 3 M KCl decreases drastically and the ability to bind pentane essentially disappears (Foster 1977), changes are observed by differential spectrophotometry and perturbation spectroscopy (Sogami 1971; Sogami and Ogura 1973), and fluorescent studies indicate a decrease in fluorescence (Noel and Hunter 1972; Halfman and Nishida 1971; Rudolph *et al.* 1975; Sogami *et al.* 1973; Ivkova *et al.* 1971). Optical rotatory dispersion studies indicate a decrease in α-helix content from $\sim 51\%$ to a plateau of $\sim 44\%$ at pH 3.6–3.9, followed by a further decrease to $\sim 35\%$ at pH 2.7 (Sogami and Foster 1968). Titration curves of bovine serum albumin in the carboxyl region suggest that as many as 40 of the carboxyl groups are masked in the N-form, possibly due to ion pair formation. A statistical mechanical model has been proposed to account for this phenomenon (Arvidsson 1972). Based on the observed changes, it has been concluded that, as the pH is lowered, the domains separate in at least three steps; N–F', F'–F, and F–E transitions (Foster 1977). The N–F' due to the breakage of ion-pair bonds results in a dumbell-like structure. The F'–F phase involves the exposure of considerable hydrophobic surface with decreasing solubility (Wilson and Foster 1971; Zurawski *et al.* 1976; Hilak *et al.* 1974). Finally, the F–E transition occurs, resulting in an extended thread with knots due to the repulsion of the cationic groups.

In addition to the acid transition, a change has been observed in the neutral range and has been designated the "N–B transition." This phenomenon has been observed in titration, dye binding, fluorescence, hydrogen exchange, nuclear magnetic resonance, and optical rotatory dispersion studies (Harmsen *et al.* 1971; Foster 1977), and resembles the N–F transition with less loss of helix content. While the evidence is inconclusive, other observations of bovine serum albumin suggest additional transitions (Nikkel and Foster 1971; Stroupe and Foster 1973; White *et al.* 1973).

β-Lactoglobulins

Extensive studies have been made of the structure and conformation of β-lactoglobulin. In the pH range from 5.2 to 7.5, all genetic variants of β-lactoglobulin investigated have been shown to exist primarily as dimers. Based on small-angle x-ray diffraction, crystallography, and hydrodynamic properties (Bell *et al.* 1970; Zimmerman *et al.* 1970; McKenzie 1971A; Gilbert and Gilbert 1973), the dimer has been shown to consist of two spheres with a radii of 17.9 Å and a distance from center to center of 33.5 Å joined so as to possess a dyad axis of symmetry. It was originally thought that hybrid dimers could not exist, but by applying moving boundary theory to free-boundary electrophoretic results, the existence of a hybrid dimer of β-lactoglobulins A and B was clearly established (Gilbert 1970).

On the acid side of the isoionic point, especially below pH 3.5, the dimer dissociates into monomers, with the extent of dissociation increasing as the pH is lowered. It was concluded that a rapid monomer–dimer equilibrium existed. The dissociation constants of the reaction varied with the genetic variant involved and increased with increasing temperature. Albright and Williams (1968) made a detailed study of the equilibrium by sedimentation equilibrium measurements (Figure 3.15) which demonstrated not only the effect of pH but also that the dissociation increased with decreasing concentration and ionic strength. While the effects of pH and ionic strength suggest that the

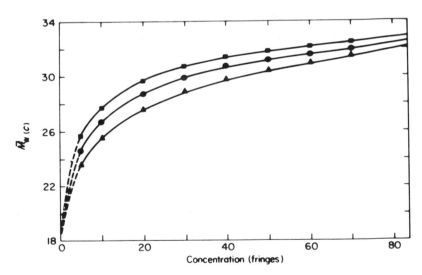

Figure 3.15. The dissociation of β-lactoglobulin B at low pH. Idealized curves for the weight-average molecular weight as a function of concentration from sedimentation equilibrium measurements. ■ = pH 2.58, ionic strength 0.15; ● = pH 2.20, ionic strength 0.15; ▲ = pH 2.58, ionic strength 0.10. 40.2 fringes · 10 g/liter. (From Albright and Williams 1968. Reprinted with permission of AVI Publishing Co., Westport, Conn.)

association is due to hydrophobic bonding, the effect of temperature does not support this postulate, and other types of noncovalent bonds such as hydrogen bonds cannot be eliminated. The dissociation no doubt results from electrostatic repulsive forces due to the increased positive charge on the monomers at lower pH.

There does not appear to be much change in monomer configuration during this transition. Little change is observed at pH values of 3.5 and below in the optical rotatory dispersion, circular dichroism, or the emission maximum of tryptophyl fluorescence (Mills and Creamer 1975). There may be a small change in the perturbation spectrum of the tryptophan residues, and their fluorescence increases significantly (Townend et al. 1969; Mills and Creamer 1975). A slight increase in the radius of the monomer at low pH has been suggested based upon hydrodynamic and x-ray crystallographic data (Swaisgood 1982). Some disagreement exists in the literature on the interpretation of optical rotatory dispersion and circular dichroism data with regard to the detailed secondary and tertiary structure in this pH range. McKenzie (1967) postulates approximately 33% α-helix, 33% β-configuration, and 33% disordered chain, while Townend et al. (1967) propose 10% α-helix, 47% β-conformation, and 43% disordered chain. While the numbers are somewhat variable, the interpretations proposed by Townend and co-workers are in general agreement with those predicted from the primary sequence (Deckmyn and Préaux 1978).

On the alkaline side of the ioionic point, dissociation of the dimer also occurs, but there is some disagreement as to its extent. McKenzie (1967) states that it becomes appreciable at pH 7.5 and above and increases with increasing pH. However, Zimmerman et al. (1970) observe that while it is appreciable in this region, the equilibrium constant is essentially unchanged in the pH range of 6.9 to 8.8. Marked reversible conformational changes accompany this dissociation, as indicated by changes observed in optical rotation and optical rotatory dispersion (Figure 3.16). The transition apparently is also associated with the titration of an abnormal carboxyl group per monomer and the increased exposure of the solvent to the tyrosyl and tryptophyl residues that is observed. The single sulfhydryl group also appears to have increased reactivity, with the C variant possessing the slowest rate of reaction.

As the pH is increased to above 8.0, not only does the dissociation of the dimer continue, but a time-dependent aggregation occurs. Addition of disodium ethylenediaminetetraacetate (EDTA) tends to slow this reaction, suggesting that this reagent, by binding copper and other ions involved in the oxidation of sulfhydryl groups prevents some of the aggregation by slowing the formation of intermolecular disulfide bonds. This aggregation is apparently accompanied by a slow

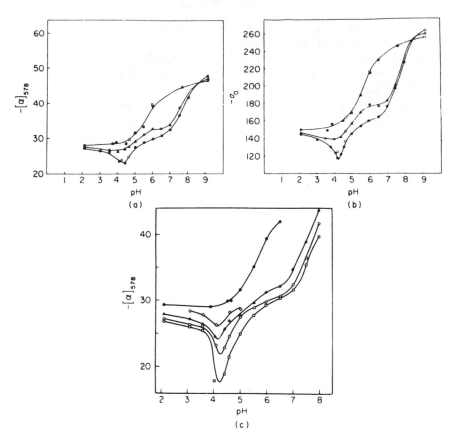

Figure 3.16. pH-dependent conformational transitions for β-lactoglobulins A, B, and C. (a) Specific rotation at 578 nm at 20°C. (b) Parameter a_0 in Moffit Yang equation at 20°C. Symbols for (a) and (b): ● = A variant, ▲ = B variant, ■ = C variant. (c) The effect of temperature on specific rotation at 578 nm for the A variant near pH 4.5: ◇ = 45°C; △ = 30°C; ○ = 20°C; □ = 10°C. The C variant at 20°C (■) is shown for comparison. (From McKenzie and Sawyer 1967; McKenzie et al. 1967. Reprinted with permission of AVI Publishing Co., Westport, Conn.)

time-dependent change in the optical rotatory dispersion after the initial transition. The rate of change is more rapid at 3°C than at 20°C and is dependent upon the genetic variant involved, with A > B > C.

In the isoelectric region, octamerization occurs in addition to the monomer–dimer association (Gilbert and Gilbert 1973). Between pH 3.5 and 5.2, the dimers of both the A and B variants associate to form octamers with maximum association at pH 4.6. The association of the A variant is much stronger than that of the B variant. These variants can also form mixed octamers. However, the other genetic variants investigated do not appear to octamerize to any detectable extent. The association is rapid, as indicated by sedimentation velocity studies, with the equilibrium constant for the reaction decreasing with increasing temperature. The model proposed for the octamer is illustrated in Figure 3.17. The octamer and the dimer were assumed to be the predominant species, and higher polymers were unlikely due to the stereo-

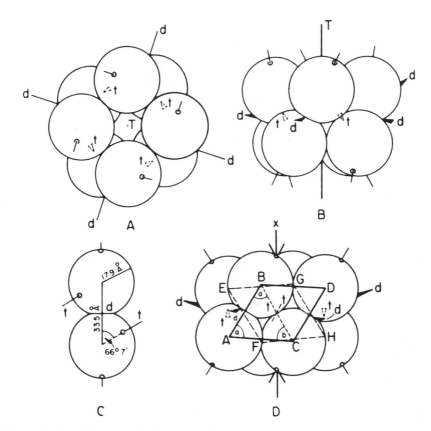

Figure 3.17. Staggered structures for the octomer of β-lactoglobulin A. (A) Top view, 422 symmetry; d = dyad axis of symmetry, t = octomer bond. (B) Side view, 422 symmetry; T = tetrad axis of symmetry. (C) Dimer structure. (D) 222 symmetry, X = overall dyad axis of symmetry. The preferred structure is 422. (From Green 1964; Timasheff and Townsend 1964. Reprinted with permission of AVI Publishing Co., Westport, Conn.)

chemistry of the polymerization (Timasheff and Townend 1964). At 0°C only small amounts of tetramer and hexamer were observed, but at higher temperatures or lower concentrations, increasing quantities of these intermediates were observed. Sedimentation equilibrium studies of β-lactoglobulin A at pH 4.6, ionic strength 0.2, and 16°C suggested that the data could be described by an indefinite association (Adams and Lewis 1968). However, column chromatography on Sephadex G-100 at pH 4.55 and 4°C resulted in two fractions: one with a higher and one with a lower association constant (Roark and Yphantis 1969). The fraction with the high association constant yielded results which were incompatible with the indefinite association model and supported the model of Timasheff and Townend (1964). The reason for the two fractions is still unknown. With regard to the mechanism of octamerization, the negative enthalpy and large negative entropy accompanying the change suggest that hydrophobic bonding is not important. The pH dependence with maximum octamerization at pH 4.6 suggests that carboxyl groups are involved, with the possible formation of hydrogen bonds between the protonated carboxyl groups. It is known that four carboxyl groups are protonated per monomer at this pH. Examination of the primary sequence of β-lactoglobulin A indicates the presence of three carboxyl groups in the vicinity of the aspartic acid residue at position 64, which is responsible for genetic variation. Replacement of this amino acid by glycine in the B variant would reduce the number of water molecules released by the monomers during octamerization if this were the site of attachment. Therefore, the change in entropy should be more negative in the octamerization of the B than the A variant. Since this is actually the case, it suggests that octamerization occurs at this site (Swaisgood 1982). It is of interest to note, however, that the Dr variant does not associate to form octamers even though it possesses an aspartic acid residue at position 64. Perhaps that carbohydrate moiety in this variant is close enough to the reactive site to prevent octamerization sterically (Bell *et al.* 1970).

Marked changes are observed in the optical rotation and optical rotatory dispersion in the isoionic region, suggesting that conformational changes occur in the β-lactoglobin molecules prior to octamer formation (Figure 3.16). The change in the optical rotatory dispersion parameter with pH suggests the binding of one proton per monomer in variants B and C but of two protons in variant A. This behavior adds additional support to the proposed involvement of the aspartic acid residue at position 64 in the octamerization. The difference in the titration curve of the C variant from that of variants A and B suggests a conformational change in the region of position 59, since a histidine is present in this location in variant C and is replaced by a glutamine residue in variants A and B.

α-Lactalbumins

Sedimentation velocity studies of α-lactalbumin on the alkaline side of the isoionic point result in sedimentation constants of 1.73 to 1.98 Svedberg units (Wetlaufer 1961; Kronman and Andreotti 1964; Szuchet-Derechin and Johnson 1965; Rawitch and Hwan 1979). From these and other measurements, it was demonstrated that α-lactalbumin exists primarily as a nearly spherical, compact globular monomer in neutral and alkaline media. Small-angle x-ray scattering studies suggested an oblate ellipsoid with the axes 2.2 × 4.4 × 5.7 nm (Krigbaum and Kugler 1970). However, conformations based on model building (Brown et al. 1969) and energy minimization (Warme et al. 1974) suggest an oblate ellipsoid of about 2.5 × 3.7 × 3.2 nm.

At pH values below the isoionic point, Kronman and Andreotti (1964) and Kronman et al. (1964) observed that α-lactalbumin associated to form dimers and trimers and aggregated to polymers with sedimentation constants in the range of 10 to 14 Svedberg units. The association was rapid, reversible, and temperature dependent, being greater at 10°C than at 25°C. In contrast, the aggregation, while reversible, was dependent upon the concentration, with little aggregation below 1% protein, and decreased with decreasing temperature, pH, and ionic strength. This association and aggregation were attributed to conformational changes in the protein molecule below pH 4 based on the observation that at pH 2 and at concentrations such that association and aggregation were absent, sedimentation velocity studies indicated an increase in hydrodynamic volume and a change in the absorption spectra. Robbins et al. (1965) amidinated the amino group of α-lactalbumin to increase its hydrophobicity without changing its charge and investigated the aggregation phenomenon under acid conditions. Since amidination did not appreciably change the configurational behavior but did increase the amount of association and aggregation, they concluded that these phenomena were due to hydrophobic bonding resulting from the conformation change or a decrease in the electrostatic barrier.

Brown et al. (1969) noted the similarity of the primary sequence of α-lactalbumin and hen's egg-white lysozyme and their functional properties and proposed a structure for α-lactalbumin based on the main-chain conformation of lysozyme. While changes in the internal side chains can generally be interrelated, there are some regions which cannot be deduced unequivocally. The surface cleft, which is the site of substrate binding in lysozyme, is shorter in α-lactalbumin. While the optical rotatory dispersion studies of Herscovits and Mescanti (1965) indicated a tightly folded molecule with ~40% α-helix, circular dichroism spectra suggest 26% α-helix, 14% β-configuration, and 60% random coil. The latter configuration is similar to the secondary structure of lysozyme (Robbins and Holmes 1970; Barel *et al.* 1972). X-ray crystallography supports this conclusion (Robbins and Holmes 1970), and other physical-chemical properties are consistent with this model but suggest that the α-lactalbumin structure is less stable and more expanded (Krigbaum and Kugler 1970; Barel *et al.* 1972). Solvent perturbation difference spectra, fluorescence investigations, photooxidation, and nitration with tetranitromethane (Habeeb and Atassi 1971; Kronman *et al.* 1972; Tamburro *et al.* 1972; Sommers *et al.* 1973) suggest that two of the four tryptophan residues are exposed at neutral pH and 25°C, while the other two residues are buried. Lowering the temperature at pH 6 produces a change in configuration which prevents large perturbants from reaching the exposed residues. From a consideration of the model, one would conclude that the tryptophans at positions 60 and 26 are buried and those at positions 104 and 118 are exposed at 25°C and neutral pH (Brown *et al.* 1969; Warme *et al.* 1974). In the model, the tryptophans at positions 60 and 104 are in the cleft region and the tryptophan at position 118 is on the surface very close to the cleft. Therefore, a slight closure of the cleft at the lower temperature would shield the tryptophans at positions 104 and 118 as well (Warme *et al.* 1974). Studies of the four tyrosine residues in α-lactalbumin indicate that they all titrate normally (Kronman *et al.* 1972), but while two residues are easily acetylated, the remaining residues are acetylated only at higher concentrations (Kronman *et al.* 1971, 1972). Three tyrosines react with cyanogen fluoride in the neutral range (Gobrinoff 1967). While all four residues react with the tetranitromethane, only two are nitrated (Habeeb and Atassi 1971; Denton and Ebner 1971). These observations are generally consistent with the proposed model (Warme *et al.* 1974). The disulfide bonds in α-lactalbumin, as predicted from the expanded model, are more rapidly reduced and, therefore, more accessible than in lysozyme (Iyer and Klee 1973). The higher rate of hydrolysis by immobilized pronase also indicates a less stable structure (Swaisgood 1982).

At alkaline pH values, even though no observable association or aggregation occurs, some changes in configuration are observed. Changes in the Cotton effect between 250 and 300 nm and in optical rotatory dispersion occur at pH 11.5. The fourth tyrosine, which was somewhat buried at neutral pH values, reacts with cyanogen fluoride at pH 10.0 and above (Gobrinoff 1967).

On the acid side of the isoelectric point, Kronman *et al.* (1965) noted that the α-lactalbumin molecule swells and yields a difference spectra with maxima at 285–286, 292–293, and 230 nm, due largely to changes in the environment of the tryptophan residues. The amplitude of the change is dependent upon pH and temperature but is practically independent of ionic strength. From solvent perturbation studies the authors demonstrated that these changes were caused by a conformational change in the environment of the buried tryptophan residues rather than by their transfer to the surface of the molecule. This conclusion is supported by the observation that, while the Moffitt and Yang b_0 value decreases from 230° to 154°, the Cotton effect at 225 nm remained the same from pH 2 to 6.

Immunoglobulins

Investigation of the structure and conformation of the immunoglobulins is complicated by their heterogeneity. The secondary and tertiary structures of the four chain unit have been investigated by enzyme and chemical modification (Dorrington and Tanford 1970). Enzyme modification indicates that the N-terminal amino acids are not on the surface of the native immunoglobin molecules. Selective reduction of the interchain disulfide bonds demonstrates that they are not involved in the specific conformation of the antigen-binding sites.

X-ray diffraction and electron microscopy studies of human IgG_1 have been employed to define the arrangement of the polypeptide chains, especially in the antigen-binding end of the molecule (Poljak et al. 1972). The Fab fragments (Fig. 3.8) have two globular structures formed from both the heavy and light chains, which is consistent with the involvement of both chains in antigen binding. Evidence secured from optical rotatory dispersion and circular dichroism spectra indirectly supports this structure by indicating a negligible amount of α-helix in the immunoglobulins (Dorrington and Tanford 1970).

Electron microscopy of IgM and IgG reveals that the four-chain units in both immunoglobulins are shaped like a Y or T with a flexible central hinge (Green 1969). In IgM, the five Y-shaped tetramers form a ring with five arms projecting from it.

In the presence of denaturing agents such as urea and guanidine-HCl, the globular antigen-binding sites are completely unfolded (Björk and Tanford 1971). The denaturation is completely reversible upon removal of the denaturing agent in neutral media. The denaturing effect of acids and alkalis is similar (Doi and Jirgensons 1970).

FRACTIONATION OF MILK PROTEINS

Numerous methods have been developed and employed in the fractionation of milk proteins; some of these are based on differential solubilities in various solvent systems, and others depend upon chromatographic or electrophoretic behavior.

Whole Casein

In the fractionation of the milk proteins, usually the first step in the process is to separate the so-called whole casein from the whey in a skim milk. A number of procedures are available (McKenzie 1971C), but the most commonly used method is based upon classical acid precipitation at the pH of minimum solubility. Several different temperatures have been employed: 2, 20, and 30°C. Except for precipitation at 2°C, where minimum solubility occurs at pH 4.3, the skim milk is adjusted to pH 4.5–4.6 with hydrochloric acid (1 M). A more recent investigation of the relationship of temperature and pH to the completeness of casein precipitation indicated that optimum yield was obtained at pH 4.3 and 35°C (Helesicová and Podrazký 1980).

Casein can also be obtained from skim milk by high-speed centrifugation at 105,000 x g at different temperatures in the presence or absence of added calcium ion (McKenzie 1971C). For example, the procedure of von Hipple and Waugh (1955) has been modified by adding $CaCl_2$ to skim milk to a final added concentration of 0.07 M at pH 6.6–6.8 and 3°C. After centrifugation, the precipitated casein is resuspended in 0.08 M NaCl and 0.07 M $CaCl_2$ and the calcium is removed by oxalate, citrate, ion-exchange resin, or exhaustive dialysis. Whole casein has also been prepared by salt precipitation with either $(NH_4)_2SO_4$, (McKenzie 1971C) or Na_2SO_4 (Wake and Baldwin 1961).

Fractionation of Casein

Differential Solubility Methods.   Numerous methods have been developed to obtain one or more of the various caseins from whole casein or directly from skim milk based on their differential solubility (Thompson 1971; Mackinlay and Wake 1971; Whitney 1977). While some early procedures indicated the possibility of fractionating whole casein into different components, it was not until the 1950s that systematic procedures were proposed for the fractionation of casein into Warner's α-, β-, and γ-caseins. Hipp *et al.* (1952) developed two procedures which have been used extensively or partially incorporated into other methods. The first is based upon the differential solubilities of the caseins in 50% alcohol in the presence of ammonium acetate by varying the pH, temperature, and ionic strength. The second procedure involves the dispersion of whole casein in 6.6 M urea and the separation of the casein fractions by dilution, pH adjustment, and, finally, the addition of $(NH_4)_2SO_4$. The order of precipitation of the caseins in both methods is α-, β-, and γ-caseins. Warner's casein fractions are now identified as follows: α-casein, a mixture of α_{s1}-, α_{s2}-, and κ-caseins; β-casein, the major β-casein component; γ-casein, a mixture of the β-caseins-1P(f 29–209), (f 106–209), and (f 108–209).

The α_s-caseins and κ-caseins have been prepared largely from either whole casein or the α-casein fraction of Hipp and co-workers (McKenzie and Wake 1961; Cheeseman 1962; Neelin *et al.* 1962; Swaisgood and Brunner 1962; Waugh *et al.* 1962; Zittle and Custer 1963; Craven and Gehrke 1967; Fox and Guiney 1972; Chiba *et al.* 1978). Crude α_s-casein is prepared from casein obtained by $CaCl_2$ precipitation at 37°C by removing the calcium with oxalate to solubilize the casein and reprecipitation with 0.25 M $CaCl_2$ at 37°C and pH 7 (Waugh *et al.* 1962). The κ-caseins are removed from the supernatant by precipitation with Na_2SO_4 followed by reprecipitation from 50% ethanol with ammonium acetate (McKenzie and Wake 1961) or by using calcium oxalate as a carrier precipitate to enhance the removal of the other caseins (Craven and Gehrke 1967). Adjustment of the pH of whole casein dispersions in urea also has been used to precipitate the α_s-caseins either by adjusting a 6.6 M urea dispersion to pH 1.3–1.5 with H_2SO_4 (Zittle and Custer 1963) or by adjusting a 3.3 M urea system to pH 4.5 (Fox and Guiney 1972). The crude κ-caseins can be obtained from the supernatant of the H_2SO_4 method by precipitation with $(NH_4)_2SO_4$ and purified by reprecipitation from aqueous ethanol (Zittle and Custer 1963). More than 90% of the α_{s1}-caseins have been recovered from acid casein by precipitation with 75 mM $CaCl_2$ at 5°C (Chiba *et al.* 1978).

Starting with the α-casein fraction of Hipp and co-workers, the α_s-caseins can be precipitated by $CaCl_2$ treatment and the κ-caseins can be removed from the supernatant by pH adjustment to 4.7 (Neelin *et al.* 1962). Swaisgood and Brunner (1962) added 12% trichloroacetic acid (TCA) to a 6.6 M urea dispersion of the same fraction at 3°C and precipitated the α_s-fraction. After removal of the urea and TCA from the supernatant, they adjusted the pH to 7.0, added $CaCl_2$ to 0.25 M, and removed the precipitate. κ-Casein was finally obtained from the supernatant at pH 4.4. In contrast, Wake (1959) prepared κ-casein from the supernatant remaining from the β-casein precipitation by the first procedure of Hipp and co-workers by adjusting the pH to 5.7. A κ-casein concentrate has been prepared from commercial casein based on the differential solubilities of the caseins in $CaCl_2$ solutions (Girdhar and Hansen 1978). A novel procedure has been developed for the isolation of α_{s1}-casein directly from skim milk, using sodium tetraphosphate (Quadrafos) (Melnychyn and Wolcott 1967).

The major β-casein component can be prepared by a simplification of the urea method of Hipp and co-workers (Aschaffenburg 1963). Whole casein is dispersed in 3.3 M urea at pH 7.5 and adjusted to pH 4.6, which precipitates the bulk of the α_{s1}-and κ-caseins. The supernatant is adjusted to pH 4.9, diluted to 1 M urea, and warmed to 30°C, precipitating the major β-casein. Owicki and Lillevik (1969) prepared the major β-casein from rennin-treated whole casein by dispersing the clot in 6.6 M urea at pH 7.8 and successively diluting and adjusting the pH of the system. After cooling to 4°C and clarifying, a precipitate rich in the β-casein was obtained by warming to 30°C.

Partition Methods.   Walter (1952) devised a countercurrent distribution procedure for the fractionation of casein in a two-phase system containing water-ethanol-phenol at pH 8.2. The concentration of β-casein increased in the phenol phase. Ellfolk (1957) employed a two-phase system consisting of collidine, ethanol, and distilled water at 20°C. The α-caseins were concentrated in the water-rich phase, while the β-caseins were concentrated in the collidine-rich phase.

Electrophoretic Methods.   Several electrophoretic procedures have been developed to fractionate or purify the various caseins (McKenzie 1971C; Thompson 1971; Whitney 1977). Wake and Baldwin (1961) fractionated whole casein by zone electrophoresis on cellulose powder in 7 M urea and 0.02 ionic strength sodium phosphate buffer at pH 7 and 5°C. Payens and co-workers employed several somewhat different electrophoretic conditions for the fractionation and purification of the caseins on cellulose columns (Payens 1961; Schmidt and Payens 1963; Schmidt 1967). Three fractions, α_s-, κ-, and β-caseins, were separated at pH 7.5 and 30°C with 4.6 M urea-barbiturate buffer. The purification of α_{s1}-casein and the separation of the genetic variants of κ-casein were accomplished by altering the electrophoretic conditions. Manson (1965) fractionated acid casein on a starch gel column stabilized by a density gradient at 25°C.

Isoelectric focusing on polyacrylamide gels containing sucrose gradients and 7 M urea effectively separated the major caseins (Josephson 1972). The isoelectric points ranged in increasing order from pH 4.9 to 6.5 for α_s-, β-, and κ-caseins and from 6.7 to 8.0 for the minor β-caseins (previously classified as γ-caseins). Pearce and Zadow (1978) modified this procedure by using 5% polyacrylamide gel containing 6 M urea and 2% ampholytes in the presence of mercaptoethanol.

Chromatographic Methods.   Many procedures have been developed for the chromatographic separation of the caseins on ion-exchange columns based upon their relative affinity for the column material in various environments (Thompson 1971; Yaguchi and Rose 1971). Their fractionation has been accomplished with both anion and cation exchangers. The elution of the proteins from anion exchangers is usually achieved by stepwise or continuous increases in chloride or phosphate concentration. Due to the tendency of the caseins to aggregate, dissociating agents such as urea are commonly used, and since the κ-caseins are present as disulfide-linked polymers, reducing agents such as mercaptoethanol are also frequently employed in the eluting system. However, some procedures have been developed without either agent (Zittle 1960; Gordin et al. 1972; Groves et al. 1962; Groves and Gordon 1969; Igarashi and Saito 1970; Schober and Heimberger 1960). Since the association of the caseins is temperature dependent, Tarrassuk et al. (1965) investigated the effect of temperature upon the chromatography of casein on DEAE-cellulose. As expected, the caseins were eluted at lower concentrations of NaCl at 4°C than at 25°C. Anion exchange columns other than DEAE-cellulose, such as triethylaminoethyl (TEAE)-cellulose and DEAE-Sephadex also have been used to fractionate the caseins without dissociating or reducing agents (Igarashi and Saito 1970; Gordin et al. 1972).

Anion-exchange chromatography in the presence of urea as a dissociating agent but without a reducing agent has been used by numerous investigators for the fractionation or purification of the caseins (Rose et al. 1969; Tripathi and Gehrke 1969, 1970; Farrell et al. 1971; Yaguchi and Rose 1971; Gordin et al. 1972; El-Negoumy 1973; Davies and Law 1977). Ribadeau-Dumas et al. (1964) chromatographed whole casein on DEAE-cellulose with urea at pH 7 with an NaCl elution gradient of 0 to 0.6 M. Comparable procedures have been used to obtain the different α_{s1}-casein variants (Thompson and Kiddy 1964) and to characterize the minor β-caseins (Tripathi and Gehrke 1969) and the κ-caseins (Tripathi and Gehrke 1970; Gordin et al. 1972). DEAE-Sephadex columns have also been used with urea-containing buffers for the fractionation of casein (Hladik and Kas 1973; Vujicic 1973), while others have used dimethylformamide as the dissociating agent, even though its vapors are toxic (Yaguchi and Rose 1971).

The combined use of dissociating agents and mercaptoethanol, which maintains the κ-caseins as monomers, has been used in a number of procedures for the anion-exchange chromatography of the caseins (Farrell et al. 1971; Hoagland et al. 1971; Yaguchi and Rose 1971; Nagasawa et al. 1973; Creamer 1974; El-Negoumy 1976). Pujolle et al. (1966) fractionated the κ-caseins on DEAE-cellulose with a buffer at pH 7.0 containing 3.3 M urea and 0.3% mercaptoethanol and a linear salt gradient from 0.02 to 0.2 M. A novel procedure was used by

Creamer (1974) to separate α_{s1}-casein-A from whole casein containing both the A and B variants. The B variant was degraded with pepsin or rennet, and the A variant was isolated from the degradation products on a DEAE-cellulose column with an NaCl gradient (0.0 to 0.5 M) in 4.5 M urea buffered at pH 5.5 containing 0.1% mercaptoethanol.

Another technique employed to facilitate the chromatographic fractionation of casein involves the reduction and subsequent alkylation of the casein prior to chromatography (Rose *et al.* 1969; Yaguchi and Rose 1971; Davies and Law 1977). Rose *et al.* (1969) reduced whole casein with mercaptoethanol, alkylated the product with iodoacetamide, and separated the components on a DEAE-cellulose column (Figure 3.18). Davies and Law (1977) modified this procedure and achieved a quantitative estimation of the major caseins.

While salt or buffer gradients at constant pH were employed for elution in all of the above procedures, Vreeman *et al.* (1977) demonstrated an improved separation of the κ-casein components on DEAE-

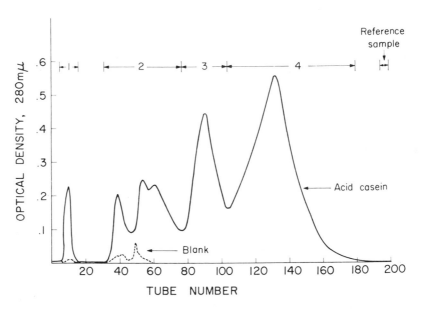

Figure 3.18. Elution pattern for a 250-mg sample of reduced and alkylated acid casein from DEAE-cellulose with NaCl gradient in buffer containing 6.6 M urea. Fraction (1), minor β-caseins and para-κ-casein-like material; (2) κ-casein and some β-casein-1P (f 29–209); (3) major β-casein; (4) α_s-caseins. (From Rose et al. *1969; Yaguchi and Rose 1971. Reprinted with permission of the American Dairy Science Association.*)

cellulose with a pH gradient elution, and Wei (1982) developed a batch procedure for the fractionation of whole casein.

Cation-exchange columns have been used effectively by some investigators for the fractionation of casein (Annan and Manson 1969; Kim et al. 1969; Kopfler et al. 1969; Snoeren et al. 1977; Saito et al. 1979). Sulfoethyl-Sephadex was used by Annan and Manson (1969) with formate buffer to fractionate the α_s-casein complex. Cellulose phosphate, carboxyl-methyl-cellulose (CMC), potassium-κ-carrageenan, and sodium Amberlite CG50 columns have also been used to fractionate the caseins (Kim et al. 1969; Kopfler et al. 1969; Snoeren et al. 1977). A batch method for the preparation of para-κ-casein from rennin-treated whole casein has been developed with CMC Sephadex (Saito et al. 1979).

Adsorption chromatography on hydroxyapatite has frequently been used to fractionate protein mixtures. The binding of proteins to hydroxyapatite is due primarily to the attraction of the positive calcium on the hydroxyapatite for the negatively charged groups of the proteins (Glueckauf and Patterson 1974). A number of investigators have used hydroxyapatite columns to fractionate caseins (Addeo et al. 1977; Donnelly 1977; Barry and Donnelly 1979, 1980). Batch methods for the large-scale separation of the minor β-caseins X-1P (f 29–209), (f 106–209), and (f 108–209) from whole casein and the purification of crude κ-casein and the major β-casein with calcium phosphate gels have been developed (Green 1969, 1971A,B; Eigel and Randolph 1974).

Gel filtration, which depends primarily on molecular size and shape, has also been used for the fractionation of caseins. However, since the various casein monomers except for the minor β-caseins are in the same size range, one must make use of their association and dissociation to secure their separation by this technique. A number of fractionations have been obtained on suitable cross-linked dextran gels (Igarashi and Saito 1970; Yaguchi and Rose 1971; Nakahori and Nakai 1972; Nakai et al. 1972). In the absence of reducing agents but under conditions which favor the dissociation of caseins, such as the presence of dissociating agents, low temperature, or high pH values, κ-casein is still largely in polymer form due to intermolecular disulfide bonds. Therefore, it is eluted largely in the void volume of Sephadex gels. Yaguchi and Tarrassuk (1967) investigated the effect of pH, NaCl concentration, and urea on the gel filtration of acid casein and skim milk on Sephadex G-100 and G-200. Highly purified κ-caseins were secured from a Sephadex G-150 column with 5 mM Tris-citrate buffer, pH 8.6, containing 6 M urea at room temperature and on Sephadex G-200 with 0.02 M phosphate buffer, pH 8.0, at 4°C (Yaguchi et al. 1968). Similar results have been obtained in the presence of SDS and EDTA on Sephadex columns (Cheeseman 1968; Nakahori and Nakai 1972). Nakai *et al.* (1972) obtained electrophoretically pure α_{s1}-, major β-, and κ-caseins directly from skim milk by employing a Sephadex G-100 column with 5 mM phosphate buffer, pH 10.8, containing 2 mM EDTA at 4°C followed by appropriate precipitation procedures (Figure 3.19).

Agarose gels have also been used as columns for the fractionation of caseins (Yamashita *et al.* 1976; Pepper and Farrell 1977). Nijhuis and

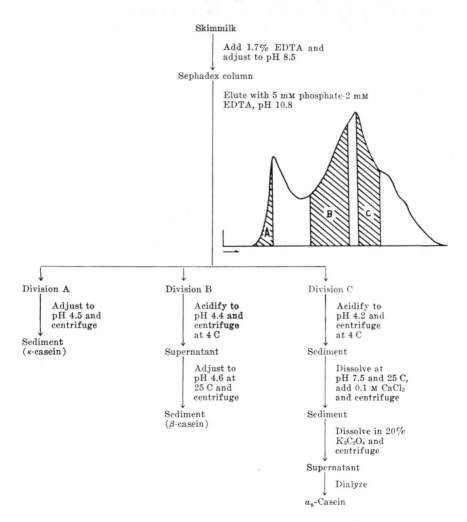

Figure 3.19. Fractonation of casein by Sephadex gel chromatography. (From Nakai et al. *1972. Reprinted with permission of the American Dairy Science Association.)*

Klostermeyer (1975) used an activated thiol-Sepharose 4B column with Tris-HCl buffer containing dithiothreitol to separate the κ-and α_{s2}-caseins from the α_{s1}-and β-caseins in whole casein. More recently, Creamer and Matheson (1981) studied the fractionation of casein by hydrophobic interaction chromatography on octyl- or phenyl-Sepharose CL-4B columns. The whole casein was adsorbed onto the column from dilute phosphate buffers. A gradient of 0 to 40% ethylene glycol followed by 6 M urea was employed to desorb the protein. Optimum separation was obtained with an increasing urea gradient. Under all conditions, the major β-casein component was eluted more readily than the α_{s1}-casein in spite of its higher hydrophobicity.

Whey Proteins

Several procedures exist for removing whey proteins from the other whey components. They can be obtained by complexing with CMC (Hansen et al. 1971). The whey is acidified to pH 3.2, diluted with an equal volume of a 0.25% solution of CMC, and the complex removed by centrifugation. Sodium hexametaphosphate (HMP) has also been used to isolate the whey protein by complex formation (Hidalgo et al. 1973). The cations in the whey are removed by elution through the Amberlite IR 120 column, the pH is adjusted to 3.0, and HMP is added (60 mg/100 ml whey) to form the complex, which is removed by centrifugation. Most of the HMP can then be removed with ion exchange on Dowex-2 or gel filtration on Biogel P-6. Protein can be recovered from cheese whey by complex formation with polyacrylic acid (Sternberg et al. 1976). Whey protein concentrates can also be prepared by complexing with Ferripolyphosphate, gel filtration on Sephadex G-25, ultrafiltration through cellulose acetate membranes, reverse osmosis, and electrodialysis (Anon. 1968; Morr et al. 1969; Fenton-May et al. 1971; Peri and Dunkley 1971; O'Sullivan 1972; Jones et al. 1972).

Fractionation of Whey Proteins

Like the caseins, the whey proteins have been isolated from whey or whey concentrates and purified by differential solubilities, electrophoresis, or chromatography.

Differential Solubility Methods. Many methods have been developed for isolating the major whey proteins based on their solubilities in different systems. Some of the methods have been employed to obtain only one protein; others can be used to secure all of the major proteins. The advantages and limitations of the various procedures have been discussed (Gordon 1971; McKenzie 1971B).

Salt fractionation with $(NH_4)_2SO_4$ has been employed in various ways to isolate blood serum albumin and β-lactoglobulin (Polis *et al.* 1950), and to obtain α-lactalbumin from the mother liquor after the crystallization of β-lactoglobulin (Gordon and Ziegler 1955). Whey prepared by adding 200 g per liter of Na_2SO_4 to milk at 40°C was used by Aschaffenburg and Drewry (1957B) in the procedure for isolating β-lactoglobulin and α-lactalbumin by $(NH_4)_2SO_4$ fractionation. Armstrong *et al.* (1967) made a thorough study of the methods available for the isolation of β-lactalbumin and α-lactalbumin and modified an earlier procedure (Robbins and Kronman 1964) to obtain more satisfactory results. A schematic representation of two of their procedures is shown in Figure 3.20. The various genetic variants of β-lactoglobulin possess different solubilities at pH 3.5 in method Ia, and the yield of the A variant by this method is low.

TCA has been used in a concentration of 34.2 g/liter to precipitate

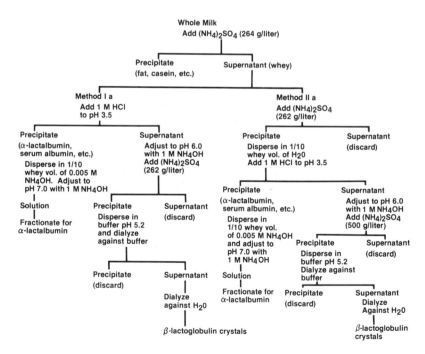

Figure 3.20. Fractionation methods of Armstrong, McKenzie, and Sawyer for the isolation of β-lactoglobulin. (From McKenzie 1967. Reprinted with permission of AVI Publishing Co., Westport, Conn.)

crude α-lactalbumin from acid whey (Fox *et al.* 1967). The supernatant is concentrated to 1/10th the original volume of the skim milk by negative pressure dialysis and either exhaustively dialyzed against water and lyophilized as β-lactoglobulin or further purified by $(NH_4)_2SO_4$ prior to dialysis. Aschaffenburg (1968B) purified the crude α-lactalbumin TCA precipitate by $(NH_4)_2SO_4$ fractionation.

Electrophoretic Methods.   Little use has been made of electrophoretic techniques for the fractionation of the whey proteins. Column isoelectric focusing has been used to fractionate further the crude immunoglobulin fraction obtained by Smith's procedure (Josephson *et al.* 1972). Two major peaks, a shoulder, and two minor peaks were obtained, but no attempt was made to identify the components in the peaks.

Butler and Maxwell (1972) employed preparative zonal electrophoresis on Pevikon C870 blocks in 0.05 M barbiturate buffer, pH 8.2, to purify bovine IgM in their procedure in isolating the various immunoglobulins from whey.

Chromatographic Methods.   Numerous procedures have been developed for the chromatographic fractionation of the whey proteins (Yaguchi and Rose 1971). Anion-exchange chromatography on DEAE-cellulose with stepwise changes in pH and/or NaCl concentrations has been used (Schober *et al.* 1959; Yaguchi *et al.* 1961). Gordin *et al.* (1972) compared a stepwise gradient and a combination of these elution techniques and concluded that gradient elution with 0.0 to 0.6 M NaCl in phosphate buffer, pH 6.8, yielded the more reproducible pattern. The immunoglobulins have been effectively separated by DEAE-cellulose chromatography in combination with other techniques (Mach *et al.* 1969; Groves and Gordon 1967). DEAE-cellulose columns have also been used to separate the genetic variants of β-lactoglobulin and to purify the various whey proteins (Yaguchi *et al.* 1961; Basch *et al.* 1965; Gordon 1971).

Other investigators used DEAE-Sephadex A-50 for the anion-exchange chromatography of the whey proteins (Yaguchi and Rose 1971). Smith *et al.* (1971) employed a DEAE-Sephadex A-50 column to separate the IgG_1 and IgG_2 immunoglobulins in the "7S" fraction obtained by Sephadex G-200 chromatography of colostral whey. Similar procedures have been used to obtain the IgG subclasses (Butler and Maxwell 1972) and to prepare a homogeneous IgM fraction (Kumar and Mikalajcik 1973). This anion-exchange material has also been used to separate the A and Dr variants of β-lactoglobulin (Bell *et al.* 1970).

Cation-exchange columns have also been used in a few instances for the fractionation of the whey proteins (Yaguchi and Rose 1971). Kiddy et al. (1965) fractionated the "albumin" fraction of whey obtained by half-saturation with $(NH_4)_2SO_4$ on a CMC column. The proteins were adsorbed at pH 4.6 and eluted by increasing the pH. The whey proteins have also been fractionated by a batch process based on protein–CMC complex formation (Hidalgo and Hansen 1971). β-Lactoglobulin and bovine serum albumin are complexed with an appropriate amount of CMC at pH 4.0 and removed by centrifugation. Adjusting the pH of the supernatant to pH 3.2 and changing the concentration of CMC results in the precipitation of an α-lactalbumin–CMC complex.

Gel filtration has been used by several investigators in their studies of the whey proteins (Yaguchi and Rose 1971). Column chromatography on either Sephadex G-75 or G-100 separates the proteins in the whey remaining after $(NH_4)_2SO_4$ precipitation of the casein into five fractions (Armstrong et al. 1970; Elfagm and Wheelock 1978A,B). In the separation on Sephadex G-100, five peaks were obtained: (1) lactoferrin and transferrin, (2) serum albumin, (3) β-lactoglobulin, (4) glyco-α-lactalbumin, and (5) α-lactalbumin. Bio-gel P-100, a porous polyacrylamide gel, has been used to fractionate the protein in cottage cheese whey (Patel and Adhikari 1973). The proteins appear to be eluted in the order of their decreasing molecular weights. Recently, Shimazaki and Sukegawa (1982) fractionated centrifugal whey and colostral whey upon a Fractogel TSKHW55F column. Residual casein polymers were observed in the void volume, followed by IgG, serum albumin, β-lactoglobulin, and α-lactalbumin. Several investigators have used gel filtration to fractionate the immunoglobulins (Mach et al. 1969; Froese 1971; Kanno et al. 1976). Starting with colostrum whey, Mach et al. (1969) separated the main classes of the immunoglobulins on Sephadex G-200. IgM appeared in the void volume and IgA on the shoulder of a large IgG peak (Figure 3.21). Large-scale fractionation of whey protein concentrates prepared by gel filtration on Sephadex G-25 or ultrafiltration was achieved by Forsum et al. (1974) on Sephadex G-75. Three fractions, which were eluted with 0.1 M phosphate buffer, pH 6.3, containing 0.2% NaN_3, were bovine serum albumin and β-lactoglobulin; β-lactoglobulin; and α-lactalbumin.

ELECTROCHEMICAL PROPERTIES OF MILK PROTEINS

All proteins are amphoteric due to the presence of acidic and basic groups on the molecule and, therefore, can possess either a net positive or a negative charge, depending upon their environment. The extent of

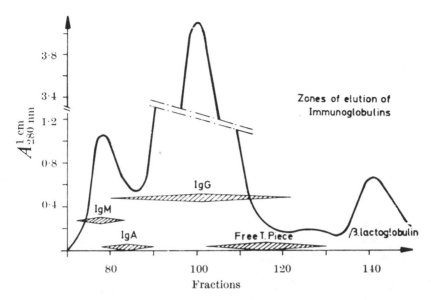

Figure 3.21. Gel filtration of total colostrum whey protein on Sephadex G-200. Column 120 cm length, 3.7 cm diameter; gel volume, 1420 cm^3, 21.6 ml/hr. (From Mack et al. *1969. Reprinted with permission of AVI Publishing Co., Westport, Conn.)*

dissociation of the protons from these groups and the character and number of any bound ions determine the charge. The electrochemical natures of the milk proteins are most commonly investigated by their titration curves and electrophoretic behavior.

Titration Curves

While the shape of the titration curve for a given milk protein is a function of its primary structure, it also reflects its configuration and any changes that occur in its conformation with pH. For example, Ho and Waugh (1965) demonstrated that, while ionic strength influences the character of coagulation below pH 6.4, the titration curve for α_s-casein indicates that all the ionizable groups are accessible and completely reversible to H$^+$ ion in all forms of α_s-casein. The agreement between the ionizable groups, as determined from the titration curves and from the primary structure for α_{s1}-casein-B-8P, is quite good. The negative logarithms of the intrinsic dissociation constants, pk_i, for the side-chain carboxyl groups were found to be 5.18 and 4.88 at ionic strengths of 0.4 and 0.05, respectively. Detailed analysis of the data in this region of the curve indicated that the apparent electrostatic interaction factor, w', varies only slightly at an ionic strength of 0.4, which is consistent with the observation that α_s-casein exists as a precipitate in this pH range. In contrast, at ionic strength 0.05, w' increased 10-fold between pH 5.3 and 3.5 and then returns to approximately the original value at lower pH values. These changes coincide with the precipitation and redispersion of the protein at this ionic strength. Thus, the titration curve is consistent not only with the primary structure but also with changes in the degree of aggregation.

In the titration of β-casein B-5P, Creamer (1972) obtained agreement between the prototropic residues detected by titration at 25°C and the primary sequence, except for the carboxyl groups and lysine (Table 3.4). The agreement between the calculated and expected pk_i for the various dissociations is very good. In the absence of urea, nonreversible or pathway-dependent behavior was observed near the isoelectric point, indicating the importance of conformation and association. In the presence of Ca^{2+}, the titration curve changes due to suppression of the protonation of phosphoserine residues and changes in configuration near the isoelectric point as Ca^{2+} is released.

A number of titration studies have been performed on the genetic variants of β-lactoglobulin (Tanford 1962; Basch and Timasheff 1967; Brignon *et al.* 1969). The titration curves as illustrated in Figure 3.22 were reversible between the acid endpoint and pH 9.7. The maximum acid-binding capacity observed in genetic variants A, B, and C indicates 20 cationic groups per monomer compared to 21 from the pri-

Table 3.4. Comparison Between the Number of Changed Residues on β-Casein in B-5P as Determined by Titration at 25°C and by Primary Sequence.

Group	Number of Residues		pK$_i$	
	Primary Sequences	Titration	Calculated from Titration	Expected
C-terminal carboxyl	1	1.0	3.6	3.6
Aspartic acid	4	19.0	4.6	4.6
Glutamic acid	18		1.5	1.5
Phosphate	5	5.0	6.4	6.6
Histidine	6	6.0	6.0	6.15
N-terminal amine	1	1.0	7.0	7.5
Tyrosine	4	4.0	9.85	9.7
Lysine	11	10.0	10.0	10.4
Arginine	5	—	—	~12.0

SOURCE: Creamer (1972). Reprinted with permission of Elsevier Science Publishing Co., Inc.

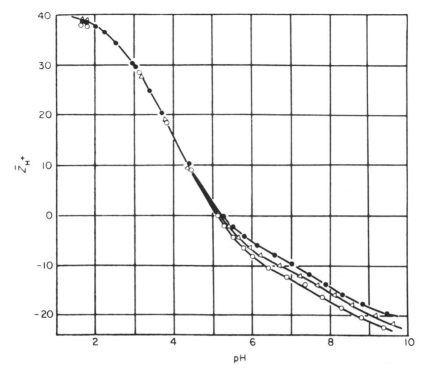

Figure 3.22. Titration curves for β-lactoglobulins, A, B, and C in 0.15 M KCl at 25°C. ○ = A variant; △ = B variant; ● = C variant; Z_H+ expressed in terms of the number of groups per dimer (36,000 daltons). (From Basch and Timasheff 1967. Reprinted with permission of AVI Publishing Co., Westport, Conn.)

mary structure of variants A and B and 22 for variant C. McKenzie (1971B) has suggested that this finding may be in error, since Ghose *et al.* (1968) observed 21 total acid-binding sites in their titration of the B variant. The additional cationic group that is undetected in the titration of the C variant has been at least partially explained by a conformational transition of a protonated histidine residue from the surface to the interior of the molecule, either with the transfer of the proton to a carboxyl group or with the formation of an ion pair with a carboxyl ion. The total number of carboxyl groups titrated per monomer for variants A, B, and C is 26, 25, and 25, respectively. These values are one less than predicted from their primary structure. In contrast, the titration curves of all of these β-lactoglobulins indicate one more histidine and α-amino group than predicted by primary structure. This anomalous behavior has been attributed to the missing carboxyl group, which is buried in the titration range of these groups, but due to a conformational change at pH 7.3 it becomes titratable, with pk_i of 7.3. If these β-lactoglobulins are denatured, the values for the number of carboxyl, imidazole, and α-amino groups obtained by titration equal those observed in the primary structure.

Brignon et al. (1969) demonstrated that the maximum acid-binding capacity of β-lactoglobulin D is the same as that of the other variants. The curves are identical at pH 4.0. At pH 6.5, one less proton is dissociated in the titration of the D variant than with the B variant, as would be predicted from the substitution of a glutamine residue for a glutamic acid residue in B. The anomalous carboxyl group observed in the other variants is also detected in the D variant.

The isoionic points of a number of the milk proteins are given in Table 3.5.

Table 3.5. The Isoionic Points of Some Milk Proteins.

Protein	Isoionic Points	
	Observed	Calculated[a]
α_{s1}-Casein A-8P	5.15	4.97
B-8P	5.05	4.96
C-8P	—	5.00
D-8P	—	4.91
α_{s2}-Casein A-10P	—	5.39
-11P	—	5.32
-12P	—	5.25
-13P	—	5.19
β-Casein A^3-5P	—	5.11
A^2-5P	—	5.19
A^1-5P	—	5.27
B-5P	—	5.35
C-4P	5.35	5.53
β-Casein X-5P (f 1–105)	—	4.55
β-Casein X-4P (f 1–28)	3.3	3.0
β-Casein X-1P (f 29–105)	—	5.2
κ-Casein A-1P	5.37	5.43
B-1P	—	5.64
Bovine serum albumin	4.71 and 4.84	—
β-Lactoglobulin A	5.14, 5.35	5.19
B	5.3, 5.41	5.28
C	— , 5.39	5.37
α-Lactalbumin B	4.8	5.4

SOURCE: Swaisgood (1982). Reprinted with permission of Elsevier Applied Science Publishers, Ltd.
[a]Calculated from primary structure by Swaisgood, (1982).

Electrophoretic Behavior

Free-boundary electrophoresis has been used extensively for the identification and characterization of the various milk proteins, but its resolving power is limited. Wake and Baldwin (1961) introduced zonal electrophoresis on starch gels in alkaline urea buffers, which greatly improved the resolution of the milk proteins. Since then, zonal electrophoresis on starch and polyacrylamide gels, with or without urea, mercaptoethanol, or SDS, has largely replaced free-boundary electrophoresis (Swaisgood 1975A). The relative mobility of the proteins on these gels is a function not only of the charge of the proteins but also of their size, configuration, and state of aggregation in the particular medium employed. For example, in the absence of SDS and mercaptoethanol, Schmidt and Both (1975) separated the genetic variants of the major α_{s1}-and β-caseins, but the κ-caseins either remained in the slot or streaked due to the presence of disulfide polymers. To resolve the κ-caseins, the system was changed to a 5 M urea–10% starch gel containing 0.03 M mercaptoethanol in the buffer, which, by forming κ-casein monomers, allowed them to be separated and identified.

Zonal electrophoresis is commonly used for phenotyping of the milk proteins, since in the majority of known cases the relative mobilities of the genetic variants of a protein are different (Figure 3.23) (Thompson 1970). The order of the α_{s1}-casein genetic variants can be explained on the basis of their relative net charge and size. Arranged in decreasing order of their net charge at pH 8.6 as calculated from their primary structure, they would be listed as follows: D > B > A = C > E. However, α_{s1}-casein-A-8P contains 13 fewer amino acids than the other variants, and would have a comparably greater charge density and less frictional resistance to its movement through the gel. Therefore, the order of relative mobilities is A > D > B > C > E. Similar explanations can be suggested for the observed relative mobilities of the genetic variants of the other milk proteins.

Electrophoresis on cellulose acetate strips has also been used for the rapid resolution of whey proteins (Bell and Stone 1979). Samples of a 10:1 concentrate of whey are applied to cellulose acetate strips which have been saturated with Tris-barbiturate buffer, pH 8.6, ionic strength 0.097, and the electrophoresis is performed at 225 V for 1 hr. This procedure separates not only the major whey proteins but also their genetic variants.

Isoelectric focusing, as previously discussed, has been used to isolate and characterize the milk proteins (Josephson *et al.* 1971; Kaplan and Foster 1971; Peterson 1971; Josephson 1972; Josephson *et al.* 1972). Greater heterogeneity was observed in the caseins than was noted in

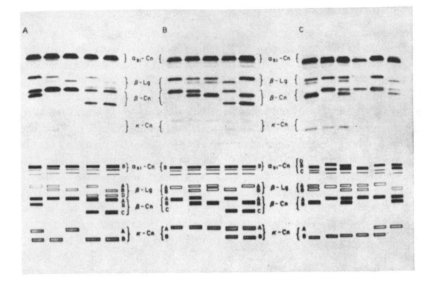

Figure 3.23. Starch-gel electrophoresis of whole milk samples in urea and mercaptoethanol. Letters on the bottom half of the photograph refer to the genetic types of the milk proteins. (Photograph courtesy Dr. W. Michalak, Warsaw, Poland). (From Thompson 1970. Reprinted with permission of the American Dairy Science Association.)

zonal electrophoresis, possibly due to ampholyte–casein complexes (Josephson 1972). The presence of urea in the medium apparently displaces the pI values of both the ampholites and the proteins (Josephson *et al.* 1971; Salaman and Williamson 1971; Ui 1971; Josephson 1972). Attempts have been made to correct for this effect.

ASSOCIATION OF MILK PROTEINS

The association that occurs between the monomers of the same milk protein as influenced by their environment has been discussed in the section "Structure and Conformation of Milk Proteins." However, in addition to this type of association, the milk proteins are known to form complexes with small ions and molecules, to bind water, and to form complexes with other macromolecules and with each other. The most important example of the last phenomenon is the formation of casein micelles, which is discussed in Chapter 9.

Association With Small Ions and Molecules

Since the proteins contain negatively charged groups such as phosphates, side-chain carboxyls, terminal carboxyls, and sulfhydryls, they bind a number of different cations, such as calcium, barium, strontium, magnesium (Dickson and Perkins 1969), copper (Dill and Simmons 1970; Aulakh and Stine 1971), thallium (Sundararajan and Whitney 1969), potassium and sodium (Ho and Waugh 1965), iron (Basch *et al.* 1974, Demott and Park 1974; Demott and Dincer 1976), cadmium (Roh *et al.* 1976), and mercury (Roh *et al.* 1975).

The binding of calcium by the milk proteins, especially the caseins, is of primary interest to milk protein chemists because of its involvement in micelle formation and its effect on the stability of the milk protein system (Thompson *et al.* 1969; Muldoon and Liska 1972; Farrell 1973; Toma and Nakai 1973; Eigel and Randolph 1976; Payens 1982). Zittle *et al.* (1958) investigated the effect of pH, temperature, and time on the binding of calcium by whole casein and observed that no calcium was bound at pH 5.0 and below, but the binding increased with increasing pH above that point. The amount of calcium bound did not appear to be appreciably dependent upon the time and temperature, but reversible aggregates form with increasing temperature. The presence of phosphates increases the amount of calcium bound. Carr and Topol (1950) and Ntailianas and Whitney (1964) observed that as the sodium ion concentration increased in the whole casein system, the amount of protein-bound calcium decreased due to the competitive binding of sodium. While the genetic variants A^2, B, and C of the major β-casein are dispersible at low temperatures over a wide range of calcium concentrations, the genetic variants of the major α_{s1}-casein are variable in their behavior (Thompson *et al.* 1969). Variants A and B are both precipitated at approximately 0.008 M, but the A variant is redispersed at 0.09 M, while the B variant requires a higher concentration. Addition of KCl to the A variant system increases the dispersibility of the protein. At temperatures of 18°C and above, both the α_{s1}-and β-caseins are precipitated at $CaCl_2$ concentrations higher than 0.008 M. Eigel and Randolph (1976) observed that β-casein A^2-1P (f 29–209) was more sensitive to calcium at low concentrations than α_{s1}-casein B-8P and β-casein A^2-5P at 30°C, but at concentrations higher than 0.008 M only 55% of the β-casein A^2-1P (f 29–209) was precipitated. The α_{s2}-casein dimer-11P–12P was found to be more sensitive to calcium than α_{s1}-casein B-8P, requiring only 0.002 M $CaCl_2$ for 80% precipitation (Toma and Nakai 1973).

The law of mass action controls the binding of calcium by proteins. The number and type of binding sites for calcium on the various caseins, and their association constants, have been investigated by a number of investigators (Demott 1969; Kramer and Lagoni 1969; Muldoon and Liska 1969; Yamauchi *et al.* 1969; Dickson and Perkins 1971; Waugh 1971; Waugh *et al.* 1971; Sundararajan and Whitney 1975; Jaynes and Whitney 1982). The variety of methods and conditions employed and the possible differences in the character of the protein preparations make it difficult to compare results or draw detailed conclusions. Most of the investigations indicate that the phosphate groups have the greatest affinity for calcium and are responsible for most of the binding at low concentrations of calcium, but there is considerable evidence that the carboxyl groups are involved as well (Figure 3.24). The α_{s1}-caseins and β-caseins have regions in their molecules containing high concentrations of polar groups which would strongly interact and, therefore, complicate the results of binding studies. Regardless of

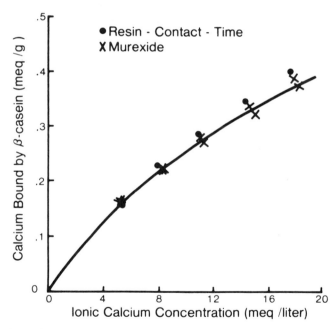

Figure 3.24. Calcium binding in 3% β-casein dispersion at $\Gamma/2 = 0.14$, pH 7.0 and 2°C, as determined by the resin contact time and murexide methods. Apparent maximum number of sites = 11.2 moles of calcium per mole. Apparent intrinsic binding constant = 76.62 liters/mole. (From Jaynes and Whitney 1982. Reprinted with permission of the American Dairy Science Association.)

the ionic strength, α_{s1}-and β-casein precipitate at the same level of bound calcium, 8 and 5.4 moles/mole, respectively (Waugh et al. 1971). Calcium may form both intra-and intermolecular bridges in the caseins. α-Lactalbumin strongly binds 1 mole of calcium per mole. (Hiraoka et al. 1980).

The adsorption of water by proteins depends upon their configuration, environment, and temperature and is due to hydrogen bonding and the structural forming properties of the exposed nonpolar groups. The various methods available for measuring the bound water of the proteins unfortunately are based on various physical properties of the bound water and, therefore, yield values dependent upon the method employed. Some studies have been made on the milk proteins. Thompson et al. (1969) determined the solvation of the A and B variants of α_{s1}-casein and the C variant of the major β-casein in different $CaCl_2$ and KCl mixtures at pH 7.0 and 37°C. They observed by centrifugal pelleting that the solvations of isoionic α_{s1}-casein A and B and β-casein C were 0.84, 0.74, and 1.50 g H_2O per gram of protein, respectively. These results are comparable to those obtained earlier by Creamer and Waugh (1966). In both studies, occluded water as well as bound water is included. Berlin et al. (1973) investigated the bound water content of the whey protein in concentrates by the calorimetric and vapor pressure equilibrium methods and observed 0.5 g of unfreezable water per gram of protein. Similar measurements on bovine serum albumin by Hasl and Pauly (1971) indicated 0.3 g caloric bound H_2O per gram of protein and 0.54 g of total bound water per gram of protein.

Numerous other small molecules are known to bind to the milk proteins, including certain antibiotics, such as dihydrostreptomycin and tetracycline (Ziv and Rasmussen 1975), antioxidants, such as the esters of gallic acid and butylated hydroxy anisole (Cornell et al. 1971) and dyes (McGann et al. 1972; Kristoffersen et al. 1974; Mickelsen and Shukri 1975). For analytical purposes, the binding of various dyes such as Amido Black, Orange G, and Acid Orange 12 are of special interest. Amido Black has been adapted for use with the Pro-Milk MK II instrument for the rapid determination of casein and whey proteins in milk (McGann et al. 1972). The noncasein protein separated from milk at pH 4.6 is measured directly on the instrument after reaction with the dye. The casein content is then calculated by the difference between the dye binding by the original milk and the dye binding by the whey protein. While the exact nature of the reaction is not known, it is generally assumed that the acid groups of the dye are bound by electrostatic forces to the basic groups on the protein. The amount of dye bound depends on the pH, the ratio of protein to dye concentration, and the character of the particular proteins involved.

Association with Macromolecules

The ability of whey proteins to form complexes with macromolecules such as CMC (Hansen *et al.* 1971; Hidalgo and Hansen 1971), and polyacrylic acid (Sternberg *et al.* 1976) has been described. CMC also has the ability to stabilize the major caseins in the pH range of 4.0 to 7.0 by soluble complex formation (Asano 1970; Asano and Ishida 1971). The behavior of the complexes with the milk proteins is not only a function of the pH but also of ionic strength, temperature, and ratio of polymer concentrations. Various sulfated polysaccharides such as agar-agar, porphyran, furcellaran, fucoidan, and carrageenan interact with the milk proteins under appropriate conditions (Lin 1977). Their action is due largely to electrostatic forces. Above the isoelectric point of the protein, polyvalent metal ions act as linkages between the negative charges on the protein and the negative sulfate on the polysaccha-

Table 3.6.   Stabilization of α_s-Casein (0.15%) by Some Hydrocolloids at pH 6.7 (Hydrocolloid/α_s-Casein = 1/4).

Groups	Glycosidic Linkages	Stabilized α_s-Casein (%)
Neutral		
Guar gum	β-1,4 (Branch α-1,6)	0
Locust bean gum	β-1,4 (Branch α-1,6)	0–0.1
Agarose	β-1,4 and β-1,3	0–1.5
Carboxylated		
CMC	β-1,4	0–0.1
Algin	β-1,4	0
Pectin	α-1,4	0–2.8
Gum arabic	β-1,3 and α-1,6	0–1.3
Hyaluronic acid	β-1,3 and β-1,4	0
Mixed carboxylated and sulfated		
Heparin	α-1,3 and α-1,4	0–6.0
Chondroitin sulfate A	β-1,3 and β-1,4	0–10.8
(4-sulfate)		
Chondroitin sulfate C	β-1,3 and β-1,4	0–1.5
(6-sulfate)		
Chondroitin sulfate D	β-1,3 and β-1,4	0–0.5
Sulfated		
Sulfated cellulose	β-1,4	6–15.4
Fucoidan	α-1,2 and β-1,4	0
Furcellaran	α-1,3 and β-1,4	40–50
λ-Carrageenan	α-1,3 and β-1,4	40.0–50.0
i-Carrageenan	α-1,3 and β-1,4	92.0–100
κ-Carrageenan	α-1,3 and β-1,4	90.0–100

SOURCE: Lin (1971). Reprinted with permission of AVI Publishing Co., Westport, Conn.

ride, while below the isoelectric point, the complex results from the association of the cationic groups on the protein and the sulfate of the polysaccharide. The number of sulfate groups and their point of attachment to the polysaccharide determine their conformation and, therefore, their ability to interact with proteins and to stabilize casein dispersons (Lin and Hansen 1970; Chakraborty and Hansen 1971; Lin 1971; O'Laughlin and Hansen 1973; Lin 1977). Lin (1971) investigated the ability of a number of hydrocolloids to stabilize the α_s-caseins at pH 6.7 and observed appreciable stabilization only when the carrageenans and furcellaran were used (Table 3.6).

REFERENCES

Aalund, O., Blakeslee, D., Butler, J. E., Duncan, J. R., Freeman, M. J., Jenness, R., Kehoe, J. M., Mach, J.-P., Rapacz, J., Vaerman, J.-P. and Winter, A. J. Proposed nomenclature for the immunoglobulins of the domesticated *Bovidae. Can. J. Comp. Med. 35*, 346–348.

Adams, E. T., Jr. and Lewis, M. S. 1968. Sedimentation equilibrium in reacting systems. VI. Some applications to indefinite self-association. Studies with β-lactoglobulin A. *Biochemistry 7*, 1044–1053.

Addeo, F., Chobert, J.-M. and Ribadeau-Dumas, B. 1977. Fractionation of whole casein on hydroxyapatite. Application of a study of buffalo kappa-casein. *J. Dairy Res. 44*, 63–68.

Aimutis, W. R. and Eigel, W. N. 1982. Identification of λ-casein as plasma-derived fragments of bovine α_{s1}-casein. *J. Dairy Sci. 65*, 175–181.

Albright, D. A. and Williams, J. W. 1968. A study of the combined sedimentation and chemical equilibrium of β-lactoglobulin B in acid solution. *Biochemistry 7*, 67–68.

Anderson, M., Cawston, T. and Cheeseman, G. C. 1974. Molecular weight estimates of milk fat globule membrane protein–sodium dodecyl sulfate complexes by electrophoresis in gradient acrylamide gels. *Biochem. J. 139*, 653–660.

Anderson, M. and Cheeseman, G. C. 1971. Some aspects of the chemical composition of the milk fat globule membrane during lactation. *J. Dairy Res. 38*, 409–417.

Anderson, L. O., Rehnstrom, A. and Eaker, D. L. 1971. Studies on nonspecific binding. The nature of the binding of fluorescein to bovine serum albumin. *Eur. J. Biochem. 20*, 371–380.

Andrews, A. L., Atkinson, D., Evans, M. T. A., Finer, E. G., Green, J. P., Phillips, M. C. and Robertson, R. N. 1979. The conformation and aggregation of bovine β-casein A. I. Molecular aspects of thermal aggregation. *Biopolymers 18*, 1105–1121.

Andrews, A. T. 1978A. The composition, structure, and origin of proteose-peptone component 5 of bovine milk. *Eur. J. Biochem. 90*, 59–65.

Andrews, A. T. 1978B. The composition, structure, and origin of proteose-peptone component 8F of bovine milk. *Eur. J. Biochem. 90*, 67–71.

Annan, W. D. and Manson, W. 1969. Fractionation of the α_s-casein complex of bovine milk. *J. Dairy Res. 36*, 259–268.

Anon. 1968. Electrodialysis leads to whey profits. *Food Eng. 40*(7), 158.

Armstrong, J. M., Hopper, K. E., McKenzie, H. A. and Murphy, W. H. 1970. On the column chromatography of bovine whey proteins. *Biochim. Biophys. Acta 214*, 419–426.

Armstrong, J. M., McKenzie, H. A. and Sawyer, W. H. 1967. On the fractionation of β-lactoglobulin and α-lactalbumin. *Biochim. Biophys. Acta 147*, 60–72.

Arvidsson, E. O. 1972. Salt binding in proteins. A model for the abnormal hydrogen ion titration and strong anion binding of serum albumin. *Biopolymers 11*, 2197–2221.

Asano, Y. 1970. Interactions between casein and carboxymethylcellulose in the acidic condition. *Agr. Biol. Chem. 34*, 102–107.

Asano, Y. and Ishida, Y. 1971. Chemical structure and protein stabilization activity of carboxymethylcellulose. *Agr. Biol. Chem. 35*, 1018–1023.

Aschaffenburg, R. 1963. Preparation of β-casein by a modified urea fractionation method. *J. Dairy Res. 30*, 259–260.

Aschaffenburg, R. 1968A. Review of the progress of dairy science. Section G. Genetics. Genetic variants of milk proteins: Their breed distribution. *J. Dairy Res. 35*, 447–460.

Aschaffenburg, R. 1968B. Preparation of α-lactalbumin from cows' or goats' milk: A method for improving the yield. *J. Dairy Sci. 51*, 1295–1296.

Aschaffenburg, R. and Drewry, J. 1957A. Genetics of the β-lactoglobulins of cows' milk. *Nature 180*, 376–378.

Aschaffenburg, R. and Drewry, J. 1957B. Improved method for the preparation of crystalline β-lactoglobulin and α-lactalbumin from cows' milk. *Biochem. J. 65*, 273–277.

Aschaffenburg, R., Sen, A. and Thompson, M. P. 1968. Genetic variants of casein in Indian and African zebu cattle. *Comp. Biochem. Physiol. 25*, 177–184.

Aulakh, J. S. and Stine, C. M. 1971. Binding of copper by certain milk proteins as measured by equilibrium dialysis. *J. Dairy Sci. 54*, 1605–1608.

Barel, A. O., Prieels, J. P., Maes, E., Looze, Y. and Leonis, J. 1972. Comparative physicochemical studies of human α-lactalbumin and human lysozyme. *Biochim. Biophys. Acta 257*, 288–296.

Barman, T. E. 1970. Purification and properties of bovine milk glyco-α-lactalbumin. *Biochim. Biophys. Acta 214*, 242–243.

Barman, T. E. 1973. The isolation of an α-lactalbumin with three disulfide bonds. *Eur. J. Biochem. 37*, 86–89.

Barry, J. G. and Donnelly, W. J. 1979. A method for the quantitative analysis of bovine casein. *Biochem. Soc. Trans. 7*, 529–531.

Barry, J. G. and Donnelly, W. J. 1980. Casein compositional studies. 1. The composition of casein from Friesian herd milk. *J. Dairy Res. 47*, 71–82.

Basch, J. J., Farrell, H. M. Jr. and Greenberg, R. 1976. Identification of the milk fat globule membrane proteins. I. Isolation and partial characterization of a glycoprotein B. *Biochim. Biophys. Acta 448*, 589–598.

Basch, J. J., Jones, S. B., Kalan, E. B. and Wondolowski, M. V. 1974. Distribution of added iron and polyphosphate phosphorus in cows' milk. *J. Dairy Sci. 57*, 545–550.

Basch, J. J., Kalan, E. B. and Thompson, M. P. 1965. Preparation of β-lactoglobulin C. *J. Dairy Sci. 48*, 604–606.

Basch, J. J. and Timasheff, S. N. 1967. Hydrogen ion equilibria of the genetic variants of β-lactoglobulin. *Arch. Biochem. Biophys. 118*, 37–47.

Becker, J. W., Ziffer, J. A., Edelman, G. M. and Cunningham, B. A. 1977. Crystallographic studies of bovine β2-microglobulin. *Proc. Natl. Acad. Sci. USA 74*, 3345–3349.

Beh, K. J. 1973. Distribution of brucella antibody among immunoglobulin classes and a low molecular weight antibody fraction in serum and whey of cattle. *Res. Vet. Sci. 14*, 381–384.

Bell, J. W. and Stone, W. K. 1979. Rapid separation of whey proteins by cellulose acetate electrophoresis. *J. Dairy Sci. 62*, 502–504.

Bell, K. 1962. One-dimensional starch-gel electrophoresis of bovine skim milk. *Nature* *195*, 705–706.

Bell, K., Hopper, K. E. and McKenzie, H. A. 1981A. Bovine α-lactalbumin C and $α_{s1}$-, β-, and κ-casein of Bali (Banteng) cattle, *Bos (Bibos) javanicus. Aust. J. Biol. Sci.* *34*, 149–159.

Bell, K. and McKenzie, H. A. 1976. The physical and chemical properties of whey proteins. *In: Milk Protein Workshop, Tanunda, South Australia.* Northfield Research Laboratories, South Australian Department of Agriculture.

Bell, K., McKenzie, H. A., Murphy, W. H. and Shaw, D. C. 1970. β-Lactoglobulin (Droughtmaster): A unique protein variant. *Biochim. Biophys. Acta 214*, 427–436.

Bell, K., McKenzie, H. A. and Shaw, D. C. 1981B. Bovine β-lactoglobulin E, F, and G of Bali (Banteng) cattle, *Bos (Bibos) javanicus. Aust. J. Biol. Sci. 34*, 133–147.

Benton, C., Floer, W. and Petzoldt, K. 1976. Research on experimental animal detection of anaphylactic antibodies of cattle in passive transport research under mixing of colostrum free fed calves and lambs. *Zbl. Vet. Med.* B *23*, 200–215 (German).

Berlin, E., Kliman, P. G., Anderson, B. A. and Pallansch, M. J. 1973. Water binding in whey protein concentrates. *J. Dairy Sci. 56*, 984–987.

Bezkorovainy, A. 1965. Comparative study of the acid glycoproteins isolated from bovine serum colostrum and milk whey. *Arch. Biochem. Biophys. 110*, 558–567.

Bezkorovainy, A. 1967. Physical and chemical properties of bovine milk and colostrum whey M-1 glycoproteins. *J. Dairy Sci. 50*, 1368.

Bezkorovainy, A. and Grohlich, D. 1969. Separation of the bovine colostrum M-1 glycoproteins into two components. *Biochem. J. 115*, 817–822.

Bhargava, H. N. and Foster, J. F. 1970. Reversible boundary spreading as a criterion of microheterogeneity of plasma albumins. *Biochemistry 9*, 1977–1983.

Bjork, I. and Tanford, C. 1971. Gross conformation of free polypeptide chains from rabbit immunoglobulin G. I. Heavy chain. *Biochemistry 10*, 1271–1280.

Bloomfield, V. A. and Mead, R. J., Jr. 1975. Structure and stability of casein micelles. *J. Dairy Sci. 58*, 592–601.

Braunitzer, G. and Chen R. 1972. The cleavage of β-lactoglobulin AB with cyanogen bromide. *Hoppe-Seyler's Z. Physiol. Chem. 353*, 674–676 (German).

Braunitzer, G., Chen, R., Schrank, B. and Stangl, A. 1972. Automatic sequential analysis of a protein (β-lactoglobulin AB). *Hoppe-Seyler's Z. Physiol. Chem. 353*, 832–834 (German).

Braunitzer, G., Chen, R., Schrank, B. and Stangl, A. 1973. The sequence analysis of β-lactoglobulin. *Hoppe-Seyler's Z. Physiol. Chem. 354*, 868–878 (German).

Brew, K., Castellino, F. J., Vanaman, T. C. and Hill, R. L. 1970. The complete amino-acid sequence of bovine α-lactalbumin. *J. Biol. Chem. 245*, 4570–4582.

Brew, K. and Hill, R. L. 1970. The isolation and characterization of the tryptic, chymotryptic, peptic, and cyanogen bromide peptides from bovine α-lactalbumin. *J. Biol. Chem. 245*, 4559–4569.

Brew, K. and Hill, R. L. 1975. Lactose biosynthesis. *Rev. Physiol. Biochem. Pharmacol. 72*, 105–158.

Brignon, G. and Ribadeau-Dumas, B. 1973. The location in the peptide chain of bovine β-lactoglobulin of the Glu/Gln substitution differentiating the genetic variants B and D. *FEBS Lett. 33*, 73–76 (French).

Brignon, G., Ribadeau-Dumas, B. and Mercier, J.-C. 1976. Primary elements of the primary structure of $α_{s2}$-bovine casein. *FEBS Lett. 71*, 111–116 (French).

Brignon, G., Ribadeau-Dumas, B., Garnier, J., Pantaloni, D., Guinand, S., Basch, J. J. and Timasheff, S. N. 1969. Chemical and physico-chemical characterization of genetic variant D of bovine β-lactoglobulin. *Arch. Biochem. Biophys. 129*, 720–727.

Brignon, G., Ribadeau-Dumas, B., Mercier, J.-C. Pelissier, J. P. and Das, B. C. 1977. Complete amino acid sequence of bovine α_{s2}-casein. *FEBS Lett. 76*, 274–279 (French).

Brown, J. R. 1975. Structure of bovine serum albumin. *Proc. Fed. Am. Soc. Exp. Biol. 34*, 591.

Brown, J. R. 1977. Serum albumin: Amino-acid sequence. In: *Albumin Structure, Function, and Uses.* V.M. Rosenoer, M. Oratz and M.A. Rothschild (Editors). Pergamon Press, New York.

Brown, J. R., Law, T., Behrens, P., Sepulveda, K., Parker, K. and Blakency, E. 1971. Amino-acid sequence of bovine and porcine serum albumin. *Fed. Proc. 30*, Pt II, 1241.

Brown, W. J., North, A. C. T., Phillips, D. C., Brew, K., Vanaman, T. C. and Hill, R. L. 1969. A possible three-dimensional structure of bovine α-lactalbumin based on that of hens' egg white lysozyme. *J. Mol. Biol. 42*, 65–86.

Brunner, J. R. 1981. Cow milk proteins: Twenty-five years of progress. *J. Dairy Sci. 64*, 1038–1054.

Buchheim, W. and Schmidt, D. G. 1979. On the size of monomers and polymers of β-casein. *J. Dairy Res. 46*, 277–280.

Burger, R. L. and Allen, R. H. 1974. Characterization of vitamin B_{12}-binding proteins isolated from human milk and saliva by affinity chromatography. *J. Biol. Chem. 249*, 7220–7227.

Butler, J. E. 1969. Bovine immunoglobulins: A review. *J. Dairy Sci. 52*, 1895–1909.

Butler, J. E. 1973. The occurrence of immunoglobulin fragments, two types of lactoferrin and a lactoferrin-IgG$_2$ complex in bovine colostral and milk whey. *Biochim. Biophys. Acta 295*, 341–351.

Butler, J. E. 1974. Immunoglobulins of the mammary secretions. In: *Lactation: A Comprehensive Treatise*, Vol. II. B. L. Larson and V. R. Smith (Editors). Academic Press, New York.

Butler, J. E., Kiddy, C. A., Pierce, C. S. and Rock, C. A. 1972. Quantitative changes associated with calving in the levels of bovine immunoglobulins and selected body fluids. I. Changes in the level of IgA, IgG, and total protein. *Can. J. Comp. Med. 36*, 234–242.

Butler, J. E. and Maxwell, C. F. 1972. Preparation of bovine immunoglobulins and free secretory component and their specific antisera. *J. Dairy Sci. 55*, 151–164.

Butler, J. E., McGivern, P. L., Conterero, L. A. and Peterson, L. 1980. Application of the amplified enzyme-linked immunosorbent assay: Comparative quantitation of bovine serum IgG$_1$, IgG$_2$, IgA, and IgM antibodies. *Am. J. Vet. Res. 41*, 1479–1491.

Butler, J. E., Winter, A. J. and Wagner, G. G. 1971. Symposium: Bovine immune systems. *J. Dairy Sci. 54*, 1309–1314.

Carr, C. W. and Topol L. 1950. The determination of sodium-ion and chloride-ion activities in protein solutions by means of permselective membranes. *J. Phys. Colloid Chem. 54*, 176–184.

Chakraborty, B. K. and Hansen, P. M. T. 1971. Electron microscopy of protein/hydrocolloid interacting systems. *J. Dairy Sci. 54*, 754.

Cheeseman, G. C. 1962. A method of preparation of kappa-casein and some observations on its nature. *J. Dairy Res. 29*, 163–171.

Cheeseman, G. C. 1968. A preliminary study by gel filtration and ultracentrifugation of the interaction of bovine milk caseins with detergents. *J. Dairy Res. 35*, 439–446.

Chiba, H., Ueda, M., Yoshikawa, M. and Sassaki, R. 1978. A simple method for fractionation of α_{s1}-casein. *Vth Int. Cong. Food Sci. Tech.* (Japan), Abst. 172.

Clark, R. F. L. and Nakai, S. 1972. Fluorescent studies of κ-casein with 8-anilinonaphthaline-1-sulfonate. *Biochim. Biophys. Acta* 257, 61–69.

Cornell, D. G., De Vilbiss, E. D. and Pallansch, M. J. 1971. Binding of antioxidants by milk proteins. *J. Dairy Sci. 54*, 634–637.

Craven, D. A. and Gehrke, C. W. 1967. Improved chemical method for κ-casein. *J. Dairy Sci. 50*, 940.

Creamer, L. K. 1972. Hydrogen ion equilibria of bovine β-casein B. *Biochim. Biophys. Acta 271*, 252–261.

Creamer, L. K. 1974. Preparation of α$_{s1}$-casein A. *J. Dairy Sci. 57*, 341–344.

Creamer, L. K. and Matheson, A. R. 1981. Separation of bovine caseins using hydrophobic interaction chromatography. *J. Chromatography* (New Zealand) *210*, 105–111.

Creamer, L. K. and Waugh, D. F. 1966. Calcium binding and precipitate solvation of Ca-α$_s$-caseinates. *J. Dairy Sci. 49*, 706.

Creamer, L. K., Zoerb, H. F., Olson, N. F. and Richardson, T. 1982. Surface hydrophobicity of α$_{s1}$-I, α$_{s1}$-casein A and B and its implication in cheese structure. *J. Dairy Sci. 65*, 902–906.

Dalgliesh, D. G. and Parker, T. G. 1979. Quantitation of α$_{s1}$-casein aggregation by the use of polyfunctional models. *J. Dairy Res. 46*, 259–263.

Dalgliesh, D. G. and Parker, T. G. 1980. Binding of calcium ions to bovine α$_{s1}$-casein and precipitability of protein–calcium ion complexes. *J. Dairy Res. 47*, 113–122.

Davies, D. T. and Law, A. J. R. 1977. An improved method for the quantitative fractionation of casein mixtures using ion-exchange chromatography. *J. Dairy Res. 44*, 213–221.

Deckmyn, H. and Préaux, G. 1978. Chain-folding prediction of the bovine β-lactoglobulin. *Arch. Intern. Physiol. Biochim. 86*, 938–939.

Demott, B. J. 1969. Calcium ion concentration in milk, whey, and β-lactoglobulin as influenced by ionic strength, added calcium, rennet concentration, and heat. *J. Dairy Sci. 52*, 1672–1675.

Demott, B. J. and Dincer, B. 1976. Binding added iron to various milk proteins. *J. Dairy Sci. 59*, 1557–1559.

Demott, B. J. and Park, J. R. 1974. Effect of processing upon association of added iron with different protein fractions. *J. Dairy Sci. 57*, 121–123.

Denton, W. L. nd Ebner, K. E. 1971. Effect of tyrosyl modification on the activity of α-lactalbumin in the lactose synthetase reaction. *J. Biol. Chem. 246*, 4053–4059.

Dickson, J. R. and Perkins, D. J. 1969. Interaction between the alkaline earth metal ions and purified bovine caseins. *Biochem. J. 113*, 7P.

Dickson, J. R. and Perkins, D. J. 1971. Studies on the interaction between bovine caseins and alkaline earth metal ions. *Biochem. J. 124*, 235–240.

Dill, C. W. and Simmons, J. 1970. Binding copper by milk proteins. *J. Dairy Sci. 53*, 641.

Doi, E. and Jirgensons, B. 1970. Circular dichroism studies on the acid denaturation of γ-immunoglobulin G and its fragments. *Biochemistry 9*, 1066–1073.

Doi, H., Ibuki, F. and Kannamori, M. 1979A. Heterogeneity of reduced bovine-κ-casein. *J. Dairy Sci. 62*, 195–203.

Doi, H., Kawaguchi, N., Ibuki, F. and Kannamori, M. 1979B. Minor components of reduced κ-casein. *J. Nutr. Sci. Vitaminol. 25*, 95–102.

Donnelly, W. J. 1977. Chromatography of milk protein on hydroxyapatite. *J. Dairy Res. 44*, 621–625.

Dorrington, K. J. and Tanford, C. 1970. Molecular size and conformation of immunoglobulins. *Adv. Immunol. 12*, 333–381.

Dosaka, S., Kaminogawa, S., Taneya, S. and Yamauchi, K. 1980A. Hydrophobic surface areas and net charge of α$_{s1}$-, κ-casein, and α$_{s1}$-casein: κ-casein complex. *J. Dairy Res. 47*, 123–129.

Dosaka, S., Kimura, T., Taneya, S., Sone, T., Kaminogawa, S. and Yamauchi, K. 1980B. Polymerization of α_{s1}-casein by calcium ions. *Agric. Biol. Chem. 44*, 2443–2448.

Dreizen, P., Noble, R. W. and Waugh, D. F. 1962. Light scattering studies of $\alpha_{s1,2}$-caseins. *J. Am Chem. Soc. 84*, 4938–4943.

Duncan, J. R., Wilkie, B. N., Heistand, F. and Winter, A. J. 1972. The serum and secretory immunoglobulins of cattle: Characterization and quantitation. *J. Immunol. 108*, 965–976.

Ebner, K. E. 1971. Biosynthesis of lactose. *J. Dairy Sci. 54*, 1229–1233.

Ebner, K. E. and Schanbacher, F. 1974. Biochemistry of lactose and related carbohydrates. *In: Lactation: A Comprehensive Treatise*, Vol. II. B. L. Larson and V. R. Smith (Editors). Academic Press, New York.

Eigel, W. N. 1977. Formation of γ_1-A^2, γ_2-A^2, and γ_3-A caseins by in *vitro* proteolysis of β-casein A^2 with bovine plasmin. *Int. J. Biochem. 8*, 187–192.

Eigel, W. N. 1981. Identification of proteose-peptone component 5 as a plasmin derived fragment of β-casein. *Int. J. Biochem. 13*, 1081–1086.

Eigel, W. N., Butler, J. E., Ernstrom, C. A., Farrell, H. M., Jr., Harwalkar, V. R., Jenness, R. and Whitney, R. McL. 1984. Nomenclature of the proteins of cows' milk: Fifth revision. *J. Dairy Sci. 67*, 1599–1631.

Eigel, W. N., Hoffmann, C. J., Chibber, B.A.K. Tomich, J. M., Keenan, T. W. and Mertz, E. T. 1979. Plasmin-mediated proteolysis of casein in milk. *Proc. Natl. Acad. Sci. USA 76*, 2244–2248.

Eigel, W. N. and Keenan, T. W. 1979. Identification of proteose-peptone component 8-slow as a plasmin-derived fragment of β-casein. *Int. J. Biochem. 10*, 529–535.

Eigel, W. N. and Randolph, H. E. 1974. Preparation of whole γ-casein by treatment with calcium phosphate gel. *J. Dairy Sci. 57*, 1444–1447.

Eigel, W. N. and Randolph, H. E. 1976. Comparison of calcium sensitivities of α_{s1}-B, β-A^2, and γ-A^2 caseins and their stabilization by κ-casein A. *J. Dairy Sci. 59*, 203–206.

Elfagm, A. A. and Wheelock, J. V. 1978A. Interaction of bovine α-lactalbumin and β-lactoglobulin during heating. *J. Dairy Sci. 61*, 28–32.

Elfagm, A. A. and Wheelock, J. V. 1978B. Heat interaction between α-lactalbumin, β-lactoglobulin, and casein in bovine milk. *J. Dairy Sci. 61*, 159–163.

Ellfolk, N. 1957. Fractionation of casein by distribution in a liquid two-phase system. *Acta Chem. Scand. 11*, 1317–1322.

El-Negoumy, A. M. 1973. Separation of lambda casein and some of its properties. *J. Dairy Sci. 56*, 1486–1491.

El-Negoumy, A. M. 1976. Two rapid and improved techniques for chromatographic fractionation of casein. *J. Dairy Sci. 59*, 153–156.

Evans, M. T. A., Irons, L., and Jones, M. 1971. Physicochemical properties of β-casein and some carboxylacyl derivatives. *Biochim. Biophys. Acta 229*, 411–422.

Farrell, H. M., Jr. 1973. Models of casein micelle formation. *J. Dairy Sci. 56*, 1195–1206.

Farrell, H. M., Jr., Thompson, M. P. and Larson, B. 1971. Verification of the occurrence of the α_{s1}-casein A allele in Red Danish cattle. *J. Dairy Sci. 54*, 423–425.

Fenton-May, R. I., Hill, C. G., Jr. and Amundson, C. H. 1971. Use of ultrafiltration-osmosis systems for the concentration and fractionation of whey. *J. Food Sci. 36*, 14–21.

Fiat, A.-M., Alais, C. and Jollès, P. 1972. The amino acid and carbohydrate sequence of a short glycopeptide isolated from bovine κ-casein. *Eur. J. Biochem. 27*, 408–412.

Forsum, E., Hambraeus, L. and Siddiqi, I. H. 1974. Large-scale fractionation of whey protein concentrates. *J. Dairy Sci. 57*, 659–664.

Foster, J. F. 1977. Some aspects of the structure and conformational properties of serum albumin. *In: Albumin Structure, Function, and Uses.* V.M. Rosenoer, M. Oratz and M.A. Rothschild (Editors). Pergamon Press, New York.

Fournet, B., Fiat, A.-M., Alais, C. and Jollès, P. 1979. Cow κ-casein: Structure of carbohydrate portion. *Biochim. Biophys. Acta 576,* 339–346.

Fournet, B., Fiat, A.-M., Montreuil, J. and Jollès, P. 1975. The sugar part of κ-casein from cow milk and colostrum and its microheterogeneity. *Biochimie 57,* 161–165.

Fox, P. F. and Guiney, J. 1972. A procedure for the partial fractionation of the α_s-casein complex. *J. Dairy Res. 39,* 49–53.

Fox, K. K., Holsinger, V. H., Posati, L. P. and Pallansch, M. J. 1967. Separation of β-lactoglobulin from other milk serum proteins by trichloroacetic acid. *J. Dairy Sci. 50,* 1363–1367.

Freeman, T. 1970. Techniques for protein separation. *In: Plasma Protein Metabolism.* M.A. Rothschild and T. Waldman (Editors). Academic Press, New York.

Freudenstein, C., Keenan, T. W., Eigel, W. N., Sasaki, M., Stadler, J. and Franke, W. W. 1979. Preparation and characterization of inner coat material associated with fat globule membranes from bovine and human milk. *Exp. Cell. Res. 118,* 277–294.

Froese, A. 1971. Isolation of dinitrophenyl-specific antibodies from bovine colostrum. *Can. J. Biochem. 49,* 522–528.

Galley, J. A. 1973. Structure of immunoglobulins. *In: The Antigens,* Vol I. M. Sela (Editor). Academic Press, New York.

Garnier, J. 1966. Conformation of β-casein in solution. Analysis of a thermal transition between 5° and 40°C. *J. Mol. Biol. 19,* 586–590 (French).

Gazia, N. and Agergoard, N. 1980. Electrophoretic studies on protein and enzyme systems in Egyptian water buffaloes. *Arsheret-K Vet. Landhohoejsk Inst. Sterileletsforsk 23,* 35–41 (Danish).

Ghose, A. C., Chaudhuri, S. and Sen, A. 1968. Hydrogen ion equilibria and sedimentation behavior of goat β-lactoglobulin. Comparison of goat and bovine β-lactoglobulin. *Arch. Biochem. Biophys. 126,* 232–243.

Gilbert, G. A. 1970. Mixed hemoglobin molecules. *Biochem. J. 119,* 32P.

Gilbert, L. M. and Gilbert, G. A. 1973. Sedimentation velocity measurements of protein association. *Methods Enzymol. 27,* Part D, 273–296.

Girdhar, B. K. and Hansen, P.M.T. 1978. Production of κ-casein concentrate from commercial casein. *J. Food Sci. 43,* 397–400.

Glueckauf, E. and Patterson, E. 1974. Adsorption of some proteins on hydroxylapatite and other adsorbents used for chromatographic separation. *Biochim. Biophys. Acta 351,* 57–76.

Gobrinoff, M. J. 1967. Exposusre of tyrosine residues in proteins. Reaction of cyanuric fluoride with ribonuclease, α-lactalbumin, and β-lactoglobulin. *Biochemistry 66,* 1606–1614.

Goodger, B. V. 1971. A low molecular weight protein in bovine serum with similar electrophoretic mobility to IgG_2. *Res. Vet. Sci. 12,* 465–468.

Gordin, S., Birk, Y. and Volcani, R. 1972. Fractionation and characterization of milk proteins by column chromatography and electrophoresis. *J. Dairy Sci. 55,* 1544–1549.

Gordon, W. G. 1971. α-Lactalbumin. *In: Milk Proteins,* Vol. II. H.A. McKenzie (Editor). Academic Press, New York.

Gordon, W. G., Groves, M. L. and Basch, J. J. 1963. Bovine milk "red protein": Amino acid composition and comparison with blood transferrin. *Biochemistry 2,* 817–820.

Gordon, W. G., Groves, M. L., Greenberg, R., Jones, S. B., Kalan, E. B., Peterson, R. F. and Townend, R. E. 1972. Probable indentification of γ-, TS-, R-, and S-casein as fragments of β-casein. *J. Dairy Sci. 55,* 261–263.

Gordon, W. G. and Ziegler, J. 1955. α-Lactalbumin. *Biochem. Prep. 4*, 16–22.

Green, D. W. 1964. Cited by S.N. Timasheff and R. Townend. Structure of the β-lacto-globulin tetramer. *Nature 203*, 517–519.

Green, M. L. 1969. Simple methods for the purification of crude κ-casein and β-casein by treatment with calcium phosphate gel. *J. Dairy Res. 36*, 353–357.

Green, M. L. 1971A. The specificity for κ-casein as the stabilizer of α_s-casein and β-casein. I. Replacement of κ-casein by other proteins. *J. Dairy Res. 38*, 9–23.

Green, M. L. 1971B. The specificity for κ-casein as the stabilizer of α_s-casein and β-casein. II. Replacement of κ-casein by detergents and water-soluble polymers. *J. Dairy Res. 38*, 25–32.

Green, N. M. 1969. Electron microscopy of immunoglobulins. *Ad. Immunol. 11*, 1–30.

Grosclaude, F., Jourdrier, P. and Mahé, M.-F. 1978. Polymorphism of α_{s2}-bovine casein: Connection of the α_{s2}-Cn locus with the loci of α_{s1}-Cn, β-Cn, and κ-Cn; evidence of a deletion in the variant α_{s2}-Cn D. *Ann Genet. Sel. Anim. 10*, 313–327 (French).

Grosclaude, F., Jourdrier, P. and Mahé, M.-F. 1979. A genetic and biochemical analysis of a polymorphism of bovine α_{s2}-casein. *J. Dairy Res. 46*, 211–213.

Grosclaude, F., Mahé, M.-F. and Mercier, J.-C. 1974A. Comparison of the genetic polymorphism of the milk proteins of zebu and bovines. *Ann. Genet. Sel. Anim. 6*, 305–329 (French).

Grosclaude, F., Mahé, M.-F., Mercier, J.-C., Bonnemarie, J. and Tessier, J. H. 1976A. Polymorphism of the milk proteins of Nepalese bovines. I. The *Yak* and biochemical characterization of two new variants: β-lactoglobulin D _{(Yak)} and α_{s1}-casein E. *Ann. Genet. Sel. Anim. 8*, 461–479 (French).

Grosclaude, F., Mahé, M.-F., Mercier, J.-C., Bonnemarie, J. and Tessier, J. H. 1976B. Polymorphism of milk proteins of Nepalese bovines. II. Polymorphism of the caseins (α_s-minors); Is locus α_{s2}-Cn linked to loci α_{s1}-Cn, β-Cn, and κ-Cn? *Ann. Genet. Sel. Anim. 8*, 481–491 (French).

Grosclaude, F., Mahé, M.-F., Mercier, J.-C. and Ribadeau-Dumas, B. 1972. Characterization of the genetic variants of bovine α_{s1}- and β-caseins. *Eur. J. Biochem. 26*, 328–337 (French).

Grosclaude, F., Mahé, M.-F., Mercier, J.-C. and Ribadeau-Dumas, B. 1973. Primary structure of α_{s1}-casein and β-casein. Correction. *Eur. J. Bichem. 40*, 323–324 (French).

Grosclaude, F., Mahé, M.-F. and Voglino, G. F. 1974B. The variant β-E and the code for the phosphorylation of bovine caseins. *FEBS Lett. 45*, 3–5 (French).

Grosclaude, F., Pujolle, J., Garnier, J. and Ribadeau-Dumas, B. 1966. Evidence of two additional variants of the proteins of cows' milk: α_{s1}-Cn D and β-Lg D. *Ann. Biol. Anim. Biochim. Biophys. 6*, 215–222 (French).

Groves, M. L. 1971. Minor milk proteins and enzymes. *In: Milk Proteins*, Vol. II. H. A. McKenzie (Editor). Academic Press, New York.

Groves, M. L. and Gordon, W. G. 1967. Isolation of a new glycoprotein-a and a γG-globulin from individual cows milks. *Biochemistry 6*, 2388–2394.

Groves, M. L. and Gordon, W. G. 1969. Evidence from amino acid analysis for a relationship in the biosynthesis of γ- and β-caseins. *Biochim. Biophys. Acta 194*, 421–432.

Groves, M. L., Gordon, W. G., Greenberg, R., Peterson, R. F. and Jenness, R. 1975. Sequencing β-casein C: Isolation of a large fragment after cleavage of thioltrifluora-cetylated β-casein C. *J. Dairy Sci. 58*, 301–305.

Groves, M. L., Gordon, W. G., Kalan, E. B. and Jones, S. B. 1972. Composition of bovine γ-casein A[1] and A[3] and further evidence for a relationship in biosynthesis of γ- and β-casein. *J. Dairy Sci. 55*, 1041–1049.

Groves, M. L., Gordon, W. G., Kalan, E. B. and Jones, S. B. 1973. TS-A², TS-B, R, and S-caseins: Their isolation, composition, and relationship to the β- and γ-casein polymorphs A² and B. *J. Dairy Sci. 56*, 558–568.

Groves, M. L. and Greenberg, R. 1977. Bovine homologue of β₂-microglobulin isolated from milk. *Biochem. Biophys. Res. Commun. 77*, 320–327.

Groves, M. L. and Greenberg, R. 1982. Complete amino acid sequence of bovine β₂-microglobulin. *J. Biol. Chem. 257*, 2619–2626.

Groves, M. L., McMeekin, T. L., Hipp, N. J. and Gordon, W. G. 1962. Preparation of β- and γ-casein by column chromatography. *Biochim. Biophys. Acta 57*, 197–203.

Groves, M. L. and Townend, R. 1970. Molecular weight of some human and cow caseins. *Arch. Biochem. Biophys. 139*, 406–409.

Guidry, A. J., Butler, J. E., Pearson, R. E. and Weinland, B. T. 1980. IgA, IgG₁, IgG₂, IgM, and BSA in serum and mammary secretions throughout lactation. *Vet. Immunol. Immunopathol. 1*, 329–341.

Habeeb, A. F. S. A. and Atassi, M. Z. 1971. Enzymic and immunochemical properties of lysozyme. IV. Demonstration of conformational differences between α-lactalbumin and lysozyme. *Biochim. Biophys. Acta 236*, 131–141.

Hagenmaier, R. D. and Foster, J. F. 1971. Preparation of bovine mercaptalbumin and an investigation of its homogeneity. *Biochemistry 10*, 637–645.

Halfman, C. J. and Nishida, T. 1971. Influence of pH and electrolyte on fluorescence of bovine serum albumin. *Biochim. Biophys. Acta 243*, 284–293.

Hammer, D. K., Kickhoefen, B. and Schmid, T. 1971. Detection of homocytotropic antibody associated with a unique immunoglobulin class in the bovine species. *Eur. J. Immunol. 1*, 249–257.

Hansen, P. M. T., Hidalgo, J. and Gould, I. A. 1971. Reclamation of whey protein with carboxymethylcellulose. *J. Dairy Sci. 54*, 830–834.

Hanson, L. A., Sammuelsson, E. G. and Halmgren, J. 1967. Detection of ceruloplasmin in bovine milk and blood serum. *J. Dairy Res. 37*, 493–504.

Harmsen, B. J. M., DeBruin, S. H., Janssen, L. H. M., Rodrigues de Miranda, J. F. and VanOs G. A. J. 1971. pK change of imidazole groups in bovine serum albumin due to conformational change at neutral pH. *Biochemistry 10*, 3217–3221.

Hasl, G. and Pauly, H. 1971. Calorimetric properties of the bound water in protein solutions. *Biophysik 7*, 283–294 (German).

Helesicová, H. and Podrazký, V. 1980. Casein coagulation by hydrochloric acid. *Prumysl. Potravin* (Czechoslovakia) *31*, 210–213 (Czechoslovakian).

Herald, C. T. and Brunner, J. R. 1957. The fat-globule membrane of normal cows' milk. I. The isolation and characteristics of two membrane-protein fractions. *J. Dairy Sci. 40*, 948–956.

Herskovits, T. T. 1966. On the conformation of caseins. Optical rotatory properties. *Biochemistry 5*, 1018–1026.

Herskovits, T. T. and Mescanti, L. 1965. Conformation of proteins and polypeptides. *J. Biol. Chem. 240*, 639–644.

Hidalgo, J. and Hansen, P. M. T. 1971. Selective precipitation of whey proteins with carboxymethylcellulose. *J. Dairy Sci. 54*, 1270–1274.

Hidalgo, J., Krusman, J. and Bahren, H. U. 1973. Recovery of whey proteins with sodium hexametaphosphate. *J. Dairy Sci. 56*, 988–993.

Hilak, M. C., Harmsen, B. J. M., Braam, W. G. M., Joordens, J. J. M. and VanOs, G. A. J. 1974. Conformational studies on large fragments of bovine serum albumin in relation to the structure of the molecule. *Int. J. Peptide Protein Res. 6*, 95–101.

Hill, R. J. and Wake, R. G. 1969. Amphiphile nature of κ-casein as the basis for its micelle stabilizing properties. *Nature 221*, 635–639.

Hill, R. L. and Brew, K. 1975. Lactose synthetase. *Adv. Enzymol. 43*, 411–490.

Hipp, N. J., Groves, M. L., Custer, J. H. and McMeekin, T. L. 1952. Separation of α-, β-, and γ-casein. *J. Dairy Sci. 35*, 272–281.

Hiraoka, Y., Segawa, T., Kuwajima, K., Sugai, S and Murai, N. 1980. α-Lactalbumin: A calcium metalloprotein. *Biochem. Biophys. Res. Commun. 95*, 1098–1104.

Hladik, J. and Kas, J. 1973. Fractionation of the whey proteins and casein of cows' milk on Sephedex DEAE A-50 and A-25. *J. Chromatography 75*, 117–121.

Ho, C. and Chen, A. H. 1967. The polymerization of bovine α_s-casein B. *J. Biol. Chem. 242*, 551–554.

Ho, C. and Waugh, D. F. 1965. Interaction of bovine α_s-casein with small ions. *J. Am. Chem. Soc. 87*, 110–117.

Hoagland, P. D., Thompson, M. P. and Kalan, E. B. 1971. Amino acid composition of α_{s3}-, α_{s4}-, α_{s5}-caseins. *J. Dairy Sci. 54*, 1103–1110.

Hoffman, C. J., Keenan, T. W. and Eigel, W. N. 1979. Association of plasminogen with bovine fat globule membrane. *Int. J. Biochem. 10*, 909–917.

Holt, C., Parker, T. G. and Dalgliesh, D. G. 1975. Thermochemistry of reactions between α_{s1}-casein and calcium chloride. *Biochim. Biophys. Acta 379*, 638–644.

Hopper, K. E. and McKenzie, H. A. 1973A. Purification and properties of bovine milk glyco-α-lactalbumin. *Biochim. Biophys. Acta 214*, 242–244.

Hopper, K. E. and McKenzie, H. A. 1973B. Minor components of bovine α-lactalbumin A and B. *Biochim. Biophys. Acta 295*, 352–363.

Horne, D. S. 1979. The kinetics of the precipitation of chemically modified α_{s1}-casein by calcium. *J. Dairy Res. 46*, 265–269.

Igarashi, Y. and Saito, Z. 1970. Some properties of temperature-sensitive casein in cows' milk. *Jpn. J. Zootech. Sci. 41*, 262–269.

Ivkova, M. N., Vendenkina, N. S. and Burshtein, E. A. 1971. Fluorescence of tryptophan residues in serum albumin. *Mol. Biol. 5*, 168–176.

Iyer, K. S. and Klee, W. A. 1973. Direct spectrophotometric measurement of the rate of reduction of disulfide bonds. Reactivity of the disulfide bonds of bovine α-lactalbumin. *J. Biol. Chem. 248*, 707–710.

Janatova, J., Fuller, J. K. and Hunter, M. J. 1968. The heterogeneity of bovine albumin with respect to sulfhydryl and dimer content. *J. Biol. Chem. 243*, 3612–3622.

Jarasch, E. D., Bruder, G., Keenan, T. W. and Franke, W. W. 1977. Redox constituents in milk fat globule membranes and rough endoplasmic reticulum from lactating mammary gland. *J. Cell. Biol. 73*, 223–241.

Jaynes, H. O. and Whitney, R. McL. 1982. Resin-contact time for the determination of protein bound calcium in milk and model systems. *J. Dairy Sci. 65*, 1074–1083.

Jenness, R. 1982. Interspecies comparison of milk proteins. *In: Developments in Dairy Chemistry*, Vol. I: *Proteins*. P.F. Fox (Editor). Applied Science Publishers, New York.

Jollès, P. and Fiat, A.-M. 1979. The carbohydrate portion of milk glycoproteins. *J. Dairy Res. 46*, 181–191.

Jollès, J., Fiat, A.-M., Alais, C. and Jollès, P. 1973. Comparative study on cow and sheep κ-casein glycopeptides: Determination of the N-terminal sequences with a sequencer and the location of the sugars. *FEBS Lett. 30*, 173–176 (French).

Jollès, P., Loucheux-Lefebvre, M. H. and Henschen, A. 1978. Structural relatedness of κ-casein and fibrinogen γ-chain. *J. Mol. Evol. 11*, 271–277.

Jollès, J., Schoentgen, F., Alais, C., Fiat, A.-M. and Jollès, P. 1972A. Studies on the primary structure of cow κ-casein. Structural features of para-κ-casein; N-terminal sequence of κ-casein-glycopeptides studied with a sequencer. *Helv. Chim. Acta 55*, 2872–2883.

Jollès, J., Schoentgen, F., Alais, C. and Jollès, P. 1972B. Studies on the primary structure of cow κ-casein: The primary sequence of cow para-κ-casein. *Chimia 26,* 645-646.

Jones, E. A. 1977. Synthesis and secretion of milk sugars. *Symp. Zool. Soc. London 41,* 77-94.

Jones, S. B., Kalan, E. B., Jones, T. B. and Hazel, J. F. 1972. "Ferripolyphosphate" as a whey protein precipitant. *J. Agric. Food Chem. 20,* 229-232.

Josephson, R. V. 1972. Isoelectric focusing of bovine milk caseins. *J. Dairy Sci. 55,* 1535-1543.

Josephson, R. V., Maheswaran, S. K., Morr, C. V., Jenness, R. and Lindorfer, R. K. 1971. Effect of urea on pI's of ampholytes and casein in isoelectric focusing. *Anal. Biochem, 40,* 476-482.

Josephson, R. V., Mikolajcik, E. M. and Singh, V. K. 1972. Isoelectric focusing of bovine colostrum immunoglobulins. *J. Dairy Sci. 55,* 1050-1057.

Kaminogawa, S., Mizobuchi, H. and Yamauchi, K. 1972. Comparison of bovine milk protease with plasmin. *Agr. Biol. Chem. 36,* 2163-2167.

Kanno, C., Emmons, D. B., Harwalker, V. R. and Elliott, J. A. 1976. Purification and characterization the agglutinating factor for lactic streptococci from bovine milk: IgM immunoglobulin. *J. Dairy Sci. 59,* 2036-2045.

Kanno, C., Shimizu, M. and Yamauchi, K. 1975. Isolation and physicochemical properties of a soluble glycoprotein fraction of milk fat globule membrane. *Agr. Biol. Chem. 39,* 1835-1842.

Kanno, C. and Yamauchi, K. 1979. Relationship of soluble glycoprotein of milk fat globule to components -3, -5, and -8 of proteose peptone. *Agr. Biol. Chem. 43,* 2105-2113.

Kaplan, L. J. and Foster, J. F. 1971. Isoelectric focusing behavior of bovine plasma albumin, mercaptalbumin, and β-lactoglobulin A and B. *Biochemistry 10,* 630-636.

Kato, A. and Nakai, S. 1980. Hydrophobicity determined by a fluorescence probe method and its correlation with surface properties of proteins. *Biochim. Biophys. Acta 624,* 13-20.

Keshavarz, E. and Nakai, S. 1979. The relationship between hydrophobicity and interfacial tension of proteins. *Biochim. Biophys. Acta 576,* 269-279.

Kiddy, C. A. 1975. Gel electrophoresis in vertical polyacrylamide beds. Procedure I and II. *In: Methods of Gel Electrophoresis of Milk Protein.* H.E. Swaisgood (Editor). American Dairy Science Association, Champaign, Ill.

Kiddy, C. A., Townend, R. E., Thatcher, W. W. and Timasheff, S. N. 1965. β-Lactoglobulin variation in milk from individual cows. *J. Dairy Res. 32,* 209.

Kim, Y. K., Yaguchi, M. and Rose, D. 1969. Isolation and amino-acid composition of para-kappa-casein. *J. Dairy Sci. 52,* 316-320.

King, N. 1955. The milk fat globule membrane. *Technical Communication No. 2.* Commonwealth Bureau of Dairy Science, Commonwealth Agriculture Bureau, Farnham Royal, Bucks, England.

King, T. P. and Spencer, E. M. 1970. Structure studies and organic ligand-binding properties of bovine plasma albumin. *J. Biol. Chem. 245,* 6134-6148.

King, T. P. and Spencer, E. M. 1972. Amino acid sequences of the amino and carboxyl terminal cyanogen bromide peptides of bovine plasma albumin. *Arch. Biochem. Biophys. 153,* 627-640.

Kitchen, B. J. 1977. Fractionation and characterization of membranes from bovine milk fat globule. *J. Dairy Res. 44,* 469-482.

Kobayashi, K., Vaerman, J. P., Bazin, H., and Lebaq-Verheyden, A. M. 1973. Identification of J-chain in polymeric immunoglobulins from a variety of species by cross-reaction with rabbit antisera to human J-chain. *J. Immunol. 111,* 1590-1594.

Kobylka, D. and Carraway, K. L. 1972. Proteins and glycoproteins of the milk fat globule membrane. *Biochim. Biophys. Acta 288*, 282–295.

Kobylka, D. and Carraway, K. L. 1973. Proteolytic digestion of proteins of the milk fat globule membrane. *Biochim. Biophys. Acta 307*, 133–140.

Kolar, C. K. and Brunner, J. R. 1969. Proteose-peptone fraction of bovine milk: Distribution in protein system. *J. Dairy Sci. 52*, 1541–1546.

Kolar, C. K. and Brunner, J. R. 1970. Proteose-peptone fraction of bovine milk: Lacteal serum components 5- and 8-casein-associated glycoproteins. *J. Dairy Sci. 53*, 997–1008.

Komar, R. and Mukkur, T. K. S. 1974. Isolation and characterization of J-chain from bovine colostral immunoglobulin M. *Can. J. Biochem. 53*, 943–949.

Kopfler, F. C., Peterson, R. F. and Kiddy, C. A. 1969. Amino acid composition of chromatographically separated β-casein. *J. Dairy Sci. 52*, 1573–1576.

Kramer, R. and Lagoni, H. 1969. Calcium selective electrode for the measurement of calcium ion activity in milk and milk products. *Milchwissenschaft 24*, 68–70 (German).

Krigbaum, W. R. and Kugler, F. R. 1970. Molecular conformation of egg white lysozyme and bovine α-lactalbumin in solution. *Biochemistry 9*, 1216–1223.

Kristofferson, T., Koo, K. H. and Slatter, W. L. 1974. Determination of casein by the dye method for estimation of cottage cheese yield. *Cult. Dairy Prod. J. 9*, 12.

Kronman, M. J. and Andreotti, R. E. 1964. Inter- and intramolecular interactions of α-lactalbumin. I. The apparent heterogeneity at acid pH. *Biochemistry 3*, 1145–1151.

Kronman, M. J., Andreotti, R. E. and Vitals, R. 1964. Inter-and intramolecular interactions of α-lactalbumin. II. Aggregation reactions at acid pH. *Biochemistry 3*, 1152–1160.

Kronman, M. J., Cerankowski, L. and Holmes, L. G. 1965. Inter- and intramolecular interactions of α-lactalbumin. III. Spectral changes at acid pH. *Biochemistry 4*, 518–525.

Kronman, M. J., Hoffman, W. B., Jeroszko, J. and Sage, G. W. 1972. Inter-and intramolecular interactions of α-lactalbumin. XI. Comparison of the exposure of tyrosyl, tryptophyl, and lysyl side chains in the goat and bovine protein. *Biochim. Biophys. Acta 285*, 124–144.

Kronman, M. J., Holmes, L. G. and Robbins, F. M. 1971. Inter- and intramolecular interactions of α-lactalbumlin X. Effect of acylation of tyrosyl and lysyl side chains on molecular conformations. *J. Biol. Chem. 246*, 1909–1921.

Kumar, S. and Mikalajcik, E. M. 1973. Selected physicochemical characteristics of bovine colostrum immunoglobulin. *J. Dairy Sci. 56*, 255–258.

Kumosinski, T. F., Brown, E. M. and Groves, M. L. 1981. Solution physicochemical properties of bovine β_2-microglobulin. Aggregate states. *J. Biol. Chem. 256*, 10949–10953.

Labib, S. R., Calvanico, N. J. and Tomasi, T. B. 1976. Bovine secretory component isolation, molecular size and shape, composition, and NH_2-terminal amino acid sequence. *J. Biol. Chem. 251*, 1969–1974.

Laemmli, U. K. 1970. Cleavage of structural proteins during assembly of the head of bacteriophage T4. *Nature 227*, 680–685.

Lascelles, A. K. 1977. Role of the mammary gland and milk in immunology. *Symp. Zool. Soc. London 41*, 241–260.

Laskowski, M., Jr. and Laskowski, M. 1950. Trypsin inhibitor in colostrum. *Fed. Proc. 9*, 194.

Laskowski, M. Jr. and Laskowski, M. 1951. Crystalline trypsin inhibitor from colostrum. *J. Biol. Chem. 190*, 563–573.

Laskowski, M., Jr., Mars, P. H. and Laskowski, M. 1952. Comparison of trypsin inhibitor from colostrum with other crystalline trypsin inhibitors. *J. Biol. Chem. 198*, 745–752.

Leach, B. E., Blalock, C. R. Y. and Pallansch, M. J. 1967. Kinin-like activity in bovine milk. *J. Dairy Sci. 50*, 763–764.

Leach, B. S., Collawn, J. F. Jr. and Fish, W. W. 1980. Behavior of glycopolypeptides with empirical molecular weight estimation methods. 2. In random coils. *Biochemistry 19*, 5741–5747.

Leger, D., Verbert, A., Loucheux, M.-H. and Spik, G. 1977. Study of the molecular weight of human lactotransferrin and serotransferrin. *Ann. Biol. Anim. Biochim. Biophys. 17*, 737–747 (French).

Lin, C. F. 1971. The casein stabilizing function of sulfated polysaccharide. Ph.D. thesis, Ohio State University, Columbus.

Lin, C. F. 1977. Interaction of sulfated polysaccharides with proteins. *In: Food Colloids.* H.D. Graham (Editor). AVI Publishing Co., Westport, Conn.

Lin, C. F. and Hansen, P. M. T. 1970. Stabilization of casein micelles by carrageenan. *Macromolecules. 3*, 269–274.

Loucheux-Lefebvre, M.-H., Aubert, J.-R. and Jollès, P. 1978. Prediction of the conformation of the cow and sheep κ-caseins. *Biophys. J. 23*, 323–336.

MacDonald, C. A. and Thomas, M. A. W. 1970. The rennin sensitive bond of bovine κ-casein. *Biochim. Biophys. Acta 207*, 139–143.

Mach, J.-P. 1970. *In vitro* combination of human and bovine-free secretory component with IgA of various species. *Nature 228*, 1278–1282.

Mach, J.-P., Pahud, J. J. and Isliker, H. 1969. IgA with "secretory piece" in bovine colostrum and saliva. *Nature 223*, 952–954.

Mackinlay, A. G. and Wake, R. G. 1971, κ-Casein and its attack by rennin (chymosin). *In: Milk Proteins*, Vol. II. H.A. McKenzie (Editor). Academic Press, New York.

Mangino, M. E. and Brunner, J. R. 1975. Molecular weight profile of fat globule membrane proteins. *J. Dairy Sci. 58*, 313–318.

Manson, W. 1965. The separation of major components of the casein of bovine milk by electrophoresis in density gradient. *Biochem. J. 94*, 452–457.

Manson, W., Annan, W. D. and Barnes, G. K. 1976. α_{s0}-Casein: Its preparation and characterization. *J. Dairy Res. 43*, 133–136.

Manson, W., Carolan, T. and Annan, W. D. 1977. Bovine α_{s0}-casein: A phosphorylated homologue of α_{s1}-casein. *Eur. J. Biochem. 78*, 411–417.

Mather, I. H. and Keenan, T. W. 1975. Studies on the structure of the milk fat globule membrane. *J. Membr. Biol. 21*, 65–85.

Mather, I. H., Weber, K. and Keenan, T. W. 1977. Membranes of mammary gland. XII. Loosely associated proteins and compositional heterogeneity of bovine milk fat globule membrane. *J. Dairy Sci. 60*, 394–402.

McGann, T. C. A., Mathiassen, A. and O'Connell, J. A. 1972. Applications of the Pro-Milk MkII. Part III. Rapid estimation of casein in milk and protein in whey. *Lab Pract. 21*, 628–631, 650.

McGuire, T. C., Musoke, A. J. and Kurtti, T. 1979. Functional properties of bovine IgG₁ and IgG₂: Interaction with complement, macrophages, neutrophils and skin. *Immunology 38*, 249–256.

McKenzie, H. A. 1967. Milk proteins. *In: Advances in Protein Chemistry*, Vol. XXII. C.B. Anfinsen, M. L. Anson and J.T. Edsall (Editors). Academic Press, New York.

McKenzie, H. A. 1970. *Milk Proteins*, Vol. I. Academic Press, New York.

McKenzie, H. A. 1971A. *Milk Proteins*, Vol. II. Academic Press, New York.

McKenzie, H. A. 1971B. β-Lactoglobulin. *In: Milk Proteins*, Vol. II. H.A. McKenzie (Editor). Academic Press, New York.

McKenzie, H. A. 1971C. Whole casein: Isolation properties and zone electrophoresis. *In:* *Milk Proteins,* Vol. II. H.A. McKenzie (Editor). Academic Press, New York.

McKenzie, H. A., Ralston, G. B. and Shaw, D. C. 1972. Location of sulfhydryl and disulfide groups in bovine β-lactoglobulin and effect of urea. *Biochemistry 11,* 4539–4547.

McKenzie, H. A. and Sawyer, W. H. 1967. Effect of pH on β-lactoglobulins. *Nature 214,* 1101–1104.

McKenzie, H. A., Sawyer, W. H. and Smith, M. B. 1967. Optical rotatory dispersion and sedimentation in the study of association-dissociation: Bovine β-lactoglobulin pH 5. *Biochim. Biophys. Acta 147,* 73–92.

McKenzie, H. A. and Wake, R. G. 1959. Studies of casein. III. The molecular size of α-, β-, and κ-casein. *Aust. J. Chem. 12,* 734–742.

McKenzie, H. A. and Wake, R. G. 1961. An improved method for the isolation of kappa-casein. *Biochim. Biophys. Acta 47,* 240–242.

Melnychyn, B. and Wolcott, J. M. 1967. Simple procedure for isolation of αs-casein. *J. Dairy Sci. 50,* 1863–1867.

Mercier, J.-C., Brignon, G. and Ribadeau-Dumas, B. 1973. Primary structure of bovine κ-casein B. Complete sequence. *Eur. J. Biochem. 35,* 222–235 (French).

Mercier, J.-C., Grosclaude, F. and Ribadeau-Dumas, B. 1971. Primary structure of αs1-casein. Complete sequence. *Eur. J. Biochem. 23,* 41–51 (French).

Mickelsen, R. and Shukri, N. A. 1975. Measuring casein by dye binding. *J. Dairy Sci. 58,* 311–312.

Mills, O. E. and Creamer, L. K. 1975. Conformational change in bovine β-lactoglobulin at low pH. *Biochim. Biophys. Acta 379,* 618–626.

Morr, C. V., Nielson, M. A. and Lin, S.H.-C. 1969. Sephadex equilibrium-diffusion technique for fractionating whey and skim milk systems. *J. Dairy Sci. 52,* 1552–1556.

Muldoon, P. J. and Liska, B. J. 1969. Comparison of a resin ion-exchange method and a liquid ion-exchange method for determination of ionized calcium in skim milk. *J. Dairy Sci. 52,* 460–464.

Muldoon, P. J. and Liska, B. J. 1972. Effect of heat treatment and subsequent storage on the concentration of ionized calcium in skim milk. *J. Dairy Sci. 55,* 35–38.

Nagasawa, T., Kiyosawa, I., Kuwahara, K. and Ganguly, N. C. 1973. Fractionation of buffalo milk casein by acrylamide gel electrophoresis and DEAE cellulose column chromatography. *J. Dairy Sci. 56,* 61–65.

Nakahori, C. and Nakai, S. 1972. Fractionation of caseins directly from skim milk by gel chromatography. I. Elution with sodium dodecylsulphate. *J. Dairy Sci. 55,* 25–29.

Nakai, S., Toma, S. J. and Nakahori, C. 1972. Fractionation of caseins directly from skim milk by gel chromatography. II. Elution with phosphate buffers. *J. Dairy Sci. 55,* 30–34.

Neelin, J. M., Rose, D. and Tessier, H. 1962. Starch gel electrophoresis of various fractions of caseins. *J. Dairy Sci. 45,* 153–158.

Nezlin, R. S. 1972. Recommendations on the nomenclature of immunoglobulins. *Mol. Biol. 6,* 639.

Nielson, C. S. and Bjerrum, O. J. 1977. Crossed immunoelectrophoresis of bovine milk fat globule membrane protein solubilized with non-ionic detergent. *Biochim. Biophys. Acta 466,* 496–509.

Nielson, K. H. 1977. Bovine reaginic antibody. III. Cross reaction of anti-human IgE and antibovine reaginic immunoglobulin antisera with sera from several species of mammals. *Can. J. Comp. Med. 41,* 345–348.

Nijhuis, H. and Klostermeyer, H. 1975. Partial fractionation of whole casein by affinity chromatography. *Milchwissenschaft 30,* 530–531 (German).

Nikkel, H. J. and Foster, J. F. 1971. A reversible sulthydryl-catalyzed structural alteration of bovine mercaptalbumin. *Biochemistry 10*, 4479–4486.

Noel, J. K. F. and Hunter, M. J. 1972. Bovine mercaptalbumin and nomercaptalbumin monomers, interconversion and structural differences. *J. Biol. Chem. 247*, 7391–7406.

Noelken, M. E. 1967. The molecular weight of α_{s1}-casein B. *Biochim. Biophys. Acta 140*, 537–539.

Noelken, M. E. and Reibstein, M. 1968. Conformation of β-casein B. *Arch. Biochem. Biophys. 123*, 397–402.

Ntailianas, H. A. and Whitney, R. McL. 1965. Dialysis equilibrium for determining binding of calcium by milk proteins. *J. Dairy Sci. 48*, 773.

O'Loughlin, K. and Hansen, P.M.T. 1973. Stabilization of rennet-treated milk protein by carrageenan. *J. Dairy Sci. 56*, 629.

O'Sullivan, A. C. 1972. Whey processing by reverse osmosis, ultrafiltration, and gel filtration. *Dairy Ind. 36*, 636, 691.

Owicki, J. C. and Lillevik, H. A. 1969. Isolation of β-casein from parasodium caseinate. *J. Dairy Sci. 52*, 902.

Parker, T. G. and Dalgleish, D. G. 1977A. The use of light-scattering and turbidity measurements to study the kinetics of extensively aggregating protein: α_s-Casein. *Biopolymers 16*, 2533–2547.

Parker, T. G. and Dalgleish, D. G. 1977B. The potential application of the theory of branching processes to the association of milk proteins. *J. Dairy Res. 44*, 79–84.

Patel, P. C. and Adhikari, H. R. 1973. Effect of gamma-irradiation on gel chromatographic patterns of cottage cheese whey. *J. Dairy Sci. 56*, 406–408.

Payens, T. A. J. 1961. Zone electrophoresis of casein in urea-buffer mixtures. *Biochim. Biophys. Acta 46*, 441–451.

Payens, T. A. J. 1982. Stable and unstable casein micelles. *J. Dairy Sci. 65*, 1863–1873.

Payens, T. A. J., Brinkhuis, J. and Van Markwijk, B. W. 1969. Self-association in nonideal systems. Combined light scattering and sedimentation measurements in B-casein solutions. *Biochim. Biophys. Acta 175*, 434–437.

Payens, T. A. J. and Heremans, K. 1969. Effect of pressure on temperature-dependent association of β-casein. *Biopolymers 8*, 335–345.

Payens, T. A. J. and Van Markwijk, B. W. 1963. Features of the association of β-casein. *Biochim. Biophys. Acta 71*, 517–530.

Payens, T. A. J. and Vreeman, H. J. 1982. Casein micelles and micelles of κ- and β-casein. *In: Solution Behavior of Surfactants*, Vol. I. K.L. Mittal and E.J. Fendler (Editors). Plenum Publishing Corp., New York.

Pearce, R. J. and Zadow, J. G. 1978. Isoelectric focusing of milk proteins. *XXth International Dairy Congress (Australia)*, Vol. E, 217–218.

Pepper, L. and Farrell, H. M., Jr. 1977. Studies of casein association by gel filtration. *Fed. Proc. 36*, 840.

Peri, C. and Dunkley, W. L. 1971. Reverse osmosis of cottage cheese whey. I. Influence of the composition of the feed. *J. Food Sci. 36*, 25–30.

Peters, T., Jr. and Feldhoff, R. C. 1975. Fragments of bovine serum albumin produced by limited proteolysis. Isolation and characterization of peptic fragments. *Biochemistry 14*, 3384–3391.

Peterson, P. A., Cunningham, B. A., Berggard, I. and Edelman, G. M. 1972. β_2-Microglobulin, a free immunoglobulin domain. *Proc. Natl. Acad. Sci. USA 69*, 1697–1701.

Peterson, R. F. 1971. Testing for purity in proteins by gel electrophoresis. *J. Agr. Food Chem. 19*, 595–599.

Polis, B. D., Shmuckler, H. W. and Custer, J. H. 1950. Isolation of a crystalline albumin from milk. *J. Biol. Chem. 187*, 349–354.

Poljak, R. J., Amzel, L. M., Avey, H. P., Becka, L. N. and Nissonoff, A. 1972. Structure of Fab' New at 6 Å resolution. *Nature New Biol. 235*, 137–140.

Polzhofer, K. P. 1972. Synthesis of a rennin-sensitive pentadecapeptide from cow κ-casein. *Tetrahedron 28*, 855–865 (German).

Porter, P. and Noakes, D. E. 1970. Immunoglobulin IgA in bovine serum and external secretions. *Biochim. Biophys. Acta 214*, 107–116.

Poulik, M. D. and Weiss, M. L. 1975. Ceruloplasmin. *In: The Plasma Proteins*, Vol. II. F.W. Putnam (Editor), Academic Press, New York.

Préaux, G., Braunitzer, G., Schrank, B. and Stangl, A. 1979. The amino acid sequence of goat β-lactoglobulin. *Hoppe-Seyler's Z. Physiol. Chem. 360S*, 1595–1604 (German).

Pujolle, J., Ribadeau-Dumas, B., Garnier, J. and Pion, R. 1966. A study of kappa-casein components. I. Preparation. Evidence of a common C-terminal sequence. *Biochem. Biophys. Res. Commun. 25*, 285–290.

Putnam, F. W. 1975. Transferrin. *In: The Plasma Proteins*, Vol. I. F. W. Putnam (Editor). Academic Press, New York.

Radl, J., Klein, F., van den Berg, P., de Bruyn, A. M. and Hijmans, W. 1971. Binding of secretory piece to polymeric IgA and IgM paraproteins *in vitro*. *Immunology 20*, 843–852.

Raj, T. and Flygare, W. H. 1974. Diffusion studies of bovine serum albumin by quasielastic light scattering. *Biochemistry 13*, 3336–3340.

Ralston, G. B. 1971. Cited by H.A. McKenzie. β-Lactoglobulin. *In: Milk Proteins*, Vol. II. H.A. McKenzie (Editor). Academic Press, New York.

Rawitch, A. B. and Hwan, R.-Y. 1979. Anilinonaphthalene sulfonate as a probe for the native structure of bovine alpha lactalbumin: Absence of binding to the native monomeric protein. *Biochim. Biophys. Res. Commun. 91*, 1383–1389.

Rawley, B. O., Lund, D. B. and Richardson, T. 1979. Reductive methylation of β-lactoglobulin. *J. Dairy Sci. 62*, 533–536.

Reed, R. G., Feldhoff, R. C., Clute, O. L. and Peters, T., Jr. 1975. Fragments of bovine serum albumin produced by limited proteolysis. Conformation and ligand bonding. *Biochemistry 14*, 4578–4583.

Reed, R. G., Putnam, F. W. and Peters, T., Jr. 1980. Sequence of residues 400–403 of bovine serum albumin. *Biochem. J. 191*, 867–868.

Ribadeau-Dumas, B., Brignon, F., Grosclaude, F. and Mercier, J.-C. 1972. Primary structure of bovine β-casein. Sequence complete. *Eur. J. Biochem. 25*, 505–514 (French).

Ribadeau-Dumas, B., Maubois, J. L., Mocquot, G. and Garnier, J. 1964. A study of casein by DEAE-cellulose column chromatography in urea. *Biochim. Biophys. Acta 82*, 494–506.

Roark, D. E. and Yphantis, D. A. 1969. Studies of self-associating systems by equilibrium ultracentrifugation. *Ann. NY Acad. Sci. 164*, 245–278.

Robbins, F. M. and Holmes, L. G. 1970. Circular dichroism spectra of α-lactalbumin. *Biochim. Biophys. Acta 221*, 234–240.

Robbins, F. M. and Kronman, M. J. 1964. A simplified method for preparing α-lactalbumin and β-lactoglobulin from cows' milk. *Biochim. Biophys. Acta 82*, 186–188.

Robbins, F. M., Kronman, M. J. and Andreotti, R. E. 1965. Inter- and intramolecular interactions of α-lactalbumin. V. The effect of amidination on association and aggregation. *Biochim. Biophys. Acta 109*, 223–233.

Roh, J. K., Bradley, R. L. Jr., Richardson, T. and Weckel, K. G. 1975. Distribution and removal of added mercury in milk. *J. Dairy Sci. 58*, 1782–1788.

Roh, J. K., Bradley, R. L. Jr., Richardson, T. and Weckel, K. G. 1976. Distribution and removal of cadmium from milk. *J. Dairy Sci. 59*, 376–381.

Rose, D., Davies, D. T. and Yaguchi, M. 1969. Quantitative determination of the major components of casein mixture by column chromatography on DEAE cellulose. *J. Dairy Sci. 52*, 8–11.

Rudolph, R., Holler, E. and Jaenicke, R. 1975. Fluorescence and stop-flow studies on N-F transition of serum albumin. *Biophys. Chem. 3*, 226–233.

Saito, T., Itoh, T. and Adachi, S. 1979. Rapid preparation method of para-kappa-casein from whole casein. *Jpn. J. Dairy Food Sci. 28*, A183–A188.

Salaman, M. R. and Williamson, A. R. 1971. Isoelectric focusing of proteins in the native and denatured states. Anomalous behavior of plasma albumin. *Biochem. J. 122*, 93–99.

Salter, D. N., Ford, J. E., Scott, K. J. and Andrews, P. 1972. Isolation of the folate-binding protein from cows' milk by the use of affinity chromatography. *FEBS Lett. 20*, 302–306.

Schewale, J. G., Sinha, S. K. and Brew, K. 1984. Evolution of α-lactalbumins. The complete amino acid sequence of the α-lactalbumin from a marsupial (*Macropus rufogriseus*) and corrections to regions of sequence in bovine and goat α-lactalbumins. *J. Biol. Chem. 259*, 4947–4956.

Schmid, K. 1975. α_1-Acid glycoprotein. *In: The Plasma Proteins*, Vol. I. F. W. Putnam (Editor). Academic Press, New York.

Schmidt, D. G. 1967. Fractionation of kappa-casein by column electrophoresis. *Protides Biol. Fluids 14*, 671–676.

Schmidt, D. G. 1970A. The association of α_{s1}-casein B at pH 6.6. *Biochim. Biophys. Acta 207*, 130–138.

Schmidt, D. G. 1970B. Differences between the association of the genetic variants B, C, and D of α_{s1}-casein. *Biochim. Biophys. Acta 221*, 140–142.

Schmidt, D. G. 1970C. The association of α_{s1}-casein at pH 2.5. *Protides Biol. Fluids 18*, 337–340.

Schmidt, D. G. 1982. Association of casein and casein micelle structure. *In: Developments in Dairy Chemistry*, Vol. I. P. F. Fox (Editor). Applied Science Publishers, New York.

Schmidt, D. G. and Both, P. 1975. Procedure III. In: *Methods of Gel Electrophoresis of Milk Proteins*. H. Swaisgood (Editor). American Dairy Science Association, Champaign, Ill.

Schmidt, D. G. and Payens, T. A. J. 1963. The purification and some properties of a calcium-sensitive α-casein. *Biochim. Biophys. Acta 78*, 492–499.

Schmidt, D. G. and Payens, T. A. J. 1972. The evaluation of positive and negative contributions to the second virial coefficient of some proteins. *J. Colloid Interface Sci. 39*, 655–662.

Schmidt, D. G., Payens, T. A. J., Van Markwijk, B. W. and Brinkhuis, J. A. 1967. On the subunit of α_{s1}-casein. *Biochm. Biophys. Res. Commun. 27*, 448–455.

Schober, R. and Heimburger, N. 1960. Fractionation of Na-caseinate by column chromatography on ion-exchange-cellulose. *Milchwissenschaft 15*, 607–609 (German).

Schober, R., Heimburger, N. and Enkelmann, D. 1959. Fractionation of milk proteins by column chromatography on ion-exchange cellulose. *Milchwissenschaft 14*, 432–435 (German).

Senft, B. and Klobasa, F. 1973. Research on the concentration of lactoferrin in colostral and ripe milk from cows. *Milchwissenschaft 28*, 750–752 (German).

Shahani, K. M., Harper, W. J., Jensen, R. G., Parry, R. M. and Zittle, C. A. 1973. Enzymes in bovine milk: A review. *J. Dairy Sci. 56*, 531–543.

Shimazaki, K.-I. and Sukegawa, K. 1982. Chromatographic profiles of bovine milk whey components by gel filtration on Fractogel TSK HW55F column. *J. Dairy Sci. 65*, 2055–2062.

Shimizu, M., Kanno, C. and Yamauchi, K. 1978. Microheterogeneity and some properties of the major glycoprotein fraction isolated from bovine milk fat globule membrane after delipidation. *Agr. Biol. Chem. 42*, 981–987.

Singer, S. J. 1974. The molecular organization of membranes. *Ann. Rev. Biochem. 43*, 805–833.

Smith, E. L. 1946. Isolation and properties of immune lactoglobulins from bovine whey. *J. Biol. Chem. 165*, 665–676.

Smith, K. L., Conrad, H. R. and Porter, R. M. 1971. Lactoferrin and IgG immunoglobulins from involuted bovine mammary glands. *J. Dairy Sci. 54*, 1427–1435.

Snoeren, T. H. M., van der Spek, C. A. and Payens, T.A.J. 1977. Preparation of κ-casein and minor α_s-caseins by electrostatic affinity chromatography. *Biochim. Biophys. Acta 490*, 255–259.

Snoeren, T. H. M., Van Markwijk, B. and Van Montfort, R. 1980. Some physicochemical properties of bovine α_{s2}-casein. *Biochim. Biophys. Acta 622*, 268–276.

Snow, L. D., Colton, D. G. and Carraway, K. L. 1977. Purification and properties of the major sialoglycoproteins of the milk fat globule membrane. *Arch. Biochem. Biophys. 179*, 690–697.

Sogami, M. 1971. Effect of salts on the N-F transition of bovine serum plasma albumin. *J. Biochem. 69*, 819–822.

Sogami, M. and Foster, J. F. 1968. Isomerization reactions of charcoal-defatted bovine plasma albumin. The N-F transition and acid expansion. *Biochemistry 7*, 2172–2182.

Sogami, M., Nagaoka, S., Itoh, K.B and Sakata, S. 1973. Fluorimetric studies on the structural transition of bovine plasma albumin in acidic solutions. *Biochim. Biophys. Acta 310*, 118–123.

Sogami, M. and Ogura, S. 1973. Structural transition in bovine plasma albumin. Location of tyrosyl and tryptophyl residues by solvent perturbation difference spectra. *J. Biochem. 73*, 323–334.

Sommers, P. B., Kronman, M. J. and Brew, K. 1973. Molecular conformation of fluorescence properties of α-lactalbumin from four animal species. *Biochm. Biophys. Res. Commun. 52*, 98–105.

Sorensen, M. and Sorensen, S.P.L. 1939. The proteins in whey. *Compt. Rend. Trav. Lab. Carlsberg. Ser. Chim. 23*, 55–59.

Spencer, E. M. 1974. Amino acid sequence of the alanyl peptide from cyanogen bromide cleavage of bovine plasma albumin. *Arch. Biochem. Biophys. 165*, 80–89.

Spencer, E. M. and King, T. P. 1971. Isoelectric heterogeneity of bovine plasma albumin. *J. Biol. Chem. 246*, 201–208.

Steklenev, E. P. and Marinchuk, G. E. 1981. Interspecies polymorphism of blood serum proteins in some representatives of a bovine subfamily (*Bovinae*) and their hybrids. *Teitol. Genet. 15*(2), 67–73 (Russian).

Sternberg, M., Chiang, J. P. and Ebert, N. J. 1976. Cheese whey protein isolated with polyacrylic acid. *J. Dairy Sci. 59*, 1042–1050.

Stroupe, S. D. and Foster, J. F. 1973. Further studies of the sulfhydryl-catalyzed structural alteration of bovine mercaptalbumin. *Biochemistry 10*, 4479–4486.

Sundararajan, N. R. and Whitney, R. McL. 1969. Binding of thallous ion by casein. *J. Dairy Sci. 52*, 1445–1448.

Sundararajan, N. R. and Whitney, R. McL. 1975. Murexide for determination of free and protein-bound calcium in model systems. *J. Dairy Sci. 58*, 1595–1608.

Swaisgood, H. E. 1973. The caseins. *In:CRC Critical Reviews in Food Technology*. Chemical Rubber Co., Cleveland.

Swaisgood, H. E. 1975A. Primary sequence of kappa-casein. *J. Dairy Sci. 58*, 583–592.

Swaisgood, H. E. 1975B. *Methods of Gel Electrophoresis of Milk Proteins*. American Dairy Science Association, Champaign, Ill.

Swaisgood, H. E. 1982. Chemistry of milk proteins. *In: Developments in Dairy Chemistry,* Vol. I. P. F. Fox (Editor). Applied Science Publishers, New York.

Swaisgood, H. E. and Brunner, J. R. 1962. Characterization of kappa-casein obtained by fractionation with trichloracetic acid in a concentrated urea solution. *J. Dairy Sci.* 45, 1–11.

Swaisgood, H. E., Brunner, J. R. and Lillevik, H. A. 1964. Physical parameters of κ-casein from cows' milk. *Biochemistry 3,* 1616–1623.

Swaisgood, H. E. and Timasheff, S. N. 1968. Association of α_{s1}-casein C in the alkaline pH range. *Arch. Biochem. Biophys. 125,* 344–361.

Szuchet-Derechin, S. and Johnson, P. 1965. The "albumin" fraction of bovine milk. I. Overall chromatographic fractionation on DEAE-cellulose. *Eur. Polymer J. 1,* 271–281.

Tamburro, A. M., Jori, G., Vidali, G., Scatturun, A. and Saccomani, G. 1972. Studies on the structure in solution of α-lactalbumin. *Biochim. Biophys. Acta 263,* 704–713.

Tanford, C. 1962. The interpretation of hydrogen ion titration curves of proteins. *In:Advances in Protein Chemistry,* Vol. 17, C. B. Anfinsen, K. Bailey and J. T. Edsall (Editors). Academic Press, New York.

Tarrassuk, N. P., Yaguchi, M. and Callis, J. B. 1965. Effect of temperature on the composition of casein fractions eluted from DEAE-cellulose column. *J. Dairy Sci.* 48, 606–609.

Taylor, R. P., Berga, S., Chau, V. and Bryner, C. 1975A. Bovine serum albumin as a catalyst. III. Conformational studies. *J. Am. Chem. Soc.* 97, 1943–1948.

Taylor, R. P., Chau, V., Bryner, C. and Berga, S. 1975B. Bovine serum albumin as a catalyst. II. Characterization of the kinetics. *J. Am. Chem. Soc.* 97, 1934–1942.

Taylor, R. P. and Vatz, J. B. 1973. Bovine serum albumin as a catalyst. Accelerated decomposition of a Meisenheimer complex. *J. Am. Chem. Soc.* 95, 5819–5820.

Teller, D. C., Swanson, E. and De Haen, C. 1979. The translational friction coefficient of proteins. *Methods Enzymol.* 61 (Part H), 103–124.

Thompson, M. P. 1970. Phenotyping milk proteins: A review. *J. Dairy Sci.* 53, 1341–1348.

Thompson, M. P. 1971. α_s-and β-Caseins. *In: Milk Proteins,* Vol. II. H. A. McKenzie (Editor). Academic Press, New York.

Thompson, M. P., Gordon, W. G., Boswell, R. T. and Farrell, H. M., Jr. 1969. Solubility, solvation, and stabilization of α_{s1}-and β-caseins. *J. Dairy Sci.* 52, 1166–1173.

Thompson, M. P. and Kiddy, C. A. 1964. Genetic polymorphism in casein of cows' milk. III. Isolation and properties of α_{s1}-casein A, B, and C. *J. Dairy Sci.* 47, 626–632.

Timasheff, S. N. and Townend. R. 1964. Structure of the β-lactoglobulin tetramer. *Nature* 203, 517–519.

Toma, S. J. and Nakai, S. 1973. Calcium sensitivity and molecular weight of α_{s5}-casein. *J. Dairy Sci.* 56, 1559–1562.

Townend, R., Herscovits, T. T. and Timasheff, S. N. 1969. The state of amino acid residues in β-lactoglobulin. *Arch. Biochem. Biophys. 129,* 567–580.

Townend, R., Kumosinski, T. F. and Timasheff, S. N. 1967. The circular dichroism of variants of β-lactoglobulin. *J. Biol. Chem.* 242, 4538–4545.

Tran, V. D. and Baker, B. E. 1970. Caseins. IX. Carbohydrate moiety of κ-casein. *J. Dairy Sci.* 53, 1009–1012.

Tripathi, K. K. and Gehrke, C. W. 1969. Chromatography and characterization of gamma-casein. *J. Chromatography 43,* 322–331.

Tripathi, K. K. and Gehrke, C. W. 1970. Chemical and chromatographic isolation of kappa-casein. *J. Chromatography 46,* 280–285.

UI, N. 1971. Isoelectric points and conformation of proteins. I. Effect of urea on the behavior of some proteins in isolectric focusing. *Biochim. Biophys. Acta 229,* 567.

Vanaman, T. C., Brew, K. and Hill, R. L. 1970. The disulfide bonds of bovine α-lactalbumin. *J. Biol. Chem. 245*, 4583–4590.

Voglino, G.-F. 1972. A new β-casein variant in Piedmont cattle. *Anim. Blood Grps. Biochem. Genet. 3*, 61–62.

Von Hipple, P. H. and Waugh, D. F. 1955. Casein: Monomers and polymers. *J. Am. Chem. Soc. 77*, 4311–4319.

Vreeman, H. J. 1979. The association of bovine SH-κ-casein at pH 7.0 *J. Dairy Res. 46*, 271–276.

Vreeman, H. J., Both, P., Brinkhuis, J. A. and van der Spek, C. 1977. Purification and some physicochemical properties of bovine kappa-casein. *Biochim. Biophys. Acta 491*, 93–103.

Vujicic, I. F. 1973. Fractionation of casein complex by ion-exchange chromatography on DEAE-Sephadex. *Milchwissenschaft 28*, 175–176 (German).

Wake, R. G. 1959. Studies of casein. IV. The isolation of kappa-casein. *Aust. J. Biol. Sci. 12*, 538–540.

Wake, R. G. and Baldwin, R. L. 1961. Analysis of casein fractions by zone electrophoresis in concentrated urea. *Biochim. Biophys. Acta 47*, 225–239.

Wallevik, K. 1973. Isoelectric focusing of bovine serum albumin. Influence of binding of carrier ampholyte. *Biochim. Biophys. Acta 322*, 75–87.

Walstra, P. and Jenness, R. 1984. *Dairy Chemistry and Physics*. John Wiley & Sons, New York.

Walter, J. C. 1952. Cited by B. Lindqvist, 1963. Casein and the action of rennin—Part I. *Dairy Sci. Abstr. 25*, 257.

Warme, P. K., Momany, F. A., Rumball, S. V., Tuttle, R. W. and Scheraga, H. A. 1974. Computation of structures of homologous proteins. α-Lactalbumin from lysozyme. *Biochemistry 13*, 768–782.

Watson, D. L. 1976. The effect of cytophilic IgG_2 on phagocytosis by polymorphonuclear leucocytes. *Immunology 31*, 159–165.

Waugh, D. F. 1971. Formation and structure of casein micelles. *In: Milk Proteins*, Vol. II. H.A. McKenzie (Editor). Academic Press, New York.

Waugh, D. F., Creamer, L. K., Slattery, C. W. and Dresdner, G. W. 1971. Core polymers of casein micelles. *Biochemistry 9*, 786–795.

Waugh, D. F., Ludwig, M. L., Gillespie, J. M., Melton, B., Foley, M. and Kleiner, E. S. 1962. The $α_s$-caseins of bovine milk. *J. Am. Chem. Soc. 84*, 4929–4938.

Wei, T.-M. 1982. Batch fractionation of casein with diethylaminoethyl cellulose. M. S. thesis, University of Illinois, Urbana.

Weiner, R. E. and Szuchel, S. 1975. The molecular weight of bovine lactoferrin. *Biochim. Biophys. Acta 393*, 143–147.

Wells, P. W. and Eyre, P. 1971. Preliminary characterization of bovine homocytotropic antibody. *Immunochemistry 9*, 88–90.

Wetlaufer, D. B. 1961. Osmometry and general characterization of α-lactalbumin. *Comp. Rend. Trav. Lab. Carlsberg, Ser. Chim. 32*, 125–138.

Wheelock, J. V. and Sinkerson, G. 1970. Carbohydrates of bovine κ-casein glycopeptides. *Biochem. J. 119*, 13P.

Wheelock, J. V. and Sinkerson, G. 1973. Carbohydrates of bovine κ-casein *J. Dairy Res. 40*, 413–420.

White, D. D., Stewart, S. and Wood, G. C. 1973. The use of reporter group circular dichroism in the study of conformational transitions in bovine serum albumin. *FEBS Lett. 33*, 305–310.

Whitney, R. McL. 1977. Milk proteins. *In: Food Colloids*. H.D. Graham (Editor). AVI Publishing Co., Westport, Conn.

Whitney, R. McL., Brunner, J. R., Ebner, K. E., Farrell, H. M., Jr., Josephson, R. V.,

Morr, C. V., and Swaisgood, H. E. 1976. Nomenclature of proteins of cows' milk: Fourth revision. *J. Dairy Sci. 59,* 785–815.

Wilson, W. D. and Foster, J. F. 1971. Conformation-dependent limited proteolysis of bovine plasma albumin by an enzyme present in commercial albumin preparations. *Biochemistry 10,* 1772–1780.

Wooding, F. P. B. 1971. The structure of the milk fat globule membrane. *J. Ultrastructure Res. 37,* 388–400.

Wyckoff, M., Rodbard, D. and Chrambach, A. 1977. Polyacrylamide gel electrophoresis in sodium dodecyl sulfate-containing buffer using multiphasic buffer systems: Properties of the stack, valid R_f measurements, and optimized procedure. *Anal. Biochem. 78,* 459–482.

Yaguchi, M., Davies, D. T. and Kim, Y. K. 1968. Preparation of κ-casein by gel filtration. *J. Dairy Sci. 51,* 473–477.

Yaguchi, M. and Rose, D. 1971. Chromatographic separation of milk proteins: A review. *J. Dairy Sci. 54,* 1725–1743.

Yaguchi, M. and Tarassuk, N. P. 1967. Gel filtration of acid casein and skim milk on Sephadex. *J. Dairy Sci. 50,* 1985–1988.

Yaguchi, M., Tarassuk, N. P. and Hunziker, H. G. 1961. Chromatography of milk proteins on anion-exchange cellulose. *J. Dairy Sci. 44,* 589–606.

Yamashita, S., Creamer, L. K. and Berry, G. P. 1976. The aggregation of bovine caseins in dilute calcium chloride solutions. *N.Z. J. Dairy Sci. Tech. 11,* 169–175.

Yamauchi, K., Yoneda, Y., Koga, Y. and Tsugo, T. 1969. Exchangeability of colloidal calcium in milk with soluble calcium. *Agr. Biol. Chem. 33,* 907–914.

Yoshida, S. 1969. Reversible transconformation of casein by heating and cooling. *J. Agr. Chem. Soc. Jpn. 43,* 514–520.

Zimmerman, J. K., Barlow, G. K. and Klotz, I. M. 1970. Dissociation of β-lactoglobulin near neutral pH. *Arch. Biochem. Biophys. 138,* 101–109.

Zittle, C. A. 1960. Column chromatography of casein on the absorbent diethylaminoethyl (DEAE)-cellulose. *J. Dairy Sci. 43,* 855.

Zittle, C. A. and Custer, J. H. 1963. Purification and some of the properties of α_s-casein and κ-casein. *J. Dairy Sci. 46,* 1183–1188.

Zittle, C. A., Della Monica, E. S., Rudd, R. K. and Custer, J. H. 1958. Binding of calcium by casein: Influence of pH and calcium and phosphate concentration. *Arch. Biochem. Biophys. 76,* 342–353.

Ziv, G. and Rasmussen, F. 1975. Distribution of labeled antibiotics in different components of milk following intramammary and intramuscular administration. *J. Dairy Sci. 58,* 938–946.

Zurawski, V. R., Jr., Kohr, W. J. and Foster, J. F. 1976. Conformational properties of bovine plasma albumin with a cleaved internal peptide bond. *Biochemistry 14,* 5579–5586.

4

Lipid Composition and Properties

Robert G. Jensen and Richard W. Clark

Milk lipids have attracted the interest of and have frustrated investigators. The lipids are readily available, for example in butter, but are exceptionally complex, both with respect to lipid classes and to component fatty acids. Furthermore, the latter have been difficult to analyze because of the shortchain fatty acids present and the large number of fatty acids in general. Jenness and Patton (1959) listed 16 fatty acids found in milk lipids. The list had grown to about 150 by 1967 (Jensen *et al.* 1967) and is now over 400 (Table 4.1).

The application of several chromatographic procedures to the separation and identification of milk lipids was mainly responsible for these endeavors. The first gas-liquid chromatographic (GLC) analysis of milk fatty acids was published by James and Martin (1956). By 1960, many laboratories were using GLC for routine analysis of fatty acids. For example, Jensen *et al.* (1962) reported the fatty acid compositions of 106 milk samples taken during 1 year. In comparison, Hansen and Shorland (1952) analyzed only six samples in a year, using distillation of methyl esters.

Column and thin-layer chromatography (TLC) came into use at about the same time as GLC, with the latter widely accepted because of its speed, ease of use, versatility, resolving power, and, probably most important, ease of visualization. Thin-layer chromatography has been particularly useful in the separation and nondestructive recovery of lipid classes. Tentative identifications can be made by comparison with known compounds, and purity can be checked. Jensen *et al.* (1961) may well have been the first group to separate milk lipid classes with TLC when they used the technique to obtain diacylglycerols from lipolyzed milk lipids.

Morrison (1970) drew attention to the many lipids found in milk during the period of intensive research which began in about 1958. Since approximately 1967, investigations on milk lipids have decreased, with relatively little activity at the present time.

Table 4.1. Fatty Acid Composition of Bovine Milk Lipids as of August 1983.

Number	Type	Identity
	Saturates	
27	Normal	2-28;
25	Monobranched	24; 13, 15, 17, 18 three or more positional isomers
16	Multibranched	16-28
	Monoenes	
62	*Cis*	10-26, except for 11:1, positional isomers of 12:1, 14:1, 16:1-18:1, and 23:1-25:1
58	*Trans*	12-14, 16-24; positional isomers of 14:1, 16:1-18:1, and 23:1-25:1
45	Dienes	14-26 evens only; cis, cis; cis, trans; or trans, cis and trans; trans, geometric isomers; unconjugated and conjugated and positional isomers
	Polyenes	
10	*Tri-*	18, 20, 22; geometric positional, conjugated and unconjugated isomers
5	*Tetra-*	18, 20, 22; positional isomers
2	*Penta-*	20, 22
1	*Hexa-*	22
	Keto (oxo)	
38	Saturated	10, 12, 14, 15-20, 22, 24; positional isomers
21	Unsaturated	14, 16, 18; positional isomers of carbonyl and double bond
	Hydroxy	
16	2-position	14:0, 16:0-26:0; 16:1, 18:1, 21:1, 24:1, 25:1
	(4- and 5-position	10:0-16:0, 12:Δ-6 and 12:1-Δ-9
60	Other positions	
	Cyclic	
1	Hexyl	11; terminal cyclohexyl

SOURCE: Compiled from Patton and Jensen (1976), Parodi (1976), and Massant-Leen *et al.* (1981).

The composition of milk lipids given in Table 4.2 (Patton and Jensen 1976) represents compilations from pooled milks. The bulk of the lipids, 97-98%, are triacylglycerols (TGs), with sterols (mostly cholesterol) and phospholipids next in quantity. The diacylglycerols (DGs) and monoacylglycerols (MGs) and free fatty acids in quantities greater than traces are the products of lipolysis (See Chapter 5). Freshly drawn milk which is promptly pasteurized contains little of these compounds. Otherwise, the spectrum of lipids found is qualitatively similar to that of lipid extracts of other biological fluids and tissues. The major components of the TGs, fatty acids, do not vary greatly because of the leveling effect of pooling, but many fatty acids find their way into milk

Table 4.2. Composition of Lipids in Whole Bovine Milk.

Lipid	Weight (%)
Hydrocarbons	Trace
Sterol esters	Trace
Triacylglycerols	97–98
Diacylglycerols	0.28–0.59
Monoacylglycerols	0.016–0.038
Free fatty acids	0.10–0.44
Free sterols	0.22–0.41
Phospholipids	0.2–1.0

SOURCE: Patton and Jensen (1976).

in minute quantities. All of these aspects and the biosynthesis of lipids in the mammary gland will be reviewed.

LIPOGENESIS

Milk is a product of metabolism and, as such, its properties and composition are alterable within the limits of our capacity to alter ruminant metabolism. To achieve this goal, it is necessary to discuss the biosynthesis or lipogenesis of milk lipids. The accumulated lipids are expelled from the secreting cell as a globule of TGs surrounded by a membrane consisting mostly of cellular proteins, phospholipids, etc. The milk fat globule membrane is discussed in Chapter 10. A portion of milk fatty acids is synthesized in the cell. These and fatty acids transported into the cell from blood are incorporated into TGs. The major components of mammary cell lipogenesis are derived from biosynthesis of fatty acids, TGs and membrane materials, and the packaging of the TGs into globules.

Effect of the Rumen on Dietary Lipids

The usual diet of ruminants consists of fresh and preserved herbage and cereals. As a result of microbial activity in the rumen, esterified dietary fatty acids are hydrolyzed, short chain fatty acids are produced by fermentation of cellulose and other polysaccharides, unsaturated fatty acids are hydrogenated and/or converted to geometric (*trans*) and positional isomers, and microbial lipids are synthesized. These activities account in part for the enormous diversity of fatty acids in milk and the unique features: short-chain and a high proportion of long chain saturated fatty acids. (Patton and Jensen, 1976; Christie, 1979B).

Synthesis of Fatty Acids

The synthesis of fatty acids for incorporation into milk fat within the mammary gland is similar to that seen in other tissues. There are two basic reactions: the conversion of acetyl-coenzyme A (CoA) to malonyl-CoA, followed by incorporation of the latter into a growing acyl chain via the action of the fatty acid–synthetase complex. However, the product of these reactions in lactating mammary tissue from many species is short and medium chain fatty acids. In most other tissues the product is palmitate. For more complete details see Moore and Christie, (1978), Bauman and Davis (1974), and Patton and Jensen (1976).

The major source of carbon for fatty acid synthesis in nonruminant mammary tissue is glucose, while in the ruminant mammary gland it is acetate and β-hydroxybutyrate (Moore and Christie, 1978; Bauman and Davis, 1974). Strong and Dills (1972) compared the rates of synthesis of fatty acids from acetate and glucose in mammary tissue from several species. In general, as the utilization of acetate increased, that of glucose decreased, with the cow primarily utilizing acetate and the rat primarily utilizing glucose. The apparent reason that the ruminant mammary cell cannot utilize glucose for fatty acid synthesis is an inability to transport acetyl units from glucose catabolism in the mitochondria to the cytosol for fatty acid synthesis (Moore and Christie 1978).

In the ruminant mammary tissue, it appears that acetate and β-hydroxybutyrate contribute almost equally as primers for fatty acid synthesis (Palmquist et al. 1969; Smith and McCarthy 1969; Luick and Kameoka 1966). In nonruminant mammary tissue there is a preference for butyryl-CoA over acetyl-CoA as a primer. This preference increases with the length of the fatty acid being synthesized (Lin and Kumar 1972; Smith and Abraham 1971). The primary source of carbons for elongation is malonyl-CoA synthesized from acetate. The acetate is derived from blood acetate or from catabolism of glucose and is activated to acetyl-CoA by the action of acetyl-CoA synthetase and then converted to malonyl-CoA via the action of acetyl-CoA carboxylase (Moore and Christie, 1978). Acetyl-CoA carboxylase requires biotin to function. While this pathway is the primary source of carbons for synthesis of fatty acids, there also appears to be a nonbiotin pathway for synthesis of fatty acids C_4, C_6, and C_8 in ruminant mammary-tissue (Kumar et al. 1965; McCarthy and Smith 1972). This nonmalonyl pathway for short chain fatty acid synthesis may be a reversal of the β-oxidation pathway (Lin and Kumar 1972).

The fatty acid–synthetase complex is located in the cytosol of the mammary cell and has a molecular weight of about 500,000 (Smith 1980). The poperties of fatty acid–synthetase isolated from mammary tissues and other tissues of several mammals were found to be very similar (Smith 1976). This complex contains a structural component with a 4'-phosphopanthetheine group and seven enzymes. During the synthesis of the fatty acids, the growing acyl chain is linked to the fatty acid–synthetase complex via a thioester linkage to the 4'-phosphopanthetheine group. In most tissues, the termination of fatty acid synthesis is by a long chain acyl thioesterase, thioesterase I (Smith 1980). Thioesterase I is covalently linked to the end of the polypeptide chain of the fatty acid–synthetase complex (Smith 1981). Thioesterase isolated from rat liver and mammary tissues exhibits identical specificity for hydrolyzing thioester bonds of acyl groups with carbon lengths of C_{16} and C_{18} (Lin and Smith 1978). This would support the observation that the primary end product of fatty acid–synthetase in most animal tissues is palmitic acid.

The milk from many species contain some short- and medium-chain fatty acids. In rabbit and rat mammary tissue there is a second thioesterase which is active toward medium- and long-chain acyl-CoA thioesters (Libertini and Smith 1978; Knudsen et al. 1976). The medium-chain thioesterase, thioesterase II, is in the cytosol and is not covalently linked to the fatty acid–synthetase complex (Libertini and Smith 1978; Knudsen et al. 1976). Thioesterase II purified from rat mammary tissue has a broad specificity for hydrolyzing thioester bonds of acyl-CoA from C_8 to at least C_{18} long (Libertini and Smith 1978). The capacity to synthesize medium chain fatty acids in the rat mammary tissue increased concurrently with thioesterase II during late pregnancy, reaching a maximum at about the time of parturition and then remaining high throughout the period of lactation (Smith and Ryan 1979). The observed increase in thioesterase II paralleled the proliferation of lobuloalveolar epithelial cells in rat mammary tissue. This led Smith and Ryan to suggest that the content of thioesterase II per mammary epithelial cell does not change during gestational development of the rat. In contrast, the thioesterase II activity in mouse mammary tissue does not reach a maximum until lactation is well established, much later than in the rat (Smith and Stern 1981). Therefore, the thioesterase II activity per mammary epithelial cell of the mouse does appear to increase. Smith and Stern concluded that thioesterase II activity is under different control mechanisms in the rat and mouse.

Both goat and cow mammary tissue synthesize medium-chain fatty acids. However, attempts to isolate thioesterase II from the cytosol of ruminant mammary tissues have not been successful (Grunnet and Knudsen 1979). In contrast to the nonruminant, the fatty acid–

synthetase from ruminant mammary tissue can synthesize medium-chain fatty acids. If there is a thioesterase II in ruminant mammary tissue, it is tightly associated with the fatty acid–synthetase complex. Further differences between ruminant and nonruminant mammary tissue are that the ruminant requires an unidentified microsomal factor and a fatty acid–removing system such as albumin, β-lactoglobulin, or methylated cyclodextrin to produce adequate quantities of medium-chain fatty acids for milk fat (Knudsen *et al.* 1981; Knudsen and Grunnet 1982). The addition of the microsomal fraction or fatty acid acceptors to nonruminant fatty acid–synthetase had no effect on medium-chain length fatty acid synthesis.

Knudsen and Grunnet (1982) have proposed an interesting system for the control of medium-chain fatty acid synthesis by ruminant mammary tissue. Their proposal is based on their observations that ruminant mammary tissue fatty acid–synthetase exhibits both medium-chain thioesterase (Grunnet and Knudsen 1978) and transacylase (Knudsen and Grunnet 1980) activity and that medium-chain fatty acids synthesized *de novo* can be incorporated into TG without an intermediate activation step (Grunnet and Knudsen 1981). They proposed that the synthesis of the medium-chain fatty acids is controlled by their incorporation into TG (Grunnet and Knudsen 1981). Further work will be needed to substantiate transacylation as a chain-termination mechanism in fatty acid synthesis by ruminant mammary tissue.

Triacylglycerol Synthesis

Two major pathways are thought to be involved in TG synthesis in mammary cells. They are the MG pathway, which utilizes an sn-2 MG as the fatty acid acceptor, and the α-glycerol phosphate pathway, which uses a glycerol phosphate as the fatty acid acceptor (Weiss and Kennedy 1956; Clark and Hübscher 1961). The relative contribution of each pathway is not known and probably varies among species. It is generally agreed that the α-glycerol phosphate pathway is the predominant pathway in ruminants (Bauman and Davis 1974; Moore and Christie 1978; Smith and Abraham 1975). However, in the mammary tissue of the pig, the MG pathway is reported to be as active as the α-glycerol phosphate pathway (Bickerstaffe and Annison 1971).

In the bovine, an interesting role has been proposed for these two pathways. In bovine milk the molecular weights of the TG exhibit a bimodal distribution. One maximum is at 38 fatty acid carbons and the second maximum is at 48 fatty acid carbons (Breckenridge and Kuksis 1967). It is postulated that the lower molecular weight TG are synthesized via the MG pathway (Barbano and Sherbon 1975; McCarthy and Coccodrilli 1975). This concept is supported by the observation that palmitate increases in the sn-2 position as the molecular weight of the TG of milk increases (Dimick *et al.* 1965). Products of lipoprotein lipase action would be the source of sn-2 Mg for the MG pathway. A major sn-2 MG produced by lipoprotein lipase would be sn-2 monopalmitate (Dimick *et al.* 1970).

The esterification of fatty acids in the mammary cell has been reported as a function of the microsomes and mitochondria (Bauman and Davis 1974; Moore and Christie 1978). While both microsomes and mitochondria may have acyltransferase activity, it has been observed to be 10 times greater in the microsomal fraction of the rat mammary cell (Tanioka *et al.* 1974). Based on autoradiographic studies, it appears that most synthesis of milk TG occurs in the rough endoplasmic reticulum of mouse mammary tissue (Stein and Stein 1971).

As can be seen in Table 4.2, the fatty acids are not randomly distributed among the three positions of the TG in bovine milk. Control of esterification is not understood, but there are several factors known to affect it. The presence of glucose is known to stimulate the synthesis of milk TG (Dimmena and Emery 1981; Rao and Abraham 1975). In the mouse, Rao and Abraham concluded that glucose was supplying factors other than NADPH or acylglycerol precursors that stimulated milk fat synthesis. The fatty acid that is esterified is known to be affected by the concentration of the acyl donors present (Marshall and Knudsen 1980; Bickerstaffe and Annison 1971). However, in studies under various conditions, palmitic acid was consistently esterified at a greater rate than other fatty acids (Bauman and Davis 1974; Moore and Christie 1978; Smith and Abraham 1975).

In many *in vitro* studies the acylation of the sn-3 position appears to be the rate-limiting step in TG synthesis. It has been suggested that the intracellular concentration of medium chain fatty acids may limit the final acylation reaction in TG synthesis (Dimmena and Emery 1981). Another theory is that the concentration of phosphatidate phosphatase, the enzyme that hydrolyzes the phosphate bond in phosphatidic acid, yielding DG, may be the limiting factor (Moore and Christie 1978). The DG acyltransferase responsible for the final acylation of milk TG has been studied in mammary tissue from lactating rats (Lin *et al.* 1976). It was observed to be specific for the sn-1,2 DG, with very little activity observed with the sn-1,3 or sn-2,3 DG. It exhibited a broad specificity for acyl donors. The acyl-CoA specificity was not affected by the type of 1,2 DG acceptor offered, which implies that the type of fatty acid introduced into the glycerol backbone was not influenced by the specificity of subsequent acylation steps. However, the concentration of acyl donors will affect the final acylation. It was observed with rat and bovine mammary tissue that incorporation of short and long chain fatty acids into the sn-3 position is dependent upon their relative acyl concentrations (Marshall and Knudsen 1980; Lin *et al.* 1976). Further work is needed to understand the factors that regulate the nonrandom distribution of fatty acids in milk TG.

LIPID CLASSES

Composition

The data in Table 4.2 are from analyses of pooled milk. As mentioned, TGs account for about 98% of the lipids; the DGs, MGs, and free fatty acids (FFA) are mostly products of lipolysis, and the cholesterol and phospholipids are cellular membrane material which accompanies the fat globule during extrusion from the secreting cell.

We will use a shorthand designation for fatty acids, i.e., 18:0, stearic acid;18:1, oleic acid; etc. The first figure is the number of carbons, the second the number of double bonds. To locate the fatty acids in acylglycerols, stereospecific numbering (sn) will be employed. If a glycerol molecule is drawn with the secondary hydroxyl to the left, the hydroxyl above is sn-1 and that below is sn-3.

Triacylglycerols. The composition of TGs refers to their structure or the identity of the fatty acids esterified to each of the three hydroxyls on glycerol and ultimately to the identity of the individual molecular species. Because there may be over 400 fatty acids in a milk sample, based on random distribution, there may be a total of 400^3 or 64×10^6 individual TGs, including all positional and enantiomeric isomers. A random distribution is defined as all possible combinations resulting from expansion of the binomial equation. If we have two fatty acids, x and y, located at random in the three positions of glycerol, the equation becomes $(x + y)^3$ or $x^3 + 3x^2y + 3xy^2 + y^3$, which, when expanded further, is $x^3 = xxx, 3x^2y = xxy, 3xy^2 = yyx, y^3 = yyy$

$$\begin{array}{cc} yxx & xyy \\ xyx & yxy \end{array}$$

Thus eight TG species are possible, including two sets of enantiomers: xxy, yxx and yyx, xyy and two monoacid or simple TGs: xxx and yyy.

Although milk fat does contain more than 400 fatty acids, Kuksis (1972) has pointed out that a more realistic figure is 20, because most of the remainder exist in trace amounts. This number still leaves the possibility of 8000 TGs, but the asymmetry of milk TGs reduces the number further. Nevertheless, the several thousand TGs undoubtedly existing in milk fat present intriguing but exasperating problems in identification. The TG structure of milk has been reviewed by Patton and Jensen (1976), Christie (1979), and Breckenridge (1978).

Analysis of the TG structure of milk usually has been done by combining various chromatographic techniques for the separation of TGs by molecular weight (GLC) or by the total number of double bonds (argentation TLC) with stereospecific analysis. In the latter, 1,2 (2,3) sn-DGs are generated from the TG by the action of pancreatic lipase and a Grignard reagent. The DGs are converted to phosphatidylphenols, which are digested by phospholipase A-2. This enzyme hydrolyzes the sn-3-phosphatidylphenol, a derivative of 1,2-sn-DGs and not the sn-1 phosphatidylphenol, thus enabling separation of the two by TLC. These procedures have been described by Patton and Jensen (1976), Breckenridge (1978), and Christie (1979). A method developed by Myher and Kuksis (1979) could provide identification of the molecular species of TGs but has not been applied to bovine milk TGs. In this analysis, the 1,2 (2,3)-sn-DGs produced as described above are derivatized to phosphatidylcholines. Phospholipase C hydrolyzes the sn-3-phosphatidylcholine to 1,2-sn-DG and phosphorylcholine in 2 min and the sn-isomer in 2 hr.

Initially, stereospecific analyses were done by Pitas et al. (1967) on whole milk fat and by Breckenridge and Kuksis (1968) on a molecular distillate of butter oil. They indicated that the short chain acids were selectively associated with the sn-3 position. In the butter oil distillate, over 90% of the TGs contained two long-chain and one short-chain fatty acids. This asymmetry has been confirmed by the observation of a small optical rotation of the TGs (Anderson et al. 1970), by proton magnetic spectroscopy (Bus et al. 1976), and by nuclear magnetic resonance spectroscopy (Pfeffer et al. 1977). Pfeffer et al. found 10.3 M% 4:0 (butyric) in the oil and determined that 97% of the acid was in the sn-3 position. It is worth noting that the analysis was done without alteration or fractionation of the oil.

Barbano and Sherbon (1975), Parodi (1979, 1982), and Christie and Clapperton (1982) have provided additional stereospecific analyses.

Barbano and Sherbon (1975) decided that the distribution of fatty acids in the high melting fractions of milk fat supported the hypothesis that at least a portion of these TGs were synthesized via the MG pathway. Trans acids were found in the fractions, indicating that these acids behave as saturates.

Parodi's (1979, Table 4.3) data again indicate the asymmetric distribution of fatty acids. The percentage of a fatty acid at position sn-3 generally decreased with an increase in the chain length of the acid, with the reverse occurring at position sn-1. There was no major vari-

Table 4.3. Composition and Stereospecific Distribution of Fatty Acids in Milk Fat Triglycerides from Bimonthly Samples of Maleny Butter.

Month	Position	Fatty Acid Composition, Mole %								
		4:0	6:0	8:0	10:0	12:0	14:0	16:0	18:0	18:1
Jan.	TG	10.1	4.4	2.0	3.4	3.4	10.1	20.9	11.1	19.9
	sn-1	0.1	0.2	0.3	1.3	2.1	7.6	27.8	18.7	23.4
	sn-2	0.1	0.7	3.0	5.4	6.1	18.9	27.0	5.3	11.6
	sn-3	30.2	12.4	2.6	3.7	2.0	3.9	7.8	9.3	24.6
Mar.	TG	10.2	4.5	2.1	3.5	3.4	9.2	18.4	11.8	21.2
	sn-1	0.2	0.2	1.1	2.1	2.5	7.6	25.8	19.6	24.5
	sn-2	0.0	0.4	3.4	5.9	6.5	18.4	25.0	5.9	14.1
	sn-3	30.2	13.1	2.0	2.5	1.2	1.7	4.4	10.1	25.1
May	TG	9.8	4.0	1.8	3.1	3.1	9.4	19.9	11.9	22.1
	sn-1	0.1	0.2	0.1	0.5	1.6	7.4	26.5	18.5	25.5
	sn-2	0.0	0.6	2.8	5.0	5.8	18.7	27.6	6.0	14.2
	sn-3	29.4	11.3	2.5	3.7	1.9	2.1	5.7	11.3	26.7
July	TG	10.6	4.1	1.8	2.9	2.8	8.6	20.0	12.0	23.2
	sn-1	0.1	0.1	0.3	0.9	1.8	6.7	26.8	20.4	26.6
	sn-2	0.1	0.5	3.1	5.1	5.5	17.4	27.3	5.7	14.5
	sn-3	31.7	11.6	2.0	2.6	1.0	1.8	5.9	10.0	28.5
Sept.	TG	10.8	4.4	1.9	3.2	3.1	9.1	20.0	11.9	22.7
	sn-1	1.0	0.8	1.0	2.0	2.4	7.9	26.2	18.2	25.0
	sn-2	0.1	0.5	3.6	6.0	6.3	18.5	26.8	5.8	14.1
	sn-3	31.4	11.9	1.1	1.7	0.7	0.9	6.9	11.7	29.0
Nov.	TG	10.2	4.4	2.1	3.6	3.5	10.0	20.3	11.7	21.1
	sn-1	1.6	0.9	0.9	2.2	2.9	9.2	28.3	18.7	20.5
	sn-2	0.1	0.7	3.5	6.3	6.5	18.9	26.5	5.7	14.0
	sn-3	28.8	11.7	1.7	2.4	1.1	1.8	6.2	10.8	28.7

SOURCE: Parodi (1979).

ation in the stereospecific distribution of fatty acids throughout the year. The variation that occurred was mainly in the medium-chain acids, which are synthesized primarily in the mammary gland. Parodi noted a decrease in the chain length of the acids esterified to sn-3, accompanied by an increase in 18:1 and 18:0 when a restricted diet (change from the normal ration to 0.25 kg hay per day at mid-lactation) was fed to one cow. The data are not shown.

Parodi (1982) obtained the positional distributions of the fatty acids in a sample of Australian butter and in the high, medium, and low molecular weight fractions. This fractionation was done by silicic acid column chromatography. Each fraction was further divided into TG classes SSS, SSM^t, SSM^c, S M^{c}M^c, and others (S = saturates, M = monoenes, t = *trans*, c - *cis*) by argentalian TLC. Once again, the exclusive location of 4:0 and 6:0 at sn-3 is confirmed. The expected fatty acids are associated with the relevant molecular weight fractions, e.g., 70.9 M% 4:0 in the low molecular weight fraction, TG class SSS and sn-3. *Trans* monoenes were incorporated into TGs in the same frashion as the *cis* isomer. The specific positional distribution of fatty acids in intact milk TGs was not always seen in the fractions or TG classes.

Earlier, Parodi (1975) detected acetodiacylglycerols in milk fat by TLC. The infrared absorption data he obtained suggested that the acetic acid was esterified mostly to the primary position. This location should be of interest to investigators studying cheese and related flavors, since the acetate ester would be quickly released by the lipolytic systems involved. Parodi (1974A) also investigated the high melting glyceride fraction of milk fat, the types and amounts of TGs, the fatty acid composition, and the TG composition. The high melting fraction (4.7% of the total fat), removed by crystallization from acetone at 20°C, contained 41.0 M% 16:0 and 29 M% 18:0. The most abundant carbon number fractions were 48, 20 M%; 50, 28 M%; and 52, 23 M%. These carbon numbers are combinations of 16:0, 18:0, and 18:1 c and t. If the data on the SSS fraction are examined, the presence of tripalmitoylglycerol and tristearoylglycerol is indicated. Parodi observed a lowering of the softening point of butter from 33.7 to 28.8°C when the high melting fraction was removed, even though the fraction amounted to only 4.7% of the total fat. Parodi has obtained additional information on the softening point which will be discussed in the section on physical characteristics. Parodi (1973A) has employed analysis of TG carbon numbers by GLC to detect adulteration of butter. When beef tallow was added to 112 samples of authentic Australian butterfat, 26.8% of the samples were not detected at the 10% level, and 4.5% were not detected at the 15% level. All were detected at the 20% level. He also discussed the analysis of adulteration with other fats.

Kuksis *et al.* (1973) have extensively analyzed the structure of milk TGs and have summarized their results as follows: there are three types of TGs. The first has acyl carbons totaling 48–54, composed of long-chain 1,2-DGs containing 18:0, 18:1, and 18:2. In type 2 the carbon numbers are 36–46 and the sn-3 position acids are 4:0, 6:0, and 8:0. These TGs are enantiomers. In type 3, the carbon numbers are 26–34, the 1,2-DGs contain medium chain fatty acids, and the 3-position acids are short and medium chain. Those TGs in type 3 that have short- or medium-chain acids in the sn-3 position that are different from those in sn-1 are also enantiomers.

The important point to remember about the TG structure of milk is the asymmetry: the sn-3 location of 4:0, 6:0, and 8:0. Physiologically, the TGs containing these acids are preferentially hydrolyzed by pancreatic and lingual lipases and the short chain acids are transported via the portal vein to the liver, where they are oxidized. The reconstituted TGs entering the chyle from the intestinal wall do not contain the short-chain acids, and some of the medium-chain acids are also missing. Another important point is that the small amounts of tripalmitoyl- and tristearoylglycerols present are likely to pass through the digestive tract untouched and probably account for much of the nonabsorbed fat in a diet containing milk fat.

Other Acylglycerols.   If some of the DGs in freshly drawn milk are involved in biosynthesis, it is possible that they are enantiomeric and are probably the sn-1,2 isomer. If so, the constituent fatty acids are long chain. Their configuration can be determined by stereospecific or other analyses, but it is difficult to accumulate enough material for analysis. Nevertheless, Lok (1979) isolated the DGs from freshly extracted cream as the trityl derivatives. Trityl chloride reacts selectively with primary hydroxyls. The stereochemical configuration of the DGs was identified as sn-1,2; therefore, these residual DGs were most likely intermediates of biosynthesis. If the DGs were products of lipolysis, they would be a mixture of 1,2/2,3 isomers in a ratio of about 1:2, since milk lipoprotein lipase preferentially attacks the sn-1 position of TGs (Jensen *et al.* 1983).

Timmen and Dimick (1972) characterized the major hydroxy compounds in milk lipids by first isolating the compounds as their pyruvic ester-2,.6-dinitrophenylhydrazones. Concentrations as weight percent of the compounds from bovine herd milk lipids were: 1,2-DGs 1.43, hydroxyacylglycerols 0.61, and sterols 0.35. Lipolysis tripled the DG content. The usual milk fatty acids were observed, except that the DGs lacked 4:0 and 6:00, again indicating that these lipids were in part intermediates in milk lipid biosynthesis. With the large hydrazone group attached to the hydroxyl, the derivatives should appreciably rotate polarized light and would therefore be detectable with a polarimeter. This was done by Lok (1979) with trityl 1,2-DGs.

Alkyl and alk-l-enyl ether diacylglycerols are also found in milk lipids (Morrison 1970). Parks *et al.* (1961) detected 0.2 μM of bound aldehyde per gram of butterfat and identified n–9 through -18 and br-11, -13, -15, -16, and -17 aldehydes. The aldehydes were derived from the alk-l-enyl diacylglycerols. Glyceryl ethers, -alkyl ether diacylglycerols, were found in milk fat at a level of 0.10% and the 16:0, 18:0, and 18:1 acyl chains were determined (Hallgren and Larsson 1962). The 1-0-alkylglycerols and 1-0-(2'-methoxy) alkylglycerols have been characterized (Hallgren *et al.* 1974). Ahrné *et al.* (1980) found glycerol ethers in the colostrum and milk of the cow and other species. The amounts found in bovine milk are presented in Table 4.4 where it can be seen that colostrum contained more of the ethers than mature milk. About 97% of the ethers were located in the neutral lipids. The acyl chains were 14, 16, and 18 carbon saturates and 14–20 unsaturates.

Phospholipids.   The phospholipids comprise approximately 1% of the total lipid in bovine milk (ca.0.3 to 0.4 g/liter). While quantitatively minor, the ability of the phospholipids to form stable colloidal suspensions or emulsions in aqueous solution cause them to be important in the formation and secretion of milk fat. (Long and Patton 1978; Patton and Keenan 1975). Their physical properties as bipolar molecules and their relatively high concentration of unsaturated fatty acids also make them an important factor to consider during the storage and

Table 4.4. Content of Glycerol Ethers in Neutral Lipids and Phospholipids Isolated from Bovine Colostrum (Means from Four Swedish Red and White Cows from Days 1 to 5) and Milk (Mean for Four Swedish Red and White Cows from 2nd Wk to 7th Mo)[a] Week 2 to Month 7.

	Colostrum (%) (wt/wt)	Milk (%) (wt/wt)
Total lipids	5.6	3.9
Neutral lipids (N) in total lipids	99.0	99.3
Phospholipids (P) in total lipids	1.0	0.7
Glycerol ethers in total lipids	0.061	0.009
Glycerol ethers in N	0.06	0.007
Glycerol ethers in P	0.16	0.25
Glycerol ethers in N of total glycerol ethers	97.4	80
Glycerol ethers in P of total glycerol ethers	2.6	20

SOURCE: Ahrne *et al.* (1980).

processing of milk. They are relatively susceptible to oxidation because of their polyunsaturated fatty acid content.

The phospholipid content of milk and milk products is given in Table 4.5 (Kurtz 1974). Total phospholipid is usually determined by measuring the lipid phosphorus content of the product and multiplying by 26 (AOCS 1975). As the total milk lipid increases in a milk product, so does the phospholipid concentration. However, the ratio of phospholipid to total lipid varies greatly. Referring to Table 4.5 skim milk contains the smallest concentration of phospholipid but the highest ratio of phospholipid to total lipid. The opposite relationship is seen in cream and butter.

Phospholipids have usually been isolated from milk lipids by silicic acid column chromatography. This is a difficult separation because the phospholipids are only 1% of the total as compared to 98% TG. Gentner et al. (1981) have not only separated the phospholipids from the remainder by TLC but have also resolved the major types of phospholipids on one plate.

Most milk lipid exists as fat globules suspended in the aqueous phase of milk. The size of the milk fat globules varies from 0.1 to 2 μ in diameter (Mulder and Walstra 1974). The core of the globule is primarily TG, which is surrounded by the milk fat globule membrane (MFGM). This membrane contains protein, glycoproteins, enzymes, phospholipids, and other polar materials. It is a major source of cholesterol and phospholipid in milk. Between the MFGM and the TG core is a dense layer 10–50 nm thick (Freudenstein et al. 1979). Based on biochemical and electron microscopic studies, it appears that the MFGM originates from the plasma membrane of the secretory cells of the mammary gland. It is believed that the apical plasma membrane of the mammary epithelial cell envelops the milk TG droplet as it is secreted from the cell. Note the similarity. For a complete review on the origin of the milk fat globules, see Chapter 10 and McPherson and Kitchen (1983).

Table 4.5. Percent Phospholipid Content of Milk and Milk Products.

Product	Phospholipids in Product	Fat in Product	Phospholipids in Fat
Whole milk	0.0337	3.88	0.87
Skim milk	0.0169	0.09	17.29
Cream	0.1816	41.13	0.442
Buttermilk	0.1819	1.94	9.378
Butter	0.1872	84.8	0.2207

SOURCE: Kurtz (1974).

Phospholipids also are found as lipoprotein complexes in skim milk. The skim milk phase may contain 30–50% of the phospholipid in milk (Bachman and Wilcox 1976; Patton and Jensen 1976); Patton and Keenan 1971). Analyzing skim milk, Cerbulis (1967) observed that 9.6% of the lipid bound to acid-precipitated casein was phospholipid, as was 21% of that bound to whey. The nature of the binding was not known but was probably of membrane origin. Based on biochemical and morphological studies, membranes from leukocytes, secretory cell debris, and components of the MFGM have been identified in skim milk (Wooding 1974; Plantz and Patton 1973; Kitchen 1974). The MFGM is a major source of membrane in skim milk. It has been proposed that as milk ages, segments of the MFGM surrounding the fat globule are lost into the skim milk by vesiculation and fragmentation (Wooding 1974; Plantz and Patton 1973). Though it has been argued that the presence of MFGM in skim milk is an artifact of handling (Baumrucker and Keenan 1973), the distributions of phospholipids in various mammary cell membranes and milk are quite similar.

The phospholipids in milk are synthesized by the mammary cell via pathways that are common to other mammalian cells. For further information on the synthesis of phospholipids in the mammary cell, see Kinsella and Infante (1978) and Patton and Jensen (1976). The major glycerophospholipids are phosphatidylethanolamine, phosphatidylcholine, phosphatidylserine, and phosphatidylinositol. A more complete composition is given in Table 4.6, Patton and Jensen (1976). The acyl and alkyl compositions will be given later. In milk, the glycerophospholipids are found predominantly in the diacyl form. However, small

Table 4.6. Phospholipid Composition of Bovine Milk.

Phospholipid	M %
Phosphatidylcholine	34.5
Phosphatidylethanolamine	31.8
Phosphatidylserine	3.1
Phosphatidylinositol	4.7
Sphingomyelin	25.2
Lysophosphatidylcholine	Trace
Lysophosphatidylethanolamine	Trace
Total choline phospholipids	59.7
Plasmalogens	3
Diphosphatidylglycerol	Trace
Ceramides	Trace
Cerebrosides	Trace
Gangliosides	Trace

SOURCE: Patton and Jensen (1976).

amounts of plasmalogens, the vinyl ether form of the glycerophospholipid, have been observed (Hay and Morrison 1971; Duin 1958). Hay and Morrison observed that 4% of the phosphatidylethanolamine was in the ether form and 1.3% of the phosphatidylcholine was in this form. Duin reported 1.3 to 2.5% of bovine phospholipids as plasmalogens.

Lysophospholipids have been found in butter serum by Cho et al. (1977). They characterized the sn-1 and -2 lysophosphatidylcholines and phosphatidylethanolamines. It is not known if these compounds are products of degradation or remnants of biosynthesis. Cho et al. (1977) searched for, but did not find, another possible product of enzymatic degradation of milk, phosphatidic acid. Phosphatidic acid can be formed by the action of phospholipase D on phosphatidylcholine, for example, but this enzymatic activity was not detected. The compound is also an important intermediate in the biosynthesis of lipids, but the concentration in tissue is always very low. The amount is also low in milk. Cho et al. (1977) found 1.2 and 0.9 (percent of total lipid P) of the lyso compounds above. The quantities of the other phospholipids were: phosphatidylethanolamine, 27.3; -choline, 29.1; -serine, 13.4; -inositol, 2.5; and sphingomyelin, 25.6.

Diphosphatidylglycerol (cardiolipin) was found in lactating mammary tissue at levels 200–300 times those found in milk (Patton et al. 1969).

Sphingo and Related Lipids.   Kayser and Patton (1970) isolated glucosyl and lactosyl ceramides (cerebrosides) from milk. They found 1.7 mg/100 ml in the globule membrane and 0.8 mg/100 ml in skim milk. The membrane-bound cerebrosides contained mainly acids of 20 to 25 carbons, and those in the skim milk contained 18 carbons or less. Morrison and Hay (1970) and Fujino and Fujishima (1972) investigated spingomyelin. Cho et al. (1977) found 25.9% (of total lipid P) sphingomyelin in milk lipids. Investigators often neglect spingomyelin as a major phospholipid in milk.

Huang (1973) found relatively high concentrations of gangliosides in buttermilk; 10–20 mg per gram of lipid. Keenan (1974) identified 5.6 nmoles of ganglioside per gram of milk. Ninety percent was associated with the globule membrane. Six were found and three identified as ceramide-glucose-galactose-sialic acid, ceramide-glucose-galactose-(sialic acid)-N-acetylgalactosamine, and ceramide-glucose-galactose-sialic acid-sialic acid. The last accounted for more than 50% of the lipid-bound sialic acid.

Sterols.   These compounds are found in the unsaponifiable fraction of milk lipids and consist mostly of cholesterol with some lanosterol. Methods for the determination of unsaponifiables using dry saponification have been given by Schwartz *et al.* (1966) and Maxwell and Schwartz (1979). The amounts were 0.33–0.36% of butter oil. Brewington *et al.* (1970) confirmed the presence of the latter sterol and identified two new constituents, dihydrolanosterol and β-sitosterol. Treiger (1979) detected β-sitosterol at levels of 0.3–0.4% of the total sterols with GLC. Mincione *et al.* (1977) isolated 17 sterols from bovine milk as trimethylsilyl ethers. Five were identified as a percent of total sterols: cholesterol, 90.5; campesterol, 1.8; stigmasterol, 0.7; β-sitosterol, 0.2; and Δ^5-avenasterol, 0.03. These analyses were done by GLC-mass spectrometry. Parodi (1973B), also using GLC, found an average of 257.6 mg cholesterol per 100 g of fat in Australian butter. Tentatively identified were 7-dehydrocholesterol, campesterol, and β-sitosterol. Keenan and Patton (1970) have reported that the cholesterol esters represent about one-tenth of the sterol content of milk. Parks (1980) found 190 nM/100 ml of cholesterol esters in skim milk, most of which were esterified with 18:1. Flanagan *et al.* (1975) isolated and identified Δ^4-0-cholesten-3-one and $\Delta^{3,5}$-cholestadiene-7-one in an hydrous milk fat and nonfat dry milk. These were probably products of cholesterol oxidation. Flanagan and Ferretti (1974) had previously found Δ^2-campestene and Δ^2-sitostene, in this product.

Reliable data on the cholesterol content of dairy products, lacking in the past, are now available (LaCroix *et al.* 1973). The amount in whole milk fat was 13.49 $\pm$ 1.01 mg per 100 g milk which contained 3.47 $\pm$ 0.74 g of fat. Data were obtained from 27 kinds of products, and from these an equation was derived for estimating the cholesterol content of dairy products with fat contents greater than those of whole milk. It is obvious that more fat is accompanied by more cholesterol, e.g., Cheddar cheese contains 102 mg/100 g, an approximate 8fold increase over whole milk.

Bachman and Wilcox (1976) found an average cholesterol content of 15.2 mg/100 ml in 356 samples (fat content 3.69%). After separation, 16.9% of the cholesterol was found in the skim milk phase. Patton *et al.* (1980) did not find any increase in the cholesterol content of skim milk obtained by 24-hr aging of milk at 2–4°C. Cholesterol was determined by nonspecific colorimetric methods in both investigations, which is acceptable since almost all of the sterols are cholesterol. Gentner and Haasemen (1979) have analyzed cholesterol in milk enzymatically with a commercially available kit, finding 13 mg/100 ml. The method is very sensitive and is more specific than colorimetric determination, but is not as good as by GLC. Determination of β-sitosterol by GLC is used to detect adulteration of butter with vegetable oils.

Lipoproteins. As pointed out previously and in Chapter 10, the bulk of the lipoprotein in milk is membranous; membrane is found around milk fat globules and as vesicles and fragments in the skim milk. Membrane is found in milk in a concentration of about 0.1% and appears to be mainly plasma membrane derived from the lactating cell.

The results of detailed analyses of the lipid composition of fat globule membranes from bovine milk have been published by Bracco *et al.* (1972). They found the high melting TGs and other lipids observed by previous investigators. Approximately 62% TGs were present, much less than in the parent milk lipid. Among the hydrocarbons isolated, squalene was positively identified with indications by GLC of odd and even alkanes, alkenes, and polyunsaturated compounds between C-31 and C-38. In addition to cholesterol, 7-dehydrocholesterol was detected; other hydrocarbons tentatively identified were carotenoids and tocophenols. Phospholipids were noted in relative quantities not greatly different from those given in Table 4.6.

Hydrocarbons.   Milk lipids contain small quantities of various hydrocarbons: carotenoids, squalene, etc. Ristow and Werner (1968) identified the C-14 to C-35 n-alkanes and some branched monolefins, but solely on the basis of GLC retention times. Flanagan and Ferretti (1973), using GLC-mass spectrometry, found 39 aliphatic hydrocarbons in the unsaponifiable fraction of anhydrous milk fat. The compounds were the C-14 to C-27 and C-29 to C-31 straight chain paraffins, their monolefin analogs, and the C-25 to C-29 branched alkanes. Phytene was identified for the first time in milk fat, and polychlorinated biphenyls (PCBs) were also present. The total hydrocarbons amounted to 30 ppm of the milk fat. Flanagan *et al.* (1975) also identified phytol and dihydrophytol in anhydrous milk fat. Urbach and Stark (1975) isolated the following hydrocarbons from butterfat: phyt-1-ene, phyt-2-ene, neophytadiene, and several other branched and n-chain compounds. The total concentration was also 30 ppm.

ACYL AND ALKYL COMPOSITION OF LIPID CLASSES: DETERMINATION OF FATTY ACIDS

Gas liquid chromatography is still the method of choice for the routine separation and tentative identification of common milk fatty acids, as well as for the resolution of the less abundant and less common acids. Although several hundred fatty acids are listed here and elsewhere as being present in milk, we remind the reader that not all of these have been rigorously identified. Some of the pitfalls in qualitative and quantitative GLC of milk fatty acids are discussed by Jensen *et al.* (1967) and those of fatty acids in general by Ackman (1980).

As always in the analysis of milk fat, the short chain fatty acids cause problems. A major difficulty has not been the GLC separation of these acids but their transfer from the esterification mixture to the GLC instrument without loss of the volatile esters. A widely used procedure is a slight modification of the method developed by Christopherson and Glass (1969) which uses sodium methoxide for transesterification. This technique can be employed with other fats, but not with those containing appreciable amounts of free fatty acids where HCl-methanol is required.

Iverson and Sheppard (1977) have compared the method above, substituting sodium butoxide for sodium methoxide to H_2SO_4 and boron trifluoride-catalyzed butyrolyses. Butyl esters were used by these and other investigators to improve the resolution of short chain esters and to reduce their volatility. They recommend the boron trifluoride method for preparation of butyl esters of milk fatty acids, although the other catalysts gave satisfactory results. Analysis of methyl esters resulted in lower values for the short chain fatty acids.

Fatty Acids in General

The number of fatty acids and related compounds in milk lipids grew from 16 in 1959 (Jenness and Patton, 1959) to 142 in 1967 (Jensen *et al.* 1967) to over 400 in 1983. However, there are only 10 fatty acids of quantitative importance. The amounts (weight percent) as butyl esters prepared by three methods of esterification were determined by Iverson and Sheppard (1977). Because of the widely differing molecular weights of the fatty acids (4:0–18:0), fatty acid compositions of ruminant milk fats are often presented as a mole percent. The nutritionist needs the data calculated in yet another manner; weight of fatty acid/100 g or 100 ml of edible portion. Analyses of food fatty acids should always be accompanied by the fat content so that the actual weights of the fatty acids and be calculated. A compilation of this type was made by Posati *et al.* (1975). Since these analyses were done with methyl esters, the contents of 4:0 are low. Data from Feeley *et al.* (1975), obtained from careful analyses, are more reliable, and USDA Handbook 8-1 (Posati and Orr 1976) has data for many milk and dairy products.

The results from the analyses of milk fatty acids as butyl esters by Jensen *et al.* (1962), Parodi (1970) and Iverson (1983) are remarkably consistent, considering that the analyses were widely separated by time and distance. The differences in quantities between June and December reflect the influence of season, that is, the availability of pasture. Determination of butyl esters as described by Iverson and Sheppard (1977) is the best method now available for the analysis of fatty acids in milk and dairy products by GLC.

Strocchi and Holman (1971), with the aid of argentation TLC and GLC-mass spectrometry, identified the fatty acids in Tables 4.7 and 4.8. We have presented all of their data because the identifications were obtained by unequivocal methods, many previously tentative identifications were confirmed, and the results were quantitative. Strocchi and Holman did not identify the positional isomers of the unsaturates but found two or three peaks for most of the carbon numbers. Iverson (1983) determined the quantities of minor and trace fatty acids, verifying the findings of other investigators.

Saturated and Branched Chain Fatty Acids

Saturated even and odd n-chain acids from 2 to 28 carbons have been found in milk (Jensen *et al.* 1967; Patton and Jensen 1976; Kurtz 1974).

Table 4.7. Fatty Acid Composition of Butter Oil as Determined by GLC-Mass Spectrometry (Weight Percent) of Total Methyl Esters

| Methyl Ester Carbons | Saturates | Monoenes | | Branched | | |
		cis	*trans*	*Iso*	*Anteiso*	*Other*
4	3.25					
6	2.32					
8	1.85					
10	4.02					
11	0.16					
12	4.15	0.03				
13	0.03			0.01	Trace	
14	11.05	0.47		0.08		
15	0.95	0.08		0.23	0.42	
16	26.15	1.25	0.03	0.32		
17	0.70	0.32	0.01	0.33	0.40	DDL pristanate, 0.01
18	9.60	20.40	5.34	0.15		
19	0.11	0.10	0.01	0.06	0.09	DDD pristanate, 0.01
20	0.19	0.15	0.01	0.04		DDL, DDD phytanates, 0.04
21	0.06	0.03	Trace	Trace	0.01	
22	0.10	0.02	Trace	Trace		
23	0.07	0.01	0.01			
24	0.06	0.02	0.01			
25	0.01					
26	0.04					

SOURCE: Adapted from Strocchi and Holman (1971).

Table 4.8. Fatty Acid Composition of Butter Oil as Determined by GLC-Mass Spectrometry (a Continuation of Table 4.7).

Methyl Ester Carbons	Weight Percent of Total Methyl Esters			
	Dienes	Trienes	Tetraenes	Pentaenes
18				
Positional	0.14	0.02		
isomers	2.30	0.60		
Conjugated				
cis, trans	0.70	di-0.03		
trans, trans	0.05	tri-0.01		
20				
Positional	0.03	0.01	0.10	
isomers	Trace	0.13	0.02	
		0.02		
22				
Positional	0.04	0.06		0.02
isomers	Trace	0.02		0.02
24				
Positional	Trace	0.01		
isomers		0.03		
		0.02		

SOURCE: Adapted from Strocchi and Holman (1971).

Most of the identifications were unequivocally confirmed by mass spectrometry. Many of these acids are present in small quantities (less than 1%) and are of little importance. See Table 4.6 for representative data.

Branched chain fatty acids are present in milk (Kurtz 1974; Patton and Jensen 1976), and the following have been identified: monomethyl 11–24; 13–19, three or more positional isomers and multimethyl 16–28. Iverson (1983) and Iverson *et al.* (1965) identified the branched and other acids with the aid of urea fractionation. Saturated n-fatty acids of longer chain length form inclusion complexes with urea more readily than acyls with functional groups. The branched chain fatty acids do not form adducts.

Massart-Leen *et al.* (1981) analyzed bovine milk fat and goat milk fat for branched chain fatty acids. They did not find the same diversity of fatty acids in bovine as in goat milk fat and as previously reported. The authors suggested that the difference—the absence of branched chain acids other than iso and anteiso in bovine milk fat—could be caused by the relative inefficiency of the incorporations of methylmalonic acid into the biosynthetic pathway.

Egge *et al.* (1972) found at least 50 branched chain fatty acids in human milk fat by identification with GLC-mass spectrometry following hydrogenation and enrichment of the acids by urea fractionation. They postulated that many of these were of bacterial origin produced in and absorbed from the intestinal tract. If so, this pathway could be a source of calories in the form of fatty acids for the mother. Ackman et al. (1972) analyzed C-15-and C-17-enriched fractions of milk fat with high-resolution open-tubular GLC, finding that only even-numbered carbons of the acyl chains bore the methyl branch. In the C-15 fraction, methyl branching occurred at the 4, 6, 8, and 10 carbons and, in the C-17 fraction, at the 4, 6, 8, 10, and 12 carbons. Most of the iso acids had been removed by prior purification. Ackman *et al.* suggested that the difficulty of interpreting mass spectra from complex mixtures may have led to assumptions concerning the existence of monomethyl branches and odd-carbon fatty acids (Egge *et al.* 1972). Conversely, several of these acids were identified by Strocchi and Holman (1971), who analyzed a fraction, obtained by TLC, with GLC-mass spectrometry, containing only n- and monomethyl branched fatty acids. Some of these differences may have been caused by the uniqueness of the individual milk fat samples. Lough (1977) found 0.7 and 13% phytanic acid in the milk and plasma from eight cows fed grass. The acid is derived from phytol, the alcohol moiety of chlorophyll.

Monounsaturated Fatty Acids

Hay and Morrison (1970) identified the monoenoic positional and geometric isomers in milk fat and determined the amounts of each total acid class and percentage of *trans* isomers. The geometric and positional isomers of the monoenes are primarily the result of biohydrogenation of polyunsaturated fatty acids in the rumen. Stearate is also produced, and *cis*-9-18:1 accounts for most of the monoenes. The several positional isomers in *trans* 16:1 and 18:1 are due to the positional isomerization of double bonds which accompanies elaidinization.

Strocchi and Holman (1971) (Table 4.6), with the aid of argentation TLC and GLC-mass spectrometry, identified several of the fatty acids observed by Hay and Morrison (1970) and more monoenes as follows: *trans* 17:1, 19:1, 20:1, 21:1, 22:1, 23:1, and 24:1. Notably missing was 11:1, either *cis* or *trans*.

Parodi (1976) determined the distribution of double bonds in *cis* and *trans* octadecenoic fatty acids from milk fat and bovine adipose tissue. About 95% of the 18:1 is the cis-9 isomer. Parodi detected the cis-12, – 13, and –14 isomers, fatty acids not observed by Hay and Morrison (1970). The 18:1 content of Australian butterfat has varied throughout the season from 17.3 to 24.9 M%, with isolated *trans* unsaturation from 4.3 to 7.6 M%.

Smith *et al.* (1978) have described a procedure for the GLC determination of *cis* and *trans* isomers of unsaturated fatty acids in butter after fractionation of the saturated, monoenoic, dienoic, and polyenoic fatty acid methyl esters by argentation TLC. Total *trans* acids were much higher, as measured by infrared spectrophotometry than by GLC, probably because some of the acids could have two or more of the *trans* bonds designated as isolated by infrared spectrophotometry. Enzymatic evaluation of methylene-interrupted *cis, cis* double bonds by lipoxidase resulted in lower values than those obtained by GLC. The authors mention that the lipoxidase method is difficult, requiring considerable skill, and suggest that their method is suitable for the determination of the principal fatty acids in complex food lipids such as bovine milk fat.

Deman and Deman (1983) have investigated the determination of *trans* unsaturation in milk fat by infrared analysis and found values of 7.4% (winter) to 9.9% (summer) when the TGs were analyzed. These are higher than the quantities found by infrared analysis of methyl esters of the fatty acids. These quantities are isolated total *trans* bonds and do not give an estimate of the positional and polyunsaturated isomers which are present. The *trans* contents obtained by Deman and Deman are higher than the 4% found by Smith *et al.* (1978).

Polyunsaturated Fatty Acids

Because of animal biohydrogenation, the content of polyunsaturated acids in milk is low, currently reported at about 5% (Smith *et al.* 1978), and is associated mostly with the phospholipids. While quantitatively unimportant, these acids are the most susceptible targets of oxidation and provide the essential fatty acids (EFA), mostly *cis, cis*-9, 12–18:2.

The requirement of humans for EFAs has been thoroughly documented (Soderhjelm *et al.* 1970; Holman 1973). Diets free from added fats or 18:2 induce the following deficiency symptoms in infants: skin lesions, inefficient weight gain, and poor wound healing (Hansen *et al.* 1958; Holman 1973; Hansen *et al.* 1963). From these and other experiments, the minimum EFA requirement has been estimated to be about 1% of the total calories (110 mg/100 kcal). Holman *et al.* (1964, 1965), noting in earlier work with animals that EFA deficiency resulted in a high ratio of triene to tetraene fatty acids in several tissues of animals while the ratio in normal animals was low, produced the same effect in infants. A diet containing less than 0.1% of the calories from 18:2 fed to infants for a month or longer resulted in a serum triene-tetraene ratio of 1.5 or more. A diet with 1.3% or more of the calories from 18:2 fed to infants of the same age changed the ratio to 0.4. Using a curve-fitting procedure, the authors concluded that the minimal 18:2 requirement for infants was about 1% of the total calories. About 4% of total calories (430 mg/100 kcal) is considered to be the optimum intake based on data obtained by the investigators mentioned above.

Cuthbertson (1976) believes that the minimum EFA requirement is too high and suggests that a daily allowance of 0.6% of calories (65 mg/100 kcal) should be sufficient. The crux of his argument is that, based on later GLC determinations of the 18:2 in bovine milk fat (mean, 2.39%; range, 1.23 to 3.7%; Jensen et al. 1962), the alkaline isomerization method used by Hansen et al. (1963) overestimated the 18:2 content as 3.3%. Cuthbertson thinks that the true EFA content of bovine milk fat is 65 to 75% of the GLC values. He combines this belief with the observation that clinical EFA deficiency symptoms have not been seen in the United Kingdom, although the baby foods used are relatively low in EFA, to arrive at his lower minimal figure for infant EFA requirements. In the United States, the question of meeting minimal EFA requirements for infants is probably academic, because the use of bovine milk as the sole source of nutrients for infants has decreased markedly in recent years.

Cuthbertson is correct in stating that GLC as done previously overestimated the true 18:2 content of bovine milk fat, partly because the acid is a minor component (thus, the error of estimation is increased), and partly because the fat contains many geometric and positional isomers of 18:2 that are included in the 18:2 peak on the chromatogram. Also, polyunsaturated fatty acids are easily oxidized, and some of the acid-catalyzed methods used to prepare methyl esters for GLC analysis destroy some of these fatty acids. However, bovine milk fat contains both 18:3 and 20:4 that have EFA activity. The data of Smith et al. (1978), apparently obviate Cuthbertson's belief because the analyses were done on double bond fractions with GLC columns of high resolving power. The sum of cis, cis-18:2, 18:3ω6, 18:3ω3, 20:3ω6, and 20:4ω6 is 5.4%, considerably higher than all earlier estimates. All of these have real or potential EFA activity. The amounts found by Smith et al. may be somewhat lower than those actually existing in whole milk, as they were done on butter, which does not contain all of the phospholipids that were originally present in the globule membrane. It appears, however, that bovine milk contains much more EFA than was previously reported.

The presence of trans isomers in partially hydrogenated food fats has aroused concern (Emken 1983). If present, trans, trans 18:2 would dilute EFA activity, as the acid must have the cis, cis configuration and the trans, trans acid does affect several enzymes involved in the metabolism of polyunsaturates and other lipids. In any case, the amount of *trans, trans* 18:2 in milk fat is almost negligible. Smith *et al.* (1978) found none.

Still to be completely isolated are the large number of isomers that could result from positional and geometric isomerization of *cis, cis*-9, 12–18:2 and the other polyunsaturates. DeJong and Van der Wel (1964) and Van der Wel and deJong (1967) presented data on the position of the double bonds in nonconjugatable 18:2 isomers of milk lipids. The total amount of these acids was about 0.02%. Parodi (1977) detected *cis, trans (trans, cis)* 9,11–18:2 and illustrated one of the pitfalls of analysis of complex fatty acid esters by GLC, i.e., overlap of retention times. He found compounds having the same equivalent chain length as conjugated *trans, trans* 18:2 and conjugated *cis, trans* 18:2 on three different columns. Smith *et al.* (1978) obtained the contents of total conjugated fatty acids by ultraviolet spectrophotometry. These amounts, which averaged about 0.7%, would probably have been included in the quantities of 18:2 obtained by GLC, (average, 5.4%) but would not contribute to EFA activity. Therefore, the total EFA would be about 4.7%, still much higher than the quantities reported earlier. The values obtained by the lipoxidase method, which is specific for *cis, cis* methylene-interrupted fatty acids, are lower than the GLC percentages. The lipoxidase procedure has been considered the test of choice for those acids which include EFA, but it is known to be difficult (Madison and Hughes 1983). At present, the GLC method as done by Smith *et al.* (1978) appears to provide the most reliable data on the EFA content of milk.

Other Acids

Milk fat contains both keto (oxo) and hydroxy fatty acids, and earlier identifications are discussed by Jensen *et al.* (1967), Morrison (1970), and Kurtz (1974). In a more recent and careful study, Weihrauch *et al.* (1974) isolated 60 oxo acids from milk fat and positively and tentatively identified 47 with the aid of mass spectrometry. These data are presented in Table 4.9. About 85% (weight) of the oxo acids were stearates, mostly the 13-isomer, and 20% were palmitates, largely the 11-isomer. Of the unsaturated oxo acids, the 9-oxo, 12-ene, and 13-oxo, 9-ene were the predominant species. Other unsaturated oxo acids which are not listed in Table 4.9 but which were possibly present, are 15:1, 16:2, 17:1, 17:2, 17:3, 18:2, 18:3, 19:1, 19:2, and 20:1.

Hydroxy acids, 10:0–16:0, with the functional group in the 4 and 5 positions, as well as 12:1Δ6, 4-OH, and 12:1Δ9, 5-OH (Dimick *et al.*

Table 4.9. N-OXO Fatty Acids in Milk Fat.

Carbon Number	Position of Carbonyl
10:0	5
12:0	4, 5, 7
14:0	5, 6, 7, 9
15:0	4, 5
16:0	4–9, 11
17:0	8
18:0	5, 8–11, 13, 16
19:0	11
20:0	9, 11, 15
22:0	11–15
24:0	14, 15
	Position of Carbonyl and Double Bond
14:1	5 (Δ9), 5 (Δ10), 9 (Δ5)
16:1	7 (Δ10), 11 (Δ7), 11 (Δ9)
18:1	9 (Δ12), 9 (Δ13), 9 (Δ15), 13 (Δ7), 13 (Δ9)

SOURCE: Weirauch *et al.* (1974).

1970; Jensen *et al.* 1962), have been found in milk fat. These isomers convert readily to lactones, some of which are flavor compounds. Schwartz (1972), in a discussion of methods for the isolation of nonlactonegenic hydroxy fatty acids (OH group on carbons other than 4 or 5), mentioned that there were at least 60 acids in this fraction.

Schwartz (1972) also noted the detection of about 70 glycerol-l-alkyl ethers in milk fat. Saturated ethers, both odd and even from C-10 through C–18, were found, with traces of ethers up to C-25 present. Fifty-five unsaturated ethers were separated, but only the Δ-9, Δ-9, 12, and Δ-9, 12,15 compounds were tentatively identified. In addition, Schwartz isolated over 50 bound aldehydes probably derived from the glycerol-l-alkenyl ethers (phosphorus free).

Ellis and Wong (1975) identified γ and Δ lactones in butter, butter oil, and margarine and showed a correlation of the lactone content with time and temperature of heating.

Cyclohexylundecanoic acid has been isolated from bovine milk and characterized (Schogt and Haverkamp Begemann 1965). Brewington *et al.* (1974) found glucuronides of 17 milk fatty acids in bovine milk. These were presumably detoxification products formed in the liver and, interestingly, included the odd-chain acids, 9:0–17:0.

FACTORS INFLUENCING FATTY ACID COMPOSITION

Except for isolated circumstances, pooling and long-distance transportation of milk have eliminated or tempered many of the dietary and environmenal effects on the fatty acid composition of milk. The amounts of fatty acids vary with the season (ultimately, the diet). This subject has been reviewed by Christie (1980), who emphasizes that the amount of milk produced must be known to determine if changes in fatty acid composition are due to actual changes in milk lipid biosynthesis. Parodi (1974B) has analyzed the variation due to the stage of lactation in the fatty acids of milk fat from seven cows as compared to a herd. Butyric acid (4:0) had a maximum value during the first month of lactation, declining thereafter and becoming minimal at the end. Hexanoic (6:0) to 14:0 all had similar variations; the values increased during the first 4 to 8 weeks of lactation, remained relatively constant until the fifth or sixth month, and then decreased again until the end of lactation. There was little variation in 16:0 throughout lactation. Stearic (18:0) and 18:1 contents were high in early lactation, decreasing until mid-lactation and increasing again to the end of lactation. Changes in 18:2 and 18:3 contents were variable.

Parodi (1973C) determined the fatty acid compositions of the milk from two dairy herds which produced milk fat with different softening points, 30.4°C as compared to 38.4°C. The herd producing the soft milk fat had higher levels of short and medium chain length fatty acids and lower levels of 18:0 than the herd producing the hard milk fat. The only major difference in the diets was that the latter herd received brewer's grains as a significant part of its diet.

PROTECTED MILK

As we have mentioned, digestion of cellulose by rumen microorganisms enables the ruminant to convert foodstuffs indigestible by humans to high-quality protein. However, this advantage is offset to some extent by inefficient utilization of proteins and lipid. Another disadvantage is the biohydrogenation of polyunsaturated fatty acids, which decreases the concentrations of these acids in milk fat to 3–5%. Australian investigators found that a polyunsaturated oil encapsulated in sodium caseinate by spray drying, followed by a denaturation treatment with formaldehyde to prevent proteolysis of the protein in the rumen, was protected against ruminal hydrogenation. These and other investigations have been reviewed by Bitman (1976), Fogerty and Johnson (1980), and Storry *et al.* (1980). For example, the 18:2 content of milk fat from a cow fed protected particles of safflower oil was 35.2% compared to 2.0% for the control animal. Protected oils are hydrolyzed in the abomasum and the fatty acids are absorbed in the small intestine, thereby avoiding hydrogenation. Scott *et al.* (1971) reported the results of feeding protected corn and peanut oils to cows on the fatty acid composition of milk fat. The 14:0, 16:0, and 18:0 contents were reduced, while the amounts of 18:2 were increased about fivefold. Similar increases were observed in plasma and depot fats. Others have confirmed the findings of the Australian workers, also noting that the 18:2 content of cow's milk fat could be increased from 3% to 35% by feeding protected safflower oil. Thus, it is possible to increase biologically the polyunsaturated fatty acid content of milk fat.

The other reason for feeding protected fat is to increase the amount of fat digested by the cow in the abomasum. As an example Wrenn *et al.* (1978) fed protected tallow in amounts providing 18% of the digestible energy.

Since the amounts of fatty acids available for acylation during the biosynthesis of milk TGs affect their placement, feeding protected oils can be expected to alter the structure of the TGs. The data of Christie and Clapperton (1982) show that total and therefore all positional 18:2s are higher than in normal milk. Palmitic acid (16:0) decreases reciprocally.

Phospholipids

Some of the earlier data tabulated by Morrison (1970) on the fatty acid compositions of milk phosphatidylcholine, phosphatidylethanolamine, and sphingomyelin are shown in Table 4.10. Included are analyses by Boatman *et al.* (1969) on phosphatidylethanolamine and phosphatidylserine and by Bracco *et al.* (1972) on phosphatidylinositol. The differences in composition between the samples of phosphatidylethanolamine and -serine can be attributed primarily to differences in metabolism.

Morrison *et al.* (1965) reported the positional distribution of the fatty acids in phosphatidylethanolamine, -serine, and -choline. In contrast to the TGs, the phospholipids had no short chain acids and many more long chain unsaturates. There were more unsaturates in phosphatidylethanolamine than in -serine or -choline. The distribution of the acids between sn-1 and sn-2 is similar to that observed in other tissues, with

Table 4.10. Fatty Acid Composition (Mole %) of Various Bovine Milk Phospholipids.[a]

Fatty Acid	Phosphatidyl Ethanolamine		Phosphatidyl Choline	Sphingomyelin	Phosphatidyl Serine		Phosphatidyl Inositol
12:0	Trace	—[b]	0.7	0.3	3.6	1.6[b]	—[c]
14:0	1.5	1.0	8.4	2.5	12.5	5.2	4.7
15:0	0.5	—	2.1	0.4	—	—	1.3
16:0	11.7	11.0	36.4	22.1	31.7	15.0	29.8
16:1	2.1	1.1	0.6	0.8	—	—	—
17:0	0.9	—	0.9	0.6	—	—	—
18:0	10.5	13.0	11.1	4.5	13.0	30.0	31.8
18:1	46.7	61.0	25.7	5.0	32.9	38.0	10.8
18:2	12.4	12.0	5.3	0.9	4.9	7.3	6.9
18:3	3.4	2.1	1.1	—	—	3.2	2.5
20:3	1.4	—	1.0	—	—	—	—
20:4	1.9	—	0.7	—	—	—	—
22:0	—	—	—	14.7	—	—	3.9
23:0	—	—	—	27.0	—	—	—
24:0	—	—	—	14.8	—	—	—

SOURCES:
[a] Adapted from Morrison (1970); minor acids omitted.
[b] Adapted from Boatman et al. (1969); minor acids omitted.
[c] From Bracco et al. (1972); 3.8% 20:0 omitted.

199

saturates at sn-1 and unsaturates at sn-2. Monoenoic acids were distributed evenly except in phosphatidylserine, where more 18:1 was present at sn-2. Morrison *et al.* isolated the phospholipids from spray-dried buttermilk, which is a convenient source.

Hay and Morrison (1971) later presented additional data on the fatty acid composition and structure of milk phosphatidylethanolamine and -choline. Additionally, phytanic acid was found only in the 1-position of the two phospholipids. The steric hindrance presented by the four methyl branches apparently prevents acylation at the 2-position. The fairly even distribution of monoenoic acids between the two positions is altered when the *trans* isomers are considered, as a marked asymmetry appears with 18:1 between the 1- and 2-positions of phosphatidylethanolamine, but not of phosphatidylcholine. Biologically, the *trans* isomers are apparently handled the same as the equivalent saturates because the latter have almost the same distribution. There are no appreciable differences in distribution of *cis* or *trans* positional isomers between positions 1 and 2 in either phospholipid. Another structural asymmetry observed is where *cis, cis* nonconjugated 18:2s are located mostly in the 2-position in both phospholipids. It appears that one or more *trans* double bonds in the 18:2s hinders the acylation of these acids to the 2-position.

Hay and Morrison (1971) did not neglect the alkyl and alkenyl ethers in milk phospholipids, finding 4% of the latter in phosphatidylethanolamine and 1.3% in phosphatidylcholine. *Trans* isomers were not found. The authors postulated that the branched chain compounds in the alkenyl ethers were derived from rumen microbial lipids.

Kitchen (1977) has analyzed the fatty acids in the phospholipids isolated from the MFGM, finding more unsaturated and less saturated acids than in the membrane TG. These findings are not unexpected.

Barbano and Sherbon (1981) found that feeding cows a protected poly unsaturated fat supplement had little influence on the fatty acid composition of the milk phospholipids. The biosynthesis of the pools of fatty acids from which milk phospholipids are synthesized is apparently independent of dietary input.

Cho *et al.* (1977) has identified the fatty acids in the lysophosphatidyl-ethanolamines and -cholines from butter serum. The compositions were similar to those of the intact phospholipids.

Sphingolipids

Morrison (1970) presented earlier data on the fatty acid composition of these lipids. Morrison and Hay (1970) described the isolation and analyses of milk sphingomyelin, glucosylceramide, and lactosylceramide. The long chain bases were similar in all compounds, consisting of normal, iso, and anteiso saturated and unsaturated dihydroxy bases. The bases present in largest quantity were 18:1, 16:1, 17:1, 16:0, 18:0, iso 18:1, and iso 17:1, with many branched chain bases occurring in smaller amounts. The major fatty acids, both normal and 2-hydroxy, were usually 22:0, 23:0, and 24:0, with some variations. Hydroxy acids were observed to comprise less than 1% of the total acids. The *trans* acid contents of total sphingolipids were 43–51%, higher than in the corresponding milk fat, with the 18:1, 22:1, 23:1, 24:1, and 25:1 isomers present in sphingomyelin, glucosylceramide, and lactosylceramide. In sphingomyelin there was a trend toward high *trans* contents in 18:1 (94.2%) to lower amounts in 25:1 (7.1%). Morrison and Hay (1970) analyzed the *cis* and *trans* 23:1, 24:1, and 25:1 acids of sphingomyelin for positional isomers. The results show that the *cis* acids were similar to the *cis* 18:1s in milk fat but not the *trans* acids, with decreased amounts of Δ-9 isomers and much larger quantities of Δ-11. The latter is unusual but might be explained by the positional isomerization known to accompany elaidinization during hydrogenation.

Morrison (1969) presented data on the composition of the long chain bases in milk sphingomyelins. In this study he did not find saturated trihydroxy bases. Later Morrison (1973) concluded, after analyzing the long chain bases in the sphingolipids of bovine milk, kidney, and other sources, that the milk and kidney sphingomyelin bases were not of dietary origin. He further decided that bovine tissues synthesize straight and branched, saturated dihydroxy and trihydroxy long chain bases.

Huang (1973) analyzed the fatty acids of milk gangliosides, finding the following amounts (%): 14:0, 4.2; 16:0, 20.2; 16:1, 2.8; 18:0, 18.1; 18:1, 36.6; 18:2, 7.8; 20:0, 3.0; and 20:4, 6.1. Hydroxy acids were not detected. The sphingosine base contents (%) were: sphinganines: C-16, 10; C-18, 5; and sphingenines: C-16, 20 and C-18, 32. Several branched bases were also noted but were not further identified. The composition of the gangliosides is quite different from that of milk sphingomyelin and other glycolipids, suggesting perhaps selectivity during biosynthesis.

Keenan (1974) isolated five gangliosides from milk and identified the acids, finding 14:0–24:0 even chain saturates, 18:1, 24:1, and 23:0. Palmitic acid (16:0) predominated.

Sterol Esters

Keenan and Patton (1970) isolated and identified the cholesterol esters from cow, sow, and goat milk and mammary tissue. The fatty acid composition of the esters from the cow is presented in Table 4.11. The au-

Table 4.11. Fatty Acid Composition of Cholesterol Esters from Bovine Milk.

Fatty Acid	Wt %	Fatty Acid	Wt %
10:0	2.9	15:1	2.6
10:1	0.3	16:0	26.9
12:0	4.1	16:1	11.9
12:1	0.2	17:0	Trace
13:0	Trace	17:1	ND
13:1	11.0	18:0	6.7
14:0	6.9	18:1	13.7
14:1	0.5	18:2	10.1
15:0	2.1		

SOURCE: Adapted from Keenan and Patton (1970).

thors commented that the concentrations of monounsaturated (other than 18:1) and odd-numbered fatty acids in the cholesterol esters were greater than those found in milk triacylglycerols. For example, only traces of 13:1 were found in the latter.

PHYSICAL PROPERTIES OF MILK FAT

Structure

Milk fat globules, most of which range from 1 to 5 μM in diameter, are covered by a loose network of bipolar compounds. These are phospholipids, proteins, diacylglycerides, and monoacylglycerides, and other surface active materials originating from the secreting cell or the milk. They are collectively designated the "milk fat globule membrane (MFGM)" and are discussed in Chapter 10. The discussion below is derived largely from the book by Mulder and Walstra (1974).

Fat in Milk Products

The fat obtained by different methods of processing varies in composition and therefore in properties (Table 4.12; Mulder and Walstra 1974). For example, the amount of phospholipid per 100 g of fat is much greater in buttermilk from 40% cream than in whole milk (21.6 vs. 0.9 g), and these phospholipids contain much more unsaturated fatty acids than the whole milk lipids.

In milk plasma, fat may be present as extremely small globules, water-soluble fatty acids and other lipids, water-dispersible lipids, or lipoprotein particles. The amount is small, 0.02–0.03%. Obviously, most of the lipid is TG in the core of the globules.

Table 4.12. Approximate Content of Lipids in Different Milk Products.

	Composition (%)			
Product	Total Fat	Phospholipids	Cholesterol	Free Fatty Acids
Milk	4	0.035	0.014	0.008
Separated milk	0.06	0.015	0.002	0.002
Cream	10	0.065	0.032	0.017
Cream	20	0.12	0.06	0.032
Cream	40	0.21	0.12	0.06
Buttermilk from 20% cream	0.4	0.07	0.007	0.002
Buttermilk from 40% cream	0.6	0.13	0.012	0.002

SOURCE: Mulder and Walstra (1974).

Summary of Physical Properties

Mulder and Walstra (1974) presented data for "liquid fat," which is synonymous with butter or the core fat of globules. Variations exist, but the causes are usually unknown (Mulder and Walstra 1974). The authors state that the thermal conductivity is about 4×10^{-4} cal/cm^{-1}/5^{1}/°9C^{-1} at room temperature and the specific heat of the liquid fat is about 0.5 cal/g^{-1}/°C^{-1}. The latter is temperature dependent. The electrical conductivity is less than 10^{-12} S/cm (mho/CM) and the dieletric constant is about 3.1.

The solubility of air in fat is 8.7 ml/100 g, of oxygen 2.8 ml/100 g; and of nitrogen 5.9 ml/100 g at room temperature and atmospheric pressure. Liquid fat in contact with air contains 0.004% oxygen. These values are related to the oxidative stability and effectiveness of the packaging of stored milk fat.

Crystallization Behavior of Milk Fat

Milk fat is liquid above 40°C and completely solid below −40°C. Between these extremes it is a mixture of crystals and oil, with the latter a continuous phase. The nature of crystallization is complex because of the large number of TGs present. The properties of milk fat are the average of the properties of the TGs, and not necessarily those of the esterified fatty acids.

Mulder and Walstra (1974) have compiled a list of the factors which influence the crystallization of milk fat. The amount of solid fat is directly affected, with considerable relevance to the isolation of milk fat, as per churning and the structure of butter.

Melting Point or Range

De Man *et al.* (1983) have reviewed the determination of melting points in fat products. They point out that although fats are customarily described as having a "melting point," it is more realistically a melting range, since fats are mixtures of mixed-acid TGs. The melting point of a fat is actually the end of the melting range.

De Man *et al.* (1983) compared several methods for determining the melting point using a variety of food fats including butter. (1) In the Mettler dropping point method a sample cup with a restricted hole in the bottom is filled with fat and, placed in an automatically heated furnace, and the falling of the first drop is detected photometrically. (2) In the falling ball (softening point) method, the fat is hardened in a test tube and a steel ball bearing is placed on the surface. The tube is heated, and the temperature at which the ball has fallen through half the height of the fat column is the softening point. (3) In the softening point or open capillary tube method, the fat is hardened in an open capillary tube, which is heated in a water bath. At a certain temperature, the fat rises in the tube. This is Method Cc 3–25 of the American Oil Chemists' Society (1960). (4) The slip point method is a capillary tube method similar to the softening point method. The authors also obtained melting points and curves by differential scanning calorimetry. They found that the reproducibilities of the Mettler dropping point and softening point were excellent, whereas that of the slip point was poor. The Mettler dropping point values were found to coincide with extra polated solid fat curves obtained with wide-line nuclear magnetic resonance for lard and margarines, but not for butter. In the case of butter, the Mettler dropping point was at a temperature where about 2.5% solid fat remained. Ideally, all fat should be liquid at the melting point.

The commercial practice of pooling milk should eliminate all but seasonal effects on the melting point of fats. However, as an example of the influence of different feeding practices on the softening point of milk fat, Parodi (1973C) found that the average softening points of the milk fats were: soft, 30.4°C and hard, 38.4°C. These were caused by changes in the fatty acid composition and their distribution in the TGs as altered by different feeds.

Parodi (1981) separated milk fat selected for softening point range into the TG classes and found that the softening point range was 31.3° to 35.0°C. The softening point correlated best with some low and high molecular weight TGs of the total fat and of the trisaturated TGs. Interesterification or randomization of the esters on the TGs raised the softening point from 31.6° to 36.3°C. by increasing the amounts of high molecular weight TGs.

CONSISTENCY

De Man (1983) has reviewed this property of fats. Consistency is defined as (1) an ill-defined and subjectively assessable characteristic of a material that depends on the complex stress–flow relation or as (2) the property by which a material resists change of shape. "Spreadability," a term used in relation to consistency, is the force required to spread the fat with a knife. The definition is similar to that for "hardness": the resistance of the surface of a body to deformation. The most widely used simple compression test in North America is the cone penetrometer method (AOCS Method Cc 16–60, 1960). More sophisticated rheological procedures are also available. Efforts have been made to calibrate instrumental tests with sensory response. With the cone penetrometer method, penetration depth is used as a measure of firmness. Hayakawa and De Man (1982) studied the hardness of fractions obtained by crystallization of milk fat. Hardness values obtained with a constant speed penetrometer reflected trends in their TG composition and solid fat content.

REFERENCES

Ackman, R. G. 1980. Potential for more efficient methods for lipid analysis. *J. Am. Oil Chem. Soc.* 57, 821A–829A.

Ackman, R. G., Hooper, S. N. and Hansen, R. P. 1972. Some monomethyl branched fatty acids: Open tubular GLC separations and indications of substitution on even numbered carbons. *Lipids* 7, 683–691.

Ahrné, L., Björck, L., Raznikiewicz, T. and Claesson, O. 1980. Glycerol ether in colostrum and milk from cow, goat, pig, and sheep. *J. Dairy Sco.* 63, 741–745.

American Oil Chemist's Society. 1960. *In: Official and Tentative Methods.* Additions and revisions Cc 4–25 and Cc 16–60, American Oil Chemist's Society, Chicago.

American Oil Chemist's Society. 1965. Official and tentative methods of the American Oil Chemist's Society. Method Co. 12–55. American Oil Chemist's Society, Chicago.

Anderson, B. A., Sutton, C. A. and Pallansch, M. J. 1970. Optical activity of butterfat and vegetable oils. *J. Am. Oil Chem. Soc.* 47, 15–16.

Association of Official Analytical Chemists. 1980. *In: Official Methods of Analysis,* 13th ed. W. Horwitz, (Editor). Association of Official Anaytical Chemists, Washington, D. C., pp. 452–456.

Bachman, K. C. and Wilcox, C. J. 1976. Factors that influence milk cholesterol and lipid phosphorus: Content and distribution. *J. Dairy Sci.* 59, 1381–1387.

Barbano, D. M. and Sherbon, J. W. 1975. Stereospecific analysis of high melting triglycerides of bovine milk fat and their biosynthetic origin. *J. Dairy Sci.* 58, 1–8.

Barbano, D. M. and Sherbon, J. W. 1981. Polyunsaturated protected lipid: Effect on milk phospholipids. *J. Dairy Sci.* 64, 2170–2174.

Bauman, D. E. and Davis, C. L. 1974. Biosynthesis of milk fat. *In: Lactation: A Comprehensive Treatise.* B.L. Larson and V.R. Smith (Editors). Academic Press, New York, pp. 31–75.

Baumrucker, C. R. and Keenan, T. W. 1973. Membranes of mammary gland. VII. Stability of milk fat globule membrane in secreted milk. *J. Dairy Sci. 56*, 1092–1094.

Bickerstaffe, R. and Annison, E. F. 1971. Triglyceride synthesis in goat and sow mammary tissue. *Int. J. Biochem. 2*, 153–162.

Bitman, J. 1976. Status report on the alteration of fatty acid and sterol composition in lipids in meat, milk, and eggs. *In: Fat Content and Composition of Animal Products.* National Academy of Sciences, Washington, D. C., pp. 200–237.

Boatman, V. E., Patton, S. and Parsons, J. G. 1969. Phosphatidyl serine of bovine milk. *J. Dairy Sci. 52*, 256–258.

Bracco, U., Hidalgo, J. and Bohren, H. 1972. Lipid composition of the milk fat globule membrane of human and bovine milk. *J. Dairy Sci. 55*, 165–172.

Breckenridge, W. C. 1978. Stereospecific analysis of triacylglycerols. *In: Handbook of Lipid Research, Vol. I: Fatty Acids and Glycerides.* A. Kuksis (Editor). Plenum Press, New York; pp. 197–232.

Breckenridge, W. C. and Kuksis, A. 1967. Molecular weight distributions of milk fat triglycerides from seven species. *J. Lipid Res. 8*, 473–478.

Breckenridge, W. C. and Kuksis, A. 1968. Specific distribution of short chain fatty acids in molecular distillates of bovine milk fat. *J. Lipid Res. 9*, 388–393.

Brewington, C. R., Caress, E. A. and Schwartz, D. P. 1970. Isolation and identification of new constituents in milk fat. *J. Lipid Res. 11*, 355–361.

Brewington, C. R., Parks, O. W. and Schwartz, D. P. 1974. Conjugated compounds in cow's milk. II. *J. Agr. Food Chem. 22*, 293–294.

Bus, J., Luk, C. M. and Gruenewegen, A. 1976. Determination of enantiomeric purity of glycerides with a chiral PMR shift reagent. *Chem. Phys. Lipids 16*, 123–132.

Cerbulis, J. 1967. Distribution of lipids in various fractions of cows' milk. *J. Agr. Food Chem. 15*, 784–786.

Chen, C.C.W., Agroudelis, C. J. and Tobias, J. 1978. Evidence for lack of phosphatidic acid and phospholipase activity in milk. *J. Dairy Sci. 61*, 1691–1695.

Cho, B.H.S., Irvine, D. M. and Rattray, J.B.M. 1977. Identification of positional isomers of lysophosphatides in butter serum. *Lipids 12*, 983–988.

Christie, W. W. 1979A. The composition, structure and function of lipids in the tissues of ruminant animals. *Prog. Lipid Res. 17*, 111–205.

Christie, W. W. 1979B. Effects of diet and other factors on the lipid composition of ruminant tissues and milk. *Prog. Lipid Res. 17*, 245–277.

Christie, W. W. 1980. The effects of diet and other factors on the lipid composition of ruminant tissues and milk. *Prog. Lipid Res. 17*, 245–277.

Christie, W. W. and Clapperton, J. L. 1982. Structures of the triglycerides of cows' milk fortified milks (including infant formulae), and human milk. *J. Soc. Dairy Technol. 35*, 22–24.

Christopherson, S. W. and Glass, R. L. 1969. Preparation of milk fat methyl esters by alcoholysis in an essentially nonalcoholic solution. *J. Dairy Sci. 52*, 1289–1290.

Clark, B. and Hübscher, G. 1961. Biosynthesis of glycerides in subcellular fractions of intestinal mucosa. *Biochim. Biophys. Acta 46*, 479–494.

Cuthbertson, W.F.J. 1976. Essential fatty acid requirements in infancy. *Am. J. Clin. Nutr. 29*, 559–568.

De Jong, K. and Van Der Wel, H. 1964. Identification of some iso-linoleic acids occurring in butter fat. *Nature 202*, 556–560.

De Man, J. M. 1983. Consistency of fats: A review. *J. Am. Oil Chem. Soc. 60*, 82–87.

De Man, L. and De Man, J. M. 1983. Trans fatty acids in milk fat. *J. Am. Oil Chem. Soc. 60*, 1095–1098.

De Man, J., De Man, L. and Blackman, B. 1983. Melting-point determination of fat products. *J. Am. Oil Chem. Soc. 60*, 91–94.

Dimenna, G. P. and Emery, R. S. 1981. Palmitate and octanoate metabolism in bovine mammary tissue. *J. Dairy Sci. 64*, 132–134.

Dimick, P. S., McCarthy, R. D. and Patton, S. 1965. Structure and synthesis of milk fat. VIII. Unique positioning of palmitic acid in milk fat triglycerides. *J. Dairy Sci. 48*, 735–737.

Dimick, P. S., McCarthy, R. D. and Patton, S. 1970. Milk fat synthesis. *In: Physiology of Digestion and Metabolism in Ruminant*, Ed. A.T. Phillipson (Editor). Oriel Press, Newcastle on Tyne, p. 534.

Duin, H. van. 1958. Investigation into the carbonyl compounds in butter. III. Phosphatide-bound aldehydes. *Neth. Milk Dairy J. 12*, 90–95.

Egge, H., Murawski, U., Ryhage, R., Gyorgy, P., Chatranon, W. and Zilliken, F. 1972. Minor constituents of human milk. IV: Analysis of the branched chain fatty acids. *Chem. Phys. Lipids 8*, 42–55.

Ellis, R. and Wong, N. P. 1975. Lactones in butter, butter oil and margarine. *J. Am. Oil Chem. Soc. 52*, 252–255.

Emken, E. A. 1983. Biochemistry of unsaturated fatty acid isomer. *J. Am. Oil Chem. Soc. 60*, 995–1004.

Feeley, R. M., Criner, P. E. and Slover, H. T. 1975. Major fatty acids and proximate composition of dairy products. *J. Am. Diet. Assn. 66*, 140–146.

Flanagan, V. P. and Ferretti, A. 1973. Hydrocarbons and polychlorinated biphenyls from the unsaponifiable fraction of anhydrous milk fat. *J. Lipid Res. 14*, 306–311.

Flanagan, V. P. and Ferretti, A. 1974. Characterization of two steroidal olefins in nonfat dry milk. *Lipids 9*, 471–475.

Flanagan, V. P., Ferretti, A., Schwartz, D. P. and Ruth, J. M. 1975. Characterization of two steroidal ketones and two isoprenoid alcohols in dairy products. *J. Lipid Res. 16*, 97–101.

Fogerty, A. C. and Johnson, A. R. 1980. Influence of nutritional factors on the yield and content of milk fat: Protected polyunsaturated fat in the diet. *Bull. Int. Dairy Fed. 125*, 96–104.

Freudenstein, C., Keenan, T. W., Eigel, W. N., Sasaki, M., Stadler, J. and Franke, W. W. 1979. Preparation and characterization of the inner coat material associated with fat globule membranes from bovine and human milk. *Exp. Cell Res. 118*, 277–294.

Fujino, Y. and Fujishima, T. 1972. Nature of ceramide in bovine milk. *J. Dairy Res. 39*, 11–14.

Gentner, P. R., Bauer, M. and Dietrich, I. 1981. Separation of major phospholipid classes of milk without previous isolation. *J. Chromatogr. 206*, 200–204.

Gentner, P. R. and Haasemen, A. 1979. Method for the determination of cholesterol in milk samples by application of a commercially available enzymatic test kit. *Milchwisseasch 34*, 344–346.

Grunnet, I. and Knudsen, J. 1978. Medium chain acyl-thioester hydrolase activity in goat and rabbit mammary fatty acid synthetase complexes. *Biochem. Biophys. Res. Commun. 80*, 745–749.

Grunnet, I. and Knudsen, J. 1979. Fatty-acid synthesis in lactating goat mammary gland. I. Medium chain fatty acid synthesis. *Eur. J. Biochem. 95*, 497–502.

Grunnet, I. and Knudsen, J. 1981. Direct transfer of fatty acids synthesized de novo from fatty acid synthetase into triacylglycerols without activation. *Biochem. Biophys. Res. Commun. 100*, 629–636.

Hallgren, B. and Larsson, S. 1962. The glyceryl ethers in man and cow. *J. Lipid Res. 3*, 39–42.

Hallgren, B., Niklasson, A., Stallberg, G. and Thorin, B. 1974. On the occurrence of 1-0-alkyglycerols and 1-0-(2-methoxyalkyl) glycerols in human colostrum, human milk, cow's milk, sheep's milk, human red bone marrow, red cells, blood plasma and a uterine carcinoma. *Acta Chem. Scand. 28B*, 1029–1034.

Hansen, A. E., Haggard, M. E., Borlsche, A. N., Adam, D.J.D. and Wiese, H. F. 1958. Essential fatty acid in infant nutrition. III. Clinical manifestations of linoleic acid deficiency. *J. Nutr. 66*, 565–576.

Hansen, R. P. and Shorland, F. B. 1962. Seasonal variations in fatty acid composition of New Zealand butter fat. *Biochem. J. 52*, 207–216.

Hansen, A. E., Wiese, H. F., Boelsche, A. N., Haggard, M. E., Adam, D.J.D. and Davis, H. 1963. Role of linoleic acid in infant nutrition. Clinical and chemical study of 428 infants fed on milk mixtures varying in kind and amount of fat. *Pediatrics 31*, 171–192.

Hay, J. D. and Morrison, W. R. 1970. Isomeric monoenoic fatty acids in bovine milk fat. *Biochim. Biophys. Acta 202*, 237–243.

Hay, J. D. and Morrison, W. R. 1971. Polar lipids in bovine milk. III. Isomeric *cis* and *trans* monoenoic and dienoic fatty acids, and alkyl and alkenyl ethers in phosphatidyl choline and phosphatidyl ethanolamine. *Biochim. Biophys. Acta 248*, 71–79.

Hayakawa, H. and De Man, J. M. 1982. Consistency of fractionated milk fat as measured by two penetration methods. *J. Dairy Sci. 65*, 1095–1101.

Holman, R. T. 1973. Essential fatty acid deficiency in humans. *In: Dietary Lipids and Postnatal Development.* C. Galli, G. Jacini, and A. Pecile (Editors). Raven Press, New York, p. 127.

Holman, R. T., Caster, W. O. and Wiese, H. F. 1964. The essential fatty acid requirement of infants and the assessment of their dietary intake of linoleate by serum fatty acid analysis. *Am. J. Clin. Nutr. 14*, 70–75.

Holman, R. T., Hayes, H. W., Rinne, A. and Soderjhelm, L. 1965. Polyunsaturated fatty acids in serum of infants fed breast milk or cow's milk. *Acta Paediatr. Scand. 54*, 573–577.

Huang, R.T.C. 1973. Isolation and characterization of the gangliosides of butter milk. *Biochim. Biophys. Acta 306*, 82–84.

Iverson, J. L. 1983. Personal communication, Washington, D.C.

Iverson, J. L., Eisner, J. and Firestone, D. 1965. Detection of trace fatty acids in fats and oils by urea fractionation and gas-liquid chromatography. *J. Am. Oil Chem. Soc. 42*, 1063–1068.

Iverson, J. L. and Sheppard, A. J. 1977. Butyl ester preparation for gas-liquid chromatographic determination of fatty acids in butter. *J. Assn. Off. Anal. Chem. 60*, 284–288.

James, A. T. and Martin, A. J. P. 1956. Gas-liquid chromatography: The separation and identification of the methyl esters of saturated and unsaturated acids from formic to *n*-octadecanoic acid. *Biochem. J. 63*, 144–152.

Jenness, R. and Patton, S. 1959. Milk lipides. *In: Principles of Dairy Chemistry.* John Wiley & Sons, New York, pp. 31–72.

Jensen, R. G., Dejong, F. A. and Clark, R. M. 1983. Determination of lipase specificity. *Lipids 18*, 239–252.

Jensen, R. G., Gander, G. W. and Sampugna, J. 1962. Fatty acid composition of the lipids from pooled raw milk. *J. Dairy Sci. 45*, 329–331.

Jensen, R. G., Quinn, J. G., Carpenter, D. L. and Sampugna, J. 1967. Gas-liquid chromatographic analysis of milk fatty acids: A Review. *J. Dairy Sci. 50*, 19–34.

Jensen, R. G., Sampugna, J. and Gander, G. W. 1961. The fatty acid composition of diglycerides from lipolyzed milk fat. *J. Dairy Sci. 44*, 1983–1988.

Kayser, S. G. and Patton, S. 1970. The function of very long chain fatty acids in membrane structure: Evidence from milk cerebrosides. *Biochem. Biophy. Res. Commun. 41*, 1572–1578.

Keenan, T. W. and Patton, S. 1970. Cholesterol esters of milk and mammary tissue. *Lipids 5*, 42–48.

Keenan, T. W. 1974. Composition and synthesis of gangliosides in mammary gland and milk of the bovine. *Biochim. Biophys. Acta 337*, 255–270.

Kinsella, J. E. and Infante, J. P. 1978. Phospholipid synthesis in the mammary gland. *In: Lactation: A Comprehensive Treatise*, Vol. 4. B.L. Larson, (Editor). Academic Press, New York, pp. 475–502.

Kitchen, B. J. 1974. A comparison of the properties of membranes isolated from bovine skim milk and cream. *Biochim. Biophys. Acta 356*, 257–269.

Kitchen, B. J. 1977. Fractionation and characterization of the membranes from bovine milk globules. *J. Dairy Res. 44*, 469–482.

Knudsen, J., Clark, S. and Dils, R. 1976. Purification and some properties of a medium chain hydrolase from lactating-rabbit mammary gland which terminates chain elongation in fatty acid synthesis. *Biochem. J. 160*, 683–691.

Knudsen, J. and Grunnet I. 1980. Primer specificity of mammalian mammary gland fatty acid synthetases. *Biochem. Biophys. Res. Commun. 95*, 1808–1814.

Knudsen, J. and Grunnet I. 1982. Transacylation as a chain-termination mechanism in fatty acid synthesis by mammalian fatty acid synthetase. *Biochem J. 202*, 139–143.

Knudsen, J., Grunnet, I. and Dils, R. 1981. Medium-chain fatty acyl-s-4'-phosphopantetheine fatty acid synthetase thioester hydrolase from lactating rabbit and goat mammary glands. *In: Methods in Enzymology*, Vol. 710. J.J. Lowenstein (Editor). Academic Press, New York, pp. 200–229.

Kuksis, A. 1972. Newer developments in determination of structure of glycerides and phosphoglycerides. *In: Progress in the Chemistry of Fats and Other Lipids*, Vol. 12. R.T. Holman (Editor). Pergamon Press, New York, p. 82.

Kuksis, A., Marai, L. and Myher, J. J. 1973. Triglyceride structure of milk fats. *J. Am. Oil Chem. Soc. 50*, 193–201.

Kumar, S., Singh, V. N. and Keren-Paz, R. 1965. Biosynthesis of short-chain fatty acids in lactating mammary supernatant. *Biochem. Biophys. Acta 98*, 221–229.

Kurtz, F. E. 1974. The lipids of milk: Composition and properties. *In: Fundamentals of Dairy Chemistry*. B.H. Webb, A. H. Johnson and J.A. Alford, (Editors). AVI Publishing Co., Westport, Conn., pp. 125–219.

LaCroix, D. E., Mattingly, W. A., Wong, N. P. and Alford, J. A. 1973. Cholesterol, fat and protein in dairy products. *J. Am. Diet. Assn. 62*, 275–279.

Libertini, L. J. and Smith S. 1978. Purification and properties of thioesterase from lactating rat mammary gland which modifies the product specificity of fatty acid synthesis. *J. Biol. Chem. 253*, 1398.

Lin, C. Y. and Kumar, S. 1972. Pathway for the synthesis of fatty acids in mammalian tissues. *J. Biol. Chem. 247*, 604–606.

Lin, C. Y. and Smith, S. 1978. Properties of the thioesterase component obtained by limited trypsinization of fatty acid synthetase multienzyme complex. *J. Biol. Chem. 253*, 1954–1962.

Lin, C. Y., Smith, S. and Abraham, S. 1976. Acyl specificity in triglyceride synthesis by lactating rat mammary gland. *J. Lipid Res. 17*, 647–656.

Lok, C. M. 1979. Identification of chiral 1,2-diacylglycerols in fresh milk fat. *Rec. Trav. Chim. 98*, 92–95.

Long, C. A. and Patton, S. 1978. Formation of intracellular fat droplets: Interrelation of newly synthesized phosphatidylcholine and triglyceride in milk. *J. Dairy Sci.* 61:1392–1399.

Lough, A. K. 1977. The phytanic acid content of the lipids of bovine tissues and milk. *Lipids 12*, 115–119.

Luick, J. R. and Kamoeka, K. K. 1966. Direct incorporation of β-hydroxybutyric acid into milk fat butyric and hexanoic acids in vivo. *J. Dairy Sci. 49*, 98–99.

Madison, B. L. and Hughes, W. J. 1983. Improved lipoxygenase method for measuring *cis, cis*-methylene interrupted polyunsaturated fatty acids in fats and oils. *J. Assoc. Anal. Chem. 66*, 81–84.

Marshall, M. O. and Knudsen, J. 1980. Factors influencing the in vitro activity of diacylglycerol acyltransferase from bovine mammary gland and liver towards butyryl-CoA and palmitoyl-CoA. *Biochim. Biophys. Acta 617*, 393–397.

Massart-Leen, A. M., DePooter, H., DeCloedt, M. and Schamp, N. 1981. Composition and variability of the branched-chain fatty acid fraction in the milk of goats and cows. *Lipids 16*, 286–292.

Maxwell, R. J. and Schwartz, D. P. 1979. A rapid, quantitative procedure for measuring the unsaponifiable matter from animal, marine, and plant oils. *J. Am. Oil chem. Soc. 56*, 634–636.

McCarthy, R. D. and Coccodrilli, G. D. 1975. Structure and synthesis of milk fat. XI. Effects of heparin on paths of incorporation of glucose and palmitic acid into milk fat. *J. Dairy Sci. 58*, 164–168.

McCarthy, S. and Smith, G. H. 1972. Synthesis of milk from β-hydroxybutyrate and acetate by ruminant mammary tissue in vitro. *Biochim. Biophys. Acta 260*, 185–196.

McPherson, A. V. and Kitchen, B. J. 1983. Review of the progress of dairy science: The bovine milk fat globule membrane—its formation, composition, structure and behavior in milk and dairy products. *J. Dairy Res. 50*, 107–133.

Mincione, B., Spagna Musso, S. and De Franciscus, G. 1977. Studies on milk from different species. Sterol content in cow's milk. *Milchwissensch 32*, 599–603.

Moore, J. H. and Christie, W. W. 1978. Lipid metabolism in the mammary gland of ruminant animals. *Prog. Lipid Res. 17*, 347–395.

Morrison, W. R. 1969. Polar lipids in bovine milk. I. Long-chain bases in sphingomyelin. *Biochim. Biophys. Acta 176*, 537–546.

Morrison, W. I. 1970. Milk lipids. *In: Topics in Lipid Chemistry*, Vol. 1. F.D. Gunstone (Editor). Logos Press, Ltd., London, pp. 51–106.

Morrison, W. R. 1973. Long-chain bases in the sphingolipids of bovine milk and kidney, rumen bacteria, rumen protozoa, hay and concentrate. *Biochim. Biophys. Acta 316*, 98–107.

Morrison, W. R. and Hay, J. D. 1970. Polar lipids in bovine milk. II. Long-chain bases, normal and 2-hydroxy fatty acids, and isomeric *cis* and *trans* monoenoic fatty acids in the sphingolipids. *Biochim. Biophys. Acta 202*, 460–467.

Morrison, W. R., Jack, E. L. and Smith, L. M. 1965. Fatty acids of bovine milk glycolipids and phospholipids and their specific distribution in the diacylglycerophospholipids. *J. Am. Oil Chem. Soc. 42*, 1142–1147.

Mulder, H. and Walstra, P. 1974. *In: The Milk Fat Globule*. Commonwealth Agricultural Bureaux, Furnham Royal, Bucks, England.

Myher, J. J. and Kuksis, A. 1979. Stereospecific analysis of triacylglycerols via racemic phosphatidylcholines and phospholipase C. *Can. J. Biochem. 57*, 117–124.

Palmquist, D. L., Davis, C. L., Brown, R. E. and Sachan, D. S. 1969. Availability and metabolism of various substrates in ruminants. V. Entry rate into the body and incorporation into milk fat of D(-)β-hydroxybutyrate. *J. Dairy Sci. 52*, 633–639.

Parks, O. W. 1980. Cholesterol esters in skim milk. *J. Dairy Sci. 63*, 295–297.

Parks, O. W., Keeney, M. and Schwartz, D. P. 1961. Bound aldehydes in butter oil. *J. Dairy Sci. 44*, 1940–1943.

Parodi, P. W. 1970. Fatty acid composition of Australian butter and milk fats. *Aust. J. Dairy Technol. 25*, 200–205.

Parodi, P. W. 1973A. Detection of synthetic and adulterated butter fat. 4. GLC trigylcer-
ide values. *Aust. J. Dairy Sci. 28*, 38–41.

Parodi, P. W. 1973B. The sterol content of milk fat, animal fats, margarines and vegetable oils. *Aust. J. Dairy Sci. 28*, 135–137.

Parodi, P. W. 1973C. The production throughout a year of soft and hard milk fat by two
dairy herds. *Aust. J. Dairy Technol. 28*, 80–83.

Parodi, P. W. 1974A. The composition of a high melting glyceride fraction from milk fat.
Aust. J. Dairy Sci. 29, 20–22.

Parodi, P. W. 1974B. Variation in the fatty acid composition of milk fat: Effect of stage
of lactation. *Aust. J. Dairy Technol. 24*, 145–148.

Parodi, P. W. 1975. Detection of aceto-diacylglycerols in milk fat lipids by thin-layer
chromatography. *J. Chromatogr. 111*, 223–226.

Parodi, P. W. 1976. Distribution of isomeric octadecenoic fatty acids in milk fat. *J. Dairy
Sci. 59*, 1870–1873.

Parodi, P. W. 1977. Conjugated octadecadienoic acids of milk fat. *J. Dairy Sci. 60*, 1550–
1553.

Parodi, P. W. 1979. Stereospecific distribution of fatty acids in bovine milk fat trigylcer-
ides. *J. Dairy Res. 46*, 75–81.

Parodi, P. W. 1981. Relationship between triglyceride structure and softening point of
milk fat. *J. Dairy Res. 48*, 131–138.

Parodi, P. W. 1982. Positional distribution of fatty acids in the triglyceride classes of
milk fat. *J. Dairy Res. 49*, 73–80.

Patton, S. and Keenan, T. W. 1971. The relationship of milk phospholipids to membranes
of secretory cell. *Lipids 6*, 58–62.

Patton S. and Keenan, T. W. 1975. The milk fat globule membrane. *Biochim. Biophys.
Acta 415*, 273–309.

Patton, S. and Jensen, R. G. 1976. *Biomedical Aspects of Lactation.* Pergamon Press,
New York.

Patton S., Hood, L. F. and Patton, J. S. 1969. Negligible release of cardiolipin during
milk secretion by the ruminant. *J. Lipid Res. 10*, 260–269.

Patton, S., Long, C. and Sokka, T. 1980. Effect of storing milk on cholesterol and phospholipid of skim milk. *J. Dairy Sci. 63*, 697–700.

Pfeffer, P. E., Sampugna, J., Schwartz, D. P. and Shoolery, J. N. 1977. Analytical [13]C
NMR: Detection, quantitation, and positional analysis of butyrate in butter oil.
Lipids 12, 869–871.

Pitas, R. E., Sampugna, J. and Jensen, R. G. 1967. Triglyceride structure of cow's milk
fat. I. Preliminary observations on the fatty acid composition of positions 1, 2, and
3. *J. Dairy Sci. 50*, 1332–1336.

Plantz, P. E. and Patton, S. 1973. Plasma membrane fragments in bovine and caprine
skim milks. *Biochem. Biophys. Acta 291*, 51–60.

Posati, L. P., Kinsella J.E. and Watt, B. K. 1975. Comprehensive evaluation of fatty
acids in foods. I. Dairy products. *J. Am. Diet. Assn. 66*, 482–489.

Posati, L. P. and Orr, M. L. 1976. *Composition of Foods, Dairy and Egg Products.* Agriculture Handbook 8-1, Agr. Res. Serv., USDA, Superintendent of Documents,
U. S. Govt. Printing Office, Washington, D.C.

Rao, G. A. and Abraham, S. 1975. Stimulatory effect of glucose upon triglyceride synthesis from acetate, decanoate, and palmitate by mammary gland slices from lactating mice. *Lipids 10*, 409–412.

Ristow, A. and Werner, H. 1968. Seasonal variation in the hydrocarbon content of milk
fat. *Fette Serifen. Anstrichm. 70*, 273–288.

Ryhage, R. 1967. Identification of fatty acids from butter fat using a combined gas
chromatograph mass spectrometer. *J. Dairy Res. 34*, 115–121.

Schogt, J. C. M. and Haverkamp Begemann, R. 1965. Isolation of II-cyclohexylundeco-noic acid from butter. J. Lipid Res. 6:466–470.

Schwartz, D. P. 1972. Methods for the isolation and characterization of trace components from milk fat. J. Am. Oil Chem. Soc. 49:312A, Abstr. 96.

Schwartz, D. P., Burgwald, L. H. and Brewington, C. R. 1966. A simple quantitative procedure for obtaining the unsaponifiable matter from butter oil. J. Am.. Oil Chem. Soc. 43, 472–473.

Scott, T. W., Cook, L. J. and Mills, S. C. 1971. Protection of dietary polyunsaturated fatty acids against microbial hydrogenation in ruminants. J. Am. Oil Chem. Soc. 48, 358–364.

Smith, G. H. and McCarthy, S. 1969. Synthesis of milk fat from β-hydroxybutyrate and acetate in mammary tissue in the cow. Biochem. Biophys. Acta 176, 664–666.

Smith, L. M., Dunkley, W. L., Franke, A. and Dairiki, T. 1978. Measurement of trans and other isomeric unsaturated fatty acids in butter and margarine. J. Am. Oil Chem. Soc. 55, 257–261.

Smith, S. 1976. Structural and functional relationships of fatty acid synthetases from various tissues and species. In: Immunochemistry of Enzymes and Their Antibodies. M.G.J. Salton, (Editor). John Wiley & Sons, New York, pp. 125–146.

Smith, S. 1980. Mechanism of chain length determination in biosynthesis of milk fatty acids. J. Dairy Sci. 63, 337–352.

Smith, S. 1981. Long-chain fatty acyl-s-4′-phosphopantetheine-fatty acid synthase thioester hydrolase from rat. In: Methods in Enzymology, Vol. 71. J.M. Lowenstein (Editor). Academic Press, New York, pp. 181–188.

Smith, S. and Abraham, S. 1971. Fatty acid synthetase from lactating rat mammary gland. Studies on the termination sequence. J. Biol. Chem. 246, 2537–2542.

Smith, S. and Abraham, S. 1975. The composition and biosynthesis of milk fat. Adv. Lipid Res. 13, 195–239.

Smith, S. and Ryan, P. 1979. Asynchronous appearance of two enzymes concerned with medium chain fatty acid synthesis in developing rat mammary gland. J. Biol. Chem. 254, 8932–8936.

Smith, S. and Stern, A. 1981. Development of the capacity of mouse mammary glands for medium chain fatty acid synthesis during pregnancy and lactation. Biochim. Biophys. Acta 664, 611–615.

Solderhjelm, L., Wiese, H. F. and Holman, R. T. 1970. Role of polyunsaturated fats in human nutrition and metabolism. Prog. Chem. Fats Lipids 9, 555–682.

Stein, O. and Stein, Y. 1971. Light and electron microscopic radioautography of lipids: Techniques and biological applications. Adv. Lipid Res. 9, 1–72.

Storry, J. E., Brumby, P. E. and Dunkley, W. L. 1980. Influence of nutritional factors on the yield and content of milk fat: Protected non-polyunsaturated fat in the diet. Bull. Int. Dairy Fed. 125, 105–125.

Strocchi, A. and Holman, R. T. 1971. Analysis of fatty acids of butter fat. Riv. Ital. Sostanze Grasse 48, 617–622.

Strong, C. R. and Dils, R. 1972. Fatty acids synthesized by mammary gland slices from lactating guinea pig and rabbit. Comp. Biochem. Physiol. 43B, 643–652.

Tanioka, H., Lin, C. Y., Smith, S. and Abraham, S. 1974. Acyl specificity in glyceride synthesis by lactating rat mammary gland. Lipids 9, 229–234.

Timmen, H. and Dimick, P. S. 1972. Structure and synthesis of milk fat. X. Characterization of the major hydroxy compounds of milk lipids. J. Dairy Sci. 55, 919–925.

Treiger, N. D. 1979. Investigation of milk fat sterols. Appl. Biochem. Microbiol. 15, 889–891.

Urbach, G. and Stark, W. 1975. The C-20 hydrocarbons of butter fat. J. Agr. Food Chem. 23, 20–24.

Van der Wel, H. and De Jong, K. 1967. Octadecadienoic acids in butter fat. II. Identification of some nonconjugated fatty acids. *Fette Seifen. Anstrichm. 64*, 277–279.

Weihrauch, J. L., Brewington, C. R. and Schwartz, D. P. 1974. Trace components in milk fat: Isolation and identification of oxofatty acids. *Lipids 9*, 883–890.

Weiss, S. B. and Kennedy, E. P. 1956. The enzymatic synthesis of triglycerides. *J. Am. Chem. Soc. 78*, 3550.

Wooding, F.B.P. 1974. Milk fat globule membrane material in skim milk. *J. Dairy Res. 41*, 331–337.

Wrenn, T. R., Bitman, J., Waterman, R. A., Weyant, J. R., Strozinski, L. L. and Hooven, N. W., Jr. 1978. Feeding protected and unprotected tallow to lactating cows. *J. Dairy Sci. 61*, 49–58.

5

Lipids of Milk: Deterioration

John L. Weihrauch

PART I. LIPOLYSIS AND RANCIDITY

Market milk and some products manufactured from milk sometimes possess a flavor described as "rancid". This term, as used in the dairy industry, denotes implicitly the flavor due to the accumulation of the proper concentrations and types of free fatty acids hydrolytically cleaved from milk fat under the catalytic influence of the lipases normally present in milk.

The development of a rancid flavor in milk and some other fluid products is usually undesirable and detracts from their market value. In contrast, the popularity of certain dairy products, notably some varieties of cheese, as well as some confectionery items containing milk as an ingredient, is thought to be partially due to the proper intensity of the rancid flavor. Hence, knowledge of the factors involved in the development of rancidity is of great practical importance to several industries.

The literature on the subject is quite large. The present review has been limited to milk lipases, but good reviews on this, other dairy products, milk esterases, and microorganisms are available (International Dairy Federation 1974, 1975, 1980; Shipe *et al.* 1978; Deeth and Fitz-Gerald 1976; Downey 1980A; Jensen and Pitas 1976; Shahani *et al.* 1980; Lawrence 1967; Kitchen 1971).

General

A "lipase" has been defined as an enzyme that hydrolyzes the esters from emulsified glycerides at an oil–water interface (Desnuelle 1961). This review adheres to this definition; as a consequence, investigations which involve water-soluble substrates or substrates containing an alcoholic moiety other than glycerol have not been included.

The flavor defect commonly referred to as "rancidity" or, more specifically, as "hydrolytic rancidity" is caused primarily by the presence in milk of a single enzyme which was proposed to be designated as "milk lipoprotein lipase" (Olivecrona 1980). There is no known physiological function for lipase in milk, and its presence has been ascribed to leakage from blood through the mammary tissues rather than to true secretion (Olivecrona 1980).

The increased use of tanks for the storage of raw milk on the farm between pickups has introduced the danger of potential off-flavor development caused by lipases that are produced by certain microorganisms (psychrotrophs) at low temperatures. The exocellular lipases of psychrotrophic bacteria are extremely heat resistant, and although the microorganisms are killed, the enzymes survive pasteurization and sterilization temperatures. Rancidity may become noticeable when cell counts exceed 10^6 or 10^7/ml. Downey (1975) has summarized the potential contribution of enzymes to the lipolysis of milk (Table 5.1).

Most, if not all, milks contain sufficient amounts of lipase to cause rancidity. However, in practice, lipolysis does not occur in milk because the substrate (triglycerides) and enzymes are well partitioned and a multiplicity of factors affect enzyme activity. Unlike most enzymatic reactions, lipolysis takes place at an oil–water interface. This rather unique situation gives rise to variables not ordinarily encountered in enzyme reactions. Factors such as the amount of surface area available, the permeability of the emulsion, the type of glyceride employed, the physical state of the substrate (complete solid, complete liquid, or liquid-solid), and the degree of agitation of the reaction medium must be taken into account for the results to be meaningful. Other variables common to all enzymatic reactions—such as pH, temperature, the presence of inhibitors and activators, the concentration of the enzyme and substrate, light, and the duration of the incubation period—will affect the activity and the subsequent interpretation of the results.

Enzymes are produced and elaborated by living cells—a fact that has prompted some investigations into the origin of milk lipases. It is only relatively recently that the synthesis of glycerides by milk lipases has been demonstrated (Koskinen *et al.* 1969; Luhtala 1969; Luhtala

Table 5.1. Contribution of Enzymes Present to Lipolysis of Milk.

Enzyme Activity	Contribution
A-type carboxylic ester hydrolaze	Negligible
Cholinesterase	Negligible
Acid lipase	Doubtful
Bacterial lipolytic enzymes	(Not critical unless counts exceed 10^6–10^7/ml)
Alkaline lipolytic enzyme(s)	Mainly responsible

SOURCE: Downey (1975).

et al. 1970A,B). Using tripalmitin isotopically labeled in both the glycerol and fatty acid moieties, Koskinen *et al.* (1969) demonstrated that glyceride synthesis occurs in freshly drawn milk and that synthesis and hydrolysis occur simultaneously (Luhtala 1969). Luhtala *et al.* (1970A) showed that intracellular enzymes isolated from homogenized somatic cells of milk are capable of synthesis and lipolysis of milk triglycerides. Downey (1980A) speculated that the synthetic activity of milk lipases may be involved in the leveling off of lipolysis over time and in the actual decrease in free fatty acid levels during the storage of lipolyzed milk. This synthetic activity is very labile, and significant loss of activity occurs in the mammary gland and on further storage at room temperature (McCarthy and Patton 1964). The effect of synthetic activity is most noticeable in fresh milk, as well as in colostrum and mastitic milk, both of which have high cell counts. In this line of investigation, it is of interest to note that Morton (1955) has shown that milk phosphatase is derived from mammary gland microsomes released into the milk during the normal secretory process.

Bovine blood serum is lipolytically active, but cows producing milk which goes rancid quickly do not have sera that are more lipolytically active than those producing normal milk. Leukocytes, which are present in large numbers in milk, are especially high in mastitic milk; they are the source of milk catalase but are apparently not the source of milk lipases (Nelson and Jezeski 1955).

The lipases of milk are apparently inactive in the udder and at the time of milking. Milk always contains relatively large proportions of unesterified fatty acids (Thomas *et al.* 1955A), but these may be left over from the metabolic pool.

Lipolysis has been classified as spontaneous or induced. This distinction is made because different measures have to be taken to correct the problem. "Induced lipolysis" is most frequently defined as lipolysis initiated in raw milk by some form of mechanical agitation. Traditionally, "spontaneous lipolysis" has been defined as lipolysis caused by the cooling of raw milk. The cooling requirement is no longer strictly adhered to, and lipolysis in raw milk is said to be spontaneous if rancidity develops without apparent mechanical agitation (Downey 1980A,B). The distinction between spontaneous and induced lipolysis is not always clear, and both may occur at the same time.

Farm Factors and Lipolysis

Spontaneous Rancidity. Studies have been undertaken to determine how widespread rancidity really is. Hemingway *et al.* (1970) examined 12 herds and reported that about 50% of the herd samples showed some initial rancidity and 21% of the samples from 15 cows were rancid. Differences in degree of rancidity were marked. Another report contended that 2 to 22% of cows in a herd produce milk which goes rancid quickly (Hileman and Courtney 1935). Milk which inherently possesses the quality of high susceptibility to rancidity has been variously termed "naturally rancid milk," "bitter milk of advanced lactation," "naturally active" or "naturally lipolytically active," "normally active," and "spontaneous" (Schwartz 1974). The last term has been more or less generally adopted in recent years. These various designations were introduced in an effort to distinguish such milk from "nonspontaneous" (normal) milk.

Lipolysis in freshly drawn milk normally proceeds at a very slow rate, even upon prolonged incubation, unless proper thermal or mechanical treatment is applied to the milk. This, of course, always occurs in practice, as raw, warm milk is never consumed in the market. It is through these necessary practices that lipolysis in normal milk is accelerated. As a consequence, milk may be made rancid either deliberately or accidentally. The so-called spontaneous type of milk needs no treatment. Cooling to 15 to 20°C when the milk is drawn or shortly afterward will hasten lipolysis (Tarassuk and Smith 1940). Once the milk has been cooled, lipolysis is not materially affected whether the milk is aged in the cold or rewarmed to 20°, 30°, or 37°C and aged at these temperatures. Lipolysis in normal milk is not accelerated to the same degree by cooling and aging.

The reason that rancidity is not more prevalent in market milk is due to the fortuitous fact that spontaneous rancidity can be prevented or reduced by mixing such milk within 1 hr after milking with four to five times its volume of normal milk (Tarassuk and Henderson 1942). Since usually only about one out of five cows in a herd produces spontaneous milk, this defect is almost automatically eliminated or reduced. It is clear, however, that farmers with only a few cows are likely to encounter spontaneously rancid milk during the lactation period.

The dilution of normal milk which has been activated by thermal or mechanical treatment does not diminish the activity of the lipases (Skean and Overcast 1961).

Feed. The cow's feed has been shown to be an important practical factor in influencing the susceptibility of the milk to rancidity. Feeding experiments and practical observations have demonstrated that green pasture decreases and dry feed increases the incidence of rancidity (Chen and Bates 1962). Rancidity is increased by feeding poor-quality rations at reduced levels (Gholson *et al.* 1966B), by abruptly lowering feed energy levels, as well as changing abruptly to normal feed levels

(Borges *et al.* 1974), and by feeding a high-carbohydrate diet (Kodgev and Rachev 1970). Astrup *et al.* (1980) observed increased blood serum and milk free fatty acid levels in cows on reduced rations. Cows receiving a 6% palmitic acid supplement had milk with increased free fatty acids and a rancid flavor. Myristic acid increased lipolysis to lesser extent, and stearic acid had no detectable effect. These researchers associated a depression of lipolytic activity with the feeding of rations containing protected rape seed oil. The reduced activity was linked to the high unsaturated fatty acid content of the oil. Abdel Hamid *et al.* (1977) reported higher lipolytic activity in buffalo milk when dry rations were fed; activity was higher in the first stage of milking than in the middle or strip phase.

Lactation. Individual cows maintained under identical conditions seem to vary markedly in the susceptibility of their milk to rancidity (Ortiz *et al.* 1970). An increased incidence of rancidity has also been associated with advanced lactation, particularly during long lactation periods (Bachmann 1961; Colmey *et al.* 1957; Dijkman and Schipper 1965). There are reports, however, which fail to show a correlation between rancidity and advanced lactation (Herrington and Krukovsky 1939; Salih and Anderson 1979A). There have been suggestions that the increased incidence of lipolysis during late lactation may be linked to the absence of pasture feeding or to other dietary changes (Jellema 1973). Ortiz *et al.* (1970) found that a negative correlation existed between the amount of milk produced and the acid degree value (ADV). They speculated that this was related to declining milk flow with advancing lactation. Murphy *et al.* (1979) found that lipase activity was higher in early than in late lactation; however, this difference did not affect free fatty acid development. They further reported higher free fatty acid levels in afternoon than in morning milk. Amounts of free fatty acids were positively related to the higher fat contents of afternoon milk.

Mastitis. Mastitis has been implicated in rancidity (Bachmann 1961; Guthrie and Herrington 1960; Tallamy and Randolph 1969; Tarassuk and Yaguchi 1958); according to Guthrie and Herrington (1960) and Tarassuk and Yaguchi (1958), it may be more important than late lactation. Luhtala and Antila (1968), however, found lower lipolytic activity in mastitic milks. They also reported that lipase activity was higher in foremilk than in strippings. Jurczak and Sciubisz (1981) observed a linear relationship between lipolysis and somatic cell counts to 1,400,000 cells/cm^3 with progressively decreasing lipolysis above this level. The highest concentration of free fatty acids (FFA) occurred when milk contained 800,000 cells/cm^3. In bulk milks, cell counts above 1,000,000 cells/cm^3 did not produce a rise in the FFA level; rather, a small depression was observed. In contrast, Salih and Anderson (1979B) observed no effect on lipase activity by high cell counts in milk. They suggested that further studies are needed to determine the relative importance and interrelationships of factors such as lipoprotein lipase activator, cell lipases, proteolytic enzymes, heparin-like substances, anions, and fat globule influence.

Estrous

The effect of the estrous period on rancidity has also been investigated. According to Wells et al. (1969), who studied lipase activity in the milk and blood of cows throughout their lactation period, the peak blood plasma lipase values occur about 24 hr before the onset of observed estrous. Changes in blood lipase activity were reflected and magnified in the milk, although it was noted that the increase in milk lipase level occurred 9 to 15 hr after it was observed in the blood. Bachmann (1961) also has indicated that hormonal disturbances are linked to rancidity. He differentiates between rancidity produced by cows in late lactation and rancidity due to hormonal disturbances on the basis of an increased in lipase concentration in the latter.

Pipeline Milkers and Farm Tanks

The increased use of pipeline milkers and farm tanks on dairy farms has coincided with a noticeable increase in rancidity (Gholson et al. 1966A; Herrington 1954; Shipe et al. 1980A; Richter 1981). About six times as much rancid milk has been reported from pipeline milkers as from nonpipeline systems (Johnson and Von Gunten 1962). The trouble has been traced to risers in the pipelines, that is, vertical sections connecting one pipeline to another at a higher level. Air leaking excessively into the milk lines primarily at the claw, teat cups, milk hose, and loose line joints causes considerable foaming of the warm, raw milk lifted in the risers under reduced pressure (Chen and Bates 1962). The formation of foam due to air agitation was found to be an important feature of the mechanism involved in the acceleration of lipolysis and the resultant appearance of a rancid flavor in milk from pipeline milkers. Optimal conditions for activation by air agitation appear to be foaming with the continuous mixing of foam and milk at temperatures that keep the milk fat liquid (Tarassuk and Frankel 1955). High inlets in holding tanks may produce excessive splashing and agitation. The addition to the tank of fresh warm milk may cause thermal activation.

A constant holding temperature of 4°C is essential, as an increase in temperature of only a few degrees may accelerate the growth of psychrotrophs (Muir et al. 1978).

Remedial measures that suppress foaming and agitation in pipeline milkers have been recommended. The use of pipeline located below the cow was reported to virtually eliminate rancidity or to significantly reduce the ADV, which is defined as ml N KOH required to neutralize the free fatty acids in 100 g fat (Gholson et al. 1966A). Shortening the main pipeline and minimizing the number of risers, joints, and sharp bends will also reduce foam formation and subsequent rancidity (Worstorff 1975; Barnard 1974, 1979A; Fleming 1980). Constant holding tank temperatures are maintained by precooling fresh milk in the piping system (Kirst 1980C). Zall and Chen (1981) have investigated the feasibility of heating raw milk to subpasteurization temperatures prior to storage in holding tanks on the farm as a measure for controlling the growth of psychrotrophs.

Distribution and Purification of Milk Lipases

Milk Lipoprotein Lipase.   Contrary to earlier reports that pointed to a multiplicity of lipases (Schwartz 1974), there is now overwhelming evidence that there is only one lipase in milk (International Dairy Federation 1974, 1975, 1980). This lipase is identical to the lipoprotein lipase in blood and represents a spillover from the mammary tissues (Downey 1975).

The milk lipoprotein lipase has been isolated from skim milk by affinity chromatography on heparin-Sepharose (Egelrud and Olivecrona 1972; Iverius and Ostlund-Lindqvist 1976; Kinnunen et al. 1976; Castberg et al. 1975A). Egelrud and Olivecrona (1972) purified the enzyme 5000- to 6,000-fold to more than 80% pure, as judged by gel electrophoresis. They reported an apparent molecular weight of 62,000 to 66,000. Kinnunen et al. (1976) reported a molecular weight of about 55,000, and Iverius and Ostlund-Lindqvist (1976) determined molecular weights of 48,300 and 50,800 under reducing conditions, and a buffer of physiological pH and ionic strength yielded a molecular weight of 96,000, which they believed was a dimer of presumably identical subunits. Molar solutions of sodium chloride inactivate the enzyme (Castberg et al. 1975A; Egelrud and Olivecrona 1973). Electrophoresis in urea or in sodium dodecyl sulfate polyacrylamide gels revealed one major component which stained for protein and carbohydrate (Egelrud and Olivecrona 1972). An antiserum against highly purified skim milk lipoprotein lipase caused total inhibition of milk lipoprotein lipase and tributyrate hydrolyzing activity in skim milk and extracts of lipid-free cream (Castberg *et al.* 1975A). Flynn and Fox (1980) have presented evidence that the enzyme purified by Fox and Tarassuk (1968) was the same enzyme purified by Egelrud and Olivecrona (1972).

Bovine milk contains 1–2 mg milk lipoprotein lipase per liter. Some milk lipoprotein lipase is in the cream fraction (Olivecrona 1980); however, practically all of the native enzyme is in the skim milk fraction, where about 90% is bound to casein micelles (Tarassuk and Frankel 1957); Downey and Andrews 1966; Downey and Murphy 1975; Gaffney and Harper. 1966; Harper *et al.* 1956A). About 10% of the enzyme is in the aqueous phase of milk (Downey 1975). The hydrophilic properties of the enzyme are confirmed by its interaction with heparin, a sulfonated polysaccharide with a highly negative charge (Olivecrona and Lindahl 1969; Olivecrona *et al.* 1971). Downey and Murphy (1975) have reviewed the literature and concluded that electrostatic interactions are mainly responsible for binding the enzyme to the casein micelles. However, compound interactions involving both electrostatic and hydrophilic interactions must be considered in explaining the binding of lipase to the various casein components of the micelles.

The binding of milk lipases to casein micelles apparently imparts some stability to the enzyme, for as purification progresses, the milk lipase becomes less stable, and more so as the concentration of casein decreases (Downey and Andrews 1966; Egelrud and Olivecrona 1972).

Lipase associated with the casein micelles in skim milk is not fully active, but both dilution and the addition of sodium chloride stimulate or restore activity, presumably by dissociating the micelle–lipase complex. Sodium chloride is an inhibitor of lipolysis, but the proper dilution and addition of this salt can elicit maximal activity (Downey and Andrews 1966).

Downey (1980) reasoned that although milk lipoprotein lipase is present in sufficient amounts to cause extensive hydrolysis and potential marked flavor impairment, this does not happen in practice for the following reasons: (1) the fat globule membrane separates the milk fat from the enzyme, whose activity is further diminished by (2) its occlusion by casein micelles (Downey and Murphy 1975) and by (3) the possible presence in milk of inhibitors of lipolysis (Deeth and Fitz-Gerald 1975). The presence in milk of activators and their relative concentration may also determine whether milk will be spontaneously rancid or not (Jellema 1975; Driessen and Stadhouders 1974A; Murphy *et al.* 1979; Anderson 1979).

Colostral Lipase. Driessen (1976) identified a lipase in bovine colostrum which is stable at pH 4.6, is bound to casein micelles, but is situated in the milk serum. Binding to heparin-Sepharose was weak, and its lipolytic activity was only partly inhibited by the antiserum against purified lipoprotein lipase from bovine milk. This colostral lipase is present only in the first three milkings after calving. From then on, only milk lipoprotein lipase is present. Driessen (1976) suggests that colostral lipase is a proenzyme of the bovine lipoprotein lipase. The work of Murphy *et al.* (1979) tends to support the report by Driessen.

Heat-Resistant Lipases.   The heat-resistant lipases and proteinases and their effects on the quality of dairy products have been reviewed (Cogan 1977, 1980). Several reports have linked the lipases from bacteria with the off-flavor development of market milk (Richter 1981; Shipe *et al.* 1980A; Barnard 1979B). The microflora developing in holding tanks at 4°C [and presumably in market milk stored at 40°F (Richter 1981)] may produce exocellular lipases and proteases that may survive ordinary pasteurization and sterilization temperatures. Rancidity of the cheese and gelation of UHT milk appear to be the major defects caused by the heat-resistant enzymes.

Muir *et al.* (1978) observed that small changes in storage temperature from 4–8°C have a significant effect on microorganism growth. They detected no lipolytic rancidity below a count of 5×10^6 colony-forming units/ml. Counts exceeding 10^6 to 10^7/ml are required in milk before microbial enzymes cause noticeable lipolysis. However, not all milks with high cell counts will develop rancidity (Muir *et al.* 1978). Milks of good microbial quality contain from 5×10^3 to $<10^5$ counts/ml (Downey 1975).

Microorganisms found in the microflora from holding tanks belong primarily to the genera *Pseudomonas, Alcaligenes, Enterobacter,* and *Achromobacter.* However, *Pseudomonas* predominates, and isolates from bulk milk show much more lipolytic and proteolytic activity than other psychrotrophs isolated (Stewart *et al.* 1975). Bacterial exocellular lipases have an optimum pH of 8.75, a relative optimum temperature at 37°C, and an absolute optimum temperature at 50°C (Driessen and Stadhouders 1974B). Kishonti (1975) reported two optimum temperatures at 30° and 55°C, respectively.

Proper sanitary procedures on the farm and in the processing plant, maintenance of a 4°C holding temperature, and reduced holding times before pasteurization have been proposed to control this problem in raw milk (Schipper 1975; Menger 1975). However, more research is needed to determine the role that lipases from microorganisms play in the flavor deterioration of market milk (Richter 1981; Cogan 1980; Stewart *et al.* 1975).

Human Milk Lipases.   Two lipases have been identified in human milk by Hernell and Olivecrona (1974A,B). One of these, lipoprotein lipase, is activated by serum; the other is stimulated by bile salts. The lipoprotein lipase has no apparent physiological function in the milk, and its presence has been ascribed to leakage from the mammary tissues (Hernell and Olivecrona 1974A). The bile-activated lipase, however, plays a significant role in the digestion of human milk fat (Hayasawa *et al.* 1974; Hernell 1975).

Studies by Hall *et al.* (1979) have shown that the bile-activated enzyme can be stimulated fifty-fold by freezing and thawing, by sonification, or by addition of bile salt; addition of glycine conjugate was four times more effective than addition of taurine conjugate. Studies on the kinetic and chemical characterization of the enzyme were performed by Wang (1981).

Goat Milk Lipoprotein Lipase. An investigation of lipolytic activity by Bjorke and Castberg (1976) has shown that goat's milk, like bovine milk, contains only one lipase, which is a lipoprotein lipase with characteristics very similar to those of bovine milk lipoprotein lipase. The extent of lipolysis is increased severalfold by homogenization, stirring, and temperature manipulation. Freezing of milk inhibits lipolysis. Marked variation in lipolysis and lipoprotein lipase activity was found among goats and among various samples from the same goat. Milks with strong goat flavor also exhibited increased lipolysis.

Activation of Lipases

Homogenization and Agitation. All methods of agitation of milk appear to increase the rate of lipolysis. The increased incidence of rancidity in pipeline milkers as opposed to conventional milking procedures due to foaming and agitation has already been discussed. Homogenization (a more violent form of agitation) of raw milk, when conducted at temperatures between 37.7° and 54.4°C, will render milk rancid within a very short time, in some cases in only a few minutes (Schwartz 1974). The length of time of homogenization as well, as the homogenization pressure (Nilsson and Willart 1960), influences subsequent lipase activity, lipolysis increasing, within limits, as the magnitude of these variables increases (Luhtala and Antila 1968; Nilsson and Willart 1960). Shipe and Senyk (1981) observed that holding times ranging from 16 to 24 sec did not affect lipolysis when the temperature was 74.4°C or higher, and varying homogenization pressures from 105 to 211 kg/cm^2 did not alter lipolysis significantly.

Other forms of agitation, including shaking raw milk containing liquid fat (Crowe 1955; Demott 1960; Sjostrom and Willart 1956), churning raw milk or cream, and pumping (Kirst 1980A,B) accelerate lipolysis. The severity of agitation and the temperature at which it is conducted are of prime importance. Kitchen and Aston (1970) observed maximum activation at an agitation temperature of 37°C. Activation declined markedly at 50°C.

Foaming due to agitation also promotes lipolysis, but the increased activity in foam is probably independent of the accelerated lipolysis due to agitation. The kind of gas entrenched in the foam is of no consequence (Fitz-Gerald 1974).

According to Tarassuk and Frankel (1955), foaming promotes lipolysis by providing (1) greatly increased surface area, (2) selective concentration of enzyme at the air–liquid interface, (3) "activation" of the substrate by surface denaturation of the membrane materials around the fat globules, and (4) intimate contact of the lipases and the "activated" substrate.

All forms of agitation, with the exception of churning, increase the surface area of the substrate, and this is the foremost reason for the increase in lipase activity. However, agitation produces other effects which are conducive to lipase action. The process of diffusion, which has been shown to be very important, is speeded up (Mattson and Volpenhein 1966). Diffusion permits the lipases to migrate more readily to the oil–water interface while simultaneously allowing the fatty acids produced in lipolysis to leave the interface. Deeth and Fitz-Gerald (1977) observed a time- and temperature-dependent redistribution of activities between the cream and the skim milk phase during the agitation of raw milk. Maximum activation was obtained in fresh milk upon agitation after 2 to 4 hr of cold storage.

Lipoprotein Lipase Activator(s). The addition of blood serum to normal milk causes lipolysis (Jellema 1975; Murphy et al. 1979). This phenomenon was explained by the presence of a thermostable cofactor, probably a phospholipoprotein, which forms a complex with lipoprotein lipase; the complex adsorbs to the fat globules, and the fat is hydrolyzed. Driessen and Stadhouders (1974A) postulated that the inclination of milk to develop spontaneous rancidity is determined by the level of phospholipid-containing substances present in milk. They further speculated that under certain physiological conditions these substances are transferred from blood to milk, where they trigger spontaneous lipolysis. Downey (1980A,B), however, cautions that without further investigations into the role of blood constituents in spontaneous lipolysis, there is the danger that too much significance will be ascribed to them.

Olivecrona et al. (1975) observed that addition of a suitable activator polypeptide caused little or no activation against tributanoylglycerol, some stimulation against trihexanoylglycerol, and a three- or four-fold stimulation against trioctanoylglycerol. These results were interpreted to indicate that the interaction between enzyme and activator takes place on the surface of the emulsified substrate.

Clegg (1980) reported that bovine serum and high-density lipoprotein (HDL) caused an increase in free fatty acid levels in unpasteurized bulk milk. Lipoprotein free serum, apo HDL, all individual HDL tested, and the unfractionated C-peptide fractions had no lipolytic effect. HDL-lipid in the presence of 2 C-peptides and the combination of HDL-lipid with unfractionated C-peptide caused a considerable stimulation of lipolysis.

Thermal Manipulation. Unlike spontaneous milk, normal (nonspontaneous) milk requires additional thermal "shocking" beyond the first cooling to activate the milk lipase system. Wang and Randolph (1978) observed a migration of lipase activity to the cream fraction upon cooling of milk to 4°C and a reversal of migration on warming. Krukovsky and Herrington (1939) were the first to demonstrate that lipolysis in normal milk could be hastened by warming cold milk to 29.4°C and then recooling it beyond the solidifying point of the fat. Most samples of milk subjected to this treatment become rancid within 24 hr. The rate of cooling apparently has no effect (Kitchen and Aston 1970). Cooling under vacuum reduces the lipolysis by 10% compared with cooling under normal pressure (Kirst 1980C). The temperature of approximately 30°C is critical, and heating below or appreciably above that point diminishes the degree of activation that can be obtained. This type of activation is of great practical importance because it can happen accidentally. For example, if warm morning milk is added to a can of milk refrigerated from the night before and all of it cooled again, the milk may be rancid by the time it is ready for processing.

Milk containing fat globules with a natural fat globule membrane can be activated, deactivated, and reactivated by proper changes in temperature. However, some loss of activity will occur upon repeated activation (Wang and Randolph 1978). The phenomenon of temperature activation is found only when the fat globules have their natural layer of adsorbed materials. Neither homogenized milk, nor emulsions of tributyrin, nor butter oil emulsified in skim milk can be activated in this manner.

Several hypotheses have been advanced to explain the peculiar phenomenon of temperature activation. These include the attainment of a favorable liquid-to-solid glyceride ratio (Henningson and Adams 1967), an increase in the permeability of the fat globule membrane to the lipases (Nilsson and Willart 1960), and reorientation of glycerides more susceptible to lipolysis toward the fat–water interface (R 1951). However, the first and last hypotheses seem to be inconsistent with the fact that homogenized milk cannot be temperature activated.

The freezing of raw milk followed by thawing to 4°C causes an increase in lipolysis compared to that of unfrozen control milk stored at 4°C, but the increase in activity varies considerably. Repeated freezing and thawing also causes a notable increase in lipolytic activity. The temperature of freezing has a marked effect, the increase in lipolysis being most pronounced when the temperature is lowered from −10 to −20°C; little further increase in activity occurs between −20° and −33°C. Slow freezing causes greater lipolysis than rapid freezing.

Chemical Activation

Downey and Andrews' (1966) experiments indicate that there is a bivalent cation requirement for full milk lipase activity. Dunkley and Smith (1951) had previously stated that small amounts of $CaCl_2$ accelerate lipolysis. These observations are in keeping with those made on lipases from other sources where Ca^{2+} was found to stimulate activity (Wills 1965; Egelrud and Olivecrona (1973).

Pitocin, a hormone, was reported to increase lipolysis (Kelly 1943, 1945), and another hormone, diethylstilbestrol, is said to increase lipase activity toward tributyrin but not toward milk fat (El-Nahta 1963).

The milk lipase system is reported to be activated by mercuric chloride. Raw milk preserved with corrosive sublimate sometimes contains a much larger concentration of free fatty acids that do unpreserved samples. Pasteurized milk preserved in a similar fashion does not show an increase in free fatty acids (Manus and Bendixen 1956).

Inhibition of Lipases

Thermal Inhibition.   Heat treatment of milk is the most important practical means of inactivating its lipases. The temperature–time relationship necessary for partial or complete inactivation has been extensively studied, but a number of discrepancies have been apparent. These are probably due to several factors, including the sensitivity of the assay procedure, the length of the incubation period following heating, the presence and concentration of fat and solids-not-fat in the milk at the time of heating, and the type and condition of the substrate. In view of these variables, references to a number of early studies on heat inactivation have been omitted.

The data of Nilsson and Willart (1960) indicate that heating at 80°C for 20 sec is sufficient to destroy all lipases in normal milk. Their studies included assays after 48 hr of incubation following heat treatment. At lower temperatures for 20 sec, some lipolysis was detected after the 48-hr incubation period after heating. Thus, 10% residual activity remained at 73°C. Below the temperature of 68°C the amount of residual activity was enough to render the milk rancid in 3 hr; temperatures below 60°C had no appreciable effect on lipolysis. With holding times of 30 min, 40°C produced only slight inactivation, and at 55°C 80% inactivation was reported.

The data of Harper and Gould (1959) are essentially in agreement with those of Nilsson and Willart. These authors also detected no inactivation until a temperature of 60°C for 17.6 sec was reached. At 87.7°C (17.6 sec) some lipase still survived.

Shipe and Senyk (1981) reinvestigated the effects of various pasteurization times and temperatures on lipolysis (Table 5.2). They concluded that processing at 76.7°C for 16 sec should be sufficient to protect most milks from lipolysis problems for 7 days after pasteurization.

Fat apparently protects the lipases to some extent from heat inactivation, 1° to 2°C higher temperatures being necessary for whole milk than for skim milk (Frankel and Tarassuk 1959; Harper and Gould 1959; Nilsson and Willart, 1960; Saito et al. 1970).

Harper and Gould (1959) indicate that besides the protective effect of fat on lipase inactivation, the solids-not-fat content is also a factor. A higher solids-not-fat concentration, within limits, affords some protection.

Inhibition by Light and Ionizing Irradiation. The milk lipoprotein

Table 5.2. Effect of Pasteurization Time and Temperature on Lipolysis[a, b]

Temperature[c] (°C)	Holding Time[d]		
	16 s	20 s	24 s
72.2	2.1	1.7	1.3
74.4	1.0	1.0	0.9
76.7	0.9	0.9	0.9
78.9	0.9	0.8	0.8
81.1	0.8	0.8	0.8

[a]Shipe and Senyk (1981).
[b]Average ADVs for six pasteurized-homogenized milk samples after storage for 7 days at 5°C. Average raw milk ADV was 0.7.
[c]ADV means for 72.2°C significantly different from all others ($P < 0.01$); values for 74.4°C differed from 81.1°C ($P < 0.05$); no significant differences between others.
[d]ADVs for different holding times were different ($P < 0.05$) at temperatures below 74.4°C.

lipase shows remarkable sensitivity to light. Kay (1946) exposed fresh milk in glass vessels to bright summer sunshine for 10 min and found that 40% of the lipolytic activity was destroyed. Exposure for 30 min resulted in a loss of 80%, and exposure to a 800-Watt quartz mercury-vapor lamp at a distance of 15 cm destroyed 75% of the activity. Kay noted, however, that if oxygen was removed from the system before exposure to sunlight, the effect of the light was greatly diminished. Kannan and Basu (1951) observed that in some cases exposure to ultraviolet light destroyed the lipase system, and diffuse daylight brought about partial inactivation. When skim milk was irradiated with UV light with a wavelength of 350 nm, marked inactivation occurred (Castberg et al. 1975A).

Frankel and Tarassuk (1959) exposed a layer of raw skim milk 1 cm thick to direct sunlight at room temperature and noted a loss in lipase activity of 84% in 5 min and of 96% after 10 min. In diffuse daylight inactivation was reduced but 71% of lipase activity was lost in 1 hr. The loss of activity by light was independent of the temperature of the milk, equal losses being observed at 0°C and at 37°C. The enzymes were markedly protected against light inactivation by the presence of fat.

Stadhouders and Mulder (1959) confirmed Kay's observation that the shorter wavelengths (about 4300 Å) of the spectrum are most destructive to milk lipases. The destructive effect of light could be repressed by the addition of reducing agents such as metol (p-methylaminophenol sulfate), hydroquinone, and especially hydrogen sulfide. Ascorbic acid and methionine had no effect, but cysteine afforded significant protection. Lipases which had inactivated by light were not reactivated by treating milk with hydrogen sulfide.

Irradiation by ionizing radiation and its effect on milk lipase activity have also been studied (Tsugo and Hayashi 1962). Irradiation doses of 6.6×10^4 rads destroyed 70% of the activity. The udders of lactating cows, when exposed to 60 Co gamma rays, gave milk with decreased lipase and esterase activity (Luick and Mazrimas 1966).

Chemical Inhibition. A large variety of chemical compounds have been added to milk or purified lipase. The conditions under which the inhibitor is studied are very important. Factors such as pH, temperature, time of addition of the chemical, sequence of addition of reactants, and the presence or absence of substrate are undoubtedly involved. The presence of substrate appears to offer some degree of protection to the enzymes. Consequently, in lipase studies, the surface area of the emulsified substrate is probably also important.

Heavy metals usually affect enzymes adversely, and milk lipases are no exception. Copper, cobalt, nickel, iron, chromium, manganese, and silver are inhibitors. Raw skim milk treated with 5 to 20 ppm Cu^{2+} for 15 min at room temperature caused 7 to 17% loss of lipolytic activity, whereas 5 ppm at 37° C for 1 hr resulted in a 69% loss. There was less inhibition in the presence of substrate (Frankel and Tarassuk 1959). Earlier, however, Krukovsky and Sharp (1940) showed that Cu^{2+} was ineffective as a lipase inhibitor in nonhomogenized milk if oxygen was absent. At the same time, they also found that oxygen alone is an active inhibitor, its effect being magnified by the presence of low percentages of copper.

A number of salts inhibit lipolysis, the most effective being sodium chloride (Gould 1941; Pijanowski et al. 1962; Willart and Sjostrom 1959; Egelrud and Olivecrona 1972). Lipolysis in cream was found to be insignificant in the presence of 4% sodium chloride and in homogenized milk containing 5 to 8% of this salt (Gould 1941).

Phosphate buffer (0.6 M) slightly inhibited lipolysis, but the same concentration of borate and barbiturate buffers was without effect. Zinc chloride, potassium cyanide, manganese sulfate, cysteine, and magnesium chloride retarded milk lapse activity to various degrees. All of these compounds were tested at pH 8.5 with tributyrin as substrate during a 30-min incubation period (Peterson et al. 1948).

N-Ethyl maleimide inhibits lipase activity in milk activated by shaking, temperature fluctuations, and homogenization, 0.02 M being completely inhibitory (Olson et al. 1956). An equimolar concentration of glutathione markedly reduces inhibition by N-ethyl maleimide. This reagent can also completely inhibit lipolysis in spontaneous milk (Tarassuk and Yaguchi 1958). It was concluded, on the basis of these experiments, that sulfhydryl groups were essential sites of activity on milk lipases. This conclusion is supported by the ability of reducing agents such as glutathione, hydroquinone, and potassium thiocyanate to stabilize the milk lipase system during storage (Frankel and Tarassuk 1959).

Other chemicals which inhibit milk lipase include hydrogen peroxide, animal cephalin, sodium arsenite, diisopropyl fluorophosphate, 2,4 dinitro-l-fluorobenzene, p-hydroxymercuribenzoate, potassium dichromate, lauryl dimethyl benzyl ammonium chloride, aureomycin, penicillin, streptomycin, and terramycin (Schwartz 1974).

An extensive study of the effects of formaldehyde in milk lipase inhibition showed that formaldehyde acts as a competitive inhibitor and, under the proper conditions, selectively inhibits the lipases of raw skim milk (Schwartz et al. 1956A). This study showed that the inhibitory effect of formaldehyde was dependent on such factors as pH, time of addition of the inhibitor, length of the incubation period, concentration and availability of the substrate, and concentration of the inhibitor. Many of the conflicting results encountered with formaldehyde (Schwartz 1974) can be explained on the basis of dependence on one or more of these factors.

Shipe *et al.* (1982) observed that the addition to raw bulk milk of 0.01 to 0.05% carrageenan reduced thermally activated lipolysis by 55 to 100% and agitation-activated lipolysis by 36 to 85%. They speculated that the inhibitory effect, at least partially, was due to interaction with lipase and did not rule out the possibility that carrageenan protected the substrate to some extent by encapsulation.

Properties of Milk Lipases

Specificity. A study of lipase specificity requires that the enzyme and substrates be virtually pure. Contamination of the lipase preparation with esterases gives rise to misleading results. Pure, synthetic substrates of known configuration are essential, and the same available surface area should be present after emulsification for meaningful data to be obtained. Since most of the earlier workers disregarded one or more of these variables, their data will not be included here.

Egelrud and Olivecrona (1973) studied the catalytic activity against several substrates of bovine milk enzyme preparations that had been purified about 7000-fold to a purity higher than 80% (Egelrud and Olivecrona 1972). The enzyme catalyzed the hydrolysis of emulsified trioleate, trioctanoate, monooleate, Tween 20 (polyoxyethylene sorbitan monolaurate), and *p*-nitrophenyl acetate. It was concluded that the enzyme had rather low substrate specificity and that the presence of activating serum factors is not needed for catalysis to occur.

pH Optimum. Enzymes usually exert their catalytic influence over a somewhat restricted pH range. Within this range the activity passes through a maximum, commonly called the "pH optimum," and then falls off again. Although the pH optimum and the pH range are generally characteristic of a given enzyme, they may sometimes be altered by such factors as type and strength of buffer, ionic strength, temperature, type of substrate employed, and, in the case of lipases, the condition of the interface where lipolysis must proceed.

Lipases are sensitive to extremes of pH, and even in the vicinity of the pH optimum, where enzymes are supposedly more stable, marked inhibition may occur (Frankel and Tarassuk 1956B). Thus, it must also be borne in mind that the length of the incubation period and the prior history of the preparation can influence the range and perhaps the shape of the pH activity curve.

Studies with purified milk lipoprotein lipase (Egelrud and Olivecrona

1973) have revealed pH maxima at 8.25 and 8.5 for hydrolysis of tributyrin and triolein, respectively. A much higher pH maximum was observed for monolaurin above 10.5, and the hydrolysis of Tween had a pH maximum of 8.5 without deoxycholate and at 9.5 with deoxycholate. Murphy *et al.* (1979) observed that a reduction of the pH from 6.7 (6.7 being the normal pH of milk) to 6.5 caused a reduction of FFA development of about one-half, while an increase in pH from 6.7 to 7.0 and 8.5 caused two- and fourfold increases in FFA development, respectively. At values below pH 6.5 little lipolysis occurred. Parry *et al.* (1966) found that the pH optimum of the lipase activity of milk ranged from 8.5 to 9.0 and was 8.6 for purified milk lipase. Driessen and Stadhouders (1974A) reported maximum lipolysis of spontaneously rancid milk presumably containing a thermostable cofactor at pH 8.0; without cofactor the pH optimum ranged from 8.8 to 9.0. These researchers further reported a maximum activity of the lipolytic enzymes of *Pseudomones fluorescence* (a psychrotroph) at pH 8.75.

The incubation of raw skim milk at pH 6.0 and at pH 8.9 for 1 hr at 37°C in the absence of substrate was subsequently shown to cause a 47% and 40% decrease, respectively, in lipase activity when the milk was later incubated with milk fat. When tributyrin was the substrate the inhibition was even more marked. Although some of the inactivation was due to temperature, the majority of it was attributable to pH exposure. Stadhouders and Mulder (1964) have also demonstrated that milk lipase subjected to incubation at pH 5.0 is almost completely destroyed.

The point on the acid side of the pH curve where milk lipase activity ceases is of considerable practical importance, but there is still controversy regarding it. Willart and Sjostrom (1962) found that milk lipase is active in the range pH 4.1 to 5.7, whereas Schwartz *et al.* (1956B) could detect no activity at pH 5.2 on butterfat. Although Peterson *et al.* (1948) found no milk lipase activity on tributyrin at pH 7.0, activity was reported on this substrate at pH 5.0 and even at pH 4.7 when 24-hr incubation periods were used (Stadhouders and Mulder 1964).

Apparent Temperature Optimum. A rise in temperature has a dual effect upon an enzyme-catalyzed reaction: it increases the rate of the reaction, but it also increases the rate of thermal inactivation of the enzyme itself. Like the pH optimum, the temperature optimum may in certain instances be altered by environmental conditions, e.g., pH, type and strength of buffer, etc. The term "temperature optimum," therefore, is useless unless the incubation time and other conditions are specified. A more enlightening term is "apparent temperature optimum," which indicates that the optimum has been obtained under a certain set of conditions and may or may not hold when these conditions are changed.

The apparent temperature optimum for the milk lipase system is reported to be around 37°C both on milk fat and on tributyrin (Frankel and Tarassuk 1956A; Roahen and Sommer 1940). This temperature has been recorded both at pH 8.9 and pH 6.6 for milk fat (Frankel and Tarassuk 1956A) and at pH 8.0 and pH 6.6 at tributyrin (Frankel and Tarassuk 1956A). Although the enzyme appears to be most active at 37°C, activity is rapidly lost at this temperature (Egelrud and Olivecrona 1973). Studying the effect of temperature on the activity of lipases of psychrotrophs, Driessen and Stadhouders (1974B) observed two optimum temperatures: a relative optimum at 37°C and an absolute optimum at 50°C; Kishonti (1975) found two temperature optima at 30 and 55°C.

Stability. Some discussion regarding stability of milk lipases was presented in the preceding section. Egelrud and Olivecrona (1973) found that the enzyme fractions from heparin-Sepharose can be stored frozen at –20°C with less than 10% loss of activity in 2 weeks. The purified enzyme had only moderate stability at 4°C; high concentrations of salt or a pH below 6.5 or above 8.5 increases the rate of inactivation.

Some Effects of Lipolysis. The most serious effect of lipolysis is the appearance of the so-called rancid flavor which becomes detectable in milk when the ADV exceeds 1.2–1.5 mEq/liter (Brathen 1980). The fatty acids and their soaps, which are thought to be implicated in the rancid flavor, have been studied in an effort to assess the role of the individual acids in the overall rancid flavor picture. Scanlan *et al.* (1965) reported that only the even-numbered fatty acids from C4 to C12 account for the contribution of fatty acids to the flavor, but that no single acid exerts a predominating influence. Another study has implicated the sodium and/or calcium salts of capric and lauric acids as major contributors to the rancid flavor (Al-Shabibi, *et al.* 1964). Butyric acid, assumed to be the compound most intimately associated with the flavor, was not singled out in either study as being especially involved.

Besides changing the natural flavor of milk, lipolysis may produce a variety of other effects. One of the most noticeable of these is the lowering of surface tension as lipolysis proceeds (Schwartz 1974). Fatty acids, especially their salts, and mono- and diglycerides, being good surface-active agents, depress the surface tension of milk (see the discussion "Methods for Determining Lipase Activity"). Milk fat obtained from milk subject to lipase action also has lower interfacial tensions with water than does milk fat obtained from nonlipolyzed milk (Bergman *et al.* 1962A; Duthie *et al.* 1961).

Rancid milk decreases the quality of cream, butter, and buttermilk made from it, and a limit on the ADV of the fat of milk from which butter is eventually to be made has been proposed.

The higher saturated fatty acids have been noted to inhibit rennet action, whereas the lower fatty acids enhance it. The inhibitory effect of the higher acids can be nullified by $CaCl_2$.

As little as 0.1% rancid milk fat proved to be a very effective foam depressant during the condensing of skim milk and whey (Brunner 1950). This effect was attributed to the mono- and diglycerides.

Lipolytic action has been observed to occur in composite samples preserved with mercuric chloride and has decreased the reading of the Babcock test as much as 0.15% (Manus and Bendixen 1956).

An inhibitory effect of rancid milk on the growth of *Streptococcus lactis* has been reported. Early reports (Schwartz 1974) claimed that rancid milk significantly inhibits the growth of bacteria in general and of *Streptococcus lactis* in particular. It has been stated that rancidity in milk may reach such a degree as to actually render the product sterile. (Schwartz 1974). Tarassuk and Smith (1940) attributed the inhibitory effect of rancid milk to changes in surface tension, but Costilow and Speck (1951) believe that the inhibition is due to the toxic effect of the individual fatty acids.

Although rancidity is a serious defect in market milk, it has also been utilized profitably. Whole milk powder made from lipase-modified milk has generally been accepted by chocolate manufacturers. It is used as a partial replacement for whole milk because it imparts a rich, distinctive flavor to milk chocolate, other chocolate products like fudge, and compound coatings, caramels, toffees, and butter creams (Ziemba 1969).

Methods for Determining Lipase Activity

A number of methods are available for following lipase activity. Although numerous modifications and variations have been introduced, the basic methods are (1) titration of the liberated fatty acids, (2) changes in surface tension, (3) colorimetric determination of the fatty acids, (4) use of gas-liquid chromatography, and (5) use of radioactive substrates. Kuzdzal-Savoie (1980) has reviewed the subject.

Titration. Titration of the fatty acids formed by the action of the milk lipase system has been the most widely used procedure. Titration has been conducted directly on the reaction medium, either manually (Gould and Trout 1936) or automatically (Parry *et al.* 1966), in the presence of added organic solvents (Dunkley and Smith 1951; Peterson *et al.* 1948), and after separation of the lipid phase by extraction (Frankel and Tarassuk 1956A; Salih *et al.* 1977A), distillation (Roahen and Sommer 1940), churning (Fouts 1940), or absorption of the medium followed by elution of the fatty acids (Harper *et al.* 1956B). All of these techniques have their shortcomings. The most widely used laboratory methods appear to be the silica gel extraction (Harper *et al.* 1956B) and pH-stat methods (Parry *et al.* 1966; Castberg *et al.* 1975B), and, in the field, the method of Thomas *et al.* (1955B), which is the basis for Bureau of Dairy Industries (BDI) method. Pillay *et al.* (1980) compared the BDI and Frankel and Tarassuk procedures in a study to detect the threshold of lipolyzed flavor. They found that the ADVs were method dependent. Brathen (1980) used an automated method to establish the upper acceptable limits for the ADVs in farm milk, retail whole milk, retail skim milk, and full fat cream (35%) at 1.0, 0.9, 0.7, and 3.0 mEq/-liter, respectively.

Surface Tension.   Efforts have been made to apply surface tension measurements to determine lipolysis in milk (Schwartz 1974). As mentioned earlier, the hydrolysis products resulting from lipase action are strongly surface active. Tarassuk and Regan (1943) have stated that the lowering of surface tension resulting from lipolysis is the most distinct change differentiating rancid from nonrancid milk. However, many variables influence the surface tension of milk, such as the elaboration of structurally different mono- and diglycerides and their concentration.

Colorimetry.   Copper (Duncombe 1963; Koops and Klomp 1977) or cobalt (Novak 1965) soaps of long-chain fatty acids ($\geq C12$) are soluble in chloroform and can be determined quantitatively by colorimetric determination of the extracted metal. Shipe *et al.* (1980B) have recently modified the original copper soap method to make it simpler, more rapid, and adaptable to automatic equipment.

Another sensitive colorimetric procedure is that of Mackenzie *et al.* (1967), which utilizes the dye Rhodamine B to form benzene-soluble complexes with fatty acids. Nakai *et al.* (1970) developed a rapid, simple method for screening rancid milk based on the foregoing procedure. The test is said to detect rancid milk with an ADV above 1.2. Like the copper or cobalt soap method, the Rhodamine B reagent is also limited to the longer-chain fatty acids. Kason *et al.* (1972) used the method employing Rhodamine 6G of Chakrabarty *et al.* (1969) to investigate the progress of rancidity in pasteurized milk during refrigerated storage.

Gas-Liquid Chromatography. Gas-liquid chromatography (GLC) affords both a qualitative and, if adequate internal standards are used, a quantitative analysis of the products of lipolysis. It is necessary, however, first to isolate the acids by a suitable method and then to inject them as free acids or as esters. The partial glycerides can be isolated by thin-layer chromatography and can also be determined by GLC of suitable derivatives. The acid(s) remaining in the partial glycerides can be identified readily by GLC following transesterification. Jensen and co-workers have utilized these techniques in their studies of lipase specificity (Jensen *et al.* 1964).

Radioactive Substrates. Koskinen *et al.* (1969), Luhtala *et al.* (1970-A,B), and Scott (1965) have used labeled triglycerides as substrates for milk lipases. This method, which is extremely sensitive, requires that the acids released by lipase action be isolated uncontaminated with any tagged glycerides. It also requires the preparation of labeled substrate and, of course, counting equipment.

Miscellaneous. A manometric technique utilizing a Warburg apparatus has been used to follow esterase activity. The carbon dioxide liberated from sodium bicarbonate by the fatty acids is measured (Willart and Sjostrom 1959). An agar diffusion procedure has been utilized for screening microorganisms for lipolyptic enzymes. The presence of lipase is indicated by clear zones in the turbid media (Lawrence *et al.* 1967).

Two assays have been developed for measuring the relative concentration of lipoprotein lipase activator in milk. Anderson (1979) developed an immunoassay and Super *et al.* (1976) used [2^3H]glycerol triolein and measured the liberated [2^3H]glycerol.

PART II. AUTOXIDATION

Lipid autoxidation in fluid milk and a number of its products has been a concern of the dairy industry for a number of years. The need for low-temperature refrigeration of butter and butter oil, and inert-gas or vacuum packing of dry whole milks to prevent or retard lipid deterioration, in addition to the loss of fluid and condensed milks as a result of oxidative deterioration, have been major problems of the industry.

The autoxidation of milk lipids is not unlike that of lipids in other edible products. However, the complex composition of dairy products, the physical state of the product (liquid, solid, emulsion, etc.), and the presence of natural anti- or pro-oxidants, as well as the processing, manufacturing, and storage conditions, tend to influence both the rate of autoxidation and the composition and percentage of autoxidation products formed.

The literature dealing with the autoxidation mechanism involved in lipid deterioration has been concerned with investigations on pure unsaturated fatty acids and their esters. The reactions involved, however, are representative of those occurring in lipids and lipid-containing food products.

Autoxidation Mechanism

The initial step in the autoxidation of unsaturated fatty acids and their esters is the formation of free radicals. Although the initiation of such radicals is not completely understood, the resulting free-radial chain reaction has been elucidated in the investigations of Farmer and Sutton (1943) and others (Bateman 1954; Bolland 1949). In the case of monounsaturated and nonconjugated polyene fatty acids—the acids of significance in milk fat—the reaction is initiated by the removal of a hydrogen atom from the methylene (-methylene) group adjacent to the double bond (I). The resulting free radical, stabilized by resonance, adds oxygen to form peroxide-containing free radicals (II); these, in turn, react with another mole of unsaturated compound to produce two isomeric hydroperoxides, in addition to free radicals (III) capable of continuing the chain reaction.

Oleic acid, having two -methylene groups, gives rise to four isomeric hydroperoxides which have been isolated in equal amounts by various workers (Farmer and Sutton 1943; Privett and Nickell 1959). The pref-

$$-CH_2-CH=CH-CH_2- \longrightarrow \overset{H}{-CH^{\cdot}};- CH=CH-CH_2- \longleftrightarrow$$
$$-CH=CH-CH^{\cdot}-CH_2-$$

$$(I)$$

$$+O_2$$

$$-CH^{\cdot}-CH=CH-CH_2- \longrightarrow -CH(OO^{\cdot})-CH=CH-CH_2-$$

$$\uparrow \qquad\qquad +O_2$$

$$-CH=CH-CH^{\cdot}-CH_2- \longrightarrow -CH=CH-CH(OO^{\cdot})-CH_2-$$

$$(II)$$

$$-CH(OO^{\cdot})-CH=CH-CH_2- \ + \ -CH_2-CH=CH-CH_2-\longrightarrow$$

$$-CH(OOH)-CH=CH-CH_2-$$
$$+$$
$$-CH^{\cdot}-CH=CH-CH_2-$$

$$-CH=CH-CH(OO^{\cdot})-CH_2- \ + \ -CH_2-CH=CH-CH_2-\longrightarrow$$

$$-CH=CH-CH(OOH) \ -CH_2-$$
$$-CH^{\cdot}-CH=CH-CH_2-$$

(III)

erential points of attack in polyene nonconjugated systems are the -methylene groups located between the double bonds. Hence the autoxidation of linoleic acid and linolenic acid can lead to the formation of three and six isomeric hydroperoxides, respectively, as a result of the attack on the C_{11} methylene group of linoleic acid and on the C_{11} and C_{14} methylene groups of linolenic acid. However, a characteristic of hydroperoxide formation is the shifting of double bonds to form the conjugated system (Cannon *et al.* 1952; Privett *et al.* 1953), and the existence of a 14-linolenate hydroperoxide has not been established (Badings 1970). The -methylene groups of polyunsaturated acids other than those located between double bonds are also subject to attack, but to a lesser degree. In all, seven hydroperoxides from linoleic acid and ten hydroperoxides from linolenic acid are theoretically possible during the autoxidation of these acids.

In addition to the formation of hydroperoxides, other reactions are known to occur simultaneously. The formation of polyperoxides, carbon-to-carbon polymerization, and the formation of epoxides and cyclic peroxides have been proposed or demonstrated in lipid oxidation.

Products of Oxidation

The hydroperoxides formed in the autoxidation of unsaturated fatty acids are unstable and readily decompose. The main products of hydroperoxide decomposition are saturated and unsaturated aldehydes. The mechanism suggested for the formation of aldehydes involves cleavage of the isomeric hydroperoxide (I) to the alkoxyl radical (II), which undergoes carbon-to-carbon fission to form the aldehyde (III) (Frankel *et al.* 1961).

$$
\begin{array}{ccc}
& ! & \\
\text{R-CH-R}^1 & \text{R-CH-R}^1 & \text{R-CHO} + \text{R}^1. \\
! & ! & \\
! & ! & \\
\text{O!OH} & \text{O}^{\cdot}! & \\
(\text{I}) & (\text{II}) & (\text{III})
\end{array}
$$

Other products, such as unsaturated ketones (Stark and Forss 1962), saturated and unsaturated alcohols (Hoffman 1962; Stark and Forss 1964, 1966), saturated and unsaturated hydrocarbons (Forss et al. 1967; Horvat et al. 1965; Khatri 1966), and semialdehydes (Frankel et al. 1961), have been observed in the decomposition of hydroperoxides of oxidized lipid systems.

A comprehensive review and study by Badings (1970) includes a listing of the carbonyls which can result from the dismutation of the theoretical hydroperoxides formed in the autoxidation of the major unsaturated acids of butterfat and those which have been observed. In addition to those carbonyls that are theoretically possible, various others have been isolated and identified in the autoxidation of pure fatty acids or their esters. Their presence suggests that migration of double bonds (Badings 1960), further oxidation of the unsaturated aldehydes initially formed (Badings 1959), and/or isomerization of the theoretical geometric form (Badings 1970) may occur during autoxidation.

In addition to the major fatty acids, milk also contains many minor polyunsaturated acids (Kurtz 1974); hence the autoxidation of dairy products can lead to a multitude of saturated and unsaturated aldehydes.

Oxidation and Off-flavors

The overwhelming consideration in regard to lipid deterioration is the resulting off-flavors. Aldehydes, both saturated and unsaturated, impart characteristic off-flavors in minute concentrations. Terms such as "painty," "nutty," "melon-like," "grassy," "tallowy," "oily," "cardboard," "fishy," "cucumber," and others have been used to characterize the flavors imparted by individual saturated and unsaturated aldehydes, as well as by mixtures of these compounds. Moreover, the concentration necessary to impart off-flavors is so low that oxidative deterioration need not progress substantially before the off-flavors are detectable. For example, Patton et al. (1959) reported that 2,4-decadienal, which imparts a deep-fried fat or oily flavor, is detectable in aqueous solution at levels approaching 0.5 ppb.

In addition to aldehydes, other secondary products of lipid oxidation, such as unsaturated ketones and alcohols, impart characteristic flavors, and their presence in oxidized milk systems has been established (Badings 1970; Stark and Forss 1962, 1964).

Generally speaking, the flavor threshold values for aldehydes are governed to varying degrees by the number of carbon atoms; degree of unsaturation; location of unsaturation in the chain; form of the geometric isomer; additive and/or antagonistic effects of mixtures of compounds; and the medium in which the flavor compounds are present (Day et al. 1963; Meijboom 1964). With respect to the last point, the flavor potency of many aldehydes identified in oxidized lipids is up to 100 times greater in an aqueous medium than in a fat or an oil. Hence, the extent of oxidative deterioration of fluid milk need not progress to the same point as that in butter oil before the onset of off-flavors in the fluid product.

The off-flavors which develop in dairy products as a result of oxidative deterioration are collectively referred to as the "oxidized flavor." However, the organoleptic properties of the off-flavor differ among products, as well as within the same product, depending on the degree of deterioration. Descriptive terms as "cappy" and "cardboard" have been used to characterize the off-flavor in fluid milk. The off-flavor in dry whole milk and in butter oil has been referred to as "oily" or "tallowy." Butter undergoes a continuous change in flavor defects during storage, defects which usually develop in an order described as "metallic," "fatty," "oily," or "trainy," and "tallowy" (Badings 1970). In an effort to standardize off-flavor nomenclature, the Committee on Flavor Nomenclature and Reference Standards of the American Dairy Science Association (Shipe et al. 1978) published an extensive bibliography and classified the descriptive terms of the oxidized flavor as "papery," "cardboard," "metallic," "oily," and fishy.

Although the conditions under which the above-mentioned products are normally stored undoubtedly influence the extent of deterioration and hence the character of the off-flavor, the lipid constituents involved in the reaction also influence the resulting flavor. The site of oxidative deterioration in fluid milk and cream is the highly unsaturated phospholipid fraction associated with the fat globule membrane material (Badings 1970; Smith and Dunkley 1959). On the other hand, in products such as butter and dry whole milk, both the phospholipids and the triglycerides are subject to oxidative deterioration (Badings 1970). The off-flavor appearing in butter oil is understandably the result of triglyceride deterioration.

Measurement of Fat Oxidation

Various methods have been employed to measure the extent of autoxidation in lipids and lipid-containing food products. For obvious reasons, such methods should be capable of detecting the autoxidation process before the onset of off-flavor. Milk and its products, which develop characteristic off-flavors at low levels of oxidation, require procedures that are extremely sensitive to oxidation. Thus methods of measuring the decrease in unsaturation (iodine number) or the increase in diene conjugation as a result of the reaction do not lend themselves to quality control procedures, although they have been used successfully in determining the extent of autoxidation in model systems (Haase and Dunkley 1969A; Pont and Holloway 1967).

Several methods have been introduced which express the degree of oxidation deterioration in terms of hydroperoxides per unit weight of fat. The modified Stamm method (Hamm et al. 1965), the most sensitive of the peroxide determinations, is based on the reaction of oxidized fat and 1,5-diphenyl-carbohydrazide to yield a red color. The Lea method (American Oil Chemists' Society 1971) depends on the liberation of iodine from potassium iodide, wherein the amount of iodine liberated by the hydroperoxides is used as the measure of the extent of oxidative deterioration. The colorimetric ferric thiocyanate procedure adapted to dairy products by Loftus Hills and Thiel (1946), with modifications by various workers (Pont 1955; Stine et al. 1954), involves conversion of the ferrous ion to the ferric state in the presence of ammonium thiocyanate, presumably by the hydroperoxides present, to yield the red pigment ferric thiocyanate. Newstead and Headifen (1981), who reexamined this method, recommend that the extraction of the fat from whole milk powder be carried out in complete darkness to avoid elevated peroxide values. Hamm and Hammond (1967) have shown that the results of these three methods can be interrelated by the use of the proper correction factors. However, those methods based on the direct or indirect determination of hydroperoxides which do not consider previous dismutations of these primary reaction products are not necessarily indicative of the extent of the reaction, nor do they correlate well with the degree of off-flavors in the product (Kliman et al. 1962).

Two variations of the thiobarbituric acid (TBA) method have been widely used to determine the degree of lipid oxidation in dairy products (Dunkley and Jennings 1951; King 1962). These methods, of approximately equal sensitivity, are based on the condensation of two molecules of thiobarbituric acid with one of malonaldehyde (Schmidt 1959), resulting in the formation of a red color complex with an absorption maximum at 532 to 540 mu. King (1962) has shown (Table 5.3) that a correlation exists between the determined TBA values and the intensity of the oxidized flavor in fluid milks. Similar observations have been reported by others in fluid milks (El-Negoumy 1965) and ultra-high-temperature creams (Downey 1969). The TBA method of Dunkley and Jennings (1951) has been reported to be more applicable than the King method in determining the extent of the off-flavor (Downey 1969). Both methods have been used extensively in studies of the autoxidation of extracted milk components and model lipid systems (Gawel and Pijanowski 1970; Haase and Dunkley 1969A). Lillard and Day (1961) reported a significant correlation between a modified TBA test and the reciprocal of the average flavor threshold of oxidized butterfat. A similar correlation also existed between the peroxide value and the reciprocal of the average flavor threshold of butterfat.

In addition to the previously mentioned chemical tests, methods based on the carbonyl content of oxidized fats have also been suggested (Henick *et al.* 1954; Lillard and Day 1961) as a measure of oxidative deterioration. The procedures determine the secondary products of autoxidation and have been reported to correlate significantly with the degree of off-flavor in butter oil (Lillard and Day 1961). The methods, however, are cumbersome and are not suited for routine analysis.

Antioxidants

The use of synthetic antioxidants in the prevention or retardation of autoxidation in lipids and lipid-containing food products has been the subject of numerous investigations. Although the present U.S. standards do not permit antioxidants in dairy products, and hence the question of their effectiveness is one of only theoretical interest, they

Table 5.3. Relation Between Organoleptic and TBA Values of Fluid Milk.

Flavor Score	Description	Range of Optical Density (532 mμ)
0	No oxidized flavor	0.010–0.023
1	Questionable to very slight	0.024–0.029
2	Slight but consistently detectable	0.030–0.040
3	Distinct or strong	0.041–0.055
4	Very strong	>0.056

SOURCE: King 1962.

are of practical interest in countries where their use is permitted. Many compounds containing two or more phenolic hydroxy groups, such as esters of gallic acid, butylated hydroxyanisole, norhydroguaiaretic acid, hydroxyquinone, and dihydroquercitin, have been employed as antioxidants in studies of dairy products (Sidhu *et al.* 1975, 1976). These compounds apparently work by interrupting the chain reaction in autoxidation by capturing the free radicals necessary for the continuation of hydroperoxide formation (Badings 1960).

Considerations, other than legal ones, that must be taken into account regarding the use of antioxidants in dairy products include off-flavors imparted by the antioxidant itself (Gelpi *et al.* 1962; Romanskaya and Valeeva 1962), ease of incorporation into the product (Hammond 1970), distribution between the water and oil phases (Cornell 1979, Cornell *et al.* 1971), and effectiveness of the antioxidant in different media. With regard to the last point, studies of the use of antioxidants in dairy products reveal variations in their antioxidative properties in different products. Norhydroguaiaretic acid is effective in preventing the development of an oxidized flavor of fluid milk but tends to increase the rate of autoxidation in milk fat (Hammond 1970). The tocopherols, while of little value in dry whole milks (Abbot and Waite 1965) and butter oil (Pont 1964), are highly effective in preventing spontaneous or copper-induced oxidation in fluid milk (Dunkley *et al.* 1967; King 1968). Compounds reported to be among the most antioxidative in specific dairy products include dodecyl gallate in spray-dried whole milks (Abbot and Waite 1962; Tamsma *et al.* 1963), ascorbyl palmitate in cold storage-cultured butter (Koops 1964B), sodium gentisate in frozen whole milk (Gelpi *et al.* 1926), and quercitin and propyl gallate in butter oil (Wyatt and Day 1965).

Synergists, such as the polybasic citric and phosphoric acids, have been used in conjunction with antioxidants. These compounds have no antioxidative value in themselves, but they increase the effectiveness of antioxidants. Their synergistic influence on antioxidants may be due to the sequestering of metallic ions (Badings 1960; Jenness and Patton 1959), inhibiting the antioxidant catalysis of peroxide decomposition (Privett and Quackenbush 1954), or regenerating the antioxidant in the system (Smith and Dunkley 1962A). It has been reported that these synergists, like the phenolic antioxidants, are capable of performing the dual role of retarding autoxidation at low levels and accelerating it at higher levels (Privett and Quackenbush 1954).

In addition to antioxidants, either alone or in the presence of synergists, metal chelating compounds, such as the various salts of ethylenediaminetetraacetic acid (Arrington and Krienke 1954; King and

Dunkley 1959A) and neocuproine (Smith and Dunkley 1962B), among others (Samuelsson 1967), have also proven their effectiveness as inhibitors of autoxidation.

Oxidative Deterioration in Fluid Milk

Fluid milks have been classified by Thurston (1937) into three categories based on their ability to undergo oxidative deterioration: (1) spontaneous, for those milks that spontaneously develop off-flavor within 48 hr after milking; (2) susceptible, for those milks that develop off-flavor within 48 hr after contamination with cupric ion; and (3) resistant, for those milks that exhibit no flavor defect, even after contamination with copper and storage for 48 hr. A similar classification has been employed by Dunkley and Franke (1967).

With the advent of noncorrodible dairy equipment, oxidative deterioration in fluid milk as a result of copper contamination has decreased significantly, although it has not been completely eliminated (Rogers and Pont 1965). However, the incidence of spontaneous oxidation remains a major problem of the dairy industry. For example, Bruhn and Franke (1971) have shown that 38% of samples produced in the Los Angeles milkshed are susceptible to spontaneous oxidation; Potter and Hankinson (1960) have reported that 23.1% of almost 3000 samples tasted were criticized for oxidized flavor after 24 to 48 hr of storage. Significantly, certain animals consistently produce milk which develops oxidized flavor spontaneously, others occasionally, and still others not at all (Parks et al. 1963). Differences have been observed in milk from the different quarters of the same animal (Lea et al. 1943).

The resistance of certain milks to oxidation, even in the presence of added copper, may be attributed to its poising action, i.e., the resistance of milk to a change in the oxidation-reduction potential (Parks 1974). That a correlation exists between the appearance of an oxidized flavor and conditions favoring milk oxidation, as measured by the oxidation-reduction potential, was shown by several researchers (Parks 1974). This apparent correlation, as well as other factors, tend to discredit theories on the role of enzymes as catalytic agents in the development of oxidized flavor. Xanthine oxidase has been proposed as the catalytic agent in the development of spontaneously oxidized milk (Astrup 1963; Aurand and Woods 1959; Aurand et al. 1967, 1977). The studies of Smith and Dunkley (1960), among others (Rajan et al. 1962), do not corroborate these findings, and the authors conclude that xanthine oxidase is itself not a limiting factor in the off-flavor. However, reports persist on the involvement of enzymes in the generation of various types of oxygen that may be involved in the autoxidation of milk lipids (Aurand *et al.* 1977; Holbrook and Hicks 1978; Hicks 1980; Hill *et al.* 1977; Gregory *et al.* 1976). Aurand *et al.* (1977) stated that the catalytic effect of the combination of light, copper, and xanthine oxidase generated singlet oxygen, which was the immediate source of the hydroperoxides that initiated lipid oxidation. Holbrook and Hicks (1978) reported that superoxide dismutase, which was believed to play an important role in the suppression of lipid oxidation, decreased the minor pro-oxidant effect of xanthine oxidase and that the concentration of superoxide dismutase in individual cow's milk did not account for the strong oxidative resistance in raw milk that had not been exposed to light. Hill and co-workers (1977), investigating factors influencing the autoxidation of high-linoleic milk, proposed a pathway whereby oxidation is induced by copper and which they believe must have the OH-radical as an intermediate and a second, probably less important, pathway which depends on superoxide O_2^- that is generated by xanthine oxidase and lactoperoxidase and which is dismutated to singlet oxygen by spontaneous, nonenzymatic dismutation. Gregory *et al.* (1976), studying the involvement of heme proteins in peroxidation of milk lipids, did not exclude their catalytic involvement.

The often conflicting reports in the literature indicate that more research is needed to clarify the role of interacting enzyme systems that control the generation and survival of active forms of oxygen and their involvement in the initiation and propagation of lipid oxidation in milk.

Despite reports of anomalous behavior in several aspects, sufficient evidence has been accumulated in recent years to establish that the susceptibility or resistance of milk to oxidative deterioration is dependent on the percentage and/or distribution of naturally occurring pro- and antioxidants.

Metals

Metal-catalyzed lipid oxidative reactions were recognized in dairy products as early as 1905 (Parks 1974). Investigations throughout the years have shown that copper and iron are the important metal catalysts in the development of oxidized flavors. Of these two metals, copper exerts the greater catalytic effect, while ferrous ion is more influential than feric ion.

Both copper and iron are normal components of milk. Murty *et al.* (1972) studied the trace mineral content of market milk from various regions of the United States for one year and found that the levels of copper and iron were highest in winter and lowest in summer. Disregarding variations due to individuality, stage of lactation, and contamination, copper is present at average levels of 20 to 105µg/liter (Horvat *et al.* 1965; Koops 1969; Murty *et al.* 1972; Lembke and Frahm 1964) and iron at average levels of 100 to 250µg/liter. Despite the greater abundance of iron in milk, copper has been shown, by the use of specific chelating agents, to be the catalytic agent in the development of oxidized fluid milk (Smith and Dunkley 1962B).

The natural copper content of milk originates in the cow's food and is transmitted to the milk via the bloodstream (Haase and Dunkley 1970). The studies of Dunkley and co-workers (1968A) and Riest *et al.* (1967) suggest that an animal's feed can influence the natural copper content of its milk—a view which is not shared by others (Mulder *et al.* 1964). Nevertheless, the total natural copper content of a milk is not the overall deciding factor in the spontaneous development of an oxidized flavor in fluid milk.

Poulsen and Jensen (1966) reported that "neither the absolute amount nor the range in content of naturally occurring copper during the lactation period has any significant influence on the tendency of the milk to acquire oxidized flavor." Samuelsson (1966) investigated milks from cows of low and high yield production ranging in copper content from 0.023 to 0.204 ppm. He concluded that oxidation may occur irrespective of the copper content, but no oxidation faults have been observed in milks with a copper content of less than 0.060 ppm. Similar results have been reported by others (King and Dunkley 1959B).

Natural copper and iron exist in milk in the form of complexes with proteins, and as such are not dialyzable at the normal pH of milk (King *et al.* 1959; Samuelsson 1970). Copper and iron added to milk are, however, slightly dialyzable, the ease of dialysis of added copper increasing with the decrease in pH (Samuelsson 1970). The latter observation suggests that the copper–protein bond of added copper is different from that of natural copper. King *et al.* (1959) reported that 10 to 35% of the natural copper and 20 to 47% of the natural iron are associated with the fat globule membrane material. Only 2 to 3% of added copper and negligible percentages of added iron, however, become associated with the fat globule membrane. Similar trends in the distribution of natural and added copper in milk have been reported by others (Parks 1974); the subject has been reviewed by Haase and Dunkley (1970).

Samuelsson (1960) observed that most of the natural copper associated with the cream phase can be removed by washing with water and that the actual fat globule membrane proteins contain approximately 4% of the total natural copper content. Nevertheless, the value represents the highest concentration of copper per gram of protein in the milk system. Koops (1969) stated that "although the amount of natural copper in early lactation may be very high, the concentration of copper (average 11.0 μg/100 g fat globules) in the membrane does not deviate substantially from that of normal uncontaminated milk." King (1958) observed that milks which developed an oxidized flavor spontaneously had a higher total copper concentration in the fat globule membrane than did milks classified as susceptible or resistant.

Samuelsson (1966) concluded, on the basis of his studies, that the close proximity of a copper–protein complex to the phospholipids which are also associated with the fat globule membrane is an important consideration in the development of an oxidized flavor in fluid milks. Haase and Dunkley (1970) stated that although "some aspects of catalysis of oxidative reactions in milk by copper still appear anomalous . . . the mechanism of oxidized flavor development with copper as catalyst involves a specific grouping of lipoprotein–metal complexes in which the spatial orientation is a critical factor."

Edmondson et al. (1971), who studied the enrichment of whole milk with iron, found that ferrous compounds normally caused a definite oxidized flavor when added before pasteurization. Aeration before addition of the iron reduced the off-flavor. The authors recommended the addition of ferric ammonium citrate followed by pasteurization at 81°C. Kurtz et al. (1973) reported that iron salts can be added in amounts equivalent to 20 mg iron per liter of skim milk with no adverse flavor effects when iron-fortified dry milk is reconstituted to skim milk or used in the preparation of 2% milk. Hegenauer et al. (1979A) reported that emulsification of milk fat prior to fortification greatly reduced lipid peroxidation by all metal complexes. These researchers (Hegenauer et al. 1979B) concluded that chelated iron and copper should be added after homogenization but before pasteurization by a high-temperature–short-time process.

Several investigators (Roh et al. 1976; Shipe et al. 1972; Gregory and Shipe 1975) investigated the removal of copper from milk. Roh et al. (1976) removed more than 90% of the copper from milk with thiosuccinylated aminoethyl cellulose. Shipe et al. (1972) used glass-bound trypsin to inhibit metal-induced peroxidation and associated vitamin A degradation. Aging milk prior to exposure to the metal catalyst increased resistance to oxidative lipid deterioration and enhanced the apparent antioxidative effect of trypsin treatment (Gregory and Shipe 1975).

Role of Ascorbic Acid

That copper, naturally occurring or present as a contaminant, accelerates the development of oxidative deterioration in fluid milk is evident.

However, its presence is not the only determinant of whether or not oxidative deterioration occurs. Olson and Brown (1942) showed that washed cream (free of ascorbic acid) from susceptible milk did not develop an oxidized flavor when contaminated with copper and stored for three days. Subsequently, the addition of ascorbic acid to washed cream, even in the absence of added copper, was observed to promote the development of an oxidized flavor (Pont 1952). Krukovsky and Guthrie (1945) and Krukovsky (1961) reported that 0.1 ppm added copper did not promote oxidative flavors in milk or butter depleted of their Vitamin C content by quick and complete oxidation of ascorbic acid to dehydroascorbic acid. Krukovsky (1955) and Krukovsky and Guthrie (1945) further showed that the oxidative reaction in ascorbic acid–free milk could be initiated by the addition of ascorbic acid to such milk. Accordingly, these workers and others have concluded that ascorbic acid is an essential link in a chain of reactions resulting in the development of an oxidized flavor in fluid milk.

Various workers (Parks 1974) have observed a correlation between the oxidation of ascorbic acid to dehydroascorbic acid and the development of an oxidized flavor. Smith and Dunkley (1962A) concluded, however, that ascorbic acid oxidation cannot be used as a criterion for lipid oxidation. Their studies showed that although ascorbic acid oxidation curves for homogenized and pasteurized milk were similar, the homogenized samples were significantly more resistant to the development of an oxidized flavor. Furthermore, whereas pasteurization caused an appreciable decrease in the rate of ascorbic acid oxidation compared to raw milk, the pasteurized samples were more susceptible to oxidation.

Haase and Dunkley (1969B,C) reported, as a result of studies on model systems of potassium linoleate, that ascorbic acid functioned as a true catalyst, i.e., it accelerated the oxidation of linoleate, but it itself was not oxidized. Hegenauer et al. (1979A,B) observed that iron catalyzed an increase in the rate of autoxidation of ascorbate to dehydroascorbate but did not alter the equilibrium concentrations of ascorbate, dehydroascorbate, and ketogluconate. When copper was added to the system, however, oxidation of ascorbic acid occurred simultaneously with oxidation of linoleate. In this connection, Smith and Dunkley (1962C) reported that a significant correlation exists between the rate of ascorbic acid oxidation and the natural copper content of milk. Furthermore, King (1963) reported a positive relation between lipid oxidation and ascorbic acid oxidation in model systems containing fat globule membrane material, the component of uncontaminated milk having the highest concentration of copper per gram of liquid. Although ascorbic acid alone in model systems of linoleate have been observed to be pro-oxidant, low concentrations of ascorbic acid in combination with copper exhibited greater catalytic activity than the additive activity of the two catalysts individually (Haase and Dunkley 1969C). Possible explanations for the enhanced catalysis include reduction of copper by ascorbic acid to the more pro-oxidative cuprous form (Bauernfeind and Pinkert 1970; Haase and Dunkley 1969C; Smith and Dunkley 1962B), an increased concentration of a semidehydroascorbic acid radical (Bauernfeind and Pinkert 1970; Haase and Dunkley 1969C), and the formation of a metal–ascorbic acid–oxygen complex (Haase and Dunkley 1969C).

The behavior of ascorbic acid in the oxidative reaction, however, is anomalous, as evidenced by the studies of several workers (Bell *et al.* 1962; Bell and Mucha 1949; Chilson 1935; Krukovsky and Guthrie 1946). Their results indicate that concentrations normal to milk (10 to 20 mg/liter) promote oxidative deterioration, while higher concentrations (50 to 200 mg/liter) inhibit the development of off-flavors.

Sidhu *et al.* (1976) added H_2O_2 just after milking, in slight excess of stochiometric amounts to delay the development of oxidized flavors in cow's milk high in linoleic acid.

Various researchers have proposed explanations for the inhibitory behavior of high concentrations of ascorbic acid in fluid milk. Chilson (1935) reported that added ascorbic acid acts as a reducing agent which oxidizes more readily than milk fat. This either prevents or prolongs the time required for fat oxidation and the development of an oxidized flavor. Bell *et al.* (1962) concluded that the addition of L-ascorbic acid to concentrated sweet cream lowers its oxidation-reduction potential and thus produces a medium less conducive to oxidation. In this connection, Campbell *et al.* (1959) reported that the oxidation-reduction potential of milk is entirely dependent on its vitamin C content, and Greenback (1948) has shown that the oxidation of ascorbic acid to dehydroascorbic acid is reflected in gradual increases in pH. Krukovsky (1961) reported that the oxidative reaction is initiated more rapidly in milk when the ratio of ascorbic to dehydroascorbic acid is approximately 1:1 or lower. He states that "an unfavorable proportion of dehydroascorbic acid could not be accumulated if the rate of its oxidation to non-reducible substances surpassed that of ascorbic acid to dehydroascorbic acid. Consequently, the protective influence of ascorbic acid added in large but variable quantities to milk could be attributed to the exhaustion of occluded oxygen prior to the establishment of a favorable equilibrium between these two forms of vitamin C." Smith and Dunkley (1962B) disputed this theory and suggested that the results were influenced by higher than normal ascorbic acid contents when the ratio of ascorbic acid to dehydroascorbic acid was greater than 1:1 in the experimental milks. In this regard, King (1958) was not able to duplicate Krukovsky's results in milks with normal ascorbic acid levels.

King (1963) theorized that when the initial concentration of ascorbic acid increases beyond that necessary to saturate the copper in the system, the oxidation of ascorbic acid becomes so rapid and the products of the reaction accumulate so rapidly that they either block the reaction involving the lipids in the system or prevent the copper from acting as a catalyst.

Haase and Dunkley (1969B) reported that although high concentrations of ascorbic acid in model systems of potassium linoleate were prooxidant, a decrease in the rate of oxidation was observed. Haase and Dunkley (1969C) further noted that certain concentrations of ascorbic acid and copper inhibited the formation of conjugated dienes, but not the oxidation of ascorbic acid, and caused a rapid loss of part of the conjugated dienes already present in the system. They theorized that certain combination concentrations of ascorbic acid and copper inhibit oxidation by the formation of free radical inhibitors which terminate free- radical chain reactions, and that the inhibitors are complexes that include the free radicals.

Role of α-Tocopherol

The literature appears to be in general agreement that the use of green feeds tends to inhibit and that of dry feeds to promote the development of oxidized flavors in dairy products (Parks 1974). Furthermore, the observation that milks produced during the winter months are more susceptible to oxidative deterioration is the result, no doubt, of differences in feeding practices.

Investigations concerned with variations in the oxidative stability of milk as a result of feeding practices have centered on the transfer to milk of natural antioxidants. Although Kanno et al. (1968) have reported the presence of γ-tocopherol, the only known natural antioxidant of consequence is α-tocopherol.

Milk fat contains, on the average, approximately 2 μg/per gram of α-tocopherol (Bruhn and Franke 1971; Erickson and Dunkley 1964; Kanno et al. 1968). Dicks (1965) has assembled a comprehensive bibliography of the literature on the α-tocopherol content of milk and its products, including data on the numerous variables which influence the vitamin E content. Foremost among these variables is the feed of the animal as influenced by the season of the year. Kanno et al. (1968) reported that milk produced from May to October on pasture feeding averaged 33.8 μg α-tocopherol per gram of fat, while that produced by dry lot feeding from November to April contained an average of 21.6 μg α-tocopherol per gram of fat. Similar results have been reported by others (King et al. 1967; Kurtz 1974; Seerless and Armstrong 1970).

Krukovsky et al. (1950) found a significant correlation between the tocopherol content of milk fat and the ability of milk to resist autoxidation. A high proportion of samples which contained less than 25 μg α-tocopherol per gram of fat were unstable and developed oxidized flavors during storage. Erickson et al. (1963) reported that the tocopherol concentration in the fat globule membrane lipids correlated more closely with oxidative stability of the milk than did the tocopherol content of the butter oil. Dunkley et al. (1968B) stated, however, that the concentration of α-tocopherol in milk is not satisfactory as a sole criterion for predicting oxidative stability and that the concentration of copper must also be considered. In this regard, King et al. (1966) found a direct relationship between the tocopherol level and the percentage of copper tolerated by milk. Spontaneous milk oxidation was reported by Bruhn and Franke (1971 to be directly proportional to the copper content and inversely proportional to the α-tocopherol content of milk.

Erickson et al. (1964) observed that, although containing only 8% of the total tocopherols in milk, the fat globule membrane contains the highest concentration of α-tocopherol per gram of fat in milk (44.0 μg/g). Erickson and co-workers (1963) had previously concluded that since "the lipids in the fat globule membrane are most susceptible to oxidants because of their unsaturation and their close association with the pro-oxidants copper and ascorbic acid, the α-tocopherol in the membrane is more important in inhibiting oxidation than the inside the fat globule." A similar conclusion has also been reached by King (1968).

Several studies have been concerned with increasing the α-tocopherol levels of milk to prevent the development of oxidized flavors when tocopherol-rich forages are not available for feed. Dunkley et al. (1966, 1967), King et al. (1966), and Merk and Crasemann (1961) have reported increases in the α-tocopherol content of milk and increased resistance to spontaneous and copper-induced oxidation when the cow's ration was supplemented with varying proportion of α-tocopherol acetate. Dunkley et al. (1966) reported that supplementing the ration of an animal with 500 mg d-α-tocopherol acetate increased the total milk tocopherol content by 28.6 μg/g lipid; and King et al. (1967) reported that supplementing the feed to achieve a total intake of 1.0 g α-tocopherol per cow per day provided an effective control against oxidation in milk containing 0.1 ppm copper contamination. Several reports (Dunkley et al. 1969B; King et al. 1966; Schingoethe et al. 1979) have shown that approximately 2% total α-tocopherol intake is transferred to milk; thus, supplementing the ration with α-tocopherol acetate is a relatively inefficient procedure. In contrast, Goering *et al.* (1976), who fed a protected safflower supplement, observed a 200% increase in the vitamin E content of the polyunsaturated milk and attributed this unusually large transfer of vitamin E to the increased amounts of lipids that were absorbed in cows fed protected lipids. King (1968) has reported that the direct addition of d-α-tocopherol acetate in an emulsified form at a concentration of 25 μg/g milk fat would prevent the development of oxidized flavor in milk containing 0.1 ppm added copper—the same α-tocopherol concentration found to be effective when the ration was supplemented with α-tocopherol acetate. Control of oxidized flavor by direct addition of emulsified α-tocopherol to milk can be achieved with only 1% of the amount required by ration supplementation.

Factors Affecting Oxidative Deterioration in Milk and Its Products

Storage Temperature. The role of storage temperature in the oxidative deterioration of dairy products is anomalous. Dunkley and Franke (1967) observed more intense oxidized flavors and higher TBA values in fluid milks stored at 0°C than at 4° and 8°C. The flavor intensity and the TBA values decreased with increasing storage temperature. Other conditions being equal, condensed milk stored at −17°C is more susceptible to the development of oxidized flavor than is condensed milk maintained at −7°C (Parks 1974).

In contrast to the above results, low storage temperatures tend to decrease the rate of light-induced oxidative deterioration (Dunkley *et al.* 1962A) and to decrease or inhibit oxidative deterioration in other dairy products. Pyenson and Tracy (1946) reported that storage temperatures of 2°C retarded the development of oxidative deterioration in dry whole milk, as determined by oxygen absorption and flavor scores, in comparison with samples stored at 38°C in an atmosphere of air. Downey (1969) reported that oxidative deterioration in UHT cream occurred 2 to 3 times more rapidly at 18°C than at 10°C, while little or no oxidation occurred at 4°C. Sattler-Dornbacher (1963) reported an increase in the oxidation-reduction potential of butter as the storage temperature increased, with a corresponding increase in the rate of flavor deterioration. Hamm *et al.* (1968) demonstrated the rates of oxidative deterioration in butter oils during storage at temperatures ranging from −10°C to +50°C. Despite dramatic differences in the rate of oxidation, increasing rates with increasing temperatures, they concluded that the same flavors were formed on storage and that the reaction sequence for flavor formation was the same at all temperatures.

Oxygen Levels.   The inhibition of oxidative deterioration in fluid milk held at higher storage temperatures has been attributed by various workers (Parks 1974) to a lowering of the oxygen content as a result of bacterial activity. In this respect, it has been noted that the increased incidence of oxidized flavor in milk has paralleled the bacteriologically improved milk supply (Jenness and Patton 1959). Collins and Dunkley (1957) have reported, however, that although large numbers of bacteria slightly retard the development of oxidized flavor, the relatively small number of bacteria normally found in market milk is of no practical consequence in determining whether or not milk will develop an off-flavor. Furthermore, Sharp et al. (1942) stated that the number of bacteria necessary to reduce the oxygen content materially would be sufficient to cause other types of deterioration.

Removal of the dissolved oxygen in fluid milk or its replacement with nitrogen was shown to inhibit the development of oxidized flavors. Sharp et al. (1941) showed the deaeration inhibits the appearance of an off-flavor even in the presence of 1 mg copper per liter of milk. Singleton et al. (1963) confirmed previous observations that oxygen was required for the development of light-induced off-flavors. Schaffer et al. (1946), applying deaeration to products other than fluid milk, concluded that to prevent the production of a tallowy flavor in butter oil, the available oxygen should be less than 0.8% of the volume of the fat. Similar storage conditions were also proposed by Lea et al. (1943). Although the deaeration of these products is of significance only from a scientific standpoint, the deaeration of dry milk products has practical applications.

Vacuum treatment or replacement of available oxygen with an inert gas has proved its reliability in preventing or retarding the onset of oxidation in dry whole milk for extended periods of storage. Greenbank et al. (1946) showed that inert gas packing to an oxygen level of 3 to 4% increased the storage life of whole milk powder two to three times over that of air-packed samples, the length of storage being dependent on the initial quality of the product. Lea et al. (1943) showed that whereas oxidation deterioration in milk powders packed at the 3 to 6% oxygen level was retarded significantly, inert gas containing 0.5 to 1.0% oxygen prevented the development of recognizable tallowy flavors for an indefinite period. Tamsma et al. (1961) showed statistically a highly significant improvement in storage stability of whole milk powders packed in inert gases containing 0.1% oxygen over those packed at 1% oxygen level. Schaffer et al. (1946) concluded that the time required for the production of a tallowy flavor is inversely proportional to the oxygen concentration.

Several deaeration techniques other than mechanical methods have been utilized to inhibit or retard the development of tallowy flavors in dry milks. Meyer and Jokay (1960) reported that milk powders packed in the presence of an oxygen scavenger (glucose oxidase-catalase) and a desiccant (calcium oxide) were comparable in flavor to samples stored in the presence of an inert gas, the enzymes demonstrating the ability to reduce oxygen levels to 0.5% in 1 week. Jackson and Loo (1959), employing an oxygen-absorbing mixture (0.5 g Na_2SO_3 and 0.75 g $CuSO_4 \cdot 5H_2O$) enclosed in porous paper pouches, demonstrated keeping qualities equal to those of dry milks stored in the presence of an inert gas. Abbot and Waite (1961) reported favorable results in the keeping quality of dry whole milk by using a mixture of 90% nitrogen and 10% hydrogen in the presence of a palladium catalyst. The metal catalyzes the formation of water from the hydrogen and residual oxygen to produce an almost oxygen-free atmosphere in the pack. Tamsma et al. (1967) reported obtaining within 24 hr a pack containing less than 0.001% oxygen by the use of an oxygen-scavenging system consisting of 95% nitrogen, 5% hydrogen, and a platinum catalyst. Marked improvements in the keeping quality of milk powders packed in the scavenging system were reported.

Heat Treatment.   Pasteurization of fluid milk leads to increased susceptibility to spontaneous (Bergman et al. 1962B), copper-induced (Parry et al. 1966; Smith and Dunkley 1962A), and light-induced oxidized flavor (Finley 1968). Heating to higher temperatures, however, reduces the susceptibility (Bergman et al. 1962B; Smith and Dunkley 1962A). A possible explanation for the increased incidence of oxidized flavor as a result of pasteurization temperatures is suggested by several studies. Sargent and Stine (1964) reported a substantial migration of added copper to the cream phase of milk at temperatures above 60°C. Duin and Brons (1967) also observed an increase in the copper content of creams prepared from pasteurized milk. Samuelsson (1967) reported that washed cream made from milk heated to 80°C for 10 min contained twice as much copper as that prepared from unheated milk. The migration of the additional copper to the cream phase, which also contains the readily oxidized phospholipids, increased the potential of the system for oxidative deterioration. Tarassuk et al. (1959) also observed that washed cream is very sensitive to the development of a trainy (fishy) flavor when heated to temperatures between 60° and 90°C. The effect of previous heat treatment on the copper content of butter was reported by Van Duin and Brons (1967). They observed that pasteurization of 78°C for 15 to 30 sec produced high copper concentrations in butter and low concentrations in the buttermilk, the reverse being true when the cream was heated to above 82°C. They recommended that creams prepared from pasteurized milks should be heated to the higher temperatures to decrease the susceptibility of butter to oxidative deterioration during storage.

The inhibitory effect of high heat treatment on oxidative deterioration of fluid milk and its products has been reported by various workers (Tamsma *et al.* 1962; Parks 1974). Gould and Sommer (1939), in conjunction with studies on the development of a cooked flavor in heated milks, noted a decrease in the oxidation-reduction potential of the product. They attributed the cooked flavor to the formation of sulfhydryl compounds and correlated the liberation of these compounds with the heat retardation and prevention of oxidized flavor. The work of Josephson and Doan (1939), conducted simultaneously with that of these workers, confirmed the relationship between sulfhydryl compounds, cooked flavor, decreased Eh, and inhibition of oxidized flavor. They further reported that most heated products do not become tallowy or oxidized until the sulfhydryls are first oxidized and the cooked flavor has disappeared. Wilson and Herreid (1969) prolonged substantially the onset of oxidative deterioration of 30% sterilized cream by increasing to 13% the solids-not-fat content of the cream prior to sterilization, presumably by increasing the potential sulfhydryl content of the finished product. Gould and Keeney (1957) showed that an oxidized flavor occurred in heated cream to which copper had been added when the active sulfhydryl compounds had decreased to a level approximating 3 mg/liter cystine HCL. Taylor and Richardson (1980A) reported that sulfhydryl groups were responsible for only part of the antioxidant activity of skim milk and that antioxidant activity resided in the proteins, principally in the casein, which had only a small amount of sulfhydryl groups. These investigators observed that sonication greatly increased the antioxidant activity of skim milk and attributed this increase to the increasing effective casein concentration that would be produced by disrupted casein micelles (Taylor and Richardson 1980B).

β-Lactoglobulin has been shown by Larsson and Jenness (1950) to be the major source of sulfhydryl groups in milk, while the fat globule membrane material contributes a minor portion of these reducing compounds. This finding was confirmed by Hutton and Patton (1952).

Time–temperature relationships have been established by various workers as being optimum for preventing or retarding the development of oxidized flavors in dairy products; cream, 88°C for 5 min; condensed milk, 76.5°C for 8 min; dry whole milk, preheated at 76.5°C for 20 min; and frozen whole milk, 76.5°C for 1 min (Parks 1974). Few, if any, instances of a tallowy flavor have been reported in evaporated milk; undoubtedly, a major reason for its stability toward oxidation is the sterilization temperatures employed in its manufacture.

Josephson (1943) reported that butterfat prepared from butter heated to 149°, 177°, and 204.5°C was extremely stable to oxidation, while that heated to 121°C oxidized readily when stored at 60°C. When butter oil itself was heated from 121 to 204.5°C, it also oxidized rapidly. However, the addition of 1% skim milk powder to butter oil prior to heating at 204.5°C for 10 min also resulted in a significant antioxidative effect, which Josephson concluded was the result of a protein–lactose reaction (carmelization). Wyatt and Day (1965) reported that the addition of 0.5% nonfat milk solids to butter oil followed by heating at 200°C and 15 mm Hg for 15 min caused the formation of antioxidants which protected the butter oil against oxidative deterioration for 1 year, surpassing the effectiveness of many synthetic antioxidants tested.

Exposure to Light. The catalytic effect of natural light in promoting off-flavor development in fluid milk has been recognized for some years. The extent of deterioration appears to be dependent on the wavelengths involved, the intensity of the source, and the length of exposure (Aurand et al. 1966; Dunkley et al. 1962A). Off-flavors have also been reported to develop in butterfat which has been exposed to the action of natural light (Parks 1974). In addition to natural light, incandescent or fluorescent light employed in storage coolers may promote deteriorative reactions (Smith and MacLeod 1955; Dimick 1973), while the development of off-flavors is the limiting factor in the preservation of dairy products by high-energy radiation (Day et al. 1957; Hoff et al. 1959). Efforts to inhibit or retard the onset of off-flavors as a result of exposure to sunlight led to the introduction of doorstep coolers and, in certain cases, of amber-colored milk bottles. A shift toward the marketing of fluid milk in plastic containers without adequate protection from light has greatly aggravated the problem of light-induced oxidized flavor (White and Bulthaus 1982).

Two distinct flavors may develop in milk exposed to light (Velander and Patton 1955): a burnt, activated, or sunlight flavor, which develops rapidly, and a typically oxidized flavor, which develops on prolonged exposure (Storgards and Ljungren 1962). It is possible that the presence of contradictory statements in the literature regarding deterioration on exposure to light may be attributed to the failure of various investigators to recognize the existence of more than one off-flavor.

Studies by Patton (1954) and Velander and Patton (1955) have shown that riboflavin plays a significant role in the development of the activated flavor. Although removal of riboflavin from milk by passing through Florisil prevented the development of an activated flavor, such treatments did not prevent the development of the oxidized flavor. The later observation does not agree with the reports of other workers (Aurand *et al.* 1966); which indicate that riboflavin plays a significant role in the development of the oxidized flavor. Ascorbic acid has also been implicated in the development of off-flavors in fluid milks exposed to light (Aurand *et al.* 1966; Dunkley *et al.* 1962B; DeMan 1980). The exact nature of its involvement, however, is not clear.

Limited studies have been conducted on the lipid components oxidized in milk exposed to sunlight. Finley (1968) and Finley and Shipe (1971) observed a decease in the oleic and linoleic acid contents of an isolated low-density lipoprotein (LDL) from milk and implicated the lipoprotein as a major substrate for the photoxidation reaction. Although previous studies (Wishner and Keeney 1963) suggested that the monoene fatty acids are important oxidizing substrates in milk exposed to sunlight, Wishner (1964) noted that photoxidation of methyllinoleate in the presence of photosensitizers produces significant percentages of the less stable 11-hydroperoxide (Khan *et al.* 1954), which on decomposition forms alk-2-enals, the significant carbonyls found in milk exposed to sunlight.

The sunlight flavor has been shown (Patton 1954) to originate in the proteins of milk. Hendrickx *et al.* (1963) concluded that the serum proteins are the main source of activated flavor in milk, with riboflavin as the photosensitizer. Similar results have been reported by Storgards and Ljungren (1962). Singelton *et al.* (1963) demonstrated a relationship between riboflavin destruction, tryptophan destruction, and the intensity of the sunlight flavor in milk, and implicated a tryptophan-containing protein rather than a single low molecular weight compound as one of the reactants. Finley (1968) reported that an LDL fraction associated with the fat globule membrane served as a carrier and a precursor for the light-induced off-flavor. Studies of the degradation of the lipoprotein on exposure to light showed that both the lipid and protein portions of the lipoprotein were degraded. In addition to tryptophan, they observed the destruction of methionine, tyrosine, cysteine, and lysine in the lipoprotein on exposure to light in the presence of riboflavin. The photoxidation of amino acids other than tryptophan has been observed in enzymes exposed to sunlight (Wishner 1964).

Methional, formed by the degradation of the amino acid methionine, has been reported (Patton 1954; Velander and Patton 1955) to be the principal contributor to the activated flavor. Samuelsson (1962) reported, in studies of dio- and tripeptides containing methionine, that irradiation did not result in any hydrolysis of the peptides, and the presence of methional in the reaction products could not be demonstrated. He concluded that methional can only occur in irradiated milks from the free methionine in the milk serum. Thiols, sulfides, and disulfides observed as products of irradiated peptides may be of greater significance in the activated flavor.

Acidity. The development of a fishy flavor in butter is well known. Cream acidities ranging from 0.20 to 0.30% appear to represent those levels at which flavor development is marginal (Parks 1974). Although the development of fishy flavors in unsalted butters is rarely encountered, it is not restricted to those products containing salt. Pont *et al.* (1960) induced the development of a fishy flavor in commercial butterfat by the addition of nordihydroguaiaretic acid and citric or lactic acid. In addition, Tarassuk *et al.* (1959) reported the development of fishy flavors in washed cream adjusted to pH 4.6.

Koops (1964A) conducted a comprehensive study of the development of the trainy (fishy) flavor which occurs in butter prepared from cultured cream (pH 4.6) during cold storage. He observed (Koops 1969) that although the acidification of milk or cream to pH 4.6 did not result in a transfer of natural copper from the plasma proteins to the fat globule membrane, 30 to 40% of added copper migrated to the membrane proteins at pH 4.6. He concluded (Koops 1964A) that the development of a trainy flavor in cultured butter is the result of the migration of the plasma-bound (contaminated) copper to the fat globule membrane and the enhanced interaction between the cephalin fraction of the membrane phospholipids, which is highly susceptible to oxygen (Koops 1963), and the copper-containing membrane protein.

Although other dairy products have not been studied extensively, reports suggest that titratable acidity as well as hydrogen ion concentration tend to influence the development of oxidative deterioration. A relationship was found between the titratable acidity and the development of an oxidized flavor in milk (Parks 1974). While milks developed an oxidized flavor at a titratable acidity of 0.19%, the deteriorative mechanism was inhibited when the milks were neutralized to acidities of 0.145% or less. An increase in pH of 0.1 was sufficient to inhibit the development of oxidized flavors in fluid milks for 24 hr (Parks 1974). In addition to fluid milk, Dahle and Folkers (1933) attributed the development of oxidized flavors in strawberry ice cream to the presence of copper and the acid content of the fruit.

Homogenization. Homogenization was found to inhibit the development of an oxidized flavor in fluid milk by Tracey *et al.* (1933). Subsequently, similar observations were reported on cream, ice cream, dry whole milk, and frozen condensed milk (Parks 1974). The inhibitory effect, however, is not absolute. Roadhouse and Henderson (1950) found that the absolute pressure required varies with different milks contaminated with the same concentration of cupric ion. The results of Smith and Dunkley (1962A) indicate that the inhibitory effect of homogenization is dependent on the degree of metallic contamination.

Various workers have proposed explanations for the inhibitory effect of homogenization on oxidative deterioration. Tracey et al. (1933) considered it to be apparent rather than actual, resulting from changes in the physical consistency of the milk, which may alter the taste. These workers based their proposal on the observation that homogenization has no apparent effect on the Eh of milk. Similar observations have been noted by others (Larson et al. 1941). Still others have proposed that the inhibition is real and is due to migration of the phospholipids into either the serum phase (Thurston et al. 1936) or the interior of the fat globule (Krukovsky 1952), to general redistribution of the phospholipids in the milk proper (Greenbank and Pallansch 1961), or to denaturation of proteins resulting in an increase in the number of available -SH groups (Forster and Sommer 1951). King (1958) proposed that homogenization produces an irreversible change in the structural configuration of the copper–protein complex in such a way that ascorbic acid is no longer able to initiate the formation of lipid free radicals. Smith and Dunkley (1962A) theorized that homogenization causes a change in the copper–protein binding by the formation of a chelate that is less active in ascorbic acid oxidation and inactive in lipid peroxidation. Tarassuk and Koops (1960) stated that "the decrease in concentration of phospholipids and the copper–protein complex per unit of newly formed fat globule surface appears to be the most important factor, if not the only one, that retards the development of oxidized flavor in homogenized milk."

Dunkley et al. (1962B) demonstrated, by the use of TBA values and a highly trained taste panel, that although homogenization inhibits light-induced lipid oxidation, the process increases the susceptibility of milk to the development of the activated flavor. An increase in the intensity of off-flavors in homogenized milks exposed to sunlight has been reported by several workers (Dahle 1938; Kelly 1942). Finley (1968) concluded, as a result of his studies, that any treatment (e.g., homogenization) which affects the fat globule membrane increases the susceptibility of milk to light-induced off-flavors. It is evident from the literature that homogenization affords a degree of protection against oxidative deterioration in fluid milks provided excessive metallic contamination and undue exposure to light are avoided.

Carbonyl Content of Oxidized Dairy Products

Considerable effort has been expended in recent years on the odorous compounds formed in autoxidized dairy products. Although some of the early identification studies lack present-day sophisticated methodology, may be incomplete, and do not differentiate between their isomeric forms of the various compounds, their contribution to the knowledge of the products of autoxidation in dairy products is invaluable.

Despite the general similarity in the qualitative carbonyl content of oxidized dairy products, flavor differences are apparent. Attempts to correlate the off-flavors with specific compounds or groups of compounds, however, are difficult for several reasons. These include (1) the multitude of compounds produced; (2) difficulties arising in the quantitative analyses of oxidized dairy products; (3) differences in threshold values of individual compounds; (4) similarity of flavors imparted by individual compounds near threshold; (5) a possible additive and/or antagonistic effect, with regard to both flavor and threshold values of mixtures of compounds; (6) the possible existence of a compound or group of compounds heretofore not identified; and (7) the difficulties involved in adding pure compounds of dairy products as a means of evaluating their flavor characteristics.

Several compounds formed by the autoxidation of milk lipids, however, have been implicated in specific off-flavors. Stark and Forss (1962) have identified 1-octen-3-one as the compound responsible for the metallic flavor which develops in dairy products. This compound has also been shown to be an integral part of other oxidized flavor defects (Badings 1970; Forss *et al.* 1960A,B). 4-*cis*-Heptenal, responsible for the creamy flavor of butter (Begeman and Koster 1964), results from autoxidation of minor isolinoleic acids in butterfat (Jong and Van der Wel 1964). At higher concentrations, this compound has also been implicated in the trainy flavor which develops in cold storage butter (Badings 1965). 6-*trans*-Nonenal has been identified as the compound responsible for the "drier" flavor (Parks *et al.* 1969) which frequently appears in freshly prepared foam spray-dried milk—an off-flavor which is peculiar to this particular product. Although the evidence suggests that it is formed in foam spray-dried milk by trace ozonolysis of minor milk lipids, it has also been identified in stored sterile milks (Parks and Allen 1972). The last observation suggests that it may also appear in dairy products as a result of autoxidation reactions (Keppler *et al.* 1967).

Shipe *et al.* (1978) summarized the principal off-flavors and standardized their reproduction primarily to serve as reference aids in training research and quality control personnel.

Other studies suggest that the preponderance of certain carbonyls or group of carbonyls is involved in the off-flavors of various dairy products. Forss *et al.* (1955A,B) reported that the C_6 to C_{11} 2-enals and the C_6 to C_{11} 2,4-dienals—and, more specifically, 2-octenal, 2-nonenal, 2,4-heptadienal, and 2,4-nonadienal—constitute a basic and characteristic factor in the copper-induced cardboard flavor in skim milk. The same workers concluded that "while these compounds in milk closely simulate the cardboard flavor, the resemblance is not complete" and that "the defect contains further subsidiary flavor elements."

Bassette and Keeney (1960) ascribed the cereal-type flavor in dry skim milk to a homologous series of saturated aldehydes resulting from lipid oxidation in conjunction with products of the browning reaction. The results of Parks and Patton (1961) suggest that saturated and unsaturated aldehydes at levels near threshold may impart an off-flavor suggestive of staleness in dry whole milk. Wishner and Keeney (1963) concluded from studies on milk exposed to sunlight that C_6 to C_{11} alk-2-enals are important contributors to the oxidized flavor in this product. Parks *et al.* (1963) concluded, as a result of quantitative carbonyl analysis and flavor studies, that alk-2-4-dienals, especially 2,4-decadienal, constitute a major portion of the off-flavor associated with spontaneously oxidized fluid milk. Forss *et al.* (1960A,B) reported that the fishy flavor in butterfat and washed cream is in reality a mixture of an oily fraction and 1-octene-3-one, the compound responsible for the metallic flavor. *n*-Heptanal, *n*-hexanal, and 2-hexanal were found to be constituents of the oily fraction in washed cream, and these three carbonyls plus heptanone-2 were constituents of the oily fraction isolated from fishy butterfat. Badings (1970) identified 40 volatile compounds in cold storage cultured butter which had a trainy (fishy) off-flavor. Included among the 14 compounds which were present at above-threshold levels were 4-*cis*-heptenal; 2-*trans*, 4-*cis*-decadienal; 2-*trans*, 6-*cis*-nonadienal; 2,2,7-decatrienal; 3-*trans*,5-*cis*-octadien-2-one; 1-octene-3-one; and 1-octen-3-ol. Keen *et al.* (1976) reported on the carbonyls in ultra-high-temperature milk. They found acetaldehyde, hexanal, heptanal, octanal, and decanal in this milk but not in the control and ascribed the presence of these aldehydes and the higher concentration of nonanal in the ultra-high-temperature milk to the heat treatment.

Comparative studies by Forss and co-workers (1960A,C) on the fishy, tallowy, and painty flavors of butterfat tended to emphasize the importance of the relative and total carbonyl contents in dairy products with different off-flavors. These researchers showed that three factors distinguished painty and tallowy butterfat from fishy butterfat. First, there was a relative increase in the *n*-heptanal, *n*-octanal, *n*-nonanal, heptanone-2, 2-heptenal, and 2 nonenal in the tallowy butterfat and a relative increase in the n-pentanal and the C_5 to C_{10} alk-2-enals in the painty butterfat. Second, 1-octen-3-one was present in such low concentrations in both the tallowy and painty butterfats as to have no effect on the flavor. Third, the total weight of the volatile carbonyl compounds was about 10 times greater in the tallowy and 100 times greater in the painty butterfat than in the fishy butterfat.

REFERENCES

Abbot, J. and Waite, R. 1961. Gas packing milk powder with a mixture of nitrogen and hydrogen in the presence of palladium catalyst. *J. Dairy Res. 28,* 285–292.

Abbot, J. and Waite, R. 1962. The effect of antioxidants on the keeping quality of whole milk powder. I. Flavones, gallates, butylhydroxyanisole and nordihydroguaiaretic acid. *J. Dairy Res. 29,* 55–61.

Abbot, J. and Waite, R. 1965. The effect of antioxidants on the keeping quality of whole milk powder. II. Tocopherols. *J. Dairy Res. 32,* 143–146.

Abdel Hamid, L. B., Mahran, G. A., Shehata, A. E. and Osman, S. G. 1977. Lipase activity in buffaloes' milk. 2. Effect of feeding system, animal age and milking phase and meal. *Egyptian J. Dairy Sci. 5,* 7–10.

Al-Shabibi, M. M. A., Langner, E. H., Tobias, J. and Tuckey, S. L. 1964. Effect of added fatty acids on the flavor of milk. *J. Dairy Sci. 47,* 295–296.

American Oil Chemists' Soc. 1971. Official and tentative methods. Official method Cd 8–53. Am. Oil Chem. Soc., Chicago, p. 122.

Anderson, M. 1979. Enzyme immunoassay for measuring lipoprotein lipase activator in milk. *J. Dairy Sci. 62,* 1380–1383.

Arrington, L. R. and Krienke, W. A. 1954. Inhibition of the oxidized flavor of milk with chelating compounds. *J. Dairy Sci. 37,* 819–824.

Astrup, H. N. 1963. Oxidized flavor in milk and the xanthine oxidase inhibitor. *J. Dairy Sci. 46,* 1425.

Astrup, H. N. 1980. Effect on milk lipolysis of restricted feeding with and without supplementation with protected rape seed oil. *J. Dairy Res. 47,* 287–294.

Astrup, H. N., Vik-Mo, L., Skrovseth, O. and Ekern, A. 1980. Milk lipolysis when feeding saturated fatty acids to the cow. *Milchwissenschaft 35,* 1–4.

Aurand, L. W., Boone, N. H. and Giddings, G. G. 1977. Superoxide and singlet oxygen in milk lipid peroxidation. *J. Dairy Sci. 60,* 363–369.

Aurand, L. W., Chu, T. M., Singleton, J. A. and Shen, R. 1967. Xanthine oxidase activity and development of spontaneously oxidized flavor in milk. *J. Dairy Sci. 50,* 465–471.

Aurand, L. W., Singleton, J. A. and Noble, B. W. 1966. Photooxidation reactions in milk. *J. Dairy Sci. 49,* 138–143.

Aurand, L. W., Woods, A. E. 1959. Role of xanthine oxidase in the development of spontaneously oxidized flavor in milk. *J. Dairy Sci. 42,* 1111–1118.

Bachmann, M. 1961. Rancidity of milk and cheese. *Schweiz. Milchztg. 87,* No. 53; *Wissenschaftl. Beilage Nr. 79,* 625–635. (German).

Badings, H. T. 1959. Isolation and identification of carbonyl compounds formed by autoxidation of ammonium linoleate. *J. Am. Oil Chemists' Soc. 36,* 648–650.

Badings, H. T. 1960. Principles of autoxidation processes in lipids with special regard to the development of autoxidation off-flavors. *Neth. Milk Dairy J. 14,* 215–242.

Badings, H. T. 1965. The flavour of fresh butter and of butter with cold-storage defects in relation to the presence of 4-*cis*-heptenal. *Neth. Milk Dairy J. 19,* 69–72.

Badings, H. T. 1970. Cold-storage defects in butter and their relation to the autoxidation of unsaturated fatty acids. Ph.D thesis, Vageningen, The Netherlands.

Barnard, S. E. 1974. Rancid flavor is causing more consumer complaints. *Hoard's Dairyman 119* 1396–1397.

Barnard, S. E. 1979A. How two rancid flavor problems were solved. *Hoard's Dairyman 124,* 936–937.

Barnard, S. E. 1979B. Quality and flavor of store purchased milk samples. *J. Dairy Sci. 62,* Suppl. 1, 34.

Bassette, R. and Keeney, M. 1960. Identification of some volatile carbonyl compounds from nonfat dry milk. *J. Dairy Sci. 43,* 1744–1750.

Bateman, L. 1954. Olefin oxidation. *Quart. Rev. 3,* 147–167.

Bauernfeind, J. C. and Pinkert, D. M. 1970. Food processing with added ascorbic acid. *In: Advances in Food Research,* Vol. 18. C. O. Chichester, E. M. Mrak and G. F. Stewart (Editors). Academic Press, New York.

Begeman, P. H. and Koster, J. C. 1964. Components of butterfat 4-*cis*-heptenal: A cream-flavoured component of butter. *Nature 202,* 552–553.

Bell, R. W., Anderson, H. A. and Tittsler, R. P. 1962. Effect of L-ascorbic acid on the flavor stability of concentrated sweetened cream. *J. Dairy Sci. 45,* 1019–1020.

Bell, R. W. and Mucha, T. J. 1949. Deferment of an oxidized flavor in frozen milk by ascorbic acid fortification and by hydrogen peroxide oxidation of the ascorbic acid of fresh milk. *J. Dairy Sci. 32,* 833–840.

Bergman, T., Beetelsen, E., Berglof, A. and Larsson, S. 1962A. The occurrence of flavour defects in milk exposed to cold storage prior to pasteurization. *Int. Dairy Congr. 4,* 579–588.

Bergman, T., Beetelsen, E., Berglof, A. and Larsson, S. 1962B. *16th Int. Dairy Cong. Proc.* Vol. A. Sect. II: 1, 579.

Bjorke, K. and Castberg, H. B. 1976. Lipolytic activity in goat's milk. *Nord. Mejeri-Tidsskift 8,* 296–304.

Bolland, J. L. 1949. Kinetics of olefin oxidation. *Quart. Rev. 3,* 1–21.

Borges, M. S., True, L. C. and Mickle, J. B. 1974. Lipase activity and milk production as related to sudden decreases in the energy of the cow's ration. Oklahoma Agr. Exp. Sta. Misc. Pub. 92. pp. 278–284.

Brathen, G. 1980. Lipolysis in milk. Automated determination of the acidity value with the autoanalyzer and assessment of results. *Meierposten 69* (13), 345–352. (Norwegian).

Bruhn, J. C. and Franke, A. A. 1971. Influence of copper and tocopherol on the susceptibility of herd milk to spontaneous oxidized flavor. *J. Dairy Sci. 54,* 761–762.

Brunner, J. R. 1950. The effectiveness of some antifoaming agents in the condensing of skim milk and whey. *J. Dairy Sci. 33,* 741–746.

Campbell, J. J. R., Phelps, R. H. and Keur, L. B. 1959. Dependence of oxidation-reduction potential of milk on its vitamin C content. *J. Milk Food Technol. 22,* 346–347.

Cannon, J. A., Zilch, K. T., Burket, S. C. and Dutton, H. J. 1952. Analysis of fat acid oxidation products by countercurrent distribution methods. IV. Methyl linoleate, *J. Am. Oil Chemists' Soc. 29,* 447–452.

Castberg, H. B., Egelrud, T., Solberg, P. and Olivecrona, T. 1975A. Lipases in bovine milk and the relationship between the lipoprotein lipase and tributyrate hydrolysing activities in cream and skim-milk. *J. Dairy Res. 42,* 255–266.

Castberg, H. B., Solberg, P. and Egelrud, T. 1975B. Tributyrate as a substrate for the determination of lipase activity in milk. *J. Dairy Res. 42*, 247–253.

Chakrabarty, M. M., Bhattacharyya, D. and Kundu, M. K. 1969. A simple photometric method for microdetermination of fatty acids in lipids. *J. Am. Oil Chemists' Soc. 46*, 473–475.

Chen, J. H. S. and Bates, C. R. 1962. Observations on the pipeline milker operation and its effect on rancidity. *J. Milk food Technol. 25*, 176–182.

Chilson, W. H. 1935. What causes most common off-flavors of market milk? *Milk Plant Monthly 24*, 24–28.

Christensen, L. J., Decker, C. W. and Ashworth, U. S. 1951. The keeping quality of whole milk powder. I. The effect of preheat temperature of the milk on the development of rancid, oxidized and stale flavors with different storage conditions. *J. Dairy Sci. 34*, 404–411.

Clegg, R. A. 1980. Activation of milk lipase by serum proteins: Possible role in the occurrence of lipolysis in raw bovine milk. *J. Dairy Res. 47*, 61–70.

Cogan, T. M. 1977. A review of heat resistant lipases and proteinases and the quality of dairy products. *Ir. J. Food Sci. Technol. 1*, 95–105.

Cogan, T. M. 1980. Heat resistant lipases and proteinases and the quality of dairy products. *Int. Dairy Fed. Bull. 118*, 26–32.

Collins, E. B. and Dunkley, W. L. 1957. Influence of bacteria on the development of oxidized flavor in milk. *J. Dairy Sci. 40*, 603.

Colmey, J. C., Demott, B. J. and Ward, G. M. 1957. The influence of the stage of lactation on rancidity in raw milk. *J. Dairy Sci. 40*, 608–609.

Cornell, D. G. 1979. Distribution of some antioxidants in dairy products. *J. Dairy Sci. 62*, 861–868.

Cornell, D. G., Devilbiss, E. D. and Pallansch, M. J. 1971. Binding of antioxidants by milk proteins. *J. Dairy Sci. 54*, 634–637.

Costilow, R. N. and Speck, M. L. 1951. Inhibition of *Streptococcus lactis* in milk by fatty acids. *J. Dairy Sci. 34*, 1104–1110.

Crowe, L. K. 1955. Some factors affecting the quantity of water insoluble fatty acids in cream. *J. Dairy Sci. 38*, 969–980.

Dahle, C. D. 1935. Tallowy flavor in milk. Pa. Agr. Exp. Sta. Bull. 320.

Dahle, C. D. 1938. Preventing the oxidized flavor in milk and milk products. *Milk Dealer 27*(5): 68–86.

Dahle, C. D., Folkers, E. C. 1933. Factors contributing to an off-flavor in ice cream. *J. Dairy Sci. 16*, 529–547.

Dahle, C. D. and Palmer, L. S. 1937. The oxidized flavor in milk from the individual cow. Pa. Agr. Exp. Sta. Bull. *347.*

Day, E. A., Forss, D. A. and Patton, S. 1957. Flavor and odor defects of gamma-irradiated skim milk. I. Preliminary observations and the role of volatile carbonyl compounds. *J. Dairy Sci. 40*, 922–931.

Day, E. A. and Lillard , D. A. 1960. Autoxidation of milk lipids. I. Identification of volatile monocarbonyl compounds from autoxidized milk fat. *J. Dairy Sci. 43*, 585–597.

Day, E. A., Lillard, D. A. and Montgomery, M. W., 1963. Autoxidization of milk lipids. III. Effect on flavor of the additive interactions of carbonyl compounds at subthreshold concentrations. *J. Dairy Sci. 46*, 291–294.

Deeth, H. C. and Fitz-Gerald, C. H. 1975. Factors governing the susceptibility of milk to spontaneous lipolysis. *Int. Dairy Fed. Doc. 86*, 24–34.

Deeth, H. C. and Fitz-Gerald, C. H. 1976. Lipolysis in dairy products: A review. *Aust. J. Dairy Technol. 31*, 53–64.

Deeth, H. C. and Fitz-Gerald, C. H. 1977. Some factors involved in milk lipase activation by agitation. *J. Dairy Res. 44*, 569–583.

Deman, J. M. 1980. Effect of fluorescent light exposure on the sensory quality of milk. *Milchwissenschaft 35*, 725–726.

Demott, B. J. J. 1960. The influence of sugars upon lipolysis in milk. *J. Dairy Sci. 43*, 436.

Desnuelle, P. 1961. Pancreatic lipase. *Adv. Enzymol. 23*. 129–161.

Dicks, M. W. 1965. Vitamin E content of foods and feeds for human and animal consumption. Wyoming Agr. Exp. Sta. Bull. *435*.

Dijkman, A. J. and Schipper, C. J. 1965. Ransheid in boerderijmelk. *Veet-en Zuivelbericht 7*, 525–531. (Dutch).

Dimick, P. S. 1973. Effect of fluorescent light on the flavor and selected nutrients of homogenized milk held in conventional containers. *J. Milk Food Technol. 36*, 383–387.

Downey, W. K. 1969. Lipid oxidation as a source of off-flavor development during the storage of dairy products. *J. Soc. Dairy Technol. 22*, 154–162.

Downey, W. K. 1975. Identity of the major lipolytic enzyme activity of bovine milk in relation to spontaneous and induced lipolysis. *Int. Dairy Fed. Doc. 86*, 80–89.

Downey, W. K. 1980A. Review of the progress of dairy science: Flavour impairment from pre- and post-manufacture lipolysis in milk and dairy products. *J. Dairy Res. 47*, 237–252.

Downey, W. K. 1980B. Flavour impairment of milk and milk products due to lipolysis. II. risks from pre- and post-manufacture lipolysis. *Int. Dairy Fed. Bull. 118*, 4–18.

Downey, W. K. and Andrews, P. 1966. Studies on the properties of cow's milk tributyrinases and their interaction with milk proteins. *Biochem. J. 101*, 651–660.

Downey, W. K. and Murphy, R. F. 1975. Partitioning of the lipolytic enzymes in bovine milk. *Int. Dairy Fed. Doc. 86*, 19–23.

Driessen, F. M. 1976. A comparative study of the lipase in bovine colostrum and in bovine milk. *Neth. Milk Dairy J. 30*, 186–196.

Driessen, F. M. and Stadhouders, J. 1974A. A study of spontaneous rancidity. *Neth. Milk Dairy J. 28*, 130–145.

Driessen, F. M. and Stadhouders, J. 1974B. Thermal activation and inactivation of extracellular lipases of some Gram-negative bacteria common in milk. *Neth. Milk Dairy J. 28*, 10–22.

Duin, H. Van and Brons, C. 1967. The effect of pasteurization on the copper content of cream and butter. *Algemeen Zuivelblad 60*, 37–41. (Dutch).

Duncombe, W. G. 1963. The colorimetric micro-determination of long-chain fatty acids. *Biochem. J. 88*, 7–10.

Dunkley, W. L. and Franke, A. A. 1967. Evaluating susceptibility of milk to oxidized flavor. *J. Dairy Sci. 50*, 1–9.

Dunkley, W. L. and Smith, L. M. 1951. Hydrolytic rancidity in milk. III. Tributyrinase determination as a measure of lipase. *J. Dairy Sci. 34*, 935–939.

Dunkley, W. L., Franke, A. A. and Robb, J. 1968B. Tocopherol concentration and oxidative stability of milk from cows fed supplements of d- or dl-o-tocopheryl acetate. *J. Dairy Sci. 51*, 531–534.

Dunkley, W. L., Franke, A. A. and Robb, J. and Ronning, M. J. 1968A. Influence of dietary copper and ethylenediaminetetraacetate on copper concentration and oxidative stability of milk. *J. Dairy Sci. 51*(6), 863–866.

Dunkley, W. L., Franklin, J. D. and Pangborn, R. M. 1962A. Effects of fluorescent light on flavor, ascorbic acid and riboflavin in milk. *Food Technol. 16*, 112–118.

Dunkley, W. L., Franklin, J. D. and Pangborn, R. M. 1962B. Influence of homogenization. Copper and ascorbic acid on light-activated flavor in milk. *J. Dairy Sci. 45,* 1040–1044.

Dunkley, W. L. and Jennings, W. G. 1951. A procedure for application of the thiobarbituric acid test to milk. *J. Dairy Sci. 34,* 1064–1069.

Dunkley, W. L., Ronning, A. A., Franke, A. A. and Robb, J. 1967. Supplementing rations with tocopherol and ethoxyquin to increase oxidative stability of milk. *J. Dairy Sci. 50,* 492–499.

Dunkley, W. L., Ronning, M. and Smith, L. M. 1966. Influence of supplemental tocopherol and carotene on oxidative stability of milk and milk fat. *17th Int. Dairy Congr. Proc. A,* 223–227.

Duthie, A. H., Jensen, R. G. and Gander, G. W. 1961. Interfacial tensions of lipolyzed milk fat–water systems. *J. Dairy Sci. 44,* 401–406.

Edmondson, L. F., Douglas, F. W., Jr. and Avants, J. K. 1971. Enrichment of pasteurized whole milk with iron. *J. Dairy Sci. 54,* 1422–1426.

Egelrud, T. and Olivecrona, T. 1972. The purification of a lipoprotein lipase from bovine skim milk. *J. Biol. Chem. 247,* 6212–6217.

Egelrud, T. and Olivecrona, T. 1973. Purified bovine milk (lipoprotein) lipase: Activity against lipid substrates in the absence of exogenous serum factors. *Biochem. Biophys. Acta 306,* 115–127.

El-Nahta, A. 1963. In-vitro studies on the effect of surface-active materials and of oestrogens on lipase activity. *Milchwiss. Berichte 13,* 139–166. (German).

El-Negoumy, A. M. 1965. Relation of composition of the aqueous phase to oxidized flavor development by dialyzed globular milk fat. *J. Dairy Sci. 48,* 1406–1412.

El-Negoumy, A. M., Miles, D. M. and Hammond, E. G. 1961. Partial characterization of the flavors of oxidized butteroil. *J. Dairy Sci. 44,* 1047–1056.

Erickson, D. R. and Dunkley, W. L. 1964. Spectrophotometric determination of tocopherol in milk and milk lipides. *Anal. Chem. 36,* 1055–1058.

Erickson, D. R., Dunkley, W. L. and Ronning, M. 1963. Effect of intravenously injected tocopherol on oxidized flavor in milk. *J. Dairy Sci. 46,* 911–915.

Erickson, D. R., Dunkley, W. L. and Smith, L. M. 1964. Tocopherol distribution in milk fractions and its relation to antioxidant activity. *J. Food Sci. 29,* 269–275.

Farmer, E. H. and Sutton, D. A. 1943. The course of autoxidation reactions in polyisoprenes and allied compounds. Part IV. The isolation and constitution of photochemically-formed methyl oleate peroxide. *J. Chem. Soc. 1943,* 119–122.

Finley, J. W. 1968. A study of chemical and physical factors affecting the development of light-induced off-flavors in milk. Ph.D thesis, Cornell University.

Finley, J. W. and Shipe, W. F. 1971. Isolation of a flavor producing fraction from light exposed milk. *J. Dairy Sci. 54,* 15–20.

Fitz-Gerald, C. H. 1974. Milk lipase activation by agitation—influence of temperature. *Aust. J. Dairy Technol. 29,* 28–32.

Fleming, M. G. 1980. Mechanical factors associated with milk lipolysis in bovine milk. *Int. Dairy Fed. Doc. 118,* 41–52.

Flynn, A. and Fox, P. A. 1980. Evidence for the identity of milk lipase and lipoprotein lipase. *Ir. J. Food Sci. Technol. 4,* 173–176.

Forss, D. A., Angelini, P., Bazinet, M. L. and Merritt, C., Jr. 1967. Volatile compounds produced by copper-catalyzed oxidation of butterfat. *J. Am. Oil Chemists' Soc. 44,* 141–143.

Forss, D. A., Dunstone, E. A. and Stark, W. 1960A. Fishy flavour in dairy products. II. The volatile compounds associated with fishy flavour in butterfat. *J. Dairy Res. 27,* 211–219.

Forss, D. A., Dunstone, E. A. and Stark, W. 1960B. Fishy flavor in dairy products. III.

The volatile compounds associated with fishy flavour in washed cream. *J. Dairy Res. 27,* 373–380.

Forss, D. A., Dunstone, E. A. and Stark, W. 1960C. The volatile compounds associated with tallowy and painty flavours in butterfat. *J. Dairy Res. 27,* 381–387.

Forss, D. A., Pont, E. G. and Stark, W. 1955A. The volatile compounds associated with oxidized flavour in skim milk. *J. Dairy Res. 22,* 91–102.

Forss, D. A., Pont, E. G. and Stark, W. 1955B. Further observations on the volatile compounds associated with oxidized flavour in skim milk. *J. Dairy Res. 22,* 345–348.

Forster, T. L. and Sommer, H. H. 1951. Manganese, trypsin, milk proteins and the susceptibility of milk to oxidized flavor development. *J. Dairy Sci. 34,* 992–1002.

Fouts, E. L. 1940. Relationship of acid number variations to the qualities and flavor defects of commercial butter. *J. Dairy Sci. 23,* 173–179.

Fox, P. F. and Tarassuk, N. P. 1968. Bovine milk lipase. I. Isolation from skim milk. *J. Dairy Sci. 51,* 826–833.

Frankel, E. N., Nowakowska, J. and Evans, C. D. 1961. Formation of methyl azelaaldehydate on autoxidation of lipids. *J. Am. Oil Chemists' Soc. 38,* 161–168.

Frankel, E. N. and Tarassuk, N. P. 1956A. The specificity of milk lipase. I. Determination of the lipolytic activity in milk toward milk fat and simpler esters. *J. Dairy Sci. 39,* 1506–1516.

Frankel, E. N. and Tarassuk, N. P. 1956B. The specificity of milk lipase. III. Differential inactivation. *J. Dairy Sci. 39,* 1523–1531.

Frankel, E. N. and Tarassuk, N. P. 1959. Inhibition of lipase and lipolysis in milk. *J. Dairy Sci. 42,* 409–419.

Gaffney, P. J., Jr. and Harper, W. J. 1966. Distribution of lipase among components of a water extract of rennet casein. *J. Dairy Sci. 49,* 921–924.

Gawel, J. and Pijanowski, E. 1970. Observations on the oxidative changes of skim milk lipids. *Nahrung 14,* 469–474. (German).

Gelpi, A. J., Rusoff, L. L. and Pineiro, E. 1962. The use of antioxidants in frozen whole milk. *J. Agr. Food Chem. 10,* 89–91.

Gholson, J. H., Gelpi, A. J., Jr. and Frye, J. B., Jr. 1966A. Effect of a high-level and a low-level milk pipeline on milk fat acid degree values. *J. Milk Food Technol. 29,* 248–250.

Gholson, J. H., Schexnailder, R. H. and Rusoff, L. L. 1966B. Influence of a poor-quality low-energy ration on lipolytic activity in milk. *J. Dairy Sci. 49,* 1136–1139.

Goering, H. K., Gordon, C. H., Wrenn, T. R., Bitman, I., King, R. L. and Douglas, F. W. 1976. Effect of feeding protected safflower oil on yield, composition, flavor and oxidative stability of milk. *J. Dairy Sci. 59,* 416–425.

Gould, I. A. 1941. Effect of certain factors upon lipolysis in homogenized raw milk and cream. *J. Dairy Sci. 24,* 779–788.

Gould, I. A. and Sommer, H. H. 1939. Effect of heat on milk with especial reference to the cooked flavor. Mich. Agri. Exp. Sta. Bull. *164.*

Gould, I. A. and Trout, G. M. 1936. The effect of homogenization on some of the characteristics of milk fat. *J. Agr. Res. 52,* 49–57.

Greenbank, G. R. 1948. The oxidized flavor in milk and dairy products: A review. *J. Dairy Sci. 31,* 913–933.

Greenbank, G. R. and Pallansch, M. J. 1961. Migration of phosphatides in processing dairy products. *J. Dairy Sci. 44,* 1597–1602.

Greenbank, G. R., Wright, P. A., Deysher, E. F. and Holm, G. E. 1946. The keeping quality of samples of commercially dried milk packed in air and in inert gas. *J. Dairy Sci. 29,* 55–61.

Gregory, J. F., Babish, J. G. and Shipe, W. F. 1976. Role of heme proteins in peroxidation of milk lipids. *J. Dairy Sci. 59*, 364–368.

Gregory, J. F. and Shipe, W. F. 1975. Oxidative stability of milk. I. The antioxidative effect of trypsin treatment and aging. *J. Dairy Sci. 58*, 1263–1271.

Guthrie, E. S. and Brueckner, H. J. 1934. The cow as a source of "oxidized" flavors of milk. N.Y. Agr. Exp. Sta. Bull. *606*.

Guthrie, E. S. and Herrington, B. L. 1960. Further studies of lipase activity in the milk of individual cows. *J. Dairy Sci. 43*, 843.

Haase, G. and Dunkley, W. L. 1969A. Ascorbic acid and copper in linoleate oxidation. I. Measurement of oxidation by ultraviolet spectrophotometry and the thiobarbituric acid test. *J. Lipid Res. 10*, 555–560.

Haase, G. and Dunkley, W. L. 1969B. Ascorbic acid and copper in linoleate oxidation. II. Ascorbic acid and copper as oxidation catalysts. *J. Lipid Res. 10*, 561–567.

Haase, G. and Dunkley, W. L. 1969C. Ascorbic acid and copper in linoleate oxidation. III. Catalysts in combination. *J. Lipid Res. 10*, 568–576.

Haase, G. and Dunkley, W. L. 1970. Copper in milk and its role in catalyzing the development of oxidized flavors. *Milchwissenschaft 25*, 656–661.

Hall, B., Muller, D. P. R. and Harries, J. T. 1979. Studies of lipase activity in human milk. *Proc. Nutr. Soc. 38*, 114A.

Hamm, D. L. and Hammond, E. G. 1967. Comparisons of the modified Stamm, iron, and iodometric peroxide determinations on milk fat. *J. Dairy Sci. 50*, 1166–1168.

Hamm, D. L., Hammond, E. G. and Hotchkiss, D. K. 1968. Effect of temperature on rate of autoxidation of milk fat. *J. Dairy Sci. 51*, 483–491.

Hamm, D. L., Hammond, E. G., Parvanah, V. and Snyder, H. E. 1965. The determination of peroxides by the Stamm method. *J. Am. Oil Chemists' Soc. 42*, 920–922.

Hammond, E. G. 1970. Stabilizing milk fat with antioxidants. *Am. Dairy Rev. 32*, 40–42.

Harper, W. J. and Gould, I. A. 1959. Some factors affecting the heat-inactivation of the milk lipase enzyme system. *15th Int. Dairy Congr. Proc. 6*, 455–462.

Harper, W. J., Gould, I. A. and Badami, M. 1956A. Separation of the major components of the milk lipase system by supercentrifugation. *J. Dairy Sci. 39*, 910.

Harper, W. J., Schwartz, D. P. and El-Hagarawy, I. S. 1956B. A rapid silica gel method for measuring total free fatty Acids in milk. *J. Dairy Sci. 39*, 46–50.

Hyasawa, H., Kiyosawa, I. and Nagasawa, T. 1974. Some observations on human milk lipase. *Proc. XIXth Int. Dairy Congr. 1E*, 559.

Hegenauer, J., Saltman, P. and Ludwig, D. 1979A. Effects of supplemental iron and copper on lipid oxidation in milk. 2. Comparison of metal complexes in heated and pasteurized milk. *J. Agr. Food Chem. 27*, 868–871.

Hegenauer, J., Saltman, P., Ludwig, D., Ripley, L. and Bajo, P. 1979B. Effects of supplemental iron and copper on lipid oxidation in milk. I. Comparison of metal complexes in emulsified and homogenized milk. *J. Agr. Food Chem. 27*, 860–867.

Hemingway, E. B., Smith, G. H., Rook, J. A. F. and O'Flanagan, N. C. 1970. Lipase taint. *J. Soc. Dairy Technol. 23*, 44–48.

Hendrickx, H., Demoor, H. and Dovogelaere, R. 1963. The mechanism of light-flavour formation in milk. II. The significance of riboflavin in the development of light-flavour. *Comite voor Wetenschappelijk en Technisch Zuivelouderzock Centrum GENT 29*, 119–140. (Dutch).

Henick, A. S., Benca, M. F. and Mitchell, J. H., Jr. 1954. Estimating carbonyl compounds in rancid fats and foods. *J. Am. Oil Chemists' Soc. 31*, 88–91.

Henningson, R. W. and Adams, J. B. 1967. Influence of the melting point of milk fat and ambient temperature on the incidence of spontaneous rancidity in cow's milk. *J. Dairy Sci. 50*, 961–962.

Hernell, O. 1975. Human milk lipases. III. Physiological implications of the bile-salt stimulated lipase. *Eur. J. Clin. Invest. 5*, 267.

Hernell, O. and Olivecrona, T. 1974A. Human milk lipases. I. Serum-stimulated lipase. *J. Lipid Res. 15*, 367-374.

Hernell, O. and Olivecrona, T. 1974B. Human milk lipases. II. Bile salt-stimulated lipase. *Biochem. Biophys. Acta 369*, 234-244.

Herrington, B. L. 1954. Lipase: A review. *J. Dairy Sci. 37*, 775-789.

Herrington, B. L. and Krukovsky, V. N. 1939. Studies of lipase action. III. Lipase action in the milk of individual cows. *J. Dairy Sci. 22*, 149-152.

Hicks, C. L. 1980. Occurrence and consequence of superoxide dismutase in milk products: A review. *J. Dairy Sci. 63*, 1199-1204.

Hileman, J. L. and Courtney, E. 1935. Seasonal variations in the lipase content of milk. *J. Dairy Sci. 18*, 247-255.

Hill, R. D., Van Leeuwen, V. and Wilkinson, R. A. 1977. Some factors influencing the autoxidation of milks rich in linoleic acid. *N.Z. J. Dairy Sci. Technol. 12*, 69-77.

Hoff, J. E., Wertheim, J. H. and Proctor, B. E. 1959. Radiation preservation of milk and milk products. V. Precursors to the radiation-induced oxidation flavor of milk fat. *J. Dairy Sci. 42*, 468-475.

Hoffman, G. 1962. 1-Octen-3-ol and its relation to other oxidative cleavage products from esters of linoleic acid. *J. Am. Oil Chemists' Soc. 39*, 439-444.

Holbrook, J. and Hicks, C. L. 1978. Variation of superoxide dismutase in bovine milk. *J. Dairy Sci. 61*, 1072-1077.

Holm, G. E., Greenbank, G. R. and Deysher, E. F. 1925. The effect of homogenization, condensation and variations in the fat content of a milk upon the keeping quality of its milk powder. *J. Dairy Sci. 8*, 515-522.

Horvat, R. J., McFadden, W. H., Ng, H., Black, D. R., Lane, W. G. and Teeter, R. M. 1965. Volatile products from mild oxidation of methyl linoleate. Analysis by combined mass spectrometry-gas chromatography. *J. Am. Oil Chemists' Soc. 42*, 1112-1115.

Hutton, J. T. and Patton, S. 1952. The origin of sulfhydryl groups in milk proteins and their contributions to "cooked" flavor. *J. Dairy Sci. 35*, 699-705.

International Dairy Federation 1974. Lipolysis in cooled bulk milk. Document No. 82.

International Dairy Federation 1975. Proceedings of the lipolysis symposium (5-7 March 1975) Document No. 86.

International Dairy Federation 1980. Flavour impairment of milk and milk products due to lipolysis. Document No. 118.

Iverius, P. H. and Ostlund-Lindqvist, A. M. 1976. Lipoprotein lipase from bovine milk, isolation procedure, chemical characterization, and molecular weight analysis. *J. Biol. Chem. 251*, 7791-7795.

Jackson, W. P. and Loo, C. C. 1959. A solid in-package oxygen absorbent and its use in dry milk products. *J. Dairy Sci. 42*, 912.

Jellema, A. 1973. Lipolysis in farm tank milk. *Meded. Neth. Inst. Dairy Res. M8*, 11.

Jellema, A. 1975. Note on susceptibility of bovine milk to lipolysis. *Neth. Milk Dairy J. 29*, 145-152.

Jenness, R. and Patton, S. 1959. Milk enzymes. *In: Principles of Dairy Chemistry.* John Wiley & Sons, New York., pp. 182-202.

Jensen, R. G. and Pitas, R. E. 1976. Milk lipoprotein lipases: A review. *J. Dairy Sci. 59*, 1203-1214.

Jensen, R. G., Sampugna, J. and Pereira, R. L. 1964. Intermolecular specificity of pancreatic lipase and the structural analysis of milk triglycerides. *J. Dairy Sci. 47*, 727-732.

Johnson, P. E. and Gunten, R. L. Von 1962. A study of factors involved in the development of rancid flavor in milk. Okla. Agr. Exp. Sta. Bull. *B-593.*

Jong, K. de and Van Der Wel, H. 1964. Identification of some iso-linoleic acids occurring in butterfat. *Nature 202,* 553-555.

Josephson, D. V. 1943. (Abst.). Ph.D. dissertation 6, Pennsylvania State College.

Josephson, D. V. and Doan, F. J. 1939. Observations on cooked flavor in milk—its source and significance. *Milk Dealer 29*(2), 35-54.

Jurczak, M. E. and Sciubisz, A. 19081. Studies on the lipolytic changes in milk from cows with mastitis. *Milchwissenschaft 36,* 217-219.

Kannan, A. and Basu, K. P. 1951. Energy of activation of hydrolysis of sodium phenyl phosphate by milk phosphatase and on the inactivation of the enzyme by heat. *Indian J. Dairy Sci. 4,* 8-15.

Kanno, C., Yamauchi, K. and Tsugo, T. 1968. Occurrence of α-tocopherol and variation of α- and β-tocopherol in bovine milk fat. *J. Dairy Sci. 51,* 1713-1719.

Kason, C. M., Pavamani, I. V. P. and Nakai, S. 1972. Simple test for milk lipolysis and changes in rancidity in refrigerated pasteurized milk. *J. Dairy Sci. 55,* 1420-1432.

Kay, H. D. 1946. A light-sensitive enzyme in cow's milk. *Nature 157,* 511.

Keen, A. R., Boon, P. M. and Walker, N. J. 1976. Off-flavour in stored whole milk powder. I. Isolation of monocarbonyl classes. *N.Z. J. Dairy Sci. Technol. 11,* 180-188.

Kelley, E. 1942. Report of Chief, Division of Market Milk Investigations. Bureau of Dairy Industries, USDA, Washington, D.C.

Kelly, P. L. 1945. The effect of pitocin on milk lipase. *J. Dairy Sci. 28,* 793-797.

Kelly, P. L. 1943. The lipolytic activity of bovine mammary gland tissue. *J. Dairy Sci. 26,* 385-399.

Kende, S. 1932. Untersuchungen uber "olig-talgige" "schmirgelige" veranderungen der milch. *Milchw. Forsch 13,* 111-143. (German).

Keppler, J. G., Horikx, M. M., Meijboom, P. W. and Feenstra, W. H. 1967. Iso-linoleic acids responsible for the formation of the hardening flavor. *J. Am. Oil Chemists' Soc. 44,* 543-544.

Khan, N. A., Lundberg, W. O. and Holman, R. T. 1954. Displacement analysis of lipids. IX. Products of the oxidation of methyl linoleate. *J. Am. Oil Chemists' Soc. 76,* 1779-1784.

Khatri, L. L. 1966. Flavor chemistry of irradiated milk fat. *Diss. Abst. 26,* 6638-6639.

King, R. L. 1958. Variation and distribution of copper in milk in relation to oxidized form. Ph.D. Thesis, University of California, Davis.

King, R. L. 1962. Oxidation of milk fat globule membrane material. I. Thiobarbituric acid reaction as a measure of oxidized flavor in milk and model systems. *J. Dairy Sci. 45,* 1165-1171.

King, R. L. 1963. Oxidation of milk fat globule membrane material. II. Relation of ascorbic acid and membrane concentrations. *J. Dairy Sci. 46,* 267-274.

King, R. L. 1968. Direct addition of tocopherol to milk for control of oxidized flavor. *J. Dairy Sci. 51,* 1705-1707.

King, R. L., Burrows, F. A., Hemken, R. W. and Bashore, D. L. 1967. Control of oxidized flavor by managed intake of vitamin E from selected forages. *J. Dairy Sci. 50,* 943-944.

King, R. L. and Dunkley, W. L. 1959A. Role of a chelating compound in the inhibition of oxidized flavor. *J. Dairy Sci. 42,* 897.

King, R. L. and Dunkley, W. L. 1959B. Relation of natural copper in milk to incidence of spontaneous oxidized flavor. *J. Dairy Sci. 42,* 420-427.

King, R. L., Luick, J. R., Litman, I. I. Jennings, W. G. and Dunkley, W. L. 1959. Distribution of natural and added copper and iron in milk. *J. Dairy Sci. 42,* 780-790.

King, R. L., Tikriti, H. H. and Oskarsson, M. 1966. Natural and supplemented tocopherol in the dairy ration and oxidized flavor. *J. Dairy Sci. 49,* 1574.

Kinnunen, P. K. J., Huttunen, J. K. and Ehnholm, C. 1976. Properties of purified bovine milk lipoprotein lipase. *Biochem. Biophys. Acta 450,* 342–351.

Kirst, E. 1980A. Lipolytic process in milk and milk products. Review of literature and study of effects of stirring and pumping on milk fat. *Die Nahrung 24,* 569–576. (German).

Kirst, E. 1980B. Lipolytic processes in milk and milk products. Literature report and studies on the effect of mixers and pumps on milk fat. *Lebensmittelindustrie 27,* 27–31. (German).

Kirst, E. 1980C. Lipolytic processes in milk and milk products. III. Effect of cooling on structure of milk fat. *Lebensmittel-industrie 27,* 464–468. (German).

Kishonti, E. 1975. Influence of heat resistant lipases and proteases in psychortrophic bacteria on product quality. Int. Dairy Fed. Doc. *86,* pp. 121–124.

Kitchen, B. J. 1971. Bovine milk esterases. *J. Dairy Res. 38,* 171–177.

Kitchen, B. J. and Aston, J. W. 1970. Milk lipase activation. *Aust. J. Dairy Technol. 25* 10–13.

Kliman, P. G., Tamsma, A. and Pallansch, M. J. 1962. Peroxide value–flavor score relationships in stored foam-dried whole milk. *J. Agr. Food Chem. 10,* 496–498.

Kodgev, A. and Rachev, R. 1070. The influence of some factors on the acidity of milkfat. *18th Int. Dairy Congr. Proc. 1E,* 200.

Koops, J. 1963. Cold storage defects of butter. *J. Verslag. Ned. Inst. Zuivelouderzoek 80.* (Dutch).

Koops, J. 1964A. Cold-storage defects of cultured butter. *Neth. Milk Dairy J. 18,* 220–225.

Koops, J. 1964B. Antioxidant activity of ascorbyl palmitate in cold stored cultured butter. *Neth. Milk Dairy J. 18,* 38–51.

Koops, J. 1969. The effect of the pH on the partition of natural and added copper in milk and cream. *Neth. Milk Dairy J. 23,* 200–213.

Koops, J. and Klomp, H. 1977. Rapid colorimetric determination of free fatty acids (lipolysis) in milk by the copper soap method. *Neth. Milk Dairy J. 31,* 56–74.

Koskinen, E. H., Luhtala, A. and Antila, M. 1969. Studies on enzymatic reactions (lipases) by means of liquid scintillation counting. *Milchwissenschaft 24,* 20–25. (German).

Krukovsky, V. N. 1952. The origin of oxidized flavors and factors responsible for their development in milk and milk products. *J. Dairy Sci. 35,* 21–29.

Krukovsky, V. N. 1955. Organoleptic study of oxygenated and copper-treated milk prior to pasteurization. *J. Dairy Sci. 38,* 595.

Krukovsky, V. N. 1961. Review of biochemical properties of milk and the lipide deterioration in milk and milk products as influenced by natural varietal factors. *J. Agr. Food Chem. 9,* 439–447.

Krukovsky, V. N. and Guthrie, E. S. 1945. Ascorbic acid oxidation, a key factor in the inhibition or promotion of the tallowy flavor in milk. *J. Dairy Sci. 28,* 565–579.

Krukovsky, V. N. and Guthrie, E. S. 1946. Vitamin C, hydrogen peroxide, copper and the tallowy flavor in milk. *J. Dairy Sci. 29,* 293–306.

Krukovsky, V. N. and and Herrington, B. L. 1939. Studies of lipase action. II. The activity of milk lipase by temperature changes. *J. Dairy Sci. 22,* 137–147.

Krukovsky, V. N. and Sharp, P. F. 1940. Inactivation of milk lipase by dissolved oxygen. *J. Dairy Sci. 23,* 1119–1122.

Krukovsky, V. N., Whiting, F. and Loosli, J. K. 1950. Tocopherol carotenoid and vitamin A content of the milk fat and the resistance of milk to the development of oxidized flavors as influenced by breed and season. *J. Dairy Sci. 33,* 791–796.

Kurtz, F. E. 1974. The lipids of milk: Composition and properties. *In: Fundamentals of Dairy Chemistry.* B.H. Webb, A. H. Johnson and J.A. Alford (Editors). AVI Publishing Co., Westport, Conn. pp. 125–210.

Kurtz, F. E., Tamsma, A. and Pallansch, J. 1973. Effect of fortification with iron on susceptibility of skim milk and nonfat dry milk to oxidation. *J. Dairy Sci. 56,* 1139–1143.

Kuzdzal-Savoie, S. 1980. Flavour impairment of milk and milk products due to lipolysis. VII. Determination of free fatty acids in milk and milk products. Int. Dairy Fed. Bull. *118,* pp. 53–66.

Larsson, B. L. and Jenness, R. 1950. The reducing capacity of milk as measured by an iodimetric titration. *J. Dairy Sci. 33,* 896–903.

Larsen, P. B., Trout, G. M. and Gould, I. A. 1941. Rancidity studies on mixtures of raw and pasturized homogenized milk. *J. Dairy Sci. 24,* 771–778.

Lawrence, R. C. 1967. Microbial lipases and related esterases. Part II. Estimation of lipase activity. Characterization of lipases. Recent work concerning their effect on dairy products. *Dairy Sci. Abstr. 29,* 59–70.

Lawrence, R. C., Fryer, T. F. and Reiter, B. 1967. Rapid method for the quantitative estimation of microbial lipases. *Nature 213,* 1264–1265.

Lea, C. H., Moran, T. and Smith, J. A. B. 1943. The gas-packing and storage of milk powder. *London J. Dairy Res. 13,* 162–215.

Lembke, A. and Frahm, H. 1964. Manufacture of butter with a long shelf life. *Kiel. Milchw. Forsch. 16,* 427–437. (German).

Lillard, D. A. and Day, E. A. 1961. Autoxidation of milk lipids. II. The relationship of sensory to chemical methods for measuring the oxidized flavor of milk fats. *J. Dairy Sci. 44,* 623–632.

Loftus Hills, G. and Thiel, C. C. 1946. The ferric thiocyanate method of estimating peroxide in fat or butter, milk and dried milk. *J. Dairy Res. 14,* 340–353.

Luhtala, A. 1969. Studies on lipase activity, lipases and glyceride synthesis in Finnish cows' milk. *Meijertiet Aikakauskirja 29,* 7–65.

Luhtala, A. and Antila, M. 1968. Lipases and lipolysis of milk. *Fette, Seifen. Anstrichmittel. 70,* 280–288. (German).

Luhtala, A., Korhonen, H., Koskinen, E. H. and Antila, M. 1970A. Glyceride synthesis and hydrolysis caused by cells in milk. *18th Int. Dairy Congr. Proc. 1E.,* 80.

Luhtala, A., Koskinen, E. H. and Antila, M. 1970B. Lipolysis in freshly drawn milk. *18th Int. Dairy Congr. Proc. 1E.,* 79.

Luick, J. R. and Mazrimas, J. A. 1966. Biological effects of ionizing radiation on milk synthesis. III. Effects on milk lipase, esterase, alkaline phosphatase, and lactoperoxidase activities. *J. Dairy Sci. 49,* 1500–1504.

McCarthy, R. D. and Patton, S. 1964. *Nature (London) 202,* 347–349.

MacKenzie, R. D., Blohm, T. R., Auxier, E. M. and Luther, A. C. 1967. Rapid colorimetric micromethod for free fatty acids. *J. Lipid Res. 8,* 589–597.

Manus, L. J. and Bendixen, H. A. 1956. Effects of lipolytic activity and of mercuric chloride on the Babcock test for fat in composite milk samples. *J. Dairy Sci. 39* 508–513.

Mattson, F. H. and Volpenhein, R. A. 1966. Enzymatic hydrolysis at an oil/water interface. *J. Am. Oil Chemists' Soc. 43,* 286–289.

Meijboom, P. W. 1964. Relationship between molecular structure and flavor perceptibility of aliphatic aldehydes. *J. Am. Oil Chemists' Soc. 41,* 326–328.

Menger, J. W. 1975. Experience with lipolytic activities in milk and dairy products. Int. Dairy Fed. Doc. *86,* pp. 108–112.

Merk, W. and Crasemann, E. 1961. The effect of tocopherol enrichment in the feed of milk cows on the amount and composition of the produced milk. *Z. Tierphysiol. Tierernahr. Futter-mittelk. 16,* 197–214.

Meyer, R. I. and Jokay, L. 1960. The effect of an oxygen scavenger packet, desiccant in package system on the stability of dry whole milk and dry ice cream mix. *J. Dairy Sci. 43,* 844.

Morton, R. K. 1955. Some properties of alkaline phosphatase of cows' milk and calf intestinal mucosa. *Biochem. J. 60*, 573–582.

Muir, D. D., Kelly, M. E. and Phillips, J. D. 1978. The effect of storage temperature on bacterial growth and lipolysis in raw milk. *J. Soc. Dairy Technol. 31*, 203–208.

Mulder, H., Menger, J. W. and Meijers, P. 1964. The copper content of cows' milk. *Neth. Milk Dairy J. 18*, 52–65.

Murphy, J. J., Connolly, J. F. and Headon, D. R. 1979. A study of factors associated with free fatty acid development in milk. *Ir. J. Food Sci. Technol. 3*, 131–149.

Murty, G. K., Rhea, U. S. and Peeler, J. T. 1972. Copper, iron, manganese, strontium and zinc content of market milk. *J. Dairy Sci. 55*, 1666–1674.

Nakai, S., Perrin, J. J. and Wright, V. 1970. Simple test for lipolytic rancidity in milk. *J. Dairy Sci. 53*, 537–540.

Nelson, H. G. and Jezeski, J. J. 1955. Milk lipase. I. The lipolytic activity of separator slime. *J. Dairy Sci. 38*, 479–486.

Newstead, D. F. and Headifen, J. M. 1981. A reappraisal of the method for estimation of the peroxide value of fat in whole milk powder. *N.Z. J. Dairy Sci. Technol. 15*, 13–18.

Nilsson, R. and Willart, S. 1960. Lipolytic activity in milk. II. The heat inactivation of fat splitting in milk. *Milk Dairy Res., Alnarp. Sweden, Rep. 64,*

Novak, M. 1965. Colorimetric ultramicro method for the determination of free fatty acids. *J. Lipid Res. 6*, 431–433.

Olivecrona, T. 1980. Flavour impairment of milk and milk products due to lipolysis. III. Biochemical aspects of lipolysis in bovine milk. Int. Dairy Fed. Bull. *118*, pp. 19–25.

Olivecrona, T., Egelrud, T., Hernell, O., Castberg, H. and Solberg, P. 1975. Is there more than one lipase in bovine milk? Doc. Ind. Dairy Fed. *86*, 61–72.

Olivecrona, T., Egelrud, T., Iverius, P. H. and Lindahl, U. 1971. *Biochem Biophys. Res. Commun. 43*, 524.

Olivecrona, T. and Lindahl, U. 1969. *Acta. Chem. Scand. 23*, 3587.

Olson, F. C. and Brown, W. C. 1942. Oxidized flavor in milk. XI. Ascorbic acid, glutathione, and hydrogen peroxide as mechanisms for the production of oxidized flavor. *J. Dairy Sci. 25*, 1027–1039.

Olson, J. C., Jr., Thomas, E. L. and Nielsen, A. J. 1956. The rancid flavor in raw milk supplies. *Am. Milk Rev. 18*, 98–102, 199.

Ortiz, M. J., Kesler, E. M., Watrous, G. H., Jr. and Cloninger, W. H. 1970. Effect of the cow's body condition and stage of lactation on development of milk rancidity. *J. Milk Food Technol. 33*, 339–342.

Parks, O. W. 1974. The lipids of milk: Deterioration, Part II. Autoxidation. *In: Fundamentals of Dairy Chemistry.* B.H. Webb, A. H. Johnson and John A. Alford (Editors). AVI Publishing Co., Westport, Conn., pp. 240–263.

Parks, O. W. and Allen, C. A. 1972. Unpublished data.

Parks, O. W., Keeney, M. and Schwartz, D. P. 1963. Carbonyl compounds associated with the off-flavor in spontaneously oxidized milk. *J. Dairy Sci. 46*, 295–301.

Parks, O. W. and Patton, S. 1961. Volatile carbonyl compounds in stored dry whole milk. *J. Dairy Sci. 44*, 1–9.

Parks, O. W., Wong, N. P., Allen, C. A. and Schwartz, D. P. 1969. 6-*trans*-Nonenal: An off-flavor component of foam spray-dried milks. *J. Dairy Sci. 52*, 953–956.

Parry, R. M., Jr., Chandan, R. C. and Shahani, K. M. 1966. rapid and sensitive assay for milk lipase. *J. Dairy Sci. 49*, 356–360.

Patton, S. 1954. The mechanism of sunlight flavor formation in milk with special reference to methionine and riboflavin. *J. Dairy Sci. 37*, 446–452.

Patton, S., Barnes, I. J. and Evans, L. E. 1959. *n*-Deca-2,4 Dienal, its origin from linoleate and flavor significance in fats. *J. Am. Oil Chemists' Soc. 36*, 280–283.

Peterson, M. H., Johnson, M. J. and Price, W. V. 1948. Determination of cheese lipase. *J. Dairy Sci. 31*, 31–38.

Pijanowski, E., Wojtowicz, M. and Lochowska, H. 1962. On the oxidative susceptibility of milk from cows of different yields. *Int. Dairy Congr. A*, 633–640.

Pillay, V. T., Myhr, A. N. and Gray, J. L. 1980. Lipolysis in milk. I. Determination of free fatty acid and threshold value for lipolyzed flavor detection. *J. Dairy Sci. 63*, 1213–1218.

Pont, E. G. 1952. Studies on the origin of oxidized flavour in whole milk. *J. Dairy Res. 19*, 316–327.

Pont, E. G. 1955. A de-emulsification technique for use in the peroxide test on the fat of milk, cream, concentrated and dried milks. *Aust. J. Dairy Technol. 10*, 72–74.

Pont, E. G. 1964. The relationship between the swift test time and the keeping quality of butterfat. *Aust. J. Dairy Technol. 19*, 108–111.

Pont, E. G., Forss, D. A., Dunstone, E. A. and Gunnis, L. F. 1960. Fishy flavour in dairy products. I. General studies on fishy butterfat. *J. Dairy Res. 27*, 205–209.

Pont, E. G. and Holloway, G. L. 1967. The effect of oxidation on the iodine values of phospholipid in milk, butter and washed-cream serum. *J. Dairy Res. 34*, 231–238.

Potter, F. E. and Hankinson, D. J. 1960. The flavor of milk from individual cows. *J. Dairy Sci. 43*, 1887.

Poulsen, P. R. and Jensen, G. K. 1966. Observations of the liability of cows to yield milk that spontaneously develops oxidized flavour. *17th Int. Dairy Cong. Proc. A2*, 229–237.

Privett, O. S. and Nickell, E. C. 1959. Determination of structure and analysis of the hydroperoxide isomers of autoxidized methyl oleate. *Fette, Seifen. Anstrichm. 61*, 842–845.

Privett, O. S. and Quackenbush, F. W. 1954. The relation of synergist to antioxidant in fats. *J. Am. Oil Chemists' Soc. 31*, 321–323.

Privett, O. S., Lundberg, W. O., Khan, N. A., Tolberg, W. E. and Wheeler, D. H. 1953. Structure of hydroperoxides obtained from autoxidized methyl linoleate. *J. Am. Oil Chemists' Soc. 30*, 61–66.

Pyenson, H. and Tracy, P. H. 1946. A spectrophotometric study of the changes in peroxide value of spray-dried whole milk powder during storage. *J. Dairy Sci. 29*, 1–12.

Rajan, T. S., Richardson, G. A. and Stein, R. W. 1962. Xanthine oxidase activity of milks in relation to stage of lactation, feed, and incidence of spontaneous oxidation. *J. Dairy Sci. 45*, 933–934.

Rao, S. R. 1951. Ph.D. thesis, University of Wisconsin.

Richter, R. 1981. Hydrolytic rancidity: Its prevalence, measurement and significance. *Am. Dairy Rev. 43*, 18DD, 18HH.

Riel, R. R. 1952. Causative factors and end-products of oxidized flavor development in milk. Ph.D. thesis, University of Wisconsin.

Riest, U., Ronning, M., Dunkley, W. L. and Franke, A. A. 1967. Oxidative stability of milk as influenced by dietary copper, molybdenum and sulfate. *Milchwissenschaft 22*, 551–554.

Roadhouse, C. L. and Henderson, J. L. 1950. *The Market Milk Industry.* McGraw-Hill Book Co., New York.

Roahen, D. C. and Sommer, H. H. 1940. Lipolytic activity in milk and cream. *J. Dairy Sci. 23*, 831–841.

Rogers, W. P. and Pont, E. G. 1965. Copper contamination in milk production and butter manufacture. *Aust. J. Dairy Technol. 20*, 200–205.

Roh, J. K., Bradley, R. L., Jr., Richardson, T. and Weckel, K. G. 1976. Removal of copper from milk. *J. Dairy Sci. 59*, 382–385.

Romanskaya, N. N. and Valeeva, A. N. 1962. Effect of synthetic antioxidants on the stability of milk fat during long storage. *Tr. Frunzensk. Politekhn. Inst.* 17–21. (Russian).

Saito, Z., Nakamura, S. and Igarashi, Y. 1970. Milk lipases. VII. Protective effect of substrate on the thermal inactivation of lipases. *Dairy Sci. Abst. 32*, 3081.

Salih, A. M. A. and Anderson, M. 1979A. Effect of diet and stage of lactation on bovine milk lipolysis. *J. Dairy Res. 46*, 623–631.

Salih, A. M. A. and Anderson, M. 1979B. Observations on the influence of high cell count on lipolysis in bovine milk. *J. Dairy Res. 46*, 453–462.

Salih, A. M. A. and Anderson, M. and Tuckley, B. 1977. The determination of short and long chain free fatty acids in milk. *J. Dairy Res. 44*, 601.

Samuelsson, E. G. 1962. Model experiments on sunlight flavour in milk di- and tripeptides of methionine. *Ind. Dairy Congr. A*, 552–560.

Samuelsson, E. G. 1966. The copper content in milk and the distribution of copper to various phases of milk. *Milchwissenschraft 21*, 335–341.

Samuelsson, E. G. 1967. The distribution of copper in milk with some aspects on oxidation reactions of the milk lipids. *Berlingska Boktryckeriet, Lund.* (English).

Samuelsson, E. G. 1970. The migration of copper in milk with change of temperature and addition of some chelating compounds. *Milk Dairy Res. (Alnarp)* Report no. 77, 5–22.

Sargent, J. S. E. and Stine, C. M. 1964. Effects of heat on the distribution of residual and added copper in whole fluid milk. *J. Dairy Sci. 47*, 662–663.

Sattler-Dornbacher, S. 1963. Studien zum Redox—Potential in butter. *Milchwiss. Ber. 13*, 53–74.

Scanlan, R. A., Sather, L. A. and Day, E. A. 1965. Contribution of free fatty acids to the flavor of rancid milk. *J. Dairy Sci. 48*, 1582–1584.

Schaffer, P. S., Greenbank, G. R. and Holm, G. E. 1946. The rate of autoxidation of milk fat in atmospheres of different oxygen concentration. *J. Dairy Sci. 29*, 145–150.

Schingoethe, D. J., Parsons, J. G., Ludens, F. C., Schaffer, L. V. and Shave, H. J. 1979. Response of lactating cows to 300 mg of supplemental vitamin E daily. *J. Dairy Sci. 62*, 333–338.

Schipper, C. J. 1975. Prevention of bacteriological lipolysis by dairy factories. Int. Dairy Fed. Doc. *86*, pp. 113–115.

Schmidt, I. H. 1959. Over the thiobarbituric acid-methyl dyestuffs. *Fette, Seifen. Anstrichmittel 61*, 881–886. (German).

Schwartz, D. P. 1974. The lipids of milk: Deterioration, Part I. Lipolysis and rancidity. *In: Fundamentals of Dairy Chemistry.* B.H. Webb, A. H. Johnson and John A. Alford (Editors). AVI Publishing Co., Westport, Conn., pp. 220–239.

Schwartz, D. P., Gould, I. A. and Harper, W. J. 1956A. The milk lipase system. II. Effect of formaldehyde. *J. Dairy Sci. 39*, 1375–1383.

Schwartz, D. P., Gould, I. A. and Harper, W. J. 1956B. The milk lipase system. I. Effect of time, pH and concentration of substrate on activity. *J. Dairy Sci. 39*, 1364–1374.

Scott, K. 1965. The measurement of esterase activity in cheddar cheese. *Aust. J. Dairy Technol. 20*, 36.

Seerless, S. K. and Armstrong, J. G. 1970. Vitamin E, vitamin A, and carotene contents of alberta butter. *J. Dairy Sci. 53*, 150–154.

Shahani, K. M., Kwan, A. J. and Friend, B. A. 1980. Role and significance of enzymes in human milk. *Am. J. Clin. Nutr. 33*, 1861–1868.

Sharp, P. F., Guthrie, E. S. and Hand, D. B. 1941. A new method of retarding oxidized flavor and preserving vitamin C-deaeration. *Int. Assoc. Milk Dealers Bull. 20*, 523–545.

Sharp, P. F., Hand, D. B. and Guthrie, E. S. 1942. Experimental work on deaeration of milk. *Int. Assoc. Milk Dealers Bull. 34*, 365–375.

Shipe, W. F. and Senyk, G. F. 1981. Effects of processing conditions on lipolysis in milk. *J. Dairy Sci. 64*, 2146-2149.

Shipe, W.F.,Bassette, R., Deane, D. D., Dunkley, W. L., Hammond, E. G., Harper, W. J., Kleyn, D. H., Morgan, M. E., Nelson, J. H. and Scanlan, R. A. 1978. Off-flavors of milk: Nomenclature, standards and bibliography. *J. Dairy Sci. 61*, 855-869.

Shipe, W. F., Senyk, G. F., Ledford, R. A., Bandler, D. K. and Wolff, E. T. et al. 1980A. Flavor and chemical evaluations of fresh and aged market milk. *J. Dairy Sci. 63*, (Suppl. 1), 43 (Abstr.).

Shipe, W. F., Senyk, G. F. and Boor, K. J. 1982. Inhibition of milk lipolysis by lambda carrageenan. *J. Dairy Sci. 65*, 24-27.

Shipe, W. F., Senyk, G. F. and Fountain, K. B. 1980B. Modified copper soap solvent extraction method for measuring free fatty acids in milk. *J. Dairy Sci. 63*, 193-198.

Shipe, W. F., Senyk, G. F. and Weetall, H. H. 1972. Inhibition of oxidized flavor development in milk by immobilized trypsin. *J. Dairy Sci. 55*, 647-648.

Sidhu, G. S., Brown, M. A. and Johnson, A. R. 1975. Autoxidation in milk rich in linoleic acid. I. An objective method for measuring autoxidation and evaluating antioxidants. *J. Dairy Res. 42*, 185-195.

Sidhu, G. S., Brown, M. A. and Johnson, A. R. 1976. Autoxidation in milk rich in linoleic acid. II. Modification of the initial system and control of oxidation. *J. Dairy Res. 43*, 239-250.

Singleton, J. A., Aurand, L. W. and Lancaster, F. W. 1963. Sunlight flavor in milk. I. A study of components involved in the flavor development. *J. Dairy Sci. 46*, 1050-1053.

Sjostrom, G. and Willart, S. 1956. Free fatty acids and lipase activity in milk. *Svenska Mejeritidn. 48*, 421-428, 435-438. (Swedish).

Skean, J. D. and Overcast, W. W. 1961. Apparent location of lipase in casein. *J. Dairy Sci. 44*, 823-832.

Smith, A. C. and MacLeod, P. 1955. The effect of artificial light on milk in cold storage. *J. Dairy Sci. 38*, 870-874.

Smith, L. M. and Dunkley, W. L. 1959. Effect of the development of oxidized flavor on the polyunsaturated fatty acids of milk lipids. *J. Dairy Sci. 42*, 896.

Smith, G. J. and Dunkley, W. L. 1960. Xanthine oxidase and incidence of spontaneous oxidized flavor in milk. *J. Dairy Sci. 43*, 278-280.

Smith, G. J. and Dunkley, W. L. 1962A. Copper binding in relation to inhibition of oxidized flavour by heat treatment and homogenization. *Int. Dairy Congr. A*, 625-632.

Smith, G. J. and Dunkley, W. L. 1962B. Pro-oxidants in spontaneous development of oxidized flavor in milk. *J. Dairy Sci. 45*, 170-181.

Smith, G. J. and Dunkley, W. L. 1962C. Ascorbic acid oxidation and lipid peroxidation in milk. *J. Food Sci. 27*, 127-134.

Stadhouders, J. and Mulder, H. 1959. The destructive effect of light on milk lipase activity. *Neth. Milk Dairy J. 13*, 122-129.

Stadhouders, J. and Mulder, H. 1964. Some observations on milk lipase. III. The effect of pH on milk lipase activity. *Neth. Milk Dairy J. 18*, 30-37.

Stark, W. and Forss, D. A. 1962. A compound responsible for metallic flavour in dairy products. I. Isolation and identification. *J. Dairy Res. 29*, 173-180.

Stark, W. and Forss, D. A. 1964. A compound responsible for mushroom flavour in dairy products. *J. Dairy Res. 31*, 253-259.

Stark, W. and Forss, D. A. 1966. *n*-Alkan-1-ols in oxidized butter. *J. Dairy Res. 33*, 31-36.

Stewart, D. B., Murray, J. G. and Neil, S. D. 1975. Lipolytic activity of organisms isolated from refrigerated bulk milk. *Int. Dairy Fed. Doc. 86*, 38-50.

Stine, C. M., Harland, H. A., Coulter, S. T. and Jenness, R. 1954. A modified peroxide test for detection of lipid oxidation in dairy products. *J. Dairy Sci. 37*, 202–208.

Storgards, T. and Ljungren, B. 1962. Some observations on the formation of light-induced oxidized flavour. *Milchwissenchaft 17*, 406–407.

Super, D. M., Palmquist, D. L. and Schanbacher, F. L. 1976. Relative activation of milk lipoprotein lipase by serum of cows fed varying amounts of fat. *J. Dairy Sci. 59*, 1409–1413.

Tallamy, P. T. and Randolph, H. E. 1969. Influence of mastitis on properties of milk. IV. Hydrolytic rancidity. *J. Dairy Sci. 52*, 1569–1572.

Tamsma, A., Kurtz, F. E. and Pallansch, M. J. 1967. Effect of oxygen removal technique on flavor stability of low-heat foam spray dried whole milk. *J. Dairy Sci. 50*, 1562–1565.

Tamsma, A., Mucha, T. J. and Pallansch, M. J. 1962. Factors related to flavor stability of foam-dried milk. II. Effect of heating milk prior to drying. *J. Dairy Sci. 45*, 1435–1439.

Tamsma, A., Mucha, T. J. and Pallansch, M. J. 1963. Factors related to the flavor stability during storage of foam-dried whole milk. III. Effect of antioxidants. *J. Dairy Sci. 46*, 114–119.

Tamsma, A., Pallansch, Mucha, T. J., M. J. and Patterson, W. I. 1961. Factors related to the flavor stability of foam-dried whole milk. I. Effect of oxygen level. *J. Dairy Sci. 44*, 1644–1649.

Tarassuk, N. P. 1942. The problem of controlling rancidity in milk. *Milk Plant Monthly 31*(4), 24–25.

Tarassuk, N. P. and Frankel, E. N. 1955. On the mechanism of activation of lipolysis and the stability of lipase systems of normal milk. *J. Dairy Sci. 38*, 438–439.

Tarassuk, N. P. and Frankel, E. N. 1957. The specificity of milk lipase. IV. Partition of the lipase system in milk. *J. Dairy Sci. 40*, 418–430.

Tarassuk, N. P. and Henderson, J. L. 1942. Prevention of development of hydrolytic rancidity in milk. *J. Dairy Sci. 25*, 801–806.

Tarassuk, N. P. and Koops, J. 1960. Inhibition of oxidized flavor in homogenized milk as related to the concentration of copper and phospholipids per unit of fat globule surface. *J. Dairy Sci. 43*, 93–94.

Tarassuk, N. P., Koops, J. and Pette, J. W. 1959. The origin and development of trainy (fishy) flavor in washed cream and butter. I. Factors affecting the development of trainy flavor in washed cream. *Neth. Milk Dairy J. 13*, 258–278.

Tarassuk, N. P. and Regan, W. M. 1943. A study of the blood carotene in relation to lipolytic activity of milk. *J. Dairy Sci. 26*, 987–996.

Tarassuk, N. P. and Smith, F. R. 1940. Relation of surface tension of rancid milk to its inhibitory effect on the growth and acid fermentation of *Streptococcus lactis*. *J. Dairy Sci. 23*, 1163–1170.

Tarassuk, N. P. and Yaguchi, M. 1958. Effect of mastitis on the susceptibility of milk to lipolysis. *West. Div. Am. Dairy Sci. Assoc. Proc. 39*, 191–196.

Taylor, M. J. and Richardson, T. 1980A. Antioxidant activity of skim milk: Effect of heat and resultant sulfhydryl groups. *J. Dairy Sci. 63*, 1783–1795.

Taylor, M. J. and Richardson, T. 1980B. Antioxidant activity of skim milk: Effect of sonification *J. Dairy Sci. 63*, 1938–1942.

Thomas, E. L., Nielsen, A. J. and Olsen, J. C., Jr. 1955B. Hydrolytic rancidity in milk— A simplified method for estimating the extent of its development. *Am. Milk Rev. 17*, 50–52, 85.

Thomas, W. R., Harper, W. J. and Gould, I. A. 1955A. Lipase activity in fresh milk as related to portions of milk drawn and fat globule size. *J. Dairy Sci. 38*, 315–316.

Thurston, L. M. 1937. Theoretical aspects of the causes of oxidized flavor particularly from the lecithin angle. *Int. Assoc. Milk Dealers Proc. 30*, Lab. Sect. 143–153.

Thurston, L. M., Brown, W. C. and Dustman, R. B. 1936. Oxidized flavor in milk. II. The effects of homogenization, agitation and freezing of milk on its subsequent susceptibility to oxidized flavor development. *J. Dairy Sci. 19*, 671–682.

Tracey, P. H., Ramsey, R. J. and Ruehe, H. A. 1933. Certain biological factors related to tallowiness in milk and cream. *Ill. Agr. Exp. Sta. Bull.* 389.

Tsugo, T. and Hayashi, T. 1962. The effect of irradiation by ionizing radiation on milk enzymes. III. Effect of irradiation on lipase and xanthine-oxydase activities in milk. *Jap. J. Zootech. Sci. 33*, 125–129.

Velander, H. J. and Patton, S. 1955. Prevention of sunlight flavor in milk by removal of riboflavin. *J. Dairy Sci. 38*, 593.

Wang, C. S. 1981. Human milk bile salt-activated lipase. *J. Biol. Chem. 256*, 10198–10203.

Wang, L. and Randolph, H. E. 1978. Activation of lipolysis. I. Distribution of lipase activity in temperature activated milk. *J. Dairy Sci. 61*, 874–880.

Wells, M. E., Pryor, O. R., Haggerty, D. M., Pickett, H. C. and Mickle, J. B. 1969. Effect of estrous cycle and lactation on lipase activity in bovine milk and blood. *J. Dairy Sci. 52*, 1110–1113.

White, C. H. and Bulthaus, M. 1982. Light activated flavor in milk. *J. Dairy Sci. 65*, 489–494.

Willart, S. and Sjostrom, G. 1959. The effect of sodium chloride on the hydrolysis of the fat in milk and cheese. *15th Int. Dairy Congr. Proc. 3*, 1482–1486.

Willart, S. and Sjostrom, G. 1962. Pasteurization of milk and its effect on the lipolytic activity measured at different pH values. *Ind. Dairy Congr. A*, 669–674.

Wills, E. D. 1965. Lipases. *In: Advances in Lipid Research.* Paoletti, Rand Kritchiosky, D. (Editors), Academic Press, New York., pp. 197–231.

Wilson, H. K. and Herreid, E. O. 1969. Controlling oxidized flavors in high-fat sterilized creams. *J. Dairy Sci. 52*, 1229–1232.

Wishner, L. A. 1964. Light-induced oxidations in milk. *J. Dairy Sci. 47*, 216–221.

Wishner, L. A. and Keeney, M. 1963. Carbonyl pattern of sunlight-exposed milk. *J. Dairy Sci. 46*, 785–788.

Worstorff, H. 1975. Mechanical factors in the milking plant affecting the level of free fatty acids in milk. *Ind. Dairy Fed. 86*, 156–161.

Wyatt, C. J. and Day, E. A. 1965. Evaluation of antioxidants in deodorized and nonde-odorized butteroil stored at 30 degrees C. *J. Dairy Sci. 48*, 682–686.

Zall, R. R. and Chen, J. H. 1981. Heating and storing milk on dairy farms before pasteurization in milk plants. *J. Dairy Sci. 64*, 1540–1544.

Ziemba, J. V. 1969. Enzymes enhance flavor of milk solids. *Food Eng. 41*, 105–106, 110.

Lactose

Virginia H. Holsinger

OCCURRENCE

The characteristic carbohydrate of milk is lactose (4-0-β-D-galactopyranosyl-D-glucopyranose), commonly referred to as "milk sugar." Practically, the milk of mammals is the sole source of lactose, a belief which is substantially correct, but with recognized exceptions. In addition to the high concentrations of lactose in milk and the mammary gland, low concentrations appear in the blood and urine, especially during pregnancy and lactation, the result of escape of lactose formed in the mammary tissues. Lactose is found in the urine in about 9% of healthy humans of either sex who consume a normal diet; its origin is probably alimentary (Flynn *et al.* 1953). Other sources are rare, e.g., as a constituent of some oligosaccharides (Trucco *et al.* 1954), in *Forsythia* flowers (Kuhn and Low 1949), and in *Sapotacea* (Reithel and Venkataraman 1956).

The literature on lactose is voluminous, and no attempt is made here to review it comprehensively. Excellent general and specialized reviews are available for those interested in additional details and references (Delmont 1983; Doner and Hicks 1982; Hobman 1984; MacBean 1979; Nickerson 1974; Paige and Bayless 1981; Renner 1983; Short 1978; Zadow 1984).

The first record of isolation of lactose was in 1633, by Bartolettus, by evaporation of whey. During the Eighteenth century, lactose became a commercial commodity, used principally in medicine. Whey had been used by physicians since the time of Hippocrates to utilize the unique biochemical functions and properties of lactose.

Lactose is present in the milk of most mammals, as shown in Table 6.1. Exceptions to this pattern are the California sea lion and other Pacific pinnipeds, which have no lactose in their milks (Johnson *et al.* 1974; Pilson 1965; Pilson and Kelly 1962; Stewart *et al.* 1983). Human milk contains one of the highest lactose contents, about 7%, of all mammalian milks (Renner 1983), whereas the average lactose content

Table 6.1. Milk Composition of Domesticated and Experimental Mammals.

Mammal	Fat (%)	Protein (%)	Lactose (%)	Total solids (%)
Cow	3.7	3.4	4.8	12.7
Man	3.8	1.0	7.0	12.4
Sheep	7.4	5.5	4.8	19.3
Goat	4.5	2.9	4.1	13.2
Water buffalo	7.4	3.8	4.8	17.2
Dromedary	4.5	3.6	5.0	13.6
Horse	1.9	2.5	6.2	11.2
Llama	2.4	7.3	6.0	16.2
Reindeer	16.9	11.5	2.8	33.1
Zak	6.5	5.8	4.6	17.3
Indian elephant	11.6	4.9	4.7	21.9
Dog	12.9	7.9	3.1	23.5
Cat	4.8	7.0	4.8	17.6
Pig	6.8	4.8	5.5	18.8
Norway rat	10.3	8.4	2.6	21.0
Golden hamster	4.9	9.4	4.9	22.6
Guinea pig	3.9	8.1	3.0	16.4
Rhesus monkey	4.0	1.6	7.0	15.4
Baboon	5.0	1.6	7.3	14.4
Rabbit	18.3	13.9	2.1	32.8
Mink	3.4	7.5	2.0	21.2

SOURCE: Jenness and Sloan (1970).
Note: Large differences within species are reported by various workers.

of normal bovine milks averages 4.8% anhydrous lactose. This usually amounts to 50 to 52% of the total solids in skim milk.

Besides lactose, small amounts of other carbohydrates are found in milk, partly in a free form and partly bound to proteins, lipids, or phosphate. In some milks, other carbohydrates occur in higher concentrations than lactose (Jenness et al. 1964). Cows' milk contains the monosaccharides glucose and galactose in concentrations of about 10 mg/10 ml (Reineccius et al. 1970); the amount of oligosaccharides is small, 100 mg/liter (Renner 1983). The review of Jenness and Sloan (1970) is an excellent reference for those interested in the evolutionary biochemistry of lactose.

Molecular Structure

Lactose is a disaccharide that yields D-glucose and D-galactose on hydrolysis. It is designated as 4-0-β-galactopyranosyl-D-glucopyranose and occurs in both alpha and beta forms. The conclusive evidence establishing this structure has been reviewed (Whittier 1944; Clamp *et al.* 1961) in detail, so only a brief description is given here.

By hydrolyzing lactosone and lactobionic acid and obtaining free D-galactose plus glucosone and gluconic acid, respectively, it has been shown that the two monosaccharides are linked through the aldehyde group of D-galactose. Thus the aldehydic portion of lactose is on the glucose residue. That the configuration of the D-galactose residue is of the beta form was shown by the use of an enzyme, β-D-galactosidase, that hydrolyzes both lactose and a methyl β-D-galactopyranoside, but not the α anomer. Conversely, it was shown that an enzyme that hydrolyzed an α-D-galactoside but not a β-D-galactoside would not hydrolyze lactose. That the D-galactose in the lactose molecule is the beta form was also shown by its synthesis from D-glucose and D-galactose. The point of union of the two monosaccharides was established through products of hydrolysis or methylated lactose. The alpha or beta configurations of lactose are easily distinguished, since the alpha designation is arbitrarily assigned to the form having the greater rotation in the dextro direction. The structural formula for lactose is represented in Figure 6.1. β-Lactose is depicted by interchanging the OH and H on the reducing group. ^{13}C nuclear magnetic resonance assignments have been investigated (Pfeffer *et al.* 1979).

A series of rare carbohydrates is known whose structures differ only slightly from that of lactose. Lactose has been prepared from epilactose, 4-0-β-D-galactopyranosyl-D-mannose, from which it differs in configuration at C-2 of the D-glucose residue. Lactulose, 4-0-β-D-galoctopyranosyl-D-fructose, does not exist in unheated cows' milk but is produced from lactose during the heat processing and storage of certain dairy products such as sterilized infant formula (Renner 1983). Its preparation, properties, presence, and significance in milk products have been reviewed (Adachi and Patton 1961; Doner and Hicks 1982). A number of oligosaccharides that may be derivatives of lactose have been found in milks of various species (Renner 1983). They may differ

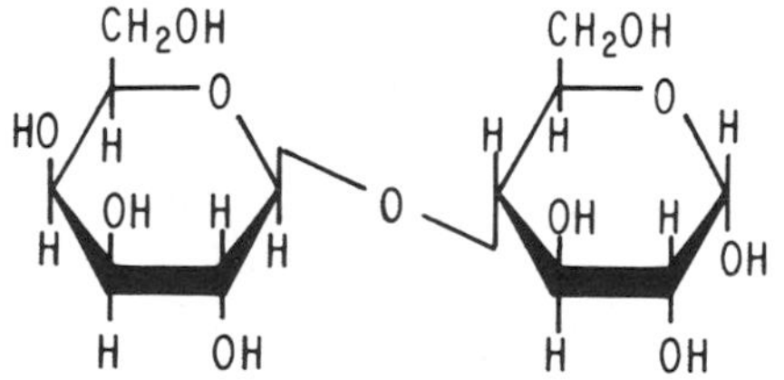

Figure 6.1. Structural formula of α-lactose. (From Herrington 1934. Reprinted with permission from the Journal of Dairy Science 17(7), 533–542.)

considerably in composition in the various milks and, on hydrolysis, yield such compounds as fucose, glucosamine, galactosamine, neuraminic acid, and D-mannose, in addition to D-glucose and D-galactose. Many exhibit bifidus growth activity for *Lactobaccillus bifidus* var. *pennsylvanicus* (Gyorgy 1953) and are of interest because of their nutritional and physiological significance (Zerban and Martin 1949). A review of the literature covers the isolation, identification, and structure of these oligosaccharides (Clamp *et al.* 1961).

BIOSYNTHESIS

Lactose biosynthesis in the mammary gland has been the subject of a considerable volume of published work which has been reviewed (Ebner and Schanbacher 1974; Jones 1978). Much of the following discussion is taken from these two references.

The biosynthesis of lactose is an unusual biochemical system in that one protein, the whey protein α-lactalbumin, acts as a protein modifier of an enzyme, a galactosyltransferase. The physiological function of this enzyme is to transfer galactose to an *N*-acetylglucoaminyl residue of the carbohydrate side chain of a glycoprotein. In the presence of α-lactalbumin, galactose may also be transferred to glucose to form lactose. The mammary gland is unique in its ability to synthesize α-lactalbumin, and this synthesis is presumably under hormonal control (Ebner and Schanbacher 1974).

The general scheme for lactose synthesis (Jones 1978) is considered to be:

Glucose	hexokinase $\longrightarrow$	glucose-6-phosphate
Glucose-6-phosphate	phosphoglucomutase $\longrightarrow$	glucose 1-phosphate
Glucose-1-phosphate	UDP glucose pyrophosphorylase $\longrightarrow$	UDP glucose
UDP glucose	UDP glucose epimerase $\longrightarrow$	UDP galactose
UDP galactose + glucose	galactosyltransferase $\longrightarrow$ α-lactalbumin	lactose + UDP

Most of these reactions take place in the cystosol of the epithelial cells surrounding the alveoli of the mammary gland, but the final step, in which the galactosyltransferase is modified by α-lactalbumin, occurs in the Golgi vesicles. The lactose thus formed is retained within the Golgi apparatus and transported to the apical surface of the cell in secretory vesicles along with the milk proteins. The vesicles then discharge their contents into the alveolar lumen.

Several possibilities existed as to how α-lactalbumin functions in the galactosyltransferase reaction. The results of kinetic and other mechanistic studies have shown that the active form of the enzyme is an α-lactalbumin-galactoslytransferase complex which forms in the presence of Mn^{2+} and uridine diphosphate (UDP) galactose or other appropriate carbohydrate acceptors (Klee and Klee 1972; Ivatt and Rosemeyer 1972; Powell and Brew 1975; Challand and Rosemeyer 1974). Ebner and Schanbacher (1974) described the main changes in substrate affinities brought about by α-lactalbumin; these changes consist mainly of an increased affinity for the carbohydrate galactosyl acceptor complicated by various types of substrate inhibition, for example, higher than optimum concentrations of α-lactalbumin.

Brew (1969) developed a hypothesis describing the role of α-lactalbumin in the control of lactose synthesis which was compatible with the kinetic properties of the enzyme system (Morrison and Ebner 1971A–C). Because of hormonal influence, α-lactalbumin is synthesized on the ribosomes, whence it passes into the tubules and vesicles of the Golgi apparatus. Here it comes into direct contact with the substrates UDP galactose and glucose and galactosyltransferase bound to the inner surface of the membrane. For this scheme to operate, the Golgi vesicles have to be impermeable to lactose but permeable to glucose and UDP galactose (Kuhn and White 1975, 1976). α-Lactalbumin disasociates from the galactosyltransferase after lactose formation but before the product is released into the milk. Although Brew originally visualized a mobile α-lactalbumin interacting with a membrane-bound galactosyltransferase, the enzyme also appears in milk, so it must also be mobile; the transit time appears to be about two days compared to 40 min for a milk protein such as α-lactalbumin (Jones 1978; Heald and Saacke 1972).

Lactose content of milk is affected by inheritance, age, stage of lactation, and interquarter differences (Walsh et al. 1968A). Such variations in lactose content constitute the major factor in variations observed in the SNF content of milk (Walsh et al. 1968B).

PHYSICAL PROPERTIES

Lactose normally occurs naturally in either of two crystalline forms— α-monohydrate and anhydrous β—or as an amorphous "glass" mixture of α- and β-lactose. Several other forms may be produced under special conditions.

α-Hydrate

Ordinary commercial lactose is α-lactose monohydrate ($C_{12}H_{22}O_{11} \cdot H_2O$). It is prepared by concentrating an aqueous lactose solution to supersaturation and allow crystallization to take place at a modern rate below 93.5°C. That α-hydrate is the stable solid form at ordinary temperatures is indicated by the fact that the other solid forms change to the hydrate in the presence of a small amount of water below 93.5°C. It has a specific optional rotation in water of $[\alpha]_D^{20} = +89.4°$ (anhydrous weight basis) and a melting point of 201.6°C. A study of the crystalline structure by x-ray diffraction (Buma and Wiegers 1967) has given the following consultants: $a = 7.98$ Å; $b = 21.68$ Å; $c = 4.836$ Å; $\beta = 109°47'$, which are in close agreement with previously reported constants (Knoop and Samhammer 1962; Seifert and Labrot 1961). These values refer to the dimensions of the unit cell and one of the axial angles. The value of ρ exp. = 1.497, indicating a z value of 2.03 molecules per unit cell.

α-Hydrate may form a number of crystal shapes, depending on the conditions of crystallization, but the most familiar forms are the prism and tomahawk shapes (Herrington 1934A; Hunziker and Nissen 1927). Since the crystals are hard and not very soluble, they feel gritty when placed in the mouth, similar to sand particles. This is the origin of the term "sandy" to describe the defect in the texture of ice cream, condensed milk, or processed cheese spread that contains perceptible α-hydrate crystals. Crystals that are 10 μ or smaller are undetectable in the mouth, but above 16 μ, fewer crystals can be tolerated without affecting the texture. When they are as large as 30 μ, only a few crystals are sufficient to cause sandiness in several products (Hunziker 1949; Nickerson 1954).

Crystalline Habit. α-Lactose hydrate crystals are observed in a wide variety of shapes, depending on conditions of crystallization. The principal factor governing the crystalline habit of lactose is the precipitation pressure, the ratio of actual concentration to solubility (Herrington 1934A). When the pressure is high and crystallization is forced rapidly, only prisms form. As precipitation pressure lessens, the dominant crystal form changes to diamond-shape plates, then to pyramids and tomahawks, and finally, in slow crystallization, to the fully developed crystal. These types of crystals are illustrated in Figure 6.2.

Detailed studies on the growth rates of the individual faces of α-lactose crystals have appreciably increased our understanding of the crystallization process. All the habits of lactose crystals found in dairy products are crystallographically equivalent to the tomahawk form; different relative growth rates on the crystal faces account for the var-

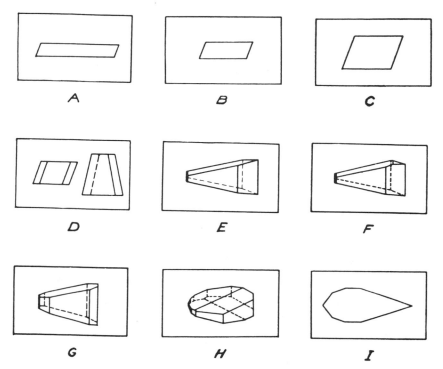

Figure 6.2. The crystalline habit of lactose α-hydrate. (A) Prism, formed when velocity of growth is very high. (B) Prism, formed more slowly than prism A. (C) Diamond-shaped plates; transition between prism and pyramid. (D) Pyramids resulting from an increase in the thickness of the diamond. (E) Tomahawk, a tall pyramid with bevel faces at the base. (F) Tomahawk, showing another face which sometimes appears. (G) The form most commonly decribed as fully developed. (H) A crystal having 13 faces. The face shown in F is not present. (I) A profile view of H with the tomahawk blade sharpened. (From van Krevald and Michaels 1965. Reprinted with permission of the Journal of Dairy Science 48(3), 259–265.)

ious shapes observed (Herrington 1934A; Van Krevald and Michaels, 1965). The axes and faces of the tomahawk crystal are depicted in Figure 6.3. Some typical lactose crystals are shown in Figure 6.4.

The rate of crystal growth increases rapidly as supersaturation (precipitation pressure) is increased. Data from several studies have shown that the growth rate increases with a supersaturation power greater than 1 (Van Krevald and Michaels 1965; Twieg and Nickerson 1968). Again, the rate is different for the different faces, altering the shape of the crystals. It is observed that the more the faces are oriented toward the b direction, the less they grow. The $(0\bar{1}0)$ face does not grow

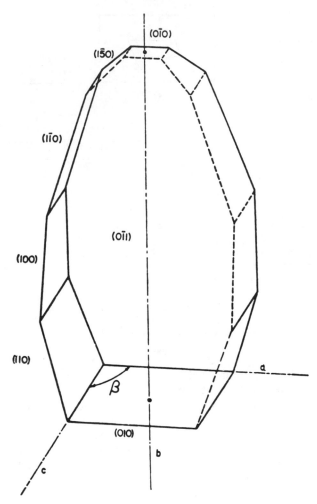

Figure 6.3. Tomahawk crystal of α-lactose monohydrate. (From Nickerson 1974.)

at all; the (011) face does not grow at low supersaturation but grows slightly at high supersaturation; the (1$\bar{1}$0) face grows slightly at low and moderately at high supersaturations; the (100) face always takes an intermediate position; and the (110) and (010) always grow fastest. Growth on the (010) face can vary enormously, whereas the (0$\bar{1}$1) and (1$\bar{1}$0) faces are fairly constant in growth rate.

Growth studies of broken crystals, as well as studies of the individual faces, have shown that lactose crystals grow only in one direction

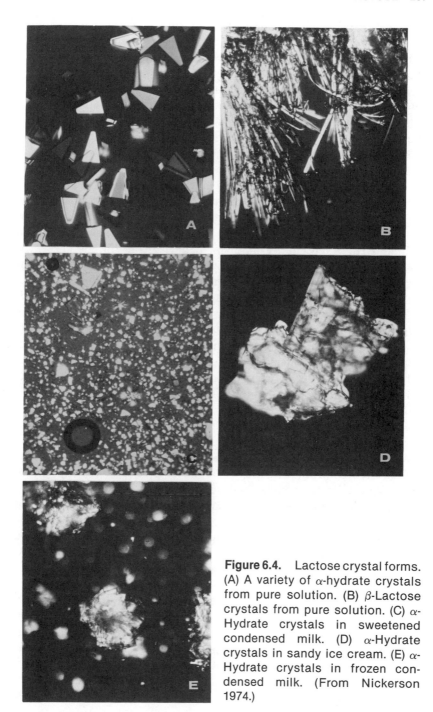

Figure 6.4. Lactose crystal forms. (A) A variety of α-hydrate crystals from pure solution. (B) β-Lactose crystals from pure solution. (C) α-Hydrate crystals in sweetened condensed milk. (D) α-Hydrate crystals in sandy ice cream. (E) α-Hydrate crystals in frozen condensed milk. (From Nickerson 1974.)

of its principal axis and therefore have their nucleus in the apex of the tomahawk (Van Krevald and Michaels 1965).

In dairy products, crystallization is more complex. The impurities (e.g., other milk components), as far as lactose is concerned, may interfere with the crystalline habit. As a result, the crystals tend to be irregularly shaped and clumped, instead of yielding the characteristic crystals obtained from simple lactose solutions. In some instances, the impurities may inhibit the formation of nuclei and thus retard or prevent lactose crystallization (Nickerson 1962).

The influence of a number of additives on growth rates has been studied; some additives resulted in marked retardation, whereas others accelerated growth on specific crystal faces (Michaels and Van Krevald 1966). Alterations in the growth process by additives are assumed to involve two opposing mechanisms: (1) acceleration of crystallization by reducing the edge energy at dislocation centers on the crystal face, thereby favoring a more rapid step generation rate by permitting a higher curvature of steps near a dislocation; and (2) inhibition of crystallization by retarding step propagation by adsorption of the additive on the crystal face. The concentration of the additive can influence the relative importance of these two reactions. For example, low concentrations of a surface-active agent, sodium dodecylbenzenesulfonate, result in "activation" of dislocation centers, thus leading to accelerated crystal growth; at higher concentrations, however, adsorption on the crystal face is rapid, resulting in inhibited growth (Michaels and Van Krevald 1966).

Although most additives that have been studied retard growth on all faces of the crystal, there are some which definitely promote growth on certain faces. For example, repeated recrystallization of lactose removes growth-promoting trace substances, so that crystal growth is much slower in supersaturated solutions of this lactose than in less purified solutions. The tendency toward spontaneous nucleation is also lowered upon repeated recrystallization.

Washing of lactose crystal clumps with distilled water increases subsequent growth (Nickerson and Moore 1974A). Jelen and Coulter (1973A) observed that partial dissolution of lactose crystals initially resulted in higher growth rates. Washing may cause an effect similar to that of breaking a face in that broken face grows much faster than normal faces until the break is healed (Van Krevald and Michaels 1965.) Washing may remove impurities accumulated on the crystal surface that inhibit growth; their removal would result in accelerated growth.

Gelatin is an example of a crystallization inhibitor that reduces the growth rate to 1/3 to 3/4 of normal even at low gelatin concentrations (Michaels and Van Krevald 1966). In highly supersaturated lactose solutions, however, gelatin cannot suppress nucleation, which explains its ineffectiveness in preventing sandiness in ice cream (Nickerson 1962).

Various marine and vegetable gums are currently in wide use in ice cream formulations. Shown to inhibit the formation of lactose crystal nuclei, they have been the principal factor responsible for the reduced incidence of sandiness in ice cream in recent years (Nickerson 1962).

Both methanol and ethanol accelerate crystallization by as much as 30 to 60% even at low (1%) concentrations, depending on which crystal face is being observed (Michaels and Van Krevald 1966). The mechanism is unexplained, but several factors seem to be involved. Although the solubility of lactose is depressed by alcohol, it does not seem to be depressed enough to account for the observed acceleration. It is more likely that the effects are due to promotion of step generation by adsorption of alcohol on the steps. Added support for the step theory was gained when it was shown that doubling the methanol concentration caused only a small increase in growth rather than an effect of increased supersaturation (Nickerson and Moore 1974B). Since alcohol promotes spontaneous nucleation, this may be another factor involved (Michaels and Van Krevald 1966).

The rate of lactose crystallization is also markedly increased at low pH (<1). Organic acids such as acetic and lactic acids are not suitable, since they do not yield the low pH necessary; lactic acid has been shown to slow crystallization (Jelen and Coulter 1973B). Sulfuric acid is an especially effective catalyst for lactose crystallization, being considerably better than hydrochloric acid even at the same pH (Nickerson and Moore 1974B). The effect has been attributed to the accelerating effect of acid on mutarotation (Nickerson and Moore 1974B). However, it has been demonstrated that mutarotation becomes limiting only when crystallization occurs rapidly on a large surface area. Since the acceleration could not be explained by mutarotation, it was suggested that the effect of low pH may be influencing the crystal-surface reaction (Twieg and Nickerson 1968).

Some carbohydrates actively inhibit the crystallization of lactose, whereas others do not. Carbohydrates that are active possess either the β-galactosyl or the 4-substituted-glucose group in common with lactose, so that adsorption can occur specifically at certain crystal faces (Van Krevald 1969). β-Lactose, which is present in all lactose solutions [see "Equilibrium in Solution (Mutarotation")], has been postulated to be principally responsible for the much slower crystallization of lactose compared with that of sucrose, which does not have an isomeric form to interfere with the crystallization process (Van Krevald 1969). Lactose solubility can be decreased substantially by the presence of sucrose (Nickerson and Moore 1972). The retarding action of β-lactose on crystallization of needles of α-hydrate lactose was ascribed to the fact that the β-galactosyl part of its molecule is the same as in α-lactose. The β-lactose molecules, along with α molecules, become attached to certain crystal faces which are acceptors of β-galactosyl groups. Once the β-lactose molecules are incorporated on the crystal, they impede further growth because of their β-glucose group, which is foreign to the crystal structure. However, β-lactose has failed to inhibit crystal growth under conditions where supersaturation was increased by its addition (Nickerson and Moore 1974B). The growth rate of regular α-hydrate crystals has been shown to depend on the amount of α-lactose in solution but to be independent of the amount of β-lactose (Nickerson and Moore 1974A).

Lactose crystal growth rates have been evaluated in the presence of certain salts and other substances found in cheese whey (Jelen and Coulter 1973B). The effect varied with species and concentration of the salt. Calcium chloride had the greatest growth-promoting effect; at the 10% impurity level, crystal growth rate was accelerated three times. Acceleration of the crystal growth rate resulted in an altered crystal shape; in the presence of calcium chloride, there was a considerable flattening of the crystal base, whereas in control solutions the crystals continued to grow in pyramid-like shapes. Jelen and Coulter (1973B) suggested that the salt effects they observed might explain the several forms of lactose crystals found in various dairy products (Van Krevald and Michaels 1965).

The retarding action of certain additives is more apparent in solutions in which crystallization is slow (low supersaturation). Under conditions of rapid growth, there is little opportunity for the additive to be adsorbed on the surface of the crystal, since the additive has greater competition with the large numbers of molecules of crystallizing material in highly supersaturated solutions.

Riboflavin also may adsorbed on growing lactose crystals and alter the crystalline habit. Since it is naturally present in the whey from which lactose hydrate is made and is present in all dairy foods, its influence on lactose crystallization may be of special interest. Adsorption is dependent upon concentration of riboflavin in solution, on degree of lactose supersaturation and on temperature (Leviton 1943, 1944; Michaels and Van Krevald 1966). No adsorption occurs below a certain minimum (critical) concentration of riboflavin (2.5 μg/ml), but adsorption increases linearly with riboflavin concentration above this critical level. Increasing the temperature of crystallization results in reduced riboflavin adsorption. Adsorption is favored at lower supersaturation levels of lactose where crystallization is slow, in keeping with the action of additives in general. By proper control of these variables, concentrations of 200 to 300 μg of riboflavin per gram of lactose are practical.

Forms of Anhydrous α-Lactose. The water of crystallization may be removed from α-hydrate crystals under various conditions to produce different types of anhydrous lactose.

Hygroscopic (Unstable) Anhydrous α-Lactose. An anhydrous form of α-lactose is produced by heating α-hydrate above 100°C *in vacuo*. The loss of moisture is negligible at 85°C, becomes significant at 90°C, and rises steadily with increasing temperature, being rapid at 120–125°C (Heinrich 1970). Its melting point is 222.8°C. Workers have had difficulty preparing this type of lactose with high purity; for example, drying at 100°C for 48 hr yielded a product of only 90 to 95% α, with the remainder β (Buma and Wiegers 1967). This form of anhydrous lactose is stable in dry air, but is highly hygroscopic and therefore unstable when exposed to normal atmospheric conditions. In the presence of water, it apparently forms the hydrate without first dissolving. In a solution saturated as to α-lactose hydrate, however, it will not dissolve. This behavior suggests little change in the crystalline structure other than removal of the water of crystallization, but the α-hydrate crystals are extensively fractured by the heat and vacuum treatment required to remove the water of crystallization, as shown by electron microscopic micrographs (Figure 6.5). The fractures remain even when the anhydrous lactose is rehydrated by absorption of 5% moisture from a moist environment (Lim and Nickerson 1973).

Stable Anhydrous α-Lactose. A stable form of anhydrous α-lactose (not hygroscopic) can be prepared by heating α-hydrate crystals in air at temperatures high enough to drive off the water of crystallization (100 to 190°C) while maintaining the atmospheric environment of the crystals at a water vapor pressure between 6 and 80 cm mercury (Sharp 1943). This environment is intermediate between rapid removal of vapor, by which the hygroscopic (unstable) anhydrous lactose is formed, and heavier vapor pressure conditions under which β-lactose is formed. This stable anhydrous α-lactose differs from the regular anhydrous form produced by heating the hydrate under vacuum in that it has greater density, is not appreciably hygroscopic, must dissolve in water before forming the hydrate, and dissolves readily in a solution that is already saturated as far as α-lactose hydrate is concerned. This solution is unstable and soon deposits crystals of α-lactose hydrate. Stable anhydrous α-lactose is more soluble in water than either

Figure 6.5. Electron Microscopic micrographs. Top left: α-hydrate crystals at 120X. Top right: regular anhydrous α-lactose (hygroscopic) at 100X. Middle left: regular anhydrous α-lactose (hygroscopic) at 2300X. Middle right: stable anhydrous α-lactose from methanol at 1150X. Bottom: β-lactose crystals at 2000X. (From Nickerson 1956. Reprinted with permission from the Journal of Dairy Science *39(10), 1342–1350.)*

α-hydrate or β-lactose. This form of lactose has been used to prepare solutions that are highly supersaturated in α but low in β (Van Krevald 1969).

A different, stable anhydrous form of α-lactose was prepared by refluxing α-lactose hydrate in absolute methanol (Lim and Nickerson 1973). At refluxing temperature, the anhydrous form was produced in a yield of 98 to 99% in 1 hr at all ratios of α-lactose hydrate to dry methanol. Other alcohols, such as ethanol, n-propanol, n-butanol, and isobutanol, were also used for this process (Nickerson and Lim 1974). It was originally assumed that the stable anhydrous α-lactose formed by these alcohol treatments was the same as that produced by heating α-lactose hydrate in air (Nickerson 1974). However, Ross (1978B), using differential scanning calorimetry, compared the stable anhydrous α-lactose obtained by heat treatment to that produced by methanol treatment of crystalline α-lactose monohydrate. The melting point of the methanol-treated anhydrous α-lactose was lower by 5.8°C, heat of fusion was higher by 33%, heat capacity was lower by 0.027 cal g^{-1} deg^{-1}, and density was higher by 0.025/g cm^{-3}. The anhydrous α-lactose crystallized from methanol has a new crystal form seen in Figure 6.5. Anhydrous α-lactoses prepared by treating α-lactose hydrate with methanol, ethanol, propanol, and n-butanol were all distinct species; each contained measurably small amounts of alcohol (Parrish et al. 1979A).

The water of crystallization can be moved from α-hydrate by refluxing it in a high-boiling organic solvent that is immiscible with water. For example, the moisture in lactose hydrate has been determined by the toluene distillation method that is often used to determine moisture in milk powder; with lactose, prolonged distillation (5 hr) is necessary to remove the hydrate moisture. The powder remaining after distillation in a stable anhydrous form (Nickerson 1974).

Another type of anhydrous lactose crystal can be prepared by shaking finely powdered α-hydrate crystals at room temperature in 10 times their weight of methanol containing 1 to 5% anhydrous hydrogen chloride (Hockett and Hudson 1931). The characteristic crystals of lactose gradually disappear, and tiny needles form. They contain a mixture of anhydrous α-and β-lactose in a ratio of 5:3. Olano et al. (1977) demonstrated that the compound having $\alpha{:}\beta = 5{:}3$ rapidly formed from α-lactose monohydrate in methanolic hydrogen chloride, provided that 1% water was present in the methanol; the crystalline anhydrous lactose had $\alpha{:}\beta = 4{:}1$ in absolute methanolic hydrogen chloride. At 27°C, the products from α-lactose monohydrate in acidic ethanol or 99 wt% aqueous ethanol were lactose with $\alpha{:}\beta = 5{:}3$ or anhydrous α-lactose, respectively; at -20°C, the product was anhydrous α-lactose in acidic, absolute or 99 wt% aqueous ethanolic or methanolic media (Simpson *et al.* 1982).

Anhydrous Lactose Glass (Amorphous Noncrystalline Glass)

When a lactose solution is dried rapidly, its viscosity increases so quickly that crystallization cannot take place. The dry lactose is essentially in the same condition as it was in solution, except for removal of the water. This is spoken of as a "concentrated syrup" or an "amorphous (noncrystalline) glass." Various workers have shown conclusively that lactose in milk powder (spray, roller, or freeze-dried) is noncrystalline and exists in the same equilibrium mixture of α- and β-lactose as existed in the milk prior to drying (Zadow 1984).

In vacuum oven methods for moisture determination, such as the official method of the Association of Official Analytical Chemists, lactose solutions are dried at about 100°C for 2 to 6 hr. The result in the dried product is amorphous lactose glass. Since lactose glass is very hygroscopic, the dried sample must be protected from moisture until final weighing. If α-hydrate crystals are present in the product to be analyzed, the sample is diluted with water to dissolve the crystals, since slow removal of the water of crystallization under the temperature and vacuum conditions of the moisture test unduly prolongs the moisture determination.

Lactose glass is stable if protected from moisture, but since it is very hygroscopic, it rapidly takes up moisture from the air and becomes sticky. When the moisture content reaches about 8% or a relative vapor pressure near 0.5, the lactose achieves a maximum weight; a discontinuity is observed in the sorption isotherm, and water is desorbed from the lactose (Berlin *et al.* 1968, 1970, 1971). α-Hydrate crystals develop at all temperatures below 93.5° C, and as they grow, the crystals bind adjacent powder particles together. Dry milk products containing lactose glass therefore tend to become lumpy or cake together during storage unless protected from moisture absorption. When moisture is absorbed, part of it is incorporated as water of crystallization in the α-hydrate and the remainder is desorbed, since crystalline α-hydrate is not hygroscopic (Supplee 1926).

β-Lactose

When lactose crystallization occurs above 93.5°C, the crystals formed are anhydrous and have a specific rotation of $[\alpha]_D^{20} = +35.0°$ and a melting point of 252.2°C. They are composed of anhydrous β-lactose,

Table 6.2. Solubilities of Lactose (g per 100 g water).

°C	Initial		Final	Super-solubility
	α	β		
0	5.0	45.1	11.9	25
10.0	5.8	—	15.1	—
15.0	7.1	—	16.9	38
25.0	8.6	—	21.6	50
30.0	9.7	—	24.8	—
39.0	(12.6)[a]	—	31.5	74
49.0	(17.8)	—	42.4	—
50.0	17.4	—	43.7	—
59.1	—	—	59.1	—
63.9	—	—	64.2	—
64.0	(26.2)	—	65.8	—
73.5	—	—	84.5	—
74.0	(34.4)	—	86.2	—
79.1	—	—	98.4	—
87.2	—	—	122.5	—
88.2	—	—	127.3	—
89.0	(55.7)	—	139.2	—
90.0	60.0	—	143.9	—
100.0	—	(94.7)	157.6	—
107.0	—	—	177.0	—
121.5	—	—	227.0	—
133.6	—	—	273.0	—
138.8	—	—	306.0	—

[a]Calculated values assuming $K = 1.50$ and solubility of one form is independent of the other.
SOURCES: Herrington (1948) and Whittier (1944).

which is sweeter and considerably more soluble than α-hydrate (Tables 6.2 and 6.6). The common form of the crystal is an uneven-sided diamond when crystallized from water and curved needle-like prisms when crystallized from alcohol. The lattice constants for the crystalline structures, as determined by x-ray diffraction (Buma and Wiegers 1967), are: $a = 10.81$ Å; $b = 13.34$ Å; $c = 4.84$ Å; $\beta = 91°15'$.

Several methods have been developed for preparation of β-lactose. α-Lactose hydrate has been converted to β-lactose in nearly quantitative yield by refluxing in methanol containing small amounts of sodium hydroxide (Olano and Rios 1978); ethanol, n-propanol and n-butanol were also effective as solvents (Olano 1978). Similarly, α-lactose hydrate was converted to β-lactose with potassium methoxide or potassium hydroxide as the base (Parrish et al. 1979B). β-Lactose was also prepared from the anhydrous forms of α-lactose if small amounts of β-lactose were present (Parrish et al. 1980A).

Itoh *et al.* (1978) have proposed an improved method for preparing β-lactose crystals by refluxing a supersaturated solution of α-lactose hydrate and seeding with β-lactose crystals. By using differential thermal analysis for characterization, a marked difference from the melting point of β-lactose previously reported was found. On the basis of their work, these authors proposed 229.5°C (decomposition) as the melting point for β-lactose. However, it has been demonstrated that mechanical treatments such as grinding or compaction significantly affect the thermic behavior of lactose; decreased melting temperatures occurred with increasing grinding and compaction times (Lerk *et al.* 1980). Mutarotation may occur in the "dry" crystals with elevated temperatures. The decreased melting point observed with β-lactose may indicate formation of decomposition products during grinding and compaction.

Equilibrium in Solution (Mutarotation)

As mentioned previously, lactose exists in two forms, α and β. By definition, α is the form with greater optical rotation in the dextro direction. The specific rotation of a substance is characteristic of that substance and is defined as the rotation in angular degrees produced by a length of 1 decimeter of a solution containing 1 g of substance per 100 ml. Therefore the specific rotation may be represented by the formula $[\alpha] = 100\ a/lc$, in which α = specific rotation, a = degrees of angular rotation, l = length of tube in decimeter and c = concentration of substance in grams per 100 ml of solution.

Also important, besides the variables of the equation, are temperature of the solution, wavelength of the light source, and concentration of the solution. The standard light source used to measure optical rotation has been the bright yellow D lines of the sodium spectrum, but the single mercury line, $\lambda = 5461$ Å, is now used frequently for precision measurements. Generally, the specific rotation is reported at 20°C and expressed as:

$$[\alpha]_D^{20} \quad or \quad [\alpha]_{Hg}^{20}$$

The following formulas (Haase and Nickerson 1966; Nickerson 1974) express variations in specific rotation in terms of these variables:

$$[\alpha]_D^t = 55.23 - 0.01688C - 0.07283\ (t - 25)$$

where C is grams of anhydrous lactose per 100 ml solution and t is degrees Centigrade;

$$[\alpha]^t_{Hg} = 61.77 - 0.007C - 0.076 (t - 20)$$

where C is grams of lactose monohydrate per 100 ml solution.

The values given earlier for specific rotations of α- and β-lactose are the initial values. When either form is dissolved in water, however, there is a gradual conversion of one form to the other until equilibrium is established. Regardless of the form used in preparing a solution, the rotation will change (mutarotation) until $[\alpha]^{20}_D = +55.3°$ at equilibrium (anhydrous weight basis). This is equivalent to 37.3% in the α form and 62.7% in the β form, since the equilibrium rotation is the sum of the individual rotations of the α and β forms. The equilibrium ratio of β to α at 20°C, therefore, is 62.7/37.3 $= 1.68$. This value is affected slightly by differences in temperature, but not by differences in pH. The proportion of lactose in the α form increases gradually and at a constant rate as the temperature rises. The equilibrium constant (β/α) consequently decreases with rising temperatures (Figure 6.6).

Mutarotation has been shown to be a first-order reaction, the velocity constant being independent of reaction time and concentration of reactants. The rate of mutarotation increases 2.8 times with a 10°C rise in temperature. By applying the law of mass action, equations have been developed to measure the rate of the reversible reaction between the α and β forms of lactose. If a dilute lactose solution at constant temperature contain a moles of α and b moles of β, then the amount of β formed (x) per unit of time is

$$\frac{dx}{dt} = k_1(a - x) - k_2 (b + x).$$

The mutarotation coefficient ($k_1 + k_2$) can be determined by the change in optical rotation with time:

$$k_1 + k_2 = \frac{1}{t} \log \frac{r_0 - r_\infty}{r_t - r_\infty}$$

where r_0 is the optical rotation at zero time, r_t is the rotation at time t, and r_∞ is the equilibrium (final) rotation. The equation expresses a first-order reaction. Plotting the difference in rotation at time t and equilibrium ($r_t - r_\infty$) against time gives a straight line with a slope equivalent to the mutarotation coefficient.

Procedures based on rotation have been used for quantitative measurement of the amounts of α- and β-lactose in fluid and dry milk products and in ice cream (Roetman (1981).

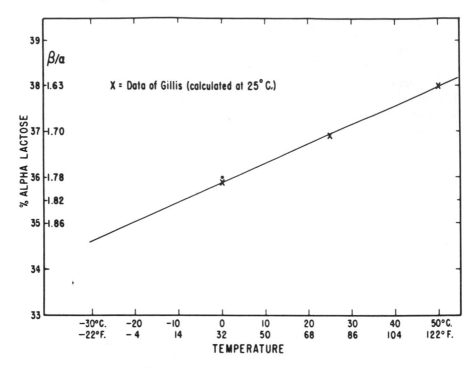

Figure 6.6. Effect of temperature upon the equilibrium ratio of β- to α-lactose. (From Troy and Sharp 1930. Reprinted with permission from the Journal of Dairy Science 13(2), 140–157.)

The rate of lactose mutarotation is influenced greatly by both temperature and pH. The rate is slow at low temperature but increases as the temperature rises, becoming almost instantaneous at about 75°C. The rate of change from α to β is given by Hudson (1908) as 51.1% complete in 1 hr at 25°C, 17.5% complete in 1 hr at 15°C, and 3.4% complete in 1 hr at 0°C. The rate of mutarotation is minimum at about pH 5.0, increasing with changes in pH on either side of this value (Figure 6.7). The rate is rapid at very low pH values but increases most rapidly in alkaline solutions, establishing equilibrium within a few minutes at pH 9.

The presence of sugars and salts can also affect the rate of mutarotation. Although the effect is small in dilute solutions, a combination of salts equal to that found in solution in milk nearly doubles the rate of mutarotation (Haase and Nickerson 1966). This catalytic effect is attributed primarily to the citrates and phosphates of milk. The presence of high levels of sucrose, on the other hand, has the opposite ef-

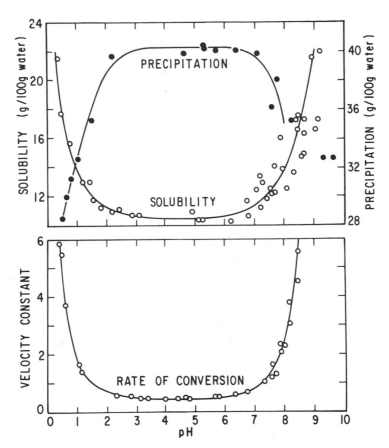

Figure 6.7. The effect of pH on the rate of change of the forms of lactose into each other as influencing the rate of solution and precipitation of lactose. (From Hunziker 1926.)

fect. The effect of sucrose is only slight at concentrations up to 40%, but as the concentration is increased above this level, mutarotation is rapidly decreased to about half the normal rate. A level of 30% sucrose or more also eliminates the catalytic effect of citrates and phosphates (Patel and Nickerson 1970). Some of the data suggest an interaction between the salts and the sugars. Some physicochemical properties have been reported for a complex of lactose and calcium chloride that has an empirical formula of α-lactose $\cdot$ CaCl$_2$ $\cdot$ 7 H$_2$O. However, a study of the calcium distribution in milk indicated that no soluble compounds between calcium and lactose exist at concentrations normally occurring in milk (Smeets 1955).

The specific rotation of lactose varies with the solvent. It is higher in glycerol than in aqueous solutions but lower in alcoholic or acetone solutions (Nickerson 1974). Not only the specific rotation but also the equilibrium ratio of α to β may be changed by the nature of the solvent. For example, upon dilution with water, concentrated solutions of lactose in methanolic calcium chloride show a high (1.3) initial to final rotation, regardless of whether the α- or β-isomer was used originally in preparing the solution (Domovs and Freund 1960).

Solubility

Mutarotation also manifests itself in the solubility behavior of lactose. When α-lactose hydrate is added in excess to water, with agitation a definite amount dissolves rapidly, after which an additional amount dissolves slowly until final solubility is attained.

Equations similar to those for mutarotation have been derived, expressing the relationship between the solubility behavior of the two forms of lactose and the equilibrium or rate constants (Hudson 1904). The constants derived by both mutarotation and solubility methods are in agreement. The solubility equations have been used to develop procedures for measuring α- and β-lactose in dry milk (Roetman 1981).

The initial solubility is the true solubility of the α form. The increasing solubility with time is due to mutarotation. As some of the α is converted to β, the solution becomes unsaturated with respect to α, and more α-hydrate dissolves. This process continues until equilibrium is established between α and β in solution and no more α-hydrate can dissolve, thus establishing the final solubility. This solution is saturated with respect to α, but a great deal of β-lactose powder can be dissolved in it because of the greater initial solubility of the β form. The solution becomes saturated with α long before the saturation point of β is reached. However, additional β dissolving in such a solution upsets the equilibrium, and mutarotation takes place. Since the solution was already saturated with α, α formed by mutarotation will crystallize to reestablish equilibrium. Since β-lactose is much more soluble and mutarotation is slow, it is possible to form more highly concentrated solutions by dissolving β- rather that α-lactose hydrate. In either case, the final solubility of the lactose in solution will be the same. Solubility values for lactose are shown in Table 6.2.

The solvent and the presence of salts or sucrose influence the solubility of lactose, as well as the rate of mutarotation. The solubility of lactose increases with increasing concentrations of several calcium salts—chloride, bromide, or nitrate—and exceedingly stable, concentrated solutions are formed (Herrington 1934B). One explanation for the increased solubility is the complex formation previously mentioned between the lactose and the salt.

It has been shown that calcium chloride also markedly increases the solubility of lactose in methanol (Domovs and Freund 1960). From the highly concentrated viscous solutions formed there slowly crystallizes a complex of β-lactose, calcium chloride, and methanol in a molecular ratio of 1:1:4. On addition of water to the concentrated solution, the complex previously described (α-lactose $\cdot$ CaCl$_1$ $\cdot$ 7H$_2$O) soon crystallizes.

The Steffen process, which uses calcium oxide for precipitation of sucrose from molasses, has been applied to the recovery of lactose from cheese whey (Cerbulis 1973). By proper control of the reaction, over 90% of the lactose can be recovered as an insoluble calcium–lactose complex. The addition of ferric chloride in combination with calcium oxide improves lactose yields. Addition of equal volumes of acetone or methanol gives almost complete precipitation of lactose and protein from whey.

Barium hydroxide has also been used to recover sucrose from molasses, but when applied to lactose, no precipitate formed on addition of alkali. This indicates that the barium–lactose complex was more soluble than the calcium–lactose complex. Even addition of acetone at levels comprising 20% of the final volume gave a much lower recovery of lactose (Nickerson 1979).

Lactose has been shown to combine with many cations in a 1:1 ratio, and equilibrium constants have been calculated (Swartz et al. 1978). No complexing could be demonstrated with K$^+$ or NH$_4$$^+$. Detailed studies with group IIA metal chlorides demonstrated 97% recovery of lactose with calcium (Quickert and Bernhard 1982).

Some studies have been made on the effects of other sugars on the solubility of lactose (Nickerson and Moore 1972). At 10 to 18°C, a 14% sucrose solution, comparable to that in ice cream mix, reduces lactose solubility only slightly. However, the data in Table 6.3 show that concentrations of 40 to 70% sucrose reduce the solubility of lactose appreciably—to 40 to 80% of normal. At temperatures near 0°C, the solubility of lactose is reduced by about one-half by saturating the solution with sucrose.

As mentioned previously, alcohol greatly reduces the solubility of lactose, but the glass or amorphous form dissolves in alcoholic solutions to form supersaturated solutions. This has been used to extract lactose from whey or skim milk powder with methanol or ethanol. A high-grade lactose subsequently crystallizes from the alcoholic solu-

Table 6.3. Relative Solubility of Lactose in Sucrose Solutions.[a]

	Temperature (°C)					
Solution	25	40	50	60	80	85
40% sucrose	74.5	76.7	75.5	81.9	89.4	80.5
50% sucrose	63.0	64.8	64.9	71.9	76.7	73.0
60% sucrose	50.9	53.5	53.3	57.8	70.2	66.4
70% sucrose	42.1	44.3	43.3	54.3	63.9	62.7

[a] Percentage of lactose solubility in distilled water at the same temperature.
SOURCE: Nickerson and Moor (1972). Reprinted with permission from the *Journal of Food Science*, 1972; *37*(1), 60–61. Copyright © by the Institute of Food Technologists.

tion. Methanol is the better solvent and allows recovery of soluble proteins in addition to the lactose (Leviton 1949; Leviton and Leighton 1938).

Ethanol and methanol (preferred with less than 3% moisture) have been used to extract lactose from skim milk or whey powders (Kyle and Henderson 1970). The dried lactose powder that crystallized from the alcoholic extract was believed to be anhydrous α-lactose, but other work indicates that the product is a mixture of anhydrous α- and β-lactose (Lim and Nickerson 1973).

Since increasing the concentration of alcohol greatly reduces the solubility of lactose, addition of alcohol accelerates crystallization and influences crystal habit (Majd and Nickerson 1976). Solubility also decreases with increasing alcohol chain length. When alcohol is added to a lactose solution, the mixture becomes milky white for a few seconds and then clears. After a few minutes, a permanent precipitate of lactose crystals appears. The composition of the precipitate may vary greatly with the percentage of alcohol added. Only α-hydrate is precipitated at low concentration, but β is also included at higher concentrations. Crystal shape changes from prisms to tomahawks as time passes or as the percentage of ethanol is decreased. Stable anhydrous α-lactose is produced when α-hydrate is treated with alcohol. Unlike α-hydrate, β-lactose is not altered by methanol, either at room or at refluxing temperatures.

Acetone also reduces the solubility of lactose, upon which a procedure to recover lactose from whey is based (Kerkkonen *et al.* 1963). Acetone is added to concentrated whey (18 to 20% lactose) in amounts sufficient to precipitate some of the impurities. After these are filtered out, the gradual addition of acetone to over 65% allows recovery of 85% of the lactose during a 3.5 hr period. The yield of lactose and rapidity of crystallization are influenced by the rate of acetone addition.

Crystallization

Solutions of lactose are capable of being highly supersaturated before spontaneous crystallization occurs. Even then, crystallization may occur only after a considerable period. In general, the supersolubility at any temperature is equal to the saturation value at a temperature 30°C higher. This is shown by the lactose solubility curves of Figure 6.8.

Ostwald in 1897 is credited with introducing the concept of supersaturation and extending it to "metastable" and "labile" areas (Mullin 1961). The metastable area occurs in the first stages of supersaturation produced by cooling a saturated solution or by continued evaporation beyond the saturation point. Crystallization does not occur readily in this supersaturation range. The labile area is found at higher levels of supersaturation, where crystallization occurs readily.

The true picture (Schoen 1961) is far more complex than indicated in Figure 6.8. In reality, a series of supersolubility curves should be pictured whose locations depend on specific seed surface, rate of supersat-

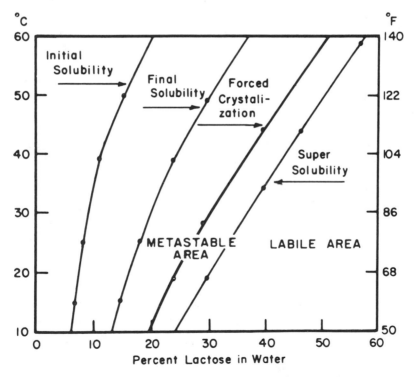

Figure 6.8.   Lactose solubility curves.

uration production (e.g., cooling rate, evaporation rate), and mechanical disturbances (e.g., agitation). The concept of "regions" of supersolubility is correct qualitatively but not quantitatively. The significant points of the concept are that (1) neither growth nor nucleation can take place in the unsaturated region; (2) growth of a crystal can take place in both the metastable and labile areas; (3) nucleation can take place in the metastable area only if seeds (centers for crystal growth) are added; and (4) spontaneous nucleation (crystallization) can take place in the labile area without addition of seeding materials.

These principles have been used to detect crystals of α-hydrate, β-lactose, or both in various products. A supersaturated solution in the metastable zone prepared with respect to the form being tested is unsaturated with respect to the other. When the product in question contains crystals, the solution will become cloudy with newly formed crystals as a result of seeding. If crystals are not present in the material, the solution will remain stable and clear.

Crystallization in general is a two-step process involving (1) nucleation and (2) growth of the nucleus to a macro size. Nucleation involves the activation of small, unstable particles with sufficient excess surface energy to form a new stable phase. This may occur in supersaturated solutions as a result of mechanical shock, the introduction of small crystals of the desired type, or the presence of certain impurities that can act as centers for growth.

A certain minimum-sized fragment is required to induce crystallization. The size of such a critical nucleus is on the order of 100 molecules (a diameter of 100 Å) (Van Hook 1961).

With increasing concentration, the probability of nucleus formation increases to a maximum and then quickly decreases to zero. The stability of lactose glass is apparently due to the small probability that nuclei will form at such high concentrations. When lactose glass absorbs moisture, as milk powder does, great numbers of nuclei form, since the concentration has been reduced to the region of maximum nucleus formation. After a nucleus forms, subsequent growth of any crystal depends on the rate of transfer of solute to the crystal surface and the rate of orientation of these molecules at the surface. Thus, the rate of crystal growth is controlled by the degree of supersaturation, the surface area available for deposition, and the diffusion rate to the crystal surface, which depends upon viscosity, agitation, and temperature of the solution. With lactose there is the additional factor of the rate of mutarotation of β to α. This rate is very rapid above 75°C but is slow at low temperatures.

It has been shown that optimum crystallization temperature (where the greatest amount of lactose will crystallize per unit time) varies with the degree of supersaturation (Twieg and Nickerson 1968). This is due to the fact that temperature influences two important aspects of the crystallization process: (1) supersaturation and (2) the crystallization rate constant, including rate of diffusion, rate of mutarotation, and rate of orientation of lactose molecules into the crystal lattice, all of which probably increase the temperature. However, supersaturation will decrease with temperature. These two factors oppose each other and in some instances cancel each other, so that changes in temperature have practically no effect on crystallization rate (Haase and Nickerson 1966). In other cases, crystallization is accelerated at higher temperatures as a result of decreased viscosity and increased kinetic activity.

Thus, the overall process of lactose crystallization can be summarized by the reaction:

$$\text{Step 1} \qquad \text{Step 2}$$

$$\beta\text{-Lactose} \underset{\leftarrow}{\rightarrow} \alpha\text{-lactose} \underset{\leftarrow}{\rightarrow} \alpha\text{-lactose hydrate crystals}$$

If mutarotation (step 1) is slower than crystallization (step 2), it will determine the overall reaction, and the α-lactose level will be lower than the mutarotatory equilibrium value (37.3% α at 20°C). Conversely, if crystallization is slower, the α- and β-lactose isomers in solution will be close to their equilibrium value. It has been shown that mutarotation occurs more rapidly under conditions normally found in milk products; thus crystallization becomes the rate-determining step (Haase and Nickerson 1966). However, under conditions of very rapid crystallization where the supersaturated α-lactose is being deposited on a large surface area of nuclei, the percentage of α in solution will drop below its equilibrium value. Under these conditions, neither mutarotation nor surface orientation appears to be completely rate-limiting (Twieg and Nickerson 1968).

Crystallization of the lactose in concentrated skim milk (40 to 54% total solids) can result in rapid increases in viscosity even during the short interval between concentration and spray drying (Baucke and Sanderson 1970).

Pallansch (1973) and his colleagues demonstrated that the rate of crystallization could be increased and viscosities held down by vigorous stirring during the holding period when working with high solids concentrates of cottage cheese whey. The crystallization of lactose at 40°C from cottage cheese whey concentrated to 69% total solids apparently followed first-order kinetics; the rate constant was $k = 3.07 \times 10^{-4}$ sec^{-1}. In this example, 70% of the lactose crystallized in the α-hydrate form after 20 min.

Buma (1980) reported viscosities for concentrated lactose solutions and concentrated cheese whey in the range of 10 to 40% total solids at temperatures of 20° to 60°C. A 40% lactose solution was considerably more viscous than a corresponding sucrose solution. The viscosity of the whey concentrate was much higher than that of the lactose; however, whey viscosity is also influenced by composition and heat treatment.

Other Physical Properties

Density. The densities of the various lactose crystals differ slightly from each other. α-Hydrate is 1.540, anhydrous β is 1.589, anhydrous α formed by dehydration under vacuum is 1.544, and anhydrous α crystallized from alcohol is 1.575. Densities of lactose solutions are not linear functions of concentration. Equations have been developed (McDonald and Turcotte 1948) relating the percentage (p) by weight of the lactose to density. The equations differ, depending on whether they are based on the hydrated or anhydrous form, temperature of solution, and range of concentration, e.g., $D_4^{20} = 0.99823 + 0.003739p + 0.00001281p^2$, where the hydrate is present between 0 and 16% and equilibrium is established. Similarly, equations have been developed relating such variables as concentration, form of lactose, and temperature to refractive index: $n_D^{20} = 1.33299 + 0.001409p + 0.00000498p^2$, where hydrate is present at less than 20% concentration and equilibrium is established. Tables are available containing precise data on the density and refractive indices of lactose solutions (McDonald and Turcotte 1948; Zerban and Martin 1949). Other physical properties of lactose are presented in Table 6.4. Berlin et al. (1971) report the heat of desorption from crystalline α-lactose hydrate and lactose glass to be 12.3 ± 0.7 and 10.8 ± 0.5 kcal mole^{-1}, respectively. The altered physical proper-

Table 6.4. Physical Properties of Lactose.

Property	α-Hydrate	β
Specific heat	0.299	0.2895
$S_{298.16}°K$ (Eu/mole)	99.1	92.3
$\Delta S_{298.16}°K$ (Eu)	−586.2	−537.2
$\Delta H°_{298.16}°K$ (cal/mole)	−592,900	−533,800
$\Delta F°_{298.16}°K$ (cal/mole)	−418,200	−373,700
Heat of combustion (cal/g)	3,761.6	3,932.7
Melting point (°C)	201.6°(d.)	252.2°(d.)

SOURCE: Anderson and Stegeman (1941), Buma and Meerstra (1969), and Whittier (1944).

ties of amorphous α-lactose produced by methanol treatment reported by Ross (1978A) and of β-lactose reported by Itoh *et al.* (1978) may have their origin in the mechanical treatment given the samples before examination by differential thermal analysis, as pointed out by Lerk *et al.* (1980).

Relative Sweetness. It has been amply demonstrated that the relative sweetness of sugars changes with the concentration. Therefore it is misleading to say that one sugar is so many times as sweet as another, because this will be true only at certain concentrations. Table 6.5 summarizes results on the relative sweetness of some common sugars. It should be noted that lactose is relatively sweeter at higher concentrations than at lower concentrations and is sweeter than is usually reported in reviews of food applications.

β-Lactose is sweeter than α-lactose (Table 6.6) but β is not appreciably sweeter than the equilibrium mixture except when the concentration of lactose solution equals or is greater than 7% (Pangborn and Gee 1961). Since there is approximately 63% β in the equilibrium mixture, a β-lactose solution differs less in sweetness from a solution in equilibrium than does α-lactose solution.

Parrish *et al.* (1981) reevaluated the sweetness of α-, β-, and equilibrium lactose. They demonstrated that β-lactose was 1.05 to 1.22 times as sweet as α-lactose, but there was no significant difference in sweet-

Table 6.5. Relative Sweetness of Sugars (Percent Concentration to Give Equivalent Sweetness).

Sucrose	Glucose	Fructose	Lactose
0.5	0.9	0.4	1.9
1.0	1.8	0.8	3.5
2.0	3.6	1.7	6.5
2.0	3.8	—	6.5
2.0	3.2	—	6.0
5.0	8.3	4.2	15.7
5.0	8.3	4.6	14.9
5.0	7.2	4.5	13.1
10.0	13.9	8.6	25.9
10.0	14.6	—	0
10.0	12.7	8.7	20.7
15.0	17.2	12.8	27.8
15.0	20.0	13.0	34.6
20.0	21.8	16.7	33.3

SOURCE: Nickerson (1974).

Table 6.6 Relative Sweetness of α- and β-Lactose.

	Concentration (%)	No. of evaluations	Sec after hydration	Percentage Response Considering		
				α Sweeter	Equilibrium sweeter	β Sweeter
α-Lactose	5.0	40	270	40.0	60.0	—
	7.0	44	222	13.6	86.4[3]	—
β-Lactose	5.0	40	124	—	45.0	50.0
	7.0	40	140	—	35.0	65.0[a]
α-Versus	5.0	40	184	25.0	—	75.0[b]
β-Lactose	7.0	40	266	12.5	—	87.5[c]

[a]Significant at $p = 0.05$.
[b]Significant at $p = 0.01$.
[c]Significant at $p = 0.001$.
SOURCES: Pangborn and Gee (1961). Reprinted by permission from *Nature 191* (4790), 810–811. Copyright © 1961, Macmillan Journals Limited.

ness between α-lactose and equilibrium lactose. The small difference in sweetness between α- and β-lactose is of no practical value in food applications such as coating sugar in baking applications.

APPLICATIONS

The crystallization principles previously discussed are applied in processing dairy products, such as sweetened condensed milk, instant milk powder, stabilized whey powders, lactose, and ice cream.

Dry Whey and Whey Permeates

In the past few years, considerable research has been devoted to whey processing and utilization. Because excellent reviews of progress in whey research are available (Whey Products Conference Proceedings, 1970, 1973, 1975, 1977, 1979, 1981, 1983, 1985; Whey Research Workshop II 1979; Evans 1980; Hobman 1984), the discussion here will be brief.

Lactose, constituting about 70% of the solids in whey, understandably plays a dominant role in determining the properties of whey and whey products. Whey is often difficult to dry by normal methods but can be dried by modifying the processes used to dry milk, i.e., roller or spray methods (Hall and Hedrick 1971). Modifications usually involve procedures to cope with the sticky hygroscopic glass that may be formed with either method. Most special processes cause a considerable portion of the lactose to crystallize by holding the product at some stage in the presence of sufficient moisture. This may be induced in the condensed product or at some stage where the product is only partially dried, or by rehumidifying the dry product.

Crystallizing conditions generally favor the production of α-hydrate crystals, but if crystallization occurs above 93.5°C, especially under pressure, anhydrous β crystals will be formed. After crystallization has occurred and the product has finally dried, the resulting whey powder is granular and free-flowing, and does not tend to become sticky and caked.

Foam-spray drying was introduced successfully for drying high-solids cottage cheese whey (Hanrahan and Webb 1961A,B). Normally, the high acidity of many such wheys causes difficulty in drying; lumps form and clog the dryer. By introducing compressed air into the whey just prior to the spray nozzle, a foam structure is produced that dries rapidly, forming free-flowing particles. Hygroscopicity problems have been overcome in this product by crystallizing lactose in the concentrate before drying (Tamsma *et al.* 1972).

Because crystallization of lactose in high-solids whey concentrates is accompanied by a rapid increase in viscosity, many whey drying operations carry out the major part of lactose crystallization after spray drying. Spray drying in this type of equipment is controlled so that sufficient moisture remains in the powder to permit lactose crystallization in the damp powder mass. When crystallization has proceeded to the desired point, the residual moisture in the powder is removed in a secondary drying system. For example, a partially crystallized concentrate is sprayed into a conical drying chamber by centrifugal atomization; the partially dried powder is collected and fed into a fluidized bed system where the lactose is crystallized and the whey is dried and cooled in sequential steps before packaging (Pallansch 1973). In the Pillsbury dryer (Young 1970), the product builds up a mat on a porous metal belt. The mat serves as a filter for the exhaust air; the porous mat can be moved to other sections of the dryer for holding, further drying, and cooling.

Other systems make use of the sticking tendency of acid whey. Partially dried whey powder coats the inner wall of the drying chamber, whence it falls when the crystalline lactose content of the powder becomes high. However, sticking of the product on the hot metal surfaces can be a problem unless sufficient moisture is present so that lactose crystallization proceeds to the point where the powder no longer adheres to the equipment (Pallansch 1973).

Hargrove *et al.* (1976) investigated spray drying of the deproteinized permeate from whey ultrafiltration. Concentrates between 40 and 50% total solids dried readily, whereas concentrates above 50% total solids failed to dry. Controlled crystallization was studied as a means of increasing the drying capability and providing a more stable product. Powders with a high level of crystalline lactose were obtained by concentrating the permeate to about 60% total solids, holding for 1 to 2 hr at 40°C, and then adjusting to 50% total solids with water and spray drying.

These researchers found no correlation between hygroscopicity of the powders and degree of lactose crystallization, but there was a direct correlation between hygroscopicity and the amount of lactic acid in the powder. Acid whey permeate dried readily, however, without many of the problems encountered during the drying of whole whey.

Other Products

The control of lactose crystallization in sweetened condensed milks has been reviewed extensively (Webb 1970; Hall and Hedrick 1971; Hunziker 1949) and will not be discussed further here. Lactose crystallization in ice cream and other frozen desserts is readily controlled by stabilizers and has been reviewed elsewhere (Keeney and Kroger 1974).

Frozen concentrated milks have always been an attractive way of preserving milk with minimal flavor change. Unfortunately, such concentrates thicken and coagulate during storage because of the crystallization of lactose followed by destabilization of the calcium caseinate-phosphate complex. Partial enzymatic hydrolysis of the lactose before concentration (3:1), followed by a postpasteurization heat treatment at 71°C for 30 min after being canned, results in samples that show only a moderate viscosity increase after 9 months of frozen storage (Holsinger 1978).

Lumping and caking in dry milk during storage is a problem if the milk powder is not protected from moisture. The absorption of moisture dilutes the lactose glass in the powder to the point where molecular orientation is possible and crystallization occurs. The α-hydrate crystals that form cement the milk powder particles together, producing lumpiness and then caking. Crystallization begins at about 7% moisture content. Part of the moisture is incorporated as water of crystallization and the remainder is desorbed (Supplee 1926; Berlin et al. 1968, 1970). The caseinate complex is insolubilized as a result of this process.

To improve the reconstitutability of spray-dried milk powder, the powder is agglomerated or instantized. This is generally carried out by wetting the surface of the milk powder particles with steam, atomized water, or a mixture of both, agglomerating the wetted powder particles by collision in turbulant air, redrying with hot air, and cooling and sizing to eliminate very large agglomerates and very small particles (Hall and Hedrick 1971). Partial lactose crystallization occurs before the particles are redried (Peebles 1956). The agglomerated powder is free-flowing and readily dispersible in water, in contrast to conventionally spray-dried powder, where the particles tend to ball up on the surface of the water when wetted. The lactose equilibrium in instantized powders is shifted to 3:2 (α:β) instead of the more usual 2:3 (Bockian et al. 1957).

Manufacture of α-Lactose

For many years, only the highly refined USP grade of lactose was marketed, but now that lactose is used in diverse products, a variety of grades are available. Specifications for some of these grades are shown in Table 6.7. In addition, small quantities of β-lactose are produced for certain uses where its higher solubility or sweetness may be an advantage.

Table 6.7. Typical Physical and Chemical Data for Various Grades of Lactose.

Analysis	Fermentation	Crude	Edible	USP	USP Spray Process
Lactose (%)	98.0	98.4	99.0	99.85	99.4
Nonhydrate moisture (%)	0.35	0.3	0.5	0.1	0.5
Protein (N × 6.38) (%)	1.0	0.8	0.1	0.01	0.05
Ash (%)	0.45	0.4	0.2	0.03	0.09
Lipids (%)	0.2	0.1	0.1	0.001	0.01
Acidity, as lactic acid (%)	0.4	0.4	0.06	0.04	0.03
Heavy metals, as Pb (ppm)	—	—	<2	<1	<2
Specific rotation $[\alpha]_5^{25}$	—	—	+52.4°	+52.4°	+52.4°
Turbidity (ppm)	—	—	<5	<5	<5
Other sugars (mg)	—	—	15	5	10
Color (ppm)	—	—	10	5	5
Bacterial estimate Standard plate count (per gm)	—	—	<100	<30	<30
Coliforms in 10 mg	—	—	Neg.	Neg.	Neg.
Sporeformers (in 10 mg)	—	—	Neg.	Neg.	Neg.
Molds (in 10 mg)	—	—	Neg.	Neg.	Neg.
Yeasts (in 10 mg)	—	—	Neg.	Neg.	Neg.

SOURCE: Nickerson (1974).

Processes for crystallization of lactose are well established, with production generally limited to a few large plants. Although a variety of cheese wheys and whey ultrafiltrates can be used for lactose production, sweet whey or ultrafiltrates are preferred (Woychik 1982). The crystallization process has three basic steps:

1. Concentration of whey to 50 to 70% total solids by evaporation.
2. Initiation of crystallization, either spontaneously or by seeding with a small quantity of lactose crystals.
3. Separation of the crystals by centrifugation.

The yield and purity of the crystals are affected by the protein and mineral content of the starting material; the highest purity and best yields are obtained from deproteinized-demineralized whey.

During centrifugation, the crystals are sprayed with clear water to remove the adhering liquor. The crystals may be dissolved for further recrystallization, or they may be dried to form "crude lactose." Refining to produce better grades of lactose may consist of redissolving in water, treating with activated carbon to decolorize the solution, filtering, concentrating, and either recrystallizing or spray-drying the solution. Commercial practice makes use of numerous modifications of this basic procedure. Several reviews are available for additional details (Short 1978; Brinkman 1976; Roetman 1972; Nickerson 1970).

Direct application of this technology to the processing of whey permeates to produce crystalline lactose is not straightforward (Hobman 1984). The permeate is saturated with calcium, so concentration by evaporation causes precipitation of calcium salts of citrate and phosphate, which can foul heat exchanger surfaces. The insoluble calcium salts will also contaminate the lactose during subsequent crystallizing operations; because of their low solubility, they are not readily removed by washing with water. It is generally accepted that the whey permeate must be pretreated either before or during the evaporation step. Suitable processes include removal of calcium by ion exchange, demineralization by electrodialysis, reducing the pH to eliminate the formation of insoluble salts, addition of food-grade calcium-chelating agents such as sodium hexametaphosphate to form insoluble complexes easily removed before crystallization, or separation of insoluble salts from hot concentrated permeate before cooling it to crystallize the lactose (Nickerson 1979).

Manufacture of crystalline lactose from permeate derived by ultrafiltration of lactic casein whey presents special problems because of the low pH, high lactate concentration, and high calcium and phosphate concentrations (Hobman 1984). Research at the New Zealand Dairy Research Institute has led to a pilot-scale process whereby calcium phosphate complexes are partially removed before evaporation by an alkali and heat treatment to precipitate them, followed by centrifugation to clarify the treated permeate. Removal of about 50% of the calcium is sufficient to avoid problems during evaporation.

The absence of protein in whey permeates has the advantage of reducing viscosity in concentrated solutions, thereby permitting concentration to higher total solids. The higher percentage of lactose in the total solids increases the yield of crystalline lactose (Brinkman 1976). Other potential advantages include shorter crystallization times and continuous crystallization (Muller 1979). The purity of the crude lactose crystals can be readily increased to more than 99% (dry weight) by slurrying with water and reseparating.

A general, recent trend has been to apply the principles of ion exchange to the purification of whey or lactose solutions. Anionic and cationic exchange resins are used to remove impurities from the solution, which can then be condensed and crystallized or spray-dried directly. Ahlgren (1977) and Delaney (1976) have reviewed developments in whey demineralization, including ion exchange and ultrafiltration. Parrish *et al.* (1979C) demonstrated an improved yield of α-lactose monohydrate, crystallized from a 50% (w/w) aqueous solution when sweet whey ultrafiltrate was demineralized with thermally regenerable ion-exchange resins. These resins are regenerable with hot water instead of acids and bases, which means reduced operating costs and decreased effluent pollutants.

Efforts are underway to develop a process based on a single-stage, continuous crystallizer with control over the nucleation rate (Muller 1979). Studies by Thurlby (1976) and Thurlby and Sitnai (1976) suggested that it might be possible to improve the economics of lactose manufacture by the use of a continuous crystallization process followed by batch crystallization operated at a low temperature to increase yields. By computer modeling, they showed that rates of crystallization could be increased in the continuous crystallizer if the crystal nuclei could be redissolved soon after formation so that existing crystals grew instead of new small crystals being formed. A 20 liter working capacity crystallizer was built to investigate the concept. When concentrated sweet whey ultrafiltrate was tested, the crystallization rates observed were between 2500 and 4000 g/hr. It was necessary to wash the crystals for improved purity. The percentage recovered is influenced by the increase in the proportion of the minerals in the mother liquor as lactose is progressively removed (Muller 1979).

The process for producing a high-quality crude lactose by methanol extraction of whey or nonfat dry milk (Leviton 1949) is presently being reevaluated (Chambers 1985). After dispersal of the powder in the solvent under controlled conditions of concentration and temperature, the resulting supersaturated solution is stable enough to permit removal of the precipitated proteins before crystallization starts. This procedure is promising, not only for producing a relatively pure lactose but for producing a lactose-free milk protein product.

The kinetics of lactose crystallization has been investigated further. Valle-Vega and Nickerson (1977) studied the effects of supersaturation and agitation with an image analyzer computer to measure changes in the distribution of crystal size during crystal growth of lactose. Valle-Vega *et al.* (1977) examined the size of lactose crystals prepared commercially and found that the major factor influencing crystal size was the cooling rate.

Further information on the kinetics of crystallization is helpful in designing improved processes for lactose recovery. Many potential food applications, such as in confectionary products, involve concentrated sugar solutions in which crystallization imparts the desired texture to the product.

Manufacture of β-Lactose

Because β-lactose has a much higher initial solubility and is sweeter than α-lactose, there is a demand for a limited amount of it. The processes for the manufacture of β-lactose are based on the fact that β is the stable form crystallized from lactose solutions above 93.5°C (Bell 1930).

New Zealand workers have produced a crude β-lactose from whey ultrafiltrate by roller or drum drying at a temperature greater than 93.5°C (Kavanagh 1975; Goldman and Short 1977). This product replaced up to 25% of the sucrose in a "high ratio" cake formula without adversely affecting cake size, tenderness, or sweetness (Goldman and Short 1977).

It has been demonstrated in the laboratory that by careful manipulation of alcohol concentration, amorphous lactose high in the β anomeric form could be precipitated (Majd and Nickerson 1976; Ross 1978A; Olano and Rios 1978). Parrish *et al.* (1980B) have formed β-lactose from stable forms of anhydrous α-lactose; and have prepared β-lactose from α-lactose monohydrate with potassium methoxide (1979B). None of these processes have been commercialized.

DETERMINATION OF LACTOSE

Harper (1979) and Doner and Hicks (1982) have reviewed the various methods for analysis of lactose and its derivatives. More recently, Roetman (1981) has described methods for the quantitative determination of crystalline lactose in milk products. The reader should consult these reviews for information on specific procedures.

Lactose and its derivatives are determined quantitatively by polarimetry, colorimetry, enzymatic procedures, cryoscopy, gas-liquid chromatography, and high-performance liquid chromatography.

Polarimetric analysis is useful for determining the anomeric form of crystalline lactose or related compounds. Quantitation by polarimetric analysis is limited to samples free of other optically active compounds.

Because of their speed and reasonable instrumental requirements, there is continuing interest in colorimetric procedures for lactose determinations. Most procedures are based on the reducing properties of lactose; samples that contain only one sugar are easily measured colorimetrically, but samples with three or four sugars may require several different colorimetric assays to determine the composition accurately.

Enzymatic methods are accurate, and specific but complex sugar samples may require multiple analyses. Instrumentation is available for the determination of glucose and lactose enzymatically. To measure lactose, the sample is first reacted with glucose oxidase to measure the glucose concentration; it is then passed through a β-galactosidase column to hydrolyze the lactose and then through a glucose oxidase column to measure the glucose in the hydrolyzed sample. Both glucose and galactose concentrations are determined. The major problem is fouling of the enzyme columns; the method is applicable to permeates but not to whey.

Several simple, accurate cryoscopic methods have been developed to measure the percentage of lactose hydrolyzed in whey and lactose solutions. Zarb and Hourigan (1979) have developed a novel method combining enzymatic treatment and cryoscopy for the measurement of lactose in milk and milk products.

The method is based on the principle that a solution of lactose specifically hydrolyzed will show a freezing point depression directly proportional to the molarity of the lactose. Doubling of the number of molecules by hydrolysis theoretically will double the effect that the carbohydrate would have on the freezing point. The method was most sensitive in the region of lactose concentration of 1.0 to 3.5% (± 0.05% lactose).

Gas-liquid chromatography is especially useful in measuring isomerization products and oligosaccharides formed during lactose hydrolysis. Several sugars can be quantitated in the same analysis. Sugars must be converted into volatile derivatives before analysis by gas-liquid chromatography, adding to analysis time and complexity.

Sugars may be directly measured by high-pressure liquid chromatography without the formation of derivatives and usually without extensive sample cleanup. Procedures have been developed for quantitating lactose in a variety of milk products. The methods require commercially available columns prepacked with high-performance silicas that have been modified with a polar amino or cyano-type bonded phase. Mobile phases in the systems may be acetonitrile–water mixtures; at flow rates of 1 to 2 ml/min, the retention time for lactose is generally 10 to 20 min (Doner and Hicks 1982). Sugars in mixtures can be conveniently quantitated by this method by comparing peak heights with those of corresponding sugars in standard solutions.

Four methods, based on different principles, are used to measure crystalline lactose (Roetman 1981):

1. Measurement of a property to which crystalline and dissolved lactose contribute to different extents. Large crystals do not contribute to properties such as refractive index and electrical conductivity, and methods based on these properties have been described.

2. Separation by removing the crystals by centrifugation or by filtration or suspensions of lactose crystals.
3. Determination of the ratio of isomers. The polarimetric determination of α- and β-lactose in solutions is well known. The fraction crystallized in dried products may be easily calculated from the shifts in the relative amounts of α-lactose on rehydration.
4. Estimation of water of crystallization. When lactose is crystallized as the α-lactose hydrate in a product, the fraction crystallized can be determined from the weight of water of crystallization lost. Water of crystallization accounts for 5% by weight of the crystalline lactose.

Other approaches use Laser-Raman spectra to differentiate five conformational states of lactose, including α-lactose monohydrate, β-lactose, and lactose glass (Susi and Ard 1974). Differential thermal analysis has also been used to measure the concentration of crystalline lactose, especially α-lactose hydrate (Ross 1978B). The specialized equipment required by these procedures may limit their use.

CHEMICAL REACTIONS

In its chemical behavior, lactose is like a number of similar carbohydrates and reacts according to the general rules of carbohydrate chemistry. Thus the reactions involve such groups as (1) glycosidic linkage between the two monosaccharides; (2) the reducing group of glucose; (3) the free hydroxyl groups; and (4) the carbon–carbon bonds. A discussion is available on the chemical properties of lactose and its early chemical history (Clamp et al. 1961).

There are three major chemical derivatives of lactose in which the β 1–4 linkage remains unbroken. They are lactitol, produced by reduction; lactulose, produced by isomerization; and lactobionic acid, produced by oxidation. The synthesis and properties of these derivatives have been reviewed (Donar and Hicks 1982).

Lactitol (4-0-β-D-Galactopyranosyl-D-Sorbitol)

Lactitol is a disaccharide sugar alcohol prepared by reduction of the glucose residue to a sorbitol group. It is prepared by hydrogenation of a lactose solution; hydrogenation at 100°C for 6 hr and 8825 kPa with a Raney nickel catalyst produces lactitol in nearly quantitative yield (van Velthuijsen 1979; Linko et al. 1980). Hydrogenation of lactose with sodium or calcium amalgam catalysts and reduction with sodium borohydride (Scholnick et al. 1975) have also been successful.

Lactitol may be crystallized as the monohydrate which melts at 94°-97°C. It is extremely soluble in water but only slightly soluble in alcohol or ether. Since it is not a reducing sugar, it does not mutarotate in solution. The absence of a potential carbonyl group confers stability on acid, base, heat, and nonenzymatic browning reactions (van Velthuijsen 1979). The monohydrate appears to be less hygroscopic than sorbitol and xylitol but more hygroscopic than mannitol. In solution, lactitol is much less hygroscopic than solutions of the other polyols.

The relative sweetness of lactitol, depending on the concentration, is about 35% of that of sucrose (van Velthuijsen 1979). For example, an 11.4% (w/w) lactitol solution has sweetness equal to a 4% (w/w) sucrose solution.

Van Velthuijsen (1979) has reviewed biological and toxicity data on lactitol. The nontoxic effect of lactitol in rats is 5% in the diet, corresponding to a noneffect level of 2.5 g/kg body weight per day. Humans can apparently consume up to 20 g in a single dose without experiencing unpleasant side effects, but knowledge of the ability of human intestinal enzymes to hydrolyze and absorb lactitol is incomplete. It has been suggested that lactitol is noncaloric and suitable for use as a sweetener by diabetics. Van Velthuijsen (1979) has reviewed studies on the caloric potential of lactitol and has concluded that a reduced caloric value can be expected. Apparently, the lactitol is neither absorbed or hydrolyzed largely in the small intestine, but is fermented by the microflora in the large intestine.

The cariogenicity of lactitol has also been investigated. Linko et al. (1980) reviewed early studies showing that lactitol was not readily fermented by Streptococcus mutans and other oral bacteria. In vivo studies, reviewed by van Velthuijsen (1979), were concerned with the reduction of pH in dental plaque; after consumption of chocolates made with lactitol, there was evidence that lactitol did not increase the incidence of dental caries.

Lactitol is potentially useful in foods as a bulking agent because of its good solubility and limited sweetening power. Its excellent chemical stability would permit its use in foods that undergo severe processing or abuse in storage. It may also have potential use in dietetic products because of its noncaloric nature, but its use in all foods would have to be carefully regulated because of its laxative effects (Doner and Hicks 1982).

Lactitol Palmitate and Other Surfactants

Lactitol palmitate may be prepared by direct esterification with fatty acids of edible fats in such a manner that formation of anhydropolyols is minimal (van Velthuijsen 1979). The reaction is carried out at a temperature of about 160°C, catalyzed by soaps of the fatty acids. These esters may be used as emulsifiers in foods or as detergents; van Velthuijsen (1979) has described the performance of lactitol palmitate under household laundering conditions.

Parrish (1977) reviewed the research and development of lactose ester-type surfactants carried out by Scholnick and his colleagues (Scholnick et al. 1974, 1975; Scholnick and Linfield 1977). Their initial attempts to form lactose esters followed the same transesterification procedures that had been used with sucrose (a fatty acid methyl ester in N,N-dimethylformamide with potassium carbonate as the catalyst). Their successful approach was the reaction of lactose in N-methyl-2-pyrrolidone as the solvent with fatty acid chlorides, resulting in yields of 88 to 95% for esters of lauric, myristic, palmitic, stearic, oleic, and tallow fatty acids. The principal product was the monoester, which is important for detergent use, since diesters and higher esters of lactose are not water soluble.

The lactose esters and their ethyoxylated (prepared by treatment with ethylene oxide) derivatives possessed surfactant properties comparable to those shown by analogous sucrose derivatives. The best detergency properties were shown by the lower fatty acid monoesters of lactose. The comparable lactitol esters were slightly better detergents than the lactose esters; no improvement was brought about in either case by the ethylene oxide adducts. The esters of both lactose and lactitol are readily biodegradable.

Lactulose (4-0-β-D-Galactopyranosyl-D-Fructose)

Lactulose is an isomer of lactose that is formed by molecular rearrangement, usually under alkaline conditions whereby the terminal aldose residue of lactose is converted into a ketose. Doner and Hicks (1982) and Zadow (1984) have described the preparation and properties of lactulose in detail. Preparation of lactulose with calcium hydroxide has long been known (Montgomery and Hudson 1930), but preparation of ketoses by this method is time-consuming, yields are less than 20%, and the keto sugar must be isolated from unreacted starting materials, alkaline degradation products, and metal salts. Hicks and Parrish (1980) have developed a method to prepare lactulose in nearly 90% yield by treatment of lactose with boric acid in an aqueous solution made basic by tertiary amines. Hicks et al. (1984) have also demonstrated that yields of lactulose exceeding 80% can be produced from sweet whey ultrafiltrate, and they describe five purification procedures to produce syrups that permit the crystallization of pure, nonhygroscopic lactulose. A high-performance liquid chromatographic procedure has been developed for analysis of mixtures of lactulose and other sugars (Parrish et al. (1980B).

Lactulose is extremely soluble in water and polar solvents such as methanol. It is difficult to crystallize, especially when traces of other sugars are present. This has led to conflicting reports about its structure in the solid state (Doner and Hicks (1982). Pfeffer et al. (1983) examined lactulose that had been crystallized from refluxing methanol (melting point, 169–171 °C) by solid-state, magic angle spinning cross-polarization ^{13}C nuclear magnetic resonance (NMR) spectroscopy. They concluded that the reducing moiety in the crystalline solid consisted of a mixture of β-fructofuranose: β-fructopyranose: α-fructofuranose forms in a ratio of about 15:3:2. This was later verified by x-ray crystallographic analysis (Jeffrey et al. 1983). In spite of varied crystallization methods, no anomerically pure crystalline form has been obtained.

Lactulose is unstable in alkaline solution, degrading by alkaline peeling and β-elimination reactions to yield galactose, isosaccharinic acids, and other acid products (Corbett and Kenner 1954). Amines can bring about dehydration and degradation reactions (Hough et al. 1953). Lactulose is similar to sucrose in humectant properties (Huhtanen et al. 1980).

Lactulose has several important uses in the food and drug industries. The two major uses are in the treatment of portal systemic encephalopathy and chronic constipation (Doner and Hicks 1982). There is much information on lactulose utilization in infant nutrition. A review by Mendez and Olano (1979) discusses its preparation, structure, properties, and metabolism. The presence of lactulose in infant feeding encourages the development of *Bifidobacterium bifidum* in the intestinal flora, imitating flora in the guts of breast-fed infants. There has been some concern about the possible laxative effects of lactulose, especially in infants; a low colonic pH might be a contributing factor to this effect (Zadow 1984). It is currently believed that lactulose cannot be digested by human alimentary enzymes, so even lactose-tolerated individuals cannot digest lactulose (Doner and Hicks 1982).

Parrish et al. (1979D) have suggested that lactulose could partially replace sucrose and corn sweeteners in intermediate-moisture foods. They studied the sweetness of lactulose over a concentration range of 5 to 35% (w/w), showing that the sweetness was 48 to 62% of that of sucrose. Because of its laxative properties, only limited amounts could be tolerated in foods.

Lactobionic Acid (4-0-β-D-Galactopyranosyl-D-Gluconic Acid

Many sugar acids can form water-soluble metal complexes because of the ability of the carboxyl and hydroxyl groups to bind cations in ring form by means of coordinate and ionic bonds. (Mehltretter *et al.* 1953). An example of such a sugar acid is lactobionic acid, which may be readily prepared from lactose under mild oxidizing conditions. This product can be used in alkaline solutions as a chelating agent for heavy metals such as iron under conditions where EDTA is not effective. However, gluconic acid, a sequestrant commonly used in the soft drink industry, is about twice as effective as lactobionic acid for this purpose (Scholnick and Pfeffer 1980). Calcium, cupric, and ferric salts of lactobionic acid have been prepared. These lactobionates may have potential for supplying heavy metals to plants (Holsinger 1979). It has been suggested that lactobionic acid may be used to prevent boiler scale and as a corrosion inhibitor (Parrish 1977).

Esters formed from lactobionic acid are not stable. However, lactobionic acid may be cyclized by dehydration to form a lactone which is reactive with amines to form stable amides (Scholnick and Pfeffer 1980). An extensive examination of the characteristics of nitrogenous derivatives such as N-dodecyl-lactobionamide or 1,6-dilactobionamido hexane was conducted, but no antimicrobial activity or other special use for these derivatives was identified.

Certain aerobic organisms, notably of the *Pseudomonas* genus but also algae and yeasts, are capable of oxidizing lactose to lactobionic acid without hydrolysis to monosaccharides (Stodola and Lockwood 1947). Lactose dehydrogenase oxidizes lactose to lactobionic-δ-lactone in the presence of a hydrogen acceptor; the lactose is then hydrolyzed to lactobionic acid by lactonase (Nishizaka and Hayaishi 1962).

Other Derivatives of Lactose

Gluconic Acid. A process to convert lactose to gluconic acid and galactose has been reported (Zadow 1984). Lactose is treated under acidic conditions with bromine, and galactose is recovered from the product concentrate by crystallization.

Other acids. Treating with nitric acid oxidizes the D-glucose and D-galactose portions of lactose to their respective dicarboxylic acids, D-glucaric (saccharic) and D-galactaric (mucic) acids. If the acid is sufficiently concentrated or hot, it may cause further oxidation to tartaric, oxalic, and carbonic acids. Complete oxidation to carbon dioxide and water can be accomplished in alkaline solution with potassium permanganate, or with such catalysts as cerous hydroxide, ferrous sulfate, or sodium sulfite. Biological oxidation is also capable of degrading lactose to carbon dioxide and water.

Stearoyl-2-lactylic acid has been prepared by reaction of benzyl lactylate with stearoyl chloride (Ellinger 1979). Ascorbic acid has also been synthesized from lactose (Danehy 1981).

Lactosylurea. Lactosylurea is formed under acid conditions from lactose and urea (McAllan *et al* 1975); conversion of 75% of the lactose was achieved. Widell (1979) described a method for its preparation from whey and urea intended for feeding to ruminants. Ruminants can utilize nonprotein nitrogen compounds for protein synthesis, but urea itself can cause difficulties through too rapid decomposition to ammonia by the rumen enzymes. Lactosylurea appears to meet the requirements for satisfactory palatability, controlled nonprotein nitrogen release, and low toxicity.

Cerbulis *et al.* (1978) studied the chemical characteristics of the reaction of lactose with urea. Maximal yields (40%) of lactosylurea were obtained at an initial pH of 2.0 (final pH, 3.0). At this pH, lactose was partially hydrolyzed to D-glucose and D-galactose, which also reacted with urea to yield minor urea-containing compounds. Lactulose was the principal secondary product formed above pH 4.0.

N-Substituted Amino Sugars. Hoagland *et al.* (1979) has reductively aminated lactose with selected alkylamines and sodium cyanoborohydride in boiling methanol in the presence of a weak organic acid. Sodium cyanoborohydride selectively reduced the imine initially formed by the condensation of the alkylamine with lactose and, as a result, minimized the formation of Amadori rearrangement products. The authors suggest that N-substituted amino sugars might have useful surface active metal-ion binding or biological growth properties.

Polymers. Polyurethane foam has been prepared from the lactose in dried whey by reaction with dimethyl sulfoxide (Hustad *et al.* 1970).

Other Reactions

Regioselective esterification and acetylation reactions of lactose were described by Thelwall (1982). These selective reactions produced partially protected derivatives which were of value in the further modification of lactose and as precursors for the synthesis of higher oligosaccharides.

Hydrolysis.   Lactose may be hydrolyzed by the enzyme β-D-galactosidase, also called "lactase," and by dilute solutions of strong acids. The initial products of acid hydrolysis are glucose and galactose, in equal proportions, but subsequent side reactions may partially consume these and oligosaccharides may form in concentrated solutions by reversion.

Many of the developments in lactose hydrolysis are described in reviews (Whey Workshop II 1979; Harju and Kreula 1980; Shukla 1975; Zadow 1984).

Acid Hydrolysis.   Lactose is resistant to acid hydrolysis compared to other disaccharides such as sucrose. In fact, organic acids, such as citric acid, that easily hydrolyze sucrose are unable to hydrolyze lactose under the same conditions. This is useful in analyzing a mixture of these two sugars, because the quantity of sucrose can be measured by the extent of these changes in the optical rotation of reducing power as a result of mild acid hydrolysis. The speed of hydrolysis of lactose varies with time, temperature, and concentration of the reactant, as shown in Table 6.8.

Coughlin and Nickerson (1975), Vujicic *et al.* (1977), and Lin and Nickerson (1977) readily hydrolyzed 5 to 40% lactose solutions (w/w) with 1 to 3 N hydrochloric acid or sulfuric acid. Ninety percent of the lactose could be hydrolyzed to the constituent monosaccharides at relatively low temperatures (60°C) and long reaction times (up to 36 hr). The authors were not able to adapt this process to whey concentrate because of degradative side reactions producing high levels of off-flavor and color. Guy and Edmondson (1978) hydrolyzed lactose with 0.1 N hydrochloric acid at short reaction times at 121°C.

Sulfonic acid-type ion-exchange resins have been used to catalyze lactose hydrolysis. The resin was equally effective on lactose solutions and acid whey permeate (Mulherin *et al.* 1979). The hydrolysis is carried out at temperatures ranging from 90° to 98°C. The advantages of this method are continuous operation, short reaction times, and no mineral acid to be removed from the hydrolyzed product. High temperature and low pH eliminate problems with microbial contamination. Best reaction rates were achieved with strong acid gelular-type cation-exchange resins with low degrees of crosslinking (MacBean 1979). The formation of oligosaccharides during acid hydrolysis seem to be much less than during enzymatic hydrolysis (Vujicic *et al.* 1977; Guy and Edmondson 1978).

Table 6.8. Hydrolysis of Lactose by Acid.

Lactose in solution (%)	HCl/1000 g Lactose solution mole	Heating conditions		Reaction after heating (pH)	Lactose hydrolyzed (%)	Velocity constant ($K^a \times 10^4$)	Calculated time to invert 99.5% (Min)
		Temp. (°C)	Time (Min)				
33.6	0.034	130	36.0	1.23	82.0	476	111.3
29.0	0.023	130	58.8	1.46	79.7	271	195.4
28.4	0.023	140	30.0	1.47	84.5	662	85.3
23.2	0.019	165	8.2	1.60	79.0	1904	27.8

[a] $K = 1/t \times 2.303 \times \log a/(a - x)$, where x is the amount of hydrolysis attained in time t and a is the initial concentration of lactose.
SOURCES: Ramsdell and Webb (1945).

Enzymatic Hydrolysis. The literature on the hydrolysis of lactose with β-galactosidase (lactase) enzymes is enormous; available reviews discuss specific aspects of their use (Zadow 1984; Harju and Kreula 1980; Shukla 1975). There has been significant progress in this field, and several processes are or almost commercially feasible.

There are three major approaches to enzymatic hydrolysis: (1) "single-use" or "throwaway" lactase systems; (2) lactase recovery systems based on membranes to retain the lactase for reuse; and (3) immobilized systems in which the enzyme is physically or chemically bound to a solid matrix.

Several lactases suitable for industrial processing of whey or lactose are available. The enzyme prepared from the yeast *Kluveromyces lactis* has a pH optimum between 6 and 7 and a temperature optimum of about 35°C. The lactase from *K. fragilis* has a pH optimum of 4.8 and a temperature optimum of about 50°C (MacBean 1979).

A batch processing operation is the simplest method of achieving enzymatic lactose hydrolysis but suffers from the disadvantage that a large amount of recoverable enzyme is needed. For small users or manufacture on an irregular basis, the single-use enzyme procedure is probably the method of choice.

Membrane reactor systems in which the enzyme is recovered by ultrafiltration of the reaction mixture after hydrolysis is complete have been developed. These systems have been pilot tested in Australia but have not been commercialized (Zadow 1984).

Lactose hydrolysis with immobilized systems is the method of choice when regular production of hydrolyzed syrups on a large scale is required. The best-known of these is the Corning immobilized system, which uses lactase from *Aspergillus niger* covalently bound to a controlled-pore silica carrier. The particle size is 0.4 to 0.8 mm, the wet bulk density is 0.6, the activity is near 500 U/g at 50°C, and the optimal pH of operation is between 3.2 and 4.3. Estimated laboratory life is 2 years (Dohan *et al.* 1980). There are at least two of these plants in commercial operation, one in the Untied States and one in the United Kingdom, each a joint venture with Corning.

The rate of hydrolysis is dependent on the mineral, lactose, and galactose concentrations, as well as on the temperature and pH. Many kinetic studies are available on lactose hydrolysis systems and enzymes (MacBean 1979). Inhibition of hydrolysis can be caused by galactose or sodium and calcium ions, so demineralization is often necessary.

Because immobilized systems are designed for long-term use, adequate techniques must be developed to ensure sanitary operations.

Common techniques use backflushing with water, acetic acid, milk alkali, and detergents with bacteriocidal activity.

At least 10 di- and oligosaccharides have been detected during β-D-galactosidase hydrolysis of lactose (Roberts and McFarren 1953). Three of the disaccharides have been identified (Pazur et al. 1958) as 3-0-D-galactopyranosyl-D-glucose, 6-0-β-D-galactopyranosyl-D-glucose, and 6-0-β-D-galactopyranosyl-D-galactose. Galactose is primarily involved in the formation of the oligosaccharides, which accounts for the lower concentration of free galactose than of free glucose during hydrolysis. Similar oligosaccharides are found in the cecal contents of rats fed a high-lactose diet (Roberts and McFarren 1953). Their formation in vivo suggests they are either of physiological importance or that they may be a means of removing excess free galactose from the system.

In a more recent study (Asp et al. 1980), six oligosaccharides from hydrolyzed lactose milk which had been treated with K. lactis lactase were isolated; from their structure, it was concluded that the enzyme had high transglycosylation activity with specificity for the formation of β-(1,6)-galactosidic bonds.

A plethora of lactase-treated products are described in the literature (Zadow 1984; Holsinger 1978). The use of hydrolyzed lactose syrups has been proposed as an alternative sweetener to corn syrup solids. Storage of syrups as concentrated liquids can be a problem due to microorganism growth or to extensive precipitation of residual lactose or galactose. Generally, both problems can be controlled by storage at −10 to −20°C if syrup total solids were around 70%. Guy (1979) has reported on the sweetness of hydrolyzed, demineralized syrups.

The most important heat-induced changes in dairy products that involve lactose are the changes associated with browning. Milk is the only important naturally occurring protein food with a high content of reducing sugar. An extensive review is available of browning and other associated changes in milk (Patton 1955). Other pertinent reviews discuss the Maillard reaction (Waller and Feather 1983; Nursten 1981) and the Amadori rearrangement (Hodge 1953). The Maillard-type browning, sugar-amino type, is the most prevalent, since it requires relatively low energy of activation and is autocatalytic. Direct caramelization, on the other hand, has a rather high energy of activation and therefore is less important.

Lactose and casein are the two principal reactants in the browning of milk products, but dried whey products containing lactose also undergo browning. Holsinger et al. (1973) studied the variation of total and available lysine in dehydrated products from cheese wheys by different processes. Roller-dried products showed significant losses in lysine content, probably because of the high temperatures encountered on the drums. Saltmarch *et al.* (1981) showed that the loss of protein quality and extent of browning were greatest in whey powders stored at $a_w = 0.44$, the point where amorphous lactose began to shift to the α-monohydrate crystalline form, with a release of water that mobilized reactants for the Maillard reaction.

The protein–carbohydrate complex or its decomposition products result in the production of reducing substances, fluorescent substances, and disagreeable flavor materials. For example, 40 compounds were isolated and identified from a model system of casein and lactose that had been stored at $80°C$ and 75% relative humidity for 8 days to accelerate browning. On the basis of gas chromatographic, infrared, and mass spectroscopic data in comparison with authentic samples, 13 furans, 9 lactones, 5 pyrazines, 2 pyridines, 2-acetylpyrrole, 2 amines, pyrrolidinone, succinamide, glutarimide, 2 carboxylic acids, acetone, 2-heptanone, and maltol were identified in the brown mixture, as well as D-galactose, D-tagatose, and lactulose (Ferretti *et al.* 1970). Nearly 40 additional compounds were found in a later study with more sensitive techniques (Ferretti and Flanagan, 1971).

A number of compounds have been shown to inhibit the browning reaction. In milk products, active sulfhydryl groups serve as natural inhibitors in retarding heat-induced browning, but the mechanism is not understood. Sodium bisulfite, sulfur dioxide, and formaldehyde also inhibit browning in milk systems as well as in simpler amino acid–sugar solutions. In actual practice, browning is controlled in dairy products by limiting heat treatments, moisture content, and time and temperature of storage.

Browning has a detrimental effect on the nutritive value of food products through interaction of the free ϵ-amino group of lysine in the proteins with carbohydrates and the resulting rearrangement products. Excellent reviews of this topic are available (Mauron 1981; Dworschak 1980). Destruction of essential amino acids, particularly lysine and probably histidine, has been shown to occur during the storage and browning of nonfat dry milk of high (7.6%) moisture content (Henry *et al.* 1948). Similar powders of low (3.0%) moisture did not deteriorate in nutritive value during storage. Reaction of β-lactoglobulin with lactose in the "dry" (10% moisture) state resulted in various degrees of lysine destruction, depending upon temperature and heating times. Neither arginine, histidine, nor the acidic and neutral amino acids were damaged by the thermal treatments (0 to $90°C$) in the presence of lactose (Freimuth and Trübsach 1969).

The reaction of sugar with protein becomes irreversible. For example, in a model system, after glucose incubation with casein, no glucose could be detected by enzymatic oxidation with glucose oxidase, nor could any glucose be regenerated by dilute acid or alkali hydrolysis (Lea and Hannan 1950).

NUTRITIONAL AND PHYSIOLOGICAL EFFECTS OF LACTOSE

The nutritional and physiological effects of lactose in the diet have become of major interest to health professionals and the public with the finding that about 70% of the world population has low levels of lactase activity in the intestine and, in many cases, an intolerance to lactose. A voluminous literature has developed (Delmont 1983; Renner 1983; Paige and Bayless 1981). Most problems with lactose digestion are attributable to the lactose molecule, but others may arise from the galactose moiety liberated on hydrolysis.

In the digestive tract, lactose may be fermented by bacteria; in the upper intestine, it may be absorbed directly or hydrolyzed by β-D-galactosidase (lactase) and its component sugars absorbed. β-D-Galactosidase is a membrane-bound enzyme in the brush border of epithelial cells of the small intestine (Paige and Bayless 1981). Hydrolysis, therefore, occurs during transport through the intestinal wall. Research on the transport and metabolism of lactose and galactose has been reviewed (Hansen and Gitzelmann 1975).

There are several forms of intolerance to lactose and galactose. Primary adult lactase deficiency is a normal age-related decrease in lactase activity seen in the majority of adults. Secondary lactase deficiency is a transient state of low enzyme activity following injury to the intestinal mucosa as a result of diseases such as celiac sprue, infectious gastroenteritis, and protein-calorie malnutrition. The last two states are common conditions (Dahlqvist 1983).

Congenital lactase deficiency is extremely rare. This condition is due to a genetic defect in which lactase enzyme is absent from birth.

There is severe lactose intolerance which is not an enzyme defect but a permeability disease. Dietary lactose that is passed into the blood through the stomach wall, seems to have toxic effects and is excreted in the urine (Dahlqvist 1983).

In malabsorption of glucose and galactose, the carrier-mediated transport is defective. In severely malnourished patients, there may be a secondary transport defect in which all carbohydrates are difficult to absorb (Dahlqvist 1983).

In the classical form of galactosemia, there is a deficiency of uridyl-transferase enzyme resulting in galactouria, cataracts, and metabolic disturbances. In galactokinase enzyme deficiency, there are cataracts and galactouria but no other symptoms. In epimerase enzyme deficiency, there appears to be an enzyme defect in the erythrocytes but not in the liver (Dahlqvist 1983).

Low lactase activity (lactase deficiency) is detected directly by a biopsy of the mucosa or by indirect methods such as the lactose tolerance test, in which the rise in blood sugar is measured after consumption of a lactose load, or the breath hydrogen test, in which the hydrogen concentration in expired air is determined by gas chromatography. Clinical signs of lactose intolerance include diarrhea, bloating, and flatulance; subjective symptoms are abdominal pain and gassiness following intake of a lactose dose. Many lactase-deficient individuals can tolerate some lactose in the diet; the amount tolerated is influenced by a number of factors, such as the form in which lactose is fed, rate of gastric emptying, age of the consumer, and intestinal transit time. For example, milk containing fat is tolerated better than skim milk and chocolate milk better than unflavored milk.

The question has arisen of whether lactase deficiency is inherited or acquired. In most mammals, lactase declines to low levels or is entirely absent after weaning. It would seem entirely feasible that it could be an acquired characteristic in people who customarily do not drink milk.

A review of medical research has led to the conclusion that ethnic differences concerning lactose intolerance are largely genetic in origin. A culture historical hypothesis has been offered to explain the present-day occurrence of various Old World groups with high and low incidences of lactose intolerance based on milk use (Simoons 1981). It is currently believed that the decline in lactase activity with age is determined by an autosomal recessive gene and is not influenced by the amount of lactose consumed.

The significance of this subject is obvious because of its implications for the suitability of milk as a food for weaning in countries where there is a high incidence of lactase deficiency in the population. As a result of these concerns, low-lactose milk has been considered as an alternative to whole milk in the treatment of protein-calorie malnourished children. No differences were found in growth, protein repletion, or nutrient absorption. There was no persistent diarrhea or abdominal pain with either supplement (Torun et al. 1983). It appears to be highly inappropriate, on the present evidence, to discourage programs aimed at improving milk supplies and increasing milk consumption among children for fear of milk intolerance.

Lactose appears to stimulate the intestinal absorption and retention of calcium. The effect may not be due to lactose but rather to its metabolic product, lactic acid, formed by microbial action in the gut. The acid pH produced by the lactic acid increases solubility of the calcium salts, thereby making more calcium available for absorption. Lactose has the ability to form soluble complexes with calcium, which may be partially responsible for the observed effects. The enhancement of calcium absorption appears to be independent of the vitamin D status and is due to increased passive diffusion. The underlying mechanism responsible for the observed effects has not yet been resolved. In addition to calcium, lactose also enhances the absorption of magnesium, phosphorus, and other essential trace elements (Renner 1983).

Since lactose is absorbed slowly, a portion usually reaches the ileum, where it is utilized by bacterial flora, with the production of lactic acid. Lactose inhibits putrefaction by promoting the growth of aciduric bacteria in the intestine.

Lactose is considered by many to be the preferred carbohydrate for modifying cow's milk for infant food formulation, which is probably its greatest single use. Lactose in the diet is necessary for the desired balance of intestinal flora. The acid conditions caused by conversion of lactose to lactic acid in the lower small intestine and the colon promote the growth of *Lactobacillus bifidus*. By promoting a more desirable flora in the lower digestive tract, it is effective in combating gastrointestinal disturbances caused by putrefactive bacteria, as well as in promoting synthesis of the B vitamins for absorption by the host. (Renner 1983; Delmont 1983).

Low lactase activity has been suggested as a factor leading to osteoporosis as a result of either reduced calcium intake or reduced calcium absorption. However, results are conflicting and further studies will be necessary to resolve this question (Paige and Bayless 1981). The relationship between irritable bowel syndrome and lactase deficiency is still unclear, but hypolactasia does not appear to be a major problem in patients with this condition (Paige and Bayless 1981).

USES OF LACTOSE

The major users of lactose are the pharmaceutical and infant formula manufacturers. With concerns about disposal of whey, a dilute solution of lactose, expanded uses for lactose are constantly being investigated.

Food Uses

Being less sweet than other commercial sugars makes lactose useful in processing many foods. It may be added to increase osmotic pressure or viscosity or to improve texture without making the product too sweet. It is added in the manufacture of beer in some instances because it is not fermented by the yeast and remains in the product to improve flavor and contribute to viscosity and mouth feel. It has similar uses in other beverages and low-calorie foods. Toppings, icings, and various types of pie fillings are examples of uses where its inclusion in the formulation can improve quality.

Lactose is a major contributor to the acceptability of milk as a beverage, and variations of 0.33% lactose are readily detected by taste test (Higgins and Lorimer 1982). Lactose may not be present at optimum levels, however, so that supplementation of milk products such as buttermilk or chocolate drinks effectively improves their acceptance, apparent richness, and smoothness. Consequently, lactose is included as an optional ingredient in standards of identity of such foods.

The candy industry uses lactose to achieve desirable characteristics in certain types of candies. It changes the crystallization habits of other sugars present and improves body, texture, chewiness, or shelf life.

Lactose excels in absorbing flavors, aromas, and coloring materials. As a result, it has found application as a carrier for flavorings or volatile aromas. It is used to trap such materials during their preparation or in filters to remove undesirable volatiles. The anhydrous forms of lactose have recently been shown to have a greater absorption capacity for certain odors than do other sugars or other forms of lactose (Nickerson 1979). To retard flavor losses, lactose may be added to various foods during processing. Likewise, it is used to carry fragrances when a gradual release of odor is desired over a period of time, such as in sachet wafers or as a carrier for seasoning. For example, wine can be incorporated into cake mixes by absorbing the wine on an anhydrous lactose. It is used in conjunction with saccharine or cyclamate to carry these sweetening agents. It gives a better color to some foods. In other cases, it is used as a carrier for colors because it dissolves slowly, releasing the color for uniform dispersion. It is used in flavoring mixtures (Nickerson 1974).

Lactose in the glass state may be used as a protective coating on certain materials, either to seal in components or to protect the material from the environment. Materials may be coated with lactose solution and dried, or a solution containing the material and lactose can be spray-dried. This latter application has been used to preserve enzymatic activity during spray drying and storage (Nickerson 1974).

The pharmaceutical industry has been using lactose for many years because its properties that aid flow characteristics and tablet or pill formation. The drug is distributed uniformly in lactose powder, which is easily molded or compressed into tablets. These have good dispersing characteristics, similar in some respects to the properties of instantized products. Other tablets are given a shell by first moistening the surface of the tablets with a small amount of coating syrup and then tumbling them in lactose powder. This process is repeated for as many coats as desired. Such a coating procedure should be very useful in producing certain food products when the coating can seal in the contents, but the product is easily handled and readily dispersible. Although the potential of this type of application seems very promising, it has not been given the consideration or study it deserves.

The desirable properties of lactose that are important to the baking industry have been reviewed (Guy 1971). Being a reducing sugar, it readily reacts with proteins by the Maillard reaction to form the highly flavored golden-brown materials commonly found in the crusts of baked goods. Caramelization by heat during baking also contributes flavor and color. Lactose is not fermented by bakers' yeast, so its functional properties are effective throughout baking and during storage. Its emulsifying properties aid in creaming and promote greater efficiency from shortening. Thus, lactose can contribute to the flavor, texture, appearance, shelf life, and toasting qualities of baked goods.

Lactase-treated products have been reviewed (Holsinger 1978) and will not be described further here.

Fermented Products

When lactose is used as a substrate for fermentation processes, a wide variety of end products are produced. Some processes are in commercial operation, for example, the Carbery, Ireland, process for the production of alcohol from whey permeate. This plant produces about 22,000 liters of alcohol from 600,000 liters of whey permeate per day. The conversion of lactose to alcohol is about 86%. Single-cell protein has been produced from whey by a Wisconsin plant for many years; this plant also produces potable alcohol.

Whey permeate may also be fermented anaerobically to fuel gas. Studies have also been reported on the production of ammonium lactate by continuous fermentation of deproteinized whey to lactic acid followed by neutralization with ammonia. Conversion of whey and whey permeate to oil and single-cell protein with strains of *Candida curvata* and *Trichosporon cutaneum* have been examined. Production of the solvents *n*-butanol and acetone by *Clostridium acetobutylicum* or *C. butyricum* is under investigation in New Zealand. Whey permeate also has potential for citric acid and acrylic acid manufacture. Extracellular microbial polysaccharide production from whey permeate has been studied. Permeate can also serve as a substrate in the manufacture of lactase enzyme.

The excellent reviews by Hobman (1984), Zadow (1984), and Short (1978) should be consulted for specific references to the above-mentioned products and processes. The expansion of biotechnology applications to agricultural areas should aid in increasing utilization of lactose as an industrial chemical.

REFERENCES

Adachi, S. and Patton, S. 1961. Presence and significance of lactulose in milk products: A review. *J. Dairy Sci. 44*, 1375-1393.

Ahlgren, R. M. 1977. Electromembrane technology for whey processing. *In: Proceedings, Whey Products Conference 1976.* Pub. No. ARS-NE-81, USDA ARS. Eastern Regional Research Center, Philadelphia.

Anderson, A. G. and Stegeman, G. 1941. The heat capacities and entropies of three disaccharides. *J. Am. Chem. Soc. 63*, 2119-2121.

Asp, N. G., Burval, A., Dahlquist, A., Hallgren, P. and Lundblad, A. 1980. Oligosaccharide formation during hydrolysis of lactose with *Saccaromyces lactis* lactase (Maxilact R). II. Oligosaccharide structures. *Food Chem. 5*, 147-153.

Baucke, A. G. and Sanderson, W. B. 1970. A study of viscosity increase in concentrated skim-milk. *19th Int. Dairy Congr. Proc. 1E* 256.

Bell, R. W. 1930. Some methods of preparing quickly soluble lactose. *Ind. Eng. Chem. 22*, 51-54.

Berlin, E., Anderson, B. A. and Pallansch, M. J. 1968. Comparison of water vapor sorption by milk powder components. *J. Dairy Sci. 51*, 1912-1915.

Berlin, E., Anderson, B. A. and Pallansch, M. J. 1970. Effect of temperature on water vapor sorption by dried milk powders. *J. Dairy Sci. 53*, 146-149.

Berlin, E., Kliman, P. G., Anderson, B. A. and Pallansch, M. J. 1971. Calorimetric measurement of the heat of desorption of water vapor from amorphous and crystalline lactose. *Therm. Acta 2*, 143-152.

Bockian, A. N., Stewart, G. F. and Tappel, A. L. 1957. Factors affecting the dispersibility of "instantly dissolving" dry milks. *Food Res. 22*, 69-75.

Brinkman, G. E. 1976. New ideas for the utilization of lactose—principles of lactose manufacture. *J. Soc. Dairy Technol. 29* 101-107.

Buma, T. J. 1980. Viscosity and density of concentrated lactose solutions and of concentrated cheese whey. *Neth. Milk Dairy J. 34*, 65-68.

Buma, T. J. and Meerstra, J. 1969. The specific heat of milk powder and of some related materials. *Neth. Milk Dairy J. 23*, 124-127.

Buma, T. J. and Wiegers, G. A. 1967. X-ray powder patterns of lactose and unit cell dimensions of β-lactose. *Neth. Milk Dairy J. 21*, 208-213.

Brew, K. 1969. Secretions of α-lactalbumin into milk and its relevance to the organization and control of lactose synthetase. *Nature (London) 222*, 671-672.

Carpenter, K. J. (with Booth, V.H.) 1973. Damage to lysine in food processing: Its measurement and its significance. *Nutr. Abst. Rev. 43*, 423-447.

Cerbulis, J. 1973. Application of Steffen process and its modifications to recovery of lactose and proteins from whey. *J. Agri. Food Chem. 21*, 255-257.

Cerbulis, J., Pfeffer, P. E. and Farrell, H. M., Jr. 1978. Reaction of lactose with urea. *Carbohydr. Res. 65*, 311–313.

Challand, G. S. and Rosemeyer, M. A. 1974. The correlation between the apparent molecular weight and the enzymic activity of lactose synthetase. *Febs Letters 47*, 94–97.

Chambers, J. V. 1985. Personal communication. West Lafayette, Ind.

Clamp, J. R., Hough, L., Hickson, J. L. and Whistler, R. L. 1961. Lactose. *In: Advances in Carbohydrate Chemistry*, Vol. 16. M.L. Wolfrom and R.S. Tyson (Editors). Academic Press, New York. pp. 159–206.

Corbett, W. M. and Kenner. J. 1954. The degradation of carbohydrates by alkali. Part V. Lactulose, maltose and maltulose. *J. Chem. Soc.* 1789–1791.

Coughlin, J. R. and Nickerson, T. A. 1975. Acid-catalyzed hydrolysis of lactose in whey and aqueous solutions. *J. Dairy Sci. 58*, 169–174.

Dahlqvist, A. 1983. Digestion of lactose. *In: Milk Intolerances and Rejection*. J. Delmont (Editor). S. Karger, Basel, pp. 11–16.

Danehy, J. P. 1981. Synthesis of ascorbic acid from lactose. U.S. Patent 4,259,443.

Delaney, R. A. M. 1976. Demineralization of whey, *Aust. J. Dairy Technol. 31*, 12–17.

Delmont, J. 1983. *Milk Intolerances and Rejections*. S. Karger, Basel.

Dohan, L. A., Baret, J.-L., Pain, S. and Delalande, P. 1980. Lactose hydrolysis by immobilized lactase:Semi-industrial experience. *Enzyme Eng. 5*, 279–293.

Domovs, K. B. and Freund, E. H. 1960. Methanol-soluble complexes of lactose and of other carbohydrates. *J. Dairy Sci. 42*, 1216–1223.

Doner, L. W. and Hicks, K. B. 1982. Lactose and the sugars of honey and maple: Reactions, properties and analysis. *In:Food Carbohydrates* D.R. Lineback and G.E. Inglett (Editors). AVI Publishing Co., Westport, Conn. pp. 74–112.

Dworschak, E. 1980. Nonenzymatic browning and its effect on protein nutrition. *CRC Crit. Rev. Food Sci. Nutr. 12*, 1–40.

Ebner, K. E. and Schanbacher, F. L. 1974. Biochemistry of lactose and related carbohydrates. *In: Lactation: A Comprehensive Treatise*, Vol. II. B. L. Larson and V. R. Smith (Editors). Academic Press, New York. pp. 77–113.

Elliger, C. A. 1979. A convenient preparation of pure stearoyl-2-lactylic acid. *J. Agri. Food Chem. 27*, 527–528.

Evans, E. W. 1980. Whey research. *J. Soc. Dairy Technol. 33*(3), 95–100.

Ferretti, A. and Flanagan, V. P. 1971. The lactose–casein (Maillard browning system. Volatile components. *J. Agri. Food Chem. 19*, 245–249.

Ferretti, A., Flanagan, V. P. and Ruth, J. M. 1970. Nonenzymatic browning in a lactose-casein model system. *J. Agri. Food Chem. 18*, 13–18.

Flynn, F. V., Harper, C. and de Mayo, P. 1953. Lactosuria and glycosuria in pregnancy and the puerperium. *Lancet 265*, 698–704.

Freimuth, V. and Trübsach, A. 1969. Studies of the Maillard reaction. 1. Determination of the reaction of β-lactoglobulin with lactose in the "dry" state. *Nahrung 13*, 199–206.

Goldman, A. and Short, J. L. 1977. Use of crude β-lactose in "high-ratio" cakes. *N.Z. J. Dairy Sci. Technol. 12*, 88–93.

Guy, E. J. 1971. Lactose:Review of its properties and uses in bakery products. *Bakers Digest 45*(2), 34–36, 38, 74.

Guy, E. J. 1979. Purification of syrups from hydrolyzed lactose in sweet whey permeate. *J. Dairy Sci. 62*, 384–391.

Guy, E. J. and Edmondson, L. F. 1978. Preparation and properties of syrups made by hydrolysis of lactose. *J. Dairy Sci. 61*, 542–549.

Gyorgy, P. 1953. Hitherto unrecognized biochemical differences between human milk and cow's milk. *Pediatrics 11*, 98–108.

Haase, G. and Nickerson, T. A. 1966. Kinetic reactions of alpha and beta lactose. II. crystallization. *J. Dairy Sci. 49*, 757–761.

Hall, C. W. and Hedrick, T. I. 1971. *Drying of Milk and Milk Products,* 2nd ed. AVI Publishing Co., Westport, Conn.

Hanrahan, F. P. and Webb, B. H. 1961A. Spray drying cottage cheese whey. *J. Dairy Sci. 44,* 1171.

Hanrahan, F. P. and Webb, B. H. 1961B. U.S. Department of Agriculture develops foam-spray drying. *Food Eng. 33*(8), 37–38.

Hansen, R. G. and Gitzelmann, R. 1975. The metabolism of lactose and galactose. *In: Physiological Effects of Food Carbohydrates.* A Jeanes and J. Hodge (Editors). ACS Symposium Series 15. American Chemical Society, Washington, D. C., pp. 100–122.

Hargrove, R. E., McDonough, F. E., LaCroix, D. E. and Alford, J. A. 1976. Production and properties of deproteinized whey powders. *J. Dairy Sci. 59,* 25–33.

Harju, M. and Kreula, M. 1980. Lactose hydrolysates. *In: Carbohydrate Sweeteners in Foods and Nutrition.* P. Koivistoinen and L. Hyvönen (Editors). Academic Press, New York. pp. 233–242.

Harper, W. J. 1979. Analytical procedures for whey and whey products. *N.Z. J. Dairy Sci. Technol. 14,* 156–171.

Heald, C. W. and Saacke, R. G. 1972. Cytological comparison of milk protein synthesis of rat mammary tissue in vivo and in vitro. *J. Dairy Sci. 55,* 621–628.

Heinrich, C. 1970. Thermogravimetric determination of moisture in milk powder. *Milchwissenschaft 25,* 387–391.

Henry, K. M., Kon, S. K., Lea, C. H. and White, J. C. D. 1948. Deterioration on storage of dried skim milk. *J. Dairy Res. 15,* 292–363.

Herrington, B. L. 1934A. Some physico-chemical properties of lactose. II. Factors influencing the crystalline habit of lactose. *J. Dairy Sci. 17,* 533–542.

Herrington, B. L. 1934B. Some physico-chemical properties of lactose. VI. The solubility of lactose in salt solutions; the isolation of a compound of lactose and calcium chloride. *J. Dairy Sci. 17,* 805–814.

Herrington, B. L. 1948. *Milk and Milk Processing.* McGraw-Hill Book Co., New York. p. 84.

Hicks, K. B. and Parrish, F. W. 1980. A new method for the preparation of lactulose from lactose. *Carbohydr. Res. 82,* 393–397.

Hicks, K. B. Raupp, D. L. and Smith, P. W. 1984. Preparation and purification of lactulose from sweet whey ultrafiltrate. *J. Agri. Food Chem. 32,* 288–292.

Higgins, J. J. and Lorimer, P. R. 1982. Flavour characteristics of crude lactose. *N.Z. J. Dairy Sci. Technol. 17,* 91–101.

Hoagland, P. D., Pfeffer, P. E. and Valentine, K. M. 1979. Reductive amination of lactose: Unusual ^{13}C-N.M.R. spectroscopic properties of N-alkyl-(1-deoxylactitol-1-yl amines. *Carbohydr. Res. 74,* 135–143.

Hobman, P. G. 1984. Review of processes and products for utilization of lactose in deproteinated milk serum. *J. Dairy Sci. 67,* 2630–2653.

Hockett, R. C. and Hudson, C. S. 1931. A novel modification of lactose. *J. Am. Chem. Soc. 53,* 4455–4456.

Hodge, J. E. 1953. Chemistry of browning reactions in model systems. *J. Agri. Food Chem. 1,* 928–943.

Holsinger, V. H. 1978. Application of lactose modified milk and whey. *Food Technol. 32,* 35–36, 38, 40.

Holsinger, V. H. 1979. Agricultural research toward increased whey utilization. *In: Proceedings—Whey Products Conference, 1978.* USDA, ARS. Eastern Regional Research Center, Philadelphia, pp. 90–110.

Holsinger, V. H., Posati, L. P., Devillbiss, E. D. and Pallansch, M. J. 1973. Variation of total and available lysine in dehydrated products made from cheese wheys by different processes *J. Dairy Sci. 56* 1498–1504.

Hough, L., Jones, J. K. N. and Richards, E. J. 1953. The reaction of amino-compounds with sugars. Part II. The action of ammonia on glucose, maltose and lactose. *J. Chem. Soc.* 2005-2009.

Hudson, C. S. 1904. The hydration of milk sugar in solution. *J. Am. Chem. Soc. 26*, 1065-1082.

Hudson, C. S. 1908. Further studies on the forms of milk sugar., *J. Am. Chem. Soc. 30*, 1767-1783.

Huhtanen, C. N., Parrish, F. W. and Hicks, K. B. 1980. Inhibition of bacteria by lactulose preparations. *Appl. Environ. Microbiol. 40*, 171-173.

Hunziker, O. F. 1949. *Condensed Milk and Milk Products*, 6th ed. Published by the author, La Grange, Illinois.

Hunziker, O. F. and Nissen, B. H. 1927. Lactose solubility and lactose crystal formation II. Lactose crystal formation. *J. Dairy Sci. 10* 139-154.

Hustad, G. O., Richardson, T. and Amundson, C. H. 1970. Polyurethane foams from dried whey. *J. Dairy Sci. 53*, 18-24.

Itoh, T., Katoh, M. and Adachi, S. 1978. An improved method for the preparation of β-lactose and observations on the melting point. *J. Dairy Res. 45*, 363-371.

Ivatt, R. J. and Rosemeyer, M. A. 1972. The complex formed between the A and B proteins of lactose synthetase. *Febs Letters 28*, 195-197.

Jeffrey, G. A., Wood, R. A., Pfeffer, P. E. and Hicks, K. B. 1983. Crystal structure and solid-state NMR analysis of lactulose. *J. Am. Chem. Soc. 105*, 2128-2133.

Jelen, P. and Coulter, S. T. 1973A. Effects of supersaturation and temperature on the growth of lactose crystals. *J. Dairy Sci.* 1182-1185.

Jelen, P. and Coulter, S. T. 1973B. Effects of certain salts and other whey substrates on the growth of lactose crystals. *J. Food Sci. 38*, 1186-1189.

Jenness, R., Regehr, E. A. and Sloan, R. E. 1964. Comparative biochemical studies of milks. II. Dialyzable carbohydrates. *Comp. Biochem. Physiol. 13*, 339-353.

Jenness, R. and Sloan, R. E. 1970. The composition of milks of various species: A review. *Dairy Sci. Abstr. 32*, 599-607.

Johnson, J. D., Kretchmer, N. and Simoons, F. J. 1974. Lactose malabsorption; its biology and history. *In: Advances in Pediatrics*, Vol. 21. I. Schulman (Editor). Yearbook Medical Publishers, Chicago, pp. 197-237.

Jones, E. A. 1978. Lactose biosynthesis. *In: Lactation: A Comprehensive Treatise*, Vol. IV. B.L. Larson (Editor). Academic Press, New York, pp. 371-385.

Kavanagh, J. A. 1975. Production of crude lactose from ultrafiltration permeate. *N.Z. J. Dairy Sci. Technol. 10*, 132.

Keeney, P. G. and Kroger, M. 1974. Frozen dairy products. *In: Fundamentals of Dairy Chemistry*, B.H. Webb, A. H. Johnson and J.A. Alford (Editors). AVI Publishing Co., Westport, Conn., pp. 873-913.

Kerkkonen, H. K., Käkkäinen, V. J. and Antila, M. 1963. On the manufacture of protein concentrate and lactose from milk. *Finnish J. Dairy Sci. 24*, 61-68.

Klee, W. A. and Klee, C. B. 1972. The interaction of α-lactalbumin and the A protein of lactose synthetase. *J. Biol. Chem. 247*, 2336-2344.

Knoop, E. and Samhammer, E. 1962. Roentgenographic studies on the crystal structure of lactose in milk powder. *Milchwissenschaft 17*, 128-131.

Krevald, A. Van 1969. Growth rates of lactose crystals in solutions of stable anhydrous α-lactose. *Neth. Milk Dairy J. 23*, 258-275.

Krevald, A. Van and Michaels, A. S. 1965. Measurement of crystal growth of α-lactose. *J. Dairy Sci. 48*, 259-265.

Kuhn, R. and Low, I. L. 1949. The occurrence of lactose in the plant kingdom. *Chem. Ber. 82*, 479-481.

Kuhn, N. J. and White, A. 1975. The topography of lactose synthesis. *Biochem. J. 148*, 77-84.

Kuhn, N. J. and White, A. 1976. Evidence for specific transport of uridine diphosphate galactose across the Golgi membrane of rat mammary gland. *Biochem. J. 154*, 243–244.

Kyle, R. C. and Henderson, R. J. 1970. Lactose manufacture. U.S. Patent 3,511,266.

Lea, C. H. and Hannan, R. S. 1950. Studies of the reaction between proteins and reducing sugars in the "dry" state. II. Further observations on the formation of the casein-glucose complex. *Biochem. Biophys. Acta 4*, 518–531.

Lerk, C. F., Buma, T. J. and Andreae, A. C. 1980. The effect of mechanical treatment on the properties of lactose as observed by differential scanning calorimetry. *Neth. Milk Dairy J. 34*, 69–73.

Leviton, A. 1943. Adsorption of riboflavin by lactose. Influence of concentration. *Ind. Eng. Chem. 35*, 589–593.

Leviton, A. 1944. Adsorption of riboflavin by lactose. Influence of temperature. *Ind. Eng. Chem. 36*, 744–747.

Leviton, A. 1949. Methanol extraction of lactose and soluble proteins from skim-milk powder. *Ind. Eng. Chem. 41*, 1351–1357.

Leviton, A. and Leighton, A. 1938. Separation of lactose and soluble proteins of whey by alcohol *extraction-extraction* from spray dried whey powder derived from sweet whey. *Ind. Eng. Chem. 30*, 1305–1311.

Lim, S. G. and Nickerson, T. A. 1973. Effect of methanol on the various forms of lactose. *J. Dairy Sci. 56*, 843–848.

Lin, A. Y. and Nickerson, T. A. 1977. Acid hydrolysis of lactose in whey versus aqueous solutions. *J. Dairy Sci. 60*, 34–39.

Linko, P., Saijonmaa, T., Heikonen, M. and Kreula, M. 1980. Lactitol. *In: Carbohydrate Sweeteners in Food and Nutrition.* P. Koivistoinen and L. Hyvönen (Editors). Academic Press, New York. pp. 243–251.

McAllan, A. B., Merry, R. J. and Smith, R. H. 1975. Glucosyl ureides in ruminant feeding. *Proc. Nutr. Soc. 34*, 90A–91A.

McDonald, E. J. and Turcotte, A. L. 1948. Density and refractive indexes of lactose solutions. *J. Res. Nat. Bur. Standards 41*, 63–68.

Majd, F. and Nickerson, T. A. 1976. Effect of alcohols on lactose solubility. *J. Dairy Sci. 59* 1025–1032.

Macbean, R. D. 1979. Lactose crystallization and lactose hydrolysis. *N.Z. J. Dairy Sci. Technol. 14*, 113–119, 128–130.

Mann, E. J. 1977. Utilization of lactose. *Dairy Ind. Int. 42*(2), 60–61.

Mauron, J. 1981. The Maillard reaction in food: A critical review from the nutritional standpoint. *Prog. Food Nutr. Sci. 5*, 5–35.

Mehltretter, C. L., Alexander, B. H. and Rist, C. E. 1953. Sequestration by sugar acids. *Ind. Eng. Chem. 45*, 2782–2784.

Mendez, A. and Olano, A. 1979. Lactulose. A review of some chemical properties and applications in infant nutrition and medicine. *Dairy Sci. Abstr. 41*, 531–535.

Michaels, A. S. and Kreveld, A. Van. 1966. Influence of additives on growth rates in lactose crystals. *Neth. Milk Dairy J. 20*, 163–181.

Montgomery, E. M. and Hudson, C. S. 1930. Relations between rotary power and structure in the sugar group. XXVII. Synthesis of a new disaccharide ketose (lactulose) from lactose. *J. Am. Chem. Soc. 52*, 2101–2106.

Morrison, J. F. and Ebner, K. E. 1971A. Studies on galactosyltransferase. Kinetic investigations with N-acetyl glucosamine as the galactosyl group acceptor. *J. Biol. Chem. 246*, 3977–3984.

Morrison, J. F. and Ebner, K. E. 1971B. Studies on galactosyltransferase. Kinetic investigations with glucose as the galactosyl group acceptor. *J. Biol. Chem. 246*, 3985–3991.

Morrison, J. F. and Ebner, K. E. 1971C. Studies on galactosyltransferase. Kinetic effects of α-lactalbumin with N-acetylglucosamine and glucose as galactosyl group acceptors. *J. Biol. Chem. 246,* 3992–3998.

Mulherin, B., Muller, T., Delaney, R. A. M. and Harper, W. J. 1979. Acid catalyzed hydrolysis of lactose with cation exchange resins. *N.Z. J. Dairy Sci. Technol. 14,* 127.

Muller, L. L. 1979. Studies on continuous crystallization of lactose. *N.Z. J. Dairy Sci. Technol. 14,* 119–121.

Mullin, J. W. 1961. *Crystallization.* Butterworths, London.

Nickerson, T. A. 1954. Lactose crystallization in ice cream. I. Control of crystal size by seeding. *J. Dairy Sci. 37,* 1099–1105.

Nickerson, T. A. 1956. Lactose crystallization in ice cream. II. Factors affecting rate and quantity. *J. Dairy Sci. 39,* 1342–1350.

Nickerson, T. A. 1962. Lactose crystallization in ice cream. IV. Factors responsible for reduced incidence of sandiness. *J. Dairy Sci. 45,* 354–359.

Nickerson, T. A. 1970. Lactose. *In: Byproducts from Milk.* B.H. Webb and E.O. Whittier (Editors). AVI Publishing Co., Westport, Conn., pp. 356–380.

Nickerson, T. A. 1974. Lactose. *In: Fundamentals of Dairy Chemistry,* 2nd ed. B.H. Webb, A. H. Johnson, and J.A. Alford (Editors). AVI Publishing Co., Westport, Conn., pp. 273–324.

Nickerson, T. A. 1979. Lactose chemistry. *J. Agri. Food Chem. 27,* 672–677.

Nickerson, T. A. and Lim, S. G. 1974. Effect of various alcohols on lactose. *J. Dairy Sci. 57,* 1320–1324.

Nickerson, T. A. and Moore, E. E. 1972. Solubility interaction of lactose and sucrose. *J. Dairy Sci. 37,* 60–61.

Nickerson, T. A. and Moore, E. E. 1974A. Alpha lactose and crystallization rate. *J. Dairy Sci. 57,* 160–164.

Nickerson, T. A. and Moore, E. E. 1974B. Lactose influencing lactose crystallization. *J. Dairy Sci. 57,* 1315–1319.

Nishizuka, Y. and Hayaishi, O. 1962. Enzymic formation of lactobionic acid from lactose. *J. Biol. Chem. 237,* 2721–2728.

Nursten, H. E. 1981. Recent developments in studies of the Maillard reaction. *Food Chem. 6,* 263–277.

Olano, A. 1978. Treatment of forms of lactose with dilute alcoholic solution of sodium hydroxide. *J. Dairy Sci. 61,* 1622–1623.

Olano, A., Bernhard, R. A. and Nickerson, T.A, 1977. Alteration in the ratio of α- to β-lactose co-crystallized from organic solvents. *J. Food Sci. 42,* 1066–1068, 1083.

Olano, O. and Rios, J. J. 1978. Treatment of lactose with alkaline methanolic solutions: Production of beta-lactose from alpha-lactose hydrate. *J. Dairy Sci. 61,* 300–302.

Paige, D. M. and Bayless, T. M. 1981. *Lactose Digestion: Clinical and Nutritional Implications.* Johns Hopkins University Press, Baltimore.

Pallansch, M. J. 1973. New methods for drying acid whey. *In: Proceedings of the Whey Products Conference, 1972,* ERRL Pub. No. 3779, U. S. USDA ARS. Eastern Regional Research Center, Philadelphia.

Pangborn, R. M. and Gee, S. C. 1961. Relative sweetness of α- and β-forms of selected sugars. *Nature 191,* 810–811.

Parrish, F. W. 1977. New uses for lactose. *In: Proceedings of the Whey Products Conference, 1976.* Bull. ARS-NE-81, USDA, Eastern Regional Research Center, Philadelphia.

Parrish, F. W., Pfeffer, P. E., Ross, K. D., Schwartz, D. P. and Valentine, K. M. 1979A. Retention of aliphatic alcohols by anhydrous lactose. *J. Agri. Food Chem. 27,* 56–59.

Parrish, F. W., Ross, K. D. and Simpson, T. D. 1979B. Formation of β-lactose from α- and β-lactose octaacetates, and from α-lactose monohydrate. *Carbohydr. Res. 71,* 322–326.

Parrish, F. W., Sharples, P. M., Hoagland, P. D. and Woychik, J. H. 1979. Demineralization of cheddar whey ultrafiltrate with thermally regenerable ion-exchange resins: Improved yield of α-lactose monohydrate. *J. Dairy Sci.* 44, 555-557.

Parrish, F. W., Talley, F. B., Ross, K. D., Clark, J. and Phillips, J. G. 1979D. Sweetness of lactulose relative to sucrose. *J. Dairy Sci.* 44, 813-815, 835.

Parrish, F. W. Hicks K.B. and Doner, L. 1980A. Analysis of lactulose preparations by spectrophotometric and high performance liquid chromatographic methods. *J. Dairy Sci.* 63, 1809-1814.

Parrish, F. W., Ross, K. D. and Valentine, K. M. 1980B. Formation of β-lactose from the stable forms of anhydrous α-lactose. *J. Dairy Sci.* 45, 68-70.

Parrish, F. W., Talley, F. B. and Phillips, J. G. 1981. Sweetness of α-, β-and equilibrium lactose relative to sucrose. *J. Food Sci.* 46, 933-935.

Patel, K. N. and Nickerson, T. A. 1970. Influence of sucrose on the mutarotation velocity of lactose. *J. Dairy Sci.* 53, 1654-1658.

Patton, S. 1955. Browning and associated changes in milk and its products: A review. *J. Dairy Sci.* 38, 457-478.

Pazur, J. H., Tipton, C. L., Budovich, T. and Marsh, J. M. 1958. Structural characterization of products of enzymatic disproprotionation of lactose. *J. Am. Chem. Soc.* 80, 119-121.

Peebles, D. D. 1956. The development of instant milk. *Food Technol.* 10, 64-65.

Pfeffer, P. E., Hicks, K. B. and Earl, W. L. 1983. Solid state structures of keto-disaccharides as probed by ^{13}C cross-polarization, "magic angle" spinning NMR spectroscopy. *Carbohydr. Res.* 11, 181-194.

Pfeffer, P. E., Valentine, K. M. and Parrish, F. W. 1979. Deuterium-induced differential isotope shift ^{13}C NMR. 1, Resonance reassignments of mono- and disaccharides. *J. Am. Chem. Soc.* 1001, 1265-1274.

Pilson, M. E. Q. 1965. Absence of lactose from the milk of the *Otarioidea*, a superfamily of marine mammals. *Am. Zool.* 5, 220-221.

Pilson, M. E. Q. and Kelly, A. L. 1962. Composition of the milk from *Zalophus californianus*, the California sea lion. *Science* 135, 104-105.

Powell, J. T. and Brew, K. 1975. On the interaction of α-lactalbumin and galactosyltransferase during lactose synthesis. *J. Biol. Chem.* 250, 6337-6343.

Quickert, S. C. and Bernhard, R. A. 1982. Recovery of lactose from aqueous solution using Group IIA metal chlorides and sodium hydroxide. *J. Food Sci.* 47, 1705-1709.

Ramsdell, G. A. and Webb, B. H. 1945. The acid hydrolysis of lactose and the preparation of hydrolyzed lactose syrup. *J. Dairy Sci.* 28, 677-686.

Reineccius, G. A., Kavanagh, T. E. and Keeney, P. G. 1970. Identification and quantitation of free neutral carbohydrates in milk products by gas-liquid chromatography and mass spectrometry. *J. Dairy Sci.* 53, 1018-1022.

Reithel, F. J. and Venkataraman, R. 1956. Lactose in the *Sapotaceae. Science* 123, 1083.

Renner, E. 1983. *Milk and Dairy Products in Human Nutrition.* W-GmbH., Volkswirtschaftlicher Verlag, Munich.

Roberts, H. R. and McFarren, E. F. 1953. The chromatographic observation of oligosaccharides formed during lactose hydrolysis of lactose. *J. Dairy Sci.* 36, 620-632.

Roetman, K. 1972. Crystallization of lactose. *Voedingsmiddelen-technoligie* 3,(43), 230-234. Cited in *Food Sci. Technol. Abstr.* 6, 2L110 (1974).

Roetman, K. 1981. Methods for the quantitative determination of crystalline lactose in milk products. *Neth. Milk Dairy J.* 35, 1-52.

Ross, K. D. 1978A. Effects of methanol on physical properties of α- and β-lactose. *J. Dairy Sci.* 61, 152-158.

Ross, K. D. 1978B. Rapid determination of α-lactose in whey powders by differential scanning calorimetry. *J. Dairy Sci.* 61, 255-259.

Saltmarch, R., Vagnini-Ferrari, M.A and Labuza, T. P. 1981. Theoretical basis and application of kinetics to browning in spray dried food systems. *Prog. Food Nutr. Sci. 5*, 331–344.

Schoen, H. M. 1961. Crystallization is a two-step process-nucleation and growth. *Ind. Eng. Chem. 53*, 607–611.

Scholnick, F., Sucharski, M. K. and Linfield, W. M. 1974. Lactose-derived surfactants (I) fatty esters of lactose. *J. Am. Oil Chemists Soc. 51*, 8–11.

Scholnick, F., Ben-et, G., Sucharski, M. K., Maurer, E. W. and Linfield, W. M. 1975. Lactose-derived surfactants. II. Fatty esters of lactitol. *J. Am. Oil Chemists' Soc. 52*, 256–258.

Scholnick, F. and Linfield, W. M. 1977. Lactose-derived surfactants. III. Fatty esters of oxyalkylated lactitol. *J. Am. Oil Chemists' Soc. 54*, 430–432.

Scholnick, F. and Pfeffer, P. E. 1980. Iron chelating ability of gluconamides and lactobionamides. *J. Dairy Sci. 63*, 471–473.

Seifert, H. and Labrot, G. 1961. About the structure of α-lactose-monohydrate (milk sugar). *Naturwissenschaft 48*, 691.

Sharp, P. F. 1943. Stable crystalline anhydrous α-lactose. U.S. Patent 2,319,562.

Short, J. L. 1978. Prospects for the utilization of deproteinated whey in New Zealand— A review. *N.Z. J. Dairy Sci. Technol. 13*, 181–194.

Shukla, T. P. 1975. Beta-galactosidase technology: A solution to the lactose problem. *CRC Crit. Rev. Food Technol. 5*(3), 325–356.

Simoons, F. J. 1981. Geographic patterns of primary adult lactose malabsorption. A further interpretation of evidence from the Old World. *In: Lactose Digestion: Clinical and Nutritional Implications.* D.M. Paige and T.M. Bayless (Editors). Johns Hopkins University Press, Baltimore, pp. 23–48.

Simpson, T. D., Parrish, F. W. and Nelson, M. L. 1982. Crystalline forms of lactose produced in acidic alcoholic media. *J. Dairy Sci. 47*, 1948–1951, 1954.

Smeets, W. T. G. M. 1955. The determination of the concentration of calcium ions in milk ultrafiltrate. *Neth. Milk Dairy J. 9*, 249–260.

Stewart, R. E. A., Webb, B. E., Lavinge, D. M. and Fletcher, F. 1983. Determining lactose content of harp seal milk. *Can. J. Zool. 61*, 1094–1100.

Stodola, F. H. and Lockwood, L. B. 1947. The oxidation of lactose and maltose to bionic acids by pseudomonas. *J. Biol. Chem. 171*, 213–221.

Supplee, G. C. 1926. Humidity equilibria of milk powders. *J. Dairy Sci. 9*, 50–61.

Susi, H. and Ard, J. S. 1974. Laser-Raman spectra of lactose. *Carbohydr. Res. 37*, 351–354.

Swartz, M. L., Bernhard, R. A. and Nickerson, T. A. 1978. Interactions of metal ions with lactose. *J. Dairy Sci. 43*, 93–97.

Tamsma, A., Kontson, A., Sutton, C. and Pallansch, M. J. 1972. Production of non-hygroscopic foam-spray dried cottage cheese whey. *J. Dairy Sci. 55*, 667.

Thelwall, L. A. W. 1982. Recent aspects of the chemistry of lactose. *J. Dairy Res. 49*, 713–724.

Thurlby, J. A. 1976. Crystallization kinetics of alpha lactose. *J. Food Sci. 41*, 38–42.

Thurlby, J. A. and Sitnai, O. 1976. Lactose crystallization: Investigation of some process alternatives. *J. Food Sci. 41*, 43–47.

Torun, B., Solomons, N. W., Caballero, B., Flores-Huerta, S., Orozco, G. and Batres, R. 1983. Intact and lactose-hydrolyzed milk to treat malnutrition in Guatemala. *In: Milk Intolerances and Milk Rejection.* J. Delmont (Editor). S. Karger, Basel, pp. 109–115.

Troy, H. C. and Sharp, P. F. 1930. Alpha and beta lactose in some milk products. *J. Dairy Sci. 13*, 140–157.

Trucco, R. E., Verdier, P. and Rega, A. 1954. New carbohydrate compounds from cow milk. *Biochem. Biophys. Acta 15*, 582–583.

Twieg, W. C. and Nickerson, T. A. 1968. Kinetics of lactose crystallization. *J. Dairy Sci. 51*, 1720–1724.

Valle-Vega, P. and Nickerson, T. A. 1977. Measurement of lactose crystal growth by image analyzer. *J. Food Sci. 42*, 1069–1072.

Valle-Vega, P., Nickerson, T. A., Moore, E. E. and Gonzenbach, M. 1977. Variability of growth of lactose crystals under commercial treatment. *J. Dairy Sci. 60*, 1544–1549.

Van Hook, A. 1961. *Crystallization: Theory and Practice.* Reinhold Publishing Corp., New York.

Velthuijsen, J. A. Van. 1979. Food additives derived from lactose: Lactitol and lactitiol palmitate. *J. Am. Chemists' Soc. 27*, 680–686.

Vujicic, I. F., Lin, A. Y. and Nickerson, T. A. 1977. Changes during hydrolysis of lactose. *J. Dairy Sci. 60*, 29–33.

Waller, G. R. and Feather, M. S. (Editors). 1983. *The Maillard Reaction in Foods and Nutrition.* ACS Symposium Series 215. American Chemical Society, Washington, D.C.

Walsh, J. P., Rook, J. A. F. and Dodd, F. H. 1968A. A new approach to the measurement of the quantitative effects of inherent and environmental factors on the composition of the milk of individual cows and of herds, with particular reference to lactose content. *J. Dairy Res. 35*, 91–105.

Walsh, J. P., Rook, J. A. F. and Dodd, F. H. 1968B. The measurement of the effects and environmental factors on the lactose content of the milk of individual cows and of the herd bulk milk in a number of commercial herds. *J. Dairy Res. 35*, 107–125.

Webb, B. H. 1970. Condensed products. *In: Byproducts from Milk,* 2nd ed. B.H. Webb and E.O. Whittier (Editors). AVI Publishing Co., Westport, Conn.

Whey Products Conference/1970. Proceedings. 1970. Pub. No. ARS-73-69. USDA, ARS. Eastern Regional Research Center, Philadelphia.

Whey Products Conference/1972. Proceedings. 1973. ERRL Pub. No. 3779. USDA, ARS. Eastern Regional Research Center, Philadelphia.

Whey Products Conference/1974. Proceedings. 1975. ERRC Pub. No. 3996. USDA, ARS. Eastern Regional Research Center, Philadelphia.

Whey Products Conference/1976. Proceedings. 1977. Pub. No. ARS-NE-81. USDA, ARS. Eastern Regional Research Center, Philadelphia.

Whey Products Conference/1978. Proceedings. 1979. USDA, ARS. Eastern Regional Research Center, Philadelphia.

Whey Products Conference/1980. Proceedings. 1981. USDA, ARS. Eastern Regional Research Center, Philadelphia.

Whey Products Conference/1982. Proceedings. 1983. USDA, ARS. Eastern Regional Research Center, Philadelphia.

Whey Products Conference/1984. Proceedings. 1985. USDA, ARS. Eastern Regional Research Center, Philadelphia.

Whey Research Workshop II. Proceedings. 1979. *N.Z. J. Dairy Sci. Technol. 14*, 73–216.

Whittier, E. O. 1944. Lactose and its utilization: A review. *J. Dairy Sci. 27*, 505–537.

Widell, S. 1979. A lactosylurea whey product for feeding to ruminants. *In: Proceedings, Whey Products Conference, 1978.* U.S. Department of Agriculture, ARS. Eastern Regional Research Center, Philadelphia, PA.

Woychik, J. W. 1982. Whey and lactose. *In: CRC Handbook of Processing and Utilization in Agriculture,* Vol. I: *Animal Products.* I. A. Wolff (Editor). CRC Press, Boca Raton, Fla.

Young, H. 1970 The drying of whey and whey products. *Proceedings, Whey Products*

Conference/1970. Pub. No. ARS-73-69, USDA, ARS. Eastern Regional Research Center, Philadelphia.

Zadow, J. G. 1984. Lactose: Properties and uses. *J. Dairy Sci. 67,* 2654-2679.

Zarb, J. M. and Hourigan, J. A. 1979. An enzymatic, cryoscopic method for the estimation of lactose in milk products. *N.Z. J. Dairy Sci. Technol. 14,* 171.

Zerban, F. W. and Martin, J. 1949. Refractive indices of lactose solutions. *J. Assoc. Agric. Chem. 32,* 709-713.

7

Nutritive Value of Dairy Foods

Lois D. McBean and Elwood W. Speckmann

The significant contributions of dairy foods in general and milk in particular to the nutrient intake and health of the American population are well recognized (Speckmann *et al.* 1981; Rechcigl 1983; Speckmann 1984). As estimated for 1984, dairy foods (excluding butter) contributed 76% of the calcium in the U.S. food supply, 36% of the phosphorus, 35% of the riboflavin, 21% of the protein, 20% of the vitamin B_{12}, 19% of the magnesium, 12% of the vitamin A, 11% of the vitamin B_6, and 10% of the energy (Table 7.1; Marston and Raper 1986). In addition, dairy foods provide a significant source of the vitamin niacin due to their content of the amino acid tryptophan, and, through fortification, fluid milk provides the majority of vitamin D in our diets.

Ideally, a food's nutritional value is assessed in the context of the total diet rather than solely from the standpoint of its components. Nutrients do not function in the body as isolated substances, but interact with each other. Dairy foods hold a key position as a separate group in basic food guides such as the "Dairy Food Guide," commonly referred to as the "Four Food Groups" or the "Basic Four (USDA, ARS 1957, 1958). Periodically, the Basic Four has been revised slightly to reflect changes in the recommended dietary allowances (RDA) (NAS, 1980A) and data from national food consumption and dietary surveys (Light and Cronin 1981). The most recent revision occurred in 1979 with publication of the U.S. Department of Agriculture's *Food. The Hassle-Free Guide to a Better Diet* (USDA, SEA 1979).

The basic food pattern is viewed as a model of moderation. In general, food selections which follow the pattern of recommended numbers and sizes of daily servings from each group provide all of the essential nutrients needed to maintain good health at an energy level of approximately 1200–1500 kcal (USDA, ARS 1957; USDA, SEA 1979). Because of their unique nutrient profile, dairy foods such as milk are designated as "protective foods," that is, foods in which the concentration of essential nutrients is high in relation to the food's energy value. In Table 7.1, the milk group can be compared with other food groups in

Table 7.1. Percentage Contribution of Major Food Groups to Nutrients Available in the United States per Capita per Day in 1984 (PRELIMINARY).

Food group	Food energy	Protein	Fat	Carbo-hydrate	Calcium	Phos-phorus	Iron	Mag-nesium	Vitamin A value	Thiamin	Ribo-flavin	Niacin	Vitamin B_6	Vitamin B_{12}	Ascorbic acid
Milk	10.3	20.9	11.7	6.0	75.8	35.8	2.3	19.1	11.6	8.9	34.7	1.6	11.5	20.1	3.1
Meat	25.0	52.9	40.4	2.3	9.4	38.1	38.8	27.9	23.1	32.8	28.8	52.4	47.5	78.3	2.0
Fruits/vegetables	8.4	6.8	0.8	16.8	8.9	10.7	17.7	25.3	48.3	15.8	8.8	14.2	30.8	0	90.4
Grain	19.5	18.4	1.3	35.5	3.6	12.7	35.7	18.4	0.3	41.9	23.0	27.9	10.1	1.7	0
Others[a]	36.6	1.1	45.7	39.4	2.3	2.7	5.5	9.2	16.6	0.5	4.8	3.9	0.1	0	4.4

[a]High-energy–low-nutrient foods such as sugars and sweeteners, fats and oils, and alcohol.
SOURCE: Marston and Raper (1986)

terms of percentage contribution to nutrients available in the United States per capita per day in 1984.

The importance of dairy foods in the U.S. diet is also demonstrated by their inclusion in federally supported child nutrition programs (Radzikowski and Gale 1984). Background information on the programs, including their original purpose and objectives, has been reviewed by Longen (1980), Owen *et al.* (1979), and the National Dairy Council (1981, 1982A). Milk is an integral component of these programs, although in recent years significant changes in the regulations regarding the choice and amount of milk served under various programs have occurred. For example, in the Special Milk Program (SMP), 449 million half-pints of milk were served in 1955. This increased to a peak of about 2.6 billion half-pints in 1973 (Longen 1980). However, due to the Omnibus Reconciliation Act of 1981, which limited this program to schools that do not participate in any other federal child nutrition programs, participation and the amount of milk served in the SMP have decreased drastically (USDA, FNS 1983A). In fiscal year 1981, 1542 million half-pints of milk were served in the SMP (USDA, FNS 1983A). This decreased to an estimated 210 million half-pints in 1983 (USDA, FNS 1983B).

NUTRIENT COMPONENTS OF MILK

Many different kinds of milk and milk products are available (USDA 1978; NDC 1976, 1983A, 1983B). Milk has two major components: fat, including fat-soluble vitamins, and milk-solids-not-fat (MSNF), which contain protein, carbohydrate, water-soluble vitamins, and minerals. The specific nutrient contribution of each of these milk products is related largely to the concentrations of milk fat and MSNF of the product. If the percentages of milk fat and MSNF are known, the values in Table 7.2 may be used to determine the nutrient contribution of a particular milk product.

By examining the major nutrient components of milk, as in the following sections, it is possible to delineate further the manner in which dairy foods contribute to human nutritional needs and consequently to overall health.

ENERGY

In terms of energy, 24g (8 fluid oz) of milk provide 90 to 150 kcal, depending on the content of fat and MSNF (USDA, CFEI 1976). Milk

Table 7.2. Nutrient Content of Milk Fat and MSNF.[a]

Nutrient	Per gram Milk fat	Per gram MSNF
Energy (kcal)[b]	8.79	3.71
Energy (kJ)	36.78	15.52
Protein, (N × 6.38) (g)		0.380
Fat (g)	1.00	
Carbohydrate, total (g)		0.536
Fiber (g)		0
Ash (g)		0.083
Minerals		
Calcium (mg)		13.8
Iron (mg)		0.006
Magnesium (mg)		1.6
Phosphorus (mg)		10.8
Potassium (mg)		17.5
Sodium (mg)		5.7
Zinc (mg)		0.044
Vitamins		
Ascorbic acid (mg)		0.108
Thiamin (mg)		0.0044
Riboflavin (mg)		0.0187
Niacin (mg)		0.0097
Niacin equivalents (mg)[c]		0.0987
Pantothenic acid (mg)		0.0362
Vitamin B_6 (mg)		0.0048
Folacin (μg)		0.6
Vitamin B_{12} (μg)		0.0412
Vitamin A (RE)[d]	9.3	
Vitamin A (IU)	37.7	
Cholesterol (mg)	4.0	

[a]The nutrient values are based on whole milk with 3.34% fat and 8.67% MSNF. For standard error of the mean, refer to USDA (1976).
[b]For rapid estimate of food energy, fat can be assumed to contribute 9 kcal/g; protein and carbohydrate, 4 kcal/g.
[c]This value includes niacin equivalents from preformed niacin and from tryptophan. A dietary intake of 60 mg tryptophan is considered equivalent to 1 mg niacin. One niacin equivalent is equal to either of these amounts.
[d]A retinol equivalent (RE) is equal to 3.33 IU retinol or 10 IU β-carotene.
SOURCE: USDA (1976). Adapted from National Dairy Council. 1983A. *Newer Knowledge of Milk and Other Fluid Dairy Products*, Rosemont, Ill. With permission.

and milk products are foods of high nutrient density because of their substantial concentration of major nutrients relative to calories. This fact is supported by data in Table 7.1, where it is shown that the milk group provides only 10% of the food energy, yet supplies appreciable quantities of other essential nutrients (Marston and Raper 1986).

Obesity is a major dietary issue for the American population (NAS 1980B; AMA, Council on Scientific Affairs 1979; USDA and

USDHEW 1980; USDHEW 1979; Vital and Health Statistics 1979; National Institutes of Health 1985; Simopoulos 1985). Dairy foods such as whole milk and cheese have been incriminated as contributing to the development of this condition (Jones 1973). However, there is no evidence that any specific food, food component, or combination of foods, when consumed in recommended amounts, leads to obesity per se (American Academy of Pediatrics 1974).

For individuals who want to either lose or maintain body weight, it is necessary to reduce energy intake and/or to increase energy expenditure. It is important to use foods of low nutrient density sparingly, as they provide primarily energy and few additional essential nutrients (NAS 1980B). Several government surveys indicate that high-energy, low-nutrient foods such as sugars and sweeteners, fats and oils, and alcohol provide approximately one-third of the energy intake, yet supply only minor or trace quantities of essential nutrients (Carroll et al. 1983; Abraham and Carroll 1981; USDA, SEA 1980). Consequently a reduction in their intake has only a minimal effect on the nutritional integrity of the meal pattern. Dairy foods of varied energy content such as whole milk, lowfat milk, and skim milk are available for persons concerned about their weight (USDA, CFEI 1976). In addition, the inclusion of dairy foods in a weight reduction regimen is appropriate because of the riboflavin (vitamin B_2) content of these foods. According to Belko et al. (1983, 1984, 1985), the need for riboflavin by women who are exercising and/or on weight reduction diets may be greater than the RDA for this vitamin.

PROTEIN

Protein in cow's milk is not only present in significant amounts but is of exceptionally high quality (Porter 1978; Milner et al. 1978; Jonas et al. 1976; Whitaker and Tannenbaum 1977; Hambraeus 1982). Milk and milk products, excluding butter, provide about 21% of the daily per capita protein available for consumption in the United States (Table 7.1).

Fluid milk contains approximately 3.5% protein (USDA, CFEI 1976). Casein, found only in milk, comprises about 82% of the total milk protein, and whey proteins, principally β-lactoglobulin and α-lactalbumin, constitute the remaining 18% (Lampert 1975; Jonas et al. 1976). Casein, because of its excellent nutritional value, is used routinely as a reference protein to evaluate the quality of protein in other foods (Jonas et al. 1976; Hambraeus 1982).

The nutritional quality of dietary protein depends largely on the pattern and concentration of essential amino acids provided for the synthesis of nitrogen-containing compounds within the body (Hambraeus 1982). The high quality of milk protein stems from the fact that it contains, in varying quantities, all of the amino acids required by humans. Moreover, the pattern of distribution of amino acids in milk protein resembles that needed by humans (Table 7.3). As shown in Table 7.4, there are numerous methods of evaluating the quality of proteins under experimental conditions, all of which indicate milk proteins to be of high quality (Hambraeus 1982).

Whey proteins are slightly superior to casein because of the limiting quantity of the total sulfur-containing amino acids (methionine plus cystine) in casein. However, because whey proteins have a relative surplus of these amino acids, casein and whey proteins, as found in milk,

Table 7.3. Amino Acid Distribution in Milk.

Amino acids	Estimated requirements of adults[a] (grams per day)	Milligrams per gram MSNF[b]	Milligrams per 100 g Fluid whole milk
Essential			
Histidine[c]		10.31	89.39
Isoleucine	0.84	22.98	199.24
Leucine	1.12	37.16	322.18
Lysine	0.84	30.12	261.14
Methionine	0.70[d]	9.50	82.36
Phenylalanine	1.12[d]	18.34	159.01
Threonine	0.56	17.12	148.43
Tryptophan	0.21	5.34	46.30
Valine	0.98	25.39	220.13
Nonessential			
Alanine		13.10	113.58
Arginine		13.76	119.30
Aspartic acid		28.79	249.61
Cystine		3.50	30.34
Glutamic acid		79.48	689.09
Glycine		8.04	69.71
Proline		36.78	318.88
Serine		20.66	179.12
Tyrosine		18.34	159.01

[a] Values calculated for a 70-kg adult male.
[b] All values have been calulated based on 8.67% MSNF for whole milk.
[c] Necessary for the growth of infants and children. No estimated requirements for adults.
[d] Value for total S-containing amino acids (methionine + cystine).
[e] Value for total aromatic amino acids (phenylalanine + tyrosine).
SOURCE: NAS (1980A) and USDA (1976). Adapted from National Dairy Council. 1983A. *Newer Knowledge of Milk and Other Fluid Dairy Products*. Rosemont, Ill. With permission.

Table 7.4. Average Measures of Protein Quality for Milk and Milk Proteins.

	BV	Digestibility	NPU	PER[a]	Chemical score
Milk	84.5	96.9	81.6	3.09	60
Casein	79.7	96.3	72.1	2.86	58
Lactalbumin	82	97	79.5[b]	3.43	—[c]
Nonfat dry milk	—	—	—	3.11	—
Biological value (BV)	=	Proportion of absorbed protein that is retained in the body for maintenance and/or growth			
Digestibility (D)	=	Proportion of food protein that is absorbed			
Net protein utilization (NPU)	=	Proportion of protein intake that is retained (calculated as BV × D)			
Protein efficiency ratio (PER)	=	Gain in body weight divided by weight of protein consumed			
Chemical score	=	The content of the most limiting amino acid expressed as a percentage of the content of the same amino acid in egg protein			

[a] PER values are often adjusted relative to casein, which may be given a value of 2.5.
[b] Calculated.
[c] Denotes no value compiled in the FAO (1970) report.
SOURCE: FAO (1970). From National Dairy Council. 1983A. *Newer Knowledge of Milk and Other Fluid Dairy Products.* Rosemont, Ill. With permission.

complement each other (Hambraeus 1982). Both casein and whey proteins have a relative surplus of the essential amino acids lysine, threonine, methionine, and isoleucine, which make milk proteins valuable in supplementing vegetable proteins, particularly those of cereals, which are limiting in these amino acids (Speckmann 1984; Rechcigl 1983). For example, the high nutritive value of milk protein makes it an integral part of food interventions in many developing countries, where protein-energy malnutrition (kwashiorkor) is found among children whose diet consists largely of cereals (Whitaker and Tannenbaum 1977; Lampert 1975; Stanley *et al.* 1981). The protein nutriture of vegetarians, as well as their nutritional status in general, can be improved by adding animal protein such as that in milk to diets containing mainly cereal and vegetable proteins (The American Dietetic Association 1980; American Academy of Pediatrics 1977). Tryptophan is also an essential amino acid abundant in milk protein (Table 7.3). It is because of its tryptophan content that milk is an excellent source of niacin equivalents (Horwitt *et al.* 1981). One niacin equivalent is defined as 1 mg niacin or about 60 mg tryptophan.

The high casein content of cow's milk is responsible for the formation of a large, firm curd which may be difficult for some infants to digest compared with the finer, soft curd formed from human milk. Consequently, cow's milk often is modified to conform more closely to the nutrient and physical requirements of infants (Fomon 1974). When cow's milk is heated, homogenized, or acidified to produce softer curd formation, the protein is used by infants as efficiently as that of human milk, which contains less casein than cow's milk (Fomon 1974).

True hypersensitivity (i.e., an immunologic reaction) to the protein component of cow's milk occurs in about 1% of the infant and child populations in industrialized countries (May and Bock 1978; Savilahti *et al.* 1981), although a range of 0.3 and 7.5% has been reported (Bahna and Gandhi 1983). The pathogenesis, incidence, manifestations, diagnosis, management, and prevention of cow's milk hypersensitivity are reviewed by several authors (Bahna and Gandhi 1983; Hill *et al.* 1984; Savilahti and Verkasalo 1984; Foucard 1985; Bock 1985; Savilahti *et al.* 1981). The condition is difficult to diagnose because of the multiplicity of symptoms (e.g., cutaneous, gastrointestinal, and respiratory manifestations), the transient nature of the reactions, and the lack of simple, reliable, objective methods to verify the diagnosis (Bahna and Heiner 1980). In general, the diagnosis depends more on clinical evaluation than on laboratory data. The tendency to ascribe any adverse reaction to cow's milk as hypersensitivity, even when no immunological basis can be demonstrated, may result in overlooking a more serious disorder or in the elimination of milk from the diet (Bock 1980). Follow-up studies in infants and young children with cow's milk sensitivity have shown that the condition often subsides within a few months and generally disappears or is greatly reduced in severity by the time the child is two to three years of age (Bock 1982, 1985). After this time, true allergenic reactions to milk in the general population are rare (American Academy of Pediatrics 1983; Bock 1982; Bahna and Gandi 1983).

FAT

Milk fat, averaging about 3.25% fat in market whole milk, exists in microscopic globules in an oil-in-water emulsion. Considered the most complex of all of the common fats, milk fat exhibits unique physcial, chemical, and biological properties not easily duplicated by other fats (Lampert 1975; Formo *et al.* 1979).

Table 7.5. Constituents of Milk Lipids.

Class of lipid	Percentages of total milk lipids	Per gram Fat	Per 100 g Fluid whole milk (3.34% Fat)
Vitamin A activity	$6-9 \times 10^{-4}$	37.7 IU (9.3 RE)	126 IU (31 RE)
Vitamin D	$8.5-21 \times 10^{-7}$	0.34–0.84 IU	1.1356–2.8056 IU
Vitamin E	2.4×10^{-3}	0.024 mg	0.080 mg
Vitamin K	1×10^{-4}	0.001 mg	0.0034 mg
Triglycerides of fatty acids	97–98	0.97–0.98 g	3.24–3.27 g
Diglycerides	$2.8-5.9 \times 10^{-1}$	0.003–0.006 g	0.01–0.02 g
Monoglycerides	$1.6-3.8 \times 10^{-2}$	0.16–0.380 g	0.53–1.27 g
Keto acid glycerides (total)	$8.5-12.8 \times 10^{-1}$	8.503–12.8 mg	28.4–42.75 mg
Ketonogenic glycerides	$3-13 \times 10^{-2}$	0.299–1.287 mg	1.00–4.3 mg
Hydroxy acid glycerides (total)	$6-7.8 \times 10^{-1}$	5.988–7.799 mg	20.0–26.05 mg
Lactonogenic glycerides	6×10^{-2}	0.599 mg	2.00 mg
Neutral glyceryl ethers	$1.6-2.0 \times 10^{-2}$	0.16–0.20 mg	0.53–0.67 mg
Neutral plasmalogens	4×10^{-2}	0.40 mg	1.34 mg
Free fatty acids	$1.0-4.4 \times 10^{-1}$	1.0–4.4 mg	3.34–14.7 mg
Phospholipids (total)	$8.0-10.0 \times 10^{-1}$	7.99–10.0 mg	26.7–33.4 mg
Sphingolipids (less sphingomyelin)	6.0×10^{-2}	0.599 mg	2.00 mg
Sterols	$2.2-4.1 \times 10^{-1}$	2.20–4.10 mg	7.35–13.69 mg
Cholesterol	4.19×10^{-1}	4.0 mg	14 mg
Squalene	7×10^{-3}	0.07 mg	0.2338 mg
Carotenoids	$7-9 \times 10^{-4}$	0.0070–0.0090 mg	0.0233–0.0301 mg

SOURCE: USDA (1976) and Webb *et al.* (1974). Adapted from National Dairy Council. 1983A. *Newer Knowledge of Milk and Other Fluid Dairy Products.* Rosemont, Ill. With permission.

In terms of composition, Table 7.5 shows that milk fat consists primarily of triglycerides with small amounts of di- and monoglycerides, phospholipids, sterols such as cholesterol, carotenoids, fat-soluble vitamins A, D, E, and K, and some traces of free fatty acids (Renner 1983; Christie 1983). The fatty acid composition of bovine milk fat is characterized by a high proportion of saturated fatty acids (60 to 70%), appreciable amounts of monounsaturated fatty acids (25 to 35%), and small amounts of polyunsaturated fatty acids (4%) (Lampert 1975). Milk fat has a relatively high content of short chain fatty acids with four to eight carbon atoms (e.g., butyric acid to caprylic), many of which are not found in other natural food fats (Lampert 1975; Renner 1983). According to Patton and Jensen (1975), over 400 different fatty acids have been detected in bovine milk lipids and still others are likely to be identified. The fatty acid composition of milk and some milk products is given in Table 7.6.

Milk fat is characterized not only by the kind and amount of fatty acids, but also by the distribution of fatty acids on the glycerol moiety. The glyceride structure appears to influence the biological properties of milk fat as well as of other fats. For example, the digestibility of fats used in infant feeding has been related to the position of individual fatty acids on the glyceride molecule (Fomon 1974; Renner 1983). The fatty acids in milk fat triglycerides are not haphazardly distributed, but are arranged so that the short chain fatty acids, mainly butyric and caproic, occur in the outer position, and long chain fatty acids, such as myristic acid, are found in position two. About 95% of the short chain fatty acid, butyric acid, occupies positions one and three, whereas 54% of the long chain fatty acid, myristic acid, is found in position 2 in the triglyceride (Renner 1983). It is thought that this arrangement contributes to the ease of digestibility of milk fat compared with some other fats, especially those with a large proportion of triglycerides, each containing three fatty acids with 18 carbons (Lampert 1975; Porter 1975; Renner 1983). Homogenization of milk, by reducing the size of fat globules and altering the physical condition of the protein, results in a milk with a lower curd tension and improved digestibility (Lampert 1975).

Milk fat, similar to other dietary fats, serves as a concentrated source of energy (Gurr 1983; Coates 1983). About 72 of the 150 kcal in 244 g (8 fluid oz) of whole milk are contributed by the fat. The short chain fatty acids present in milk fat, but virtually absent in most vegetable oils, are absorbed through the intestinal wall without being resynthesized to glycerides and are transported in the portal vein directly to the liver, where they are immediately converted to utilizable forms of energy. They serve, therefore, as a quick source of energy which may be important, especially in early life. In contrast, the long chain fatty acids undergo a much more complex chain of reactions during digestion, absorption, and transport.

Whole cow's milk is recommended for infant feeding when the infant over six months of age is eating approximately 20 g beikost (foods other than milk or formula) daily (i.e., equivalent to about one and one half 4.75 oz jars (135 g) of strained foods commercially prepared for infants) (Anderson *et al.* 1985; American Academy of Pediatrics 1983,

1986; Fomon *et al.* 1979). Milks with reduced fat content such as skim and lowfat milks are not recommended during infancy (Fomon *et al.* 1977, 1979; Fomon 1974; American Academy of Pediatrics 1983). Infants fed skim and lowfat milks receive insufficient energy to support maintenance requirements. Growth is achieved, albeit at a reduced rate, and energy is obtained by mobilization of body fat, clinically evidenced by a substantial reduction in triceps and subscapular skinfold thicknesses. Furthermore, regular consumption of large volumes of calorically dilute food may not promote the development of sound eating habits (Fomon *et al.* 1979).

Cholesterol, which is synthesized in the body at a rate sufficient to meet body need (0.5–1.0 g daily), has several important functions in the body (Sabine 1977; Renner 1983). It serves as a structural element of cell membranes and as a precursor of bile acids, steroid hormones, and vitamin D (Sabine 1977; Renner 1983). It has been suggested from animal studies that cholesterol intake (e.g., as in cow's milk) early in life may influence specific enzyme systems which, in turn, enable one to maintain a serum cholesterol concentration in adulthood within a normal range regardless of cholesterol ingestion (Reiser and Sidelman 1972; Hahn and Kirby 1973). However, studies in humans (Glueck *et al.* 1972; Hodgson *et al.* 1976; Friedman and Goldberg 1975), as well as further work with animals (Kris-Etherton *et al.* 1979; Green *et al.* 1981), have failed to substantiate this hypothesis. More research, in particular studies of longer duration, is necessary to determine whether early exposure to dietary cholesterol contributes to adult cholesterol homeostatic mechanisms (Kris-Etherton *et al.* 1979; Reiser 1975).

Concern has been expressed regarding the content of fat in the diet in general and of cholesterol and saturated fatty acids in particular (Segall 1977). In the 1950s, scientists postulated that an increased dietary intake of cholesterol and saturated fatty acids (e.g., as found in animal fats) elevated blood cholesterol levels and increased the risk of coronary heart disease. This hypothesis, called the "lipid hypothesis," has been subjected to much research (McGill 1979A, B; Glueck 1979; Coates 1983). However, despite extensive epidemiological data, direct proof that the level of cholesterol and/or saturated fatty acids as currently consumed in the United States may predispose the individual to coronary heart disease morbidity and mortality is lacking (American Academy of Pediatrics 1986; Council for Agricultural Science and Technology 1985; Coates 1983; Samuel *et al.* 1983; Harper 1983; Oliver 1982; McNamara 1982; Ahrens 1976, 1979, 1982, 1985; American Council on Science and Health 1982; NAS 1982, 1980B; McGill 1979A,B).

Table 7.6. Fatty Acid Composition of Milk and Selected Milk Products (AMOUNT in 100 g, EDIBLE PORTIONS).

Fatty Acids (g)

Milk Product	Total Lipid	Total	Saturated 4:0 Butyric	6:0 Caproic	8:0 Caprylic	10:0 Capril	12:0 Lauric[a]	14:0 Myristic[a]	16:0 Palmitic[a]	18:0 Stearic[a]	Total[b]	Unsaturated 16:1 Palmitoleic	18:1 Oleic	18:2 Linoleic	18:3 Linolenic
Whole Milk	3.34	2.08	0.11	0.06	0.04	0.08	0.09	0.34	0.88	0.40	1.08	0.08	0.84	0.08	0.05
Lowfat milk, 2% fat	1.92	1.20	0.06	0.04	0.02	0.05	0.05	0.19	0.50	0.23	0.63	0.04	0.48	0.04	0.03
Lowfat milk, 1% fat	1.06	0.66	0.03	0.02	0.01	0.03	0.03	0.11	0.28	0.13	0.35	0.02	0.27	0.02	0.02
Skim milk	0.18	0.117	0.009	0.001	0.002	0.004	0.003	0.017	0.053	0.019	0.054	0.007	0.038	0.005	0.002
Chocolate milk	3.39	2.10	0.10	0.06	0.04	0.08	0.09	0.32	0.89	0.47	1.11	0.07	0.87	0.08	0.05
Evaporated whole milk	7.56	4.59	0.20	0.13	0.05	0.11	0.16	0.73	2.03	0.92	2.58	0.16	2.10	0.17	0.08
Evaporated skim milk	0.20	0.121	0.005	0.004	0.001	0.003	0.004	0.019	0.054	0.024	0.068	0.004	0.056	0.004	0.002
Sweetened condensed whole milk	8.70	5.49	0.28	0.17	0.10	0.07	0.18	0.78	2.40	1.21	2.77	0.14	2.19	0.22	0.12
Nonfat dry milk, regular	0.77	0.50	0.03	0.01	0.01	0.02	0.01	0.08	0.24	0.08	0.23	0.02	0.17	0.02	0.01
Nonfat dry milk, instant	0.72	0.47	0.03	0.01	0.01	0.02	0.01	0.08	0.22	0.08	0.22	0.02	0.16	0.02	0.01
Dry whole milk	26.71	16.74	0.87	0.24	0.27	0.60	0.61	2.82	7.52	2.85	8.58	1.20	6.19	0.46	0.20
Buttermilk, cultured, fluid	0.88	0.55	0.30	0.02	0.01	0.02	0.02	0.09	0.23	0.11	0.28	0.02	0.22	0.02	0.01
Dry butter milk, sweet cream, from manufacture of butter	5.78	3.60	0.19	0.11	0.06	0.14	0.16	0.58	1.52	0.70	1.89	0.13	1.45	0.13	0.08
Sour cream	20.96	13.05	0.68	0.40	0.23	0.53	0.59	2.11	5.51	2.54	6.83	0.47	5.27	0.47	0.30
Sour half-and-half	12.00	7.47	0.39	0.23	0.13	0.30	0.34	1.21	3.16	1.45	3.92	0.27	3.02	0.27	0.18
Yogurt Plain, whole milk (8 g protein/8 fl oz)	3.25	2.10	0.10	0.07	0.04	0.09	0.11	0.34	0.89	0.32	0.98	0.07	0.74	0.06	0.03
Plain, lowfat (12 g protein/8 fl oz)	1.55	1.00	0.05	0.03	0.02	0.04	0.05	0.16	0.42	0.15	0.47	0.03	0.35	0.03	0.01

Product														
Plain, skim milk (13 g protein/8 fl oz)	0.18	0.116	0.005	0.004	0.002	0.005	0.006	0.019	0.049	0.018	0.054	0.004	0.041	0.004	0.001
Fruit, lowfat (10 g protein/8 fl oz)	1.08	0.70	0.03	0.02	0.01	0.03	0.04	0.11	0.29	0.10	0.33	0.02	0.25	0.02	0.01
Half-and-half	11.50	7.16	0.37	0.22	0.13	0.29	0.32	1.16	3.02	1.39	3.75	0.26	2.89	0.26	0.17
Light cream	19.31	12.02	0.63	0.37	0.22	0.48	0.54	1.94	5.08	2.34	6.30	0.43	4.86	0.44	0.28
Medium cream, 25% fat	25.00	15.56	0.81	0.48	0.28	0.63	0.70	2.51	6.58	3.03	8.15	0.56	6.29	0.56	0.36
Light whipping cream	30.91	19.34	1.08	0.30	0.31	0.63	0.37	3.29	8.84	3.37	9.97	1.01	7.66	0.62	0.27
Heavy whipping cream	37.00	23.03	1.20	0.71	0.41	0.93	1.04	3.72	9.73	4.48	12.06	0.83	9.31	0.84	0.54
Whipped cream topping, pressurized	22.22	13.83	0.72	0.43	0.25	0.56	0.62	2.24	5.84	2.69	7.24	0.50	5.59	0.50	0.32
Cheddar cheese	32.8	20.2	1.05	0.46	0.28	0.63	0.74	3.30	8.94	3.78	10.7	1.01	8.24	0.51	0.42
Cottage cheese, creamed	4.0	2.6	0.13	0.03	0.03	0.06	0.09	0.43	1.20	0.44	1.2	0.12	0.91	0.09	0.03
Cottage cheese, uncreamed	0.4	0.2	0.02	—	0.01	0.01	—	0.04	0.11	0.04	0.1	0.01	0.08	0.01	—
Gouda	27.4	17.6	1.00	0.64	0.43	0.92	1.21	3.04	6.85	2.92	8.4	0.89	6.39	0.26	0.39
Mozzarella, low-moisture, part skim	16.4	10.2	0.53	0.10	0.12	0.26	0.17	1.64	4.99	1.99	5.1	0.45	3.98	0.34	0.14
Swiss	27.6	17.6	1.13	0.50	0.35	0.71	0.79	3.0	7.85	2.78	8.7	0.73	6.53	0.49	0.55
Pasteurized process cheese, American	28.9	18.0	0.94	0.29	0.32	0.59	0.54	2.71	8.33	3.52	9.5	0.94	7.21	0.64	0.33
Pasteurized process cheese spread, plain	20.8	13.1	0.66	0.40	0.31	0.50	0.62	2.14	5.8	2.39	6.7	0.55	5.13	0.39	0.22

Figures shown represent average nutrient values. The *Handbook 8-1* (USDA 1976) includes the standard error of the mean.

[a] Considered by FDA to be saturated fatty acids (for labeling purposes) and to have an effect on plasma lipids and cholesterol.

[b] Traces of additional fatty acids (i.e., 14:1, 20:1, 22:1, 18:4, 20:4, 22:5, 22:6) are also present; thus, totals do not always add up. The fatty acid composition of these milk products with different lipid levels can be calculated if the lipid level is known. Assuming product 1 is listed in this table and product 2 is the same product with a different lipid level:

$$\frac{(\text{g fatty acid in product 1})\,(\text{g total lipid in product 2})}{(\text{g total lipid in product 1})} = \text{g fatty acid in product 2}$$

The gram amounts of fatty acid per serving size can be calculated as follows:

$$\frac{(\text{serving size in g})\ \text{g fatty acid}}{100\ \text{g}} = \text{g fatty acid per serving}$$

SOURCE: USDA (1976) and Posati *et al.* (1975). Adapted from National Dairy Council. 1983A. *Newer Knowledge of Milk and Other Fluid Dairy Products.* Rosemont, Ill. With permission.

Coronary heart disease is a condition influenced by a number of risk factors (e.g., advancing age, male sex, hypertension, cigarette smoking, elevated blood cholesterol, diabetes, stress, obesity, lack of exercise) (Coates 1983; Ahrens 1976; Renner 1983). When a group of men aged 35–57 years at high risk of developing coronary heart disease (high blood cholesterol, high blood pressure, smoking habit) were subjected to a multifactor intervention program to reduce all three risk factors for an average period of seven years, their mortality rate did not differ from that of men not subjected to the intervention (Multiple Risk Factor Intervention Trial Research Group 1982). Similarly, when cholestyramine was fed to men with blood cholesterol levels in the top five percent of the population, blood cholesterol and coronary heart disease morbidity and mortality decreased, but there was no statistically significant change in total mortality (Lipid Research Clinics Program 1984). It is well known that individuals vary greatly in their response to dietary cholesterol (Samuel et al. 1983; Ahrens 1982, 1979; McNamara 1982).

Regardless of the above, the general healthy population, irrespective of their blood cholesterol level (mean for men 18–74 years old is about 215 mg/100 ml, according to Abraham et al. 1977), has been advised by various medical experts and organizations to consume a diet low in cholesterol and saturated fatty acids, specifically by reducing their consumption of meat, eggs, and dairy foods, in an effort to lessen the risk of coronary heart disease (Zilversmit 1982; Weidman et al. 1983; American Heart Association 1982; USDA, USDHEW 1980; Stamler 1981). In contrast, other organizations, such as the American Academy of Pediatrics (1986), are more moderate in their recommendations for dietary change. In its position statement, "Prudent Lifestyle for Children: Dietary Fat and Cholesterol," the American Academy of Pediatrics (1986) concluded that "it would seem prudent not to recommend changes in current dietary patterns in the United States for the first two decades of life without first assessing the effects on growth, development, and such measures of nutritional adequacy as the status of iron." The Academy specifically recommended that meat and dairy foods not be restricted in the diet of adolescents and children because these foods are important sources of iron and calcium, respectively (American Academy of Pediatrics 1986).

Dairy foods are not particularly high in either total fat or cholesterol (Renner 1983). Moreover, a wide variety of dairy foods of varied fat and cholesterol content is available to meet consumer needs (Renner 1983). For example, individuals on fat-restricted diets can choose from such lowfat dairy foods as 2% milk, 1% milk, skim milk, lowfat yogurt, lowfat cheeses, and buttermilk. Some foods in the milk group, such as skim milk, provide less than 1 g fat per serving, and a serving of whole milk contains only 8 g fat, which is less than that of a broiled hamburger (17 g fat). With respect to cholesterol, 244 g (8 fluid oz) of whole milk, lowfat milk (2%), and skim milk contain 33, 18, and 4 mg, respectively, amounts less than those found in most other animal foods (Feeley et al. 1972B; LaCroix et al. 1973; Renner 1983). For example, one egg contains 274 mg cholesterol and a 3-oz serving of liver, pork and fish fillets has 372, 76, and 34–75 mg cholesterol, respectively (Feeley et al. 1972B; LaCroix et al. 1973). If a reduction in cholesterol intake is warranted on an individual basis, this can best be accomplished by moderating the intake of foods high in this substance, such as organ meats, eggs, and shellfish (Feeley et al. 1972B).

The hypothesis that dairy foods contain a cholesterol-lowering "milk factor" evolved from the observation that Maasai tribesmen of East Africa have low serum concentrations of cholesterol and a low incidence of cardiovascular disease in spite of their consumption of 4 to 5 liters/day of fermented whole milk (Mann and Spoerry 1974). Subsequently, sufficient amounts of yogurt, as well as unfermented milk (whole, lowfat, skim), were reported to exhibit a hypocholesterolemic effect both in humans and in laboratory animals in several studies (Mann 1977; Howard and Marks 1977, 1979; Nair and Mann 1977; Kritchevsky et al. 1979; Richardson 1978; Hussi et al. 1981). However, not all investigators have observed a hypocholesterolemic effect of milk. This inconsistency may be explained in part by differences in the experimental design and in the specific type of dairy food used by investigators.

Furthermore, the identity of the "milk factor(s)," if present, is unknown (Renner 1983). Hypocholesterolemic activity has been ascribed to several substances, including hydroxymethylglutarate (HMG), orotic acid, lactose, calcium, and factors in the milk fat globule (Mann and Spoerry 1974; Mann 1977; Howard and Marks 1977, 1979; Richardson 1978; Howard 1977; Mitchell et al. 1968; Thakur and Jha 1981; Keim et al. 1981; Ahmed et al. 1979). HMG is known to inhibit HMG-CoA reductase, the rate-limiting enzyme in cholesterol biosynthesis, but whether HMG is present in sufficient concentration in cow's milk to elicit a hypocholesterolemic effect remains to be determined (Richardson 1978; Howard 1977; McNamara et al. 1972; Boguslawski and Wrobel 1974). According to Ahmed et al. (1979), bovine milk contains not only orotic acid, which inhibits hepatic cholesterol biosynthesis by suppressing the conversion of acetate to mevalonate, but also a second inhibitor of unknown identity and physiological significance. The suggestion that calcium influences plasma cholesterol levels remains to be substantiated (Howard and Marks 1977; Howard 1977; Mitchell *et al.* 1968; Thakur and Jha 1981; Keim *et al.* 1981).

Not only is the hypocholesterolemic effect of fresh milk unresolved, but there is some suggestion that the cholesterol-lowering activity is greater in fermented than in nonfermented milk (Mann 1977; Thakur and Jha 1981; Rao *et al.* 1981; Hussi *et al.* 1981; Sinha *et al.* 1979). Yogurt has been shown to be hypocholesterolemic in humans (Mann 1977; Thakur and Jha 1981) and in animals (Nair and Mann 1977; Thakur and Jha 1981). Rao *et al.* (1981) report that milk fermented with *Streptococcus thermophilus* (one of the lactic cultures used in the preparation of yogurt) is more hypocholesterolemic than unfermented milk when fed to rats. In contrast, Thompson *et al.* (1982) observed that one liter supplements of cultured buttermilk, yogurt, and acidophilus milk did not affect blood cholesterol levels in young healthy adults. It is apparent that more study is necessary to substantiate the hypocholesterolemic activity of specific dairy foods and to determine the nature of the hypocholesterolemic agent, if present.

The wisdom of lowering blood cholesterol levels in the healthy U.S. population in an effort to decrease the risk of coronary heart disease is being challenged by evidence that low blood cholesterol levels (<190 mg/dl) are associated with an increased risk of colon cancer, at least in men (National Heart, Lung, and Blood Institute 1981). However, the association between blood cholesterol levels and cancer risk is inconsistent (Sidney and Farquhar 1983; Committee on Diet, Nutrition and Cancer 1982). Likewise, despite findings from epidemiological and experimental studies indicating a relationship between total fat intake and the occurrence of cancer at certain sites, particularly the breast and colon (Committee on Diet, Nutrition and Cancer 1982), evidence of a causal relationship is lacking (Council for Agricultural Science and Technology 1982). Furthermore, Pariza (1984) considers it inappropriate to recommend that the general population reduce its fat intake in an effort to decrease the risk of cancer when evidence of any benefit is lacking.

The suggestion that *homogenized* cow's milk might be the culprit in promoting heart disease in humans was first made in 1971 (Oster 1971). Specifically, it was proposed that xanthine oxidase (XO), an enzyme occurring naturally in cow's milk, as well as in other animal, plant, and human tissues (the greatest activity of XO in humans is found in the liver and the intestinal mucosa), is absorbed intact from the gastrointestinal tract into the bloodstream. This absorbed XO then is alleged to deplete plasmalogen (an important structural component of cell membranes in arterial and myocardial tissues), thereby intiating the atherosclerotic process, eventually resulting in heart disease. The homogenization process is claimed to increase the biological availability of XO by trapping XO within liposomes or membrane-bound structures which protect it from the action of digestive acids and enzymes.

Recently, increased publicity has been given to this XO hypothesis, in particular by promoters of imitation milk products. Imitation milk is alleged to be superior to real homogenized milk because it lacks the enzyme XO. It is important to keep in mind, however, that the XO hypothesis has never been proven (Deeth 1983; Clifford *et al.* 1983; American Heart Association 1981; Carr *et al.* 1975; Bierman and Shank 1975).

In 1975, the Life Sciences Research Office of the Federation of American Societies for Experimental Biology (Carr *et al.* 1975), upon an extensive review of the available evidence, concluded that it was doubtful whether XO in homogenized cow's milk was a causal or risk factor for heart disease. More recently, Clifford *et al.* (1983) and Deeth (1983), in critical reviews of the homogenized cow's milk XO hypothesis, have arrived at a similar, if not more definitive, conclusion. As stated by Clifford *et al.* (1983), "experimental evidence has failed to substantiate, and in many cases has refuted, the hypothesis that homogenized bovine milk xanthine oxidase intake or plasmalogen depletion are causal factors in the development of atherosclerosis." And, according to Deeth (1983), "there appears to be no unequivocal evidence that the absorbed enzyme has any pathological effects that may contribute to development of atherosclerotic heart disease."

It is well established that pasteurization and homogenization of milk reduce its XO activity (Cerbulis and Farrell 1980; Zikakis and Wooters 1980). In a study in which the activity of XO was assayed in 195 commercially processed dairy foods, it was found that commercial processing (homogenization, pasteurization) destroyed about 82% of the XO activity in raw milk (Zikakis and Wooters 1980).

With regard to the absorption of dietary XO, research findings indicate that the XO molecule is too large to be absorbed into the bloodstream from the intestine (Bierman and Shank 1975). The XO molecule has a molecular weight of 300,000, whereas the largest observed compound to be absorbed from the intestine has a molecular weight of 80,000 (Bierman and Shank 1975). Several investigators (Ho and Clifford 1976; Mangino and Brunner 1976; Volp and Lage 1977; Zikakis *et al.* 1977) have examined the stability of XO (purified XO, fresh raw milk XO, and homogenized cow's milk XO) in the environment of the digestive tract. In all studies, XO was found to be inactivated in an environment similar to the acidity of the stomach's juices. Other researchers directly measured the absorption of XO in laboratory rats (Zikakis *et al.* 1977; Clark and Pratt 1976; Ho *et al.* 1978), in miniature pigs (McCarthy and Long 1976; Doughterty *et al.* 1977), and in humans (McCarthy and Long 1976). Likewise, these studies were unable to demonstrate that XO is absorbed intact. Moreover, there is no evidence that liposomes are formed during the homogenization of milk or that substances (e.g., XO) trapped in liposomes are absorbed from the gastrointestinal tract (Clifford *et al.* 1983; Patel and Ryman 1981).

Studies which have correlated the intake of homogenized dairy foods with levels of blood XO activity by measuring antibody titers to XO (Rzucidlo and Zikakis 1979) or which have measured antibody titers to XO in atherosclerotic and healthy normal humans (Oster *et al.* 1974) suffer from several inadequacies, the most serious being the lack of specificity of the method (hemagglutination assay) used to detect the antibody (Clifford *et al.* 1983). This method is ineffective in identifying the source of the XO which brought about the antibody response (Clifford *et al.* 1983). That is, XO from human liver is not distinguishable from that of cow's milk. As such, no XO activity specifically of cow's milk origin has been found in the blood of humans who drink homogenized cow's milk who have died from heart disease.

The few published articles (Oster 1971; Oster and Hope-Ross 1966) that have suggested plasmalogen depletion as a cause of heart disease have failed to provide quantitative data or controls (Clifford *et al.* 1983). In addition, the method used to detect plasmalogen has been demonstrated to have serious limitations (Clifford *et al.* 1983; Rapport and Norton 1962). In a study designed to investigate directly the question of whether plasmalogen depletion causes cadiovascular lesions, Ho and Clifford (1977) showed that neither arterial nor coronary tissue plasmalogens were depleted, nor was plague formed when rabbits received large intravenous doses of cow's milk XO. This study thus fails to provide evidence that XO in the body depletes plasmalogens or induces arterial plaque formation.

Claims that high doses of folic acid (up to 200 times the RDA for this vitamin) are beneficial in the prevention and treatment of myocardial infarction and chest pain, allegedly because of the vitamin's ability to block the action of XO and help rebuild plasmalogen, are based on the observation that folic acid in high concentrations inhibits XO activity *in vitro* (Clifford *et al.* 1983; DeRenzo 1956). However, several investigators (Clifford *et al.* 1983; Ho and Clifford 1976; Kaplan 1980) have provided direct experimental data to the contrary—that is, indicating that folic acid does not inhibit XO activity in the body.

CARBOHYDRATE

Lactose, a disaccharide, is the predominant carbohydrate in milk, accounting for about 54% of the total SNF content and about 45 of the 150 kcal (i.e., 30%) of energy supplied in 244 g (8 fluid oz) of whole milk (Lampert 1975). In addition to lactose, minor quantities of glucose, galatose, and oligosaccharides are present. For all practical purposes, the milk group is the sole source of lactose in the diet. The lactose content of milk and some milk products is listed in Table 7.7. Most varieties of cheese contain an insignificant quantity of lactose. Ripened cheeses generally have no measurable amounts of lactose, whereas unripened cheeses such as cottage or cream cheese may contain less than 1% lactose (NDC 1983B; Kosikowski 1982). A wide range in the lactose content of cottage cheese has been reported due to the addition of lactose as an optional ingredient to the creaming mixture.

Cow's milk contains about 4.8% lactose, whereas human milk contains 7% (Fomon 1974). Lactose is used in the preparation of modified cow's milk formulas for infant feeding to duplicate as closely as possible the lactose content of human milk. Although a specific need for

Table 7.7. Lactose Content of Milk and Selected Milk Products.

	Weight/Unit		Lactose (g/unit)	Lactose (%)	Total Carbo- hydrate (g/unit)
Milk					
Whole	224g	1 c	11	4.5	11.4
Lowfat, 2%	244g	1 c	9–13	3.7– 5.3	11.7– 13.5
Skim	244g	1 c	12–14	4.9– 5.7	11.9– 13.7
Buttermilk	245g	1 c	9–11	3.7– 4.5	11.7
Chocolate	244g	1 c	10–12	4.1– 4.9	25.9
Sweetened condensed whole	306g	1 c	35	11.4	166.5
Cream					
Half-and-half	15g	1 T	0.6	4	0.64
Light	15g	1 T	0.6	4	0.55
Whipped cream topping	3g	1 T	0.4	13	0.38
Yogurt					
Lowfat	227g– 228g	8 oz	11–15	4.8–6.6	16.0– 43.2

SOURCE: USDA (1976) and Welsh (1978).

lactose or its component simple sugar, galactose, in the body has not been demonstrated, a few roles for this carbohydrate have been suggested. Lactose may serve as a substrate for acid-forming bacteria present in the intestine, with the result that organic acids such as lactic acid are formed and the growth of undesirable putrefactive bacteria is inhibited (Lampert 1975; Renner 1983). Lactose, therefore, may be said to be effective in combatting gastrointestinal disturbances. Lactose has been shown to increase calcium absorption in experimental animals (Armbrecht and Wasserman 1976; Schaafsma and Visser 1980), although in humans the effect of lactose on calcium absorption has been less consistent (Condon *et al.* 1970; Kocian *et al.* 1973; Kobayashi *et al.*, 1975). However, Ziegler and Fomon (1983) recently demonstrated that in infants the absorption of calcium, as well as other minerals, is significantly increased by lactose. There is some suggestion that galactose, one of the monosaccharides in lactose, has a role in the early development of the infant's brain and spinal column (Lampert 1975; Renner 1983). Galactose is also derived from glucose in the liver (Renner 1983).

To be utilized by the body, lactose must first be hydrolyzed by the enzyme lactase in the brush border of the small intestine to its component simple sugars, glucose and galactose (Lampert 1975). Low levels of the enzyme lactase in the intestine may render it difficult for some individuals to metabolize lactose completely, a condition called "lactose malabsorption." "Lactose intolerance" is defined as clinical symptoms (e.g., abdominal pain, bloating, flatulence, diarrhea) following a lactose tolerance test. The latter consists of administering a standard dose of lactose (50 to 100 g/m^2 or 2 g/kg or less) mixed in water to a person with proven lactose malabsorption. Comprehensive reviews of lactose malabsorption and lactose intolerance, including their diagnosis, incidence, etiology, clinical and nutritional consequences, and management, are provided by the National Dairy Council (1985), Torun *et al.* (1979), Paige and Bayless (1981), Delmont (1983), Renner (1983), and Newcomer and McGill (1984A).

Lactose intolerance, as determined by the above lactose tolerance test, was demonstrated in 70% of black and 6 to 12% of white persons in the United States (Bayless and Rosensweig 1966). The majority of nonwhite individuals demonstrated symptomatology (intolerance) to a single large clinical test dose of lactose (50–100 g), the amount in one to two quarts of milk. Worldwide, lactose intolerance is relatively high among nonwhite populations (Torun *et al.* 1979; Paige and Bayless 1981). However, research has revealed that lactose intolerance is not synonymous with milk intolerance and that most individuals diagnosed as being lactose intolerant can safely consume the usual amounts of milk without discomfort. That is, while symptoms of lactose intolerance may occur in individuals given 50 to 100 g lactose in one feeding, most of these individuals can tolerate the amount of lactose contained in typical servings of dairy foods (Paige and Bayless 1981).

"Milk intolerance," generally defined as the inability to digest completely the amount of lactose in an 8-oz (244 g) serving of milk, is a rare phenomenon even among populations with a high prevalence of lactose intolerance (Torun et al. 1979). According to Haverberg et al. (1980) and Kwon et al. (1980), the true prevalence of milk intolerance due to lactose malabsorption can be determined only by double-blind studies. If a double-blind procedure including a placebo (i.e., lactose-free milk) is not employed, it is not known whether any symptoms following milk intake are caused by the lactose in milk or by other chemical, physiological, or psychological factors.

Torun et al. (1979), reviewing 195 publications on this subject, concluded that "the poor correlation between lactose malabsorption and intolerance to the amounts of milk ordinarily ingested in a meal indicates that the assumption of milk intolerance by many populations is exaggerated." Several professional groups support this conclusion and state that there is no reason to discourage supplemental milk-feeding programs targeted at children on the basis of primary lactose intolerance (Protein Advisory Group of the United Nations 1972; Food and Nutrition Board, NAS 1972; American Academy of Pediatrics 1974, 1978A). For the few individuals who truly are milk intolerant, suitable alternatives are available. These include consumption of milk in smaller quantities more frequently throughout the day, most cheeses, many cultured and culture-containing dairy foods, and lactose-hydrolyzed milk and milk products (Paige and Bayless 1981).

There is overwhelming scientific evidence that dietary carbohydrates, and sugars in particular, contribute to dental caries (Newbrun, 1982). However, recent findings from both laboratory animal and human dental plaque investigations suggest that dairy foods, including milk and certain cheeses, may protect against caries (Silva et al. 1986; Morrissey et al. 1984; Schachtele and Harlander 1984; Jensen and Schachtele 1983; Edgar et al. 1982). In laboratory rats, for example, large reductions in the incidence of dental caries occurred when cheese intake followed selected exposure to a high-sucrose diet (Edgar et al. 1982). Likewise, human dental plaque studies have shown that at least seven cheeses—aged Cheddar, Swiss, Blue, Monterey Jack, Mozzarella, Brie, and Gouda—produce little or no plaque acid (Schachtele and Harlander 1984). Moreover, three of the cheeses—aged Cheddar, Monterey Jack, and Swiss—effectively prevented sucrose-induced changes in plaque pH when consumed 30 minutes before sucrose intake. The precise mechanism or protective factor responsible for the cariostatic action of cheese and milk is yet to be established.

VITAMINS

Water-Soluble Vitamins

All of the vitamins known to be essential for humans have been detected in milk, the water-soluble vitamins being present in the nonfat portion (Lampert 1975). In general, the concentration of water-soluble vitamins in milk is relatively constant and is little affected by the diet of the cow. For a comprehensive discussion of the factors (e.g., breed, season) influencing the vitamin content of milk and milk products, refer to Gregory (1975), Hartman and Dryden (1965), and Webb *et al.* (1974). Table 7.8 lists the average amounts of water-soluble vitamins in milk. In cheeses, the vitamin content varies widely, depending on such factors as the vitamin content of the milk used, the manufacturing process, the cultures or microorganisms used, and the conditions and duration of the curing period. A portion of the water-soluble vitamins of milk is lost in the whey which is removed during the cheese-making operation. However, more of the water-soluble riboflavin and thiamin of milk remains in the curd than might be expected. In general,

Table 7.8. Water-Soluble Vitamins in Milk[a]

Vitamin	Per gram MSNF	Per 100 g Fluid whole milk
Riboflavin (mg)	0.0187	0.162
Vitamin B_{12} (μg)	0.0412	0.357
Niacin equivalents (mg)[b]	0.0987	0.856
Thiamin (mg)	0.0044	0.038
Vitamin B_6 (mg)	0.0048	0.042
Ascorbic acid (mg)	0.108	0.94
Folacin (μg)	0.6	5
Pantothenic acid (mg)	0.0362	0.314

[a]The nutrient values are based on whole milk with 8.67% MSNF. The *Handbook 8-1* (USDA 1976) includes the standard error of the mean.
[b]This value includes niacin equivalents from preformed niacin and from tryptophan. A dietary intake of 60 mg tryptophan is considered equivalent to 1 mg niacin. One niacin equivalent is equal to either to those amounts.
SOURCE: USDA (1976). Adapted from National Dairy Council. 1983A. *New Knowledge of Milk and Other Fluid Dairy Products.* Rosemont, Ill. With permission.

the more whey retained in the cheese, the greater the content of these water-soluble vitamins in cheese. Generally, any cheese variety high in one B vitamin is high in most of the other B vitamins.

The bacterial surface-ripened and mold-ripened cheese varieties (e.g., Limburger, Camembert, blue, and Roquefort cheeses) may contain a higher concentration of the B-complex vitamins than the hard and semihard types of cheese (e.g., Cheddar, Swiss, Mozzarella). In the bacterial surface-ripened varieties of cheese, the B-complex vitamins can be synthesized by the surface-ripening microorganisms during curing. In soft-ripened cheeses such as Brie and Camembert, the outer layers can show an increase in several of the B-complex vitamins, whereas little change in vitamin content is evident in the center portion. Cheeses in which proteolysis is extensive (soft-ripened and semisoft types) can have a higher content of the B-complex vitamins than hard and soft unripened types (NDC 1983B). Values for the vitamin content of specific cheeses and cheese products are provided by the Consumer and Food Economics Institute, USDA (1976).

Of all of the water-soluble vitamins in milk, riboflavin (vitamin B_2) is present in greatest concentration. As a component of two coenzymes, flavin mononucleotide (FMN) and flavin adenine dinucleotide (FAD), riboflavin is involved in the oxidation of glucose, fatty acids, amino acids, and purines in the body. Basically, riboflavin functions as the reactive portion of these flavoproteins that serve essentially as electron carriers. Milk and milk products are excellent sources of riboflavin. As shown in Table 7.1, the milk group contributes 34.7% of the riboflavin available for civilian consumption in the United States. The average content of riboflavin in milk is 0.16 mg/100 g. Riboflavin is relatively heat-stable and is only slightly, if at all, affected by the heat of pasteurization of milk. This vitamin, however, is sensitive to light, and appreciable losses can occur depending upon light intensity, illuminance and wavelength, distance from the light source, amount of exposed surface area, surface-to-volume ratio, duration of exposure, temperature and fat content of the milk, and the packaging material (Bradley 1980; Senyk and Shipe 1981; DeMan 1981; Hedrick and Glass, 1975).

In general, a greater potential for riboflavin loss in milk occurs with more intense light, longer exposure, higher temperature during storage, and a lower fat content. Of the common packaging materials used, paperboard or fiberboard and opaque plastic afford greater protection against light than clear glass or translucent plastic containers. The riboflavin content decreased by 14% in plastic containers and by 2% in paperboard containers under experimental conditions when skim milk was exposed for 24 hr to fluorescent light at a high intensity (2000 lumens/m^2) (Senyk and Shipe 1981). In another study, riboflavin in whole milk packaged in clear plastic containers and held for 10 hr under fluorescent light similar to that in the dairy case declined by 7%, whereas that in milk packaged in paperboard containers declined by 4% (Hedrick and Glass 1975). The greater destruction of riboflavin in clear plastic containers in the first study (14% loss versus 7% loss in the second study) can be explained by the increased duration and intensity of light exposure and the lower fat content of the milk.

Under experimental or laboratory conditions, the intensity and duration of exposure to fluorescent light may be more extreme than those generally encountered in the retail dairy case. The real question is whether there is a meaningful nutritional difference in the riboflavin content of milk packaged in clear plastic versus paperboard at the retail level. Reif *et al.* (1983) analyzed the riboflavin content of milk in paperboard or plastic containers purchased in California retail outlets. They showed that the average riboflavin content (1.5 mg/liter) was similar to that reported in the literature and was unaffected by the type of packaging material.

Milk and milk products are good sources of vitamin B_{12}, a vitamin necessary for growth, maintenance of nerve tissues, and the formation and development of red blood cells. The milk group contributes 20.1% of the vitamin B_{12} available for civilian consumption in the United States (Table 7.1). The average content of vitamin B_{12} in milk is 0.36 $\mu g/100$ g. Because vitamin B_{12} is found almost exclusively in foods of animal origin, vegetarians who consume no animal products (i.e., vegans) are at risk of developing vitamin B_{12} deficiency. On the other hand, for lacto-ovo vegetarians who include dairy foods and eggs in their diet, milk and dairy foods generally supply most of their need for this vitamin (The American Dietetic Association 1980). Pasteurization causes only a slight, if any, destruction of vitamin B_{12} in milk.

Niacin, a water-soluble vitamin vital for oxidation by living cells, functions in the body as a component of two important coenzymes: nicotinamide adenine dinucleotide (NAD) and nicotinamide adenine dinucleotide phosphate (NADP). NAD and NADP are involved in the release of energy from carbohydrate, fat, and protein, and in the synthesis of protein, fat, and pentoses for nucleic acid formation. Milk is a poor source of preformed niacin, containing about 0.08 mg per 100 g. However, milk's niacin value is considerably greater than indicated by its niacin content (Horwitt *et al.* 1981). Not only is the niacin in milk fully available, but the amino acid tryptophan in milk can be used by the body for the synthesis of niacin. For every 60 mg of tryptophan consumed, the body synthesizes 1 mg of niacin. Therefore, the niacin equivalents in 100 g milk equal 0.856 mg including that from preformed niacin and that from tryptophan. Pellagra, a disease caused by niacin deficiency, can be prevented or cured by consuming milk. Niacin is stable in foods, and its content in milk is not reduced by storage, exposure to light, or various heat treatments such as pasteurization.

Thiamin (vitamin B_1), as part of the coenzyme thiamin pyrophosphate, takes part in biochemical reactions involving the metabolism of carbohydrate. Related roles of thiamin include maintenance of neurological function, normal appetite, good digestion, muscle tone, and growth. An average of 0.04 mg thiamin is found in 100 g milk. Pasteurization results in about a 10% loss of the thiamin originally present in milk. The amount of destruction of thiamin in milk increases directly with the frequency and severity of the heat treatment, independent of the presence of oxygen.

"Vitamin B_6" is a collective term for three naturally occurring pyridines: pyridoxine, pyridoxal, and pyridoxamine. Phosphates of pyridoxal and pyridoxamine serve as coenyzmes involved in amino acid and protein synthesis. Vitamin B_6 is also needed for the metabolism of unsaturated fatty acids, especially linoleic and arachidonic acids. The content of vitamin B_6 in milk varies considerably, but on the average, 100 g milk contain 0.04 mg vitamin B_6. No significant decreases in vitamin B_6 occur during the pasteurization of milk, although prolonged exposure to sunlight and ultraviolet irradiation may significantly destroy vitamin B_6 activity.

Ascorbic acid (vitamin C), the antiscurvy vitamin, is used by the body to form and maintain intercellular and skeletal material such as the collagen of fibrous tissue and the matrix of bone, dentin, and cartilage. Milk and milk products are not considered a significant source of this vitamin and should not be relied upon as such. Freshly drawn cow's milk contains about 2 mg ascorbic acid per 100 g milk, but as vitamin C is heat labile and easily destroyed by oxidation, the vitamin C content of pasteurized milk is reduced to about 0.94 mg/100 g.

"Folacin," a generic term comprising folic acid and its derivatives, functions as a coenzyme for the transfer of one-carbon units in nucleic acid and amino acid metablism (Anderson and Talbot 1981). Milk contains an average of 5 μg folacin per 100 g. There appears to be a protein in milk that combines with folate to enhance the absorption of the vitamin, at least in *in vitro* studies (Colman *et al.* 1981; Anon. 1982A). Research is needed to identify this specific folic acid–binding protein and to determine its functional significance. Pasteurization can result in a 0 to 12% loss of folacin in milk; the amount of heat-induced destruction can be minimized by excluding oxygen (Webb *et al.* 1974).

Pantothenic acid, as a component of coenzyme A, is involved with the release of energy during gluconeogenesis, in the synthesis and degradation of fatty acids, and in the synthesis of sterols, steroid hormones, and acetylcholine, among other compounds. Pantothenic acid occurs in milk at a substantial level, averaging 0.31 mg/100 g. Pasteurization of milk has little or no effect on its pantothenic acid content.

Biotin is involved in many carboxylation and decarboxylation reactions in carbohydrate, fatty acid, protein, and nucleic acid metabolism. Milk is a fairly good source of this vitamin, generally providing about $3\mu g/100$ g. Pasteurization has a minimal effect on the biotin content of milk.

In addition to the above water-soluble vitamins, other compounds are present in milk, such as choline, myo-inositol, and para-aminobenzoic acid, compounds for which there is no proof of a dietary requirement for humans (NAS 1980A).

Fat-Soluble Vitamins

Vitamins A, D, E, and K are associated with the fat component of milk. Milk is considered to be a major source of vitamin A and, as a result of fortification, of vitamin D, but contains only small amounts of vitamin E and K. Unlike water-soluble vitamins, which are relatively stable in concentration in milk, the fat-soluble vitamin content is greatly influenced by the dietary intake of the cow and, to a lesser extent, by other factors such as breed (Lampert 1975). Table 7.5 includes the fat-soluble vitamin content of fluid whole milk. In cheesemaking, most of the fat in milk is retained in the curd. Hence, the cheese product contains most of the fat-soluble vitamins of the milk employed in its manufacture. For example, cheese such as Cheddar, which is made from whole milk, is a good source of vitamin A activity (303 retinol equivalents or 1059 IU/100 g), whereas cheese made with skim milk, such as cottage cheese (dry curd), contains a comparatively smaller amount of this vitamin (8 retinol equivalents or 30 IU/100 g). During ripening or storage, there is very little change in the vitamin A content of cheese (NDC 1983B).

Vitamin A is necessary for growth and reproduction, resistance to infection, maintenance and differentiation of epithelial tissues, stability and integrity of membrane structures, and the process of vision. In terms of the last function, vitamin A is a component of rhodopsin or visual purple, a photosensitive pigment in the eye that is needed for vision in dim light. An early mild clinical symptom of vitamin A deficiency is night blindness; a severe deficiency of this fat-soluble vitamin results in xerophthalmia, an eye condition leading to blindness.

In food, vitamin A exits in two forms: preformed vitamin A (retinol and retinyl esters) in animal products and pro-vitamin A carotenoids in plants. The latter are converted primarily to vitamin A (retinol) in the intestine. Of the pro-vitamin A carotenoids, β-carotene has the greatest biological activity. Both vitamin A (retinol) and carotenoids are present in high but variable quantities in the fat portion of milk. The carotenoids are the yellow pigments that give milk its characteristic creamy color. The proportion of carotenoids in milk varies from 11 to 50% of the vitamin A (retinol) value, depending upon the breed and ration of the cow and the season of the year, among other factors (Hartman and Dryden 1965). For example, the cow's intake of carotenoids, which are abundant in plant feed, generally is greater during summer months than during winter months; thus the fraction of total vitamin A activity due to carotenoids is greater in summer milk.

Milk and milk products are an important dietary source of vitamin A, providing about 11.6% of the vitamin A in the U.S. food supply (Table 7.1). Milk contains about 126 IU (31 retinol equivalents) vitamin A per 100 g (Table 7.5). Pasteurization does not affect the vitamin A content of milk (Lampert 1975). However, exposure to sunlight or artificial light can result in losses of vitamin A, especially vitamin A (retinyl palmitate) added to lowfat and skim milks. As with riboflavin, many factors influence light-induced changes in milk's vitamin A content. These include light intensity, illuminance and wavelength, distance between the light source and the milk container, container material, amount of exposed surface area, surface-to-volume ratio, temperature and duration of exposure, and fat content of the milk (Senyk and Shipe 1981; DeMan 1981).

Under scientific conditions simulating in-store conditions, losses of vitamin A occurred in lowfat milk packaged in conventional containers (Senyk and Shipe 1981). For example, after 24 hr of exposure to high-intensity fluorescent lighting, 90% of added vitamin A was lost from skim milk in clear plastic containers compared with 15% from milk in paperboard containers (Senyk and Shipe 1981). However, under more realistic conditions, such as those in the dairy case (i.e., 4–6 hr of storage at a lower fluorescent light intensity), little change in the vitamin A content of lowfat milk was evident, regardless of the type of container (Bruhn et al. 1984).

As vitamin A and carotene are in the fat portion of milk, the vitamin A activity is removed with the milk fat during separation into cream and lowfat and skim milks. Consequently, standards of identity established by the U.S. Food and Drug Administration (FDA) mandate the addition of vitamin A (e.g., retinyl palmitate) to fluid lowfat and skim milks and to nonfat dry milk to a level approximating that found in whole milk from cows on summer pasture. That is, at least 2000 IU of vitamin A must be present in each quart of lowfat and skim milk (FDA

1973). The fortification of these milk products with vitamin A is endorsed by the American Medical Association, with the concurrence of the Food and Nutrition Board, National Academy of Sciences, National Research Council and the Expert Panel on Food Safety and Nutrition of the Institute of Food Technologists (AMA 1982). The fortification of dried skim milk with vitamin A is viewed by the World Health Organization and the Food and Agricultural Organization (WHO 1977) as an important measure to combat vitamin A deficiency in developing countries, where 20,000 to 100,000 children yearly develop blindness from a lack of vitamin A in their diets (DeLuca et al. 1979).

Vitamin D, by current definitions, can be considered both a vitamin and a hormone (DeLuca 1981). In terms of function, vitamin D plays a central role in calcium and phosphorus homeostasis in the body by stimulating the absorption of these minerals in the intestine and promoting normal mineralization of newly forming bone (Parfitt et al. 1982). Unfortified cow's milk traditionally has been regarded as a poor source of vitamin D, supplying 5 to 35 IU/liter (Leerbeck and Sondergaard 1980). Generally, the amount of vitamin D in milk is greater during summer months than during winter months (Lampert 1975). The suggestion that the vitamin D content of milk might have been substantially underestimated appeared in the late 1960s and mid-1970s with reports that a large quantity of water-soluble vitamin D sulfate, a conjugated form of vitamin D, was present in the whey fraction of milk (i.e., 204 μg/liter) (Leerbeck and Sondergaard 1980). According to several recent investigations, however, it is evident that the amount of vitamin D sulfate in bovine milk whey is insignificant; furthermore, synthesized vitamin D sulfate has negligible biological activity (Hollis et al. 1981, 1982; Anon. 1982B; Reeve et al. 1982).

Although unfortified cow's milk contains only small amounts of vitamin D, milk lends itself well to vitamin D fortification (NAS 1980A). While fortification is optional, approximately 98% of fluid milk marketed in the United States is fortified with vitamin D to obtain standardized amounts of 400 IU or 10 μg/quart (FDA 1973). Vitamin D fortification of milk has been largely responsible for the virtual elimination of rickets in the United States (AMA 1955; Gallagher and Riggs 1978; DeLuca 1978). Moreover, vitamin D fortification of fluid whole milk, as well as other milk products such as lowfat and skim milks and nonfat dry milk, is endorsed by the American Academy of Pediatrics, Committee on Nutrition (1967), and the American Medical Association, with the concurrence of the Food and Nutrition Board, National Academy of Sciences, National Research Council and the Expert Panel on Food Safety and Nutrition of the Institute of Food Technologists

(AMA 1982). For persons susceptible to hypovitaminosis D, such as those who lack exposure to sunlight (e.g., the elderly), individuals consuming lowfat diets, and vegans, inclusion of fortified vitamin D milk in the diet is particularly important.

"Vitamin E" is a generic term for a number of compounds made by plants, the tocopherols and tocotrienols. Of the eight naturally occurring tocopherols, α-tocopherol has the greatest biological activity for humans (DeLuca 1978; Machlin 1980; Scott 1980). It is generally accepted that vitamin E functions in cellular and subcellular membranes as a biological antioxidant or free radical scavenger, thus stabilizing cellular membranes. Polyunsaturated fatty acids, which are predominant in cellular membranes, are liable to interact with active oxygen via a free radical pathway, ultimately leading to tissue damage. Vitamin E is viewed as the first line of defense against peroxidation of lipids within membranes. Vitamin E protects vitamine A, carotene, and polyunsaturated fatty acids from oxidation in the body.

In cow's milk, nearly all of the vitamin E is α-tocopherol and the level can vary with the cow's feed and the season of the year (Lampert 1975). For example, summer milk can contain five times more vitamin E (1.1 mg α-tocopherol per quart) than winter milk (0.2 mg/quart) (Hertig and Drury 1969; McLaughlin and Weihrauch 1979). It is suggested that vitamin E, due to its antioxidant properties, may have some effect in retarding the development of oxidized flavor in milk (Lampert 1975).

Vitamin K is required for the synthesis of blood clotting factors, prothrombine (factor II), and factors VII, and IX, and X. Identification of vitamin K–dependent proteins in tissues such as bone (e.g., "osteocalcin") implies that vitamin K may have functions in addition to that in blood clotting (Suttie 1980). The vitamin K_1 (phylloquinone) content of pasteurized Friesian (Holstein) cow's milk has been reported to range from 3.6 to 8.9 μg/liter (mean, 4.9 μg/liter), as measured by high-performance liquid chromatography (Haroon et al. 1982). This level is significantly higher than that in human milk (mean, 2.1 μg/liter). The vitamin K_1 content of cow's milk is influenced by the breed of the cow but not by boiling of the milk (Haroon et al. 1982).

MINERALS

Minerals can be classified into two groups according to the amounts needed in the daily diet. The first group is composed of macrominerals (i.e., those needed at levels of 100 mg or more daily), of which calcium, phosphorus, and magnesium are of importance in milk. The second group consists of trace elements which are neded in much smaller quantities, generally a few milligrams or less daily (NAS 1980A).

Macrominerals

Calcium, which is essential for bone mineralization, as well as for other vital physiological processes, is the most abundant mineral in the body, comprising 1.5 to 2.0% of an adult's body weight (NDC 1984A,B; Albanese et al. 1978; Albanese 1977). About 99% of the body's calcium is found in bones—skeletal bones supporting the body and alveolar bones in the jaw supporting the teeth. The remaining 1% of calcium that exists outside bone in extracellular fluids and soft tissues is necessary for the transmission of nerve impulses, contraction of muscles, blood coagulation, and several enzymatic and secretory processes (Schaafsma 1983).

Throughout life, bone is constantly being formed and resorbed. The process, called "bone remodeling," occurs more rapidly during early life and at a declining rate with advancing age (Albanese et al. 1978; Albanese 1977; Chinn 1981; Heaney et al. 1982; Spencer et al. 1982A). Thus, an adequate intake of calcium is necessary not only during the years of skeletal growth and bone consolidation but also thereafter to maintain optimal bone integrity. Milk and other dairy foods are the major sources of calcium in the diet (Feeley et al. 1972A). As shown in Table 7.1, the milk group furnishes 75.8% of the calcium available for civilian consumption in the United States. Table 7.9 shows the macromineral content of selected milk and milk products.

Milk is one of the best dietary sources of calcium, not only because of the significant quantity of the mineral present but also because of (1) its calcium-to-phosphorus ratio (1.3:1), which is conducive to optimal skeletal growth, and (2) the presence of nutrients such as lactose and vitamin D in vitamin D–fortified milk, which promote calcium absorption (Ziegler and Fomon 1983; Schaafsma 1983; Renner 1983).

The bioavailability of calcium from dairy foods is considered to be excellent (Schaafsma 1983). Evidence from animal studies suggests that the form of calcium in dairy foods may influence the bioavailability of this mineral (Wong and LaCroix 1980). For example, dairy foods that contain colloidal calcium phosphate or calcium caseinate (e.g., as in Cheddar cheese) appear to be somewhat better sources of calcium than foods that contain ionic calcium (e.g., yogurt, buttermilk). However, calcium in milk and other milk products is of greater bioavailability to humans than calcium found in other food sources. According to Renner (1983), calcium utilization from skim milk powder is 85% compared with 22–74% from vegetables. Dietary fiber in plant cell

Table 7.9. Macrominerals in Milk and Milk Products.

Product	Macrominerals (mg/100 g)				
	Calcium	Phosphorus	Magnesium	Sodium	Potassium
Whole milk	119	93	13	49	152
Lowfat (2%) milk	122	95	14	50	154
Half-and-half	105	95	10	41	130
Light cream	96	80	9	40	122
Buttermilk, cultured, fluid	116	89	11	105	151
Sour cream	116	85	11	53	144
Yogurt, plain, lowfat	183	144	17	70	234
Cheddar cheese	728	518	29	628	100
American, pasteurized process	621	753	21	1449	164
Gouda	707	553	29	828	121
Mozzarella, low-moisture, part skim	739	532	25	535	96
Swiss	971	610	36	264	111
Cottage, creamed	60	131	5	402	84
Cottage, uncreamed	32	104	4	12	33

SOURCE: USDA (1976).

walls such as uronic acids or sodium alginate impair calcium bioavailability (Allen 1982). Oxalic acid in certain foods such as chocolate, rhubarb, spinach, and chard binds calcium, thereby limiting the bioavailability of this mineral. However, if calcium intake is adequate, the formation of insoluble oxalate salts is of little nutritional significance (Allen 1982). In particular, concern regarding the reduced availability of calcium in chocolate milk is unfounded (Mitchell and Smith 1945; Bricker et al. 1949). Calculations show that only about 2 mg of the 280 mg calcium in an 8-oz (250-g) serving of chocolate milk are rendered unavailable by the oxalic acid present in the chocolate.

Several government surveys reveal that calcium is one nutrient likely to be consumed in less than recommended amounts (i.e., 800 mg RDA for nonpregnant, nonlactating women) by a substantial percentage of Americans, particularly females 12 years of age and older (Carroll et al. 1983; Heaney et al. 1982; Chinn 1981; USDA, SEA 1980; Radzikowski 1983). As shown in Figures 7.1 and 7.2, daily calcium intake values reported in the Second National Health and Nutrition Examination Survey, 1976–1980 (HANES II), have been plotted as a function of age for males and females, respectively, and are compared to the RDA. Although the RDA is the same for both sexes, it is clear

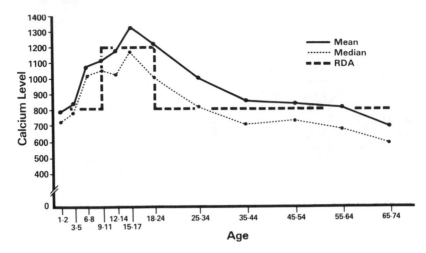

Figure 7.1. Daily calcium intake (mg) for males (U.S. Population 1976–1980). (Carroll *et al.* 1983.)

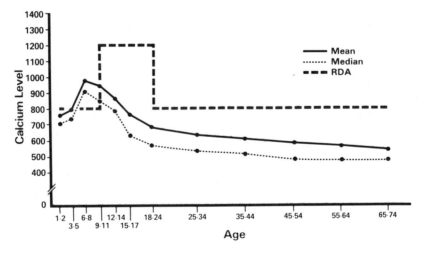

Figure 7.2. Daily calcium intake (mg) for females (U.S. population, 1976–1980). (Carroll *et al.* 1983.)

that the average U.S. male consumes up to twice as much calcium as the U.S. female of the same age, the greatest difference being between ages 15 and 50 (Heaney *et al.* 1982). The mean daily calcium intake for males is at or above the RDA from years 12 through 64, and 50–75% of men between the ages of 18 and 34 years consume the recommended amounts of calcium. In sharp contrast, the mean daily calcium intake for females does not exceed 85% of the RDA after age 12. During the years of peak bone mass development (18 to 30), more than 66% of all U.S. women fail to consume the recommended amounts of calcium on any given day; after age 35, this percentage increases to over 75% (Heaney et al. 1982; Carroll et al. 1983).

The consequence of inadequate calcium intake, particularly over prolonged periods of time, is speculative. However, there is considerable scientific evidence to support the view that diet, particularly a sustained low-calcium intake, is one of several contributing factors to osteoporosis (Allen 1986; Consensus Development Panel 1984; NDC 1984A,B; Heaney et al. 1982; National Institute of Arthritis and Musculoskeletal and Skin Diseases 1986; The American Society for Bone and Mineral Research 1982; Spencer et al. 1984). Osteoporosis is regarded as one of the most common, but most poorly understood, debilitating disorders of the elderly, especially postmenopausal women (Allen 1986; Consensus Development Panel 1984; NDC 1984A,B, 1982B; DeLuca et al. 1981; Albanese 1977). Age-related bone loss is characterized by a reduction in the amount of bone present in the skeleton, leading in many cases to bone fractures and breaks (Chinn 1981; DeLuca 1981). A calcium intake of 800 mg, the RDA for most adults, may not be adequate, particularly for individuals who do not adapt readily to varying calcium intakes (Spencer et al. 1984) or for those who ingest high-protein diets which lead to increased urinary calcium excretion (Heaney et al. 1982; Allen 1982; Marcus 1982). Also, some medications negatively influence calcium absorption and utilization (Spencer et al. 1982A,B). A calcium intake of 1000 to 1500 mg may be necessary to ensure a favorable calcium balance and optimal bone density in practically all healthy persons in the United States (National Institute of Arthritis and Musculoskeletal and Skin Diseases 1986; Consensus Development Panel 1984; The American Society for Bone and Mineral Research 1982; Heaney et al. 1977, 1982; Chinn 1981; Allen 1982; Marcus 1982; Spencer et al. 1984).

It is hypothesized that a prolonged low dietary intake of calcium may be one of several factors contributing to loss of alveolar (jaw) bone which supports the teeth (NDC 1984B; Rogoff et al. 1984; Albanese 1983; Daniell 1983). Loss of alveolar bone may accelerate periodontal disease and therefore contribute to tooth loss (Rogoff et al. 1984; Albanese 1983). Moreover, continued loss of alveolar bone after tooth loss can lead to unstable or poor-fitting dentures, which not only adversely affects nutrient intake but may also exacerbate alveolar bone loss (Rogoff et al. 1984).

Some researchers have suggested that a relationship exists between osteoporosis and alveolar bone density (Albanese 1983; Daniell 1983; Kribbs et al. 1983). According to Albanese (1983), alveolar bone density may serve as an early indicator of osteoporosis. A recent study indicates that osteoporosis may contribute to adult tooth loss. In this study, women with osteoporosis in their sixties required dentures three times as frequently after age 50 as nonosteoporotic women (Daniell 1983). The author suggests that therapeutic measures to prevent or reduce osteoporosis, such as an increase in calcium intake, may be beneficial in preserving the bone that anchors the teeth as well. Kribbs et al. (1983) also have shown a significant correlation between skeletal osteopenia and mandibular density in postmenopausal women. These authors suggest that a calcium intake greater than 800 mg/day may be required to prevent bone loss. Although the relationship of calcium nutrition, alveolar bone loss, and periodontal disease remains unclear, it appears that progressive loss of bone from the alveolar ridge may be a manifestation of osteoporosis resulting from a diet low in calcium (Rogoff et al. 1984; Daniell 1983). Maintaining a long-term positive calcium balance may be expected to prevent or slow the rate of alveolar bone loss, thereby minimizing periodontal disease and subsequent tooth loss.

A relatively new and promising area of research concerns the role of inadequate dietary calcium in the development of essential hypertension or high blood pressure (Villar et al. 1986; Karanja and McCarron 1986; Resnick 1985; NDC 1984A,B; McCarron 1985, 1983, 1982; McCarron et al. 1982). While most reports relating diet to hypertension have emphasized sodium, it appears that only a small proportion of the U.S. population is genetically sodium sensitive and that for the majority, dietary sodium intake has little effect on blood pressure. As discussed below, inadequate calcium intake, either alone or in combination with other factors, appears to predispose to high blood pressure by a mechanism(s) as yet unknown.

A number of observations both in experimental animals and in humans are suggestive of a crucial role for dietary calcium in regulating blood pressure (Villar et al. 1986; Karanja and McCarron 1986; McCarron et al. 1984, 1982; McCarron 1985, 1983, 1982; Belizan et al. 1983A,B). For example, laboratory rats fed insufficient calcium experience a rise in blood pressure. Increasing the calcium intake either decreases blood pressure or reduces the development of high blood pressure in animals predisposed to hypertension (Karanja and McCarron 1986). Furthermore, disturbances in calcium metabolism, including a depression in serum ionized or free calcium, have been reported in spontaneously hypertensive rats, as well as in untreated patients with hypertension.

Data from several epidemiological studies indicate that hypertension is more prevalent in populations characterized by a low calcium than a high calcium intake (Villar *et al.* 1986; Karanja and McCarron 1986). In a pilot nutritional survey, the calcium intake of hypertensive individuals (668 ± 55 mg) was signficantly less (22%) than that reported by normotensive controls (886 ± 89 mg) (McCarron *et al.* 1982). Except for calcium, the diets of the two groups were similar. Likewise, data from HANES I show that hypertensive persons consume 18% less dietary calcium than normotensive individuals (McCarron 1983; McCarron *et al.* 1984). According to information collected from this survey, a reduction in the intake of dairy foods is the food behavior related most closely to high blood pressure in the United States (McCarron *et al.* 1984).

Support for the animal and epidemiological evidence of an inverse relationship between calcium intake and blood pressure comes from clinical trials in which blood pressure is reduced by adding calcium to the diet (Johnson *et al.* 1985; McCarron and Morris 1985; Belizan *et al.* 1983A,B). For example, McCarron and Morris (1985) found that systolic blood pressure decreased by 10 mm Hg or more in 44% of hypertensive and 19% of normotensive persons who had received 1000 mg calcium per day for eight weeks in a randomized, double-blind, placebo-controlled crossover trial. Similarly, in another study, systolic blood pressure decreased by 13 mm Hg in hypertensive women receiving 1500 mg calcium per day, whereas it increased by 7 mm Hg in women who did not receive this extra calcium (Johnson *et al.* 1985). While much remains to be learned about the relationship between calcium and blood pressure, and in particular about the possible therapeutic effect of increased calcium intake for hypertensives, data to date suggest that persons at risk for developing high blood pressure should consume enough calcium to at least meet the current RDA (McCarron 1983).

The observed association between calcium intake and hypertension leads to an important consideration with respect to current dietary recommendations emphasizing a restriction of dietary sodium to protect against hypertension and its sequelae (McCarron *et al.* 1982; Engstrom and Tobelmann 1983). A risk in reducing the intake of foods contributing sodium to the diet (e.g., foods in the meat, grain, and milk groups) is that the intake of other essential nutrients such as calcium, iron, magnesium, and vitamin B_6, nutrients which are already consumed at levels below the RDA by a significant proportion of the population, may be further reduced (Engstrom and Tobelmann 1983). If such a restriction in sodium intake decreases the dietary intake of calcium, hypertension may be aggravated instead of alleviated (McCarron *et al.* 1982).

Calcium and other factors in dairy foods may also play a role in protecting against colorectal cancer (Garland and Garland 1986; Garland *et al.* 1985; Lipkin and Newmark 1985). Garland *et al.* (1985) observed a strong inverse correlation between intake of vitamin D and cancer and the later development of colorectal cancer in a 19-year prospective study of men in the Chicago area. Moreover, findings of a recent clinical trial showed that increasing dietary calcium by 1250 mg/day for two-to-three months suppressed epithelial cell proliferation in the colonic mucosa of subjects at high risk of developing colon cancer (Lipkin and Newmark 1985). Abnormal cellular proliferation is a hallmark of neoplasia. While further investigation is necessary to substantiate the protective effect of calcium against colorectal cancer and to determine the mechanism involved, the findings to date are important, considering that colorectal cancer affects about 6% of the U.S. population.

Phosphorus, in combination with calcium, is important for bone mineralization, as well as for many chemical reactions in the body (NAS 1980A). Like calcium, most of the body's phosphorus (80–90%) is found in bones. The remaining 10 to 20% exists as soluble phosphate in blood, cells, lipids, proteins, carbohydrates, and energy-transfer enzymes (NAS 1980A).

A dietary deficiency of phosphorus is unlikely, as this mineral occurs in nearly all foods and many food additives (NAS 1980A; Greger and Krystofiak 1982). Foods in the milk group, for example, contribute 35.8% of the phosphorus available for civilian consumption (Table 7.1). Adults generally consume about twice as much phosphorus as the RDA of 800 mg (NAS 1980A; Greger and Krystofiak 1982).

A potential concern has been the dietary ratio of calcium to phosphorus (Ca:P) in relation to bone health. Based on data from animal studies and on the relative calcium content in bone, a dietary Ca:P ratio of 1:1 to 2:1 is recommended as beneficial for bone mineralization in humans (NAS 1980A; Chinn 1981; Linkswiler and Zemel 1979). The Ca:P ratio in cow's milk (1.3:1) closely approximates that found in bones. The average American diet is estimated to contain a Ca:P ratio of 1:1.6 (Chinn 1981; Greger and Krystofiak 1982) to 1:3 (Linkswiler and Zemel 1979), and if no dairy foods are consumed, it may be as low as 1:4.

There is some evidence, mostly from animal studies, to suggest that high dietary levels of phosphorus, especially if dietary levels of calcium are low, may adversely affect bone mass and calcium metabolism (Greger and Krystofiak 1982). However, in humans there is little direct evidence to indicate that large variations in dietary phosphorus or in the Ca:P ratio have any significant influence on calcium utilization or balance (Heaney *et al.* 1982). Some preliminary findings, however, suggest that the form of phosphorus may influence calcium absorption (Zemel *et al.* 1982). Hexametaphosphate, as compared with orthophosphate, decreased calcium absorption when human subjects consumed either low- or high-calcium diets (Zemel *et al.* 1982).

Magnesium, an essential mineral for humans, is closely related in location and function to both calcium and phosphorus. Similar to calcium, a large fraction of the body's magnesium is located in bones. Magnesium is involved in phosphate transfer systems and is essential for energy-requiring biological functions such as membrane transport, generation and transmission of nerve impulses, contraction of muscles, and oxidative phosphorylation. As part of many enzyme systems, magnesium participates in the synthesis of protein from amino acids, as well as in lipid and carbohydrate metabolism (NAS 1980A). A deficiency of magnesium is rare, as this mineral is widely distributed in foods (NAS 1980A; Greger *et al.* 1978). Milk is considered a good source of magnesium, containing 13 mg/100 g (Table 7.9) or providing 19.1% of the magnesium available in the U.S. diet (Table 7.1).

Trace Elements

"Trace elements" are defined as those elements occurring in "trace" concentrations (microgram per gram or parts per million) in biological materials (Mertz 1981; Underwood 1977; Prasad 1978). Those shown to be either essential or beneficial for higher animals include arsenic, chromium, cobalt, copper, fluorine (or fluoride), iodine, iron, manganese, molybdenum, nickel, silicon, selenium, vanadium, and zinc. An RDA for humans has been determined for iodine, iron, and zinc, and ranges of "estimated safe and adequate daily dietary intakes" are established for chromium, copper, fluoride, manganese, molybdenum, and selenium (NAS 1980A). Safe and adequate ranges of intake are amounts considered at the lower end of the range to meet nutrient needs or prevent deficiencies and at the upper end to be below known toxic doses (NAS 1980A). The range concept takes into account not only incomplete knowledge of requirements but also nutrient interactions which influence dietary requirements. Pronounced deficiencies in humans have been described for iron, iodine, and zinc and to a lesser degree for chromium, copper, and selenium in various areas of the world. In the United States, marginal deficiencies and toxicities of trace elements are of particular concern, although difficult to assess due to the lack of sensitive diagnostic tests.

Although new analytical techniques with greater sensitivity have been employed to determine the levels of trace elements in dairy foods, the values obtained must be considered approximate. The trace element content of milk and other dairy foods can vary as a result of the stage of lactation, season, milk yield, amount of trace element in the cow's ration, handling of the milk following pasteurization, storage conditions, and methods and accuracy of analysis (Wong *et al.* 1978; Murthy 1974; Lonnerdal *et al.* 1981; Hegarty 1981). Representative concentrations of trace elements in milk are shown in Table 7.10. In general, the levels of trace elements in milk and milk products are low (Jarrett 1979). However, iron, iodine, and zinc in milk deserve special comment.

Table 7.10. Trace Elements in Milk.

Trace elements	μg per gram MSNF[a]	μg per 100 g Fluid whole milk
Aluminum	5.35	46.0
Arsenic[b]	0.58	5.0
Barium	—[c]	—[c]
Boron[b]	3.14	27.0
Bromine	6.98	60.0
Bromine (coastal area)	32.56	280.0
Cadmium	0.30	2.6
Chromium[d]	0.17	1.5
Cobalt[b, d]	0.007	0.06
Copper[d]	1.51	13.0
Fluoride[b, d]	1.74	15.0
Iodine[b, d]	0.50	4.3
Iron[b, d]	0.006	0.0492
Lead[b]	0.47	4.0
Lithium[b, d]	—[c]	—[c]
Manganese[b, d]	0.26	2.2
Molybdenum[b, d]	0.85	7.3
Nickel	0.31	2.7
Rubidium[b]	23.26	200.0
Selenium (nonseleniferous area)[b, d]	0.47	4.0
Selenium (seleniferous area)[b, d]	14.77	up to 127.0
Silicon	16.63	143.0
Silver	0.55	4.7
Strontium	1.99	17.1
Tin	—[c]	—[c]
Titanium	—[c]	—[c]
Vanadium	0.0011	0.0092
Zinc	44.0	381.1

[a]Calculated for fluid whole milk containing 8.67% SNF.
[b]Effect of feed supplement.
[c]Dashes denote qualitative data; therefore, it is difficult to assign a specific value.
[d]Trace elements for which the present state of knowledge allows an evaluation for human nutrition. An RDA has been established for iron, iodine, and zinc, and an "estimated safe and adequate daily dietary intake" has been recommended for chromium, copper, fluoride, manganese, molybdenum, and selenium (NAS 1980A).
SOURCE: USDA (1976) and Webb *et al.* (1974). Adapted from National Dairy Council. 1983A. *Newer Knowledge of Milk and Other Fluid Dairy Products.* Rosemont, Ill. With permission.

Iron, an essential element for humans, is a constituent of hemoglobin, myoglobin, and numerous enzymes important in oxygen, carbon dioxide, and electron transport (NAS 1980A; Bothwell *et al.* 1979). A deficiency of iron results in anemia, a condition most commonly found among young children, adolescents, and women of menstrual age (Dallman *et al.* 1984). Milk, which provides 10 to 90 μg iron per 100 g, is a poor source of dietary iron. However, milk is viewed as an appropriate vehicle for enrichment with iron (Hegenauer *et al.* 1979; Douglas *et al.* 1981; Cook and Reusser 1983). An important consideration is the choice of the iron compound for fortification, as iron can catalyze the oxidation of milk fat, leading to unacceptable flavor and color changes.

Ingestion of cow's milk has been implicated as a factor contributing to iron deficiency anemia in infants and young children (Wilson *et al.* 1974; Woodruff 1977, 1978). Cow's milk contains only a small amount of iron, and excessive consumption of milk at the expense of iron-rich foods may result indirectly in iron deficiency anemia. Insidious loss of blood in the gastrointestinal tract, leading to iron deficiency anemia, has been reported in some infants who have ingested excessive amounts of pasteurized, homogenized cow's milk prematurely, that is, during the first few months of life. Although not a clinically significant problem (Fomon *et al.* 1979), the condition appears to occur when relatively large amounts of pasteurized milk are fed to very young infants. When milk is heated beyond pasteurization temperatures, as is done with evaporated or ultra-high-temperature milk, this problem does not occur. Moreover, when cow's milk is introduced as recommended (i.e., when the infant over six months of age is receiving at least 200 g beikost daily), it is unlikely that milk will provoke occult loss of blood from the gastrointestinal tract or iron deficiency anemia (Anderson *et al.* 1985; Fomon 1974; Fomon *et al.* 1979; American Academy of Pediatrics 1983). The magnitude, frequency, etiology, and functional consequence of enteric bleeding associated with cow's milk have not been firmly established (American Academy of Pediatrics 1978B, 1983; Anon. 1974). However, the vast majority of normal infants over six months of age experience no such problems when consuming recommended amounts of milk.

Iodine is necessary for the production of the thyroid hormones, thyroxine and triiodothyroxine, which are important in regulating energy metabolism. A deficiency of iodine leads to thyroid enlargement (goiter) and cretinism. Worldwide, endemic goiter remains a problem, but in the United States iodine-deficiency goiter is rare as a result of iodine-fortified table salt and the use of iodine in certain food processing techniques. During the period 1974–1978, there was an apparent upward trend in iodine consumption in the United States (Talbot *et al.* 1976;

Mertz 1981; Hemken 1980; Crocco and White 1981; Park *et al.* 1981), although in 1979 and 1980 the iodine intake decreased substantially (Pennington 1980; Allegrini *et al.* 1983). Adventitious sources of iodine in various compounds used in modern farming and food processing have contributed greatly to the average iodine intake, which presently exeeds the RDA (Park *et al.* 1981). However, there is no evidence that this increase in iodine intake has resulted in a corresponding increase in the incidence of iodine toxicity or hypersensitivity in the human population or has had an adverse effect on general health (Talbot *et al.* 1976; Crocco and White 1981).

Dairy foods can be a major contributor of iodine to the diet at all ages (Park *et al.* 1981; Bruhn *et al.* 1981, 1983; Swanson 1981; Allegrini *et al.* 1983; Bruhn and Franke 1985). Between 1974 and 1978, dairy foods accounted for 38 to 56% of the iodine in adult diets and from 56 to 85% in infant and toddler diets (Pennington 1980; Park *et al.* 1981). The concentration of iodine in milk varies widely, from 6 to 500 μ/100 g. However, analyses of milk samples from numerous states have shown that the majority have iodine concentrations below 50 μg/100 g (Bruhn and Franke 1985; Hemken 1980). LaCroix and Wong (1980), using a specific ion electrode method, reported the average iodide values for raw milk and commercially processed milks to be 22 and 62 μg/100 g, respectively. Possible sources of iodine in milk include iodine in the rations of dairy cows, in certain veterinary medications, and in disinfectants and sanitizers used in the dairy industry (Bruhn and Franke 1985; Bruhn *et al.* 1981, 1983; Swanson 1981; Park *et al.* 1981; Crocco and White 1981; Allegrini *et al.* 1983). The greatest potential for reducing the iodine content of milk appears to be improved management practices in the use of substances containing iodine (Bruhn and Franke 1985). In particular, iodine-containing veterinary medications intended for systemic use should be evaluated critically for efficacy and suitability (Crocco and White 1981). In California, the iodine in raw milk was reduced substantially in 1981 to 26 $\pm$ 23 μg/100 g compared with that in 1980 of 47 $\pm$ 30 μg/100 g by discontinuing iodine supplementation of dairy feeds (Bruhn *et al.* 1983). As mentioned above, the concentration of iodine in cow's milk can vary widely (Bruhn and Franke 1985). However, because milk from several farms is mixed before being processed for retail sale, it is "unlikely that any population group purchasing processed dairy foods at the retail level in the USA will be exposed to excessive concentrations of iodine in the dairy products they consume" (Bruhn and Franke 1985).

Zinc is essential for the function of more than 100 enzymes (e.g., thymidine kinase, carbonic anhydrase, lactic dehydrogenase, alkaline phosphatase) involved in a variety of metabolic activities in the body, including protein and nucleic acid function (Prasad 1979). Among its many roles, zinc is important for immune mechanisms, sexual development, taste acuity, hormone metabolism, and tissue repair (collagen synthesis). Marginal states of zinc nutriture have been shown in segments of the U.S. population.

Zinc in cow's milk varies from 0.3 to 0.6 mg/100 g, with an average of 0.38 mg/100 g. According to Jarrett (1979), cow's milk is considered to be low in zinc, but compared to the concentration of many other trace elements in other foods, this trace element exists in relatively large amounts in milk. Results of gel filtration chromatography reveal that zinc in cow's milk is associated with high molecular weight fractions, whereas in human milk this trace element is associated with low molecular weight fractions (Eckhert et al. 1977; Lonnerdal et al. 1981). Furthermore, it is postulated that zinc is of higher bioavailability in human milk than in cow's milk due to the low molecular weight zinc-binding ligands and that, as such, human milk may be of therapeutic value in acrodermatis enteropathica, a genetic disorder of zinc deficiency (Eckhert et al. 1977; Lonnerdal et al. 1981; Sandström et al. 1983). Cow's milk, however, is still a better source of zinc than many other foods such as plant products, particularly soy. Isolated soybean protein has been shown in experimental animal studies to inhibit zinc availability (Solomons 1982). And in a study carried out by Sandström et al. (1983), the absorption of zinc from soy formula (14 ± 4%) was significantly lower than that from cow's milk (28 ± 15%). Much remains to be learned regarding both zinc binding to specific milk components and the bioavailability of zinc from human and cow's milk (Cousins and Smith 1980).

ELECTROLYTES

Sodium, potassium, and chloride are electrolytes found in cow's milk for which the Food and Nutrition Board has estimated safe and adequate daily dietary intakes for infants, children and adolescents, and adults (NAS 1980A). Sodium functions in the body to maintain blood volume and cellular osmotic pressure and to transmit nerve impulses (NAS 1980A). The estimated safe and adequate daily dietary intake of sodium is 1100–3300 mg (2.8–8.4 g sodium chloride) for healthy adults (NAS 1980A). The American Medical Association, Council on Scientific Affairs (1979), suggested 4800 mg sodium per day as a tentative definition of moderation in sodium intake.

The average daily intake of sodium from all sources is 3900–4700 mg (10–12 g sodium chloride)—1200 mg (3 g sodium chloride) occurring naturally in foods, 1300–2500 mg (3–6 g sodium chloride) added by the cook or at the table (i.e., discretionary), and 1600–2300 mg (4–6 g sodium chloride) added during the commercial processing of food (Select Committee on GRAS Substances 1979). Discretionary or consumer-controlled use of sodium accounts for about one-fourth to one-half of total intake; the rest is non-discretionary, either commercially controlled or occurring naturally (Select Committee on GRAS Substances 1979). Sodium in the form of common table salt (sodium chloride) or as sodium-containing ingredients (e.g., monosodium glutamate) plays an essential role in the processing of many foods (AMA 1983). For example, in cheesemaking, sodium chloride controls the moisture content of the final cheese, aids in controlling the fermentation process, and contributes to the flavor and texture of the product (AMA 1983; Shank et al. 1982; Kosikowski 1982). In processed cheeses, cheese foods, and cheese spreads, sodium phosphates and citrates dissolve milk protein and aid in the formation of desired texture and firmness, as well as decrease bacterial growth (AMA 1983).

The relationship of sodium intake to hypertension or high blood pressure, a primary risk factor for coronary heart disease and stroke, has become an issue of increasing concern in the United States (Nicholls 1984; Select Committee on GRAS Substances 1979; White and Crocco 1980; Shank et al. 1982; Crocco 1982). In fact, FDA (1984) recently issued a regulation to make information on the sodium content of foods a mandatory item in nutrition labeling. Although the results of some epidemiological studies and animal experiments reveal an association between increased sodium intake and elevated blood pressure, absolute proof of a cause-and-effect relationship is lacking (NAS 1980A; Shank et al. 1982; McCarron 1983; McCarron et al. 1982, 1984). In fact, an analysis of the HANES I (1971–1974) data reveals that a low sodium intake, as opposed to a high sodium intake, is associated with higher blood pressures (McCarron et al. 1984). An individual's blood pressure is influenced by many factors, including genetics, age, race, associated medical problems, and environmental determinants such as psychological stress and nutrition (Van Itallie 1982).

According to the FDA Total Diet Study, in which dietary sodium and potassium intakes of three age groups of Americans were estimated from 1977 through 1980, cow's milk supplied 32 to 39% of the total sodium intake for infants, whereas the percentage for toddlers was much lower, 12 to 14% (Shank et al. 1982). Dairy products contributed about 10% of the sodium in the total diet consumed by adults. Data obtained from the National Health and Nutrition Examination Survey (HANES II) 1976–1980 (Carroll et al. 1983) reveal that the median daily consumption of sodium found naturally in food and added during processing was 2922 mg for males (6 months-74 years) and 2060 mg for females (6 months–74 years). Discretionary sodium intake was not included in this study. The milk group contributed 12.8% of the sodium, compared with 19.1% for the meat group, 10.6% for the fruit-vegetable group, 24.1% for the grain group, and 14.8% for the "others" category, which includes many high-energy–low-nutrient foods. A final category of combination foods including soups, gravies, and mixed protein dishes contributed a total of 20%. Whole milk contains 49 mg sodium per 100 g (Table 7.9).

Potassium is the principal cation in intracellular fluid in the body. Variations in the sodium-to-potassium ratio in the diet can affect blood pressure under certain circumstances (Shank et al. 1982; NAS 1980A). In fact, preliminary evidence suggests that potassium may protect against a sodium-induced increase in blood pressure (Langford 1983; Tannen 1983). Milk and milk products are shown in the FDA Total Diet Study to be the largest contributors of potassium in the diet (Shank et al. 1982). Milk contributed about 55% and 40% of the total potassium in the diet of infants and toddlers, respectively, and dairy foods accounted for 26% of the potassium in the adult diet. Whole milk contains 152 mg potassium per 100 g.

Chloride, a normal constituent of extracellular fluid in the body, is an important anion in the maintenance of fluid and electrolyte balance, as well as a necessary component of gastric juice. Whole milk contains 103 mg chloride per 100 g (NDC 1983A).

CULTURED AND SPECIALTY MILK PRODUCTS

A review of the nutritive value of milk would be incomplete without a consideration of some other milk products. In the following discussion, the nutritional value of cultured and culture-containing dairy foods, ultra-high-temperature dairy foods, and imitation and substitute dairy products is examined briefly.

CULTURED AND CULTURE-CONTAINING DAIRY FOODS

Milk can be converted easily by lactic acid starter cultures into various cultured and culture-containing milk products. Within the last ten years, consumption of these products (e.g., yogurt, sour cream, and acidophilus milk) has increased appreciably in the United States. (Rasic and Kurmann 1978; Shahani and Chandan 1979; Helferich and

Westhoff 1980; Sellars 1981). This may be explained in part by the nutritive and therapeutic qualities ascribed specifically to cultured and culture-containing products. Several reviews on this subject are available (Sellars 1981; Speck and Katz 1980; Ayebo and Shahani 1980; Shahani and Chandan 1979; Rasic and Kurmann 1978; Chandan and Shahani 1982; International Dairy Federation Group F20 1983; Renner 1983; Deeth and Tamine 1981; Shahani 1983; NDC 1984C).

In terms of their favorable nutritive characteristics, cultured milk products are digested and absorbed more easily than milk due to the partial hydrolysis of milk constituents such as protein, carbohydrate, and fat during the fermentation process (International Dairy Federation Group F20 1983; Renner 1983). In addition, folic acid may be increased in fermented milk products, while other vitamins either are slightly decreased (e.g., vitamin B_{12}) or unchanged (International Dairy Federation Group F20 1983; Alm 1982; Speckmann 1984; Renner 1983). Fermentation of milk has little influence on its mineral content, although lactic acid fermentation improves calcium, phosphorus, and iron utilization (Rasic and Kurmann 1978; Renner 1983).

A hypocholesterolemic effect of fermented dairy foods such as yogurt has been observed by some investigators (Mann 1977; Mann and Spoerry 1974; Thakur and Jha 1981; Hepner et al. 1979). Culture-containing milk products also may be beneficial for individuals with lactose malabsorption (International Dairy Federation Group F20 1983; Gallagher et al. 1974; Kolars et al. 1984; Newcomer and McGill 1984B). Gallagher et al. (1974) reported that lactase-deficient individuals tolerated fermented dairy foods without symptoms of intolerance. This benefit is attributed to the lower lactose content of many fermented dairy foods, in addition to the starter culture, which may contain lactase, the enzyme necessary for the metabolism of lactose. For example, the yogurt cultures, Streptococcus thermophilus and Lactobacillus bulgaricus, are shown to contain lactase (Kilara and Shahani 1976; Friend et al. 1983), whereas L. acidophilus, S. lactis, and S. cremoris have little, if any, lactase (Newcomer et al. 1983; Farrow 1980). More recently, it was shown that 10 healthy, lactase-deficient individuals absorbed the lactose in yogurt better than that in milk (Kolars et al. 1984; Newcomer and McGill 1984B).

Cultured and culture-containing dairy foods may influence growth and metabolism (International Dairy Federation Group F20 1983; Speckmann 1983; Wong et al. 1983). For example, Hargrove and Alford (1978, 1980) found that weanling rats fed yogurt gained significantly more weight than rats fed unfermented milk or other types of fermented milks. Stimulation of rat growth was associated with improved feed efficiency. Hargrove and Alford (1978, 1980) postulate that the growth stimulatory effect may be due to improved bioavailability of protein in the fermented dairy food. However, more carefully controlled studies in which energy intake and expenditure are measured need to be conducted before any conclusions can be drawn regarding the positive effect of cultured dairy foods on weight gain and feed efficiency in animals and humans (International Dairy Federation Group F20 1983).

During the fermentation of milk, by-products or metabolites are formed which can have beneficial functions in the body. For example, ingestion of yogurt or acidophilus milk and their constituent organisms may inhibit pathogenic and food spoilage organisms in the intestine which cause intestinal infections, diarrhea, flatulence, and other digestive problems (Shahani and Chandan 1979; Renner 1983). This antagonistic action to unwanted bacteria is explained by the production of natural antibiotics by lactic organisms and by the increase in the acidity of the intestine as a result of lactic acid, acetic acid, and hydrogen peroxide, which are formed during the fermentation of milk. Intake of cultured and culture-containing milk products has been credited with reestablishing a desirable microfloral balance following antibiotic therapy, which destroys the intestinal flora and upsets the established biological balance (International Dairy Federation Group F20 1983).

In addition, there is interest in the antitumor qualities of cultured dairy foods such as yogurt, although the antitumor component(s) has not been identified (Ayebo et al. 1981). Results of some studies (Goldin and Gorbach 1976, 1977, 1980, 1984; Goldin et al. 1978, 1980) indicate that when L. acidophilus organisms are consumed in milk, there is a reduction in fecal bacterial enzymes (e.g., β-glucuronidase, azoreductase, and nitroreductase) associated with the risk of colon cancer. Furthermore, tumorigenesis is delayed and its severity is reduced when yogurt or yogurt-containing L. acidophilus and L. Bulgaricus organisms are fed with known carcinogens to laboratory animals (Reddy et al. 1973, 1983). However, at this time it is premature to conclude that factors altering gut microfloral enzymes have an effect on cancer risk and tumor formation in the large bowel (Goldin and Gorbach 1984).

ULTRA-HIGH-TEMPERATURE DAIRY FOODS

Advances in the processing and packaging of dairy foods have made possible the production of milk and other dairy foods that can be stored at room temperature for up to six months (Miller 1985). Ultra-high-temperature (UHT) processing of milk, combined with aseptic filling techniques and hermetically sealed packaging, has been used in some foreign countries to produce shelf-stable milk for a number of years. Not until 1981, however, did the FDA approve the aseptic packaging procedure for use in the United States.

Basically, UHT milk is heat treated to 138 to 150°C (280 to 302°F) for one or two seconds by a process that renders the milk commercially sterile. The product is then placed in containers by an aseptic filling process and packaged in hermetically sealed containers. This combination of heat treatment and packaging results in milk that can be stored unrefrigerated for extended periods of time (Miller 1985; Arnold and Roberts 1982).

For practical purposes, the nutritional quality of UHT milk is similar to that of conventionally pasteurized milk (Miller 1985; Arnold and Roberts 1982; Burton 1980, 1982; Renner 1980; Kosaric et al. 1981; Katz et al. 1981; Ford and Thompson 1981). As with conventionally pasteurized milk, the degree of heat treatment and the conditions of storage can affect the nutrient content of the milk (Burton 1980, 1982; Renner 1980; Kosaric et al. 1981; Ford and Thompson 1981; Mehta 1980). However, careful controls on UHT processing, along with proper packaging, can ensure that the nutritional value of milk with regard to protein, fat, carbohydrate, fat-soluble vitamins, most water-soluble vitamins, and minerals is retained (Ford and Thompson 1981).

For example, the UHT heat treatment results in some protein denaturation, but the biological value and net utilization of milk proteins remain essentially unaffected (Katz et al. 1981; Ford and Thompson 1981; Mehta 1980). In fact, UHT milk may be more digestible than raw or conventionally pasteurized milk because of the greater protein denaturation. The results of both animal feeding studies and studies of human infants indicate that the protein utilization of UHT milk is at least as high as that of conventionally pasteurized milk (Renner 1980; Katz et al. 1981).

No adverse effects of either UHT processing or storage on the nutritional properties of milk fat have been demonstrated. Although increases in the milk's free fatty acid content have been noted when UHT milk is stored at room temperature rather than refrigerated, these changes do not appear to affect the nutritional value of the milk. No changes of nutritional importance have been noted in the carbohydrate components of UHT milk (Ford and Thompson 1981).

In general, vitamins appear to be at least as stable during UHT processing as during conventional pasteurization (Mehta 1980). Levels of the fat-soluble vitamins A, D, and E, as well as those of the water-soluble vitamins, riboflavin, nicotinic acid, pantothenic acid, and biotin in milk, are not decreased by UHT processing. Furthermore, no loss of these nutrients occurs during storage, provided that the packaging of the UHT milk protects the food from light. Losses of heat-sensitive or oxygen-sensitive vitamins in milk such as thiamin, vitamin B_6, vitamin B_{12}, folacin, and ascorbic acid may vary considerably, depending upon such factors as the degree of heat treatment during processing, oxygen content of the milk storage, exposure of the milk to light, and the temperature and length of time of storage (Burton 1980; Renner 1980; Kosaric *et al.* 1981; Ford and Thompson 1981; Mehta 1980). Properly controlled UHT heat treatment, like conventional pasteurization, results in no nutritionally significant losses of these five nutrients.

UHT processing and storage have no effect on the total calcium content or calcium bioavailability. Calcium, phosphorus, and magnesium are shown to be equally bioavailable to rats from UHT milk, raw milk, and traditionally processed milk (Katz *et al.* 1981). Also, human infants retain similar amounts of calcium, potassium, and phosphorus whether fed UHT milk or conventionally pasteurized milk (Renner 1980; Mehta 1980). Data to date indicate no significant changes in the nutritional value of UHT milk under controlled heat treatment and subsequent storage.

IMITATION AND SUBSTITUTE DAIRY PRODUCTS

Products which imitate and substitute for dairy foods such as nondairy coffee creamers, margarine, nondairy whipped toppings, imitation milk, and imitation and substitute cheeses have attained a sizable share of the market for traditional dairy foods (NDC 1983C). Although these products may be used by the consumer in place of traditional foods, they are not necessarily the same in nutritional value.

Acording to the FDA (1982A), an imitation food is a product that substitutes for and resembles a traditional food but is nutritionally inferior to that food. That is, the term "imitation" denotes nutritional inferiority. On the other hand, FDA has determined that if a food substitutes for and resembles another food but is not nutritionally inferior to that food, it need not be called an imitation. Rather, by FDA definition, such a substitute food may be called "nutritionally equivalent." This term implies that the nutrients in the substitute food are identical in quantity, biological activity, and bioavailability to those found in the traditional food. However, examination of FDA's nutrient profile for nutritional equivalency and consideration of the differences in the biological activity and bioavailability of nutrients, as well as the possibility that traditional foods may contain beneficial factors, reveal serious weaknesses in FDA's definition of nutritional equivalence (NDC 1983C).

In 1978 FDA published proposed standards of identity for substitutes for milk, cream and cheese (FDA 1978). Although the proposal has been withdrawn (FDA 1983), the underlying basis for it remains in force (FDA 1982B). FDA most likely will continue to interpret this regulation when applied to dairy food substitutes in much the same manner as in the 1978 proposal.

In the proposed regulation, nutrient profiles for substitute dairy products were established to serve as the basis for determining nutritional equivalence between the substitute dairy food and its traditional counterpart. The FDA yardstick for nutritional equivalence is the U.S. RDA (FDA 1982B). Although at least 42 nutrients are known to be essential for humans, FDA has established U.S. RDAs for only 20 of these nutrients and indirectly has considered nine essential amino acids. Furthermore, only those U.S. RDA nutrients present in a "measurable amount" (i.e., 2% of the U.S. RDA per serving) are considered in establishing nutrient equivalency levels. That is, nutrients below 2% of the U.S. RDA per serving are disregarded.

In defining the nutritional equivalence of dairy foods, FDA considered only 11 to 15 nutrients for milk substitutes, 1 nutrient for cream substitutes, and 4 to 9 nutrients for cheese substitutes (FDA 1978). Yet, data from the Consumer and Food Economics Institute, USDA (1976), reveal that traditional milk, cream, and cheese contain an array of nutrients including protein, fat, carbohydrate, and at least 15 minerals and vitamins and 18 amino acids. Thus, under FDA's proposal (FDA 1978), which has been withdrawn (FDA 1983) but, as mentioned above, may in effect be applied, a substitute dairy product could be declared nutritionally equivalent to its traditional counterpart and yet (1) not contain all of the nutrients in the traditional food, or (2) contain some or all of these other nutrients but in lesser quantities, or (3) contain some of the nutrients such as sodium in excessive amounts, or (4) contain more or less energy (NDC 1983C).

Substitute milk products may not be equivalent to cow's milk in terms of the quantity and in some cases the quality of fat, carbohydrate, vitamins, and minerals. Fat, carbohydrate, sodium, fiber, and energy, as well as the nutrients for which no U.S. RDA has been established, were not considered by FDA in its proposed definition of nutritional equivalency. In terms of quality, coconut oil, the primary and in most instances the sole fat used in substitute as well as imitation milk products, is a more saturated fat than milk fat and lacks linoleic acid, an essential fatty acid. Thus, a substitute dairy product formulated with hydrogenated coconut oil and sucrose and containing more sodium than the traditional product but meeting FDA's proposed criteria for nutritional equivalency could be considered nutritionally equivalent to that product by FDA.

The nutrient content of imitation and substitute milk products can vary widely, depending on the product formula used. For example, imitation milk products generally contain about 1 to 5% protein, 3 to 4% vegetable fat (hydrogenated coconut oil, soybean oil, or cottonseed oils), 6 to 10% carbohydrate (corn syrup solids, sucrose), and various additives including stabilizers and emulsifiers. Not only can some nutrients be found in lower concentrations in an imitation or substitute product than its traditional counterpart, but other nutrients may be present in higher amounts (e.g., sodium) (NDC 1983C). Because of their variability in composition, imitation and substitute milk products cannot uniformly be depended upon to supply nutrients consistently at specified levels. Biological tests using animal feeding studies have shown that imitation milk and imitation cheese do not support the growth and well-being of young rats as well as their traditional counterparts (NDC 1983C; Lowe *et al.* 1983; Kotula *et al.* 1983, 1984).

Various government agencies and professional organizations have recommended against using imitation milk as a milk alternative or substitute in child nutrition programs, in place of infant formula, in meal programs for the elderly, and for individuals with milk allergy, lactose intolerance, or those on fat- or sodium-controlled diets (NDC 1983C; American Academy of Pediatrics 1984).

REFERENCES

Abraham, S. and Carroll, M. D. 1981. Fats, cholesterol, and sodium intake in the diet of persons 1–74 years; United States, 1971–1974. Advance data No. 54 (revised). U.S. Department of Health and Human Services, Public Health Service, Office of Health Research, Statistics, and Technology, Hyattsville, Md.

Abraham, S., Johnson, C. L. and Carroll, M. D. 1977. A comparison of levels of serum cholesterol of adults 18–74 years of age in the United States in 1960–62 and 1971–74. Advance data No. 5. Vital and Health Statistics, National Center for Health Statistics, Hyattsville, Md.

Ahmed, A. A., McCarthy, R. D. and Porter, G. A. 1979. Effect of milk constituents on hepatic cholesterol-genesis. *Atherosclerosis 32*, 347–357.

Ahrens, E. H., Jr. 1976. The management of hyperlipidemia: Whether, rather than how. *Ann. Intern. Med. 85*, 87–93.

Ahrens, E. H., Jr. 1979. Dietary fats and coronary heart disease: Unfinished business. *Lancet 2*, 1345–1348.

Ahrens, E. H., Jr. 1982. Diet and heart disease: Shaping public perceptions when proof is lacking. *Arteriosclerosis 2*, 85–86.

Ahrens, E. H., Jr. 1985. The diet–heart question in 1985: Has it really been settled? *Lancet 1*, 1085–1087.

Albanese, A. A. 1977. *Bone Loss: Causes, Detection, and Therapy.* Alan R. Liss, New York.

Albanese, A. A. 1983. Calcium nutrition throughout the life cycle. *Bibl. Nutr. Dieta 33,* 80–99.

Albanese, A. A., Edelson, A. H., Lorenze, E. J., Wein, E. H. and McBean, L. D. 1978. *Calcium Throughout the Life Cycle.* National Dairy Council, Rosemont, Ill.

Allegrini, M., Pennington, J. A. T. and Tanner, J. T. 1983. Total diet study: Determination of iodine intake by neutron activation analysis. *J. Am. Diet. Assoc. 83,* 18–24.

Allen, L. H. 1982. Calcium bioavailability and absorption: A review. *Am. J. Clin. Nutr. 35,* 783–808.

Allen, L. H. 1986. Calcium and age-related bone loss. *Clin. Nutr. 5,* 147–152.

Alm, L. 1982. Effect of fermentation on B-vitamin content of milk in Sweden. *J. Dairy Sci. 65,* 353–359.

American Academy of Pediatrics, Committee on Nutrition. 1967. The relation between infantile hypercalcemia and vitamin D—public health implications in North America. *Pediatrics 40,* 1050–1061.

American Academy of Pediatrics, Committee on Nutrition. 1974. Should milk drinking by children be discouraged? *Pediatrics 53,* 576–582.

American Academy of Pediatrics, Committee on Nutrition. 1977. Nutritional aspects of vegetarianism, health foods, and fad diets. *Pediatrics 59,* 460–464.

American Academy of Pediatrics, Committee on Nutrition. 1978A. The practical significance of lactose intolerance in children. *Pediatrics 62,* 240–245.

American Academy of Pediatrics, Committee on Nutrition. 1978B. Relationship between iron status and incidence of infection in infancy. *Pediatrics 62,* 246–250.

American Academy of Pediatrics, Committee on Nutrition. 1983. The use of whole cow's milk in infancy. *Pediatrics 72,* 253–255.

American Academy of Pediatrics, Committee on Nutrition. 1984. Imitation and substitute milks. *Pediatrics 73,* 876.

American Academy of Pediatrics, Committee on Nutrition. 1986. Prudent lifestyle for children: Dietary fat and cholesterol. *Pediatrics 78,* 521–525.

American Council on Science and Health. 1982. *Diet Modification: Can It Reduce the Risk of Heart Disease?* American Council on Science and Health, New York.

American Heart Association, Nutrition Program Committee. 1981. Advisory statement concerning the claims that consumption of homogenized milk increases the risk of heart disease. Supplement to guidelines for the development of nutrition programs. American Heart Association.

American Heart Association, Report of Nutrition Committee. 1982. Rationale of the diet–heart statement of the American Heart Association. *Circulation 65,* 839A–854A.

American Medical Association. 1982. The nutritive quality of processed foods: General policies for nutrient additions. *Nutr. Rev. 40,* 93–96.

American Medical Association, Council on Foods and Nutrition. 1955. Importance of vitamin D milk. *JAMA 159,* 1018–1019.

American Medical Association, Council on Scientific Affairs. 1979. American Medical Association concepts of nutrition and health. *JAMA 242,* 2335–2338.

American Medical Association, Council on Scientific Affairs. 1983. Sodium in processed foods. *JAMA 249,* 784–789.

Anderson, G. H., Morson-Pasut, L. A., Bryan, H., Cleghorn, G., Tanaka, P., Yeung, D. and Zimmerman, B. 1985. Age of introduction of cow's milk to infants. *J. Pediatr. Gastroenterol. Nutr. 4,* 692–698.

Anderson, S. A. and Talbot, J. M. 1981. A review of folate intake, methodology, and status. Life Sciences Research Office, Federation of American Societies for Experimental Biology, Bethesda, Md.

Anon. 1974. Fresh cows' milk and iron deficiency in infants. *Nutr. Rev. 31*, 318–320.

Anon. 1982A. Folate binder in milk may facilitate folate absorption. *Nutr. Rev. 40*, 90–92.

Anon. 1982B. The vitamin D activity of milk. *Nutr. Rev. 40*, 27–28.

Armbrecht, H. J. and Wasserman, R. H. 1976. Enhancement of Ca^{++} uptake by lactose in the rat small intestine. *J. Nutr. 106*, 1265–1271.

Arnold, S. and Roberts, T. 1982. UHT milk: Nutrition, safety, and convenience. *National Food Rev. 18*, 2–5.

Ayebo, A. D. and Shahani, K. M. 1980. Role of cultured dairy products in the diet. *Cultured Dairy Products J. 15*(4), 21–29.

Ayebo, A. D., Shahani, K. M. and Dam, R. 1981. Antitumor component(s) of yogurt: Fractionation. *J. Dairy Sci. 64*, 2318–2323.

Bahna, S. L. and Gandhi, M. D. 1983. Milk hypersensitivity. I. Pathogenesis and symptomatology. *An. Allergy 50*, 218–223.

Bahna, S. L. and Heiner, D. C. 1980. *Allergies to Milk*. Grune and Stratton, New York.

Bayless, T. M. and Rosensweig. 1966. A racial difference in incidence of lactase deficiency. A survey of milk intolerance and lactase deficiency in healthy adult males. *JAMA 197*, 968–972.

Belizan, J. M., Pineda, O., Sainz, E., Menendez, L. A. and Villar, J. 1981. Rise of blood pressure in calcium-deprived pregnant rats. *Am. J. Obstet. Gynecol. 141*, 163–169.

Belizan, J. M., Villar, J., Pineda, O., Gonzales, A. E., Sainz, E., Garrera, G. and Sibrian, R. 1983A. Reduction of blood pressure with calcium supplementation in young adults. *JAMA 249*, 1161–1165.

Belizan, J. M., Villar, J., Zalazar, A., Rojas, L., Chan, D. and Bryce, G. F. 1983B. Preliminary evidence of the effect of calcium supplementation on blood pressure in normal pregnant women. *Am. J. Obstet. Gynecol. 146*, 175–180.

Belko, A. Z., Meredith, M. P., Kalkwarf, H. J., Obarzanek, E., Weinberg, S., Roach, R., McKeon, G. and Roe, D. A. 1985. Effects of exercise on riboflavin requirements: Biological validation in weight reducing women. *Am J. Clin. Nutr. 41*, 270–277.

Belko, A. Z., Obarzanek, E., Kalkwarf, H. J., Rotter, M. A., Bogusz, S., Miller, D., Haas, J. D. and Roe, D. A. 1983. Effects of exercise on riboflavin requirements of young women. *Am. J. Clin. Nutr. 37*, 509–517.

Belko, A. Z., Obarzanek, E., Roach, R., Rotter, M., Urban, G., Weinberg, S. and Roe, D. A. 1984. Effects of aerobic exercise and weight loss on riboflavin requirements of moderately obese, marginally deficient young women. *Am. J. Clin. Nutr. 40*, 553–561.

Bierman, E. L. and Shank, R. E. 1975. Editorial: Homogenized milk and coronary artery disease: Theory, not fact. *JAMA 234*, 630–631.

Bock, S. A. 1980. Food sensitivity. A critical review and practical approach. *Am. J. Dis. Child. 134*, 973–982.

Bock, S. A. 1982. The natural history of food sensitivity. *J. Allergy Clin. Immunol 69*, 173–1977.

Bock, S. A. 1985. Natural history of severe reactions to foods in young children. *J. Pediatr. 107*, 676–680.

Boguslawski, W. and Wrobel, J. 1974. An inhibitor of sterol biosynthesis present in cow's milk. *Nature 247*, 210–211.

Bothwell, T. H., Charlton, R. W., Cook, J. D. and Finch, C. A. 1979. *Iron Metabolism in Man*. Blackwell Scientific Publications, Oxford.

Bradley, R. L., Jr. 1980. Effect of light on alteration of nutritional value and flavor of milk: A review. *J. Food Protection 43*, 314–320.

Bricker, M. L., Smith, J. M., Hamilton, T. S. and Mitchell, H. H. 1949. The effect of cocoa upon calcium utilization and requirements, nitrogen retention and fecal composition of women. *J. Nutr. 39*, 445–461.

Bruhn, J. C. and Franke, A. A. 1985. Iodine in cow's milk produced in the USA in 1980–1981. *J. Food Protection 48*, 397–399.

Bruhn, J. C., Franke, A. A. and Amirhosseini, D. S. 1981. Iodine in raw milk. *J. Dairy Sci. 64*(suppl. 1), 56.

Bruhn, J. C., Franke, A. A., Reif, G. D. and Frazeur, D. R. 1984. Milk quality surveys in California. Submitted to *J. of Food Protection.*

Bruhn, J. C., Franke, A. A., Bushnell, R. B. Weisheit, H., Hutton, G. H. and Gurtle, G. C. 1983. Sources and content of iodine in California milk and dairy products. *J. Food Protection 46*, 41–46.

Burton, H. 1980. An introduction to the ultra-high temperature processing of milk and milk products. *In: Proceedings of the International Conference on UHT Processing and Aseptic Packaging of Milk and Milk Products.* Department of Food Science, North Carolina State University, Raleigh, N.C.

Burton, H. 1982. Sterilized milk and milk products. *In: CRC Handbook of Processing and Utilization in Agriculture,* Vol. I, *Animal Products.* I. A. Wolff (Editor). CRC Press, Boca Raton, Fla., pp. 379–387.

Carr, C. J., Talbot, J. M. and Fisher, K. D. 1975. A review of the significance of bovine milk xanthine oxidase in the etiology of atherosclerosis. Life Science Research Office, Federation of American Societies for Experimental Biology, Bethesda, Md. (Prepared for the Food and Drug Administration, Washington, D. C., Contract No. FDA 223-75-2090.)

Carroll, M. D., Abraham, S. and Dresser, C. M. 1983. Dietary intake source data: United States, 1976–80. Data from the National Health Survey. Series II, No. 231, DHHS Pub. No. (PHS) 83-1681. U.S. Department of Health and Human Services, Public Health Service, National Center for Health Statistics, Hyattsville, Md.

Cerbulis, J. and Farrell, H. M. 1977. Xanthine oxidase activity in dairy products. *J. Dairy Sci. 60*, 170–176.

Chandan, R. C. and Shahani, K. M. 1982. Cultured milk products *In: CRC Handbook of Processing and Utilization in Agriculture,* Vol. I, *Animal Products.* I. A. Wolff (Editor). CRC Press, Boca Raton, Fla., pp. 365–377.

Chinn, H. I. 1981. *Effects of Dietary Factors on Skeletal Integrity in Adults: Calcium, Phosphorus, Vitamin D, and Protein.* Life Sciences Research Office, Federation of American Societies for Experimental Biology, Bethesda, Md.

Christie, W. M. W. 1983. The composition and structure of milk lipids. *In: Developments in Dairy Chemistry,* Vol. 2., *Lipids.* P. F. Fox (Editor). Applied Science Publishers, New York, pp. 1–35.

Clark, A. J. and Pratt, D. E. 1976. Xanthine oxidase activity in rat serum after administration of homogenized bovine cream preparation. *Life Sci. 19*, 887–892.

Clifford, A. J., Ho, C. Y. and Swenerton, H. 1983. Homogenized bovine milk xanthine oxidase: A critique of the hypothesis relating to plasmalogen depletion and cardiovascular disease. *Am J. Clin. Nutr. 38*, 327–332.

Coates, M. E. 1983. Reviews of the progress of dairy science: Dietary lipids and ischaemic heart disease. *J. Dairy Res. 50*, 541–557.

Colman, N., Hettiarachchy, N. and Herbert, V. 1981. Detection of a milk factor that facilitates folate uptake by intestinal cells. *Science 211*, 1427–1429.

Committee on Diet, Nutrition, and Cancer, Assembly of Life Sciences, National Research Council. 1982. *Diet, Nutrition, and Cancer.* National Academy Press, Washington, D.C.

Condon, J. R., Nassim, J. R., Hilbe, A., Millard, F. J. C. and Stainthorpe, E. M. 1970.

Calcium and phosphorus metabolism in relation to lactose tolerance. *Lancet 1*, 1027–1029.

Consensus Development Panel, Office of Medical Applications of Research, National Institutes of Health. 1984. Osteoporosis—Consensus Conference. *JAMA 252*, 799–802.

Cook, J. D. and Reusser, M. E. 1983. Iron fortification: An update. *Am. J. Clin. Nutr. 38*, 648–659.

Council for Agricultural Science and Technology. 1982. *Diet, Nutrition, and Cancer: A Critique.* Special Pub. No. 13. Council for Agricultural Science and Technology, Ames, Iowa.

Council for Agricultural Science and Technology. 1985. *Diet and Coronary Heart Disease.* Report No. 107. Council for Agricultural Science and Technology, Ames, Iowa.

Cousins, R. J. and Smith, K. J. 1980. Zinc-binding properties of bovine and human milk in vitro: Influence of changes in zinc content. *Am. J. Clin. Nutr. 33*, 1083–1087.

Crocco, S. C. 1982. The role of sodium in food processing. *J. Am. Diet. Assoc. 80*, 36–39.

Crocco, S. C. and White, P. L. 1981. Iodine: Fifty years after goiter. *In: Stokely–Van Camp Annual Symposium. Food in Contemporary Society. Emerging Patterns.* University of Tennessee Press, Knoxville, Tenn., pp. 149–164.

Dallman, P. R., Yip, R. and Johnson, C. 1984. Prevalence and causes of anemia in the United States, 1976 to 1980. *Am. J. Clin. Nutr. 39*, 437–445.

Daniell, H. W. 1983. Postmenopausal tooth loss. Contributions to edentulism by osteoporosis and cigarette smoking. *Arch. Intern. Med. 143*, 1678–1682.

Deeth, H. C. 1983. Homogenized milk and atherosclerotic disease: A review. *J. Dairy Sci. 66*, 1419–1435.

Deeth, H. C. and Tamine, A. Y. 1981. Yogurt: Nutritive and therapeutic aspects. *J. Food Protection 44*, 78–86.

Delmont, J. (Editor). 1983. *Milk Intolerances and Rejection.* Karger, New York.

DeLuca. H.F. 1978. *The Fat-Soluble Vitamins.* Plenum Press, New York.

DeLuca, H. F. 1981. The vitamin D system: A view from basic science to the clinic. *Clin. Biochem. 14*, 213–222.

DeLuca, H. F., Frost, H. M., Jee, W. S. S., Johnston, C. C., Jr. and Parfitt, A. M. 1981. *Osteoporosis: Recent Advances in Pathogenesis and Treatment.* University Park Press, Baltimore.

DeLuca, L. M., Glover, J., Heller, J., Olson, J. A. and Underwood, B. 1979. *Guidelines for the Eradication of Vitamin A Deficiency and Xerophthalmia.* VI. *Recent Advances in the Metabolism and Function of Vitamin A and Their Relationship to Applied Nutrition.* Nutrition Foundation, New York.

DeMan, J. M. 1981. Light-induced destruction of vitamin A in milk. *J. Dairy Sci. 64*, 2031–2032.

De Renzo, E. C. 1956. Chemistry and biochemitry of xanthine oxidase. *Adv. Enzymol. 17*, 293–328.

Dougherty, T. M., Zikakis, J. P. and Rzucidlo, S. J. 1977. Serum xanthine oxidase studies on miniature pigs. *Nutr. Rep. Int. 16*, 241–248.

Douglas, F. W., Jr., Rainey, N. H., Wong, N. P., Edmondson, L. F. and Lacroix, D. E. 1981. Color, flavor, and iron bioavailability in iron-fortified chocolate milk. *J. Dairy Sci. 64*, 1785–1793.

Eckhert, C. D., Sloan, M. V., Duncan, J. R. and Hurley, L. S. 1977. Zinc binding: A difference between human and bovine milk. *Science 195*, 789–790.

Edgar, W. M., Bowen, W. H., Amsbaugh, S., Monell-Torrens, E. and Brunelle, J. 1982. Effects of different eating patterns on dental caries in the rat. *Caries Res. 16*, 384–389.

Engstrom, A. M. and Tobelmann, R. C. 1983. Nutritional consequences of reducing sodium intake. *Ann. Intern. Med. 98*(Part 2), 870–872.

Farrow, J. A. E. 1980. Lactose hydrolysing enzymes in *Str. lactis* and *Str cremoris* and also in some other species of streptococci. *J. Appl. Bact.* 49, 493–503.

Feeley, R. M., Criner, P. E. and Watt, B. K. 1972B. Cholesterol content of foods. *J. Am. Diet. Assoc. 61*, 134–149.

Feeley, R. M., Criner, P. E., Murphy, E. W. and Toepfer, E. W. 1972A. Major mineral elements in dairy products. *J. Am. Diet. Assoc. 61*, 505–510.

Fomon, S. J. 1974. *Infant Nutrition,* 2nd ed. W.B. Saunders Co., Philadelphia.

Fomon, S. J., Filer, L. J., Jr. Anderson, T. A. and Ziegler, E. E. 1979. Recommendations for feeding normal infants. *Pediatrics 63,* 52–59.

Fomon, S. J., Filer, L. J., Jr., Ziegler, E. E., Bergmann, K. E. and Bergmann, R. L. 1977. Skim milk in infant feeding. *Acta Paediatr. Scand. 66,* 17–30.

Food and Agriculture Organization of the United Nations. Food Policy and Food Science Service Nutrition Division. 1970. Amino acid content of foods and biological data on proteins. FAO Nutritional Studies No. 24, Rome.

FDA, Department of Health, Education, and Welfare. 1973. Milk and Cream. Title 21, Part 18. *Federal Register* 38, 27924–27929.

FDA. 1978. Substitutes for milk, cream and cheese. Standards of Identity. Proposed Rule. *Federal Register 43,* 42118–42141.

FDA, Department of Health and Human Services. 1981. Indirect food additives: Adjuvants, production aids, and sanitizers; hydrogen peroxide. Final Rule. *Federal Register 46,* 2341–2343.

FDA, Department of Health and Human Services. 1982A. Code of Federal Regulations. Title 21, Chapter 1, Part 131, Milk and Cream, Section 131–110, Milk. Revised as of April 1. *Federal Register.*

FDA, Department of Health and Human Services. 1982B. Code of Federal Regulations. Title 21, Chapter 1, Part 101, Section 101.3. Identity of labeling of food in packaged form. Revised as of April 1. *Federal Register.*

FDA. 1983. Substitutes for milk, cream and cheese; withdrawal of proposed standards of identity. *Federal Register 48,* 37666–37668.

FDA. 1984. Food labeling; declaration of sodium content of foods and label claims for foods on the basis of sodium content. *Federal Register 49,* 15510–15535.

Food and Nutrition Board, National Academy of Sciences, National Research Council. 1972. Statement on milk intolerance. *J. Am. Diet Assoc. 61,* 241–242.

Ford, J. E. and Thompson, S. Y. 1981. The nutritive value of UHT milk. *In: New Monograph on UHT Milk.* International Dairy Federation Bulletin, Document 133. pp. 65–70.

Formo, M. W., Jungermann, E., Norris, F. A. and Sonntag, N. O. V. 1979. *Bailey's Industrial Oil and Fat Products,* 4th ed., Volume 1. John Wiley & Sons, New York.

Foucard, T. 1985. Development of food allergies with special reference to cow's milk allergy. *Pediatrics 75,* 177–181.

Friedman, G. and Goldberg, S. J. 1975. Concurrent and subsequent serum cholesterol of breast-and formula-fed infants. *Am. J. Clin. Nutr. 28,* 42–45.

Friend, B. A., Fiedler, J. M. and Shahani, K. M. 1983. Influence of culture selection on the flavor, antimicrobial activity, β-galactosidase and B-vitamins of yogurt. *Milchwissenschaft 38,* 133–136.

Gallagher, C. R., Molleson, A. L. and Caldwell, J. H. 1974. Lactose intolerance and fermented dairy products. *J. Am. Diet. Assoc. 65,* 418–419.

Gallagher, J. C. and Riggs, B. L. 1978. Nutrition and bone disease. *N. Engl. J. Med. 298,* 193–195.

Garland, C., Barrett-Connor, E., Rossof, A. H., Shekelle, R. B., Criqui, M. H. and Paul, O. 1985. Dietary vitamin D and calcium and risk of colorectal cancer: A 19-year prospective study in men. *Lancet 1*, 307–309.

Garland, C. F. and Garland, F. C. 1986. Calcium and colon cancer. *Clin. Nutr. 5*, 161–166.

Glueck, C. J. 1979. Appraisal of dietary fat as a causative factor in atherogenesis. *Am. J. Clin. Nutr. 32*, 2637–2643.

Glueck, C. J., Tsang, R., Balistreri, W. and Fallat, R. 1972. Plasma and dietary cholesterol in infancy: Effects of early low or moderate dietary cholesterol intake on subsequent response to increased dietary cholesterol. *Metabolism 21*, 1181–1192.

Goldin, B. R., Dwyer, J., Gorbach, S. L., Gordon, W. and Swenson, L. 1978. Influence of diet and age on fecal bacterial enzymes. *Am. J. Clin. Nutr. 31* (suppl.), 5136–5140.

Goldin, B. R. and Gorbach, S. L. 1976. The relationship between diet and rat fecal bacterial enzymes implicated in colon cancer. *J. Natl. Cancer Inst. 57*, 371–375.

Goldin, B. R. and Gorbach, S. L. 1977. Alterations in fecal microflora enzymes related to diet, age, *Lactobacillus* supplements, and dimethylhydrazine. *Cancer 40*, 2421–2426.

Goldin, B. R. and Gorbach, S. L. 1980. Effect of *Lactobacillus acidophilus* dietary supplements on 1,2-dimethylhydrazine dihydrochloride–induced intestinal cancer in rats. *J. Natl. Cancer Inst. 64*, 263–265.

Goldin, B. R. and Gorbach, S. L. 1984. The effect of milk and lactobacillus feeding on human intestinal bacterial enzyme activity. *Am. J. Clin. Nutr. 39*, 756–761.

Goldin, B. R., Swenson, L., Dwyer, J., Sexton, M. and Gorbach, S. L. 1980. Effect of diet and *Lactobacillus acidophilus* supplements on human fecal bacterial enzymes. *J. Natl. Cancer Inst. 64*, 255–261.

Green, M. H., Dohner, E. L. and Green, J. B. 1981. Influence of dietary fat and cholesterol on milk-lipids and on cholesterol-metabolism in the rat. *J. Nutr. 111*, 276–286.

Greenwood, M. R. C. (Editor). 1983 *Obesity*. Churchill Livingstone, New York.

Greger, J. L. and Krystofiak, M. 1982. Phosphorus intake of Americans. *Food Technol. 34*, 78–84.

Greger, J. L., Marhefka, S. and Geissler, A. H. 1978. Magnesium content of selected foods. *J. Food Sci. 43*, 1610–1612.

Gregory, M. E. 1975. Water-soluble vitamins in milk and milk products. *J. Dairy Res. 42*, 197–216.

Gurr, M. I. 1983. The nutritional significance of lipids. *In: Developments in Dairy Chemistry*, Vol. 2, *Lipids*. P. F. Fox (Editor). Applied Science Publishers, New York, pp. 365–417.

Hahn, P. and Kirby, L. 1973. Immediate and late effects of premature weaning and of feeding a high fat or high carbohydrate diet to weanling rats. *J. Nutr. 103*, 690–696.

Hambraeus, L. 1982. Nutritional aspects of milk proteins. *In: Developments in Dairy Chemistry*, Vol. 1, *Proteins*. P. F. Fox (Editor). Applied Science Publishers, New York, pp. 289–313.

Hargrove, R. E. and Alford, J. A. 1978. Growth rate and feed efficiency of rats fed yogurt and other fermented milks. *J. Dairy Sci. 61*, 11–19.

Hargrove, R. E. and Alford, J. A. 1980. Growth response of weanling rats to heated, aged, fractionated, and chemically treated yogurts. *J. Dairy Sci. 63*, 1065–1072.

Haroon, Y., Shearer, M. J., Rahim, S., Gunn, W. G., Mcenery, G. and Barkhan, P. 1982. The content of phylloquinone (vitamin K_1) in human milk, cows' milk and infant formula foods determined by high-performance liquid chromatography. *J. Nutr. 112*, 1105–1117.

Harper, A. E. 1983. Coronary heart disease—an epidemic related to diet? *Am. J. Clin. Nutr. 37*, 669–681.

Hartman, A. M. and Dryden, L. P. 1965. *Vitamins in Milk and Milk Products.* American Dairy Science Association, Champaign, Ill.

Haverberg, L., Kwon, P. H. and Scrimshaw, N. S. 1980. Comparative tolerance of adolescents of differing ethnic backgrounds to lactose-containing and lactose-free dairy drinks. I. Initial experience with a double blind procedure. *Am. J. Clin. Nutr. 33*, 17–21.

Heaney, R. P., Gallagher, J. C., Johnston, C. C., Neer, R., Parfitt, A. M. and Whedon, G. D. 1982. Calcium nutrition and bone health in the elderly. *Am. J. Clin. Nutr. 36*, 986–1013.

Heaney, R. P., Recker, R. R. and Saville, P. D. 1977. Calcium balance and requirements in middle-aged women. *Am. J. Clin. Nutr. 30*, 1603–1611.

Hedrick, T. I. and Glass, L. 1975. Chemical changes in milk during exposure to fluorescent light. *J. Milk Food Technol. 38*, 129–131.

Hegarty, P. V. J. 1981. Some practical considerations in the nutritional evaluation of the mineral content of dairy products. *Ir. J. Food Sci. Technol. 5*, 157–163.

Hegenauer, J., Saltman, P., Ludwig, D., Ripley, L. and Ley, A. 1979. Iron-supplemented cow milk. Identification and spectral properties of iron bound to casein micelles. *J. Agr. Food Chem. 27*, 1294–1301.

Helferich, W. and Westhoff, D. C. 1980. *All About Yogurt.* Prentice-Hall, Englewood Cliffs, NJ.

Hemken, R. W. 1980. Milk and meat iodine content: Relation to human health. *J. Am. Vet. Med. Assoc. 176*, 1119–1121.

Hepner, G., Fried, R., St. Jeor, S., Fusetti, L. and Morin, R. 1979. Hypocholesterolemic effect of yogurt and milk. *Am. J. Clin. Nutr. 32*, 19–24.

Hertig, D. C. and Drury, E. E. 1969. Vitamin E content of milk, milk products, and simulated milks; relevance to infant nutrition. *Am. J. Clin. Nutr. 22*, 147–155.

Hill, D. J., Ford, R. P. K., Shelton, M. J. and Hosking, C. S. 1984. A study of 100 infants and young children with cow's milk allergy. *Clin. Rev. Allergy 2*, 125–142.

Ho, C. Y. and Clifford, A. J. 1976. Digestion and absorption of bovine milk xanthine oxidase and its role as an aldehyde oxidase. *J. Nutr. 106*, 1600–1609.

Ho, C. Y. and Clifford, A. J. 1977. Bovine milk xanthine oxidase blood lipids and coronary plaques in rabbits. *J. Nutr. 107*, 758–766.

Ho, C. Y., Crane, R. T. and Clifford, A. J. 1978. Studies on lymphatic absorption of and the availability of riboflavin from bovine milk xanthine oxidase. *J. Nutr. 108*, 55–60.

Hodgson, P. A., Ellefson, R. D., Eiveback, L. R., Harris, L. E., Nelson, R. A. and Weidman, W. H. 1976. Comparison of serum cholesterol in children fed high, moderate, or low cholesterol milk diets during neonatal period. *Metabolism 25*, 739–746.

Holdren, R. A., Ostfeld, A. M., Freeman, D. H., Hellenbrand, K. G. and D'Atri, D. A. 1983. Dietary salt intake and blood pressure. *JAMA 250*, 365–369.

Hollis, B. W., Roos, B. A., Draper, H. H. and Lambert, P. W. 1981. Occurrence of vitamin D sulfate in human milk whey. *J. Nutr. 111*, 384–390.

Hollis, B. W., Roos, B. A. and Lambert, P. W. 1982. Vitamin D compounds in human and bovine milk. *In: Advances in Nutritional Research*, Vol 4. H.H. Draper (Editor). Plenum Press, New York, pp. 59–75.

Horwitt, M. K., Harper, A. E. and Henderson, L. M. 1981. Niacin–tryptophan relationships for evaluating niacin equivalents. *Am. J. Clin. Nutr. 34*, 423–427.

Howard, A. N. 1977. The Masai, milk and the yogurt factor: An alternative explanation (letter). *Atherosclerosis 27*, 383–385.

Howard, A. N. and Marks, J. 1977. Hypocholesterolaemic effect of milk (letter). *Lancet* 2, 255-256.

Howard, A. N. and Marks, J. 1979. Effect of milk products on serum cholesterol. *Lancet* 2, 957.

Hussi, E., Miettinen, T. A., Ollus, A., Kostiainen, E., Ehnholm, C., Haglund, B., Huttunen, J. K. and Manninen, V. 1981. Lack of serum cholesterol-lowering effect of skimmed milk and buttermilk under controlled conditions. *Atherosclerosis 39*, 267-272.

International Dairy Federation. Group F20. 1983. Cultured dairy foods in human nutrition. FIL-IDF Document 159. Brussels, Belgium.

Jakobsson, I. and Lindberg, T. 1979. A prospective study of cow's milk protein intolerance in Swedish infants. *Acta Paediatr. Scand. 68*, 853-859.

Jarrett, W. D. 1979. A review of the important trace elements in dairy products. *Aust. J. Dairy Technol. 34*, 28-34.

Jensen, M. E. and Schachtele, C. F. 1983. The acidogenic potential of reference foods and snacks at interproximal sites in the human dentition. *J. Dent. Res. 62*, 889-892.

Johnson, N. E., Smith, E. L. and Freudenheim, J. L. 1985. Effects on blood pressure of calcium supplementation of women. *Am. J. Clin. Nutr. 42*, 12-17.

Jonas, J. J., Craig, T. W., Huston, R. L., Marth, E. H., Speckmann, E. W., Steiner, T. F. and Weisberg, S. M. 1976. Dairy products as food protein resources. *J. Milk Food Technol. 39*, 778-795.

Jones, J. L. 1973. *Homemakers' Opinion About Dairy Products and Imitations; a Nationwide Survey.* United States Department of Agriculture Market Research Report 995. Washington, D.C.

Kaplan, H. G. 1980. Inhibition of buttermilk xanthine oxidase by folate analogues and derivatives. *Biochem. Pharmacol. 29*, 2135-2141.

Karanja, N. and McCarron, D. A. 1986. Calcium and hypertension. *Ann. Rev. Nutr. 6*, 475-494.

Katz, R. S., Lofgren, P. A., Speckmann, E. W., Derse, P. H. and Robaidek, E. S. 1981. Nutritional evaluation of raw, pasteurized, and UHT milks. *J. Dairy Sci. 64*(Suppl. 1), 43.

Keim, N. L., Marlett, J. A. and Amundson, C. H. 1981. The cholesteremic effect of skim milk in young men consuming controlled diets. *Nutr. Res. 1*, 429-442.

Kilara, A. and Shanani, K. M. 1976. Lactase activity of cultured and acidified dairy products. *J. Dairy Sci. 59*, 2031-2035.

Kobayashi, A., Kawai, S., Ohbe, Y. and Nagashima, Y. 1975. Effects of dietary lactose and a lactase preparation on the intestinal absorption of calcium and magnesium in normal infants. *Am J. Clin. Nutr. 28*, 681-683.

Kocian, J., Skala, I. and Bakos, K. 1973. Calcium absorption from milk and lactose-free milk in healthy subjects and patients with lactose intolerance. *Digestion 9*, 317-324.

Kolars, J. C., Levitt, M. D., Aouji, M. and Savaiano, D. A. 1984. Yogurt—an autodigesting source of lactose. *N. Engl. J. Med. 310*, 1-3.

Kosaric, N., Kitchen, B., Panchal, C. J., Sheppard, J. D., Kennedy, K. and Sargant, A. 1981. UHT milk: Production, quality, and economics. *CRC Crit. Rev. Food Sci. Nutr. 14*, 153-199.

Kosikowski, F. 1982. *Cheese and Fermented Milk Foods*, 2nd ed. F. V. Kosikowski and Associates, Brooktondale, N.Y.

Kotula, K. T., Nikazy, J. N., McGinnis, M. and Briggs, G. M. 1983. Development of a rat model to test the nutritional equivalency of traditional vs. fabricated foods: Cheddar cheese vs. fabricated cheddar cheese. *J. Food Sci. 48*, 1674-1677, 1704.

Kotula, K. T., Nikazy, J. N., McGinnis, N. and Briggs, G. M. 1984. Protein quality of cheddar cheese compared with casein and fabricated cheese in the rat. Submitted to *J. Food Sci.*

Kribbs, P. J., Smith, D. E. and Chesnut, C. H., III. 1983. Oral findings in osteoporosis. Part II: Relationship between residual ridge and alveolar bone resorption and generalized skeletal ostopenia. *J. Prosthet. Dent. 5,* 719–724.

Kris-Etherton, P. M., Layman, D. K., York, P. V. and Frantz, I. D., Jr. 1979. The influence of early nutrition on the serum cholesterol of the adult rat. *J. Nutr. 109,* 1244–1257.

Kritchevsky, D., Tepper, S. A., Morrissey, R. B., Czarnecki, S. K. and Klurfeld, D. M. 1979. Influence of whole or skim milk on cholesterol metabolism in rats. *Am. J. Clin. Nutr. 32,* 597–600.

Kwon, P. H., Jr., Rorick, M. H. and Scrimshaw, N. S. 1980. Comparative tolerance of adolescents of differing ethnic background to lactose-containing and lactose-free dairy drinks. II. Improvement of a double-blind test. *Am. J. Clin. Nutr. 33,* 22–26.

Lacroix, D. E., Mattingly, W. A., Wong, N. P. and Alford, J. A. 1973. Cholesterol, fat and protein in dairy products. *J. Am. Diet. Assoc. 62,* 275–279.

Lacroix, D. E. and Wong, N. P. 1980. Determination of iodide in milk using the iodide specific ion electrode and its application to market milk samples. *J. Food Protection 43,* 672–674.

Lampert, L. M. 1975. *Modern Dairy Products,* 3rd edition, Chemical Publ. Co., New York.

Langford, H. G. 1983. Dietary potassium and hypertension: Epidemiologic data. *Ann. Intern. Med. 98*(Part 2), 770–772.

Leerbeck, E. and Sondergaard, H. 1980. The total content of vitamin D in human milk and cow's milk. *Br. J. Nutr. 44,* 7–12.

Light, L. and Cronin, F. J. 1981. Food guidance revisited. *J. Nutr. Educ. 13,* 57–62.

Linkswiler, H. M. and Zemel, M. B. 1979. Calcium to phosphorus ratios. Contemp. Nutr. 4(5), 1–2.

Lipid Research Clinics Program. 1984. The lipid research clinics coronary primary prevention trial results. I. Reduction in incidence of coronary heart disease. *JAMA 251,* 351–364.

Lipkin, M. and Newmark, H. 1985. Effect of added dietary calcium on colonic epithelial-cell proliferation in subjects at high risk for familial colonic cancer. *N. Engl. J. Med. 313,* 1381–1384.

Longden, K. 1980. *Domestic Food Programs: An Overview.* USDA Economics, Statistics and Cooperative Service. ESCS-81. Washington, D.C.

Lonnerdal, B., Keen, C. L. and Hurley, L. S. 1981. Iron, copper, zinc, and manganese in milk. *Ann. Rev. Nutr. 1,* 149–174.

Lowe, C. M., Kotula, K. T. and Briggs, G. M. 1983. Nutrition studies of real nonfat and fabricated dry milks with respect to the Food and Drug Administration's 1978 proposal for the nutritional equivalence (NE) of low-fat milk substitutes in rats. *J. Dairy Sci.* (Suppl. 1) *66,* 90.

Machlin, L. J. 1980. *Vitamin E. A Comprehensive Treatise.* Marcel Dekker, New York.

Mangino, M. E. and Brunner, J. R. 1976. Homogenized milk: Is it really the culprit in dietary-induced atherosclerosis? *J. Dairy Sci. 59,* 1511–1512.

Mann, G. V. 1977. A factor in yogurt which lowers cholesteremia in man. *Atherosclerosis 26,* 335–340.

Mann, G. V. and Spoerry, A. 1974. Studies of a surfactant and cholesteremia in the Maasai. *Am. J. Clin. Nutr. 27,* 464–469.

Marcus, R. 1982. The relationship of dietary calcium to the maintenance of skeletal integrity in man—an interface of endocrinology and nutrition. *Metabolism 31,* 93–102.

Marston, R. and Raper, N. 1986. Nutrient content of the food supply. *National Food Rev. 32*, 6–12.

May, C. D. and Bock, S. A. 1978. Adverse reactions to food due to hypersensitivity. *In: Allergy. Principles and Practice*, Vol. 2. E. Middleton, Jr., C. E. Reed and E.F. Ellis (Editors). C.V. Mosby Co., St. Louis, Mo. pp. 1159–1171.

McCarron, D. A. 1982. Low serum concentrations of ionized calcium in patients with hypertension. *N. Engl. J. Med. 307*, 226–228.

McCarron, D. A. 1983. Calcium and magnesium nutrition in human hypertension. *Ann. Intern. Med. 98*(Part 2), 800–805.

McCarron, D. A. 1985. Is calcium more important than sodium in the pathogenesis of essential hypertension? *Hypertension 7*, 607–627.

McCarron, D. A., and Morris, C. D. 1985. Blood pressure response to oral calcium in persons with mild to moderate hypertension. *Ann. Intern. Med. 103*, 825–831.

McCarron, D. A., Morris, C. D. and Cole, C. 1982. Dietary calcium in human hypertension. *Science 217*, 267–269.

McCarron, D. A., Morris, C. D., Henry, H. J. and Stanton, J. L. 1984. Blood pressure and nutrient intake in the United States. *Science 244*, 1392–1398.

McCarthy, R. D. and Long, C. A. 1976. Bovine milk intake and xanthine oxidase activity in blood serum. *J. Dairy Sci. 59*, 1059–1062.

McGill, H. C., Jr. 1979A. Appraisal of cholesterol as a causative factor in atherogenesis. *Am. J. Clin. Nutr. 32*, 2632–2636.

McGill, H. C., Jr. 1979B. The relationship of dietary cholesterol to serum cholesterol concentration and to atherosclerosis in man. *Am. J. Clin. Nutr. 32*, 2644–2702.

McLaughlin, P. J. and Weihrauch, J. L. 1979. Vitamin E content of foods. *J. Am. Diet. Assoc. 75*, 647–665.

McNamara, D. J. 1982. Diet and hyperlipidemia: A justifiable debate. *Arch. Intern. Med. 142*, 1121–1124.

McNamara, D. J., Quackenbush, F. W. and Rodwell, V. W. 1972. Regulation of hepatic 3-hydroxy-3-methylglutaryl coenzyme A reductase. Developmental pattern. *J. Biol. Chem. 247*, 5805–5810.

Mehta, R. S. 1980. Milk processed at ultra-high temperatures—a review. *J. Food Protection 43*, 212–225.

Mertz, W. 1981. The essential trace elements. *Science 213, 1332–1338.*

Miller, J. J. 1985. Familiar product, new form describes UHT milk. *National Food Rev. 28*, 10–14.

Milner, M., Scrimshaw, N. S. and Wang, D. I. C. 1978. *Protein Resources and Technology: Status and Research Needs.* AVI Publishing Co., Westport, Conn.

Mitchell, H. H. and Smith, J. M. 1945. The effect of cocoa on the utilization of dietary calcium. *JAMA 129*, 871–873.

Mitchell, W. D., Fyfe, T. and Smith, D. A. 1968. The effect of oral calcium on cholesterol metabolism. *J. Atheroscler. Res. 8*, 913–922.

Morrissey, R. B., Burkholder, B. D. and Tarka, S. M., Jr. 1984. The cariogenic potential of several snack foods. *J. Am. Dent. Assoc. 109*, 589–591.

Multiple Risk Factor Intervention Trial Research Group. 1982. Multiple risk factor intervention trial. Risk factor changes and mortality results. *JAMA 248*, 1465–1477.

Murthy, G. K. 1974. Trace elements in milk. *CRC Crit. Rev. Environ. Control 4*, 1–37.

Nair, C. R. and Mann, G. V. 1977. A factor in milk which influences cholesterolemia in rats. *Atherosclerosis 26*, 363–367.

National Academy of Sciences. 1980A. *Recommended Dietary Allowances*, 9th rev. ed. National Academy of Sciences, National Research Council, Food and Nutrition Board, Washington, D.C.

National Academy of Sciences. 1980B. *Toward Healthful Diets.* National Academy of Sciences, National Research Council, Food and Nutrition Board, Washington, D.C.

National Academy of Sciences. 1982. *Outlook for Science and Technology. The Next Five Years.* W.H. Freeman, San Francisco.

National Dairy Council. 1976. Composition and nutritional value of dairy foods. *Dairy Council Digest 47*(5), 25–30.

National Dairy Council. 1981. Child nutrition programs. *Dairy Council Digest 52*(1), 1–6.

National Dairy Council. 1982A. Child nutrition program update. *Dairy Council Digest 53*, 31–36.

National Dairy Council. 1982B. Diet and bone health. *Dairy Council Digest 53*(5), 25–30.

National Dairy Council. 1983A. *Newer Knowledge of Milk and Other Fluid Dairy Products.* National Dairy Council, Rosemont, Ill.

National Dairy Council. 1983B. *Newer Knowledge of Cheese and Other Cheese Products.* National Dairy Council, Rosemont, Ill.

National Dairy Council. 1983C. Imitation and substitute dairy foods. *Dairy Council Digest 54*, 1–6.

National Dairy Council. 1984A. The role of calcium in health. *Dairy Council Digest 55*, 1–8.

National Dairy Council. 1984B. *Calcium: A Summary of Current Research for the Health Professional.* National Dairy Council, Rosemont, Ill.

National Dairy Council. 1984C. Cultured and culture-containing dairy foods. *Dairy Council Digest 55*, 15–20.

National Dairy Council. 1985. Nutritional implications of lactose and lactase activity. *Dairy Council Digest 56*, 25–30.

National Heart, Lung, and Blood Institute and National Cancer Institute. 1981. *Summary. Workshop of Cholesterol and Non-Cardiovascular Disease Mortality, May 11–12.* Bethesda, Md.

National Institute of Arthritis and Musculoskeletal and Skin Diseases. 1986. Osteoporosis. Cause, treatment, prevention. NIH Pub. No. 86–2226. U.S. Department of Health and Human Services, Public Health Service, National Institutes of Health. Bethesda. Md.

National Institutes of Health Consensus Development Conference Statement. 1985. Health implications of obesity. February 11–13. Bethesda, Md.

Newbrun, E. 1982. Sugar and dental caries: A review of human studies. *Science 217*, 418–423.

Newcomer, A. D. and McGill, D. B. 1984A. Clinical consequences of lactase deficiency. *Clin. Nutr. 3*, 53–58.

Newcomer, A. D. and McGill, D. B. 1984B. Clinical importance of lactase deficiency. *N. Engl. J. Med. 310*, 42–43.

Newcomer, A. D., Park, H. S., O'Brien, P. C. and McGill, D. B. 1983. Response of patients with irritable bowel syndrome and lactase deficiency using unfermented acidophilus milk. *Am. J. Clin. Nutr. 38*, 257–263.

Nicholls, M. G. 1984. Reduction of dietary sodium in Western society. Benefit or risk? *Hypertension 6*, 795–801.

Oliver, M. F. 1982. Diet and coronary heart disease. *Human Nutr. Clin. Nutr. 36C*, 413–427.

Oster, K. A. 1971. Plasmalogen diseases: A new concept of the etiology of the atherosclerotic process. *Am. J. Clin. Res. 2*, 30–35.

Oster, K. A. and Hope-Ross, P. 1966. Plasmal reaction in a case of recent myocardial infarction. *Am. J. Cardiol. 17*, 83–85.

Oster, K. A., Oster, J. B. and Ross, D. J. 1974. Immune response to bovine xanthine oxidase in atherosclerotic patients. *Am. Lab. 7*, 41–47.

Owen, A. L., Owen, G. M. and Lanna, G. 1979. Health and nutritional benefits of federal food assistance programs. *In: Costs and Benefits of Nutritional Care, Phase 1.* American Dietetic Association, Chicago, pp. 67–79.

Paige, D. M. and Bayless, T. M. (Editors). 1981. *Lactose Digestion: Clinical and Nutritional Implications.* Johns Hopkins University Press, Baltimore.

Parfitt, A. M., Gallagher, J. C., Heaney, R. P., Johnston, C. C., Neer, R. and Whedon, G. D. 1982. Vitamin D and bone health in the elderly. *Am. J. Clin. Nutr. 36*, 1014–1031.

Pariza, M. W. 1984. A perspective on diet, nutrition, and cancer. *JAMA 251*, 1455–1458.

Park, Y. K., Harland, B. F., Vanderveen, J. E., Shank, F. R. and Prosky, L. 1981. Estimation of dietary iodine intake of Americans in recent years. *J. Am. Diet. Assoc. 79*, 17–24.

Patel, H. M. and Ryman, B. E. 1981. Systemic and oral administration of liposomes. *In: Liposomes. From Physical Structure to Therapeutic Applications.* C. G. Knight (Editor). Elsevier/North-Holland Biomedical Press, Amsterdam, pp. 409–441.

Patton, S. and Jensen, R. G. 1975. Lipid metabolism and membrane functions of the mammary gland. *In: Progress in the Chemistry of Fats and Other Lipids.* R. G. Holman (Editor). Pergamon Press, Oxford, pp. 163–277.

Pennington, J. A. T. 1980. Total diet study—results and plans for selected minerals in foods. *FDA Bylines 10*, 179–188.

Porter, J. W. G. 1975. *Milk and Dairy Foods.* Oxford University Press, London.

Porter, J. W. G. 1978. The present nutritional status of milk protein. *J. Soc. Dairy Technol. 31*, 199–202.

Posati, L. P., Kinsella, J. E. and Watt, B. K. 1975. Comprehensive evaluation of fatty acids in foods. *J. Am. Diet. Assoc. 66*, 482–488.

Prasad, A. S. 1978. *Trace Elements and Iron in Human Metabolism.* Plenum Medical Book Co., New York.

Prasad, A. S. 1979. *Zinc in Human Nutrition.* CRC Press, Boca Raton, Fla.

Protein Advisory Group of the United Nations. 1972. PAG statement 17 on low lactase activity and milk intake. *PAG Bull 2*(2), 9–11.

Radzikowski, J. 1983. *The National Evaluation of School Nutrition Programs: Final Report—Executive Summary.* U.S. Department of Agriculture, Washington, D.C.

Radzikowski, J. and Gale, S. 1984. Requirements for the national evaluation of school nutrition programs. *Am. J. Clin. Nutr. 40*(Suppl), 365–367.

Rao, D. R., Chawan, C. B. and Pulusani, S. R. 1981. Influence of milks and *S. thermophilus* milk on plasma cholesterol levels and hepatic cholesterogenesis in rats. *J. Food Sci. 46*, 1339–1341.

Rapport, M. M. and Norton, W. T. 1962. Chemistry of the lipids. *Ann. Rev. Biochem. 31*, 103–138.

Rasic, J. L. and Kurmann, J. A. 1978. *Yogurt. Scientific Grounds, Technology, Manufacture and Preparations. Fermented Fresh Milk Products*, Vol. 1. Rasic and Kurmann, Copenhagen.

Rechcigl, M., Jr. (Editor). 1983. *CRC Handbook of Nutritional Supplements*, Vol. 1, *Human Use.* CRC Press, Boca Raton, Fla., pp. 133–252.

Reddy, G. V., Friend, B. A., Shahani, K. M. and Farmer, R. E. 1983. Antitumor activity of yogurt components. *J. Food Protection 46*, 8–11.

Reddy, G. V., Shahani, K. M. and Banerjee, M. R. 1973. Inhibitory effect of yogurt on Ehrlich asites tumor-cell proliferation. *J. Natl. Cancer Inst. 50*, 815–817.

Reeve, L. E., Jorgensen, N. A. and DeLuca, H. F. 1982. Vitamin D compounds in cow's milk. *J. Nutr. 112*, 667–672.

Reif, G. D., Franke, A. A. and Bruhn, J. C. 1983. Retail dairy foods quality—an asessment of the incidence of off-flavors in California milk. *Dairy and Food Sanitation 3,* 44–46.

Reiser, R. 1975. Letter: experimentation with human subjects. *Am. J. Clin. Nutr. 28,* 2.

Reiser, R. and Sidelman, Z. 1972. Control of serum cholesterol homeostasis by cholesterol in the milk of the suckling rat. *Am. J. Nutr. 102,* 1009–1016.

Renner, E. 1980. Nutritional and biochemical characteristics of UHT milk. *In: Proceedings of the International Conference on UHT Processing and Aseptic Packaging of Milk and Milk Products.* Department of Food Science, North Carolina State University, Raleigh, N.C.

Renner, E. 1983. *Milk and Dairy Products in Human Nutrition.* W. Gmbh. Volkswirtschaftlicher Verlag, Munich, W. Germany.

Resnick, L. M. 1985. Calcium and hypertension: The emerging connection. *Ann. Intern. Med. 103,* 944–945.

Richardson, T. J. 1978. The hypocholesterolemic effect of milk—a review. *J. Food Protection 41,* 226–235.

Rogoff, G. S., Galburt, R. B. and Nizel, A. E. 1984. Role of dietary calcium and vitamin D in alveolar bone health. Literature Review Update. *In: Calcium in Biological Systems.* R.P. Rubin, G. B. Weiss and J.W. Putney, Jr. (Editors). Plenum Pub. Co., New York, pp. 591–595.

Rzucidlo, S. J. and Zikakis, J. P. 1979. Correlation of dairy food intake with human antibody to bovine milk xanthine oxidase. *Proc. Soc. Exp. Biol. Med. 160,* 477–482.

Sabine, J. R. 1977. *Cholesterol.* Marcel Dekker, New York.

Sanuel, P., McNamara, D. J., and Shapiro, J. 1983. The role of diet in the etiology and treatment of atherosclerosis. *Ann. Rev. Med. 34,* 179–194.

Sandström, B., Cederblad, A. and Lonnerdal, B. 1983. Zinc absorption from human milk, cow's milk and infant formulas. *Am. J. Dis. Child. 137,* 726–729.

Savilahti, E., Kuitunen, P. and Visakorpi, J. K. 1981. Cow's milk allergy. *In: Textbook of Gastroenterology and Nutrition in Infancy.* E. Lebenthal (Editor). Raven Press, New York, pp. 689–708.

Savilahti, E. and Verkasalo, M. 1984. Intestinal cow's milk allergy: Pathogenesis and clinical presentation. *Clin. Rev. Allergy 2,* 7–23.

Schaafsma, G. 1983. The significance of milk as a source of dietary calcium. *In: Nutrition and Metabolism.* International Dairy Federation Bulletin, Document 166, pp. 19–30.

Schaafsma, G. and Visser, R. 1980. Nutritional interrelationships between calcium, phosphorus and lactose in rats. *J. Nutr. 110,* 1101–1111.

Schachtele, C. F. and Harlander, S. K. 1984. Will the diets of the future be less cariogenic? *J. Can. Dent. Assoc. 3,* 213–219.

Scott, M. L. 1980. Advances in our understanding of vitamin E. *Fed. Proc. 39,* 2736–2739.

Segall, J. J. 1977. Is milk a coronary health hazard? *Br. J. Prev. Soc. Med. 31*(2), 81–85.

Select Committee on GRAS Substances. 1979. *Evaluation of the Health Aspects of Sodium Chloride and Potassium Chloride as Food Ingredients.* SCOGS–102. Federation of American Societies for Experimental Biology, Life Sciences Research Office, Bethesda, Md.

Sellars, R. L. 1981. Fermented dairy foods. *J. Dairy Sci. 65,* 1070–1076.

Senyk, G. F. and Shipe, W. F. 1981. Protecting your milk from nutrient losses. *Dairy Field 164,* 81–85.

Shahani, K. M. 1983. Nutritional impact of lactobacillic fermented foods. *In: Nutrition and the Intestinal Flora.* Symposia of the Swedish Nutrition Foundation XV. B. Hallgren (Editor). Almqvist & Wiksell International, Stockholm, pp. 103–111.

Shahani, K. M. and Chandan, R. C. 1979. Nutritional and healthful aspects of cultured and culture-containing dairy foods. *J. Dairy Sci. 62,* 1685-1694.

Shank, F. R., Park, Y. K., Harland, B. F., Vanderveen, J. E., Forbes, A. L. and Prosky, L. 1982. Perspective of Food and Drug Administration on dietary sodium. *J. Am. Diet. Assoc. 80,* 29-39.

Sidney, S. and Farquhar, J. W. 1983. Cholesterol, cancer, and public health policy. *Am. J. Med. 75,* 494-508.

Silva, M. F. de A., Jenkins, G. N., Burgess, R. C. and Sandham, H. J. 1986. Effects of cheese on experimental caries in human subjects. *Caries Res. 20,* 263-269.

Simopoulos, A. P. 1985. The health implications of overweight and obesity. *Nutr. Rev. 43,* 33-40.

Sinha, D. K., Dam, R. and Shahani, K. M. 1979. Evaluation of the properties of a nonfermented acidophilus milk. *J. Dairy Sci. 62*(Suppl. I), 52.

Solomons, N. W. 1982. Factors affecting the bioavailability of zinc. *J. Am. Diet. Assoc. 80,* 115-121.

Speck, M. L. and Katz, R. S. 1980. ACDPI status paper. Nutritive and health values of cultured dairy foods. *Cultured Dairy Products J. 15*(4), 10-11.

Speckmann, E. W. 1984. Nutritional characteristics of dairy products. *In: Dairy Products for the central Processing Industry.* J. L. Vetter (Editor). American Association of Cereal Chemists. St. Paul, Minn., pp. 55-82.

Speckmann, E. W., Brink, M. F. and McBean, L. D. 1981. Dairy foods in nutrition and health. *J. Dairy Sci. 64,* 1008-1016.

Spencer, H., Kramer, L. and Osis, D. 1982A. Factors contributing to calcium loss in aging. *Am. J. Clin. Nutr. 36,* 776-787.

Spencer, H., Kramer, L., Lesniak, M., Debartolo, M., Norris, C. and Osis, D. 1984. Calcium requirements in humans. Report of original data and a review. *Clin. Orthop. 184,* 270-279.

Spencer, H., Kramer, L., Norris, C. and Osis, D. 1982B. Effect of small doses of aluminum-containing antacids on calcium and phosphorus metabolism. *Am. J. Clin. Nutr. 36,* 32-40.

Stamler, J. 1981. Primary prevention of coronary heart disease. *Am. J. Cardiol. 47,* 722-735.

Stanley, D. W., Murray, E. D. and Lees, D. H. 1981. *Utilization of Protein Resources.* Food and Nutrition Press, Westport, Conn.

Strazzullo, P., Nunziata, V., Cirillo, M., Giannattasio, R., Ferrara, L. A., Mattioli, P. L. and Mancini, M. 1983. Abnormalities of calcium metabolism in essential hypertension. *Clin. Sci. 65,* 137-141.

Suttie, J. W. 1980. *Vitamin K Metabolism and Vitamin K-Dependent Proteins.* University Park Press, Baltimore.

Swanson, E. W. 1981. Investigating iodine in milk and beef—working for FDA. *In: Stokely-Van Camp Annual Symposium. Food in Contemporary Society. Emerging Patterns.* University of Tennessee Press, Knoxville, Tenn., pp. 165-169.

Talbot, J. M., Fisher, K. D. and Carr, C. J. 1976. *A Review of the Effects of Dietary Iodine on Certain Thyroid Disorders.* Federation of American Societies for Experimental Biology, Life Sciences Research Office, Bethesda, Md.

Tannen, R. L. 1983. Effects of potassium on blood pressure control. *Ann. Intern. Med. 98*(Part 2), 773-780.

Thakur, C. P. and Jha, A. N. 1981. Influence of milk, yoghurt and calcium on cholesterol-induced atherosclerosis in rabbits. *Atherosclerosis, 39,* 211-215.

The American Dietetic Association. 1980. Position paper on the vegetarian approach to eating. *J. Am. Diet. Assoc. 77,* 61-69.

The American Society for Bone and Mineral Research. 1982. *Osteoporosis.* Kelseyville, Calif.

Thompson, L. U., Jenkins, D. J. A., Amer, V., Reichert, R., Jenkins, A. and Kamulsky, J. 1982. The effect of fermented and unfermented milks on serum cholesterol. *Am. J. Clin. Nutr. 36*, 1106–1111.

Torun, B., Solomons, N. W. and Viteri, F. E. 1979. Lactose malabsorption and lactose intolerance: Implications for general milk consumption. *Archivos Latinoamericanos de Nutricion 29*, 445–494.

Underwood, E. J. 1977. *Trace Elements in Human and Animal Nutrition*, 4th ed. Academic Press, New York.

USDA. 1978. *Cheese Varieties and Descriptions*. Agriculture Handbook No. 54. Dairy Laboratory, Eastern Regional Research Center, Agriculture Research Service, Philadelphia.

USDA Agricultural Research Service. 1957. *Essentials of an Adequate Diet*. Home Economics Research Report No. 3. U.S. Government Printing Office, Washington, D.C.

USDA Agricultural Research Service. 1958. *Food for Fitness—A Daily Food Guide*, 3rd rev. ed. USDA Leaflet No. 424. U.S. Government Printing Office, Washington, D.C.

United States Department of Agriculture and the United States Department of Health, Education and Welfare. 1980. *Nutrition and Your Health: Dietary Guidelines for Americans*. Home and Garden Bull. 232. U.S. Department of Agriculture and U.S. Department of Health, Education and Welfare, Washington, D.C.

USDA, Consumer and Food Economics Institute. 1976. *Composition of Foods, Dairy and Egg Products. Raw, Processed, Prepared*. Agriculture Handbook No. 8-1. U.S. Department of Agriculture, Washington, D.C.

USDA, Food and Nutrition Service, Management Information Division. 1983A. *Annual Historical Review of FNS Programs: Fiscal Year 1982*. U.S. Department of Agriculture, Washington, D.C.

USDA, Food and Nutrition Service, Office of Analysis and Evaluation. 1983B. Personal communication. Rosemont, Ill.

USDA, Science and Education Administration. 1979. *Food. The Hassle-Free Guide to a Better Diet*. Home and Garden Bull. 228. U.S. Government Printing Office, Washington, D.C.

USDA, Science and Education Administration. 1980. *Food and Nutrient Intakes of Individuals in One Day in the United States, Spring 1977*. Nationwide Food Consumption Survey 1977–78, Preliminary Report No. 2. Consumer Nutrition Center, Hyattsville, Md.

United States Department of Health, Education and Welfare. 1979. *Healthy People: The Surgeon General's Report on Health Promotion and Disease Prevention*. DHEW Pub. No. 79-55071. U.S. Department of Health, Education, and Welfare. Public Health Service. Office of the Assistant Secretary for Health and Surgeon General, Washington, D.C.

Van Itallie, T. B. 1982. Symposium on current perspectives in hypertension: summary. *Hypertension 4*(Suppl. III), III-177–III-183.

Villar, J., Repke, J. and Belizan, J. M. 1986. Calcium and blood pressure. *Clin. Nutr. 5*, 153–160.

Vital and Health Statistics of the National Center for Health Statistics. 1979. Overweight adults in the United States. Advance data. DHEW Pub. No. 79-1250. Department of Health, Education and Welfare, Washington, D.C.

Volp, R. F. and Lage, G. L. 1977. Studies on the intestinal absorption of bovine xanthine oxidase. *Proc. Soc. Exp. Biol. Med. 154*, 488–492.

Webb, B. H., Johnson, A. H. and Alford, J. A. 1974. *Fundamentals of Dairy Chemistry*, 2nd ed. AVI Publishing Co., Westport, Conn.

Weidman, W., Kwiterovich, P., Jr., Jesse, M. J. and Nugent, E. 1983. Diet in the healthy child. *Circulation 67,* 1411A–1414A.

Welsh, J. D. 1978. Diet therapy in adult lactose malabsorption: Present practices. *Am. J. Clin. Nutr. 31,* 592–596.

Whitaker, J. R. and Tannenbaum, S. R. 1977. *Food Proteins.* AVI Publishing Co., Westport, Conn.

White, P. L. and Crocco, S. C. 1980. *Sodium and Potassium in Foods and Drugs.* American Medical Association, Chicago.

Wilson, J. F., Lahey, M. E. and Heiner, D. C. 1974. Studies on iron metabolism. V. Further observations on cow's milk-induced gastrointestinal bleeding in infants with iron-deficiency anemia. *J. Pediatr. 84,* 335–344.

Wong, N. P. and LaCroix, D. E. 1980. Biological availability of calcium in dairy products. *Nutr. Rep. Int. 21,* 673–680.

Wong, N. P., LaCroix, D. E. and Alford, J. A. 1978. Mineral content of dairy products. *J. Am. Diet. Assoc. 72,* 288–291.

Wong, N. P., McDonough, F. E. and Hitchins, A. D. 1983. Contribution of *Streptococcus thermophilus* to growth-stimulating effect of yogurt on rats. *J. Dairy Sci. 66,* 444–449.

Woodruff, C. W. 1977. Iron deficiency in infancy and childhood. *Pediatr. Clin. North Am. 24,* 85–94.

Woodruff, C. W. 1978. The science of nutrition and the art of infant feeding. *JAMA 240,* 657–661.

World Health Organization. 1977. Enrichment of dried skim milk. *Food Nutr. 3,* 2–7.

Zemel, M. B., Soullier, B. A. and Steinhardt, N. J. 1982. Effects of calcium, ortho- and polyphosphates on calcium, zinc, iron, and copper bioavailability in man. *Fed. Proc. 42,* 397.

Ziegler, E. E. and Fomon, S. J. 1983. Lactose enhances mineral absorption in infancy. *J. Pediatr. Gastroenterol. Nutr. 2,* 288–294.

Zilversmit, D. B. 1982. Diet and heart disease: Prudence, probability and proof. *Artherosclerosis 2,* 83.

Zikakis, J. P., Rzucidlo, S. J. and Biasotto, N. O. 1977. Persistence of bovine milk xanthine oxidase activity after gastric digestion in vivo and in vitro. *J. Dairy Sci. 60,* 533–541.

Zikakis, J. P. and Wooters, S. C. 1980. Activity of xanthine oxidase in dairy products. *J. Dairy Sci. 63,* 893–904.

8

Physical Properties of Milk

John W. Sherbon

Physically, milk is a rather dilute emulsion combined with a colloidal dispersion in which the continuous phase is a solution. Its physical properties are similar to those of water but are modified by the concentration of solutes and by the state of dispersion of the other components.

In the dairy industry, measurements of the physical properties of milk and dairy products are made to secure data necessary for the design of dairy equipment (e.g., heat conductivity and viscosity), to determine the concentration of a component or group of components (e.g., specific gravity to estimate the solids-not-fat or freezing point to determine added water), or to assess the extent of a chemical or physical change (e.g., titratable acidity to follow bacterial action or viscosity to assess the aggregation of protein miscelles or fat globules). The great advantage of physical measurements for such purposes is their speed and simplicity, as well as their potentiality for automation.

The use of a physical property to measure concentrations or changes in the degree of dispersion demands knowledge of the contribution of the several components to that property. Furthermore, the natural range of variation of the property in milks or products is of major interest. The precision and suitability of possible methods of measurement are also of prime importance.

In this chapter, several physical properties will be discussed in terms of (1) general physical principles, (2) objectives of study in the dairy field, (3) methods of measurement, (4) contributions of milk components, (5) normal range of values and extent of natural variations, and (6) effects of processing treatments.

This chapter deals primarily with the physical properties of milk itself (and is confined to milk of the bovine species). Most of the principles discussed are, however, applicable in some degree to the physical properties of various milk products.

ACID-BASE EQUILIBRIA

The equilibria involving protons and the substances which bind them are among the most important in dairy chemistry. The ionized and ionizable components of milk are in a state of rather delicate physical balance. Certain treatments which alter the state of dispersion of proteins and salts are reflected in the status of the protons. Thus the intensity (pH) and capacity (buffer power) factors of the acid-base equilibria have come to be widely used in processing control.

The principles involved in these equilibria are presented in detail in numerous works and will not be repeated here. For a thorough treatment, including the definition of pH scales and methods of measurement, the reader is referred to such works as those of Edsall and Wyman (1958) and Bates (1964). Applications to milk are discussed by Walstra and Jenness (1984). It should suffice here to present some of the basic relationships in equation form.

$$pH = log_{10} \frac{1}{a_{H_3O}} = log \frac{1}{f_H [H_3O]}$$

where a_{H3O} = activity of the hydronium ion, $[H_3O^+]$ = concentration of the hydronium ion in moles/liter, and f_H = activity coefficient of the hydronium ion.

For many purposes, it is sufficiently accurate to use $[H_3O^+]$ instead of ^aH_3O. It is often written simply $[H^+]$. For a weak acid, HA, dissociating into H^+ and A^-, the dissociation constant, K_a, is given by the expression

$$K_a = \frac{[H^+][A^-]}{[HA]}$$

and hence

$$pH = pk_a + log \frac{[A^-]}{[HA]}$$

where pK_a is defined in an analogous manner to pH.

The measure of buffer capacity [dB/d(pH)] is the slope of the titration curve (pH plotted against increments of the base added) at any point.

$$\frac{dB}{dpH} = 2.303 \left[\frac{K_a C[H^+]}{(K_a + [H^+])^2} + [H^+] + [OH^-] \right]$$

where C = total concentration of the weak acid; for a close approxima-
tion between pH values of 3 to 11 and values of C from 0.01 to 0.10
M, the last two terms may be neglected.

Maximum buffering occurs when pH = pK_a; thus:

$$\left(\frac{dB}{dpH} \right)_{max} = \frac{2.303C}{4} = 0.576C$$

In applications to milk, dB/d(pH) is evaluated experimentally, since
calculations from the concentrations of buffer salts present are extremely involved.

The pH of cow's milk is commonly stated as falling between 6.5 and
6.7, with 6.6 being the most usual value. It should be emphasized, however, that this value applies only at temperatures of measurement near
25°C. The pH of milk exhibits a greater dependence upon temperature
than that of buffers such as phosphate, which is the principal buffer
component of milk at pH 6.6. Miller and Sommer (1940) reported a
specimen with a pH of 6.64 at 20°C, decreasing to 6.23 at 60°C. Over
the same temperature range, a phosphate buffer decreases only from
pH 6.88 to 6.84 (Bates 1964). Likewise, Dixon (1963) observed that the
pH of milk decreases by about 0.01 unit per degree Celsius between 30
and 10°C, and emphasized the importance of careful temperature control in making pH measurements. The marked temperature dependence of the pH of milk probably is attributable to insolubilization of
calcium phosphate as the temperature is raised and its solution as the
temperature is lowered.

The temperature compensator on pH meters does not account for
the effect of temperature on milk pH or on buffer pH. Its only function
is to adjust for the effect of temperature on the electrical characteristics of the electrodes.

Differences in pH and buffering capacity among individual lots of
fresh milk reflect compositional variations arising from the functions
of the mammary gland. In general the pH is lower in colostrum (down
to pH 6.0; McIntyre et al. 1952) and higher in cases of mastitis (up to
pH 7.5; Prouty 1940) than in normal milk of mid-lactation. As discussed in Chapter 1, colostrum and mastitis milks are known to differ
radically in their proportions of the proteins and certain salts. Milks
of lower phosphorus, casein, and Ca^{2+} tend to be low in titratable acidity, while excessive acidity is related to hyperketonemia, inadequate
calcium and excessive concentrates in the ration (Bonomi 1978).

Titration curves for milk have been published by Buchanan and Peterson (1927), Clark (1934), McIntyre et al. (1952), Watson (1931), Whit-

tier (1933A), and Wiley (1935). Most of these deal with the range between pH 4 and 9. Over this range, milk exhibits a pronounced maximum buffering between pH 5 and 6, the position of this maximum depending on the titration method used (Wiley 1935). The low buffer capacity in the region of the phenolphthalein endpoint (pH 8.3) contributes to the practicality of the well-known procedures for determination of the titratable acidity.

Table 8.1 presents data for the pH and titratable acidity observed in milks of different breeds, as well as for pooled milk. There appears to be a reduction in the mean titratable acidity over the 28 years spanned by the data. The drop in maximum values is consistent with an improvement in the microbiological quality of the milk supply.

In principle, it would be logical to combine plots of the buffer index curves of each of the buffer components of milk and thus obtain a plot which could be compared with that actually found for milk. It is not difficult, of course, to conclude that the principal buffer components are phosphate, citrate, bicarbonate, and proteins, but quantitative assignment of the buffer capacity to these components proves to be rather difficult. This problem arises primarily from the presence of calcium and magnesium in the system. These alkaline earths are present as free ions; as soluble, undissociated complexes with phosphates, citrate, and casein; and as colloidal phosphates associated with casein. Thus precise definition of the ionic equilibria in milk becomes rather complicated. It is difficult to obtain ratios for the various physical states of some of the components, even in simple systems. Some concentrations must be calculated from the dissociation constants, whose

Table 8.1. Titratable Acidity and pH of Milk.

Breed	No. of samples	Titratable Acidity		Mean pH
		Range	Mean	
		— % lactic acid —		
Ayrshire	229[a]	0.08–0.24	0.160	—
Holstein	297[a]	0.10–0.28	0.161	—
	606[b]	—	0.133	6.71
Guernsey	153[a]	0.12–0.30	0.172	—
	384[b]	—	0.151	6.65
Jersey	132[a]	0.10–0.24	0.179	—
	1062[b]	—	0.149	6.66
Pooled	361[c]	0.12–0.21	0.134	6.66[d]

[a] Caufield and Riddell (1936).
[b] Wilcox and Krienke (1964).
[c] Herrington et al. (1972).
[d] $n = 850$.

values in turn depend upon the ionic strength of the system. (See Table 1.2 and the associated discussion.)

Calcium and magnesium influence the titration curves of milk because as the pH is raised they precipitate as colloidal phosphates, and as the pH is lowered, colloidal calcium and magnesium phosphates are solubilized. Since these changes in state are sluggish and the composition of the precipitates depends on the conditions (Boulet and Marier 1961), the slope of the titration curves and the position of the maximum buffering depend upon the speed of the titration.

Three approaches have been used in attempting to account for the buffer behavior of milk in terms of the properties of its components. These are calculation, fractionation, and titration of artificial mixtures. Whittier (1933A,B) derived equations for dB/dpH in calcium phosphate and calcium citrate solutions, taking into account available data on dissociation constants and solubility products. Presumably this approach could be extended to calculate the entire buffer curve. It demands precise knowledge of the dissociation constants of the several buffers, the dissociation of the calcium and magnesium complexes, and the solubility products of the calcium and magnesium phosphates under the conditions of a titration of milk.

Fractionation of milk and titration of the fractions have been of considerable value. Rice and Markley (1924) made an attempt to assign contributions of the various milk components to titratable acidity. One scheme utilizes oxalate to precipitate calcium and rennet to remove the calcium caseinate phosphate micelles (Horst 1947; Ling 1936; Pyne and Ryan 1950). As formulated by Ling, the scheme involves titrations of milk, oxalated milk, rennet whey, and oxalated rennet whey to the phenolphthalein endpoint. From such titrations, Ling calculated that the caseinate contributed about 0.8 mEq of the total titer of 2.2 mEq/100 ml (0.19% lactic acid) in certain milks that he analyzed. These data are consistent with calculations based on the concentrations of phosphate and proteins present (Walstra and Jenness 1984). The casein, serum proteins, colloidal inorganic phosphorus, and dissolved inorganic phosphorus were accounted for by van der Have et al. (1979) in their equation relating the titratable acidity of individual cow's milks to the composition. The casein and phosphates account for the major part of the titratable acidity of fresh milk.

Titrations of artificially prepared mixtures containing phosphate, calcium, citrate, and sometimes proteins have been employed to study the precipitation of calcium phosphate and the inhibitory effect of citrate thereon (Boulet and Rose 1954; Eilers et al. 1947; Wiley 1935). The technique is valuable for basic studies because the composition of the system can be controlled.

In dairy processing operations, the pH and buffering power of milk are influenced (aside from the action of microorganisms) by heat treatments that may be applied. Moderate heating such as pasteurization produces small shifts in pH and buffering by expulsion of CO_2 and by precipitation of calcium phosphate with release of hydrogen ions (Pyne 1962). The drastic heat treatments used in sterilization produce acids by degradation of lactose (Gould 1945; Gould and Frantz 1945; Walstra and Jenness 1984). The rate is slow below 90°C but increases markedly above 100°C (Whittier and Benton 1927).

Concentration of milk lowers the pH. At concentrations of 30 and 60 g solids/100 g water, the pH values are about 6.2 and 6.0, respectively (Eilers et al. 1947; Howat and Wright 1934).

During slow freezing, the pH of milk has been observed to fall to values as low as 5.8, whereas little change in pH occurs during fast freezing (van den Berg, 1961). During storage of frozen milk at −7 or −12°C, the pH decreases to a minimum of about pH 6.0 and increases gradually thereafter. These effects of freezing and frozen storage are considered to be caused by insolubilization of salt constituents.

It is generally accepted that the higher the developed acidity, the lower the heat stability of the milk. The many factors involved are reviewed by Fox and Morrisey (1977). Heat stability of acid milks is improved by anion exchange (Tikhomirova et al. 1979). However, increased acidity resulting from increased protein, phosphorus, or calcium does not cause as serious a problem in normal milks (Sebela and Klicnik 1977).

OXIDATION-REDUCTION EQUILIBRIA

Whether a reversible oxidation-reduction reaction involves a transfer of oxygen, hydrogen, both, or neither, there is a transfer of electrons between atoms or molecules. Reduction is the addition of electrons and oxidation is the withdrawal of electrons from a molecule. On this basis, and the law of mass action, the following basic equation can be derived (Clark 1960):

$$E_h = E_o - \frac{RT}{nF} \ln \frac{[Red]}{[Ox]}$$

where E_h = oxidation reduction potential, E_o = standard oxidation-reduction potential of the system, R = gas constant, T = absolute temperature, n = number of electrons transferred per molecule, F = the Faraday constant, [Red] = molar concentration of the reduced form, and [Ox] = molar concentration of the oxidized form. At 25°C and one electron transfer the equation becomes

$$E_h = E_o + 0.06 \, log \, \frac{[Ox]}{[Red]}$$

E_h is considered by biochemists to be more positive if the oxidized form predominates.

The standard potential, E_o, obviously is the value of the potential at equal concentrations of the oxidized and reduced forms. Its value is an index of the relative position of the system on the scale of potential.

In considering oxidation-reduction equilibria in milk, the principal interest is in the potentials of the system relative to one another and to those that may be superimposed. Hence, it does not seem necessary to discuss the effects of differences in n on the slopes of curves of E_h plotted against percentage reductions or of the relations between pH and E_h. A discussion of these relationships is found in the monograph by Clark (1960). In the following discussion, the symbol E_0 is used to designate the potential of a system containing equal concentrations of oxidant and reductant at a *specified pH* value. The curves in Figure 8.1 are for systems present in milk and for indicators added to milk for the purpose of measuring bacterial activity. These curves indicate the relationship among these systems at different pH values.

In a fluid such as milk, which contains several oxidation-reduction systems, the effect of each system on the potential depends on several factors. These include the reversibility of the system, its E_0 value or position on the scale of potential, the ratio of oxidant to reductant, and the concentration of active components of the system. Only a reversible system gives a potential at a noble metal electrode, and this measured potential is an intensity factor analogous to the potential measured on a hydrogen electrode in determining hydrogen ion concentrations.

The quantity factor in oxidation-reduction is the overall concentration of active substance, [Ox] + [Red]. Two solutions of the same system having the same ratio of reductant to oxidant have the same potential but may have different quantity factors, such as [0.1]/[0.2] and [0.8]/[1.6], in which case the second will be able to oxidize nearly eight times as much reduced substance in a system of low potential as the first. If two reversible systems are combined, their potentials change to a common value intermediate between those of the two initial potentials. Part of the oxidant of the system that was initially more positive will be reduced and part of the system that was initially more negative

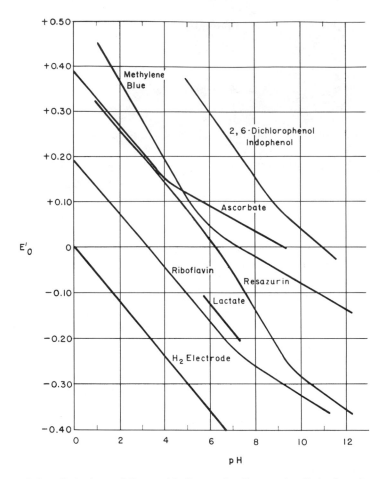

Figure 8.1. Relation of the oxidation-reduction potential of various systems to pH.

will be oxidzed. The value of the final potential will depend upon the relative concentrations of the two systems.

It is also important to consider the kinetics of the reactions, as well as the position of the equilibrium, which is where E_h values should be measured. Not all of the reactions occur at the same rate, and back reactions may differ in rate from forward reactions. Thus it becomes difficult to be sure that the several systems of milk have reached equilibrium prior to making a measurement.

Oxidation-reduction systems exhibit resistance to change of potential when the concentrations of the oxidant and reductant are close to being equal. This phenomenon, analogous to buffer action in acid-base equilibria, is known as "poising."

Fresh milk, as ordinarily produced, exhibits a potential at a gold or platinum electrode of between +0.20 and +0.30 V. That dissolved oxygen is a major factor in the establishment of this potential has been shown in several ways. Milk drawn from the udder anaerobically reduces methylene blue, indicating that its potential is more negative than that of the methylene blue system (Jackson 1936). When such a milk is exposed to oxygen, it becomes more positive than the methylene blue system. Washing oxygen-containing milk with oxygen-free gas, or allowing growing S. lactis or lactobacilli to remove free oxygen from milk, causes the potential to change in the negative direction (Eilers et al. 1947; Harland et al. 1952). Bubbling air or oxygen through these milks will restore the positive potential. This leaves the systems involving ascorbate, lactate, and riboflavin as those that may be responsible for the values of the oxidation-reduction in oxygen-free milk and may participate in the stabilization of the potential of oxygen-containing milk. The relative positions of these systems are shown in Figure 8.1. The concentration of ascorbic acid in milk is sufficient for it to exert an appreciable effect, and the system is reversible. The oxidized form of the system, dehydroascorbate, readily undergoes further oxidation, but irreversibly; hence the ratio of the concentration of ascorbate to that of dehydroascorbate will remain large until the system disappears from the milk, and this system will tend to stabilize the potential at approximately 0.0 V. The lactate-pyruvate system is irreversible, activatable by enzymes and mediators and of a highly negative normal potential (Barron and Hastings 1934), but it is present in fresh milk in such minute quantities that even if it were activated, its effect would be very slight. The riboflavin system is an active, reversible one and of highly negative normal potential; but since its concentration is low and it is present in fresh milk entirely in the oxidized form, its influence would be slight and would not be exerted in the direction of negative potentials. It seems logical to conclude that that the ascorbate–dehydroascorbate system is the principal one stabilizing the potential of oxygen-free milk at a value near 0.0 V and is the system that functions, along with the oxygen system, to stablize the potential of oxygen-containing milk in the zone of +0.20 to +0.30 V.

When milk undergoes fermentation by S. lactis, the oxidation-reduction potential of the milk changes with time (Rangappa 1948B), typically as shown in Figure 8.2. This pattern has been observed and recorded by a number of investigators. Curves characteristic of fermentation by other organisms that may be present in milk differ somewhat, but the potential tends to be changed in a negative direction.

The rapid change of potential shown in Figure 8.2 occurs only after the dissolved oxygen has been consumed by the bacteria and may be identified by the change in color of certain dyes added to the milk. These dyes are oxidizers of a redox system. Since the time elapsing before these dyes are reduced to the colorless reductant form is roughly proportional to the number of bacteria present, this "reduction time" is an index of the degree of bacterial contamination.

Using potentiometric titration at 37°C with sodium hydrosulfite ($Na_2S_2O_4$), Nilsson and co-workers (1970) demonstrated large variations in the poising effect of milk specimens from individual cows and bulked supplies. The poising index, defined as equivalents of $Na_2S_2O_4$ per liter of milk required to attain the fully reduced state, varied from 0.00488 to 0.0229 equiv. liter^{-1}·V^{-1} for 58 specimens. The highest poising indexes were found in specimens taken early in lactation, but no correlation of poising with the concentration of any natural component was made. Feed was found to affect E_h; cows on pasture only produced milk with an E_h some 20 mV less than the −180 mV observed for milks

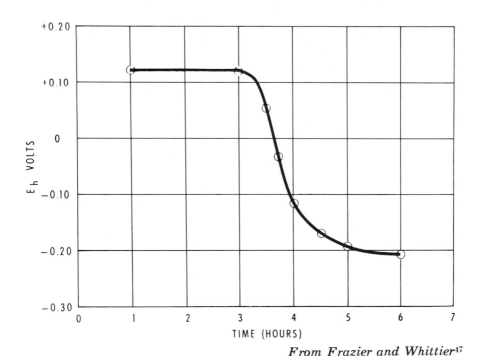

Figure 8.2. Decrease in the oxidation-reduction potential of milk during incubation with a strain of *S. lactis* at 25°C. (Frazier and Whittier 1931.)

from cows receiving both pasture and silage (Zlabinger and Stock 1978).

The effects of heat treatment of milk on the oxidation-reduction potential have been studied to a considerable extent (Eilers et al. 1947; Gould and Sommer 1939; Harland et al. 1952; Josephson and Doan 1939). A sharp decrease in the potential coincides with the liberation of sulfhydryl groups by denaturation of the protein, primarily β-lactoglobulin. Minimum potentials are attainable by deaeration and high-temperature–short-time heat treatments (Higginbottom and Taylor 1960). Such treatments also produce dried milks of superior stability against oxidative flavor deterioration (Harland et al. 1952).

A discussion of redox reactions and photooxidation is presented in the text by Walstra and Jenness (1984). This topic is especially important in view of the long light exposures given to fluid milk in transparent packages during the marketing process.

DENSITY

Interest in milk density is twofold. It has been used, along with the fat test, to estimate total solids contents. Raw milk is purchased by weight, but processed milk is sold by volume.

It should be remembered that "density" refers to the weight per unit volume of product and "specific gravity" is the ratio of the density of a product to that of water at some specified temperature. The coefficient of thermal expansion is the effect of temperature on density, and each substance has its own coefficient. Thus, when speaking of specific gravity, it is desirable to state both the sample and water temperatures; frequently, they are the same.

The density of milk is the resultant of the densities of the various components. It is complicated by changes related to the liquid-solid fat ratio and to the degree of hydration of the proteins. Thus the density of a given specimen of milk is determined by its previous temperature history, as well as by its composition.

Empirical equations of the form $T = aF + bD + c$, expressing the relation between total solids (T), fat (F), and density (D), have been used for years. Such derivations assume constant values for the density of the fat and of the mixture of solids-not-fat which enter into the calculation of the coefficients $(a, b, \text{ and } c)$. Since milk fat has a high coefficient of expansion and contracts as it solidifies (note that the solid-liquid equilibrium is established slowly), the temperature of measurement and the previous history of the product must be controlled carefully (see Sharp and Hart 1936). Variations in the composition of fat and in the proportions of lactose, proteins, and salts probably influence the equations much less than do variations due to the physical state of the fat. There have been many comparisons of total solids calculated by such equations with gravimetric results (Jenness and Patton 1959; Rowland and Wagstaff 1959). Modifications of the lactometer equations have been proposed in order to reduce the differences encountered. This amounts, of course, to including in the equation factors appropriate to the series of specimens analyzed and also to compensating for systematic errors in the determination of fat, total solids, and density.

Just as an increase in solids-not-fat increases milk density, so does the removal of water by processing. If there were no changes in physical state or chemical activity coefficients (e.g., hydration of proteins or insolubilization of salts), the density of the concentrated milk could be calculated from an equation derived by Jenness (1962) and presented in the second edition of this book. Data presented by Mojonnier and Troy (1922) conform to the equation but lack sufficient precision to indicate the small changes associated with some of the changes in physical state.

There appears to be no effect of environmental conditions, stage of lactation, lactation number, or nutritional level of the animal on milk density, aside from the effects of these parameters on milk composition.

Densities of liquid dairy products as varied as milk, whey, evaporated milk, sweetened condensed milk, and freshly frozen ice cream have been measured (1) by weighing a given volume, as in pycnometry; (2) by determining the extent to which an object sinks, as with hydrometry; (3) by hydrostatic weighing of an immersed bulb, as with a Westphal balance (McKennell 1960); (4) by measuring the volume of a given weight of product, as in a dilatometer (Short 1955); and (5) by measuring the distance that a drop of product falls in a density gradient column (Stull, et al. 1965). The most frequently used method, hydrometry, utilizes lactometers or a series of beads of graded densities (Golding 1959) to measure densities of fluid and condensed milks. The choice of method for a given purpose requires a balance between precision and speed.

It is difficult to summarize the data in the literature, especially when trying to compare milk, its fractions, and its products because the measurements were made at different temperatures, not always clearly specified. Much of the older data is given as specific gravity at 15.5° C/15.5°C, where the value of fresh whole mixed herd milk seldom lies outside the range 1.030–1.035, and 1.032 is often quoted as an average value. Skim milk at this temperature has a specific gravity of about 1.036 and evaporated whole milk of about 1.0660.

Table 8.2 gives the results at 20°C/20°C for milk from cows of various breeds.

The density of milk decreases as the temperature is raised to about 40°C (Short 1955, 1956; Wegener 1953). This is recognized in precautions for temperature control during density determinations and in factors for converting readings taken at one temperature to equivalents at another. Whitnah *et al.* (1957) used hydrostatic weighing and supercooling to find the temperature of maximum density, −5.2°C. This temperature of maximum density increased linearly to +4°C with dilution with water but decreased as the sample was held at low temperatures. Milk density increased during refrigerated storage, due to slow crystallization of the fat and a change in the hydration of the globule membrane.

For the temperature range 1° to 10°C, Watson and Tittsler (1961) derived the relations:

$$\text{Density} = 1.003073 - 0.000179t - 0.000368F + 0.003744N$$

where t = temperature in °C, F = percent fat, and N = percent nonfat solids. Data obtained by Short (1955) in the range 10° to 45°C for whole milk of 3% fat and 8.7% solids-not-fat and skim milk of 0.02% fat and 8.9% solids-not-fat were fitted to empirical equations of the form density $= a + bt + ct^2 + dt^3$, where t = temperature in °C. Average effects of varying the percentages of fat and solids-not-fat were also determined. All of these data are presented in Table 8.3. The relative importance of the linear term is less and that of the cubic term is greater for skim milk than for whole milk. The downward deviations from linearity as SNF is increased indicates a relationship between ex-

Table 8.2. Specific Gravity of Milk from Cows of Various Breeds.

| Breed | No. of cows | No. of samples | Specific Gravity at 20°C/20°C | | |
			Range	Mean	(SD)
Ayrshire	14	208	1.0231–1.0357	1.0317	0.0022
Brown Swiss	17	428	1.0270–1.0366	1.0318	0.0016
Guernsey	16	321	1.0274–1.0398	1.0336	0.0018
Holstein	19	268	1.0268–1.0385	1.0330	0.0024
Jersey	15	199	1.0240–1.0369	1.0330	0.0024

SOURCE: Data from Overman *et al.* (1939).

Table 8.3. Coefficients in the Equation $D - 1 = a + bt + ct^2 + dt^3$ where D = Density (g/ml) and t = Temperature (°C).

	a	b	c	d
Whole milk	3.50×10^{-2}	-3.58×10^{-4}	4.9×10^{-6}	-1.0×10^{-7}
Skim milk	3.66×10^{-2}	-1.46×10^{-4}	2.3×10^{-6}	-1.6×10^{-7}
Ave. difference per 1% fat[a]	4.8×10^{-4}	-3.9×10^{-5}	6.1×10^{-7}	$-2. \times 10^{-9}$
Ave. difference per 1% solids-not-fat[a]	3.8×10^{-4}	$-8. \times 10^{-6}$	-1.0×10^{-6}	$+4. \times 10^{-9}$

[a] Interpolations on mixtures of cream and skim milk from a single sample were apparently used to determine these averages. Such mixtures would have a constant ratio of each constituent of solids-not-fat and would not apply to other samples (e.g., with a different protein/lactose ratio).
SOURCE: Data from Short (1955).

tent of hydration and concentration of solids. It should be emphasized that these coefficients are valid only up to about 45°C. In this range, not only the density but also the specific gravity decreases (Rishoi and Sharp 1938; Wegener 1953). This phenomenon occurs with skim milk and whole milk, but not with a 5% lactose solution; it thus appears to represent changes (perhaps degree of hydration) in the proteins. The decrease in specific gravity is of the order of 5×10^{-5} per °C in the 10 to 40°C range.

Data in the literature are not in agreement as to the effect of temperature above 40°C on the specific gravity of milk. Of course, the absolute density decreases, but whether this decrease is as great as that of water in the same temperature range is unclear. Some pycnometric data (Wegener 1953) indicate that the specific gravity remains virtually constant, while some dilatometric data (Short 1956) show a pronounced increase between 40° and 90°C. This increase is on the order of 5×10^{-5} per °C and thus is of significance. Density measurements may be of use in detecting changes in the state of milk proteins.

Fat content and temperature have been related to the density of creams. Phipps (1969) devised a nomograph covering up to 50% fat and temperatures from 40 to 80°C. Homogenization slightly increases the density of whole milk but not of skim milk, and sterilization decreases the density of both milks (Short 1956). These changes are very small and the sample-to-sample variation is large; thus, they are essentially negligible.

The specific gravities of fluid milk products at various temperatures were measured throughout the United States by Herrington for a committee representing 13 Federal Milk Marketing orders (USDA 1965). A unique method was developed whereby the weight of the sample required to fill a Babcock bottle from the 0% to the 4% mark was compared to the weight of distilled water required to fill the same bottle to the same mark at that same temperature. Glassware and thermometers were carefully calibrated before use. Fat and total solids were measured by Babcock and by gravimetric methods. Over 8000 raw and processed samples were analyzed over the course of a year. The results are summarized in Table 8.4. After adjustment to a constant composition, the specific gravity of homogenized milk was not affected significantly by geography. Regression equations were developed for relating the fat and SNF contents of the milk to the weight per gallon at 40, 50, and 68°F; these equations were followed closely even though extra milk solids were added to some products.

Table 8.4A. Density of Various Fluid Dairy Products.

Product	Product composition		Pounds per gallon at:			
	Fat (%)	SNF (%)	40°F	50°F	68°F	102°F
Producer milk	4.00	8.95	8.625	8.614	8.589	8.525
Homogenized milk	3.60	8.60	8.613	8.604	8.581	9.518
Skim milk, packaged	0.15	8.90	8.635	8.629	8.612	8.557
Fortified skim milk	0.15	10.15	8.677	8.671	8.652	8.597
Half and half	12.25	7.75	8.559	8.544	8.502	8.420
Half and half, fort.	11.30	8.90	8.593	8.584	8.537	8.456
Light cream	20.00	7.20	8.510	8.488	8.433	8.333
Heavy cream	36.60	5.55	8.406	8.376	8.288	8.154

Table 8.4B. Specific Gravity of Various Fluid Dairy Products

Product	Product composition		Specific gravity[a] at:			
	Fat (%)	SNF (%)	4.4°C	10°C	20°C	38.9°C
Producer milk	4.00	8.95	1.0346	1.0336	1.0321	1.0302
Homogenized milk	3.60	8.60	1.0332	1.0324	1.0312	1.0293
Skim milk, packaged	0.02	8.90	1.0358	1.0354	1.0349	1.0341
Fortified skim milk	0.02	10.15	1.0409	1.0404	1.0397	1.0389
Half and half	12.25	7.75	1.0267	1.0252	1.0217	1.0175
Half and half, fort.	11.30	8.90	1.0308	1.0300	1.0259	1.0218
Light cream	20.00	7.20	1.0208	1.0185	1.0134	1.0070
Heavy cream	36.60	5.55	1.0083	1.0050	0.9960	0.9854

[a]Compared to water at that same temperature.
SOURCE: Data from Marketing Research Report 701. USDA (1965).

VISCOSITY

A review by Swindells *et al.* (1959) discusses the theoretical basis of viscosity and methods of measurement. Viscosity may be defined by the following equation:

$$\eta = F/(dv/dx)$$

where η is the coefficient of viscosity, F = force in dynes cm^{-2} necessary to maintain a unit velocity gradient between two parallel planes separated by unit distance, and dV/dx = velocity gradient in sec^{-1} perpendicular to the planes.

The unit of viscosity, the poise, is defined as the force in dynes cm^{-2} required to maintain a relative velocity of 1 cm/sec between two parallel planes 1 cm apart. The unit commonly used for milk is the centipoise (10^{-2} poise). A useful quantity in fluid flow calculations is the kinematic viscosity, or viscosity divided by density.

In dealing with solutions and colloidal dispersions, the following quantities are often used:

Relative viscosity: $\eta_{rel} = \eta_{soln}/\eta_{solv}$
Specific viscosity: $\eta_{sp} = \eta_{rel} - 1$
Reduced viscosity: $\eta_{red} = \eta_{sp}/c$ where c is the concentration of the solute
Intrinsic viscosity: $[\eta] = \lim (\eta_{sp}/c)$ as c goes to zero.

Fluids for which the viscosity coefficient η depends only on temperature and pressure and is independent of the rate of shear are called "Newtonian." A plot of shear stress versus rate of shear for such fluids is a straight line passing through the origin. Behavior of this type is exhibited by gases, pure liquids, and solutions of materials of low molecular weight. On the other hand, many colloidal dispersions and solutions of high polymers in which the molecular species is large show marked deviation from Newtonian behavior, the rate of shear depending upon the shear stress (plasticity) or the duration of shear (thixotropy) or both. Many such materials also exhibit hysteresis, whereby the coefficient of viscosity at a particular shear rate depends upon whether the shear rate is being decreased or increased.

Skim milk and whole milk do not differ appreciably from Newtonian behavior, but cream, concentrated milks, butter, and cheese exhibit varying degrees of non-Newtonian behavior. The literature on these products is summarized and reviewed in the monograph edited by Scott-Blair (1953).

The three types of viscometers that have been used in most studies of the viscosities of dairy products are coaxial cylinder (e.g., McMichael, Couette, and Brookfield), falling spheres (e.g., Hoeppler) and capillary tubes (e.g., Ostwald). McKennell (1960) considers falling sphere and capillary tube viscometers unsuitable for measurements of non-Newtonian fluids because corrections for nonuniform shear rates are tedious and/or not sufficiently exact. The two conditions proposed for minimizing uncertainty are uniform shear rate and a consistent procedure regarding rate and duration of shear. These points are also emphasized by Potter et al. (1949) in a study of the applicability of coaxial cylinder viscometers to various concentrated milk products. A general discussion of falling-sphere viscometers (Weber 1956) presents conditions for and limitations of their use. A "mobilometer" (Maxcy and Sommer 1954), which has some features of both the coaxial cylinder and falling-sphere types, has been used to measure viscosities of evaporated milk. A sealed microviscometer of the falling-sphere type in which the specimen can be sterilized has been proposed for studies of changes occurring in sterilization and storage of concentrated milk products (Leviton and Pallansch 1960).

The value of η of 1.0019 ± 0.0003 for water at 20°C (Swindells et al. 1952 and Cragoe's equation (Coe and Godfrey 1944)

$$\log(\eta_t/\eta_{20}) = [1.2348)20 - t) - 0.001467(t - 20)^2]/(t + 96)$$

for other temperatures seem slow in receiving the recognition they deserve for their careful determinations. Calibration of nearly all viscometers is based on similar earlier determinations.

The viscosity of milk and dairy products depends upon the temperature and on the amount and state of dispersion of the solid components. Representative values at 20°C are: whey 1.2 centipoise (cp), skim milk 1.5 cp, and whole milk 2.0 cp. From these values, it is evident that the caseinate micelles and the fat globules are the most important contributors to the viscosity. Specific data are given in Table 8.5.

Many workers (Andrade 1952) have proposed equations of the following form to relate viscosity of fluids to temperature: $\ln_e \eta = a + b/T$, where a and b are constants and T is the absolute temperature. Caffyn (1951), in a careful study of the viscosity of homogenized milk, found that plots of $\ln \eta$ versus $1/T$ were not linear over the range 20°–80°C but exhibited a sharp break at about 40°C, which he attributed to the melting of the fat. (Actually, the fat becomes more fluid throughout the range 20–30°C.) Jenness replotted data of Eilers et al. (1947) and of Whitaker et al. (1927) for skim milk and found distinct breaks

Table 8.5.   Viscosity of Milk, Skim Milk, and Whey.

No. of samples	Temp.	Viscosity of:		
		Whole milk	Skim Milk	Whey
	°C	c.p.	c.p.	c.p.
180[a]	27	1.45		
14[b]	25		1.42	
62[c]	24		1.47	
9[c]	24			1.16

[a]Puri *et al.* (1963).
[b]Eilers *et al.* (1947).
[c]Whitaker *et al.* (1927)

at 30° and 65°C. Therefore it appears that equations of this sort do not apply to the milk system over such a wide temperature range. Cox (1952) examined available data, including Caffyn's (1951), and fitted an empirical equation of the form

$$d\eta = b_1(dt) + b_2(dt)^2 + b_3(dt)^3$$

The average coefficients were -5×10^{-4}, -10×10^{-6}, and -7×10^{-8}, respectively, but the coefficients differed for different samples.

Cox, *et al.* (1959) surveyed available literature on the relation between the composition and viscosity of cow's milk. Although data from any one study could be fitted to an empirical equation of the form

$$\eta = A + B_1p + B_2q + B_3p^2$$

where p and q are the respective percentages of fat and solids-not-fat, the coefficients differed markedly from one set of data to another. Viscosity increased with increasing concentrations of both fat and solids-not-fat, of course, but a consistent general relationship was not obtained.

Whitnah (1962) found high positive correlations between viscosity at 4°C of milks from individual cows and their fat and protein contents. Surprisingly, the correlation between viscosity and the content of solids other than fat or protein was negative, which he attributed to an inverse effect of lactose on viscosity.

The viscosity of colloidal systems depends upon the volume occupied by the colloidal particles. The simple equation of Einstein,

$$\eta_{rel} = 1 + 2.5\theta$$

where θ is the fraction of the total volume occupied by the dispersed phase, was derived on the assumption of rigid spherical particles of equal size and appears not to fit the situation for skim milk or whole milk. Eilers *et al.* (1947) applied the empirical equation

$$\eta_{\text{rel}} = \left[1 + \frac{1.25\ VCv}{1 - 1.35VC_V} \right]^2$$

to skim milk. When the viscosity of raw skim milk relative to that of rennet whey was used, this equation gave a value for the "rheological concentration" or fractional volume occupied, VCv, of 0.88 for the caseinate particles. Cv is the concentration of caseinate on a dry basis and V is the "voluminosity," which is a composite factor including both hydration and electrical effects. Since Cv was 2.9 g/100 ml in the skim milk dealt with, the voluminosity of the caseinate particles was 3 ml/g at 25°C (i.e., 1 g dry caseinate appears to occupy about 3 ml). Whitnah and Rutz (1959) applied an equation furnished by Ford to determine voluminosity at various temperatures of casein fractions obtained by fractional centrifugation of skim milk. This yielded values close to 3.9 ml/g at 25°C, but voluminosity appeared to increase sharply below and decrease moderately above this temperature. A plot of the viscosity of skim milk relative to that of whey at various temperatures (Eilers *et al.* 1947) shows a sharp decline from 5 to 30°C, reflecting a decrease in the voluminosity of the caseinate micelles. Above 30°C, the decrease is less marked until about 65°C, where whey proteins begin to be denatured. From this point on, the viscosity of whey relative to that of water increases, but that of skim milk relative to whey remains relatively constant.

Changes in the caseinate micelles produced by either raising or lowering the pH result in an increase in viscosity (Eilers *et al.* 1947; Puri and Gupta 1955). For example, the viscosity is approximately doubled by the addition of 10 ml 1.4 to 3.8 N ammonia to 90 ml milk. Addition of alkali (to pH's up to 11.7), urea (up to 4.8 M), and calcium complexing agents to concentrated (22.7% solids) skim milk causes a marked transient increase of severalfold in viscosity, followed by a sharp decline (Beeby and Kumetat 1959; Beeby and Lee 1959). This was interpreted as resulting from swelling of the miscelles, followed by their disintegration.

Various measurements of viscosity as a function of solids in diluted and concentrated milk reveal a curvilinear relationship (Bateman and Sharp 1928; Deysher, *et al.* 1944; Leighton and Kurtz 1930). Eilers *et al.* (1947) calculated voluminosity at several solids concentrations (skim milk), showing that the voluminosity of the caseinate particles does not change but that the apparent voluminosity of the total solids decreases as the concentration is raised. This may merely signify that the equations used, which had been derived for particles of equal size, do not fit the situation of diverse particle sizes found in skim milk.

Torssell et al. (1949) attempted to develop a mathematical relationship among the viscosity, temperature, and total solids content of skim milk and whole milk. After showing that Walther's equation

$$m = \frac{\log \log (v_{t1} + 0.8) - \log \log (v_{t2} + 0.8)}{\log T^2 - \log T^1}$$

applies, in that a straight line was obtained by plotting log log (v + 0.8) versus log T (i.e., that m is a constant), they found that a plot of m versus the solids percentage is also a straight line. These relationships held for skim milk up to 36.5% solids and for whole milk up to 41.7% solids. Undoubtedly the specific plots would differ from milk to milk.

Phipps (1969) was able to relate the viscosity of creams up to 50% fat with the fat content and the temperature, and the viscosity of the skim milk phase to its temperature alone.

Quick cooling of either skim milk or whole milk that had been heated to 65°C results in a temporary decrease of a few hundredths of a centipoise in the viscosity. The data of Whitnah, et al. (1956) and Eilers et al. (1947) agree in indicating that this effect is due to a reversible change in the caseinate micelles. The relaxation of this change was exponential with time. This change could be related to the migration of β-casein from the micelle.

The increase in viscosity when fluid milk or concentrated milk is heated sufficiently to aggregate the proteins is discussed in Chapter 11. The effects of homogenization and clustering of the fat globules on their contribution to the viscosity of the product are dealt with in Chapter 10.

SURFACE AND INTERFACIAL TENSION

The area of contact between two phases is called the "interface," or (especially if one of the phases is gaseous) the "surface." The properties of interfaces and surfaces are determined by the number, kind, and orientation of the molecules located in them. A widely used measure is energy per unit area. It is the work required to extend the surface by unit area or force required per unit length, is expressed as ergs cm^{-2} or dynes cm^{-1}, and is often symbolized by the Greek letter gamma (γ).

Surface-active solutes accumulate in the interface between two phases in accordance with their concentration and their ability to reduce the interfacial tension. The relation may be expressed by the Gibbs equation as follows:

$$\Gamma = \frac{-a}{RT} \cdot \frac{d\gamma}{da}$$

where Γ = excess concentration of solute in the interface over that in the bulk solution and $d\gamma/da$ = the rate of change in interfacial tension with the change in the activity of the solute in the bulk phase.

This reversible and ideal relationship predicts that the more effective depressants of interfacial tension tend to accumulate in the interface to the exclusion of others. Actually, in many cases the amount of material concentrated at the interface is greater than would be predicted by the Gibbs equation, and the system is not reversible or only sluggishly so.

In milk, the important interfaces are those between the liquid product and air and between the milk plasma and the fat globules contained therein. Studies of the surface tension (liquid/air) have been made to ascertain the relative effectiveness of the milk components as depressants; to follow changes in surface-active components as a result of processing; to follow the release of free fatty acids during lipolysis; and to attempt to explain the foaming phenomenon so characteristic of milk. Interfacial tensions between milk fat and solutions of milk components have been measured in studies of the stabilization of fat globules in natural and processed milks.

The chief methods for measuring surface and interfacial tension are described and critically evaluated by Harkins and Alexander (1959). They may be classified as dynamic and static. In the former, measurements are made on freshly formed surfaces during the period required for equilibrium to be established. Such methods enable the rate of orientation of molecules in the interface to be followed. The method of vibrating jets is a dynamic method (Harkins and Alexander 1959). It has been applied to milk in two studies, which agree that the surface tension did not change during the observed age of the jet. In the first study (Leviton and Leighton 1935), it was concluded that the surface tension in the jet was equal to that of water and thus that surface orientation had not occurred. The second study (Whitnah et al. 1949) concluded that the surface tension equaled the static value for milk, i.e., that the change had been completed during the early part of the first wave. The large decrease in surface tension during the interval 0.001–0.01 sec for diluted milk, or for milk from which the proteins had been precipitated, seems to justify the second conclusion. Orientation in newly formed surfaces and interfaces in milk may thus be considered extremely rapid in comparison to time intervals involved in dairy-processing operations.

The large majority of the information on surface and interfacial phenomena in milk has been obtained with various static methods (Whitnah 1959). Five of the principal types of data involve the determination of (1) height of rise of liquid in a capillary; (2) weight or volume of drops formed by liquid flowing from a capillary tip (sometimes considered "semidynamic"); (3) force required to pull a ring or plate out of a surface; (4) maximum pressure required to force a bubble of gas through a nozzle immersed in the liquid; and (5) shape of a drop hanging from a capillary. Each of these parameters may be related theoretically and practically to the surface tension (Harkins and Alexander 1959). The method involving pulling of a ring or plate from the surface is undoubtedly the most widely employed. It is rapid, simple, and capable of accuracy of ± 0.25% or better when the force is measured with an analytical balance (Harkins and Alexander 1959). Apparatus employing tension balances with various degrees of sensitivity (DuNoüy balances) are also available. The critical discussion of Herrington (1954) of certain aspects of the measurement of surface tension emphasizes some of the variables and necessary precautions.

The surface tension of milk is on the order of 50 dynes cm^{-1} at 20°C, compared to that of water, which is 72.75 dynes cm^{-1} at the same temperature. The milk proteins, milk fat, phospholipids, and free fatty acids are the principal surface-active components determining the surface properties of milk.

Using the ring method at 20°C, Sharma (1963) reported values of 42.3 to 52.1 dynes cm^{-1} (mean, 46.8 ± 2.3) for 51 specimens of milk from individual Indian cows, and Parkash (1963) found an average of 46.02 ± 1.14 dynes cm^{-1} for 100 specimens. Values of 47.5 to 48.0 dynes cm^{-1} at 20°C were found for individual cow specimens in France by Calandron and Grillet (1964). Mohr and Brockmann (1930) in Germany reported 51.0 dynes cm^{-1} for skim milk, 46.7 for whole milk and 44.8 for cream of 0.04, 2.4, and 34.0% fat, respectively, from a single original lot. During renneting, the surface tension increases by about 10% at flocculation and by another 2–3% at cutting (Tambat and Srinivasan 1979).

The effectiveness of surface and interfacial tension depressants can be compared by plots of concentration versus tension. Various dilution studies of milk, skim milk, wheys, and solutions of milk proteins reveal that casein and the proteins of the lactalbumin fraction (β-lactoglobulin, α-lactalbumin, and bovine serum albumin) are powerful depressants, while the proteins of the immunoglobulin fraction are somewhat less so (Aschaffenburg 1945; El-Rafey and Richardson 1944; Jackson and Pallansch 1961; Johnston 1927; Spremulli 1942; Igarashi and Saito 1972). In skim milk and rennet whey, the concentrations of proteins are far above the levels at which their effects are additive; both products have nearly the same surface tension, but that of whey drops more markedly with dilution (Aschaffenburg 1945). Minor protein fractions that were especially effective depressants but have not been fully characterized were prepared by Ansbacher *et al.* (1934) from casein by elution with salt solution and by Aschaffenburg (1945) from the serum from heated skim milk. The protein and protein–phospholipid complex from the surface of the milk fat globules is one of the most powerful and significant depressants of tension at both the milk-air and plasma–fat interfaces (Jackson and Pallansch 1961; Mohr and Brockmann 1930; Palmer 1944). The fact that the surface tension of whole milk lies a few dynes cm^{-1} below that of skim milk may be due to the presence of these substances, as well as to traces of free milk fat in the surface. Surface tension decreases with increasing fat content up to about 4% fat but does not decline to any extent with a further increase (Watson 1958). The protein–phospholipid complex is undoubtedly largely responsible for the very low surface tension of sweet cream buttermilk.

As temperature is raised in the range 10 to 60°C, the surface tension of skim milk and whole milk decreases (Mohr and Brockmann 1930; Watson 1958). This decrease is comparable in magnitude to that observed in the surface tension of water, which decreases about 10 dynes cm^{-1} over this range.

It has frequently been observed that the surface tension of milk that has been held at 5°C and brought to 20°C is lower (2 or 3 dynes cm^{-1} than that of milk cooled to 20°C and measured immediately (Mohr and Brockmann 1930; Sharp and Krukovsky 1939). The latter authors also demonstrated that skim milk separated at 60°C has a higher (2 to 3 dynes cm^{-1}) surface tension than that separated at 5°C. It thus appears as though a surface-active substance is released from fat globules at low temperatures. This behavior is the opposite of that of the fat globule "agglutinin," also studied by Sharp and Krukovsky (1939), which is adsorbed on solid fat globules. The peculiar phenomenon regarding surface tension has not been explained satisfactorily.

Free fatty acids released by lipolysis of milk fat greatly depress the surface tension of milk. In fact, surface tension has been used to some extent as an objective index of the development of hydrolytic rancidity (Dunkley 1951; Herrington 1954; Hetrick and Tracy 1948; Tarassuk and Smith 1940). Its value for this purpose is somewhat limited by natural variations due to other causes. Of course, it must also be recognized that the shorter-chain fatty acids contribute more to rancid flavor and the longer ones are more effective surface-tension depressants.

Homogenization of raw whole milk or cream stimulates lipolysis and thus leads to a decrease in surface tension, but if the product has been previously pasteurized, the effect of homogenization is an increase in surface tension (Trout *et al.* 1935; Watson 1958; Webb 1933). The reason for such an increase is not known, but suggestions have been made that it results from denaturation or other changes in the lipoprotein complex or from a reduction in the amount of protein available to the milk–air interface because of adsorption on the extended fat surface. The latter explanation seems unlikely in view of the very slight effect of fivefold dilution on the surface tension of skim milk. Another possible suggestion is that homogenization reduces the amount of free fat in the product.

Heat treatment of milk has little effect on surface tension except that sterilizing treatments cause an increase of a few dynes cm^{-1} coinciding with grain formation (Nelson 1949). This effect undoubtedly results from denaturation and coagulation of the proteins so that they are no longer effective surface-active agents.

FREEZING POINT

The freezing point of milk, like that of any aqueous system, depends on the concentration of water-soluble components. The mathematical relationship between depression of the freezing point and concentration of the solute was determined by Raoult (1884) and is expressed in the equation

$$T_f = K_f M$$

where T_f = the difference between the freezing points of the solvent and the solution, K_f the molal depression constant (1.86°C for water), and M = the molal concentration of the solute. As Raoult pointed out, this relationship is valid only for dilute solutions of undissociated solutes. The freezing point is a property controlled by the number of particles, rather than the kinds of particles, in the solvent. The other colligative properties are boiling point elevation and osmotic pressure.

In chemical research, the freezing point of a solution composed of known weights of solute and solvent affords a means of determining M, the molal concentration, and hence the molecular weight of the solute. In the dairy field, however, the objective of freezing point measurements is virtually restricted to determination of the water content of the product in order to detect the illegal addition of water. Its value for this purpose rests on the fact that the freezing point of authentic bovine milk varies within very narrow limits. Since the depression of the freezing point is directly proportional to the number of particles in solution, it is obvious that it is primarily determined by the major constituents of low molecular weight, the lactose and the salts, and is nearly independent of variations in the concentrations of colloidal micelles and fat globules. Furthermore, a complementary relationship exists between lactose and sodium chloride in milk such that the osmotic pressure and hence the freezing point is maintained within a narrow range.

The determination of the freezing point demands meticulous attention to detail. The general principle employed is to supercool a sample slightly, to induce crystallization, and then to observe the maximum temperature attained. The temperature of the cooling bath must be controlled; otherwise the rate of heat loss will be greater than the rate of heat transfer to the bath by the heat of fusion of the solution, and the observed freezing point will be too low. Control of the temperature of supercooling and seeding techniques is extremely important; if it varies, the amount of solvent that crystallizes out, and consequently the observed freezing point, also varies. Attempts have been made to apply a correction factor to the observed freezing point in order to enable calculation of the "true freezing point." A review of the factors involved is given in *Richmond's Dairy Chemistry* (Davis and MacDonald 1953). However, from a practical standpoint, it is not necessary to determine the true freezing point; rather, a high degree of reproducibility between samples and analysts is sought.

In 1921, Hortvet published a method and a description of apparatus for freezing point determinations, feeling that earlier methods were not sufficiently standardized. Although the Hortvet method has been universally accepted as official, a number of modifications have been developed, such as replacing the ether cooling system with mechanical refrigeration and adding mechanical stirring and tapping devices (Shipe *et al.* 1953; Temple 1937). Furthermore, cryoscopes employing thermistors in place of mercury-in-glass thermometers have been developed (Shipe 1956, 1958). In 1960, the Association of Official Agricultural Chemists approved the thermistor-type cryoscope as an alternative official apparatus. In 1978 this association approved a method based on dew point depression in the vapor phase (Richardson *et al.* 1978). This method uses equipment similar to the thermistor cryoscopes, and because the measured parameter is based on a colligative property (osmotic pressure), the results are directly comparable. The thermistor cryoscopes have become more popular than the Hortvet apparatus because of their speed and ease of operation.

Regardless of the method used, determination of the freezing points of solutions is empirical. In recognition of this fact, Hortvet (1921) emphasized the need to use standardized equipment and techniques. His procedure involves measurement of the difference between the freezing point of a standard solution and the freezing point of milk. Any systematic errors should be reflected in both observed values and should thus be eliminated. This emphasis on the calibration procedure apparently has not been recognized by all analysts. Of course, it is presumed that the systematic errors affect both the standard solutions and the milk in the same way. It is difficult to obtain reproducible results with sucrose solutions, and they readily undergo microbial decomposition. Salt (NaCl) solutions have been approved as secondary freezing point standards on the basis of a collaborative study (Shipe 1960). Salt standards sterilized in sealed ampules kept for over a year without a change in freezing point. Henningson (1966) studied the factors affecting results obtained with thermistor-type cryoscopes. His studies led to the development of specific directions pertaining to cooling, seeding, and reading (Henningson 1966).

The freezing points of the two standards used by Hortvet were thought to be -0.422 and -0.621 on the Celsius scale; however, they have been shown to be -0.406 and $-0.598°C$ (Prentice 1978). Although the data reported between 1921 and about 1980 were expressed in degrees Celsius, they should have been expressed in degrees Hortvet. Absolute values in degrees Hortvet are about 3.7% lower than they are in degrees Celsius. Added water calculations based on an arbitrary freezing point would be in error by the same amount, while added water calculations based on the *difference* between the sample and the calibrating standard remain directly comparable, within practical limits. Since 1980, the Association of Official Analytical Chemists has recommended that cryoscopes be calibrated to degrees Celsius and that results be so reported (AOAC 1980). Results in this section have been converted to degrees Celsius. [The conversion formula published by the AOAC (1984) does not convert the standards exactly, and the one published by the International Dairy Federation (Harding 1983) has been used.]

The freezing point of bovine milk is usually within the range -0.512 to $-0.550°C$. The average value is close to $-0.522°C$ (Dahlberg *et al.* 1953; Davis and MacDonald 1953; Henningson 1969; Robertson 1957; Shipe 1959; Eisses and Zee 1980). It is easy to determine that lactose and chloride are the principal constituents responsible for this depression. Thus, in a milk containing 12.5% solids, 4.75% lactose, and 0.1% chloride, the molal concentration of lactose is (4.75 $\times$ 1000) (342 $\times$ 87.5) = 0.159, and the corresponding depression of the freezing point, assuming ideal behavior, is 0.159 $\times$ 1.86 = 0.296°C. Actually, sugars do not behave entirely ideally even at these concentrations. For example, data of Whittier (1933B) indicate a molal depression by lactose of 2.02°C instead of 1.86°C. This difference is attributable to solvation of the sugar, and division of the predicted freezing point depression by (1 − 0.1023 M) will give agreement with empirical data at concentrations up to 15% sucrose (Prentice 1978). The concentration of chloride is 0.032 M, and assuming that each chloride ion is accompanied by a monovalent ion of opposite charge (i.e., Na^+ or K^+), the depression expected is 1.86 $\times$ 2 $\times$ 0.032 = 0.119°C. These values are in close agreement with those of Cole and Mead (1955) and Cole et al. (1957), who measured the effects of adding increments of lactose and chloride (0.415°C in the example given) representing 75 to 80% of the entire depression. Have et al. (1980) noted that the relative contributions of lactose, chloride, and phosphate to freezing point depression varied from cow to cow, but the total of the three did not. The contributions of the other solutes cannot be readily determined, since information is lacking on the distribution of calcium, magnesium, phosphate, and citrate, among various ions and complexes. The degree of departure from ideal behavior is also not known.

Variations in the freezing point of milk as it is drawn from the udder must reflect the variability in the physiological limits of that gland. The exact magnitude of these limits has not been established. Wheelock et al. (1965) found that milk is in osmotic equilibrium with blood flowing through the udder. They noted that freezing point values for milk agreed more closely with those for mammary-venous blood than with those for jugular-venous blood. Tucker (1970B) also found that the freezing points of milk and mammary-venous blood were highly correlated, whereas Peterson and Freeman (1966) found that there was no significant correlation between them. Little is known about whether variability in freezing points reflects primarily deviations from the complementary relationship between lactose and chloride or variations in the content of other osmotically active components. Rees (1952) has presented some evidence in favor of the latter view, indicating that the concentration of the nonchloride fraction, principally the soluble acid phosphates, was the primary cause of variations.

Environmental factors associated with variations in freezing point have been studied by several investigators. Articles published prior to 1960 have been reviewed (Davis and MacDonald 1953; Robertson 1957; Shipe 1959), but since then, several additional articles (Demott

1966; Demott et al. 1967, 1968, 1969; Freeman et al. 1971; Henningson 1963; Schoenemann et al. 1964; Brathen 1983) have been published.

Some of the variations have been attributed to seasonal effects, feed, water intake, stage of lactation, breed of cow, heat stress, and time of day (i.e., morning versus evening milk). In some cases, the effects of these factors have been shown to be interrelated. For example, the differences between the freezing point of morning and evening milk have been shown to be affected by the time of feeding and watering. Geographical differences that have been reported may be due primarily to differences in breed of cattle and in feeding practices. Although variations have been shown to occur, results obtained in England (Stubbs and Elsdon 1934), Australia (Tucker 1963, 1970B), India (Dastur et al. 1952), and the United States (Bailey 1922; Henningson 1969; Kleyn and Shipe 1957) all exhibit approximately the same range. Undernourished cows produce milk with a higher freezing point than adequately nourished animals (Bartsch and Wickes 1979). Although the mineral content of milk is known to be responsive to minerals in the diet, the higher sodium content induced by sodium in the drinking water is compensated for by other changes and does not affect the freezing point (Mussenden et al. 1977). Heat stress increases the freezing point depression of evening milk over that of morning milk from the same animals (Eley et al. 1978). The freezing point of the milk decreases in response to water restriction to around $-0.570\,°C$ (Evans and Johnson 1978).

Handling treatments of milk between the time of drawing and the freezing point determination may be expected to alter the freezing point if they change the net number of osmotically active particles in solution. Effects of some such treatments are reviewed by Shipe (1959) and by Harding (1983). Of course, microbial decomposition with the production of such water-soluble components as lactic acid from lactose will lower the freezing point. Storing samples at low temperatures of freezing them has been reported to raise freezing points slightly (Demott and Burch 1966; England and Neff 1963; Henderson 1963). Likewise, heating has been observed to raise the freezing point by some authors, but not by others. Undoubtedly, chilling or heating may produce aggregation of dissolved salts or transfer of dissolved materials to the colloidal caseinate micelles or the fat globules. Such effects would raise the freezing point. Since the change may be slowly reversible after treatment, the magnitude of change in the freezing point would vary with the length of time between treatment and determination of the freezing point. However, it would be difficult to follow such sluggishly reversible equilibria by means of the freezing point because the changes are so small. Vacuum treatment of milk has been shown

(Demott 1967; Shipe 1964) to raise its freezing point, presumably as a result of the removal of carbon dioxide. If water is lost during the vacuum treatment, the freezing point will be lowered, thereby partially compensating for the loss of carbon dioxide. Bottled milk should have a freezing point very close to that of the raw milk from which it was made, and usually the same standards apply. However, the occurrence of freezing points above the maximum is much higher in retail milk than in raw milk (Watrous *et al.* 1976).

There has been considerable controversy over the interpretation of freezing point values. In 1970 the Association of Official Analytical Chemists adopted an interpretation (Henningson 1970) which specified that milk with a freezing point of $-0.525\,^\circ$H ($-0.505\,^\circ$C) or below may be presumed to be water-free. Detailed procedures have been given for confirming the absence or presence of added water. The choice of the $-0.525\,^\circ$H value as an upper limit is based on a statistical evaluation of data from a 1968 cooperative North American survey (Henningson 1969) of freezing points of authentic samples. Eisses and Zee (1980) suggest that the upper limit, above which the milk would be considered adulterated, should not be fixed but that it vary according to the type of milk, season, region, and other factors.

ELECTRICAL CONDUCTIVITY

The specific electrical resistance of an electrolyte solution is defined as the resistance of a cube 1 cm in length and 1 cm^2 in cross-sectional area.

$$\rho = \frac{\alpha R}{l}$$

where ρ = specific resistance in ohm centimeters, α = cross-sectional area in square centimeters, l = length in centimeters, and R = measured resistance in ohms ($R = E/I$). Specific conductance, K, is the reciprocal of specific resistance.

$$K = \frac{1}{\rho} = \frac{l}{\alpha R}$$

Measurements of conductance are made with glass cells in which the solution is contained between platinum electrodes. The resistance between the electrodes is measured. If the cell were of uniform and measurable length and cross section, the conductivity could be computed directly. In practice, however, the resistance of a solution of known conductivity (a KCl solution, for example) is measured in the cell and a cell constant is computed as the product, KR. The specific conductance of the unknown electrolyte can then be computed:

$$K = \frac{cell\ constant}{R}$$

K is expressed in units of reciprocal ohms (mhos) per centimeter (i.e., $ohm^{-1}\ cm^{-1}$). Methods of measurement, cell selection and use, and the appropriate electrical circuitry are discussed by Shedlovsky (1959).

A chronological summary of the literature up to 1954 on the conductivity of milk has been made by Schulz (1956). Conductivity has been considered as a possible index of mastitic infections, of added water, of added neutralizers, and as a means of monitoring changes in the concentration and composition of solids during dairy processing. The specific conductance of cow's milk, reflecting its concentration and activity of ions, is on the order of $0.005\ ohm^{-1}cm^{-1}$ at 25°C. Most normal samples fall within the range 0.0040 to $0.0055\ ohm^{-1}cm^{-1}$. Higher values usually represent mastitic infections which increase the concentration of sodium and chloride in the milk (Fredholm 1942; Pinkerton and Peters 1958; Schulz 1956).

Specific conductivity continues to attract interest as a convenient, objective means of detecting mastitis at the subclinical level. Although somatic cell counts do correlate with specific conductivity, the cell counts appear to be more sensitive for mastitis detection (Kozanecki et al. 1982). Fernando et al. (1982) found that the ratio of specific conductivities of fore-and postmilking strippings was an effective index of mastitis due to the sharp rise in the conductivity of the postmilk from infected quarters. Introduction of a foreign body, sterile polyethylene, into the udder caused changes in somatic cell counts and conductivity similar to those of mastitis (Jaster et al. 1982).

Temperature control is important in conductivity measurements, since the conductivity of milk increases by about $0.0001\ ohm^{-1}cm^{-1}$ per degree Celsius rise in temperature (Gerber 1927; Muller 1931; Pinkerton and Peters 1958). Increased dissociation of the electrolytes and decreasing viscosity of the medium with increasing temperature are undoubtedly responsible for this effect. An investigation (Sudheendranath and Rao 1970) of the viscosity and electrical conductivity of skim milk from cows and buffaloes failed to reveal a simple relationship. The authors attributed the lack of linear correlations to variations in casein structure and its hydration.

The sodium, potassium, and chloride ions of milk are the greatest contributors to its electrical conductivity, since they are present in the highest concentration. Schulz and Sydow (1957) proposed that "chloride-free conductivity" may be a more sensitive index of certain changes and adulterations than conductivity itself. Chloride-free conductivity is the difference between total conductivity and that of a sodium chloride solution having the same chloride content as the milk sample. In tests on 41 specimens of mixed raw milk, conductivity averaged 0.00485, chloride conductivity 0.00305, and chloride-free conductivity 0.00180 ohm^{-1}cm^{-1}. The correlation between chloride content and conductivity has been confirmed by Puri and Parkash (1963). These workers reported that there was no significant difference between the conductivity of cow and buffalo milk, whereas Pal (1963) claimed that adulteration of buffalo milk with cow milk causes a detectable change in conductivity. Rao et al. (1970) found cows' milk to have a higher conductivity than buffaloes' milk; by measuring the conductivity of a formic acid extract, they detected 5–25% added water.

The fat globules of milk reduce the conductivity by occupying volume and by impeding the mobility of ions. Thus the conductivity of whole milk is less than that of skim milk by about 10%, and that of cream varies with the fat content (Gerber 1927; Muller 1931; Prentice 1962). Homogenization of milk does not measurably influence conductivity (Prentice 1962). The conductivity of whey and ultrafiltrate is slightly greater than that of skim milk (Schulz 1956; Schulz and Sydow 1957). A possible relationship between the electrical conductivity and physical stability of evaporated milk and concentrated infant milk products has been reported (Hansson 1957). Samples of poor physical stability tended to have relatively low conductivity values compared to those of the more stable products.

The production of acidity by bacterial action, of course, increases the conductivity of milk. An increase of about 0.00001 ohm^{-1} cm^{-1} per Soxhlet-Henkel degree has been noted (Ruge-Lenartowicz 1955; Tillmans and Obermeier 1920). (The Soxhlet-Henkel degree, °SH, is the number of milliliters of N/4 NaOH required to titrate 100 ml milk to the phenolphthalein pink.) Conductivity can be used to detect added neutralizers.

The influence of dilution and concentration on the conductivity of milk is complicated by their respective effects in promoting and repressing dissociation of salt complexes and solubilization of colloidal salts. Data of various workers (Coste and Shelbourn 1919; Schulz and Sydow 1957; Sorokin 1955; Torssell et al. 1949) indicate that as milk is concentrated, a maximum of conductivity is reached. With skim milk the maximum is about 0.0078 ohm^{-1}cm^{-1} and occurs with a solids content of about 28%. Concentration beyond this point results in a decrease in conductivity.

Direct conductivity measurements do not provide a satisfactory index of added water in milk. However, it has been reported (Rao *et al.* 1970) that measurement of conductivity in nonaqueous solvents can be useful in detecting adulteration. The conductivities of extracts using two different solvent systems were correlated with the percentage of added water in the original milk. One solvent system consisted of 10 ml acetone and 90 ml methanol plus 3 g sodium chloride, and the other contained 2.65 g formic acid in 100 ml acetone.

HEAT CAPACITY AND THERMAL CONDUCTIVITY

The heat capacity of a substance is the quantity of heat required to raise the temperature of a unit mass through a unit range. Heat capacity at any temperature, T, is the limiting value of dQ/dT as dT approaches zero, where dQ is the amount of heat required to raise the temperature from T to $T + dT$. Heat capacity is normally evaluated over a temperature range (dT) of several degrees. It is usually expressed in terms of cal g^{-1}C^{-1}, although the current custom is to use the units J mol^{-1}K^{-1} (or 4.186 × MW × cal g^{-1}C^{-1}). The term "specific heat" is used almost interchangeably with "heat capacity." It is the ratio of heat capacity to that of water at 15°C (0.99976 cal g^{-1}C^{-1}), and thus is dimensionless. The numerical value of specific heat is nearly the same as that of heat capacity. The heat capacity of air-free water at 1 atm pressure is within 1% of 1 cal g^{-1}C^{-1} over the range 0° to 100°C (Overman *et al.* 1939).

Heat capacity is best determined with a calorimeter incorporating an electric heater. The net energy input and the resultant temperature rise are both measured. Procedures and precautions for such direct calorimetry are discussed thoroughly by Sturtevant (1959). Differential scanning calorimetry is convenient to use for the determination of heat capacity (Watson *et al.* 1964).

The heat capacity of skim milk has been carefully measured by Phipps (1957), who compared his results with those of earlier workers. Skim milk exhibits a small but definite linear increase in heat capacity between 1 and 50°C from about 0.933 to 0.954 cal g^{-1}C^{-1}. Bertsch (1982) used a continuous-flow calorimeter to measure heat capacities at temperatures up to 80°C. Since the total time in the calorimeter was 10 sec, the values of 0.968 (skim) and 0.939 cal g^{-1}C^{-1} (whole milk) at

80°C should apply to the undenatured milk system. There is a marked decrease in heat capacity as the total solids content of the sample is increased (Rambke and Konrad 1970) with some discontinuities around 70 and 80°C (Agarwala and Ojha 1973). Dried skim milk products have heat capacities of 0.28 to 0.32 cal $g^{-1}C^{-1}$ in the 18 to 30°C temperature range (Buma and Meerstra 1969).

The heat capacity of milk fat in either the solid or the liquid state is about 0.52 cal $g^{-1}C^{-1}$, and its latent heat of fusion is about 20 cal/g (Yoncoskie 1969). Thus the heat capacities of milk and cream depend strongly upon the fat content. Furthermore, in temperature ranges at which the melting of fat occurs, the apparent heat capacity is the sum of the "true" heat capacity and the energy absorbed by melting of the fat. Thus, the results will vary widely, depending upon the proportion of the fat that was solid at the start of the determination, which in turn depends upon the composition of the fat and the temperature history of the sample. Many workers in the field have observed these effects (Jack and Brunner 1943; Norris, et al. 1971; Phipps 1957; Rishoi and Sharp 1938; Sherbon and Coulter 1966). The apparent heat capacity of fat-containing dairy products has a maximum at 15 to 20°C and often shows a second inflection at about 35°C (Phipps 1957; Sherbon 1968).

Thermal conductivity is the rate of heat transfer by conduction through a unit thickness across a unit area of substance for a unit difference of temperature:

$$\lambda = \frac{Q\,d}{A\,t\,(T_2 - T_1)}$$

where Q is amount of heat transferred through the sample of cross-sectional area A and thickness d in time t, with a temperature differential of $T_2 - T_1$.

Thermal conductivty can be determined using either equilibrium or dynamic methods. Equilibrium methods involve a heated surface, a thin layer of sample, and a cooled surface. The energy required to maintain a steady state for a given temperature difference is measured and used in the calculations. Dynamic methods are based on thermal diffusivity, which is obtained from the curvatures of heating or cooling plots at various depths within the product. Procedures and applications of thermal conductivity measurements to foods have been reviewed (Peeples 1962; Reidy 1968; Woodams and Nowrey 1968).

Thermal conductivity, λ, is expressed in cal cm^{-1} $sec^{-1}C^{-1}$ or in kcal $m^{-1}hr^{-1}C^{-1}$. The value of λ for water increases from about 0.48 to 0.58 kcal $m^{-1}hr^{-1}C^{-1}$ between 0 and 100°C. The thermal conductivity of milk decreases slightly between 0° and about 37° C and then increases, but assuming a linear increase over the temperature range 0° to 100°C is usually sufficient (Bogdanov and Gochiyaev 1962; Fernandez-Martin and Montes 1970; Reidy 1968; Woodams and Nowrey 1968). Typical values for λ are 0.46 kcal $m^{-1}hr^{-1}C^{-1}$ at 37°C and 0.53 kcal $m^{-1}hr^{-1}C^{-1}$ at 80°C (Peeples 1962; Reidy 1968; Woodams and Nowrey 1968). There are marked decreases in λ with increases in fat, total solids, or concentration (Leidenfrost 1959; Lepilkin and Borisov 1966; Spells 1960; Fernandez-Martin and Montes 1972; Sweat and Parmelee 1978), but the magnitude of the change is temperature dependent (Fernandez-Martin and Montes, 1970, 1977). Thermal conductivity of dried dairy products depends upon bulk density as well as composition (Farrall *et al.* 1970; Norris *et al.* 1971; Reidy 1968).

REFRACTIVE INDEX

The refractive index of a substance is defined as the ratio of the speed of light in a vacuum to its speed in that substance. One consequence of refraction is to change the direction of a light ray as it enters or leaves the substance. Measurement of this bending gives a direct measure of the refractive index, n. Specifically $n = \sin i/\sin r$, where i is the angle of the ray to the surface as it approaches (incidence) and r is the exit angle (refraction). The principles involved and a detailed critique of the methods of measurement are presented by Bauer and co-workers (1959). Since the refractive index varies with the sample temperature and the wavelength of the light, these must be controlled and specified. Thus n_D^{20} refers to the index at 20°C with the D line of the sodium spectrum (589.0 and 589.6 nm).

The refractive index of water is $n_D^{20} = 1.33299$. The value of n_D^{20} for cow's milk generally falls in the range 1.3440 to 1.3485. Buffalo milk is similar to cow's milk (Hofi *et al.* 1966), while human, goat, and ewe milks appear to have higher (Rangappa 1964) refractive index values. Since the refractive increments contributed by each solute in a solution are additive, much consideration has been given to the possible use of refractive index as a means of determining total solids or added water in milk. The refractive index of milk itself is somewhat difficult to determine because of the opacity, but by using a refractometer such as the Abbe instrument, which employs a thin layer of sample, it is possible to make satisfactory measurements, particularly with skim milk products and sweetened condensed milk (Ludington and Bird 1941; Rice and Miscall 1926).

The relation between solids content (on the basis of weight per unit volume) and refractive index is linear, and the contributions of the several components are additive (Goulden 1963). However, the individual components of milk differ in specific refractive increment, $\Delta n(\rho c)$, where ρ is the density of the sample and c is the weight/weight concentration of the component. Thus the relation between percent solids and refractive index will vary between lots of milk. Goulden (1963) reported the following specific refractive increments (mn g^{-1}): casein complex 0.207, soluble proteins 0.187, and lactose 0.140. The total contribution to the refractive index for a milk containing 2.34% casein complex, 0.83% soluble proteins, and 4.83% lactose becomes 0.00500 + 0.00159 + 0.00695, or 0.01354. The residue of 0.95% contributes 0.00166 to the total difference between the refractive indices of water and the milk. Similar data have been reported by Rangappa (1947, 1948B). The refractive index of milk fat is 1.4537 to 1.4552 at 40°C; it is the same in bulk and in globules (Walstra 1965). The fat does not contribute to the refractive index of whole milk because refraction occurs at the interface of air and the continuous phase (Goulden 1963). Sterilization does not alter the refractive index (Armandola and Brezzi 1964), nor does subsequent storage (Chiofalo and Iannuzzi 1963).

Clarification by removal of casein with such agents as calcium chloride, acetic acid, cooper sulfate, or rennin has often been employed to obtain a serum more suitable for refractometric measurements. Obviously the composition, and hence the refractive index, of such sera will depend on the method of preparation. Furthermore, some of the serum proteins may be precipitated with the casein by some of the agents used, particularly if the milk has been heated. Refractive index measurements of such sera are not generally considered as satisfactory as freezing point measurements for detection of added water (David and MacDonald 1953; Munchberg and Narbutas 1937; Schuler 1938; Tellmann 1933; Vleeschauwer and Waeyenberge 1941). Menefee and Overman (1939) reported a close relation between total solids in evaporated and condensed products and the refractive index of serum prepared therefrom by the copper sulfate method. Of course, a different proportionality constant would hold for each type of product.

The estimation of casein in milk by refractometric techniques appears to hold some promise. The casein may be precipitated, washed, and redispersed to yield a solution suitable for refractometry (Brereton and Sharp 1942; Schober et al. 1954). Another method involves computation from the difference between the refractive indices of two samples, one made alkaline to dissolve the casein and the other treated with copper sulfate to precipitate it (Hansson 1957). Heating the milk would cause the serum proteins to precipitate with the casein. The total solids of co-precipitate preparations can be determined by refractive index using 2.5 N NaOH as a dispersant (Dunkley 1970).

The refractive constant or specific refractive index computed by the Lorenz-Lorentz formula,

$$K = \frac{n^2 - 1}{n^2 + 2} \times \frac{1}{\rho}$$

where n is the refractive index and ρ is the density, has sometimes been used for milk or sera (Ramakrishnan and Banerjee 1952; Rangappa 1948B). It is independent of the temperature but not entirely independent of the concentration. Milk has a K value of about 0.2075.

LIGHT ABSORPTION AND SCATTERING

Absorption of electromagnetic radiation by a substance occurs when the radiation has the same energy content as some transition at the molecular level and when the molecule has either a permanent or an induced dipole. At short wavelengths, such as the ultraviolet, the energy is absorbed by the transformation of electrons to higher energy levels. At long wavelengths, such as the infrared, the transitions are in the vibrational and rotational states of molecules. Transitions associated with various regions of the spectrum are shown in Table 8.6. In all regions, the amount of radiation absorbed is proportional to the number of absorption centers as well as the kind. This is commonly expressed in Beer's law,

$$\log_{10} I_0/I = A = abc$$

Table 8.6. Summary of Transitions Interacting with Radiation.

Spectral region	Wavelength	Molecular transitions involved
X-ray	0.001–10 nm	Inner electrons
Ultraviolet-(UV)–vacuum	10–200 nm	Sigma electrons
UV-far	200–290 nm	n and pi electrons
UV-near	290–400 nm	Conjugated systems
Visible	400–800 nm	Highly conjugated systems
Infrared	0.7–60 μm	Vibrational and rotational
Microwave	cm	Rotational
Nuclear magnetic resonance	(1–100 megacycles/sec)	

where I_0 and I = the powers of the incident and emergent beams, respectively, A = the absorbance (formerly called the "optical density"), a = the proportionality constant "absorptivity," b = the sample thickness through which the radiation is passed, and c = the concentration of sample in solvent (w/v). If more than one species is present which absorbs radiation of a given wavelength, the absorbances (A's) are additive. It should be obvious that absorptivity, a, depends upon the units of b and c and the wavelength of the radiation used. It is usually sensitive to the optical bandwidth of the measuring instrument, as is the apparent wavelength of maximum absorption.

Fluorescence and phosphorescence are the reemission at longer wavelengths of absorbed radiation with shorter or longer delays, respectively. The intensity of the emitted radiation follows Beer's law, in addition to being proportional to the amount of light absorbed originally. The wavelengths of the reemitted light are controlled by the structure of the molecule.

The text by Pomeranz and Meloan (1971) contains a good introduction to spectroscopy, and Volume 9 the Weissberger (1956) series contains full information on all aspects (West 1956).

The foregoing discussion applies to substances in true solution or continuous phases. Dispersed particles also scatter light if the particle size and the wavelength of the radiation are of the same order of magnitude. Colloids and emulsions scatter ultraviolet and visible radiation quite effectively. In contrast, electrons scatter x-rays. The preferred angles of scattering may be observed, as in x-ray diffraction, or the attenuation of the incident radiation may be studied, as in turbidimetry. Scattering of visible light by emulsions has been described adequately by Goulden (1961) and by Walstra (1965).

Milk is a colloidal dispersion of proteins and an emulsion of fat in an aqueous solution of lactose, salts, and other compounds. Thus it not only absorbs light at many wavelengths, because of the large number of compounds present, but also scatters it as a result of the presence of particles of various sizes. The well-known absorption by proteins in the 220- to 380-nm region can be distinguished, as can absorption in the 400- to 520-nm region, by fat pigments. Scattering is decreased as the wavelength increases (Goulden 1963); thus, in the infrared region, most of the attenuation is due to absorption. Various specific absorptions can be seen in the near infrared and infrared, most notably those by OH groups near 2.84 μm, CH_2 groups near 3.45 μm, $C=O$ groups at 5.74 μm, and NH_2 groups at 6.56 μm. Since water strongly absorbs infrared radiation, milk is opaque to a major portion of the infrared region (Goulden 1961).

The recent interest in light absorption, fluorescence, and scattering by milk is largely quantitative rather than qualitative. Direct analysis of milk by spectrophotometric techniques offers definite advantages of speed, simplicity, and capabilities of automation. Simple reactions with specific milk components can be used in many cases where direct spectrophotometry is impossible.

The most obvious fluorescent compound in milk is riboflavin, which absorbs strongly at 440–500 nm and emits fluorescent radiation with a maximum at 530 nm. Riboflavin in whey is measured easily by fluorescence (Amer. Assoc. Vitamin Chemists 1951). Proteins also fluoresce because of their content of aromatic amino acids. Part of the ultraviolet radiation absorbed at 280 nm is emitted at longer wavelengths as fluorescent radiation. A prominent maximum near 340 nm is attributable to tryptophan residues in the protein. Use of fluorescence for quantitation of milk proteins was proposed by Konev and Kozunin (1961), and the technique has been modified and evaluated by several groups (Bakalor 1965; Fox et al. 1963; Koops and Wijnand 1961; Porter 1965). It seems to be somewhat less accurate than desired because of difficulties in disaggregating the caseinate particles and in standardizing instruments. It also involves a basic uncertainty due to natural variations in the proportions of individual proteins which differ in tryptophan content.

Goulden (1961; Goulden et al. 1964) described a method for fat, protein, and lactose in milk based on absorption of infrared energy at specific wavelengths. Originally, the difference in absorbances of a homogenized sample of milk and pure water was measured at 5.8 ($C=O$ stretch), 6.5 (Amide II), and 9.6 μm for fat, protein, and lactose, respectively. Good results have been obtained by using the CH stretch (3.4 μm) rather than the $C=O$ stretch for the measurement of fat (Clemmensen 1980; Mills and van de Voort 1982; Gecks 1981). Homogenization is used to reduce fat globule sizes to less than 2 μm to eliminate light-scattering rather than to produce a specific size distribution; therefore the nature of the homogenization is less critical than with light-scattering techniques. The infrared methods correlate well with chemical methods for the various components (Adda et al. 1968; Briggs 1964, 1978) if care is taken for proper calibration and operation. Besides the use of the CH stretch at 3.4 μm, more minor modifications of the method have been tried. van de Voort (1980) found a single-cell, dual-wavelength instrument capable of meeting AOAC specifications.

Both the fat and protein of milk scatter light, the amount of scattering depending upon the number and size of the particles, the wavelength of the incident radiation, and the difference in refractive index between the different kinds of particles and the solvent (Ashworth

1969; Flux *et al* 1982; Goulden and Sherman 1962; Haugaard 1966; Jeunet and Grappin 1970; Walstra 1967). Measurement of fat in milk by light scattering was first described by Haugaard and Pettinati (1959). Homogenization is used to achieve uniform fat globule size distributions in different samples. Protein particles are solubilized at high pH with disodium ethylenediaminetetraacetate (EDTA), which also serves to dilute the milk sufficiently to avoid disturbing multiple scattering effects. The commercial version of this method, the "Milko-Tester," utilizes white light in a special photometric system to determine attenuation due to scattering, and thus, by correlation, the fat content of the milk. Dilution of milk with the EDTA solution before homogenization was found to improve the results (Aegiduo 1969; Grappin and Juenet 1970). With proper attention to instrument calibration and operation, this method compared favorably to more traditional methods of measuring the fat content of milk (Grappin and Jeunet 1970; Shipe 1969; Shipe and Senyk 1973, 1975, 1980) and cream (Packard *et al* 1973; Szijarto and van de Voort 1982).

Nakai and Le (1970) have used a different approach to the measurement of fat by light scattering. They dissolved both the fat and protein particles with acetic acid, measured the protein content by the absorbance at 280 nm, and then reformed a fat emulsion by adding a solution of urea and imidazole. The turbidity was measured at 400 nm and was found to be independent of the initial fat globule size distribution.

Both the light scattering and infrared techniques are widely used, and there are many reports of the correlation between one of these methods and a reference method, usually Babcock or Gerber in the case of fat analysis and total nitrogen by Kjeldahl for proteins. One source of disparity between the various correlations has been the choice of a reference method. The realization that different methods for fat analysis measure different chemicals and thus can be expected to give different results has been slow in reaching the industry.

The correlation between the instrumental method and the chosen reference method also varies with the nature of the sample. Factors associated with the milk that affect this correlation for fat include fat composition and sample condition. The molecular weight of the fat is affected drastically by protected lipid feeding, which in turn changes the correlation between the Babcock test and an Infrared Milk Analyzer (Franke *et al* 1975). This same effect of molecular weight can appear as seasonality (Mogot *et al* 1982) or as a result of lipolysis (Robertston *et al* 1981; van Reusel 1975). Sample condition effects noted have been compositing and/or preserving (Ng-Kwai-Hang and Hayes

1982; Packard et al. 1973; Robertson et al. 1981), cooling (Dill et al. 1979), and aging (Ng-Kwai-Hang and Hayes 1982). Most of these factors are known to be or suspected of being fat test depressants using a given method as compared to values on a fresh aliquot (Packard and Ginn 1973; Tomaszewski and Dill 1978).

Factors affecting the instrument calibrations when testing for protein content have not been as well explored, but it is apparent that amino acid composition is important. Thus, anything affecting the ratios of the various proteins can be expected to be significant, as can the protein phenotype when testing milks from individual animals. The hydration and charge states of the proteins will probably have detectable effects. This brings pH, osmotic pressure, and ionic composition into the picture.

The relationship between Milko-Tester and Gerber results is curvilinear, and the difference between the two methods correlates well with the refractive index of the fat (Flux et al. 1982). Variable results have been reported on the effect of sample aging and sample preservative on the difference between Milko-Tester and reference method results (Minzner and Kroger 1974). This indicates the importance of calibration with the same type of samples as to be tested.

Addition of specific compounds to milk has been used to allow spectrophotometric measurement of lactose as the osazone (Wahba 1965) and fat by fluorescence (Bakhiren and Butov 1968; Konev and Kozlova 1970). The dye-binding method for measuring protein in milk is based on the ability of sulfonic acid dyes to complex with the basic amino acid residues of milk proteins at low pH (Fraenkel-Conrat and Cooper 1944). Dye binding correlates well with Kjeldahl (Sherbon 1970) and infrared (Mogot et al. 1982; Grappin et al. 1980) results, but variations are caused by the different compositions of the different milk proteins (Ashworth 1966; Vanderzant and Tennison 1961).

It should be pointed out that the use of the various instrumental methods on milks of other species can be successful, but special calibrations are required (Grappin and Jeunet 1979; Grappin et al. 1979).

Lin et al. (1971) used inelastic scattering of plane-polarized light of 632.8-nm wavelength from a He-Ne laser to determine the diffusion coefficient and thereby the hydrodynamic radii of monodisperse caseinate micelle fractions from milk. The cumulative distribution curve of the weight fraction of micelles revealed that about 80% of the casein occurs in micelles with radii of 50 to 100 nm and 95% between 40 and 220 nm, with the most probable radius at about 80 nm. This method has the advantage that the micelles are examined in their natural medium.

REFERENCES

Adda, J., Blane-Platin, E., Jeuenet, R., Grappin, R., Mocquot, G., Paujardieu, B. and Ricordeau, G. 1968. Trial of the infrared milk analyzer. *Lait 48*, 145–154.

Aegiduo, P. E. 1969. Fat content determination. U.S. Patent 3,442, 623.

Agarwala, S. P. and Ojha, T. P. 1973. Specific heat of concentrated whole milk at higher temperatures. *Ind. J. Dairy Sci. 26*, 83–87.

Amer. Assoc. Vitamin Chemists. 1951. *Methods of Vitamin Assay*, 2nd ed. John Wiley & Sons, New York.

Andrade, E. N. 1952. Viscosity of liquids. *Proc. Royal Soc. (London) 215A*, 36–43.

Ansbacher, S., Flanigan, G. E. and Supplee, G. C. 1934. Certain foam producing substances of milk. *J. Dairy Sci. 17*, 723–731.

Armandola, P. and Brezzi, G. 1964. Effect of uperization of milk on its f.p., t. s., refractive index and density of serum. *Latte 38*, 1013–1017.

Aschaffenburg, R. 1945. Surface activity and proteins of milk. *J. Dairy Res. 14*, 316–329.

Ashworth, U. S. 1966. Determination of protein in dairy products by dye-binding. *J. Dairy Sci. 49*, 113–137.

Ashworth, U. S. 1969. Turbimetric methods for measuring fat content of homogenized milk. *J. Dairy Sci. 52*, 262–263.

Assoc. Official Analytical Chemists. 1980. *Methods of Analysis*, 13th ed. Washington, D.C.

Assoc. Official Analytical Chemists. 1984. *Methods of Analysis*, 14th ed. Washington, D.C.

Bailey, E. M. 1922. Cryoscopy of milk. *J. Assoc. Official Agr. Chem. 5*, 484–497.

Bakalor, S. 1965. The estimation of protein in milk from its fluorescence in the ultraviolet region. *Aust. J. Dairy Technol. 20*, 151–153.

Bakhiren, N. F. and Butov, G. P. 1968. Fluorescence method for determination of milk fat. *Nauchno-tekk Byul Elecktref, selsk. Khoz* (2)34–39. In *DSA 414*, 1970.

Barron, E. S. G. and Hastings, A. B. 1934. Studies on biological oxidations. III. The oxidation-reduction potential of the system lactate-enzyme pyruvate. *J. Biol. Chem. 107*, 567–578.

Bartsch, B. D. and Wickes, R. B. 1979. The freezing point of milk as influenced by nutrition of the cow. *Aust. J. Dairy Technol. 34*, 154–158.

Bateman, G. F. and Sharp, P. F. 1928. A study of the apparent viscosity of milk as influenced by some physical factors. *J. Agr. Res. 36*, 647–674.

Bates, R. G. 1964. *Determination of pH, Theory and Practice*. John Wiley & Sons, New York, p. 435.

Bauer, N., Fajans, K. and Lewin, S. Z. 1959. Refractometry. *In: Techniques of Organic Chemistry*, 3rd ed., Vol. 1. A. Weissberger (Editor). Interscience, New York.

Beeby, R. and Kumetat, K. J. 1959. Viscosity changes in concentrated skim milk treated with alkali, urea, and calcium complexing agents. I. The importance of the casein micelle. *J. Dairy Res. 26*, 248–257.

Beeby, R. and Lee, J. W. 1959. Viscosity changes in concentrated skim milk treated with alkali, urea, and calcium complexing agents. II. The influence of concentration, temperature, and rate of shear. *J. Dairy Res. 26*, 258.

Berg, L. Vanden. 1961. Changes in pH of milk during freezing and frozen storage. *J. Dairy Sci. 44*, 26–31.

Bertsch, A. J. 1982. Specific heat capacity of whole and skim milk between 50 and 140°C. *Lait 62*, 265–275.

Biggs, D. A. 1964. Infra-red analysis of milk for fat, protein, lactose and solids-not fat. *Conv. Proc. Milk Industry Foundation 1964*, 28–34.

Biggs, D. A. 1978. Instrumental infrared estimation of fat, protein and lactose in milk: A collaborative study. *J. Assoc. Official Anal. Chem. 61*,1015–1034.

Bogdanov, S. and Gochiyaev B. 1962. Study of thermal and physical properties of milk. *Mol. Prom. 22*(6), 16–20.

Bonomi, A. 1978. Relationship between feeding and the characteristics of milk intended for Parmigiano-Reggiano cheesemaking. *Sci. Tech. Lattiero-Casearia 29*, 397–418. In *DSA 41*, 4645, 1979.

Boulet, M. and Marier, J. R. 1961. Precipitation of calcium phosphates from solution at near physiological concentrations. *Arch Biochem. Biophys. 93*, 157–165.

Boulet, M. and Rose, D. 1954. Titration curves of whey constituents. *J. Dairy Res. 21*, 227–237.

Brereton, J. G. and Sharp, P. F. 1942. Refractometric determination of casein in skim milk. *Ind. Eng. Chem., Anal. Ed. 14*, 872–874.

Brathen, G. 1983. Factors affecting the freezing point of genuine cows' milk. *In: Measurement of Extraneous Water by the Freezing Point Test.* F. Harding (Editor). Bull. 154 FIL/IDF. Brussels, pp. 6–11.

Buchanan, J. H. and Peterson, E. E. 1927. Buffers of milk and buffer value. *J. Dairy Sci. 10*, 224–231.

Buma, T. J. and Meerstra, J. 1969. The specific heat of milk powder and of some related materials. *Neth. Milk Dairy J. 23*, 124–127.

Caffyn, J. E. 1951. The viscosity temperature coefficient of homogenized milk. *J. Dairy Res. 18*, 95–105.

Calandron, A. and Grillet, L. 1964. Measurement of the surface tension of certain milks with a Nouy tensiometer. *Lait 44*, 505–509.

Caulfield, W. J. and Riddell, W. H. 1936. Some factors influencing the acidity of freshly drawn cows' milk. *J. Dairy Sci. 19*, 235–242.

Chiofalo, L. and Iannuzzi, L. 1963. Variations in some properties of sterilized milk during storage. *Zootec e Vita 6*, 32–55. In *DSA 27*, 1287, 1965.

Clark, W. M. 1934. The acid-base and oxidation-reduction equilibria of milk. *In: Fundamentals of Dairy Science*, 2nd ed. Assoc. of Rogers Reinhold Pulishing Corp, New York, pp. 137–154.

Clark, W. M. 1960. *Oxidation-Reduction Potentials of Organic Systems*. Williams & Wilkins Co., Baltimore.

Clemmensen, K. 1980. Modified fat determination. *Dairy Field 163*(12), 51–52, 54.

Coe, J. R. and Godfrey, T. B. 1944. Viscosity of water. *J. Appl. Phys. 15*, 625.

Cole, E. R., Douglas, J. B. and Mead, M. 1957. The lactose-chloride contribution to the freezing point depression of milk. II. Examination of partial contribution over the full lactation period of two cows. *J. Dairy Res. 24*, 33–47.

Cole, E. R. and Mead, M. 1955. The lactose-choloride contribution to the freezing point depression of milk. *J.Dairy Res. 22*, 340–344.

Coste, J. H. and Shelbourne, E. T. 1919. The electrical conductivity of milk. *Analyst 44*, 158–165.

Cox, C. P. 1952. Changes with temperature in the viscosity of whole milk. *J. Dairy Res. 19*,72–82.

Cox, C. P., Hasking, Z. D. and Posener, L. N. 1959. Relation between composition and viscosity of cow's milk. *J. Dairy Res. 26*, 182–189.

Dahlberg, A. C., Adams, H. S. and Held, M. E. 1953. Sanitary milk control and its relation to the sanitary nutrition of milk. *Natl. Res. Council 250*, 174.

Dastur, N. N., Dharmarajan, C. S. and Rao, R. V. 1952. Composition of milk of Indian animals. III. Freezing point, lactose and chloride content of milk samples from different farms in India. *Ind. J. Vet. Sci. 22*, 123-133.

Davis, J. G. and MacDonald, F. J. 1953. *Richmond's Dairy Chemistry*, 5th ed. Charles Griffin & Co., London.

Demott, B. J. 1966. The freezing point of milk produced in four markets in Tennessee. *Milk Food Technol. 29*, 319-322.

Demott, B. J. 1967. The influence of vacuum pasteurization upon the freezing point and specific gravity of milk. *Milk Food Technol. 30*, 253-255.

Demott, B. J. and Burch, T. A. 1966. Influence of storage upon the freezing point of milk. *J. Dairy Sci. 49*, 317-318.

Demott, B. J., Hinton, S. A. and Montgomery, N. J. 1967. Influence of some management practices and season upon freezing point of milk. *J. Dairy Sci. 50*, 151-154.

Demott, B. J., Hinton, S. A., Swanson, E. W. and Miles, T. J. 1968. Influence of added sodium chloride in grain ration on the freezing point of milk. *J. Dairy Sci. 51*, 1363-1365.

Demott, B. J., Montgomery, M. J. and Hinton, S. A. 1969. Influence of changing from dry lot feeding to pasture on the freezing point of milk. *J. Milk Food Technol. 32*, 210-212.

Deysher, E. F., Webb, B. H. and Holm, G. E. 1944. The viscosity of evaporated milks of different solids concentration. *J. Dairy Sci. 27*, 345-355.

Dill, C. W., Herlick, S. A., Richter, R. L., and Davis, J. W. 1979. Fat test depression during chilled storage of milk samples in plastic containers for analysis by the Milko-Tester J. *Food Protection 42*, 314-316.

Dixon, B. 1963. The effect of temperature on the pH of dairy products. *Aust. J. Dairy Technol. 18*, 141-144.

Dunkley, J. 1970. Total solids determination in coprecipitate solutions. *Proc. 18th Int. Dairy Congr. IE*, 430.

Dunkley, W. L. 1951. Hydrolytic rancidity in milk. I. Surface tension and fat acidity as measures of rancidity. *J. Dairy Sci. 34*, 515-520.

Edsall, J. T. and Wyman, J. 1958. *Biophysical Chemistry*, Vol. 1. Academic Press, New York.

Eilers, H., Saal, R. H. J. and Waarden, M. van den. 1947. *Chemical and Physical Investigations on Dairy Products*. Elsevier Publishing Co., New York.

Eisses, J. and Zee, B. 1980. The freezing point of authentic cow's milk and farm tank milk in the Netherlands. *Neth. Milk Dairy J. 34*, 162-180.

Eley, R. M., Collier, R. J., Bruss, M. L., Horn, H. H. van and Wilcox, C. J. 1978. Interrelationships between heat stress parameters and milk composition and yield in dairy cattle. *J. Dairy Sci. 61*(Suppl. 1), 147.

El-Rafey, M. S. and Richardson, G. A. 1944. The role of surface-active constituents involved in the foaming of milk and certain milk products. II. Whey, skimmed milk and their counterparts. *J. Dairy Sci. 27*, 19-31.

England, C. W. and Neff, M. J. 1963. The accuracy of cryoscope methods. *J. Assoc. Offic. Agr. Chem. 46*, 1043-1049.

Evans, E. W. and Johnson, V. W. 1978. Effect on the freezing point of milk of restricting and providing water to the cow. *Proc. 20th Int. Dairy Congr. E.*, pp. 212-213.

Farrall, A. W., Heldman, D. R., Wang, P. Y., Ojha, T. P. and Chen, A. C. 1970. Thermal conductivity of dry milk. *Proc. 18th Int. Dairy Congr. 1E*, 269.

Fernandez-Martin, F. and Montes, F. 1970. Thermal properties of milk and milk products. III. Thermal conductivity, its correlation with temperature and composition. *Proc. 18th Int. Dairy Congr. 1E*, 471.

Fernandez-Martin, F. and Montes, F. 1972. Influence of temperature and composition of some physical properties of milk and milk concentrates. III. Thermal conductivity. *Milchwissenschaft 27*, 772-776.

Fernandez-Martin, F. and Montes, F. 1977. Thermal conductivity of creams. *J.Dairy Res. 44*, 103-109.

Fernando, R. S., Rindsig, R. B. and Spahr, S. L. 1982. Electrical conductivity of milk for detection of mastitis. *J. Dairy Sci. 65*, 659-664.

Flux, D. S., Raven, J. A. and Gray, I. K. 1982. Accuracy of the Milko-Tester over a wide range of milk fat concentrations. *N.Z. J. Dairy Sci. Technol. 17*, 15-25.

Fox, K. K., Holsinger, V. H. and Pallansch, M. J. 1963. Fluorimetry as a method of determining protein content of milk. *J. Dairy Sci. 46*, 302-309.

Fox, P. F. and Morrissey, P. A. 1977. Reviews of the progress of dairy science: The heat stability of milk. *J. Dairy Res. 44*, 627-646.

Fraenkel-Conrat, H. and Cooper M.J. 1944. The use of dyes for the determination of acid and basic groups in proteins. *J. Biol. Chem. 154*, 239-246.

Franke, A. A., Dunkley, W. L. and Smith, L. M. 1975. Comparison of Babcock and infrared milk analyzer methods for determining fat in milk from cows fed protected lipid supplement. *J. Dairy Sci. 58*, 791.

Frazier, W. C. and Whittier, E. O. 1931. Studies on the influence of bacteria on the oxidation-reduction potential of milk. I. Influence of pure cultures of milk organisms. *J. Bact. 21*, 239-262.

Fredholm, H. 1942. The specific electrical conductivity of Swedish cow milk, with special reference to the diagnosis of udder disease. *Nord. Jordburgsforskning 1942*, 195-213.

Freeman, T. R., Bucy, J. L. and Kratzer, D. D. 1971. The freezing point of herd milk produced in Kentucky. *J. Milk Food Technol. 34*, 212-214.

Gecks, E. 1981. Improvement of fat determination in the infrared measurement procedures. *Deutsche Molkerei-Zeitung 102*.

Gerber, V. 1927. The significance of the specific electrical conductivity of milk and a new practical procedure for its determination. *Z. Untersuch Lebensm. 54*, 257-270.

Golding, N. S. 1959. A solids-not-fat test for milk using density plastic beads as hydrometers. *J. Dairy Sci. 42*, 899.

Gould, I. A. 1945. The formation of volatile acids in milk by high-temperature heat treatment. *J. Dairy Sci. 28*, 379-386.

Gould, I. A. and Frantz, R. S. 1945. Some relationships between pH, titrable acidity, and the formol titration in milk heated to high temperatures. *J. Dairy Sci. 28*, 387-399.

Gould, I. A. and Sommer, H. H. 1939. Effect of heat on milk with especial reference to the cooked flavor. *Mich. Agr. Exp. Sta. Tech. Bull. 164*, pp. 48.

Goulden, J. D. S. 1961. Quantitative analysis of milk and other emulsions by infra-red absorption. *Nature 191*, 905-906.

Goulden, J. D. S. 1963. Determination of SNF in milk and unsweetened condensed milk from refractive index measurements. *J. Dairy Res. 30*, 411-417.

Goulden, J. D. S. and Sherman, P. 1962. A simple spectroturbimetric method for the determination of the fat content of homogenized ice cream mixes. *J. Dairy Res. 29*, 47-53.

Goulden, J. D. S., Shields, J. and Haswell, R. 1964. The infrared milk analyzer. *J. Soc. Dairy Technol. 17*, 28-33.

Grappin, R. and Jeunet, R. 1970. The Milko-testor automatic for routine determination of fat in milk. *Lait 50*, 233-256.

Grappin, R. and Jeunet, R. 1979. Routine methods for measuring fat and protein in goats milk. *Lait 59*, 345-360.

Grappin, R., Jeunet, R. and Le Dore, A. 1979. Determination of the protein content of cow's and goat's milk by dye binding and infrared methods. *J. Dairy Sci. 62*, 38–39.

Grappin, R., Packard, V. S. and Ginn, R. E. 1980. Repeatability and accuracy of dye binding and infrared methods for analyzing protein and other milk components. *J. Food Protection 43*, 374–375.

Hansson, E. 1957. Estimation of casein with aid of the refractometer. *Svenska Mejereitid 49*, 277–279.

Harding, F. 1983. The effect of processing. *In: Extraneous Water.* F. Harding (Editor). Bull. 154. FIL/IDF. Brussels, pp. 11.

Harkins, W. D. and Alexander, A. E. 1959. Determination of Surface and Interfacial Tension. *In: Technique of Organic Chemistry*, 3rd ed., Vol. 1. A Weissberger (Editor) Interscience Publishers, New York, pp. 757–814.

Harland, H. A., Coulter, S. T. and Jenness, R. 1952. The interrelationship of processing treatments and oxidation-reduction systems as factors affecting the keeping quality of dry whole milk. *J. Dairy Sci. 35*, 643–654.

Haugaard, G. 1966. Photometric determination of fat in milk. *J. Dairy Sci. 49*, 1185–1189.

Haugaard, G. and Pettinati, J. D. 1959. Photometric milk fat determination. *J. Dairy Sci. 42*, 1255–1275.

Have, A. J. van der, Deen, J. R. and Mulder, H., 1979. The composition of cow's milk. IV. Calculation of the titratable acidity studied with separate milkings of individual cows. *Neth. Milk Dairy J. 33*, 164–171.

Have, A. J., van der, Deen, J. R. and Mulder, H. 1980. The composition of cow's milk. V. The contribution of some milk constituents to the freezing point depression studies with separate milkings of individual cows. *Neth. Milk Dairy J. 34*, 1–8.

Henderson, J. L. 1963. The effect of handling and processing on the freezing point of milk. *J. Assoc. Offic. Agr. Chem. 46*, 1030–1035.

Henningson, R. W. 1963. The variability of the freezing point of fresh raw milk. *J. Assoc. Offic. Agr. Chem. 46*, 1036–1042.

Henningson, R. W. 1966. Cryoscopy of milk: Effect of variation in the method. *J. Assoc. Offic. Anal. Chem. 49*, 511–515.

Henningson, R. W. 1967. Determination of the freezing point of milk by thermistor cryoscopy. *J. Assoc. Offic. Anal. Chem. 50*, 533–537.

Henningson, R. W. 1969. Thermistor cryoscopic determination of the freezing point value of milk produced in North America. *J. Assoc. Offic. Anal. Chem. 52*, 142–151.

Henningson, R. W. 1970. Regulatory agency acceptance of the interpretation of the freezing point value of milk as part of the official cryoscopic method. *J. Assoc. Offic. Anal. Chem. 53*, 539–542.

Herrington, B. L. 1954. Lipase: A review. *J. Dairy Sci. 37*, 775–789.

Herrington, B. L., Sherbon, J. W., Ledford, R. A. and Houghton, G. E. 1972. Composition of milk in New York State. *NY Food Life Sci. Bull.* 18.

Hetrick, J. H. and Tracy, P. H. 1948. Effect of high-temperature short-time heat treatment on some properties of milk. II. Inactivation of the lipase enzyme. *J. Dairy Sci. 31*, 881–887.

Higginbottom, C. and Taylor, M. M. 1960. The oxidation-reduction potential of sterilized milk. *J. Dairy Res. 27*, 245–257.

Hofi, A. A., Riffat, I. D. and Khorshid, M. A. 1966. Studies on some physical and physico-chemical properties of Egyptian buffalo's and cow's milk. *Ind. J. Dairy Sci. 19*, 118–121.

Horst, M. G. ter. 1947. The condition and mutual relationship of calcium caseinate and calcium phosphate in milk. *Neth. Milk Dairy J. 1*, 137–151.

Hortvet, J. 1921. The cryoscopy of milk. *Ind. Eng. Chem. 13*, 198–208.

Howat, G. R. and Wright, N. C. 1937. Factors affecting the solubility of milk powders. III. Some physico-chemical properties of concentrated solutions of milk solids. *J. Dairy Res. 5*, 236–244.

Igarashi, Y. and Saito, Z. 1972. Milk components affecting the surface tension of bovine milk. *Bull. Faculty Agr. Hirosaki Univ. 18*, 43–48.

Jack, E. L. and Brunner, J. R. 1943. The relation between the degree of solidification of fat in cream and its churning time. I. Measurement of the degree of solidification. *J. Dairy Sci. 26*, 169–178.

Jackson, J. 1936. Factors in the reduction of methylene blue in milk. *J. Dairy Res. 7*, 31–40.

Jackson, R. H. and Pallansch, M. J. 1961. Influence of milk proteins on interfacial tension between butteroil and various aqueous phases. *J. Agr. Food Chem. 9*, 424–427.

Jaster, E. H., Smith, A. R., McPherron, T. A. and Pedersen, D. K. 1982. Effect of an intramammary polyethylene device in primiparous dairy cows. *Am. J. Vet. Res. 43*, 1587–1589.

Jenness, R. 1962. Unpublished derivation.

Jenness, R. and Patton, S. 1959. *Principles of Dairy Chemistry.* John Wiley & Sons, New York.

Jeunet, R. and Grappin, R. 1970. A note on the relationship between the refractive index of milk fat and the precision of determination of fat by means of the Milko-tester. *Lait 50*, 654–657.

Johnston, J. H. St. 1927. Surface tension of protein solutions. III. *Biochem. J. 21*, 1314–1328.

Josephson, D. V. and Doan, F. J. 1939. Observation on cooked flavor in milk—its source and signficance. *Milk Dealer 29*(2), 35–36, 54, 56, 58–60, 62.

Kleyn, D. H. and Shipe, W. F. 1957. Has water been added to milk? *Am. Milk Rev. 19*, 26.

Konev, S. V. and Kozlova, G. G. 1970. Application of secondary luminescence for fat analysis in milk and some dairy products. *Proc. 18th Int. Dairy Congr. IE*, 84.

Konev. S.V. and Kozunin, I. I. 1961. Fluorescence method for the determination of protein in milk. *DSA 23*, 103–105.

Koops, J. and Wijnand, H. P. 1961. Determination of protein in milk by fluorescence. *Neth. Milk Dairy J. 15*, 333–357.

Kozanecki, M., Sciubisz, A. and Kasperwicz, A. 1982. Interrelationships between the somatic cell number and lactose level and conductivity in cow's milk and their diagnostic significance in detection of mastitis. Proc. 12th World Cong. Diseases of Cattle 2, 1054–1058.

Leidenfrost, W. 1959. Measurement of heat conductivity of milk of different water content in a temperature range of 20–100°C. *Fette Seifen. Anstrichmitt 61*, 1005–1010.

Leighton, A. and Kurtz, F. 1930. The pseudo-plasticity of skim milk. *Agri. Eng. 11*, 22–23.

Lepilkin, A. and Borisov, V. 1966. Thermal coefficient of cream. *Mol. Prom. 27*, 12–13. In *DSA 28*, 3090, 1966.

Leviton, A. and Leighton, A. 1935. The action of milk fat as a foam depressant. *J. Dairy Sci. 18*, 105–112.

Leviton, A. and Pallansch, M. J. 1960. Laboratory studies on high temperature–short time sterilized evaporated milk. I. Easily constructed eccentric falling ball type bomb microviscometers. *J. Dairy Sci. 43*, 1389–1395.

Lin, S. H. C., Dewan, R. K., Bloomfield, V. A. and Morr, C. V. 1971. Inelastic light-scattering study of the size distribution of bovine milk casein micelles. *Biochemistry 10*, 4788–4793.

Ling, E. R. 1936. The titration of milk and whey as a means of estimating the colloidal calcium phosphate of milk. *J. Dairy Res. 7*, 145-155.

Ludington, V. D. and Bird, E. W. 1941. Application of the refractometer to determination of total solids in milk products. *Food Res. 6*(4), 421-434.

Maxcy, R. B. and Sommer, H. H. 1954. Fat separation in evaporated milk. I. Homogenization, separation and viscosity tests. *J. Dairy Sci. 37*, 60-71.

McIntyre, R. T., Parrish, D. B. and Fountain, F. C. 1952. Properties of the colostrum of the dairy cow. VII. pH, buffer capacity and osmotic pressure. *J. Dairy Sci. 35*, 356-362.

McKennell, R. 1960. Influence of viscometer design on non-Newtonian measurements. *Anal. Chem. 32*, 1458.

Menefee, S. G. and Overman, O. R. 1939. The relation of the refractive index of evaporated and condensed milk serum to the total solids content. *J. Dairy Sci. 22*, 831-840.

Miller, P. G. and Sommer, H. H. 1940. The coagulation temperature of milk as affected by pH, salts, evaporation and previous heat treatment. *J. Dairy Sci. 23*, 405-422.

Mills, B. L. and van de Voort, F. R. 1982. Evaporation of CH stretch measurement for estimation of fat in aqueous emulsions using infrared spectroscopy. *J. Assoc. Offic. Anal. Chem. 65*, 1357-1361.

Minzner, R. A., Jr. and Kroger, M. 1974. Physicochemical and bacteriological aspects of preserved milk samples and their effect on fat percentages as determined with the Milko-tester. *J. Milk Food Technol. 37*, 123-128.

Magot, M. F. K., Koops, J., Neeter, R., Slangen, K. J., van Hemert, H., Kooyman, O. and Wooldrik, H. Routine testing of farm tank milk with the Milko-Scan 203. II. Fat and protein contents of individual supplies compared with those obtained by Gerber (fat) and dye-binding (protein). *Neth. Milk Dairy J. 36*, 195-210.

Mohr, W. and Brockmann, C. 1930. Surface tension measurements of milk. *Milchwiss. Forsch 10*, 72-95.

Mojonnier, T. S. and Troy, H. C. 1922. *Technical Control of Dairy Products*. Mojonnier Brothers, Chicago.

Muller, W. 1931. Contribution to the electrical conductivity of milk. *Milchwiss. Forsch. 11*, 243-251.

Munchberg, F. and Narbutas, J. 1937. Contribution to the refractometric investigation of protein-free milk serum. *Milchwiss. Forsch. 19*, 114-121.

Mussenden, S., Hodges, J. and Hiley, P. G. 1977. Sodium and chloride in cows' drinking water and freezing point of milk. *J. Dairy Sci. 60*, 1554-1558.

Nakai, S. and Le, A. C. 1970. Spectrophotometric determination of protein and fat in milk simultaneously. *J. Dairy Sci. 53*, 276-278.

Nelson, V. 1949. The physical properties of evaporated milk with respect to surface tension, grain formation and color. *J. Dairy Sci. 32*, 775-785.

Ng-Kwai-Hang, K. F. and Hayes, J. F. 1982. Effect of potassium dichromate and sample storage time on fat and protein by Milko-Scan and on protein and casein by a modified Pro-Milk MK II method. *J. Dairy Sci. 65*, 1895-1899.

Nilsson, G., Carlson, C. and Lau-Eriksson, A. 1970. Studies on the poising effect of milk. *Lantbruks-Hogskolans Ann. 36*, 211-234.

Norris, R., Gray, I. K., McDowell, A. K. R. and Dolby, R. M. 1971. The chemical composition and physical properties of fractions of milk fat obtained by a commercial fractionation process. *J. Dairy Res. 38*, 179-191.

Overman, O. R., Garrett, O. F., Wright, K. E. and Sanmann, F. D. 1939. Composition of milk of Brown Swiss cows. Ill. Agr. Exp. Sta. Bull. 457, 575-623.

Packard, V. S., Jr. and Ginn, R. E. 1973. The influence of previous treatment on accuracy of milkfat analyses determined in a Mark II Milko-tester. *J. Milk Food Technol. 36*, 28-30.

Packard, V. S., Jr., Ginn, R. E. and Rosenau, J. R. 1973. A comparison of Babcock, Mojonnier, and Milko tester Mark III methods in the analysis of milkfat in cream. *J. Milk Food Technol. 36,* 523–525.

Pal, R. N. 1963. Electrical conductivity to determine adulteration of milk. *Ind. J. Dairy Sci. 16,* 92–97.

Palmer, L. S. 1944. The structure and properties of the natural fat globule "membrane." *J. Dairy Sci. 27,* 471–481.

Parkash, S. 1963. Studies in physico-chemical properties of milk. XIV. Surface tension of milk. *Ind. J. Dairy Sci. 16,* 98–100.

Peeples, M. L. 1962. Forced convection heat transfer characteristics of fluid milk products. A review. *J. Dairy Sci. 45,* 297–302.

Peterson, R. W. and Freeman, T. R. 1966. Effect of ration on freezing point of milk and blood serum of the dairy cow. *J. Dairy Sci. 49,* 806–810.

Phipps, L. W. 1957. A calorimetric study of milk, cream and the fat in cream. *J. Dairy Res. 24,* 51–67.

Phipps, L. W. 1969. The interrelationship of the viscosity, fat content and temperature on cream between 40 and 80°C. *J. Dairy Res. 36,* 417–426.

Pinkerton, F. and Peters, I. I. 1958. Conductivity, percent lactose and freezing point of milk. *J. Dairy Sci. 41,* 392.

Pomerantz, V. and Meloan, C. E. 1971. *Food Analysis: Theory and Practice.* AVI Publishing Co., Westport, Conn.

Porter, R. M. 1965. Fluorometric determination of protein in whole milk, skim milk and milk serum. *J. Dairy Sci. 48,* 99–100.

Potter, F. E., Deysher, E. F. and Webb, B. H. 1949. A comparison of torsion pendulum type viscosimeters for measurement of viscosity in dairy products. *J. Dairy Sci. 32,* 452–457.

Prentice, J. H. 1962. The conductivity of milk—the effect of the volume and degree of dispersion of the fat. *J. Dairy Res. 29,* 131–139.

Prentice, J. H. 1978. Freezing point data on aqueous solutions of sucrose and sodium chloride and the Hortvet test: A reappraisal. *Analyst 103,* 1269–1273.

Prouty, C. C. 1940. Observations on the growth response of *Streptococcus lactis* in mastitis milk. *J. Dairy Sci. 23,* 899–904.

Puri, B. R. and Gupta, H. L. 1955. Studies in physico-chemical properties of milk. V. Viscosity of milk. *Ind. J. Dairy Sci. 8,* 78–82.

Puri, B. R. and Parkash, S. 1963. Studies in physico-chemical properties of milk. XIII. Electrical conductivity of milk. *Ind. J. Dairy Sci. 16,* 47–50.

Puri, B. R., Parkash, S. and Totaja, K. K. 1963. Studies in physico-chemical properties of milk. XVI. Effect of composition and various treatments on viscosity of milk. *Ind. J. Dairy Sci. 17,* 181–189.

Pyne, G. T. 1962. Review of the progress of dairy science. C. Dairy Chemistry. Some aspects of the physical chemistry of the salts of milk. *J. Dairy Res. 29,* 101–130.

Pyne, G. T. and Ryan, J. J. 1950. The colloidal phosphate of milk. I. Composition and titrimetric estimation. *J. Dairy Res. 17,* 200–205.

Ramakrishnan, C. V. and Bannerjee, J. N. 1952. Studies on the refractive index of milk. *Ind. J. Dairy Sci. 5,* 25–31.

Rambke, K. and Konrad, H. 1970. Physical properties of fluid milk products. Specific heat of milk, cream and milk concentrates. *Nahrung 14,* 475–485.

Rangappa, K. S. 1947. Contribution of the major constituents to the refractive index of milk. *Nature 160,* 179.

Rangappa, K. S. 1948A. Cryoscopy and refraction in milk. *Biochim. Biophys. Acta. 2,* 207–209.

Rangappa, K. S. 1948B. Contribution of the major constituents to the total refraction in milk. *Biochim. Biophys. Acta 2,* 210–216.

Rangappa, K. S. 1964. Refractive index of human, goat, and sheep milk. *Ind. J. Dairy Sci. 7*, 137–138.

Rao, D. S., Sudheendranath, C. S., Rao, M. B. and Anantakrishan, C. P. 1970. Studies on the electrical conductivity of milk in non-aqueous mixed solvents. *Proc. 18th Int. Dairy Congr. IE*, 88.

Raoult, F. M. 1884. The general law on the freezing of solvents. *Ann. Chem. Phys. 2*, 66–93.

Rees, H. V. 1952. *A Study of the Mechanism of Solids Not Fat and Freezing Point Variation with Progression of the Lactation Period of the Dairy Cow.* Research Service Bulletin, Tasmanian Department of Agriculture, Hobart.

Reidy, G. A. 1968. I. *Methods for Determining Thermal Conductivity and Thermal Diffusivity of Foods. II. Values for Thermal Properties of Food Gathered from the Literature.* Department of Food Science, Michigan State University, Lansing, Michigan, p. 77.

Reusel, A. van 1975. Influence of lipolysis on milko-tester analyses. *Ann. Bull. IDF* No. 86, 185–186.

Rice, F. E. and Markley, A. L. 1924. The relation of natural acidity in milk to composition and physical properties. *J. Dairy Sci. 7*, 468–483.

Rice, F. E. and Miscall, J. 1926. Sweetened condensed milk. IV. A refractometric method for determining total solids. *J. Dairy Sci. 9*, 140–152.

Richardson, G. H., Mortensen, M. S. and Crockett, R. G. 1978. Quantitation of added water in milk using vapor pressure osmometry. *J. Assoc. Offic. Anal. Chem. 61*, 1038–1040.

Rishoi, A. H. and Sharp, P. F. 1938. Specific heat and physical state of the fat in cream. *J. Dairy Sci. 21*, 399–405.

Robertson, A. H. 1957. Cryoscopy of milk, a 1954–1956 survey. I–IV. *J. Assoc. Offic. Agr. Chem. 40*, 618–662.

Robertson, N. H., Dixon, A., Nowers, J. H. and Brink, D. P. S. 1981. The influence of lipolysis, pH and homogenization on infrared readings for fat, protein, and lactose. *S. African J. Dairy Technnol. 13*, 3–7.

Rowland, S. J. and Wagstaff, A. W. 1959. The estimation of the total solids and solids-not-fat of milk from the density and fat content. *J. Dairy Res. 26*, 83–87.

Ruge-Lenartowiz, R. 1955. The influence of acidity on the electrical conductivity of milk. *Roczn. Zalk. Hig Warsz 5*, 91–102. In *DSA 17*, 613, 1955.

Schober, R., Christ, W. and Niclause, W. 1954. The refractometric estimation of casein in milk. *Lebensm. Untersuch Forsch 99*, 299–302.

Schoenemann, D. R., Finnegan, E. J. and Sheuring, J. J. 1964. Statistical analysis of the freezing point of milk and associated factors in surveys of Florida and Georgia milk. *J. Dairy Sci. 47*, 683.

Schuler, A. 1938. The significance of the refraction-chlorine number for the detection of the watering of milk. *Milchwiss. Forsch 19*, 373–384.

Schulz, M. E. 1956. Measuring electrical conductivity: A help to the chemist in analyzing milk. *Kieler Michwiss. Forschb. 8*, 641–652.

Schulz, M. E. and Sydow, G. 1957. The electrical conductivity (chloride-free) of milk and dairy products. *Milchwissenschaft 12*, 174–184.

Scott-Blair, G. W. 1953. *Foodstuffs: Their Plasticity, Fluidity and Consistency.* North-Holland Publishing Co., Amsterdam.

Sebela, F. and Klicnik, V. 1977. Characteristics of fresh milk of increased acidity. *Prumysl Potravin 28*, 208–210. In *DSA 39*, 7595, 1977.

Sharma, R. R. 1963. Determination of surface tension of milk by the drop method and the ring method. *Ind. J. Dairy Sci. 16*, 101–108.

Sharp, P. F. and Hart, R. G. 1936. The influence of the physical state of the fat on the calculation of solids from the specific gravity of milk. *J. Dairy Sci. 19*, 683–695.

Sharp, P. F. and Krukovsky, V. N. 1939. Differences in adsorption of solid and liquid fat globules as influencing the surface tension and creaming of milk. *J. Dairy Sci. 22,* 743-751.

Shedlovsky, T. 1959. *Conductometry. In: Technique of Organic Chemistry,* 3rd ed. A. Weissberger (Editor). Interscience Publishers, New York.

Sherbon, J. W. 1968. Thermal studies of milk fat. *In: Analytical Calorimetry.* R.S. Porter and J.F. Johnson (Editors). Plenum Press, New York, pp. 173-180.

Sherbon, J. W. 1970. Dye binding method for protein content of dairy products. *J. Assoc. Offic. Anal. Chem. 53,* 862-864.

Sherbon, J. W. and Coulter, S. T. 1966. Relation between thermal properties of butter and its hardness. J. Dairy Sci. 49, 1376-1380.

Shipe, W. F. 1956. The use of thermistors for freezing point determinations. *J. Dairy Sci. 39,* 916.

Shipe, W. F. 1958. Report on cryoscopy of milk. *J. Assoc. Offic. Agr. Chem. 41,* 262-267.

Shipe, W. F. 1959. The freezing point of milk. A review. *J. Dairy Sci. 42,* 1745-1762.

Shipe, W. F. 1960. Cryoscopy of milk. *J. Assoc. Offic. Agr. Chem. 43,* 411-413.

Shipe, W. F. 1964. Effect of vacuum treatment on freezing point of milk. *J. Assoc. Offic. Agr. Chem. 47,* 570-572.

Shipe, W. F. 1969. Collaborative study of the Babcock and Foss Milko-Tester methods for measuring fat in raw milk. *J. Assoc. Offic. Anal. Chem. 52,* 131-138.

Shipe, W. F., Dahlberg, A. C. and Herrington, B. L. 1953. A semi-automatic cryoscope for determining the freezing point of milk. *J. Dairy Sci. 36,* 916-923.

Shipe, W. F. and Senyk, G. F. 1973. Collaborative study of the Foss Milko-Tester method for measuring fat in milk. *J. Assoc. Offic. Anal. Chem. 56,* 538-540.

Shipe, W. F. and Senyk, G. F. 1975. Collaborative study of the Milko-Tester method for measuring fat in homogenized and unhomogenized milk. *J. Assoc. Offic. Anal. Chem. 58,* 572-575.

Shipe, W. F. and Senyk, G. F. 1980. Evaluation of Milko-Tester Minor for determining fat in milk. *J. Assoc. Offic. Anal. Chem. 63,* 716-719.

Short, A. L. 1955. The temperature coefficient of expansion of raw milk. *J. Dairy Res. 22,* 69-73.

Short, A. L. 1956. The density of processed milks. *J. Soc. Dairy Technol. 9,* 81-86.

Shugliashvili, G. V., Charuev, N. G. and Abram, V. I. 1967. Spectrophotometric study of milk. *Protesov Prom. 4,* 91-106. In *DSA 32,* 1312, 1970.

Sorokin, Yu. 1955. Automatic control of the concentration of milk. *Mol. Prom. 16,* 38-39. In *DSA 17,* 741, 1955.

Spells, K. E. 1960. The thermal conductivity of some biological fluids. *Phys. Med. Biol. 5,* 139-153.

Spremulli, G. H. 1942. A study of the effects of time, buffer, composition, specifications, and ionic strength on the surface tension of solutions of β-lactoglobulin. Pub. 510. University of Michigan Microfilms, Ann Arbor, Michigan, p. 130.

Stubbs, J. R. and Elsdon, G. D. 1934. The examination of one thousand milks by the Hortvet freezing-point process. *Analyst. 59,* 146-152.

Stull, J. W., Taylor, R. R. and Ghlander, A. M. 1965. Gradient balance method for specific gravity determination in milk. *J. Dairy Sci. 48,* 1019-1022.

Sturtevant, J. M. 1959. Calorimetry. In *Technique of Organic Chemistry,* 3rd ed. Vol. 1, Part I. A. Weissberger (Editor). Interscience Publishers, New York.

Subheendranath, C. S. and Rao, M. B. 1970. The relationship between relative viscosity and electrical conductivity of skim milk. *Proc. 18th Int. Dairy Congr. IE,* 89.

Sweat, V. E. and Parmelee, C. E. 1978. Measurement of thermal conductivity of dairy products and margarines. *J. Food Proc. Eng. 2,* 187-197.

Swindells, J. F., Coe, J. R. and Godfrey, T. B. 1952. Absolute viscosity of water at 20°C. *J. Res. Nat. Bur. Stand. 48,* 1-31.

Swindells, J. F., Ullman, R. and Mark H. 1959. *Viscosity In: Technique of Organic Chemistry,* 3rd ed., Vol. I, Part I., A. Weissberger (Editor). Interscience Publishers, New York.

Szijarto, L. and van de Voort, F. R. 1982. Evaluation of the Foss Mark III Milko-Tester for payment of farm separated cream. *J. Dairy Sci. 65,* 1900–1904.

Tambat R.V. and Srinivasan, M. R. 1979. Changes in surface tension, viscosity and curd tension of buffalo and cow milk during Cheddar cheese manufacture. *Ind. J. Dairy Sci. 32,* 173–176.

Tarassuk, N. P. and Smith, F. R. 1940. Relation of surface tension of rancid milk to its inhibitory effect on the growth and acid fermentation of *Streptococcus lactis. J. Dairy Sci. 23,* 1163–1170.

Tellmann, E. 1933. A contribution to Rothenfusser's *Refractometry of the Protein-Free (Pb) Serum of Milk. Milchwiss. Forsch 15,* 294–314.

Temple, P. L. 1937. A new apparatus for the rapid and economical determination of the freezing-point of milk. *Analyst 62,* 709–712.

Tikhomirova, G. P., Donskeya, G. A., Kuzmin, V. M., Kararynskaya, R. K., Gorshkov, A. I., Koznetsov, V. D. and Kasyanov, V. F. 1979. Effect of ion-exchange treatment of raw milk on biological value of pasteurization and sterilized milk. *Inst. Molochroi Prom. 1979,* 7–10.

Tillmans, J. and Obermeier, W. 1920. The hydrogen-ion concentration of milk. *Z. Untersuch Nahr u. Genussm 40,* 23–34.

Tomaszewski, M. A. and Dill, C. W. 1978. Fat test fluctuations on DHI milk samples shipped to a central laboratory. *J. Dairy Sci. 61,* 223.

Torssell, H., Sandberg, V. and Thureson, L. E. 1949. Changes in viscosity and conductivity during concentration of milk. *Proc. 12th Int. Dairy Congr. 2,* 246–258.

Trout, G. M., Halloran, C. D. and Gould, I. A. 1935. Effect of homogenization on some of the physical and chemical properties of milk. *Mich. Agr. Exp. Sta. Bull. 145,* pp. 3–34.

Tucker, V. C. 1963. Variation in the freezing point of genuine farm milks in Queensland. *Queensl. J. Agr. Sci. 20,* 161–171.

Tucker, V. C. 1970A. Variation in the freezing point of factory milk. *Aust. J. Dairy Technol. 25,* 126–127.

Tucker, V. C. 1970B. Effect of nutrition on the freezing point of milk. *Aust. J. Dairy Technol. 25,* 137–139.

United States Depatment of Agriculture. 1965. Volume-weight conversion factors for milk. Consumer and Marketing Services, Dairy Division Marketing Research Rep.701 and supplement. Washington, D.C.

van de Voort, F. R. 1980. Evaluation of Milko-Scan 104 Analyzer. *J. Assoc. Offic Anal. Chem. 63,* 973–980.

Vanderzant, C. and Tennison, W. R. 1961. Estimation of the protein content of milk by dye binding with buffalo black. *Food Technol. 15,* 63–66.

Vleeschauwer, A. de, and van Waeyenberge, K. 1941. Investigations about the addition of water to milk and buttermilk. *Meded. Land. Hoogesch. Opzoeksta Gent. 9,* 56–70.

Wahba, N. 1965. A simple micro colorimetric method for the determination of lactose in milk. *Analyst 90,* 432–434.

Walstra, P. 1965. Light scattering by milk fat globules. *Neth. Milk Dairy J. 19,* 93–109.

Walstra, P. 1967. Turbidimetric method for milk fat determination. *J. Dairy Sci. 50,* 1839–1840.

Walstra, P. and Jenness, R. 1984. *Dairy Chemistry and Physics.* Wiley-Interscience, New York.

Watrous, G. H., Jr., Barnard, S. E. and Coleman, W. W. II. 1976. Freezing points of raw and pasteurized milks. *J. Milk Food Technol. 39,* 462–463.

Watson, E. S., O'Neill, M. J., Justin, J. and Brenner, N. 1964. A differential scanning calorimeter for quantitative differential thermal analysis. *Anal. Chem. 36,* 1233–1237.

Watson, P. D. 1931. Variations in the buffer value of herd milk. *J. Dairy Sci. 14,* 50–58.

Watson, P. D. 1958. Effect of variations in fat and temperature on the surface tension of various milks. *J. Dairy Sci. 41,* 1693–1698.

Watson, P. D. and Tittsler, R. P. 1961. The density of milk at low temperatures. *J. Dairy Sci. 44,* 416–424.

Webb, B. H. 1933. A note on the surface tension of homogenized cream. *J. Dairy Sci. 16,* 369–373.

Weber, W. 1956. Systematic investigation of falling ball viscometers with inclined tubes. Kolloid Z. 147, 14–28.

Wegener, H. 1953. Viscosity measurements of whole and skim milk at different temperatures. *Milchwissenschaft 8,* 433–434.

West, W. 1956. *Chemical Applications of Spectroscopy. In: Techniques of Organic Chemistry,* Vol. IX. A. Weissberger (Editor). Interscience Publishers, New York, p. 787.

Wheelock, J. V., Rook, J. A. F. and Dodd, F. H. 1965. The relationship in the cow between the osmotic pressure of milk and of blood. *J. Dairy Res. 32,* 79–88.

Whitaker, R., Sherman, J. M. and Sharp, P. F. 1927. Effect of temperatures on the viscosity of skim milk. *J. Dairy Sci. 10,* 361–371.

Whitnah, C. H. 1959. The surface tension of milk. A review. *J. Dairy Sci. 42,* 1437–1449.

Whitnah, C. H. 1962. The viscosity of milk in relation to the concentration of major constituents and to seasonal differences in the voluminosity of complexes of sedimentable nitrogen. *J. Agr. Food Chem. 10,* 295–296.

Whitnah, C. H., Conrad, R. M. and Cook, G. L. 1949. Milk surfaces. I. The surface tension of fresh surfaces of milk and certain derivatives. *J. Dairy Sci. 32,* 406–417.

Whitnah, C. H., Medved, T. M. and Rutz, W. D. 1957. Some physcial properties of milk. IV. Maximum density of milk. *J.Dairy Sci. 40,* 856.

Whitnah, C. H. and Rutz, W. D. 1959. Some physcial properties of milk. V. Effects of age on the viscosity of pasteurized fractions of milk. *J. Dairy Sci. 42,* 227.

Whitnah, C. H., Rutz, W. D. and Fryer, H. C. 1956. Some physcial properties of milk. II. Effects of age upon the viscosity of pasteurized whole milk. *J. Dairy Sci. 39,* 356.

Whittier, E. O. 1933A. Buffer intensities of milk and milk constituents. III. Buffer action of calcium phosphate. *J. Biol. Chem. 102,* 733–747.

Whittier, E. O. 1933B. Freezing points and osmotic pressures of lactose solutions. *J. Phys. Chem. 37,* 847–849.

Whittier, E. O. and Benton, A. G. 1927. The formation of acid in milk by heating. *J. Dairy Sci. 10,* 126–138.

Wiegner, G. Z. 1910. The physical chemistry of the calcium chloride milk serum. *Z. Nahr. Genussm. 20,* 70–86. In *CA 4,* 3258. 1910.

Wilcox, C. J. and Krienke, W. A. 1964. Variability and interrelationships of composition and yield of dairy milk samples. *J. Dairy Sci. 77,* 638.

Wiley, W. J. 1935. A study of the titratable acidity of milk. I. The influence of the various milk buffers on the titration curves of fresh and sour milk. *J. Dairy Res. 6,* 71–85.

Woodams, E. E. and Nowrey, J. E. 1968. Literature values of thermal conductivities of foods. *Food Technol. 22,* 150–158.

Yoncoskie, R. A. 1969. The determination of heat capacities of milk fat by differential thermal analysis. *J. Am. Oil. Chemists' Soc. 46,* 49–55.

Zlabinger, K. and Stock, H. 1978. The oxidation-reduction potential of milk from cows fed with silage and other feeds. *Proc. 20th Int. Dairy Cong. E,* 325–326.

9

Physical Equilibria: Proteins

Harold M. Farrell, Jr.

INTRODUCTION

Milk is a complex biological fluid secreted by mammals explicitly for the nourishment of their young. Through the centuries, evolution has produced this stable, fluid, concentrated source of lipid, protein, minerals, and carbohydrate. Because of its unusual stability (for a biological fluid), milk has become a valuable foodstuff, a commodity, yet many of the problems which arise in the processing of milk stem from the biochemical nature of its components. In dealing with skim milk, the retention of the unique properties of the milk proteins during processing is of the utmost importance.

Chemically, the skim milk system can be classified as a lyophilic colloid (Payens 1979) because the protein complexes of skim milk, which constitute the dispersed phase, are in the correct size range, interact with and are stabilized by the solvent, and do not coagulate spontaneously. The milk–protein complex is stable to the earth's gravitational field, yet can be separated by centrifugation. Milk, then, can be considered to be a biocolloid; the properties of the dispersed phase (the casein–protein complex) and the dispersion medium (the milk serum) will be discussed.

THE CASEIN MICELLE

For better or worse, the term "micelle' has been applied to the dispersed phase of milk, the casein–protein complex. The electron micrograph of Figure 9.1 shows a number of typical casein micelles. The nature of the casein micelle has been investigated in many laboratories, and with good reason, for this complex is the essence of many problems encountered in dairy technology, whether it be the preservation of the stability of fluid milk, the curd tension of cheese, or the production of

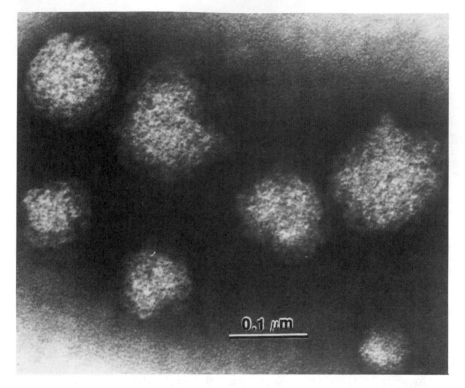

Figure 9.1   Casein micelles of bovine skim milk, fixed in 1% glutaraldehyde and negatively stained with phosphotungstic acid. (Courtesy of R. J. Carroll.)

a synthetic engineered food. Hence, understanding those forces which hold the casein micelle together, and, a posteriori, those forces which cause disruption of the casein–protein complex, is of paramount importance.

Protein Components of the Casein Micelle

Until the 1930s, the principal protein of cow's milk was considered to be the rather "homogeneous" protein, casein. Then Linderstrøm-Lang (1929) and Mellander (1939) demonstrated the heterogeneity of bovine casein. The latter worker termed the electrophoretically distinct fractions α-, β-, and γ-caseins. From that time until the late 1950s, many methods of fractionation were developed and various casein fractions were isolated and characterized (see Chapter 3). The most significant fractionation was accomplished by Waugh and Von Hippel (1956) when they discovered that the α-casein fraction was a mixture of α_{s1}-casein and κ-casein. Indeed, a sample of casein from pooled milk, subjected to gel electrophoresis in urea and mercaptoethanol, yields up to 20 casein components. The demonstration of genetic polymorphism in the β-casein fraction by Aschaffenburg (1961), followed by the work of Thompson $et\ al.$ (1962) on the genetics of the α_{s1}-fraction, began to introduce a unifying concept to the field. The contention of Gordon and co-workers (1972) that the γ-, R-, S-, and TS-fractions are but degradation products of β-casein led us (Farrell and Thompson 1974) to speculate that the other minor casein fractions, which were reported in the literature, such as m- and λ-caseins and the proteose-peptone fraction, were also degradation products of one or more of the major casein components. Work by Eigel and his co-workers (1979) has demonstrated the existence of plasmin and plasminogen in milk. This protease leads to the formation of the γ-casein fragments and also, most likely, to components of the proteose-peptone fraction, which Andrews (1978A,B) has shown to be the N-terminal fragments of β-casein. Aimutis and Eigel (1982) have also shown that components of the λ-fraction may be plasmin-generated fragments of α_{s1}-casein. Variations in the degree of posttranslational modification (phosphorylation and glycosylation) produce other bands (Chapter 3; Eigel $et\ al.$ 1984).

From all of the work on the characterization of casein, four major components of the casein–protein complex have been described, namely, α_{s1}-, α_{s2}-, β-, and κ-casein. The major protein of the casein complex is the α_{s1}-fraction; the exact margin by which this fraction exceeds β- and κ-casein seems to be open to debate, depending in part on the method of quantitation. However, the best present estimate is: α_{s1}— 38%, α_{s2}—10%, β—36% and κ—13% (Davies and Law 1980).

The names of the various fractions used here are in accord with those of the American Dairy Science Association Committee, whose reports have done much to order the field of milk protein nomenclature (Eigel $et\ al.$ 1984; Whitney $et.\ al.$ 1976). Thompson (1970) has reviewed the methods available for the detection of the various known genetic polymorphs of the milk proteins, and Farrell and Thompson (1971) have reviewed their occurrence in various breeds and the possible biological significance of milk protein genetic polymorphism.

α_{s1}-Casein B is a single-chain polypeptide, of known sequence (Mercièr $et\ al.$ 1971), with 199 amino acid residues and a molecular weight of 23,619 (for the complete sequence, see Chapter 3). The α_{s1}-B molecule contains eight phosphate residues, all of which exist as phosphomonoesters of serine. Seven of these phosphoserine residues are clustered in an acidic portion of the molecule bounded by residues 43 and

80 (the second fifth of the molecule from the N → C terminal end). This highly acidic segment contains 12 carboxylic acid residues as well as 7 of the phosphate residues, as postulated in 1970 by Waugh and co-workers. It also contains the largest segment of the molecule's net negative charge, as well as a large segment of the predicted α-helix (Creamer *et al.* 1981). Theoretically, from a knowledge of the complete sequence, one can calculate the charge frequency, net charge, and hydrophobicity for various segments of the molecule. These data for α_{s1}-casein are presented in Table 9.1.

The hydrophobicity shown in Table 9.1 was calculated using the method of Bigelow (1967) and can be taken as a quantitative measure of the apolarity of a segment of a molecule or of the molecule itself. The data in Table 9.1 reveal noncoincidence of high charge frequency and apolarity in the segments shown; however, there exist local areas of charge surrounded by an apolar environment. The proline content of α_{s1}-casein is high and these residues are rather evenly distributed, and proline residues are known to disrupt helical and β-structures. Thus, the sequence data confirm the physical-chemical data (Creamer *et al.* 1981; Herskovits 1966), which indicated that the α_{s1}-molecule has nearly 70% unordered structure, with only a small amount of secondary structure such as α-helix or β-structure. The high degree of hydrophobicity and the small amount of structural content exhibited by the segment containing residues 100–199 is probably responsible, in part, for the pronounced self-association of the α_{s1}-casein monomer in aqueous solution (Schmidt 1970; Swaisgood and Timasheff 1968; Waugh *et al.* 1970). This self-association approaches a limiting size under conditions of lowered ionic strength; the highly charged phosphopeptide region can readily account for this phenomenon through charge repulsions. However, at ionic strengths ≥ 0.5, α_{s1}-casein precipitates from solution at 37°C. It is noteworthy that while the self-

Table 9.1.   Profile of the α_{s1}-Casein Molecule Derived from Its Primary Structure.

Residues considered	Net charge[a]	Charge frequency[b]	Average hydrophobicity[b]
1–40	+4	0.30	1340
41–80	−23 ½	0.58	641
81–120	−2	0.38	1310
121–160	−1	0.25	1264
161–199	−4	0.15	1164

[a]Serine phosphate = −2, histidine = +½.
[b]Calculated as described by Bigelow (1967).
SOURCE: Adapted from Merciér *et al.* (1971).

association of α_{s1}-casein is mostly hydrophobic, and hence temperature dependent (Schmidt and Payens 1976), some ionic bonding, as postulated by Schmidt (1970), must also occur in the reaction. The solubility of α_{s1}-casein in aqueous Ca^{2+} solutions has been studied by Waugh *et al.* (1971) and by Thompson *et al.* (1969C). At calcium ion concentrations of 5 to 10 mM, α_{s1}-casein forms an insoluble precipitate. With the exception of the rare genetic variant α_{s1}-A, the calcium solubility of α_{s1}-casein is temperature independent. Thus, the major protein component of milk is insoluble under the conditions of pH, ionic strength, and temperature which normally occur in milk.

β-Casein is the second most abundant milk protein. The β-A^2 molecule is a single chain of known sequence (Ribadeau-Dumas *et al.* 1972) with five phosphoserine residues and a molecular weight of 23,980 (for the complete sequence, see Chapter 3). The charge frequency, hydrophobicity, and net charge for the various segments of β-casein are presented in Table 9.2. Analysis of these data indicates that β-casein is a linear amphiphile and thus is much more "soaplike" than α_{s1}-casein. The N-terminal portion of the β-casein molecule (residues 1–40) contains the phosphoserine residues and carries essentially all of the protein's net charge, as well as most of the protein's potential α-helical residues (Creamer *et al.* 1981). The C-terminal half of the molecule (actually, residues 136–209) contains many apolar residues (as demonstrated by its high hydrophobicity) and only two short stretches of potential β-structure. β-Casein undergoes an endothermic self-association which reaches a maximum or limiting size depending upon the ionic strength (Waugh *et al.* 1970; Schmidt and Payens 1976). The N-terminal concentration of charge and the highly hydrophobic and random C-terminal may account for the temperature dependence of this self-association, since hydrophobic interactions are temperature dependent. In fact, the self-association of β-casein can be fitted best to

Table 9.2. Profile of the β-Casein Molecule Derived from Its Primary Structure.

Residues considered	Net charge[a]	Charge frequency[b]	Average hydrophobicity[b]
1–43	−16	0.55	783
44–92	−1½	0.10	1429
93–135	+1	0.28	1173
136–177	+2	0.12	1467
178–209	0	0.13	1738

[a]Serine phosphate = −2, histidine = +½.
[b]Calculated as described by Bigelow (1967).
SOURCE: Derived from the data of Ribadeau-Dumas *et al.* (1972).

a model which describes the association as going through a critical micelle concentration and achieving a limiting size (Schmidt and Payens 1976). In aqueous solution, β-casein has been characterized as a random coil with little or no secondary structure (Herskovits 1966; Noelken and Reibstein 1968). From the complete amino acid sequence of β-casein, the proline content of the molecule is seen to be rather evenly distributed, explaining, in part, why the protein lacks any appreciable secondary structure (Creamer et al. 1981). Like α_{s1}-casein, β-casein is insoluble at room temperature in the presence of Ca^{2+} at concentrations below those encounted in milk. However, β-casein's precipitation from solution is temperature dependent and the calcium–β-caseinate complex is soluble at $1°C$ at concentrations of up to 400 mM Ca^{2+} (Thompson et al. 1969C). Again, this temperature dependence is probably due to the charge distribution of the β-casein monomer.

α_{s2}-Casein was the last of the bovine series to be sequenced; it also has the most unique primary structure of all of the caseins (Brignon et al. 1977). The molecular weight from sequence is 25,150 for the 10-phosphate form (see Chapter 3). The protein contains no known carbohydrate but does have two cysteine residues. Unlike κ-casein, which occurs as a large disulfide-linked aggregate, the α_{s2}-casein forms only dimers or may have some intrachain disulfide (Snöeren et al. 1980). Brignon et al. (1977) pointed out that this protein has two very large segments of 76 residues (50–123 and 133–207) which display 38% sequence homology and stated that these may have arisen by gene duplication. Additionally, the N-terminal portion of the whole molecule displays some homology with the first half of each of the large sequences. This being the case, the overall molecule (Table 9.3) has five distinct areas composed of only two repeating structures. It would appear that these structures, in turn, dictate the physical properties of α_{s2}-.

Table 9.3. Profile of the α_{s2}-Casein Molecule Derived from Its Primary Structure.

Residues considered	Net charge[a]	Charge frequency[b]	Average hydrophobicity[b]
1–41	$-7\frac{1}{2}$	0.39	860
42–80	$-14\frac{1}{2}$	0.51	780
81–125	$+1$	0.15	1460
126–170	-5	0.49	861
171–207	$+6\frac{1}{2}$	0.24	1674

[a]For the form of the molecule with 11 phosphorylated residues; serine phosphate $= -2$, histidine $= +\frac{1}{2}$.
[b]Calculated as described by Bigelow (1967).
SOURCE: Derived from Brignon et al. (1977).

Snoeren *et al.* (1980) concluded, from a combination of light scattering and viscosity measurements, that its self-association is isodesmic and produces spherical polymers of ~ 4 nm at 0.6 M NaCl. The alternating negatively charged and hydrophobic areas (which contain net positive charges) make this casein the one least susceptible to aggregation phenomena.

κ-Casein, the fourth major component of the milk–protein complex, differs from α_{s1}-, α_{s2}-, and β-caseins in that it is soluble over a very broad range of calcium ion concentrations (Waugh and Von Hipple 1956). It is this calcium solubility which led these workers, upon discovering the κ-fraction, to assign to it the role of casein micelle stabilization. It is also the κ-casein fraction which is most readily cleaved by chymosin (rennin) (Kalan and Woychik 1965); the resulting products are termed "para-κ-" and the "macropeptide." It would appear that κ-casein is the key to micelle structure in that it stabilizes the calcium-insoluble α_s- and β-caseins and is the primary site of attack by the enzyme chymosin. κ-Casein, like α_{s2}-, contains cystine (or possibly cysteine). The occurrence of free sulfhydryl groups in the milk–protein complex has been reported by Beeby (1964), but not by others (Jollès *et al.* 1962); hence, the degree of disulfide bonding which occurs in κ-casein is uncertain. Swaisgood and Brunner (1962, 1963) reported a molecular weight of 19,000 for reduced κ-casein. These authors also reported (1964) a nonreduced molecular weight on the order of 60,000 for the lightest component, indicating at least a disulfide-linked trimer of the isolated κ-casein; the weight average molecular weight of their preparation was ~ 110,000. Cheeseman (1968) studied the effect of the binding of the detergent sodium dodecyl sulfate on casein by gel filtration. He concluded that the majority of the κ-casein occurs as disulfide-linked aggregates, which elute at the void volume of Sephadex G-200, even in the presence of the detergent. Recently, Pepper and Farrell (1982), using agarose zonal gel chromatography without denaturants, confirmed these observations. κ-Casein was shown to occur as a high molecular weight mixture of polymers. After reduction with dithiothreitol, the elution profiles of κ-casein exhibited concentration-dependent behavior like those of the other caseins. Vreeman *et al.* (1981) studied the aggregation of reduced (SH-) κ-casein and found that the protein self-associates into a limiting polymer which is spherical and contains ~ 30 monomers, with an average diameter of ~ 11 nm. The degree of polymerization was not dependent upon ionic strength between 0.1 and 1 M; as with β-casein, the best model for the association of κ-casein appears to be one describing the association as proceeding through a critical micelle concentration to a limiting polymer size.

κ-Casein is the only major component of the casein complex which contains carbohydrate. Nearly all of the carbohydrate associated with the κ-casein is thought to be bound to the macropeptide (Whitney *et al.* 1976; Eigel *et al.* 1984), which is the highly soluble portion formed by chymosin hydrolysis. In addition to being a glycoprotein, κ-casein contains one or two phosphate residues per reduced monomer and, as noted before, is soluble in Ca^{2+}, although it binds this ion (Dickson and Perkins 1971).

The total primary structure of κ-casein has been determined (Merciér *et al.* 1973), and the molecular weight of the amino acid sequence for the unglycosylated form is 19,012 (see Chapter 3). Earlier speculation (Hill and Wake 1969) that κ-casein might be a linear amphiphile seems to be only partially true. The N-terminal fifth of the molecule (Table 9.4) has a relatively high charge frequency (0.34), but its net charge is zero and it is relatively hydrophobic. Residues 35–68 represent an exceptionally hydrophobic area with almost no charge. The molecule becomes progressively more hydrophilic and culminates in the negatively charged macropeptide. It is easy to see from Table 9.4 why cleavage of the Phe–Met bond (residues 105–106) produces the cationic para- κ-and the anionic macropeptide (Delfour *et al.* 1965; Kalan and Woychik 1965). The folded κ-casein molecule may have some interesting properties, as shown by the work of Loucheux-Lefebvre and coworkers (1978) using theoretical calculations. The C-terminal portion is predicted to have two runs of α-helix, while the highly hydrophobic area from 35 to 80 is predicted to contain three possible β-bends as well as three long stretches of β-structure. The hydrophilic N-terminal tail may be near the surface, and it is interesting to note that Cys-11 is located between the two predicted α-helical segments and Cys-88 is located in the predicted β-bend. If both of these residues are located near the surface of the molecule, this would account for the ability of the molecule to form the interchain disulfide-bonded polymers men-

Table 9.4. Profile of the κ-Casein Molecule Derived from Its Primary Structure.

Residues considered	Net charge[a]	Charge frequency[b]	Average hydrophobicity[b]
1–34	0	0.34	1170
35–68	+3	0.09	1640
69–105	+2½	0.16	1110
106–137	−1	0.22	1190
138–169	−8	0.22	870

[a]Serine phosphate = −2, histidine = +½.
[b]Calculated as described by Bigelow (1967).
SOURCE: Adapted from Merciér *et al.* (1973).

tioned above. Loucheux-Lefebvre *et al.* (1978) also found, from circular dichroism studies, that κ-casein has approximately 30% β-structure and 14% α-helix. A somewhat different structure was predicted by Raap *et al.* (1983), who indicated that the chymosin-sensitive Phe–Met bond may be contained within a stretch of β-structure rather than the α-helix predicted by Loucheux-Lefebvre *et al.* For comparison, the α-helix and β-structure contents calculated for κ-, α_{s1}-, and β-caseins are given in Table 9.5.

FORCES RESPONSIBLE FOR THE STABILITY OF THE CASEIN MICELLE

In 1929 Linderstrøm-Lang, as a result of his studies on casein, postulated that the colloidal milk complex is composed of a mixture of calcium-insoluble proteins stablized by a calcium-soluble protein. The latter protein would be readily split by chymosin, destabilizing the colloid and allowing coagulation to occur. As we have seen, such fractions do exist. The α_{s1}-, α_{s2}-, and β- caseins are indeed calcium insoluble, while κ-casein is not only soluble in the presence of calcium ions but is readily split by chymosin. In addition, Waugh *et al.* (1971) have demonstrated that α_{s1}- and κ-casein complexes can be reformed from the isolated fractions, as measured by sedimentation velocity experiments. Pepper (1972) demonstrated the interaction of α_{s1}-and κ-casein by gel filtration and studied the concentration dependence of the interaction. The complexes formed by the interaction of the isolated α_{s1}-and κ-caseins aggregate to form simulated casein micelles upon the addition of Ca^{2+} in 0.01 M imidazole buffer, pH 6.7. As viewed by electron microscopy (Bingham *et al.* 1972), these synthetic micelles are virtually identical to fresh milk micelles except for their increased size. However, Schmidt *et al.* (1977,1979) and Schmidt and Koops (1977) have produced altered size distributions by incorporating β-casein and manipulating the

Table 9.5. Contents of α-Helix, β-Sheet, and Random Coil Structures as Predicted from Primary Structure.

	α-Helix (%)	β-(%)	Random (%)
α^{s1}-Casein[a]	22	8	70
β-Casein[a]	12	11	77
κ-Casein[b]	16	17	67

[a]Calculated by Creamer *et al.* (1981) (Chou-Fasman method).
[b]Calculated by Raap *et al.* (1983) (Lim method).

κ-casein content of the micelles. The precise mechanism of formation of the natural casein micelles is still uncertain, although several theories will be reviewed later. In the course of the discussion of casein micelle structure, a brief summary of the types of bonding forces which are responsible for the stabilization of protein structure will be given.

Hydrophobic Interactions

One of the most significant contributions to our understanding of protein stability was made by Kauzmann (1959), who elucidated the nature of hydrophobic interactions in proteins. These interactions come about primarily because water exhibits decreased entropy as a result of the occurrence of apolar amino acid residues within the solvent. If these residues are forced out of the water and into the interior of a protein molecule, where they can interact with other apolar groups, the associated water becomes more disordered (increased entropy) and a small quantity of stabilization energy is gained per residue transferred from the solvent. Several model systems based on the energy of transfer of amino acids from water to ethanol have been studied and yield confirmatory results. These hydrophobic interactions are highly temperature sensitive, being minimal below 5°C and maximal at higher temperatures. It should be pointed out that for proteins whose crystallographic structure is known, many apolar side chains do exist fully or partially exposed to the solvent, and therefore many proteins may exhibit hydrophobic surface patches which are available for interactions with small molecules and with other protein molecules. The role of the "hydrophobic effect" in biological systems has been reviewed by Tanford (1978).

From the amino acid sequences of the α_{s1}-, β-, and κ-caseins, it is quite apparent that large numbers of apolar residues occur in these proteins, and it is clear that these hydrophobic residues are somewhat clustered for the α_{s1}- and β-caseins, as well as for κ-casein (Tables 9.1–9.4). According to the calculations of Hill and Wake (1969), the caseins rank among the most hydrophobic proteins of those tabulated by Bigelow (1967). It is not unexpected, then, that the casein micelle should be stablized, in part, by hydrophobic bonding. In the absence of calcium, the self-association of β-casein is highly temperature dependent. However, both β-casein and SH-κ-casein display concentration-dependent associations which appear to fit the equations for a critical micelle-like model (Schmidt 1982). Berry and Creamer (1975) removed the two C-terminal amino acids (Ile–Val) of β-casein and found that this reduced the ability of the protein to self-associate. Removal of the 20 C-terminal amino acids (whose hydrophobicity is ~1800; see Table 9.2) destroyed completely the ability of β-casein to form polymers. Several investigators (Rose 1968; Downey and Murphy 1970; Ali *et al.* 1980; Davies and Law 1983) have noted that β- and κ-caseins, and α_{s1}-casein to some extent, diffuse out of the micelle at low temperatures. As the temperature decreases, hydrophobic stabilization energy decreases, and the caseins are able to dissociate from the micelle. These observations are consistent with the analyses of the known primary structures of β-casein (Table 9.2) and κ-casein (Table 9.4). While all of the authors cited above agree that β-, and to a lesser extent κ- and α_{s1}-caseins, can be removed from the casein micelle at 1°C, some question arises as to the exact amount released. Rose (1968) reported high values for β-casein (up to 30%), while Downey and Murphy's values (1970) (up to 15%) are lower. The latter workers, however, pointed out that the stage of lactation and the health of the animal play a role in the amount of cold-soluble casein present. There is also some disagreement on the quantity of κ- which is released (Davies and Law 1983). All of the authors cited above concur that the α_{s1}- and α_{s2}-fractions do not diffuse from the micelle to as great an extent as the other caseins.

The rare α_{s1}-A genetic variant, however, does exhibit interactions which are highly temperature dependent. This variant is the result of the sequential deletion of 13 amino acid residues bounded by residues 13 and 27; the majority of these deleted amino acids are apolar (Whitney *et al.* 1976). In a peculiar way, loss of these hydrophobic residues increases the importance of hydrophobic interactions, as the net result of this deletion is to bring the charged phosphorus-rich area closer to the N-terminal region. This makes the α_{s1}-A genetic variant more β-casein-like in its charge distribution, and the physical and solubility properties of this variant mirror those of β-casein (Thompson *et al.* 1969A). As a result, micelles containing this protein are less stable in the cold and under certain processing conditions.

Increased pressure, which is thought to act primarily on hydrophobic interactions, tends to dissociate caseins and disrupt casein micelle structure, as evidenced by light scattering (Schmidt and Payens 1976) and by electron microscopy (Schmidt and Buchheim 1970). The pressure apparently reduces the micelle to small subunits of 10 to 20 nm in diameter. The temperature- and pressure-dependent properties of the hydrophobic interactions may also explain why skim milk can withstand moderate to high temperatures but is destabilized by extremely low temperatures (such as freezing) and can be affected by high pressures.

Electrostatic Interactions

It has been pointed out that the majority of the ionic side chains in the proteins, whose crystallographic structure is known, are exposed to the solvent (Tanford 1978; Perutz 1978). Ionic bonding, then, between negatively charged carboxylic acid residues and positively charged groups, contributes very little to the stability of a monomeric protein. Notable exceptions to this rule may occur when a salt bridge or an ion pair can be formed, especially within a hydrophobic environment (Tanford 1961). Certain interactions of subunits of multimeric proteins may provide just such an environment. Physical-chemical evidence for the role of ionic bonding in subunit interactions is abundant, while detailed crystallographic evidence is limited to hemoglobin and tobacco mosaic virus. Conversely, electrostatic interactions between amino acid side chains and metal ions can impart reasonable structural stability to a protein. Many metalloenzymes, especially calcium-binding proteins, derive a good deal of their stabilization from specific metal coordination complexes (Anfinsen and Scheraga 1975; Privalov 1982).

The role of inter- and intramolecular ionic bonds among the α_s-, β-, and κ-caseins in stabilization of micelle structure is difficult to assess. Many potential sites for strong ion pair bonds within an apolar environment exist, as deduced from consideration of the known sequences, and such bonds may play a role in micelle subunit interactions. Pepper et al. (1970) demonstrated that carbamylation of five of the nine lysine residues of κ-casein abolished the ability of κ-casein to stabilize α_{s1}-casein, thus demonstrating that ionic interactions may play a role in micelle structure. Furthermore, Hill (1970) modified the arginine side chains of the caseins and found major differences in micelle stability and in coagulation by chymosin.

The estimated calcium content of milk is around 30 mM, far above the concentrations of Ca^{2+} required to precipitate the isolated α_s- and β-caseins at room temperature (Thompson et al. 1969C). When the data of Dickson and Perkins (1971) are reanalyzed on a molar basis, their binding studies on the isolated β- and κ-caseins show an approximate 1:1 correlation between calcium ions bound and current data on the number of phosphate residues. The role of the phosphate residues in calcium binding has been investigated by the enzymatic dephosphorylation of α_{s1}-casein. Bingham et al. (1972) demonstrated that dephosphorylated α_{s1}-casein was still precipitated by calcium and showed decreased stabilization by κ-casein. The latter authors postulated that two nonphosphate calcium-binding sites ocurr in α_{s1}-casein, and that it is the binding to these sites which induces percipitation of the dephosphorylated casein. Their investigation of the κ-casein stabilized, dephosphorylated α_{s1}-casein by electron microscopy showed larger but fewer micelle-like structures. In milks containing α_{s1}-A (Thompson et al. 1969B,C) such large micelles are poorly solvated and less stable. Thus, the formation of micelle-like structures is not totally dependent upon the formation of calcium-phosphate bonds between caseins; however, the resulting micelles may be less stable.

Although structures resembling native casein micelles have been formed by mixtures of calicum and casein alone, more recent evidence points to a crucial role for inorganic phosphate in the structure of the casein micelle. Schmidt and co-workers (1977, 1979) and Schmidt and Koops (1977) found that synthetic micelles formed in the presence of both calcium and phosphate exhibit physical stabilities more nearly like those of native casein micelles. Slattery (1979) also found that phosphate promoted more regular growth of the micelle and produced aggregates with increased stability in the cold. Finally, Visser *et al.* (1979), using ^{31}P nuclear magnetic resonance, demonstrated a specific downfield shift which they concluded was the result of the binding of inorganic phosphate to casein. However, since the shift occurs only in the presence of both calcium and phosphate, these two ions must associate cooperatively with the casein. These workers suggested that the phosphate binds to amino groups on the casein (as mentioned above, chemical modification of lysine and arginine residues does decrease micelle stability). Thus, binding of inorganic phosphate, as well as calcium, to caseins must be important to casein micelle structure.

The total number of charged groups of the casein monomers (Tables 9.1–9.4) reveals that in the formation of a casein micelle, not all of these ionic groups can occupy a surface position. This would indicate either that much energy is used to bury these groups or that the structure is porous and available to the solvent, water. The latter proposition appears to fit the experimental evidence. Ribadeau-Dumas and Garnier (1970) noted that carboxypeptidase A is able to remove ~70% the carboxyl-terminal residues from the α_{s1}-, β-, and κ-caseins of native micelles; they hypothesized that this enzyme (MW ~40,000) is able to penetrate the center of the casein micelle. However, an alternative interpretation is that rapid exchange of surface, internal, and serum caseins may occur, and the removal of the hydrophobic C-terminal residues of soluble casein could destabilize the micelle in part (Berry and Creamer 1975 showed decreased polymerization of β-casein treated with carboxypeptidase), leading to further accessibility of the monomers. Thompson *et al.* (1969B,C), however, have shown that the casein micelle is a highly solvated structure with an average of 1.9 g of water per gram of protein, and other workers have reported higher values (Dewan *et al.* 1973; Holt 1975). The absolute degree of solvation of the micelle is somewhat controversial, but it does depend upon a variety of factors including the urea content and the calcium:phosphate ratio of the milk (Holt and Muir 1978). However, all data indicate that the micelle is more highly solvated than most globular proteins and therefore has a rather porous structure. Substances which tend to decrease solvent interaction, such as ethanol, lower the stability of the micelle and in turn destabilize the milk. These interactions relate back to the proposition that the majority of ionic residues of the individual casein monomers cannot be totally buried but must be exposed to solvent.

The better early measurements of the monomer molecular weights of the isolated casein fractions were obtained at pH 11 to 12 (Waugh and Von Hipple 1956). At these pH levels, the positively charged lysine residues and a portion of the arginine residues are neutralized, thus increasing the charge repulsions of the carboxyl and phosphate residues. However, prolonged exposure to high pH may produce degradation, as pointed out by Noelken (1967). These same effects operate in the casein micelle; as the pH of milk is brought to 11 to 12, the micelle structure is disrupted, with accompanying changes in turbidity and viscosity. Presumably, exposure to high pH for long periods of time in the production of sodium caseinate may cause degradation and, hence, alter the characteristics of the product.

Secondary and Tertiary Structure of Hydrogen Bonding

Many globular proteins, such as myoglobin, are stabilized by a high degree of α-helical structure. In addition to the fibrous proteins, the so-called β- or pleated sheet structure has been detected by x-ray crystal-lography in globular proteins, notably lactate dehydrogenase and β-lactoglobulin (Papiz et al. 1986). These structures are stabilized by hydrogen bonds along the polypeptide backbone (Anfinsen and Scheraga 1975). Many proteins have been shown to contain significant amounts of secondary structure, as determined by spectral methods such as circular dichroism, optical rotatory dispersion, and infrared spectroscopy (Timasheff and Gorbunoff 1967). These spectral methods can provide a good estimate of the amount of secondary structure (Chen et al. 1974), but they are subject to error. In many cases, then, some degree of stabilization is achieved by the formation of α-helical or β-structure, but not all stable proteins contain considerable amounts of these conformations. Other bonding forces (noted above), and perhaps even "sterically restricted" random structures, may contribute significantly to the stabilization of a protein.

Spectral investigations of the isolated caseins have shown that these proteins possess little secondary structure. Herskovits (1966) demonstrated by optical rotatory dispersion, using Moffit-Yang, Drude, and Shechter-Blout analyses, that in aqueous solutions, neither the individual casein components (α_{s1}-, β-, or κ-) nor whole sodium caseinate exhibit an appreciable degree of α-helical content. The methods of Chou-

Fasman and Lim, which predict the secondary structure of a protein from its primary structure, have been applied to α_{s1}-, β-, and κ-caseins (Loucheux-Lefebvre *et al.* 1978; Creamer *et al.* 1981; Raap *et al.* 1983). The predicted structures are given in Table 9.5. The Chou-Fasman method predicts a low degree of secondary structure for the α_{s1}-and β-caseins and a higher degree for κ-casein; the method of Lim predicts a substantially lower secondary structure for κ-casein and is more in agreement with spectroscopic data for κ-casein (Raap *et al.* 1983). Comparison of these predictions with the values obtained by spectroscopy showed fair agreement for the α_{s1}- and β-caseins (Creamer *et al.* 1981) and for the κ-casein structure predicted by the method of Lim (Raap *et al.* 1983). Thus, for each of the three major caseins, 70 to 75% of the protein exists in aperiodic conformations. Noelken and Reibstein (1968) concluded that β-casein exhibits a random coil conformation in both aqueous solution and 6 M guanindine-HC1. The observed properties of the caseins mentioned above are in good agreement with the high incidence of proline scattered throughout the α_{s1}- and β-caseins, as derived from analyses of their sequences. Since little periodic structure occurs in the individual casein components, one would expect that the degree of stabilization contributed to the casein micelle by α-helix or β-structure would be quite low.

Theoretically, hydrogen bonds between ionizable side chains accessible to the solvent, water, contribute to a limited degree to the stabilization of monomeric proteins. These groups are already hydrogen bonded to water, and the water–residue hydrogen bond would have to be broken before a residue–residue hydrogen bond could be formed. Nevertheless, once two subunits of a protein begin to interact, these surface groups may not longer be totally hydrated and hydrogen bonds could form between monomers as a result of the altered environment. Hydrogen bonding between casein monomers in the casein micelle may occur. Subunit interactions, at present, have not been sufficiently detailed by crystallographic evidence to support or rule out these types of bonds, but some intrachain hydrogen bonds do occur in monomeric proteins. It is also possible that some hydrogen bonding may occur in the self-association or α_{s1}-casein (Schmidt 1970). Certainly, in the formation of the highly aggregated casein micelle, such bonds between the various casein components would be possible.

The Role of Disulfide Bonds

The folding of helical segments, pleated sheet areas, and unordered structures of a polypeptide chain is referred to as "tertiary structure." The tertiary structure of proteins can be locked in place by the formation of disulfide bonds between distal cysteine residues. In fact, nonidentical polypeptide chains can be held together by disulfide bonding, as in the case of immunoglobulins (quaternary structure). Evidence has been presented that for several proteins, the disulfide bridges do not cause the formation of secondary, tertiary, and quaternary structures but tend to stabilize the preformed conformations (Anfinsen and Scheraga 1975). Proteins such as lysozyme and ribonuclease, with a relatively high degree of disulfide bonding, are quite stable, but not all stable proteins necessarily contain disulfide bonds.

As noted above, κ-casein and α_{s2}-casein both contain cystine (or cysteine), and the degree of disulfide cross-linkages, which normally occur in the casein micelle, is controversial. Although Swaisgood and Brunner (1962, 1963) and Swaisgood *et al.* (1964) reported that purified κ-casein contains a good deal of S-S cross-linking, Woychik *et al.* (1966) demonstrated that reduced and alkylated κ-casein stabilized α_{s1}-casein against calcium precipitation as well as native κ-casein. It appears that while the disulfide bridges of the casein micelle may contribute to the overall stability of the casein micelle, they are not the driving force for micelle formation. Recent evidence has shown that the degree of disulfide interactions which occurs within the casein micelle may act as a natural bifunctional probe for casein micelle structure. Pepper and Farrell (1982) found that for soluble whole casein, in the absence of Ca^{2+}, κ-casein occurs as a high molecular weight polydisperse complex. At low protein concentrations, this complex can be separated from the other caseins on gel chromatography in the absence of urea. The addition of reducing agents converts the κ-casein to the SH-form, which does exhibit concentration-dependent associations, both with itself and with other caseins. Thus, although κ-represents only 13% of the casein, many κ-molecules must be somewhat contiguous with each other in the micelle in order to form these disulfide-linked aggregates. Additionally, Slattery (1978) isolated several different size classes of micelles and showed that the larger the micelles, the higher the apparent molecular weight of the κ-casein disulfide-bonded polymers. This too indicates that the nearest neighbors of κ-casein monomers are often other κ-caseins. In contrast, α_{s2}-casein forms dimers at the most and some intermolecular disulfides. The occurrence in milk of sulfhydryl oxidase (Janolino and Swaisgood 1975) may account for the high degree of disulfide bonding in the casein micelle (Pepper and Farrell 1982).

Colloidal Calcium Phosphate

The total calcium content of milk has been estimated to be 31 mM (Cerbulis and Farrell 1976), but the calcium ion content of serum, prepared by ultrafiltration or centrifugation of skim milk, is only ~2.9 mM. Specific ion electrode studies give a value of 2.5 mM calcium (II) for skim milk (Demott 1968). Thus, more than 90% of the calcium content of skim milk is either associated in some way with other ions or found within the casein micelles (Holt et al. 1981). Washing of the micelles with water removes only a small portion of the calcium and other salts. The mineral content of washed micelles, prepared by centrifugation (Davies and White 1960), and "primary micelles," prepared by gel filtration (Boulet et al. 1970), are compared in Table 9.6; both methods appear to yield similar calcium and phosphate contents. The existence of this so-called colloidal calcium phosphate was postulated as early as 1915 by Van Slyke and Bosworth, who concluded that the nonprotein-bound calcium phosphate was present within the colloidal phase of skim milk with a molar ratio which approximates that of dicalcium phosphate. Other workers have calculated that the colloidal calcium phosphate more closely resembles tricalcium phosphate, with a $Ca:PO_4$ molar ratio of 1.5. Calculation of such a ratio after subtracting the casein-bound calcium is subject to some inherent error, since the concentrations of the caseins and their phosphorylated forms are variable. Assuming a 1:1 correlation between bound calcium ions and casein phosphate residues, and an average distribution of the various caseins, a $Ca:PO_4$ molar ratio can be calculated from Table 9.6. The ratios obtained for washed micelles and micelles prepared by gel filtration are 1.6 and 1.8, respectively. The latter value differs from that calculated by Boulet et al. (1970) because these authors assumed a 2:1 casein phosphate:calcium ion ratio. These calculations also neglect phosphate binding to casein (Slattery 1979); this bound phosphate may be dead-

Table 9.6.   Total Mineral Composition of Casein Micelles (mMoles/23.3 g Casein).[a]

	Washed micelles[b] by centrifugation	Micelles by repeated gel filtration[c]	Unwashed Micelles by:	
			Centrifugation[b]	Gel filtration[c]
Calcium	16.2	16.0	16.5	18.4
	14.9			
Magnesium	0.98	0.77	1.04	1.61
Sodium	1.04			
Potassium	1.44	1.44		
Casein (PO_4)[d]	5.17	6.57		
	5.41			
Inorganic (PO_4)	6.73	5.08	11.1	10.1
	6.48			
Citrate	0.37	0.0	1.44	1.10

[a]Casein N × 6.4 (23.3 g = hypothetical millimolar molecular weight of casein monomers in grams).
[b]Adapted from McMeekin and Groves, Fundamentals of Dairy Chemistry," Chapter 9, Table 70.
[c]From Boulet et al. (1970).
[d]Theoretical value as in footnote (a) would be 6.4 moles P per mole of casein.

ended or may cooperatively bind more calcium (Visser et al. 1979). In addition, the role of citrate in the mineralized complex must be taken into account (Posner et al. 1983). Thus, attempting an exact assignment of calcium to either the casein fraction or the colloidal calcium phosphate fraction can cause discrepancies in the calculated ratio. It must be realized that these data are average values based on average distributions of the caseins and minerals. Not only does the mineral content and the casein distribution vary from one milk to another (Cerbulis and Farrell 1975, 1976), but the various micelle fractions within a single sample are probably not of uniform composition with respect to their casein and calcium and phosphate contents.

It is clear, from Table 9.6, that there are two distinct forms of ions associated with the casein micelle—an outer system, perhaps in the form of a charged double layer, and an inner system not easily washed away. As noted above, the casein micelle is a highly porous, well-solvated system, and the occlusion of ions within this network is not unexpected; however, some actual complex formation between occluded salts and the casein cannot be ruled out. If one examines the pK's of phosphoric acid, it would seem most likely that the associating species of phosphate would be $(HPO_4)^{2-}$. Termine and Posner (1970) studied the in vitro formation of calcium phosphate at pH 7.4 and concluded that an amorphous calcium phosphate phase (with a Ca:PO$_4$ molar ratio of 1.5) formed prior to the transition to crystalline apatite. In a subsequent study (Termine et al. 1970), it was shown that casein and some other macromolecules enhanced the stability of the amorphous calcium phosphate and, in fact, retarded the amorphous to crystalline transition for the apatite. It would appear, then, that conditions should favor the formation of an amorphous calcium phosphate–caseinate complex in milk. Direct observations of the mineral components of the casein micelle have been limited in number. Knoop et al. (1979) studied the reconstitution of casein micelles by electron microscopy and found mineral inclusions resembling those observed in native micelles only when calcium, phosphate, and citrate were present, indicating that the mineral complex must include all three components as well as perhaps magnesium. Holt and co-workers (1982) used extended x-ray absorption fine structure (EXAFS) to study freeze-dried micelles, as well as a calcium phosphate gel made by proteolysis of casein. They compared these spectra with model compounds and found the closest match to be brushite, i.e., $CaHPO_4 \cdot 2H_2O$. This however, means that the complex is phosphate poor. These workers did not include citrate in their interpretations of possible models. Posner et al. (1983), however, isolated a complex from deproteinized skim milk which contains 10% citrate and has an x-ray radial distribution function similar to that of other phosphate-poor cellular mineral inclusions. Thus the complex in milk might well be a brushite-type complex which is limited in growth by the presence of citrate or possibly casein phosphate. Although the exact nature of this complex (or occlusion) is still undetermined, its role in casein micelle stabilization is well documented.

Pyne and McGann (1960) demonstrated that the colloidal calcium phosphate content of milk decreases as the pH is lowered from 6.7 to 5.0 at 5°C. If a small sample of this pH 5.0 milk is then dialyzed at 5°C against several large volumes of the original milk, the pH returns to 6.7 but the colloidal calcium phosphate is no longer present. Milk brought to a colloidal calcium phosphate concentration of essentially zero at pH 5, and dialyzed back to 6.7 in this manner, has been termed "colloidal calcium phosphate-free milk (CPF milk)." In a later study, McGann and Pyne (1960) investigated the properties of CPF milk in comparison to those of the original nontreated milk. The CPF milk is more translucent and has a greatly increased viscosity. Addition of Ca^{2+} up to ~1 M has little effect on normal milk at 25°C, provided that the increase in pH is compensated for. CPF milks, however, are precipitated at added calcium ion concentrations of only 25 mM. There is no apparent difference between the CPF and normal milks with regard to the primary phase of chymosin attack, as measured by release of soluble nitrogen, but, interestingly, the CPF milks are slightly more heat stable. Rose (1968) noted that while the addition of Ca^{2+} generally decreases the serum casein content of milk, lowering the pH of milk to 5.3 and the subsequent release of Ca^{2+} actually increases the serum casein content. This result led Rose to speculate that the colloidal calcium phosphate aids in maintaining micelle stability. CPF milks and normal milks were compared by Downey and Murphy (1970) with respect to their elution volumes on gel chromatography (Sepharose 2 B) in a synthetic milk serum. The normal casein micelles eluted at V_0 yielded a molecular weight of $>10^8$, but CPF micelles eluted at a volume consistent with a molecular weight of $\sim 2 \times 10^6$. However, this result could also be explained by a marked change in shape.

The preceding discussion indicates that colloidal calcium phosphate is involved in maintaining the structural integrity of the casein micelle. Occlusion of amorphous calcium phosphate as apatite or brushite or possible complexation of the minerals with citrate must occur, but the exact mechanism by which stabilization is achieved is still unknown.

CASEIN MICELLE STRUCTURE

From the above discussion, one might conclude that the individual caseins have been studied in sufficient detail to yield an intimate knowledge of the structure of the casein micelle. This is not the case, and nearly as many models have been proposed as there are investigators. Let us briefly consider why this situation exists. Analysis of the casein micelles of bovine milk by electron microscopy (Figure 9.2) indicates an average diameter of ~140 nm for the spherically shaped micelles (Carroll *et al.* 1968). Thus, the volume occupied by a micelle would be on the order of ~1.4 × 10⁶ nm³. For comparison, the β-lactoglobulin monomer occupies a volumes of ~2.4 nm³. Theoretically, more than 50,000 β-lactoglobulin-like monomers could be arranged into a sphere the size of a casein micelle. Molecular weight measurements by light scattering for the micelles range from 10^8 to 10^9 (Holt 1975; Lin *et al.* 1971). A speculative calculation, based on an average molecular weight of 23,000 for the casein monomers [(3 α_{s1}-+ 3 β-+ 1 α_{s2}-+ 1 κ-)/8] and employing only 25,000 monomers, yields a micelle molecular weight of 6 × 10⁸. This would indicate a low density packing of the casein monomers, which is consistent with the high hydrations, the random structures, and the high negative charge densities of the caseins, as compared to those of β-lactoglobulin. It is therefore understandable that the mechanism of assembly of this aggregate of around

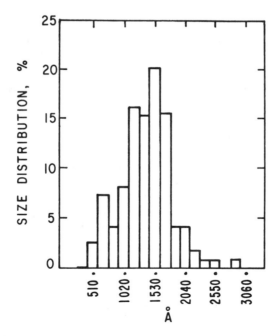

Figure 9.2 Determination of the size distribution of glutaraldehyde-fixed casein micelles from skim milk. (From Carroll, *et al.* 1968.)

25,000 monomers has not been fully elucidated. For the purpose of discussion, we shall group the various proposed models into three classes.

Coat-Core Models

The first class of models to be discussed actually contains two diametrically opposed theories. The model proposed by Waugh and his co-workers (1970) is based primarily upon their studies of the solubilities of the caseins in Ca^{2+} solutions. The model, in essence, describes the formation of low weight ratio complexes of α_{s1}- and κ-casein in the absence of calcium. Upon addition of calcium ions, the α_{s1}- or β-caseins, depicted as monomers with a charged phosphate loop in Figure 9.3A, begin to aggregate to a limiting size (the caseinate core). Precipitation of the caseinate is prevented by the formation of a monolayer of the low weight α-$_{s1}$-κ-complexes. This coat has the κ-casein monomers spread out entirely on the surface, and the micelle size is therefore dictated by the amount of κ-casein available. In the absence of κ-casein, the α_{s1}- and β-cores agglutinate and precipitate from solution. Waugh's model, as presented in Figure 9.3A, has a good deal of appeal since it explains the lyophilic nature of the colloidal casein complex, as well as the ready accessibility of κ-casein to the enzymes chymosin.

Parry and Carroll (1969) attempted to locate the surface κ-casein proposed by Waugh by the use of electron microscopy. Using ferritin-labeled anti-κ-casein immunoglobulins, they investigated the possibility of surface κ-casein on fixed whole micelles and found little or no concentration of κ-casein on the surface of the casein micelles. Based on these results and on the size of the isolated κ-casein complex, Parry and Carroll concluded that the κ-casein might serve as a point of nucleation about which the calcium-insoluble caseins might cluster and subsequently be stablized by colloidal calcium phosphate (Figure 9.3B). The action of rennin on the micelles was accounted for by demonstrating that serum κ-casein can participate in the coagulation reaction and may be involved in the formation of bridges between micelles. Convincing evidence against this model was presented by Green (1972), who demonstrated that pretreatment of milk serum with chymosin did not alter the clotting time, nor did washing the micelles at 23°C (whereas Parry and Carroll had washed the micelles at 16°C).

The models of Parry and Carroll and Waugh both predict a rather precise distribution of κ-casein and in a sense are based upon nucleation about a core (Parry and Carroll's core = κ-casein; Waugh's core = α_{s1}, β-calcium caseinate). It is important to note that neither model predicts submicellar structures. Waugh's model has been revived by the invocation of the "hairy micelle" concept (Walstra 1979).

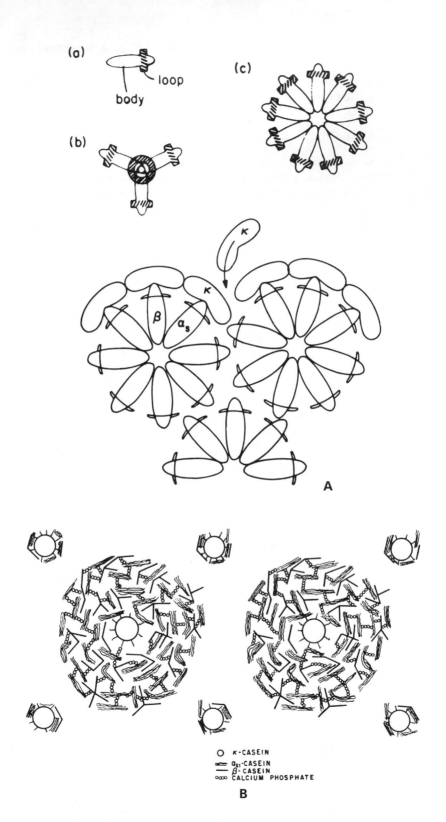

Ashoor *et al.* (1971) demonstrated that papain, which had been cross-linked by glutaraldehyde into a large, insoluble polymer, caused proteolysis of all three major components of isolated casein micelles. The α_{s1}-, β-, and κ-caseins were all cleaved proportionately by the enzyme superpolymer. Therefore, all three components must occupy some surface positions on the micelle. This result apparently rules out any totally preferential localization of κ-casein on the surface or at the core of the casein micelles.

Internal Structure Models

The second class of models to be discussed are based upon the properties of the isolated casein components, which in turn cause or direct the formation of the internal structure of the casein micelle.

Garnier and Ribadeau-Dumas (1970) have proposed a model for the casein micelle which places a good deal of emphasis on κ-casein as the keystone of micelle structure. Trimers of κ-casein are linked to three chains of α_{s1}- and β-casein which radiate from the κ-casein node (a Y-like structure), as shown in Figure 9.4A. These chains of α_{s1}- and β-casein may connect with other κ-nodes to form a loosely packed network. Garnier and Ribadeau-Dumas favored this type of network because it yields an open, porous structure, and they have demonstrated that carboxypeptidase A, with a molecular wieght of ~ 40,000, is able to remove the C-terminal amino acids of all of the casein components. The model provides the demonstrated porosity but places great steric restraints upon κ-casein, which possesses few prominent secondary structures, and calls for a uniform distribution of κ-casein regardless of micelle size. In addition, studies by Pepper and Farrell (1982), Cheeseman (1968), and Swaisgood and Brunner (1962, 1963) indicate that, while disulfide-linked trimers of κ-casein do occur, the majority of the κ-casein may form aggregates of higher orders. Finally, the model assigns no definite role to calcium caseinate interactions and ignores the possibility of colloidal calcium phosphate involvement in stabilization of the micelle.

Figure 9.3 (A) Waugh's proposed model for the casein micelle. (a) Monomer model of α_{s1}- or β-casein with charged loop. (b) a tetramer of α_{s1}-casein monomers. (c) Planar model of a core polymer of α_{s1}- and β-caseins. The lower portion shows how κ-casein might coat core polymers. (Adapted from Rose 1969.) (B) Casein micelle model proposed by Parry and Carroll (1969), depicting the location of κ-casein in the micelle.

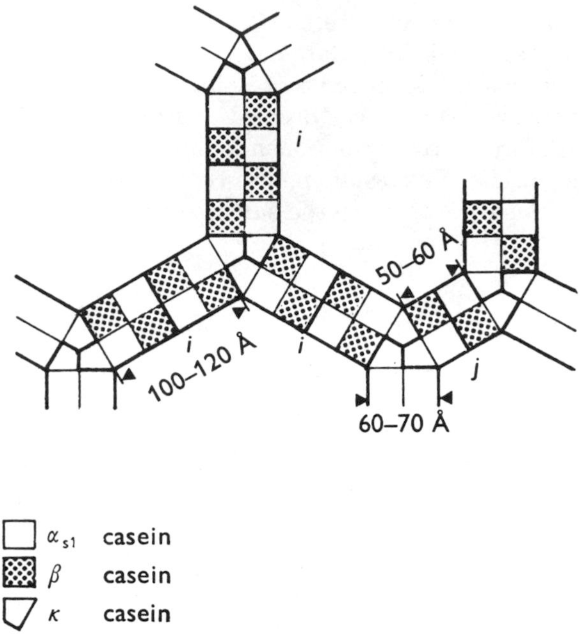

Figure 9.4 (A) Structure of the repeating unit of the casein micelle. (From Garnier and Ribadeau-Dumas 1970.)

Rose (1969) used the known endothermic polymerization of β-casein as the basis for his micelle model. In this model, β-casein monomers begin to self-associate into chain-like polymers to which α_{s1}-monomers become attached (Figure 9.4B) and κ-casein, in turn, interacts with the α_{s1}-monomers. The β-casein of the thread is directed inward, the κ-outward, but as these segments coalesce, a small amount of κ-casein is inevitably placed in an internal position. As the micelle is formed, colloidal calcium phosphate is incorporated into the network as a stabilizing agent. The model is appealing in that it accounts for the occurrence of some overall stoichiometry of the various casein components while demonstrating the role of colloidal calcium phosphate in micelle stabilization. The choice of β-casein as the basis of micelle formation is, however, questionable, since Waugh *et al.* (1970) have shown that the α_{s1}-and β-caseins tend to form mixed polymers randomly; in addition, β-casein is quite structureless in solution and, as noted above (Schmidt and Payens 1976), tends to form micellar-like complexes rather than linear polymers. Finally, synthetic micelles can be formed

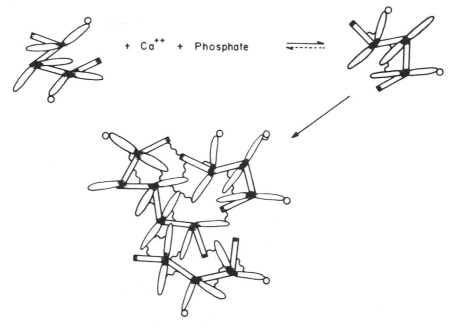

+ Ca^{++} + Phosphate

Figure 9.4 (B) Schematic representation of the formation of a small casein micelle. The rods represent β-casein, the more eliptical rods represent α_{s1}-casein, and the S-shaped lines depict apatite chain formation. The circles represent κ-casein. (Adapted from Rose 1969.)

from simple α_{s1}- and κ-casein complexes in the complete absence of β-casein.

Submicellar Models

The final class of models to be discussed is those which propose a submicellar structure for the casein micelle. Shimmin and Hill (1964) first postulated such a model based upon their study, by electron microscopy, of ultrathin cross sections of embedded casein micelles. They predicted a diameter of 10 nm for the submicelles.

Morr (1967) studied the disruption of casein micelles and proposed that the α_{s1}-, β-, and κ-monomers formed small uniform submicelles in much the same fashion as Waugh (1971) had proposed for the entire micelle. Morr's submicelles, as estimated by sedimentation velocity studies, have a diameter of ~30 nm. The submicelles are stabilized by hydrophobic bonding and calcium caseinate bridges and, in turn, are aggregated into micellar structure by colloidal calcium phosphate.

Morr's model is summarized in Figure 9.5. The average submicellar size, postulated by Morr, is somewhat larger than that of Shimmin and Hill. Several other researchers have attempted to study the nature of the submicelle by using various methods for disruption of the casein micelles. Carroll *et al.* (1971A) used EDTA, urea, sodium lauryl sulfate, and sodium fluoride to disrupt micelles and found particles of ~10 ± 2 nm diameter. Schmidt and Buchhein (1970) dialyzed milk free of calcium in the cold and used high pressure to disrupt casein micelles; in both cases, they obtained submicelles of 10 nm diameter; these results were confirmed by Buchheim and Welsch (1973). Subsequently, Pepper and Farrell (1982) employed gel chromatography in studies of the concentration-dependent interactions of soluble whole casein, prepared by dissociation of casein micelles with EDTA. With increasing protein concentration at pH 6.6 and 37°C, the components of whole casein associated to polymers which approached molecular radii with an apparent upper limit of 9.4 ± 0.4 nm. Analysis of the concentration-dependent elution profiles showed that casein submicelles could be formed by the interaction of SH-κ-casein monomers with those of α_s- and β-caseins.

The hypothesis of Shimmin and Hill (1964) that sections of the casein micelles contain submicellar particles ~10 nm in diameter was invoked by Carroll *et al.* (1970) and Farrell and Thompson (1971), who observed, by electron microscopy, particles of ~10 nm diameter in the Golgi vacuoles of lactating rat mammary gland. These particles were

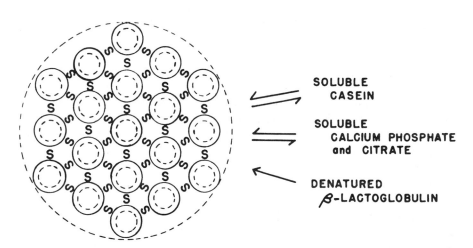

SOLUBLE
CASEIN

SOLUBLE
CALCIUM PHOSPHATE
and CITRATE

DENATURED
β-LACTOGLOBULIN

Figure 9.5 Structure of the casein micelle proposed by Morr (1967). The S-shaped lines represent calcium phosphate linkages between small spherical complexes of the α_{s1}-, β-, and κ-casein.

uniform in size and appeared to coalesce into the spherically shaped casein micelle (Figure 9.6A,B).

Slattery and Evard (1973), based upon studies of casein interactions, proposed an interesting approach which appears to combine the best features of most micelle models. They suggested that casein monomers interact to form submicelles, but that the submicelles might be of variable composition with respect to their casein content. In this model the κ-casein-rich submicelles are found predominantly on the surface of the casein micelle, where they contribute a stabilizing force. Conversely, submicelles poor in κ-casein would more likely be internalized. Limitations on the growth of the micelles are also included; the Slattery and Evard model is depicted in Figure 9.7. Since its publication this model has received much support, as it reconciles some of the conflicting data in the literature. For example, the model accommodates the inverse relationship found between micelle size and κ-casein content (Sullivan et al. 1959; Slattery 1978; McGann et al. 1980). It predicts that in large (κ-casein-poor) micelles, κ-will occupy a surface position, while in smaller (κ-rich) micelles, κ-will be more uniformly distributed. Carroll and Farrell (1983), using a ferritin-labeled double-antibody technique coupled with electron microscopy, determined on ultrathin sections that the location of the κ-casein is indeed related to casein micelle size (Figure 9.8). Casein micelles with the κ-casein located predominantly at the periphery of the micelle were found to have diameters of 142 $\pm$ 32 nm, while those with a diameter of 92 $\pm$ 22 nm had a more uniform distribution of κ-casein. These results are in accord with the inverse relationship between micelle size and κ-casein content and with the finding that larger casein micelles contain higher polymers of κ-casein, indicating that κ–κ interactions are greater in κ-poor micelles (see the discussion above on the role of disulfide bonds). These results contradict the models of Waugh, Parry, Garnier and Ribadeau-Dumas, and Morr, as cited above, since the κ-casein is not totally precisely localized in the micelles and supports the more flexible model of Slattery and Evard.

Although the submicelle model is widely accepted, it may not represent the final answer to the structure of the casein micelle. It is important to distinguish between the concepts of casein micelle formation and disruption and those concerning the casein micelle at some steady state. In vivo assembly of the casein micelle from submicelles through calcium accretion in the Golgi apparatus appears to be the best model for formation. However, once formed, new contact points between submicelles must inevitably occur, and after secretion from the Golgi vesicles or deposition into the cistern, changes may occur. That there may not be a distinct submicellar structure in the micelle has been sug-

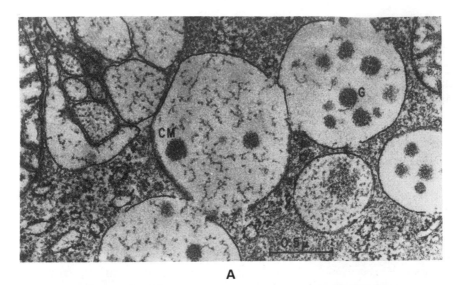

A

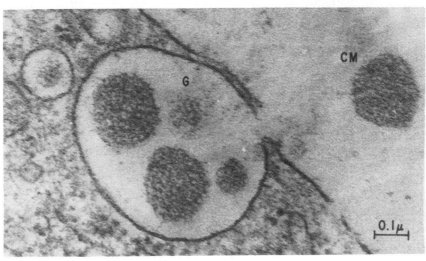

B

Figure 9.6 (A) Formation of casein micelles (CM) within Golgi Vacuoles (G) of lactating rat mammary gland. Initially, submicellar structures appear; then more compact micelles seem to occur. Sections of the gland were fixed in buffered OsO_4, Epon embedded, and stained with uranyl acetate and lead citrate. (From Carroll *et al.* 1970.) (B) A Golgi vacuole about to discharge its contents into the alveolar lumen. The Golgi vacuole shown appears to impinge upon the plasma membrane. A casein micelle is already present in the lumen. (From Carroll *et al.* 1970.)

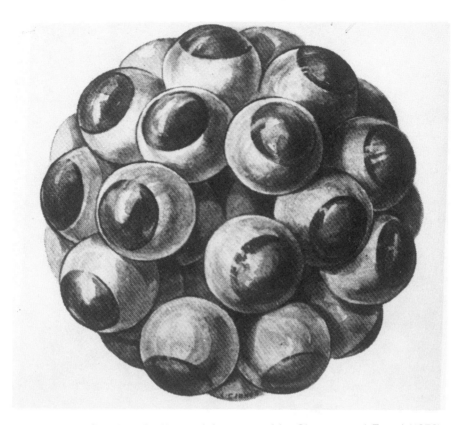

Figure 9.7   Casein micelle model proposed by Slattery and Evard (1973). The micelle is composed of about 40 submicelles. The lighter portions of the submicelles represent the α_{s1}- and β-casein, while the darker portions represent the κ-casein carbohydrate-containing molecules. Interior submicelles have similar structures.

gested by Holt (1975), and a possible gel-like structure for the micelle has been proposed. The steady-state condition of the micelle as a loose gel cannot be discounted by currently available data. However, perturbation of the micelle structure by EDTA, pressure, etc., will cause dissociation into submicellar structures by cleavage of appropriate structural elements. In essence, submicelle integrity may be lost on formation of the micelle but can reappear on dissociation; the dissociated products may not necessarily reflect the original submicelles. The interactions of κ-casein within whole casein observed by Pepper and Farrell (1982) are an example of this point in that S-S and SH-κ-casein behave quite differently in their interactions with the other caseins, although both can stabilize the calcium-sensitive caseins (Woychik *et al.* 1966).

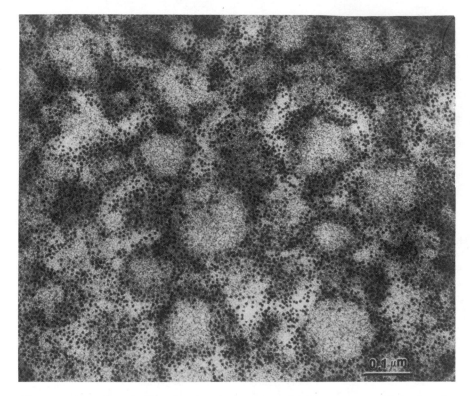

Figure 9.8 κ-Casein localization in casein micelles. Immunohistochemical localization of κ-casein in casein micelles. Ultrathin-sectioned casein micelles (underfixed) treated with a primary antibody (rabbit anti-κ-casein), followed by exposure of the sections to the secondary antibody (ferritin-conjugated goat anti-rabbit IgG). The electron-dense ferritin label indicates the position of the κ-casein on the casein micelles. (From Carroll and Farrell 1983.)

From the biosynthetic point of view, the buildup of the micelle from submicelles is quite attractive (Farrell 1973). Protein–protein interactions among the casein components can occur within the lumen of the endomembrane system. Phosphorylation in the Golgi apparatus and addition of calcium ions could cause the polymerization of casein submicelles into micellar spheres by the deposition of colloidal calcium phosphate. The assembly of the casein micelle from preformed submicelles is not nearly as specific as that of tobacco mosaic virus, but the analogy is worthy of consideration. In the latter case, its structured RNA core (Butler 1977) plays a vital role in directing correctly the assembly of the virus particle from its preformed subunits, whereas in the case of the micelle, amorphous calcium phosphocitrate may serve this function. The viral particles are of constant size and composition, while the casein micelles, though roughly spherical, vary within limits in size and in composition. In attempting to solve the problem of casein micelle structure, it should be borne in mind that the biological function of the micelle is efficient nutrition. Hence, the interactions which yield this product, the casein micelle, need not be as specific as those which result in the formation of a virus or an enzyme, and perhaps must be quite flexible so as to secrete milk salts and proteins under a variety of dietary and environmental conditions.

Proteolytic Action and Micelle Models

The action of chymosin on the casein micelle is primarily the hydrolysis of the highly sensitive Phe–Met peptide bond of κ-casein. Sequence data (Merciér et al. 1973) shows that three to eight residues from this bond in either direction comprise the unusual sequence proline–proline, which perhaps accounts for the high susceptibility of this specific bond. The splitting of the bond (Kalan and Woychik 1965) causes the formation of the rather insoluble para-κ-casein and the highly hydrophilic (and carbohydrate-rich) peptide termed the "glycomacropeptide (GMP)" or "macropeptide (MP)." As a result of the action of chymosin, the micelles coagulate or clot. All of the casein micelle models must account for this phenomenon.

Waugh's model solves this problem most readily; since all of the κ-casein is on the surface, the cleavage of the GMP results in the loss of charge repulsions which normally prevent coagulation. Parry and Caroll (1969) pointed out the possible role of serum κ-casein in the rennin reaction, though Green's data (1972) discount this, and Garnier and Ribadeau-Dumas (1970) noted the possibility that chymosin could penetrate a highly porous system. Ashoor et al. (1971) showed that papain which had been insolubilized by glutaraldehyde cross-linkages caused proteolysis of all three major components of casein micelles, demonstrating that not only κ- but α_{s1}- and β-casein occupy surface positions. Carroll and Farrell (1983) pointed out that, based on their studies, the larger micelles have κ-casein predominantly on their periphery, and the average size of these micelles has been determined to be ~ 142 nm. Since the average size of the casein micelle is ~ 140 nm, roughly half of the micelle population would contain κ-casein only at the periphery, while smaller micelles of 92 ± 22 nm may have κ-casein, and hence α_{s1}-and β- more evenly distributed and available to proteases. These conclusions are more in accord with the model of Slattery and Evard (1973) and its recent elaboration by Schmidt (1982) (Figure 9.9). Because of the high incidence of apolar residues in all of the caseins and the large number of monomers, it is inconceivable that all of the apolar chains would be buried; therefore, many hydrophobic patches probably exist on the surface of the micelle. Normally, coagulation is prevented by the charged surface groups contributed by all three components. When a sufficient number of the κ-casein–GMP fragments have been removed by rennin action, coagulation or clotting could occur, with the possible addition of some serum para-κ-bridges, as postulated by Parry and Carroll (1969). Internal hydrolysis of all casein components (and this too occurs) would cause the porous micelle to dehydrate on loss of internal MP, leading to additional destabilization of the system. Finally, as will be discussed later, equilibrations of serum and micellar caseins will alter susceptibility of protease activity and consequently influence micelle structure.

What has been noted above in regard to the enzyme chymosin would also be true of any protease which would be added to or occur naturally in milk. Chen and Ledford (1971) have studied the "milk protease" and found it to be trypsin-like. Its identity with plasminogen has been established by Eigel and co-workers (1979). Limited proteolysis by this or other as yet uncharacterized proteases or their reactivation upon

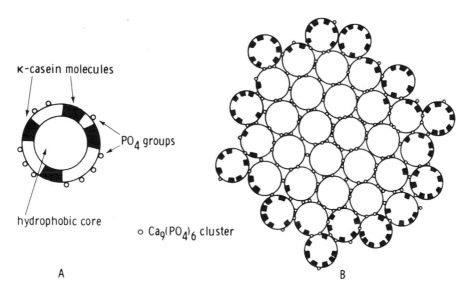

κ-casein molecules

PO_4 groups

hydrophobic core

o $Ca_9(PO_4)_6$ cluster

A   B

Figure 9.9   Casein micelle model proposed by Schmidt (1982). Schematic representation of a submicelle (A) and a casein micelle composed of submicelles (B).

storage may be the cause of many of the problems encountered in the storage of processed whole or concentrated milks and dairy products.

THE MILK SERUM PHASE

The Serum Proteins

The proteins of the dispersed phase of milk, the casein micelles, account for up to 74% of the total protein of skim milk (Table 9.7). The serum or whey proteins are those which remain in solution after the micelles are removed; these proteins are not incorporated into colloidal complexes. Depending upon the method of removal of the casein micelles, varying amounts of serum casein will be included in this fraction. The major noncasein serum proteins are β-lactoglobulin, α-lactalbumin, bovine serum albumin, and the milk immunoglobulins (Chapter 3; Eigel *et al.* 1984).

β-**Lactoglobulin.** β-Lactoglobulin is the major serum protein; it accounts for up to 50% of the noncasein protein of skim milk (Cerbulis and Farrell 1975, 1976) and exhibits unique structural features. β-Lactoglobulin has a monomer molecular weight of 18,360. The complete amino acid sequence for the molecule (see Chapter 3) has been completed and its x-ray structure elucidated (Papiz *et al.* 1986). From physical data and chemical modification experiments, much has been deduced concerning the topography of the molecule (Townend *et al.* 1969), as well as its rather unique modes of self-association. The 18,360-dalton monomer exists only below pH 3.5 and above pH 7.5. Between these pH values, β-lactoglobulin occurs as a dimer with a mo-

Table 9.7. **Average composition of Warm Skim Milk Protein.**

	Grams per 100 g milk	% Total protein[a]
Colloidal casein	2.36	74
Serum casein	0.26	8
β-Lactoglobulin	0.29	9
α-Lactalbumin	0.13	4
Bovine serum albumin	0.03	1
Total immunoglobulins	0.06	2
Other proteins	0.06	2

[a] All values normalized to 3.2 g total protein/100 g milk.
SOURCE: Averaged from data of several sources (Farrell and Thompson 1974).

lecular weight of 36,720, and its conformation (Townend *et al.* 1967) has been well characterized. At reduced temperatures, between pH 3.7 and 5.1, the dimers of the β-lactoglobulin A genetic variant form a specific octamer with 422 symmetry. From the symmetry of the octamer, one would predict a central cavity containing nonbulk water; this water has been detected and quantified by nuclear magnetic relaxation techniques in D_2O (Kumosinski and Passen 1982). β-Lactoglobulin B does self-associate to an appreciable extent, and the C genetic variant apparently does not octamerize at all (Townend *et al.* 1967). The conformational transitions, which bring about these associations, have been well documented and the analogy of these interactions to allosteric control mechanisms has been noted (Timasheff and Townend 1969). The precise biological function of this molecule is unknown, though several hypotheses have been advanced (Farrell and Thompson 1971, Papiz *et al.* 1986).

Because of its ready availability and unique properties, β-lactoglobulin has been used as a model protein in many studies. It also plays an important role during processing as the result of two of its features. First, β-lactoglobulin has well-defined secondary, tertiary, and quaternary structures which are susceptible to denaturation. Second, one free sulfhydryl group occurs per 18,360-dalton monomer (Green *et al.* 1979). Thus, in the case of fluid products, the integrity of the β-lactoglobulin molecule must be partially retained to prevent irreversible denaturation, which may result in coagulation; additionally, conditions which affect free sulfhydryl groups or cause disulfide interchange with other proteins must be avoided.

Protein denaturation has been reviewed by Tanford (1970) and Privalov (1979, 1982). Environmental influences which cause reversible denaturation, when carried to an extreme, generally produce some irreversible denaturation (see Chapter 11). In discussing denaturation, Tanford outlined the contributions of conformational, interchain, protein solvent, and electrostatic interactions to the free energy of the transition from the native (N) to a denatured (D) state, as well as the influence of environmental factors such as temperature, pressure, and binding effects, on the N → D transition. Increases in temperature and pressure, as well as the binding of denaturants (e.g., guanidine, urea, and detergents), the binding of inorganic ions, and the binding of hydrogen ions (pH) may facilitate the transition to the denatured state. Tanford (1970) further notes—and this is important for milk proteins—that each type of denaturing agent may lead to a particular denatured state. In each case, the resulting polypeptide chain has more exposed apolar groups, relative to the native state, whcih increases the probability of aggregation unless these areas are stabilized by the solvent or cosolutes (e.g., urea and guanidine). Denaturation in aqueous solution often leads to aggregation and, finally, to precipitation or coagulation. It is this latter event which one wishes to prevent in the processing of fluid dairy products, since it is generally thought that excessive denaturation of the whey proteins leads to problems in these products. Carroll *et al.* (1971B) suggested that the denaturation and subsequent precipitation of whey proteins was one basis for the gelation of sterilized, concentrated milk. Hence, it would appear that environmental influences, which tend to preserve the native structure of β-lactoglobulin, and the other whey proteins favor stability, while denaturation and subsequent coagulation lead to instability. Conversely, in products such as pasta, denatured whey protein is actually preferable (Schoppet *et al.* 1976). Factors which influence the denaturation of β-lactoglobulin have been reviewed by Tilley (1960) and McKenzie (1971). Generally speaking, heat, pH above 8.6, and increased calcium ion concentration (Zittle *et al.* 1957) tend to increase the N $\rightarrow$ D transition of β-lactoglobulin. Urea denaturation of β-lactoglobulin has been studied in detail by Span and Lapanje (1973) and Greene and Pace (1974). The process occurs in two reversible stages which result, at high concentrations of urea (8 M), in a randomly coiled polypeptide with one molecule of urea bound per three amino acid residues. Preferential interactions occur at all concentrations, and similar results have been obtained for guanidine hydrochloride. Like other proteins, alkyl ureas at lower concentrations (Pace and Marshall 1980; Lapanje and Kranjc 1982) are effective denaturants of β-lactoglobulin. However, the alkyl ureas most likely result in denatured states which differ from that of the 8 M urea-unfolded protein.

The mechanism of thermal denaturation of β-lactoglobulin has been studied by fluorescence spectroscopy (Mills 1976) and differential scanning calorimetry (DeWit and Swinkels 1980). At neutral pH, both methods indicate that at temperatures up to 70°C the denaturation is reversible, but above this temperature irreversible denaturation occurs, presumably due to aggregations. At pH 2.5, however, Harwalkar (1980A,B) found that heating produced a partially denatured β-lactoglobulin with some degree of unfolding and altered functional properties.

β-Lactoglobulin, in its native state, binds aromatic hydrophobes at low molar ratios (Robillard and Wishnia 1972); these include retinol (Fugate and Song 1980; Futterman and Heller 1972), long chain fatty acids (Spector and Fletcher 1970), and detergents (Seibles 1969) such as sodium dodecyl suflate (SDS). The binding occurs in stoichiometric amounts (1 mole per monomer). The association constants for this class of sites is one the order of 10^5 to 10^8. At more elevated detergent concentrations, very strong complexes of 30 moles per dimer can be formed (Jones and Wilkinson 1976). These complexes are partially ionic and have enthalpies on the order of those associated with salt bridges formed in a hydrophobic environment (Kresheck *et al.* 1977); their formation is accompanied by rather large conformational changes. At very high levels of SDS, binding increases up to roughly 1.4 g SDS per gram of protein (~ 90:1 molar ratio); this class of binding sites is weaker and more characteristic of micellar (hydrophobic) interactions (Kresheck *et al.* 1977). Exhaustive dialysis (Basch and Farrell 1979) of reduced β-lactoglobulin treated with very high levels of SDS yielded a protein-detergent complex of 21% by weight SDS (~ 20 moles SDS per monomer). Su and Jirgensons (1977) studied the conformation of β-lactoglobulin treated at very high SDS:protein ratios by circular dichroism; they found that at acid pH the SDS caused a moderate amount of α-helix to be formed, but this effect was reduced at alkaline pH, so that little ordered structure exists for the highest SDS–protein complexes.

Upon partial denaturation, β-lactoglobulin as well as α-lactalbumin have been shown to interact strongly with phospholipids. The nature of these complexes, the structural basis for them, and their possible practical applications have been reviewed by Brown (1984).

A potential source of instability in β-lactoglobulin is its unique sequence, –Cys–Gln–Cys– (residues 119, 120, 121). It has been reported that either residue 119 or 121 is the free sulfhydryl group of β-lactoglobulin (Eigel *et al.* 1984), while the other residue is therefore involved in a disulfide bond. Above pH 6.5, β-lactoglobulin undergoes a distinct conformational change with a transition centering at pH 7.5 (Tanford *et al.* 1959; Waissbluth and Grieger 1974). This change results in the alteration of the environment of several residues, most importantly in the sulfhydryl group (Dunnill and Green 1965; Brown and Farrell 1978). The SH-group, according to the most recent x-ray data (Papiz *et al.* 1986), does exist near the surface of the molecule; thus, upon denaturation, the sulfhydryl group becomes more reactive. This phenomenon was documented in the case of alkaline denaturation of β-lactoglobulin (Brown and Farrell 1978). In the latter study, it was observed that β-lactoglobulin formed rapidly reversible ionic complexes with cytochrome c. These complexes were observed by gel chromatography and analytical ultracentrifugation. At pH's above 8, time-dependent changes in difference spectra indicative of the conversion of ferri- to ferro-cytochrome c were observed. Reduction was shown by a derivatization reaction to be due to the free sulfhydryl group of β-lactoglobulin. The Arrhenius plot for the rate of reduction displayed a minimum at 15°C, which directly parallels the rate of alkaline denaturation of β-lactoglobulin (Waissbluth and Grieger 1974). Since model compounds reduced the cytochrome c iron rapidly, it was concluded that denaturation of the β-lactoglobulin to expose the SH-group was the rate-determining step. In summary, a fast protein–protein interaction is followed by a slow denaturation and then by a rapid reaction to reduce the cytochrome c.

Douglas *et al.* (1981) studied the effects of UHT-treatment of skim milk (148°C for 2.5 sec). They found a 56% denaturation of whey protein, accompanied by the formation of a β-lactoglobulin–casein complex. The complex coprecipitated with the caseins, altering their functional properties. In this instance, the heat treatment of the milk causes unfolding of the β-lactoglobulin, which may have already been bound to or may subsequently react with the κ-caseins of the casein micelles. Euber and Brunner (1982) demonstrated the interaction of immobolized β-lactoglobulin with κ-casein, and recent evidence by immunohistochemistry has shown that the κ-casein molecules which contain disulfide bonds are readily available at the surface of the casein micelles (Figure 9.8). κ-Casein may thus rapidly undergo disulfide interchange reactions with the denatured β-lactoglobulin, causing a deposit to be formed on the surface of the micelles. Evidence of this surface deposition has been obtained by electron microscopy (Smits and van Brouwershaven 1980). Hence, exposure to a strong base or to heat may promote disulfide interchange and lead to denaturation of β-lactoglobulin. McKenzie (1971) has summarized the work of several groups who have demonstrated in model systems that the sulfhydryl group of β-lactoglobulin has been implicated in the formation of complexes with κ-casein upon heating. Since molecules such as carboxypeptidase can penetrate the casein micelle (Ribadeau-Dumas and Garnier 1970), β-lactoglobulin may likewise do so, thereby inducing the formation of internal disulfide bonded complexes with κ-casein upon heating. Such complexes do alter the stability of the casein micelle (Farrell and Douglas 1983; Smits and van Brouwershaven 1980).

α-**Lactalbumin.** α-Lactalbumin accounts for up to 25% of the whey protein and ~4% of the total milk protein (Table 9.7), and has been assigned a unique biochemical role as the specifier protein of the lactose synthetase system (Ebner and Schanbacher 1974). The complete sequence of α-lactalbumin is known (Vanaman *et al.* 1970), and the molecule exhibits a strong structural relationship to lysozyme. A computer-generated model based upon the crystal structure of lysozyme is available (Warme *et al.* 1974). When compared to β-lactoglobulin, α-lactalbumin is found to exhibit far more structural stability. The protein contains no free sulfhydryl groups and has four disulfide bonds.

α-Lactalbumin has a good deal of predicted secondary structure and is therefore quite compact, with a monomer weight of 14,174. Kuwajima and co-workers have studied the denaturation of α-lactalbumin by guanidine, and their evidence for a three-state process has been reviewed by Kuwajima (1977). Kronman and colleagues (Kronman et al. 1964, have studied in detail the acid denaturation of the protein, and an excellent summary of the denaturation of α-lactalbumin by guanidine, acid, and salts has been given by Contaxis and Bigelow (1981). Brown (1984) has reviewed the interaction of lipids with partially denatured α-lactalbumin. However, all of these studies now need to be reexamined in light of the discovery by the Japanese workers (Hiraoka et al. 1980) that α-lactalbumin is a calcium-binding protein. Indeed Kronman et al. (1981) and Permyakov et al. (1981) have shown that removal of calcium from α-lactalbumin produces profound conformational changes equivalent to those occurring upon acid denaturation. Murakami et al. (1982) have estimated that the association constant for Ca^{2+} is 10^{10} to 10^{12} M, indicating one strong binding site, whose structure has been elucidated by x-ray data (Stuart et al. 1986). Therefore calcium and other metal ions may influence strongly the stability of α-lactalbumin. Previous work has shown that α-lactalbumin is denatured at low pH's and then undergoes an association reaction which requires a somewhat elevated protein concentration and may thus not occur in milk products (Kronman et al. 1964). However, in whey concentrates, both acid denaturation and calcium binding by α-lactalbumin may play an important role in the retention or loss of the proteins' functional properties, and such products also need to be studied in light of this new information. Hunziker and Tarassuk (1965) have also reported that the free sulfhydryl of β-lactoglobulin can promote complex formation with α-lactalbumin through disulfide reactions at elevated temperatures.

Other Serum Proteins. Serum albumin and immunoglobulins (Eigel et al. 1984) occur in skim milk to a limited extent and may, in conjunction with various enzymes (Shahani et al. 1973), account for up to 4% of the total milk protein. All of these proteins contain a significant amount of native structure and are also susceptible to various forms of denaturation. In fact, limited protein denaturation is a desired result in terms of the required inactivation of many of the enzymes which have been noted to occur in milk. The reactivation of phosphatases and of proteolytic and other enzymes upon storage of processed dairy products is undoubtedly the source of many problems. Hence, it appears that conditions must be controlled to inactivate enzymes adequately and to reduce bacterial contamination without causing severe denaturation and, consequently, coagulation of the major whey proteins.

The Serum Caseins

Not all of the casein secreted by the lactating mammary cells is incorporated into casein micelles. Rose (1968) and Downey and Murphy (1970) have studied the occurrence and distribution of the serum caseins. Rose found that in warm milk the serum caseins, on the average, account for around 10% of the total casein, and the serum casein contains all of the major components of the micelle in varying amounts, but always in the order β- > κ- > α_{s1}-casein. In cold milks the relative amounts of β- and κ-casein in the serum increase as the total serum casein increases. Rose concluded that the serum casein does not appear to be in true equilibrium with the micellar casein. However, recent data accumulated by Creamer et al. (1977), Davies and Law (1983), and Ali et al. (1980) argue strongly against this position and in favor of an equilibrium between the serum and micellar caseins. Serum caseins may play some role in the clotting of casein micelles by chymosin; however, the experiments of Green (1972) argue against an important role, as previously proposed (Parry and Carroll 1969).

Salt Content of Milk Serum

The salt content of milk serum, separated by several methods, has been determined by Davies and White (1960). Their average results for two milks, obtained by separating milk serum by diffusion, centrifugation, and clotting with rennet, are compared in Table 9.8. These values are in essential agreement with each other and with the very extensive previous results on the composition of the nonprotein aqueous phase of milk by the same authors (White and Davies 1958). The values for Jenness-Koops (1962) buffer in terms of total concentrations are given in Table 9.8 for comparison. This simulated milk salt formulation, although in use for over 20 years, still represents the best artifical milk serum available. It has been widely used under a variety of names, including the most popular, simulated milk ultrafiltrate (SMUF). The ionic equilibria among the main milk serum salt constituents has been calculated by computer modeling (Holt et al. 1981) and found to compare favorably with experimental results. Such computer programs now provide a basis for a broader understanding of the equilibria which occur in milk.

Table 9.8. Distribution of Salts (in mg/100 g Milk) between Serum and Colloidal States in Milk.[a]

	Total content (milk)	Serum				Colloidal State (Average of Methods)[c]
		Diffusate (20°C)	Rennet whey	Centrifuged serum	Jenness-Koops buffer[b]	
Total calcium	114.2	38.1	39.9	40.9	35.9	74.6
Magnesium	11.0	7.4	7.8	8.1	7.8	3.3
Sodium	50	46	47	47	42.0	3.3
Potassium	148	137	143	141	154	8.0
Total phosphorus	84.8	37.7	37.4	37.9	36.0	47.1
Inorganic phosphorus	—	31.8	30.8	31.8		—
Citric acid	166	156	152	154	185	12.0
Chloride	106.3	106.5	106.2	105.6	115	—
Total nitrogen	—	20.7	124.6	110.7		—
Casein nitrogen	364	0	21.6	6.8		—
Lactose[c]	4800	4800	4800	4800		—

[a] Average of two separated milks, corrected for bound water.
[b] Jenness and Koops (1962).
[c] Obtained by averaging diffusate, rennet whey, and centrifuged serum and subtracting their average value from the total milk content.
SOURCE: Adapted from Davies and White (1960).

Equilibria Between the Colloidal and the Serum Phases

Because of the complex nature of the milk system, the question of whether or not true equilibria occur is difficult to assess. The *Handbook of Chemistry and Physics* defines equilibrium as "the state of affairs in which a reaction and its reverse reaction are taking place at equal velocities, so that the concentration of reactants is constant." Such conditions, in terms of physical equilibria, apparently exist for milk components. For the sake of discussion, let us assume that the casein micelle is a porous, highly hydrated complex with some degree of subunit structure, that the micelle contains an amorphous form of calcium phosphocitrate, and that it is surrounded by a double layer of ions. Clearly, the outer layers of the ions can be in equilibrium with those of the solvent, and these ions are removed readily by gel filtration. If one considers the voluminosity (cubic centimeters per gram) of the casein micelles and calculates the volume fraction occupied by them, there appears to be an appropriate distribution between the serum and colloidal phases for sodium, potassium, chloride, and lactose. This distribution can be accounted for by simple equilibrium exchange. Thus, the bulk of the inorganic ions of milk is most likely in equilibrium between the phases. Calcium, magnesium, phosphate, and to a lesser extent citrate distribute disproportionately between the two phases because of their involvement in colloidal calcium phosphate. Wiechen and Knoop (1978) added ^{45}Ca to milk and found that upon equilibration overnight the radioactive calcium had distributed between the micellar and serum phases in nearly the same percentages found for total calcium (Table 9.8), indicating that the calcium in milk, and probably the phosphate and citrate as well, are in a slower but dynamic equilibrium. At lower temperatures, β-casein readily dissociates from the micelle and enters the serum phase, along with κ-casein and some α_{s1}-casein. These fractions were thought to be in equilibrium because, as the milk warmed, the caseins return to the colloidal phase (whether true microscopic reversibility occurs or not is questionable, as the opportunity for hysteretic effects in the milk system is enormous). Rose (1968), however, interpreted his data based on dilutions of micelles to mean that at room temperature and above, the serum and micellar caseins were not in equilibrium. Creamer and co-workers (1977), however, added ^{14}C-labeled β-casein to milk and found that both in the cold and at room temperature the protein equilibrated proportionately between the micellar and serum caeins. These data, along with the ^{45}Ca data, point to the fact that most components of the casein micelle are in a dynamic state in milk. That is, the micelle is the result of a series of multiple competing equilibria. Under the conditions normally found in milk, the equilibrium lies far to the right with respect to micelle formation. Added calcium tends to decrease serum casein (Rose 1968), but it also reduces heat stability (Fox 1982). Conversely, dilution with or dialysis against milk or milk serum does not appreciably alter micelle stability (Parker and Horne 1980). It has been shown repeatedly in model systems that when the calcium ion content drops below 3 to 5 mM, dissociation of the micelles occurs. Apparently an appropriate Ca:P ratio maintains the system's integrity by keeping the equilibrium toward micelle formation. Those cases in which investigators have proposed that serum and micellar components are not in equilibrium may have occurred because dilutions sufficient to cause dissociation had not been made. Serum proteins, such as β-lactoglobulin, must also be in equilibrium with the ionic environment, and added salts appear to affect their stability (Zittle et al. 1957). If the micelle is sufficiently porous to admit carboxypeptidase, β-lactoglobulin and other serum proteins should equilibriate within the micellar phase as well. The water of the milk system, as well as the lactose, should also be in equilibrium between the dispersed and serum phases. The occluded calcium phosphate, regardless of its structure, undoubtedly affects the hydration and, hence, the heat stability of the micelle system; its components too may be in dynamic equilibrium, as demonstrated by ^{45}Ca experiments noted above. The ability to model the ionic components of the milk serum by computer, and to predict free ion concentrations accurately, indicates the reliability of the equilibrium constants now available for these components. These experiments (Holt et al. 1981) represent a first excellent step toward the ultimate goal of a understanding of the skim milk system as a series of competing multiple equilibria.

Generally speaking, multiple equilibria do exist in skim milk. The innate stability of this biological fluid may depend on the correct balance among these states; however, a good deal of flexibility is apparently inherent in the system. Conditions which perturb the various equilibria affect the milk system through the formation of new steady states with some altered properties. However, environmental influences such as irreversible denaturation, which disrupt these equilibria, tend to cause a greater destabilization of milk and lead to potential problems in fluid milks. Conversely, in the manufacture of dairy products such as cheese or yogurts, these equilibria need to be overcome in order to bring about good yields of these products.

REFERENCES

Aimutis, W. R. and Eigel, W. N. 1982. Identification of λ-casein as plasmin-derived fragments of bovine α_{s1}-casein. J. Dairy Sci. 65, 175–181.

Ali, A. E., Andrews, A. T. and Cheeseman, G. C. 1980. Influence of storage of milk on casein distribution between micellar and soluble phases. *J. Dairy Res. 47*, 371–382.

Andrews, A. T. 1978A. The composition, structure, and origin of proteose-peptone component 5 of bovine milk. *Eur. J. Biochem. 90*, 59–65.

Andrews, A. T. 1978B. The composition, structure, and origin of proteose-peptone component 8F of bovine milk. *Eur. J. Biochem. 90*, 67–71.

Anfinsen, C. B. and Scheraga, H. A. 1975. Experimental and theoretical aspects of protein folding. *Ad. Protein Chem. 29*, 205–300.

Aschaffenburg, R. 1961. Inherited casein variants in cow's milk. *Nature 192*, 431–432.

Ashoor, S. H., Sair, R. A., Olson, N. F. and Richardson, T. 1971. Use of a papain superpolymer to elucidate the structure of bovine casein micelles. *Biochim. Biophys. Acta 229*, 423–430.

Basch, J. J. and Farrell, H. M., Jr. 1979. Charge separation of proteins complexed with sodium dodecyl sulfate by acid gel electrophoresis in the presence of cetyltrimethylammonium bromide. *Biochim. Biophys. Acta 577*, 125–131.

Beeby, R. 1964. The presence of sulfhydryl groups in κ-casein. *Biochim. Biophys. Acta 82*, 418–419.

Berry, G. P. and Creamer, L. K. 1975. Association of bovine β-casein. Importance of the C-terminal region. *Biochemistry 14*, 3542–3545.

Bigelow, C. C. 1967. On the average hydrophobicity of proteins and the relation between it and protein structure. *J. Theoret. Biol. 16*, 187–211.

Bingham, E. W., Farrell, H. M., Jr. and Carroll, R. J. 1972. Properties of dephosphorylated α_{s1}-casein. Precipitation by calcium ions and micelle formation. *Biochemistry 11*, 2450–2454.

Boulet, M., Yang, A. and Riel, R. R. 1970. Examination of the mineral composition of the micelle of milk by gel filtration. *Can. J. Biochem. 48*, 816–822.

Braunitzer, G., Chen, R., Schrank, B. and Stangl, A. 1972. Automatic sequence analysis of a protein (β-lactoglobulin AB). *Hoppe-Seyler's Z. Physiol. Chem. 353*, 832–834.

Brignon, G., Ribadeau-Dumas, B., Mercier, J. C., Pelissier, J. P. and Das, B. C. 1977. Complete amino acid sequence of bovine α_{s2}-casein. *Febs Letters 76*, 274–279.

Brown, E. M. 1984. Interaction of β-lactoglobulin and α-lactalbumin with lipids: A review. *J. Dairy Sci. 67*, 713–722.

Brown, E. M. and Farrell, H. M., Jr. 1978. Interaction of β-lactoglobulin and cytochrome c: Complex formation and iron reduction. *Arch. Biochem. Biophys. 185*, 156–164.

Buchheim, W. and Welsch, U. 1973. Evidence for the submicellar composition of casein micelles on the basis of electron microscopy. *Neth. Milk Dairy J. 27*, 163–180.

Butler, P. J. G. 1977. Tobacco mosaic virus, protein aggregation and virus assembly. *Ad. Protein Chem. 31*, 187–251.

Carroll, R. J. and Farrell, H. M., Jr. 1983. Immunological approach to location of κ-casein in the casein micelle using electron microscopy. *J. Dairy Sci. 66*, 679–686.

Carroll, R. J., Farrell, H. M., Jr. and Thompson, M. P. 1971A. Electron microscopy of the casein micelle—forces contributing to its integrity. *J. Dairy Sci. 54*, 752.

Carroll, R. J., Thompson, M. P. and Farrell, H. M., Jr. 1970. Formation and structure of casein micelles in lactating mammary tissue. *28th Annual EMSA Proc.*, pp. 150–151.

Carroll, R. J., Thompson, M. P. and Melnychyn, P. 1971B. Gelation of concentrated skim milk: Electron microscopic study. *J. Dairy Sci. 54*, 1245–1252.

Carroll, R. J., Thompson, M. P. and Nutting, G. C. 1968. Glutaraldehyde fixation of casein micelles for electron microscopy. *J. Dairy Sci. 51*, 1903–1908.

Cerbulis, J. and Farrell, H. M., Jr. 1975. Composition of milks of dairy cattle. I. Protein, lactose, and fat contents and distribution of protein fraction. *J. Dairy Sci. 58*, 817–827.

Cerbulis, J. and Farrell, H. M., Jr. 1976. Composition of milks of dairy cattle. II. Ash, calcium, magnesium, and phosphorus. *J. Dairy Sci. 59,* 589–593.

Cheeseman, G. C. 1968. A preliminary study by gel filtration and ultra-centrifugation of the interaction of bovine milk caseins with detergents. *J. Dairy Res. 35,* 439–446.

Chen, J. H. and Ledford, R. A. 1971. Purification and characterization of milk protease. *J. Dairy Sci. 54,* 763.

Chen, Y.-H, Yang, J. T. and Chau, K. H. 1974. Determination of the helix and β-form of proteins in aqueous solution by circular dichroism. *Biochemistry 13,* 3350–3359.

Contaxis, C. C. and Bigelow, C. C. 1981. Free energy changes in α-lactalbumin denaturation. *Biochemistry 20,* 1618–1622.

Creamer, L. K., Berry, G. P. and Mills, O. E. 1977. A study of the dissociation of β-casein from the bovine casein micelle at low temperature. *N.Z. J. Dairy Sci. Technol. 12,* 58–66.

Creamer, L. K., Richrdson, T. and Parry, D. A. D. 1981. Secondary structure of bovine α_{s1}-and β-casein in solution. *Arch. Biochem. Biophys. 211,* 689–696.

Davies, D. T. and Law, A. J. R. 1980. Content and composition of protein in creamery milks in South-West Scotland. *J. Dairy Res. 47,* 83–90.

Davies, D. T. and Law, A. J. R. 1983. Variation of the protein composition of bovine casein micelles and serum casein in relation to micellar size and milk temperature. *J. Dairy Res. 50,* 67–75.

Davies, D. T. and White, J. C. D. 1960. The use of ultrafiltration and dialysis in isolating the aqueous phase of milk and in determining the partition of milk constituents between the aqueous and disperse phases. *J. Dairy Res. 27,* 171–190.

Delfour, A., Jollès, J., Alais, C. and Jollès, P. 1965. Caseino-glycopeptides: Characterization of a methionine residue at the N-terminal sequencve. *Biochem. Biophys. Res. Commun. 19,* 452–455.

DeMott, B. J. 1968. Ionic calcium in milk and whey. *J. Dairy Sci. 51,* 1008–1012.

Dewan, R. K., Bloomfield, V. A., Chudgar, A. and Morr, C. V. 1973. Voluminosity and viscosity of bovine milk casein micelles. *J. Dairy Sci. 56,* 699–705.

DeWit, J. N. and Swinkels, G. A. M. 1980. A differential scanning calorimetric study of the thermal denaturation of bovine β-lactoglobulin. *Biochim. Biophys. Acta 624,* 40–50.

Dickson, I. R. and Perkins, D. J. 1971. Studies on interactions between purified bovine casein and alkali earth metal ions. *Biochem. J. 124,* 235–240.

Douglas, F. W., Jr., Greenberg, R., Farrell, H. M., Jr. and Edmondson, L. F. 1981. Effects of ultra-high-temperature pasteurization on milk proteins. *J. Agr. Food Chem. 29,* 11–15.

Downey, W. K. and Murphy, R. F. 1970. The temperature dependent dissociation of β-casein from bovine casein micelles. *J. Dairy Res. 37,* 361–372.

Dunnill, P. and Green, D. W. 1965. Sulfhydryl groups and the N ⇆ R conformational change in β-lactoglobulin. *J. Mol. Biol. 15,* 147–151.

Ebner, K. E. and Schanbacher, F. 1974. Biochemistry of lactose and related carbohydrates. *In: Lactation: A Comprehensive Treatise,* Vol. II. B. L. Larson and V.R. Smith (Editors). Academic Press, New York, pp. 77–113.

Eigel, W. N., Butler, J. E., Ernstrom, C. A., Farrell, H. M., Jr., Harwalkar, V. R., Jenness, R. and Whitney, R. McL. 1984. Nomenclature of the proteins of cow's milk: Fifth revision. *J. Dairy Sci. 67,* 1599–1631.

Eigel, W. N., Hoffmann, C. J., Chibber, B. A. K., Tomich, J. M., Keenan, T. and Mertz, E. T. 1979. Plasmin mediated proteolysis of casein in bovine milk. *Proc. Natl. Acad. Sci. USA 76,* 2244–2248.

Euber, J. R. and Brunner, J. R. 1982. Interaction of κ-casein with immoblized β-lactoglobulin. *J. Dairy Sci. 65,* 2384–2387.

Farrell, H. M., Jr. 1973. Models for casein micelle formation. *J. Dairy Sci. 56*, 1195–1206.

Farrell, H. M., Jr. and Douglas, F. W., Jr. 1983. Effects of UHT pasteurization on the functional and nutritional properties of the milk proteins. Symposium on Role of Milk Proteins in Human Nutrition. *Kieler Milch. Forsch. 35*, 239–464.

Farrell, H. M., Jr. and Thompson, M. P. 1971. Biological significance of milk protein polymorphism. *J. Dairy Sci. 54*, 1219–1228.

Farrell, H. M., Jr. and Thompson, M. P. 1974. Physical equilibria: Proteins. *In: Fundamentals of Dairy Chemistry*, 2nd ed. B. Webb, A. H. Johnson and J. A. Alford (Editors). AVI Publishing Co., Westport, Conn., pp. 442–471.

Fox, P. F. 1982. Heat induced coagulation of milk. *In: Developments in Dairy Chemistry*, Vol. 1. P.F. Fox (Editor). Applied Science Publishers, London, pp. 189–223.

Fugate, R. D. and Song, P. S. 1980. Spectroscopic characterization of β-lactoglobulin-relinol complex. *Biochim. Biophys. Acta 625*, 28–42.

Futterman, S. and Heller, J. 1972. Enhancement of fluorescence and the decreased susceptibility to enzymatic oxidation of retinol complexed with bovine serum albumin, β-lactoglobulin and retinol binding protein of human plasma. *J. Biol. Chem. 247*, 5168–5172.

Garnier, J. and Ribadeau-Dumas, B. 1970. Structure of the casein micelle: A proposed model. *J. Dairy Res. 37*, 493–504.

Gordon, W. G., Groves, M. L., Greenberg, R., Jones, S. B., Kalan, E. B., Peterson, R. F. and Townend, R. E. 1972. Probable identification of γ-, TS-, R-, and S-caseins as fragments of β-casein. *J. Dairy Sci. 55*, 261–263.

Green, D. W., Aschaffenburg, R., Camerman, A., Coppola, J. C., Dunhill, P., Simmons, R. M. Komorowski, E. S., Sawyer, L., Turner, E. M. and Woods, K. F. 1979. Structure of bovine β-lactoglobulin at 6A resolution. *J. Mol. Biol. 131*, 375–397.

Green, M. L. 1972. On the mechanism of milk clotting by rennin. *J. Dairy Res. 39*, 55–63.

Greene, R. F., Jr. and Pace, C. N. 1974. Urea and guanidine hydrochloride denaturation of ribonuclease, lysozyme, α-chymotrypsin and β-lactoglobulin. *J. Biol. Chem. 249*. 5388–5393.

Harwalkar, V. R. 1980A. Measurements of thermal denaturation of β-lactoglobulin at pH 2.5. *J. Dairy Sci. 63*, 1043–1051.

Harwalkar, V. R. 1980B. Kinetics of thermal denaturation of β-lactoglobulin at pH 2.5. *J. Dairy Sci. 63*, 1052–1057.

Herskovits, T. T. 1966. On the conformation of caseins. Optical rotatory properties. *Biochemistry 5*, 1018–1026.

Hill, R. D. 1970. Effect of modification of arginine side chains on the coagulation of rennin-altered casein. *J. Dairy Res. 37*, 187–192.

Hill, R. J. and Wake, R. G. 1969. Amphiphile nature of κ-casein as the basis for its micelle stabilizing property. *Nature 221*, 635–639.

Hiraoka, Y., Segawa, T., Kuwajima, K., Sugai, S. and Murai, N. 1980. α-lactalbumin: A calcium metalloprotein. *Biochem. Biophys. Res. Commun. 95*, 1098–1104.

Holt, C. 1975. Casein micelle size from elastic and quasi-elastic light scattering measurements. *Biochim. Biophys. Acta 400*, 293–301.

Holt, C., Dalgleish, D. G. and Jenness, R. 1981. Calculation of the ion equilibria in milk diffusate and comparison with experiment. *Anal. Biochem. 113*, 154–163.

Holt, C., Hasnain, S. S. and Hukins, D. W. L. 1982. Structure of bovine milk calcium phosphate determined by X-ray absorption spectroscopy. *Biochim. Biophys. Acta 719*, 299–303.

Holt, C. and Muir, D. D. 1978. Natural variation in the size of bovine casein micelles. *J. Dairy Res. 45*, 347–353.

Hunziker, H. G. and Tarassuk, N. P. 1965. Chromatographic evidence for heat-induced interaction of α-lactalbumin and β-lactoglobulin. *J. Dairy Sci. 48*, 733–734.

Janolino, V. G. and Swaisgood, H. E. 1975. Isolation and characterization of sulfhydryl oxidase from bovine milk. *J. Biol. Chem. 250*, 2532–2538.

Jenness, R. and Koops, J. 1962. Preparation and properties of a salt solution which simulates milk ultrafiltrate. *Neth. Milk Dairy J. 16*, 153–164.

Jollès, P., Alais, C. and Jollès, J. 1962. Amino acid composition of κ-casein and terminal amino acids of κ- and para κ-casein. *Arch. Biochem. Biophys. 98*, 56–57.

Jones, M. N. and Wilkinson, A. 1976. The interaction between β-lactoglobulin and sodium dodecylsulfate. *Biochem. J. 153*, 713–718.

Kalan, E. B. and Woychik, J. H. 1965. Action of rennin on κ-casein, the amino acid composition of para κ-casein. *J. Dairy Sci. 48*, 1423–1428.

Kauzmann, W. 1959. Some factors in the interpretation of protein denaturation. *Adv. Protein Chem. 14*, 1–63.

Klotz, I. M. 1970. Comparison of molecular structure of proteins: α-Helicies and apolar residues. *Arch. Biochem. Biophys. 138*, 704–706.

Knoop, A.-M., Knoop, E. and Wiechen, A. 1979. Sub-structure of synthetic casein micelles. *J. Dairy Res. 46*, 347–350.

Kresheck, G. C., Hargraves, W. A. and Mann, D. C. 1977. Thermometric titration studies of ligand binding to macromolecules: sodium dodecyl sulfate to β-lactoglobulin. *J. Phys. Chem. 81*, 532–537.

Kronman, M. J., Andreotti, R. E. and Vitols, R. 1964. Inter- and intramolecular interactions of α-lactalbumin. II. Aggregation reactions at acid pH. *Biochemistry 3*, 1152–1160.

Kronman, M. J., Sinha, S. K. and Brew, K. 1981. Characteristics of the binding of Ca^{2+} and other metal ions to bovine α-lactalbumin. *J. Biol. Chem. 256*, 8582–8587.

Kumosinski, T. F. and Pessen, H. 1982. A deuteron and proton magnetic resonance study of β-lactoglobulin A association: Sonac approaches to the scatchard hydration of globular proteins. *Arch. Biochem. Biophys. 218*, 286–302.

Kuwajima, K. 1977. A folding model of α-lactalbumin deduced from the three-state denaturation mechanism. *J. Mol. Biol. 114*, 241–258.

Lapanje, S. and Kranjc, Z. 1982. Interaction of β-lactoglobulin with some alkyl areas. *Biochim. Biophys. Acta 705*, 111–116.

Lin, S. H. C., Dewan, R. K., Bloomfield, V. A. and Morr, C. V. 1971. Inelastic light scattering study of bovine casein micelles. *Biochemistry 10*, 4788–4793.

Linderstrøm-Lang, K. 1929. Casein. III. The fractionation of casein. *Compt. Rend. Trav. lab. Carlsberg 17*(9), 1–116.

Loucheux-Lefebvre, M.-H. Aubert, J.-P. and Jollès, P. 1978. Prediction of the conformation of cow and sheep κ-casein. *Biophys. J. 23*, 323–336.

McGann, T. C. A., Donnelly, W. J., Kearney, R. D. and Buchheim, W. 1980. Composition and size distribution of casein micelles. *Biochim. Biophys. Acta 630*, 261–270.

McGann, T. C. A. and Pyne, G. T. 1960. The colloidal phosphate of milk. III. Nature of its association with casein. *J. Dairy Res. 27*, 403–417.

McKenzie, H. A. 1971. β-Lactoglobulin. *In: Milk Proteins*, Vol 2. H.A. McKenzie (Editor). Academic Press, New York, p. 257.

Mellander, O. 1939. Electrophoresis of casein. *Biochem. Z. 300*, 240–245.

Merciér, J. C., Brignon, G. and Ribadeau-Dumas, B. 1973. Primary structure of κ-casein B bovine. Complete sequence. *Eur. J. Biochem. 35*, 222–235.

Merciér, J. C., Grosclaude, F. and Ribadeau-Dumas, B. 1971. Primary structure of α_{s1}-casein bovine. Complete sequence. *Eur. J. Biochem. 23*, 41–51.

Mills, O. E. 1976. Effect of temperature on tryptophan fluorescence of β-lactoglobulin B. *Biochim. Biophys. Acta 434*, 324–332.

Morr, C. V. 1967. Effect of oxalate and urea on the ultracentrifugation properties of raw and heated skim milk casein micelles. *J. Dairy Sci. 50*, 1744–1751.

Murakami, K., Andree, J. P. and Berliner, L. J. 1982. Metal ion binding to α-lactalbumin species. *Biochemistry 21*, 5488-5494.

Noelken, M. 1967. Molecular weight of α_{s1}-casein B. *Biochim. Biophys. Acta 140*, 537-539.

Noelken, M. and Reibstein, M. 1968. The conformation of β-casein B. *Arch. Biochem. Biophys. 123*, 397-402.

Pace, C. N. and Marshall, H. F., Jr. 1980. A comparison of protein denaturants of β-lactoglobulin and ribonuclease. *Arch. Biochem. Biophys. 199*, 270-276.

Papiz, M. Z., Sawyer, L., Eiopoulos, E. E., North, A. C. T., Findlay, J. B. C., Sivaprasadarao, R., Jones, T. A., Newcomer, M. E., and Kraulis, P. J. 1986. The structure of β-lactoglobulin and its similarity to plasma retinol binding protein. *Nature 324*, 383-385.

Parker, T. G. and Horne, D. S. 1980. Light scattering investigation of the stability of bovine casein micelles to dilution. *J. Dairy Res. 47*, 343-350.

Parry, R. M., Jr. and Carroll, R. J. 1969. Lactation of κ-casein in milk micelles. *Biochim. Biophys. Acta 194*, 138-150.

Payens, T. A. J. 1979. Casein micelles: the colloidal-chemical approach. *J. Dairy Res. 46*, 291-306.

Pepper, L. 1972. Casein interactions as studied by gel chromatography and ultracentrifugation. *Biochim. Biophys. Acta 278*, 147-154.

Pepper, L. and Farrell, H. M., Jr. 1982. Interactions leading to formation of casein submicelles, *J. Dairy Sci. 65*, 2259-2266.

Pepper, L., Hipp, N.J. and Gordon, W. G. 1970. Effects of modification of ε-amino groups on interactions of κ-and α_{s1}-caseins. *Biochim. Biophys. Acta 207*, 340-346.

Permyakov, E. A., Yarmolenko, V. V., Kalinichenko, L. P., Morozova, L. A. and Burstein, E. A. 1981. Calcium binding to α-lactalbumin: Structural rearrangements and association constant. *Biochem. Biophys. Res. Commun. 100*, 191-197.

Perutz, M. F. 1978. Electrostatic effects in proteins. *Science 201*, 1187-1191.

Posner, A. S., Betts, F., Blumenthal, N. C., McGann, T. C. A., Kearney, R. D. and Buchheim, W. 1983. Amorphous calcium phosphate in biological systems: Milk micelles and mitochondria. XVIIth European Symposium on Calcified Tissue, Pavos, Switzerland, April 11-14.

Privalov, P. L. 1979. Stability of proteins. Small globular proteins. *Ad. Protein Chem. 33*, 167-241.

Privalov. P. L. 1982. Stability of proteins. Proteins which do not present a single cooperative system. *Ad. Protein Chem. 35*, 1-104.

Pyne, G. T. and McGann, T. C. A. 1960. The colloidal phosphate of milk. II. Influence of citrate. *J. Dairy Res. 27*, 9-17.

Raap, J., Kerling, K. E. T., Vreeman, H. J. and Visser, S. 1983. Peptide substrates for chymosin: Conformational studies of κ-casein and related peptides by circular dichroism and structure prediction. *Arch. Biochem. Biophys. 221*, 117-124.

Ribadeau-Dumas, B., Brignon, G., Grosclaude, F. and Merciér, J. C. 1972. Primary structure of β-casein bovine. Complete sequence. *Eur. J. Biochem. 25*, 505-514.

Ribadeau-Dumas, B. and Garnier J. 1970. Structure of casein micelles. Accessibiity of subunits to various reagents. *J. Dairy Res. 37*, 269-278.

Robillard, K. A., Jr. and Wishnia, A. 1972. Aromatic hydrophobes and β-lactoglobulin A. Thermodynmics of binding. *Biochemistry 11*, 3835-3840.

Rose, D. 1968. Relation between micellar and serum casein in bovine milk. *J. Dairy Sci. 51*, 1897-1902.

Rose, D. 1969. A proposed model of micelle structure in bovine milk. Review Article No. 150. *Dairy Sci. Abstr. 31*, 171-175.

Schmidt, D. G. 1970. The association of α_{s1}-casein B at pH 6.6. *Biochim. Biophys. Acta 207*, 130-138.

Schmidt, D. G. 1982. Association of caseins and casein micelle structure. *In: Developments in Dairy Chemistry*, Vol. 1. P.F. Fox (Editor). Applied Science Pub., Essex, England, pp. 61–82.

Schmidt, D. G., Both, P. and Koops, J. 1979. Properties of artifical casein micelles. 3. Relationship between salt composition, size, and stability towards ethanol, dialysis, and heat. *Neth. Milk Dairy J. 33*, 40–48.

Schmidt, D. G. and Buchheim, W. 1970. Electron microscopy of the fine structure of the casein micelles of cows milk. *Milchwissenshaft 25*, 596–600.

Schmidt, D. G. and Koops, J. 1977. Properties of artificial casein micelles. 2. Stability towards ethanol, dialysis, pressure, and heat in relation to casein composition. *Neth. Milk Dairy J. 31*, 342–357.

Schmidt, D. G., Koops, J. and Westerbeek, D. 1977. Properties of artificial casein micelles. 1. Preparation, size distribution, and composition. *Neth. Milk Dairy J. 31*, 328–341.

Schmidt, D. G. and Payens, T. A. J. 1976. *Colloidal Aspects of Casein in Surface and Colloid Science*, Vol. 9. E. Matijevic (Editor). John Wiley & Sons. New York, pp. 165–229.

Schoppet, E. F., Sinnamon, H. I., Talley, F. B., Panzer, C. C. and Aceto, N. C. 1976. Enrichment of pasta with cottage cheese whey proteins. *J. Food Sci. 41*, 1297–1300.

Seibles, T. S. 1969. Interaction of dodecylsulfate with native and modified β-lactoglobulin. *Biochemistry 8*, 2949–2953.

Shanani, K. M., Harper, W. J., Jensen, R. G., Parry, R. M. and Zittle, C. A. 1973. Enzymes in bovine milk: A review. *J. Dairy Sci. 56*, 531–543.

Shimmin, P. D. and Hill, R. D. 1964. An electron microscope study of the internal structure of casein micelles. *J. Dairy Res. 31*, 121–123.

Slattery, C. W. 1978. Variation in the glycosylation pattern of bovine κ-casein with micelle size and its relationship to a micelle model. *Biochemistry 17*, 1100–1104.

Slattery, C. W. 1979. A phosphate-induced submicelle-micelle equilibrium in reconstituted casein micelle systems. *J. Dairy Res. 46*, 253–258.

Slattery, C. W. and Evard, R. 1973. A model for the formation and structure of casein micelles from subunits of variable composition. *Biochim. Biophys. Acta 317*, 529–538.

Smits, P. and Van Brouwershaven, J.H. 1980. Heat induced association of β-lactoglobulin and casein micelles. *J. Dairy Res. 47*, 313–325.

Snöeren, T. H. M., Van Markwijk, B. and Van Montfort, R. 1980. Some physicochemical properties of bovine α_{s2}-casein. *Biochim. Biophys. Acta 622*, 268–276.

Span, J. and Lapanje, S. 1973. Solvation of β-lactoglobulin and chymotrypsinogen A in aqueous urea solutions. *Biochim. Biophys. Acta 295*, 371–378.

Spector, A. A. and Fletcher, J. E. 1970. Binding of long chain fatty acids to β-lactoglobulin. *Lipids 5*, 403–411.

Stuart, D. I., Acharya, K. R., Walker, N. P. C., Smith, S. G., Lewis, M. and Phillips, D. C. 1986. Lactalbumin posses a novel calcium binding loop. *Nature 324*, 84–87.

Su, Y. T. and Jirgensons, A. 1977. Further studies on detergent-induced conformational changes in proteins. Circular dichroism of ovalbumin, bacterial α-amylase, papain, and β-lactoglobulin. *Arch. Biochem. Biophys. 181*, 137–146.

Sullivan, R. A., Fitzpatrick, M. M. and Stanton, E. K. 1959. Distribution of κ-casein in skim milk. *Nature 183*, 616–617.

Swaisgood, H. E. and Brunner, J. R. 1962. Characterization of κ-casein. *J. Dairy Sci. 45*, 1–11.

Swaisgood, H. E. and Brunner, J. R. 1963. Characteristics of κ-casein in the presence of various dissociating reagents. *Biochem. Biophys. Res. Commun. 12*, 148–151.

Swaisgood, H. E., Brunner, J. R. and Lillevik, H. A. 1964. Physcial parameters of κ-casein from cow's milk. *Biochemistry 3*, 1616-1623.

Swaisgood, H. E. and Timasheff, S. N. 1968. Association of α_{s1}-casein C in the alkaline pH range. *Arch. Biochem. Biophys. 125*, 344-361.

Tanford, C. 1961. In: *Physcial Chemistry of Macromolecules*, John Wiley & Sons, New York, p. 131.

Tanford, C. 1970. Protein denaturation, part C. *Adv. Protein Chem. 24*, 1-95.

Tanford, C. 1978. The hydrophobic effect and the organization of living matter. *Science 200*, 1012-1018.

Tanford, C., Bunville, L. G. and Nozaki, Y. 1959. The reversible transformation of β-lactoglobulin at pH 7.5. *J. Am. Chem. Soc. 81*, 4032-4036.

Termine, J. D., Peckauskas, R. A. and Posner, A. S. 1970. Calcium phosphate formation *in vitro*. II. Effects of environment on amorphous-crystalline transformation. *Arch. Biochem. Biophys. 140*, 318-325.

Termaine, J. D. and Posner, A. S. 1970. Calcium phosphate formation *in vitro*. I. Factors affecting initial phase separation. *Arch Biochem. Biophys. 140*, 307-317.

Thompson, M. P. 1970. Phenotyping milk proteins: A review. *J. Dairy Sci. 53*, 1341-1348.

Thompson, M. P., Boswell, R. T., Jenness, R. and Kiddy, C.A., Jr. 1969B. Casein-pellet-solvation and heat stability of individual cow's milk. *J. Dairy Sci. 52*, 796-798.

Thompson, M. P., Farrell, H. M., Jr. and Greenberg, R. 1969A. α_{s1}-Casein A (*Bos taurus*): A probable sequential deletion of eight amino acid residues and its effect on physical properties. *Comp. Biochem. Physiol. 28*, 471-475.

Thompson, M. P., Gordon, W. G., Boswell, R. T. and Farrell, H. M., Jr. 1969C. Solubility, solvation, and stabilization of α_{s1}- and β-caseins. *J. Dairy Sci. 52*, 1166-1173.

Thompson, M. P., Kiddy, C. A., Pepper, L. and Zittle, C. A. 1962. Variations in the α_s-casein fraction of individual cow's milk. *Nature 195*, 1001-1002.

Tilley, J. M. A. 1960. Chemical and physical properties of bovine β-lactoglobulin. Review Article No. 86. *Dairy Sci. Abstr. 22*, 111-125.

Timasheff, S. N. and Gorbunoff, M. J. 1967. Conformation of proteins. *Ann. Rev. Biochem. 36*, 13-54.

Timasheff, S. N. and Townend, R. 1969. β-Lactoglobulin as a model of subunit enzymes In: *Protides of the Biological Fluids: Proceedings of the 16th Colloquium, Bruges, 1968*. H. Peeters (Editor). Pergamon Press, New York, pp. 33-40.

Townend, R., Herskovits, T. T., Timasheff, S. N. and Gorbunoff, M. J. 1969. State of amino acid residues in β-lactoglobulin. *Arch. Biochem. Biophys. 129*, 567-580.

Townend, R., Kumosinski, T. F. and Timasheff, S. N. 1967. Circular dichroism of variants of β-lactoglobulin. *J. Biol. Chem. 242*, 4538-4545.

Vanaman, T. C., Brew, K. and Hill, R. L. 1970. The disulfide bonds of α-lactalbumin. *J. Biol. Chem. 245*, 4583-4590.

Van Slyke, L. L. and Bosworth, A. W. 1915. Condition of casein and salts in milk. *J. Biol. Chem. 20*, 135-152.

Visser, J., Schaier, R. W. and Van Gorkom, M. 1979. Role of calcium, phosphate and citrate ions in the stabilization of casein micelles. *J. Dairy Res. 46*, 333-335.

Vreeman, H. J., Brinkhuis, J. A. and Van Der Spek, C. A. 1981. Some associaton properties of bovine SH-κ-casein. *Biophys. Chem. 14*, 184-193.

Waissbluth, M. D. and Grieger, R. A. 1974. Alkaline denaturation of β-lactoglobulins. Activation parameters and effect of dye binding site. *Biochemistry 13*, 1285-1288.

Walstra, P. 1979. The voluminosity of bovine casein micelles and some of its implications. *J. Dairy Res. 46*, 317-323.

Warme, P. K., Momany, F. A., Rumball, S. V., Tuttle, R. W. and Scheraga, H. A. 1974.

Computation of structures of homologues proteins: α-Lactalbumin from lysozyme. *Biochemistry 13,* 768–782.

Waugh, D. F. 1971. Formation and structure of casein micelles. *In: Milk Proteins,* Vol. II. H.A. McKenzie (Editor). Academic Press, New York, p. 3.

Waugh, D. F., Creamer, L. K., Slattery, C. W. and Dresdner, G. W. 1970. Core polymers of casein micelles. *Biochemistry 9,* 786–795.

Waugh, D. F., Slattery, C. W. and Creamer, L. K. 1971. Binding of cations to caseins: Site binding, donanan binding, and system characteristics. *Biochemistry 10,* 817–823.

Waugh, D. F. and Von Hippel, P. H. 1956, κ-Casein and the stabilization of casein micelles. *J. Am. Chem. Soc. 78,* 4576–4582.

White, J. C. D. and Davies, D. T. 1958. The relation between the chemical composition of milk and the stability of the caseinate complex. I. General introduction, description of samples, methods and chemical composition of samples. *J. Dairy Res. 25,* 236–255.

Whitney, R. McL. Brunner, J. R., Ebner, K. E., Farrell, H. M., Jr., Josephson, R. V., Morr, C. V. and Swaisgood, H. E. 1976. Nomenclature of the proteins of cow's milk: Fourth revision. *J. Dairy Sci. 59,* 795–815.

Wiechen, V. A. and Knoop, A. M. 1978. Investigations on Ca distribution between serum and casein by means of the radioisotope Ca^{45} in low cooled and pasteurized milks. *Milchwissenschaft 33,* 213–215.

Woychik, J. H., Kalan, E. B. and Noelken, M. E. 1966. Chromatographic isolation and partial characterization of reduced κ-casein components. *Biochemistry 5,* 2276–2288.

Zittle, C. A., Dellamonica, E. S., Rudd, R. K. and Custer, J. H. 1957. The binding of calcium ions by β-lactoglobulin both before and after aggregation by heating in the presence of calcium ions. *J. Am. Chem. Soc. 79,* 4661–4666.

10

Physical Equilibria: Lipid Phase

Thomas W. Keenan, Ian H. Mather, and Daniel P. Dylewski

Milk contains a complex mixture of lipids in terms of fatty acid composition and in the distribution of these acids in neutral lipids and phosphoglycerides. In addition, there are major variations among species in both the amount and fatty acid composition of milk lipids. In terms of amount, certain seals and whales produce milk which is over 50% lipid by weight, while the milk of certain rhinoceri contains less than 0.1% lipid (Jenness 1974). As discussed by Jenness in this volume (Chapter 1), relative to the number of mammalian species extant, we have but a rudimentary knowledge of the composition of milk. Most present-day knowledge of the organization of milk lipids has come from studies of milk from cows, although in recent years there has been increased interest in extending these studies to other species, particularly *Homo sapiens* (Jensen *et al.* 1980; Blanc 1981). Most of the discussion in this chapter will deal with bovine milk lipids. Where instructive, comparisons with other species will be made.

Lipids in cow's milk occur in two physically separable forms: those which float on centrifugation of milk and those which sediment in a centrifugal field. The floating lipid fraction is composed of spherical globules, while the sedimentable lipids are associated primarily with membrane fragments, cells, and cell fragments (Patton and Jensen 1976). More than 96% of the total milk lipids are recovered in the globule fraction, and these globules are composed primarily of triglycerides (98 to 99%). Globules also contain phospholipids, glycolipids, free and esterified cholesterol, partial glycerides (di- and monoglycerides), free fatty acids, hydrocarbons, and lipid-soluble vitamins (Huang and Kuksis 1967; Jenness 1974; Patton and Jensen 1976; Blanc 1981). With other species, the bulk of the milk lipids has also been found to consist of triglycerides, but one must add that milk from remarkably few species has been analyzed for the presence of globules. To our knowledge, there is no report of a lack of lipid globules in milk from any species, but it may be instructive to perform such studies with milks which contain only traces of lipids. The composition of the lipid fraction from milk serum (the terms "serum" and "milk serum" will be used throughout to refer to that phase of milk prepared by centrifugal removal of lipid globules) has been determined for very few species. Huang and Kuksis (1967) found that cow's milk serum lipids were composed principally of phospholipids (30 to 45%) and triglycerides (40 to 55%), with smaller amounts of partial glycerides, free fatty acids, and cholesterol. Similar results were obtained by others for both cow and goat milks (Patton and Keenan 1971; Plantz et al. 1973; Kitchen 1974).

MILK LIPID GLOBULES

Historical Background

Scientific inquiry into the nature of the lipids in milk has persisted over the past 300 years. The first recorded observation of lipid globules in milk appears to have been a paper published by van Leeuwenhoek in 1674 (Brunner 1974). Over 160 years later, Ascherson recognized that a substance must surround fat globules which would stabilize the lipid emulsion in the aqueous phase of milk; he believed this substance to be a milk protein which condensed and aggregated at the lipid droplet surface. In 1897 Storch detected an envelope material which was stained by ammoniacal picrocarmine on the surfaces of globules in cream washed to remove extraneous proteins (Brunner 1969). For a period of about 50 years following Storch's publication, progress was made in establishing the composition of the material surrounding milk lipid globules. However, the fact that this material is a biological membrane originating from milk-secreting mammary epithelial cells has been recognized only in the past 25 years. These developments will be discussed in a subsequent section of this Chapter.

Globule Size Distribution

Over the past 100 years or more, numerous investigators have studied the size distribution of lipid globules in milk. Brunner (1974) presented a thorough review of this subject, and Mulder and Walstra (1974) also reviewed studies on globule size distribution. Since these reviews, there have been few new studies in this area, and only a summary of the available information will be given here. Early investigators of globule size distribution were limited in techniques to use of the light microscope. However, when this instrument was used to enumerate numbers of globules within various size ranges, much was learned about globule volume and size distribution, and it was also recognized that the average globule diameter varied both with breed of cow and with stage of lactation. As other methods for determination of size distribution became available, it was found that very small lipid globules had not been accurately enumerated in many of the earlier microscopic studies. Walstra and his colleagues (Walstra 1968; Walstra and Oortwijn 1969; Walstra *et al.* 1969) evaluated the techniques of fluorescence microscopy, spectroturbidimetry, and conductimetry (the last using a commercially available electronic particle counter known as a "Coulter counter") for determination of size distribution and concluded that use of all three methods was necessary for accurate determinations. Using these methods, they made an extensive study of globule size distribution in the milk of (mainly) Holstein cows (Walstra *et al.* 1969; Walstra 1969A,B); a summary of these observations is given below.

Lipid globules in milk range in diameter from less than 0.2 μm to 20 μm or more. Small globules (below 1 μm) are most numerous, comprising about 80% of the total number of globules, but they account for only a small percentage of the total lipid. Intermediate globules, with diameters ranging from about 1 to 8 μm, comprise 90% or more of the total lipid. Larger globules are few in number but, because of their volume, account for 1% or more of the total lipid (Figure 10.1).

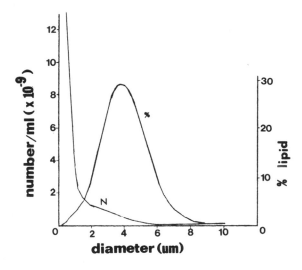

Figure 10.1 Size distribution of lipid globules in milk of a Holstein cow. The number of globules (N) of various diameters and the percentage of the total lipid present in globules of indicated diameters are plotted. (Redrawn from Mulder and Walstra 1974, p. 55, with permission of PUDOC, Centre for Agricultural Publishing and Documentation.)

Walstra (1969A) found average diameters of 3.4 and 4.5 μm for globules in milks from Holstein and Jersey cows, respectively. Variations observed in globule size, and in lipid content within globule size classes, for the common dairy breeds are given in Table 10.1. In addition to variation in size distribution between different breeds, there are differences in size distribution profiles of globules within breeds and in individual animals at different stages of lactation. As lactation progresses, there is a decrease in the average diameter of milk lipid globules, as illustrated for Guernsey and Holstein cows in Figure 10.2. Several factors such as interval between milkings, feeding regimen, and whether or not the sample is of fore or hind milk have been reported to influence size distribution; however, studies purporting to show this are inconclusive.

There are quantitative data allowing comparison of size distribution profiles of milk lipid globules for only a few species. Rüegg and Blanc (1981), using a conductimetric method, found that the ranges of globule sizes and average globule diameter in human milk are similar to those of cow's milk. They observed that globules less than 1 μm in diameter accounted for 70 to 90% of the total number of globules. Limited studies with goat, ewe, sow, and buffalo suggest that ranges of globule size are similar to those of cows and humans (Whittlestone 1952; Fahmi et al. 1956; Puri et al. 1961; Kuzdzal-Savoie 1979). With other species there is a lack of quantitative data, although published electron micrographs suggest that milk lipid globules of the about 20 species examined fall within the size range of cow's milk lipid globules.

Table 10.1. Distribution of the Fat Phase in the Milk of Four Bovine Breeds According to the Size of Fat Globules.

Breed	Average Size of Fat Globules in Distribution Classes (μm)					
	0–2.4	2.4–4.8	4.8–7.2	7.2–9.6	9.6–12.0	12–14.4
	Percentage of Fat Globules in Each Group[a]					
Jersey	8.1	38.3	32.1	18.1	5.3	1.1
Guernsey	6.5	38.9	35.0	14.4	4.4	0.7
Ayrshire	14.6	54.0	23.4	6.2	1.6	0.2
Holstein	14.5	54.6	24.5	5.1	1.1	0.2
	Percentage of Total Fat in Each Group					
Jersey	0.1	11.3	26.1	30.7	23.9	7.9
Guernsey	0.1	11.3	33.2	29.7	25.7	—
Ayrshire	0.3	34.0	41.6	17.8	6.3	—
Holstein	0.3	38.3	50.1	11.3	—	—

[a]Approximately 1000 globules counted.
SOURCE: Reproduced from Brunner (1974, p. 480).

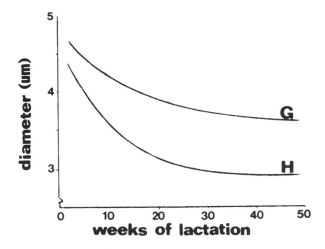

Figure 10.2 Stage of lactation and breed differences in size distribution of milk lipid globules. The average globule diameter (in micrometers) versus the stage of lactation is shown for Guernsey (G) and Holstein (H) cows. (Redrawn from Mulder and Walstra 1974, p. 59, with permission of PUDOC, Centre for Agricultural Publishing and Documentation.)

Recently, Tedman (1983) has studied the ultrastructure of lactating mammary tissue from the Weddell seal, a cold-water pinniped which produces milk of over 40% lipid (Jenness 1974). Lipid globules in this species fall within the size range of those of cow's milk, although in the seal there appears to be a higher proportion of large lipid globules and few small globules.

Since lipid globules do not appear to coalesce in milk, their ultimate size must be determined within the epithelial cells from which they originate. We still know little about the cellular mechanisms involved in determining or controlling lipid globule size. The origin, growth, and secretion of lipid globules from cells will be considered in a later section of this chapter. It may be that control of globule size is one mechanism by which milk lipid content is regulated. With quantitative data for a range of species, it may be possible to formulate working hypotheses regarding the control of globule size distribution.

Composition of the Globule Core

As mentioned previously, more than 95% of the total milk lipid is found in the globule fraction, and the globule is composed largely of glycerides (98 to 99%). Of the total lipids in the globule, it can be calculated that, on the average, nearly 99% are in the globule core and about 1% are associated with the surface membrane (Huang and Kuksis 1967; Morrison 1970; Mulder and Walstra 1974; Patton and Jensen 1976). Lipids in the globule core are largely triglycerides and diglycerides; triglycerides predominate in all cases, but the triglyceride to diglyceride ratio varies (Mulder and Walstra 1974). Whether variation in diglyceride content is either caused by degradation of triglycerides by, e.g., lipoprotein lipase (Hohe et al. 1985) or is due to variation in the completion of triglyceride synthesis from animal to animal remains undetermined (Patton and McCarthy 1963).

Very small amounts of sterols and sterol esters, phospholipids, certain glycolipids, and fat-soluble vitamins have been detected in the core lipid fraction of milk lipid globules prepared by removal of the membrane (Huang and Kuksis 1967; Keenan et al. 1972D; Keenan 1974A; Patton and Jensen 1976). Whether or not these constituents are indigenous to the core fraction or originate in the membrane and simply partition into the core lipid is problematic. That phospholipids reported in globule cores may originate, in part, from membrane is suggested by the work of Huang and Kuksis (1967), who prepared core lipid which was devoid of phospholipid. However both sterols and gangliosides can still be detected in core lipid fractions in the absence of phospholipid (Huang and Kuksis 1967; Keenan 1974A). Other work indicates that during the intracellular formation of globules, phospholipids, gangliosides, and cholesterol are present in or on the globule surface and associate with the membrane which subsequently envelops these globules. Lipid droplets isolated from homogenates of mammary tissue which was minced and washed to remove entrained milk were found to contain phospholipids, gangliosides, cholesterol, and proteins (Keenan et al. 1970; 1972D; Hood and Patton 1973; Dylewski et al. 1984; Deeney et al. 1985; Keenan and Dylewski 1985). The presence of cholesterol on intracellular lipid droplets in lactating rat mammary tissue was also shown by the use of electron microscopy and the antibiotic filipin, which forms morphologically distinct complexes with 3-β-hydroxysterols (Montesano et al. 1983). Many of these constituents appear to be membrane components derived from the endoplasmic reticulum which associate with the surfaces of intracellular globules during lipid droplet formation.

The structure of milk triglycerides is considered in Chapter 4. Here we emphasize only that many different triglyceride molecules are present in lipid globules. Whether these triglycerides are randomly distributed throughout the core or are concentrated into discrete zones or shells according to molecular species is unknown. One could envision that different molecular species of triglycerides are synthesized and added to growing lipid droplets at different rates or times. In fact, it has been suggested that lipid droplet growth may be terminated by accumulation of triglycerides with high melting points at the droplet surface (Patton 1973). However, even though there may be selectivity in deposition, the lipids would also be expected to equilibrate rapidly throughout the droplets. As discussed elsewhere (Patton 1973), fatty acids are arranged in triglycerides in such a manner as to ensure that they are liquid at body temperature. In a fluid lipid environment, one would expect rapid diffusion and equilibration to abolish any initial selectivity in the deposition of triglycerides.

Milk lipid is liquid at temperatures near $40\,^{\circ}$C and is completely solidified at temperatures below about $-40\,^{\circ}$C (Mulder and Walstra 1974). Thus, when milk is withdrawn from the animal and cooled, crystallization of triglycerides in lipid globules begins. The rate and extent of crystallization are dependent on the rate of cooling and the final temperature to which the milk is cooled. Brunner (1974) and Mulder and Walstra (1974) have reviewed this subject. Whether such temperature-induced crystallization occurs at random throughout the core lipid or in localized regions is uncertain. It is conceivable that crystal growth could occur in such a manner that zones or shells of triglycerides with similar melting ranges would form.

SURFACE OF THE MILK LIPID GLOBULE

Historical Background

Ascherson (Brunner 1969, 1974) appears to have been the first to recognize the presence of an emulsion-stabilizing substance on the surface of milk lipid globules. Since his 1840 description of what he termed a "haptogenic membrane," the nature and origin of this surface material have occupied the attention and time of many investigators. Much of the effort has focused on visualization of this material, development of methods for its isolation, and studies on its origin and nature. While research in the first two areas was largely successful, little information which has stood the test of time came from early studies of the origin and nature of the membrane material. Many of the earlier investigators were severely hampered by the techniques available at the time. In retrospect, it appears that many scientists were additionally hampered by acceptance of erroneous views which were widely held over the first hundred or more years after Ascherson's description of what is now known as the "milk lipid globule membrane." During this period, it was believed that proteins from milk serum formed the emulsion-stabilizing substance by adsorption onto the globule surface. It was not until 1924, when the presence of phospholipids in this membrane material was recognized, that investigators realized that substances other than proteins were also present (Palmer and Samuelson 1924). It required an additional 35 years to demonstrate that the globule membrane was a true biological membrane which originated from the milk-secreting epithelial cell (Bargmann and Knoop 1959). Brunner (1974) reviewed research in this area through 1971, and this presentation is recommended as an excellent summary of earlier work. Later work was reviewed by Patton (1973), Anderson and Cawston (1975), Patton and Keenan (1975), Patton and Jensen (1976), Keenan et al. (1978), Mather and Keenan (1983), and McPherson and Kitchen (1983).

Research on milk lipid globule membranes can be divided into two periods: the early era, which persisted up to the recognition that this membrane was a true membrane of cellular origin, and the present era, which dates from the biochemical studies of Morton (1954) and Bailie and Morton (1958) and the pioneering electron microscopic studies of milk lipid secretion by Bargmann and others (Bargmann and Knoop 1959; Bargmann et al. 1961; Bargmann and Welsch 1969). Morton's biochemical studies established the presence of several enzymes in the milk lipid globule membrane, including xanthine oxidase and alkaline phosphatase (Morton 1954). A comparison of milk lipid globule membranes with microsomal membranes from mammary gland led to the conclusion that lipid globule membrane is of cellular origin (Bailie and Morton 1958). The term "milk microsomes" was used to describe this membrane (Morton 1954), although the precise origin of the globule membrane within mammary secretory cells was never established.

Prior to this time, the morphology of the lactating mammary gland had been thoroughly studied with the light microscope, and it was known that mammary epithelial cells secreted lipid droplets which were large in relation to the dimensions of the cell. In fact, 50 years ago, Jeffers (1935) described milk lipid globule secretion as involving enmeshing of the globule in the apical cytoplasm and cell surface membrane. It is not clear how early investigators envisioned passage of lipid globules out of cells without acquisition of a membrane. Perhaps they held the view, prevalent for a time, that milk was discharged by holocrine secretion, that is, by rupture of the cell and the discharge of cellular contents (Kurosumi et al. 1968; Linzell and Peaker 1971). Whatever the reason, much effort was spent in attempting to show that the membrane was formed by orientation of milk constituents on the surface of a triglyceride droplet. As late as 1955, this view was expounded by King (1955) in his widely circulated monograph.

Origin of the Milk Lipid Globule Membrane

Development of the electron microscope and of fixation and sectioning techniques for biological materials was crucial in establishing the origin of the milk lipid globule membrane. Early applications of these techniques by Porter, Sjöstrand, Palade, and Claude, among others, were with liver and pancreas; nevertheless, knowledge gained through the efforts of these pioneers heavily influenced those who made early ultrastructural studies of milk formation (for a historical review of the developments in biological electron microscopy, see Pease and Porter 1981). In the period from 1959 to 1961, electron micrographs of the lactating epithelium of the rat (Bargmann and Knoop 1959), mouse (Wellings et al. 1960A,B; Bargmann et al. 1961; Hollmann 1974), hamster (Bargmann et al. 1961), and cow (Feldman 1961) were published, which gave clear evidence that lipid droplets were extruded from mammary epithelial cells by progressive envelopment in apical regions of the plasma membrane. This process, which is illustrated in Figure 10.3, was subsequently confirmed by a number of investigators for the above and several other species (Wooding 1977; Pitelka and Hamamoto 1977). Fat droplets appear to originate as small precursor "lipovesicles" in the endoplasmic reticulum and to migrate through the cytoplasm to apical regions of the cell (Stein and Stein 1967; Dylewski et al. 1984; Deeney et al. 1985; Keenan and Dylewski 1985). These droplets appear to grow during basal to apical transit by the fusion of lipovesicles with larger droplets, and evidence suggests that growth continues in the apical cytoplasm and especially during secretion (Stemberger and Patton 1981; Stemberger et al. 1984).

It is now widely, but not universally, accepted that lipid droplets acquire an outer coating of membrane by budding directly from the apical surface (Figure 10.3). This was first documented by Bargmann and Knoop (1959) and has been repeatedly observed in later studies by transmission (Kurosumi et al. 1968; Saacke and Heald 1974), freeze fracture (Peixoto de Menezes and Pinto da Silva 1978), and scanning electron microscopy (Nemanic and Pitelka 1971). In contrast to this widely held view, Wooding (1971A, 1973) has suggested that lipid droplets become surrounded by secretory vesicles in such a manner that intracytoplasmic vacuoles containing lipid droplets partially coated by membrane are formed. Release of lipid droplets from the cell is then proposed to occur by exocytotic fusion of secretory vesicles

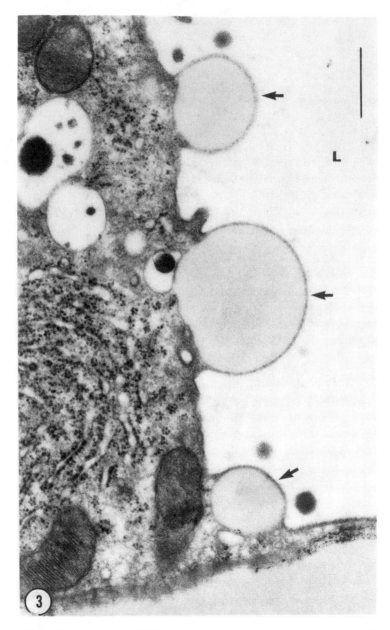

Figure 10.3 Secretion of lipid globules from bovine mammary epithelial cells. At the time of fixation, three lipid globules (arrows) were in the process of budding from this cell. These globules are partially extruded into the alveolar lumen (L), and the extruded portions are enveloped by a specialized region of apical plasma membrane. Bar = 0.5 μm; magnification $\times$ 39,000.

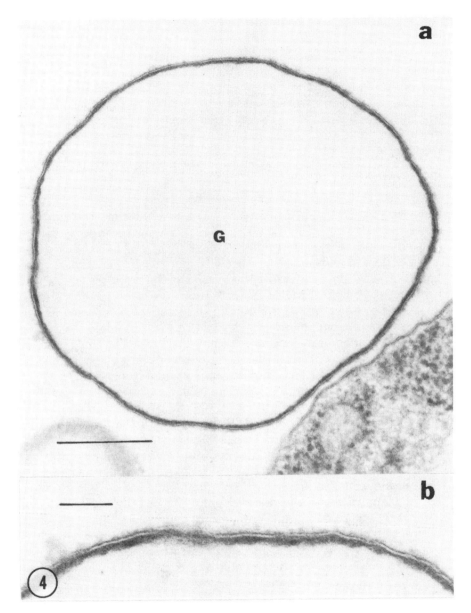

Figure 10.4 Lipid globule in human milk surrounded by a membrane which has a typical unit or bilayer appearance. This membrane is separated from the globule core by a layer of intensely stained coat material. The material on the outer face of the membrane has an appearance typical of the glycocalyx observed on plasma membranes. (b) A higher magnification of a region of membrane of the lipid globule shown in (a). Note that in certain regions a bilayer-like membrane is indiscernible. Bars = 0.1μm, magnification (a) × 75,000, (b) × 137,000. Prints are of a micrograph published previously (Freudenstein *et al.* 1979) and are included with permission of Academic Press, Inc.

521

with apical plasma membrane. Wooding believes that some droplets are secreted in this manner and that others are entirely enveloped in apical plasma membrane. Available morphological evidence is insufficient for resolution of this question. Micrographs showing globules partially enveloped in apical plasma membrane (Bargmann and Knoop 1959; Kurosumi *et al.* 1968; Patton 1973; Patton and Keenan 1975; Wooding 1977), as well as micrographs showing the association of secretory vesicles with intracytoplasmic lipid droplets (Wooding 1971A, 1973, 1977; Morré 1977; Franke and Keenan 1979), have been published. Thus there is evidence for both processes. Of concern is that electron microscopy cannot be used to determine the *rate* and therefore the extent of lipid globule secretion by either process.

Results of biochemical studies conducted to establish the origin of the milk lipid globule membrane are also equivocal. Detection of enzymatic activities normally associated with the Golgi apparatus in membranes of milk lipid globules (Martel-Pradal and Got 1972; Martel *et al.* 1973; Powell *et al.* 1977) has been used to argue the case for the involvement of Golgi apparatus–derived secretory vesicles in the formation of the globule membrane (Powell *et al.* 1977). The compositional similarity between milk lipid globule membrane and plasma membrane fractions isolated from lactating mammary gland has been used as evidence for an apical plasma membrane origin of the milk lipid globule membrane (Keenan *et al.* 1970, 1971, 1974B, 1978; Patton and Trams 1971; Patton and Keenan 1975). With respect to the presence of enzymes normally associated with the Golgi apparatus, interpretation is complicated by the possibility that cytoplasmic membranes, entrained during globule secretion (to be discussed subsequently), contaminated the fraction analyzed. Comparison of globule membrane with isolated plasma membrane is not necessarily valid because the isolated plasma membrane is derived in part from lateral and basal cell surfaces (i.e., the fractions isolated are enriched in junctional complexes) and there is no assurance that the composition of these membranes reflects that of the *apical* cell surface. It is possible, for example, that the apical cell surface contains constituents or enzymes which are also present in Golgi apparatus membranes but which are largely absent from basolateral regions of plasma membrane.

In an attempt to gain further insight into the origin of the milk lipid globule membrane, the composition of this membrane has been compared to that of secretory vesicle membranes (Keenan *et al.* 1979). The vesicle membrane was found to be compositionally intermediate between Golgi apparatus membranes and lipid globule membrane. However, the isolation procedure yielded a fraction enriched in immature secretory vesicles and mature secretory vesicles may more closely resemble lipid globule membrane in composition.

Biochemical analysis of the globule membrane is also complicated by the observation that intracellular fat droplets acquire an amorphous layer of sterols, phospholipid, gangliosides, and proteins from the rough endoplasmic reticulum, and possibly the cytoplasm, before envelopment in apical plasma membrane (Hood and Patton 1973; Keenan et al. 1983A; Dylewski et al. 1984; Deeney et al. 1985; Keenan and Dylewski 1985). This material includes two polypeptides of approximately relative molecular mass (M_r) 44,000 and a fraction of the glycoprotein butyrophilin (Deeney et al. 1985), which is also located on apical membranes (Franke et al. 1981; Johnson and Mather 1985). Material acquired by fat droplets within the cell almost certainly contributes to an inner coating of protein and lipid sandwiched between the core fat and the outer phospholipid bilayer.

This coat material, which has a paracrystalline appearance in freeze fracture, is seen along the inner face of the membrane surrounding secreted milk lipid globules (Figure 10.4) (Wooding 1971A,B, 1973; Freudenstein et al. 1979; Buchheim 1982) and remains associated with the membrane when displaced and isolated from milk lipid globules by various methods (Keenan et al. 1971; Wooding and Kemp 1975A,B; Jarasch et al. 1977; Freudenstein et al. 1979). Originally this coat was believed to be composed of triglycerides with high melting points (Keenan et al. 1971; Bauer 1972), but subsequent studies have shown this material to be primarily protein (Wooding and Kemp 1975A,B; Freudenstein et al. 1979). Major proteins of this coat material are butyrophilin (Freudenstein et al. 1979; Franke et al. 1981; Deeney et al. 1985), a hydrophobic glycoprotein with an M_r of approximately 67,000, and xanthine oxidase (Mangino and Brunner 1977A; Mather et al. 1977; Freudenstein et al. 1979; Jarasch et al. 1981; Deeney et al. 1985), a complex redox enzyme containing iron and molybdenum with a monomeric M_r of about 155,000. The properties of these proteins will be discussed in a subsequent section. While butyrophilin and xanthine oxidase may have specific functions in the recognition and envelopment of lipid droplets in apical plasma membrane (Freudenstein et al. 1979; Franke et al. 1981; Jarasch et al. 1981; Keenan et al. 1982), we still have no knowledge, in molecular terms, of how these proteins function in secretion.

The mechanism whereby fat droplets are directed to the apical cytoplasm and become coated with membrane has been the subject of much speculation. There is some evidence that microtubules and microfilaments, elements of the cytoskeleton, may be involved. During lactation a majority of the microtubules in the apical cytoplasm are oriented parallel to the lateral cell surfaces (Nickerson and Keenan 1979; Nickerson et al. 1982). These may act as guides for the directional transfer of lipid droplets to the apical plasma membrane. However, much of the evidence for the involvement of microtubules in secretion is based on the use of inhibitors (Knudson et al. 1978; Sasaki and Keenan 1978; Nickerson et al. 1980A,B), and in some cases results have been contradictory.

Patton and Fowkes (1967) proposed that London van der Waals forces may be involved in the attraction of apical plasma membrane around lipid droplets. They calculated that forces of about 1 atm would be generated when lipid droplets approached within 2 nm of the membrane and considered that these forces would be sufficient to expel water from the space between the lipid droplet surface and the membrane. However, formation of the inner coat material ensures that intracellular lipid droplets are separated from plasma membrane by a distance of at least 10 nm at the closest approach (Wooding 1971A,B) and because of this, the hypothesis of Patton and Fowkes has been questioned (Wooding 1971A, 1977). More recent considerations, however, indicate that this may still remain a plausible hypothesis. In calculating the magnitude of the attractive forces, Patton and Fowkes (1967) estimated the surface tension of the plasma membrane to be intermediate between those of glycerides and proteins. Evidence has recently been obtained that butyrophilin and xanthine oxidase contain tightly, perhaps covalently, bound fatty acids (Keenan et al. 1982), which would be expected to result in surface tension properties intermediate between those of pure protein and pure glycerides. Thus van der Waals forces of the magnitude calculated by Patton and Fowkes may be generated when lipid droplet surfaces approach within 2 nm of the electron-dense coat material on the inner face of the apical plasma membrane.

To our knowledge, the mechanism used by mammary epithelial cells to secrete lipid is unique; we know of no other cell which secretes lipids in this manner. However, there appear to be similarities between milk lipid globule secretion and the cellular discharge of enveloped viruses. It is known that certain virally coded proteins are selectively inserted into regions of plasma membrane which ultimately envelop the nucleocapsid (Richardson and Vance 1976; Katz et al. 1977), and several viruses bud selectively from the polarized domains of epithelial cells (Rodriguez Boulan and Sabatini 1978). Preliminary evidence suggests that a portion of butyrophilin and xanthine oxidase (Franke et al. 1981; Jarasch et al. 1981) is selectively added to the apical plasma membrane of milk-secreting epithelial cells. Certain virally coded envelope proteins are known to contain covalently bound fatty acids (Schlesinger 1981); butyrophilin and xanthine oxidase also contain tightly bound fatty acids (Keenan et al. 1982). Envelopment of nucleocapsids in plasma membrane appears to involve the formation of microfilament cleavage furrows behind the nucleocapsid (for a discussion of the ultrastructure of viral secretion, see Dalton and Haguenau 1973). Whether actin filaments are involved in lipid droplet envelopment is not known. In view of the apparent similarities in envelopment mechanisms, this possibility should be considered in a search for the molecular mechanisms involved in milk lipid globule secretion. It is known that actin is present in only trace amounts, if at all, in preparations of the milk lipid globule membrane (Keenan et al. 1977A). However, this does not exclude an involvement of actin filaments in the formation of the lipid globule membrane. Antibodies to actin show strong positive staining, in immunofluorescence microscopy, of basal regions of budding milk lipid globules, suggesting a concentration of actin in this region of apical cytoplasm (Franke et al. 1981). In tangential sections through apical cell regions where budding lipid globules were present, we have observed structures resembling actin filament cleavage furrows (Franke and Keenan, unpublished). Moreover, Amato and Loizzi (1981) have shown actin microfilaments to be abundant in apical cytoplasmic regions just under the plasma membrane in mammary secretory cells of guinea pig.

In spite of the fact that the driving forces remain speculative, there is now little doubt that, at secretion, lipid droplets are enveloped in apical plasma membrane, with perhaps some contribution from secretory vesicle membrane. Many questions remain, however, regarding the nature and origin of the inner coat material which lies between the triacylglycerol core and the outer bilayer membrane. To what extent is this material derived from the amorphous surface material seen on lipovesicles within the cell (Dylewski et al. 1984; Deeney et al. 1985; Keenan and Dylewski 1985) and the electron-dense coat on the cytoplasmic face of the apical plasma membrane (Franke et al. 1981)? Also to be considered is the clathrin-like coat observed on the outer surface of secretory vesicles (Franke et al. 1976; Mather and Keenan 1983), which may contribute a substantial quantity of material to the lipid droplets (Franke and Keenan 1979), if secretory vesicles do indeed contribute to the formation of the milk lipid globule membrane.

Some questions also remain as to whether cytoplasmic materials, including cellular organelles, are entrained between the membranes surrounding the globule and the triglyceride core. In alveolar spaces and in expressed milk, lipid globules entirely surrounded by a unit-like membrane are routinely observed (Figure 10.4). However, some globules appear to have cytoplasmic material trapped between the limiting membrane and the globule core (Figure 10.5). Electron micrographs showing extracellular lipid globules with entrained cytoplasmic materials were first published by Kurosumi et al. (1968) and by Helminen

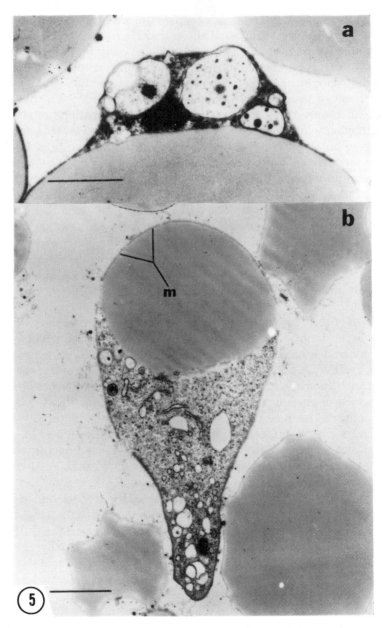

Figure 10.5 Electron micrographs of bovine lipid globules containing cytoplasmic material entrained between the core lipid and the membrane surrounding the globule [indicated by m in (b)]. (a) Secretory vesicle-containing casein micelles are present in the entrained material. (b) Vesicles of various sizes and fragments of rough endoplasmic reticulum are present. Bars = 1.0 (a) and 2.0 (b) μm; magnification ×22,000 (a), ×9000 (b).

and Ericsson (1968). Subsequently Wooding *et al.* (1970) confirmed and extended these observations and used the term "cytoplasmic crescents" and "signets" to describe these cytoplasmic inclusions. These observations revived the theory that milk is secreted in part by an apocrine mechanism, that is, that milk secretion involves detachment of the cell apex and that milk consists in part of cellular debris. The extent of cytoplasmic entrainment has not yet been fully explored, although Wooding *et al.* (1970) estimated that, in freshly expressed goat milk, 1 to 5% of the globules contain cytoplasmic crescents. Further study suggested that the extent of cytoplasmic entrainment varies among species (Wooding 1977; Janssen and Walstra 1982). In our experience, cytoplasmic inclusions are seen infrequently in lipid globules in alveolar lumina and in expressed milk from the cow (unpublished). Janssen and Walstra (1982) came to similar conclusions using fluorescence microscopy to estimate the occurrence of cytoplasmic crescents in fat globules from several species. Membrane material associated with milk lipid globules from cows and goats contains little or no cardiolipin, a lipid characteristic of mitochondria (Patton *et al.* 1969). Results from quantitative morphometry suggest that most membrane material associated with cow's milk lipid globules is of plasma membrane (or perhaps mature secretory vesicle) origin (Jarasch *et al.* 1977). In contrast, rat's milk lipid globules frequently contain cytoplasmic inclusions, and these lipid globules appear to contain only about 70% of membrane which is of plasma membrane origin (Jarasch *et al.* 1977). Entrainment of cytoplasmic material could perhaps explain why certain workers have reported the presence of constituents and enzymic activities of intracellular membranes in the membrane material from milk lipid globules of some species (Martel-Pradal and Got 1972; Martel *et al.* 1973). It must be emphasized that methods commonly used for the collection of milk lipid globule membranes result in the simultaneous collection of cytoplasmic membranes entrained between globule cores and the limiting membrane. Such cytoplasmic membrane contamination must therefore be taken into account when using preparations of milk lipid globule membrane as a source of apically derived membrane from mammary secretory cells.

The extent to which the membrane surrounding lipid globules is lost from the surface is a matter of controversy, and estimates of the extent of this loss vary widely (Wooding 1971B; Bauer 1972; Baumrucker and Keenan 1973; Patton and Keenan 1975; Jarasch *et al.* 1977; Freudenstein *et al.* 1979). In electron microscopic studies, the electron-dense coat material on the inner face of the milk lipid globule membrane was observed to thicken in localized areas, and patches of membrane with associated coat material appeared to vesiculate and be lost from globule surfaces (Wooding 1971B). The extent of this loss

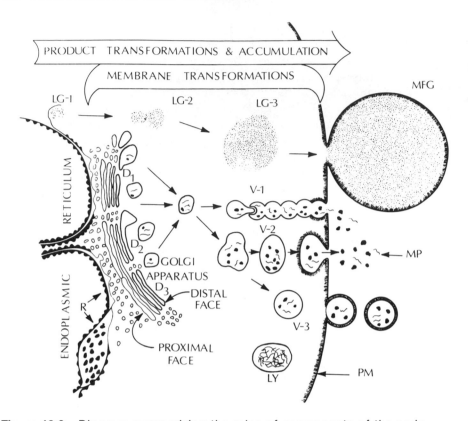

Figure 10.6 Diagram summarizing the roles of components of the endo-membrane system of mammary epithelial cells in the synthesis and secretion of the constituents of milk. Intracellular lipid globules (LG-1, LG-3) appear in the cytoplasm, but it is probable that these globules originate from or in association with endoplasmic reticulum (LG-1). At least some lipid globules appear to increase in size as they move to the apical region of the cell; this growth may occur by fusion of triglyceride-containing vesicles with the globules (LG-2). Lipid globules are discharged from the cell by progressive envelopment in regions of apical plasma membrane which have coat material along the inner membrane face. MFG denotes a lipid globule being enveloped in plasma membrane. Milk proteins (MP) are synthesized on polysomes of endoplasmic reticulum and are transported, perhaps in small vesicles which bleb from endoplasmic reticulum, to dictyosomes (D_1, D_2, D_3) of the Golgi apparatus. These small vesicles may fuse to form the proximal cisterna of Golgi apparatus dictyosomes. Milk proteins are incorporated into secretory vesicles formed from cisternal membranes on the distal face of dictyosomes. Lactose is synthesized within cisternal luminae of the Golgi apparatus and is incorporated into secretory vesicles. Lactose appears to pull water into secretory vesicles, causing them to swell. Evidence that certain of the ions of milk are also present in

appears to depend to some extent on the conditions used to fix lipid globule specimens for electron microscopic examination. Under some fixation conditions, nearly all of the globules examined are entirely surrounded by a unit-like membrane and associated electron-dense coat material (Freudenstein *et al.* 1979). Biochemical studies have shown that there is little loss of membrane constituents under appropriate storage conditions for up to 24 hr after withdrawal of milk from cows (Baumrucker and Keenan 1973; Patton *et al.* 1980). This suggests that the membrane loss which occurs happens before withdrawal of milk from the animal.

Secretion of milk fat globules involves a net loss of membrane from the cell. Obviously, if the cell is to be maintained intact and functional, this membrane must be replaced. It has been calculated that an amount of membrane equivalent to the entire apical surface of milk-secreting cells of the cow must be replaced every 8 to 10 hr (Franke *et al.* 1976). A thorough discussion of current concepts on membrane regeneration is beyond the scope of this presentation; they are discussed elsewhere (Keenan *et al.* 1974B, 1978; Mather and Keenan 1983); for general reviews of endomembrane differentiation and flow, see Morré (1977) and Morré *et al.* (1979). In brief, the secretory vesicle membrane is believed to be integrated into the apical plasma membrane, replenishing that membrane expended during lipid globule discharge. Secretory vesicles contain lactose and the major milk proteins (Sasaki *et al.* 1978; Keenan *et al.* 1979), and probably contain many of the ions and much of the water of milk as well (Peaker 1978, 1983). Thus, secretion of the lipid and nonlipid phases of milk appears to be interrelated through membrane flow. These concepts are summarized diagramatically in Figure 10.6.

---

secretory vesicles has been obtained, and it is currently believed that the content of secretory vesicles is effectively the nonfat phase of milk. Three different mechanisms for exocytotic interaction of secretory vesicle with apical plasma membrane have been described (V-1, V-2, V-3). One way in which vesicle contents are secreted is through the formation of a chain of fused vesicles (V-1) (Dylewski and Keenan 1983). Another is by fusion of individual vesicles with apical plasma membrane (V-2), with integration of vesicle membrane into plasma membrane. This may be the mechanism by which plasma membrane expended in envelopment of lipid globules is replenished. A third, and perhaps minor, mechanism is by direct envelopment of secretory vesicles in apical plasma membrane (V-3) (Franke *et al.* 1976; Dylewski and Keenan 1983). Lysosomes (LY) may function in the degradation of excess secretory vesicle membrane. (Modified from Keenan and Huang 1972B by permission of the American Dairy Science Association.)

ISOLATION OF MILK LIPID
GLOBULE MEMBRANE

Palmer and Samuelson (1924) were the first to report the isolation and partial characterization of membrane material from milk lipid globules. The methods they used are similar to those used for this isolation today. Brunner (1974), who himself was instrumental in development of isolation methods (Brunner *et al.* 1953), provided a detailed review of the history of isolation methods. In this section we will summarize the currently used methods and indicate their advantages and disadvantages.

Whatever the method selected for displacement and recovery of lipid globule membrane, the first step is recovery of lipid globules from milk. This is most conveniently done by centrifugal flotation, as the lipid globules, being less dense than the aqueous phase of milk, can be rapidly separated by centrifugation for brief periods at relatively low gravity forces. Small volumes of milk can be easily processed in nearly all commercially available laboratory centrifuges; large volumes are more conveniently handled in mechanical cream separators which are essentially specialized flow-through centrifuges. Once lipid globules are obtained, they are usually washed to remove entrained or adsorbed components of milk serum. This is normally accomplished by resuspension and reflotation of lipid globules in buffered or unbuffered water made isotonic with milk serum by addition of, for example, sucrose or sodium chloride. This washing cycle can be repeated many times. The extent to which globules must be washed is determined by the purpose for which the membrane will be used. Washing is more efficient when the milk, globules, and wash solution are not cooled. By maintaining the temperature at or above 25°C, we have found two wash cycles for globules prepared in a laboratory centrifuge, using a total volume of wash solution equal to the starting volume of milk, to be sufficient to remove caseins and major whey proteins to a level below detection when proteins associated with globules were examined by electrophoretic separation and by immunodiffusion analysis (Mather and Keenan 1975). In our experience with a mechanical cream separator, three or four washes are necessary to achieve the same degree of freedom from caseins and whey proteins. Visual inspection for proteins on electrophoretic gels stained with coomassie blue is not a sensitive detection method, and it is possible that small amounts of milk serum proteins could remain associated with globules over multiple washing cycles. If it is crucial to remove all traces of milk serum constituents, one could develop radioimmunoassays, preferably for at least one major casein and for at least one major whey protein, and use these assays to monitor the efficiency of washing; to our knowledge, this has not yet been done.

As discussed by Brunner (1974), there are potential problems with washing procedures. One is that components loosely associated with the globule surface can be washed away. This is a valid criticism but, as with fractionation of any biological material, something must be compromised. Washing would not be expected to remove any integral membrane constituents selectively. Another potential problem is that the ionic strength and composition of the wash solution frequently differ from those of milk, and this difference may enhance or diminish the strength of the interaction of constituents ionically bound to the membrane. Documentation of this conclusion is lacking, but one could attempt to determine if such changes occur by using, as the washing medium, a salt solution which closely duplicates the ionic composition of milk (Jenness and Koops 1962). Of more concern is that, as demonstrated by Swope and Brunner (1968) and confirmed by Anderson and Brooker (1975), repeated washing causes erosion of membrane material. Some loss of membrane material from centrifugally floated lipid globules may be due to selective loss of very small lipid globules which, because of their higher density, are not floated readily. Some loss may also be due to rupture of some globules, with release of membrane into the aqueous phase. Brunner (1974) cited examples from his experience where lipid globules exhibited pronounced instability after three washes, while lipid globules from other batches of milk remained stable for up to 20 wash cycles. In our experience, lipid globules from fresh uncooled milk are more stable than those from cooled or aged milk. In order to circumvent some of the above problems, McPherson et al. (1984A) recently described a method by which membranes were separated from milk serum proteins by sucrose-density gradient centrifugation, using unwashed cream as the starting material. This method allowed the preparation of globule membrane from milk samples containing cream of unknown or altered stability, such as pasteurized, homogenized, or ultra-heat-treated milks (McPherson et al. 1984B,C).

For some studies, washed lipid globules can be used as such. More commonly, it is necessary to dissociate membranes from globules by chemical or physical methods. Chemical methods normally involve direct extraction of constituents from the globules. Sodium dodecyl sulfate solutions have been used to recover membrane proteins from washed lipid globules for subsequent electrophoretic characterization (Kobylka and Carraway 1972; Mather and Keenan 1975). Other workers have used solutions of detergents such as deoxycholate (Hayashi and Smith 1965) and Triton X–100 (Patton 1982). With these detergents, portions of the membrane material are solubilized, while other portions remain insoluble and can be sedimented by centrifugation. Membrane phospholipids have been recovered by direct extraction of lipid globules with mixtures of chloroform and methanol (Patton and Keenan 1971). The bulk of the membrane proteins are insoluble in chloroform-methanol and can be recovered as an insoluble residue by filtration or centrifugation of extracts. Solutions with high concentrations of salt have been used to extract selectively certain salt-soluble proteins directly from washed lipid globules (Mather and Keenan 1975).

Physical methods commonly used to dissociate membrane from globules include mechanical agitation (churning, rapid mixing, or stirring), slow freezing and thawing, and exposure to ultrasound. Theories on the mechanism involved in agitation-induced inversion of cream into butter and buttermilk (i.e., inversion from an oil-in-water to a water-in-oil emulsion) are discussed elsewhere (Jenness and Patton 1959; Mulder and Walstra 1974). Churning proceeds more rapidly at temperatures between about 15 and 25°C (Jenness and Patton 1959). In our experience, yields of membrane are highest when globules are diluted with aqueous solution to 30 to 50% lipid content before churning. Use of dilute buffers or distilled water gives higher yields of membrane than isotonic solutions. By whatever means chosen to accomplish it, phase inversion results in a butter phase containing the bulk of the triglycerides and an aqueous phase in which membrane fragments are suspended. Phase inversion is usually accomplished while maintaining cream at temperatures below the apparent solidification point of the triglyceride mixture. Ultimate yields of recovered membrane can be improved by repeated washing, with vigorous agitation or blending, of the semi-solid butter phase or by melting this phase and separating the aqueous phase from the oil.

Membrane fragments can be collected from aqueous phases by precipitation with salts or by pH adjustment (Herald and Brunner 1957; Kitchen 1974), collection of the precipitate being accomplished by centrifugation or filtration. Alternatively, membrane fragments can be collected by ultracentrifugation of the aqueous phase without prior treatment. Ultracentrifugation conditions for optimum yields vary in their seeming dependence on the degree of fragmentation of membrane caused by the method used to effect phase inversion. After churning in a Waring blender or other high-speed spinning knife device (a severe method), we have found centrifugation at about 20×10^6 g-min (i.e., about 110,000 g for 3 hr) to be adequate; longer centrifugation times or increased speeds result in collection of little additional membrane material.

It must be emphasized that, depending on the choice of preparative conditions, not only will the yield of membrane vary but the composition of the material obtained will differ. For certain constituents, wide ranges in composition have been reported. In particular, there are large variations in reported protein-to-lipid ratios and in relative amounts of triglycerides and neutral lipids (Patton and Keenan 1975). Enzymatic activities of membrane preparations also vary when different preparative procedures are used (Bhavadasan and Ganguli 1976). While not necessarily recognized at the time, much of this variation can be ascribed to differences in isolation methods or conditions. For example, if after churning, the aqueous phase is collected without melting the butter, the membrane pellets obtained by ultracentrifugation are brown and have a lipid-to-protein ratio considerably lower than that of the predominantly white membrane pellets obtained when butter is melted after churning and the entire aqueous phase is centrifuged (Brunner 1969, 1974; Anderson and Cawston 1975; Patton and Keenan 1975). We have observed the sediment collected after melting butter to be enriched in a material which has extensive amounts of lipid bound to glycoproteins of M_r about 44,000 and 48,000 (Heid and Keenan unpublished; Kitchen 1977; McCarthy and Headon 1979). Another factor which can markedly affect lipid-protein ratios and protein distribution is the choice between precipitation or ultracentrifugation for recovery of membrane material. When ultracentrifugation alone is used, the supernatant retains soluble proteins, primarily xanthine oxidase (M_r 155,000), and glycoproteins (M_r about 44,000 and 48,000) (Mather *et al.* 1977). Depending on the conditions of precipitation, these proteins may be recovered along with membrane in the insoluble fraction, thus increasing the protein-to-lipid ratio and quantitatively changing the amounts of individual proteins in the sample. The temperature of churning can also alter the composition of the membranes recovered. For example, Vasic and De Man (1966) and Walstra (1974) noted that high melting triglycerides were much lower in membranes prepared from cream held at about 40°C than in membranes prepared from cooled cream. These examples illustrate the apparent compositional differences which can arise from variation in preparative method. The choice of isolation method should be based on the purpose for which the membrane material is to be used. A recent comparison of six commonly used procedures is summarized in Eigel *et al.* (1984).

The extent to which membranes isolated from washed lipid globules originate from the apical plasma membrane, with perhaps some contribution from mature secretory vesicle membrane, can be an important consideration. Any intracellular membranes, contained in cytoplasmic crescents, will be present in preparations obtained by any of the methods outlined above (Patton and Keenan 1975). While entrained cytoplasmic material does not appear to be abundant in lipid globules from cow's milk, for certain studies even a very small contamination with intracellular membranes may be intolerable. For purification, membrane preparations can be fractionated by density-gradient centrifugation. Since milk lipid globule membrane is not homogeneous with respect to buoyant density, in using this approach some caution must be exercised. Mather et al. (1977) collected seven fractions of milk lipid globule membrane, ranging in density from less than 1.13 to more than 1.19 g/cm³, from sucrose step gradients. All fractions were essentially homogeneous with respect to polypeptide profiles and specific activities of xanthine oxidase and 5'-nucleotidase. However, the lipid content of these fractions varied inversely with the density. Kobylka and Carraway (1972) and Kitchen (1977) also obtained fractions of milk lipid globule membranes from sucrose density gradients, which all contained the same major polypeptides. However, the fractions isolated by Kitchen differed considerably in the relative amounts of each protein in the individual fractions. In particular, the fractions of highest density were enriched in a glycoprotein with an apparent M_r of 53,000, probably glycoprotein B of Basch et al. (1976). Morphological studies have shown the cytoplasmic membrane contaminants of globule membrane preparations to be concentrated primarily in a fraction banding at more than 1.22 g/cm³ on sucrose density gradients; in confirmation of this finding, the cytoplasmic protein actin was observed only in this heavy fraction in trace amounts and was undetectable in fractions of lower density (Keenan et al. 1977A).

COMPOSITION OF THE MILK LIPID GLOBULE MEMBRANE

Gross Composition

While there has been an upsurge of interest in recent years in comparative studies of the composition of milk from different species, the lipid globule membrane of bovine milk remains the only thoroughly characterized milk membrane system. Values for the gross composition of globule membranes from cow's milk are summarized in Table 10.2. Protein has been reported to account for 25 to 60% of the dry weight of the membrane, and total lipids have been reported to range from 0.5 to 1.1 mg per milligram of protein. While the proportions of protein and lipid vary widely, together they account for over 90% of the membrane dry weight (Patton and Keenan 1975). Variations in composition among different samples can be due to the method of preparation (dis-

Table 10.2. Gross Composition of Milk Lipid Globule Membranes.[a]

Constituent Class	Amount
Protein	25 to 60% of dry weight
Total lipid	0.5 to 1.1 mg/mg protein[a,b]
Phospholipid	0.13 to 0.34 mg/mg protein[a-c]
Neutral lipid	56 to 80% of total lipid
Hydrocarbons	1.2% of total lipid
Sterols	0.2 to 5.2% of total lipid
Sterol esters	0.1 to 0.8% of total lipid
Glycerides	53 to 74% of total lipid
Free fatty acids	0.6 to 6.3% of total lipid
Cerebrosides	3.5 nmoles/mg protein
Ganglioside sialic acid	6 to 7.4 nmoles/mg protein[a,b]
Total sialic acids	63 nmoles/mg protein
Hexoses	0.6 μmole/mg protein
Hexosamines	0.3 μmole/mg protein
Cytochromes b_5 + P-420	30 pmoles/mg protein[d,e]
RNA	20 μg/mg protein[d]
Uronic acids	99 ng/mg protein[f]

[a] Unless indicated otherwise, values are from a review by Patton and Keenan (1975).
[b] Keenan et al. (1979).
[c] Kitchen (1974).
[d] Jarasch et al. (1977).
[e] Bruder et al. (1978).
[f] Shimizu et al. (1981); the value calculated from Lis and Monis (1978), 58 μg/mg protein, appears to be unrealistic.

cussed earlier) and to factors such as breed of the animal, stage of lactation, season of the year, age, and treatment of the milk. What differences are caused by each of these factors is largely unknown. The most variable portion of the membrane appears to be the neutral lipid content. Since triglycerides account for some 95% of the total mass of lipid globules, it is probable that the amount of triglyceride which adheres to the membrane during isolation greatly influences the amount of total lipid found in the membrane. Whether triglycerides are true constituents of the globule membrane, or are simply contaminants adsorbed on membrane faces or entrained in membrane vesicles, is unknown. Plasma membranes from liver and other tissues contain only small amounts of triglycerides (Pfleger et al. 1968; Ray et al. 1969; Keenan and Morré 1970). Phospholipids appear to be more constant in amount per unit of protein in the globule membrane, averaging about 0.25 mg per milligram of protein.

Both protein- and lipid-bound carbohydrates are present in milk lipid globule membrane (Table 10.2). Lipid-bound carbohydrates, present predominantly in glucosyl and lactosyl ceramides and in gangliosides, are glucose, galactose, N-acetylgalactosamine, and N-acetylneuraminic acid (sialic acid). Lis and Monis (1978) have identified hyaluronic acid, chondroitin sulfate, and heparan sulfate in a glycosaminoglycan fraction isolated from milk lipid globule membrane. Shimizu *et al.* (1981) confirmed the presence of glycosaminoglycans in preparations of lipid globule membranes and identified chondroitin sulfate and heparan sulfate as constituents of this fraction. Shimizu *et al.* (1981) found 5 to 10 times higher levels of glycosaminoglycans in human milk lipid globule membranes than in bovine lipid globule membranes.

RNA was first detected in bovine milk lipid globule membrane by Swope and Brunner (1965) and was subsequently detected in human milk lipid globule membrane (Martel *et al.* 1973). In bovine globule membranes, RNA is present at a level of about 20 μg per milligram of protein; high salt extraction of this membrane reduces RNA levels to about 10 μg per milligram of protein (Jarasch *et al.* 1977). RNA associated with the globule membrane remains uncharacterized with regard to type. Whether this small amount of RNA is associated with the globule membrane or is contributed by endoplasmic reticulum fragments or ribosomes on the surface of intracellular lipovesicles (Dylewski *et al.* 1984) or from the material in cytoplasmic crescents is unknown. DNA appears to be absent from preparations of bovine, rat, and human lipid globule membranes (Martel *et al.* 1973; Jarasch *et al.* 1977). In addition to the constituents listed in Table 10.2, Brunner and co-workers (Swope and Brunner 1968; Brunner 1974) have identified several elements, notably calcium, copper, iron, magnesium, manganese, molybdenum, phosphorus, potassium, sodium, sulfur, and zinc, in globule membranes from cow's milk.

Lipid Composition

When total lipids extracted from lipid globule membrane preparations are fractionated, the neutral lipid fraction (composed of glycerides, sterols, free fatty acids, and hydrocarbons) is invariably found to account for more than half of the total lipid (Thompson *et al.* 1961; Huang and Kuksis 1967; Anderson 1974). As noted earlier, the neutral lipid content of the membrane is variable in a manner seemingly dependent on the membrane isolation method. Glycerides are by far the most abundant constituent class in membrane lipids, and triglycerides constitute 90% or more of the glyceride fraction (Thompson *et al.* 1961; Huang and Kuksis 1967). Whether these glycerides are associated with the membrane before globule envelopment or simply are adsorbed from the globule core is unknown, as discussed above. Results of the microelectrophoretic characterization of milk lipid globules led to the conclusion that the outer surface of the membrane contains little neutral lipid (Newman and Harrison 1973). Localization of triglycerides in or on milk globule membrane needs further study.

Sterols, which constitute a significant proportion of membrane lipids, are principally cholesterol and cholesterol esters (Anderson and Cawston 1975; Patton and Keenan 1975; McPherson and Kitchen 1983). Lanosterol and dihydrolanosterol have been identified, in unesterified form only, and occur at less that 2% of the amount of cholesterol (Schwartz et al. 1968). Relative to plasma membranes from serveral different tissues, milk lipid globule membrane contains abnormally low amounts of cholesterol on a phospholipid basis; for example, in one study, the cholesterol-to-phospholipid molar ratio was 0.11 (Keenan et al. 1979), while the values usually found for plasma membrane are up to four times this amount (Pfleger et al. 1968; Ray et al. 1969: Keenan and Morré 1970). In contrast, cholesterol-to-phospholipid ratios of intracellular endomembranes such as the Golgi apparatus and secretory vesicles from mammary gland are comparable to those of these membranes from other tissues (Keenan and Morré 1970; Keenan et al. 1979). It is possible that the low cholesterol levels in lipid globule membranes are a true reflection of the cholesterol content of the apical surfaces of mammary secretory cells. Until methods are devised for isolating apical plasma membrane from mammary tissue, this question will remain unresolved. Another possibility is that some cholesterol may be extracted from the globule membrane into the core lipid during and after the formation of milk lipid globules (Keenan et al. 1979). Core lipid does contain cholesterol (Huang and Kuksis 1967), although some of this may derive from lipovesicle precursors during the formation of intracellular lipid droplets (Deeney et al. 1985). A recent preliminary report indicates that cytoplasmic lipid droplets are labeled with the antibiotic filipin (Montesano et al. 1983), which binds to cholesterol and can be used as a cytochemical probe for 3-β-hydroxysterols (Elias et al. 1979). Interestingly, the filipin labeling was polar, suggesting that cholesterol is asymmetrically distributed on the surface of intracellular droplets.

Numerous free fatty acids have been identified in globule membrane lipids, as have the hydrocarbons β-carotene and squalene (Thompson et al. 1961; Huang and Kuksis 1967). Brunner (1974) has summarized the fatty acid composition of various milk lipid globule membrane neutral lipid constituents, and these values will not be reproduced here.

Phospholipids are found in the milk lipid globule membrane and in milk serum; in the latter, they appear to be primarily, if not exclusively, associated with a membrane fraction which sediments when milk is centrifuged at high speeds (the nature of this membrane material will be discussed in a subsequent section of this chapter). Milk lipid globule membranes contain about 60% of the total milk phospholipid. Major phospholipids of the globule membrane are sphingomyelin and the phosphatides of choline, ethanolamine, inositol, and serine. Other phospholipids, such as lyso-derivatives of the major phosphatides, have been detected in milk and globule membrane, but these are relatively minor constituents when freshly obtained samples handled in a manner which minimizes lipid degradation are used for analysis. An identical distribution pattern for the major phospholipids is found for whole milk, globule membrane, and milk serum (Patton et al. 1964; Huang and Kuksis 1967; Patton and Keenan 1971). Representative values of phospholipid distribution in milk lipid globule membrane are included in Table 10.3, along with values for phospholipid distribution in plasma membrane, endoplasmic reticulum, and intracellular lipid droplet fractions from lactating bovine mammary gland. This comparison illustrates the similarity of plasma membrane and milk lipid globule membrane in phospholipid distribution. However, this phospholipid distribution pattern is distinct from that of the intracellular membranes from mammary gland which have been analyzed, including endoplasmic reticulum (Table 10.3) and nuclear membrane, Golgi apparatus, secretory vesicles, and mitochondria, in that the sphingomyelin-to-phosphatidyl choline ratio is high (Patton and Keenan 1975; Keenan

Table 10.3   Distribution of Major Phopholipids in Endoplasmic Reticulum, Intracellular Lipid Droplets, Plasma Membrane, and Milk Lipid Globule Membrane Fractions from Bovine Mammary Tissue.

Tissue Fraction	Percent of Total Lipid Phosphorus				
	SP	PC	PE	PI	PS
Endoplasmic reticulum[a]	4.1	55.2	28.9	9.1	2.7
Intracellular lipid droplets[a]					
Cytoplasmic lipid droplets	14.6	45.8	20.9	11.9	6.8
"Heavy" microlipid droplets	12.8	49.0	24.9	8.5	4.8
"Light" microlipid droplets	11.5	53.6	22.7	7.7	4.5
Plasma membrane[b]	24.5	28.9	25.4	12.7	8.4
Milk lipid globule membrane[a]	21.9	36.2	27.5	10.2	4.1

For calculation it was assumed that these five phospholipids together accounted for all of the phospholipids of the membrane. In the actual analyses, other phospholipids (primarily lysophosphatidyl choline and lysophosphatidyl ethanolamine) were found to account for less than 5% of the total. The "heavy" and "light" microlipid droplet fractions were obtained by sucrose density gradient centrifugation. Heavy fractions were collected between 0.5 and 1.0 M sucrose and light fractions banded at the 0.5 M sucrose–buffer interface.
Abbreviations: SP, sphingomyelin; PC, phosphatidyl choline; PE, phosphatidyl ethanolamine; PI, phosphatidyl inositol; PS, phosphatidyl serine.
SOURCE: Data from [a]Dylewski et al. (1984), [b]Keenan et al. 1970.

et al. 1974B; Keenan *et al.* 1979). Various fractions of intracellular lipid droplets, either cytoplasmic lipid droplets with diameters more than 1 μm in diameter or microlipid droplets with diameters between 0.05 and 0.25 μm, also contain phospholipids (Dylewski *et al.* 1984; Keenan and Dylewski 1985). Sphingomyelin values in these fractions are intermediate between the levels in endoplasmic reticulum and milk lipid globule membrane (Table 10.3). The fate of the phospholipids associated with intracellular lipid droplets remain unknown, although they could contribute either to the surface membrane on secreted fat droplets or to the triacylglycerol core. This latter possibility seems unlikely, however, since previous analyses have indicated that core fat is devoid of phospholipid (Huang and Kuksis 1967; Patton and Keenan 1971). With the proviso that some phospholipid in the globule membrane may be contributed from intracellular lipid droplets, the distinctive distribution of phospholipids shared by plasma membrane and globule membrane provides biochemical support for a plasma membrane origin of the outer surface of the lipid globule membrane.

The same five major phospholipids, with a similar distribution pattern, are present in milk or globule membranes of other species including sheep, Indian buffalo, camel, ass, pig, human (Morrison 1968), rat (Keenan *et al.* 1972C), goat (Patton and Keenan 1971), and mouse (Calberg-Bacq *et al.* 1976). However, guinea-pig lipid globule membrane has comparatively low levels of sphingomyelin, which is apparently a consequence of low levels of this polar lipid in intracellular membranes of lactating guinea-pig mammary tissue (Greenwalt and Mather, unpublished).

Alkyl and alkenyl ethers have been identified in phosphatidyl choline and phosphatidyl ethanolamine fractions recovered from whole milk (Hay and Morrison 1971). To our knowledge, levels of alkyl and alkenyl ethers in phospholipids of milk lipid globule membrane have not been measured.

Corresponding classes of phospholipids in intracellular lipid droplets, globule membrane, and milk serum are virtually identical in fatty acid composition (Huang and Kuksis 1967; Dylewski *et al.* 1984). Morrison (1970) has compiled data on the fatty acid composition of phospholipid classes from milks of a number of species, including the cow. Phosphoglycerides of milk have higher levels of di- and polyunsaturated fatty acids and much lower levels of short chain fatty acids than do milk triglycerides. Sphingomyelin contains predominantly saturated fatty acids, including appreciable amounts of behenic (22:0, number of carbons:number of double bonds), *n*-tricosanoic (23:0), and lignoceric (24:0) acids. The glycosphingolipids of milk lipid globule membrane (cerebrosides and gangliosides) are similar to sphingomyelin in fatty acid composition (Morrison *et al.* 1965; Hladik and Michalec 1966; Kayser and Patton 1970; Morrison 1970; Keenan 1974A; Bushway and Keenan 1978).

Milk lipid globule membrane, intracellular lipid droplets, and endomembrane fractions to contain a number of carbohydrate-containing sphingolipids which are members of glycolipid classes known as neutral glycosphingolipids (cerebrosides) and gangliosides (sialic acid–containing glycophingolipids). Morrison and Smith (1964) first identified mono- and dihexosylceramides in bovine milk lipids. These were subsequently shown to be concentrated in milk lipid globule membrane (Hladik and Michalec 1966; Kayser and Patton 1970), although they are also found in intracellular lipid droplets (Dylewski *et al.* 1984) and in milk serum (Kayser and Patton 1970), where they are probably associated with skim milk membranes. These neutral glycosphingolipids have the structures β-glucosyl-(1→1)-N-acylsphingosine (glucosylceramide) and β-galactosyl-(1→4)-β-glucosyl-(1→1)-N-acylsphingosine (lactosylceramide) (Fujino *et al.* 1970). Glucosyl- and lactosylceramides occur in nearly equal proportions in milk lipid globule membrane and, on a protein basis, at about four times the level in cytoplasmic lipid droplets (Dylewski *et al.* 1984). Free ceramide (N-acylsphingosine) has been identified in milk lipids but has not yet been demonstrated in the globule membrane (Fujino and Fujishima 1972). Neutral glycosphingolipids more complex than lactosylceramide have not been detected in milk lipid globule membrane. Human lipid globule membrane also contains glucosyl- and lactosylceramides, but in this membrane galactosylceramide is the most abundant cerebroside (Bouhours and Bouhours 1979).

Keenan *et al.* (1972A,D) noted the presence of six chromatographically distinguishable gangliosides in bovine milk lipid globule membrane and mammary endomembrane fractions. These observations have been confirmed, and the carbohydrate sequences of these gangliosides have been elucidated (Huang 1973; Keenan 1974A; Bushway and Keenan 1978). The structures of these six gangliosides are given in Table 10.4. On a sialic acid basis, these gangliosides together occur in milk lipid globule membrane at a level of about 6 nmoles per milligram of protein (Keenan 1974A; Keenan *et al.* 1972A, 1979), and the disialoganglioside GD_3 is the major constituent. Fractions of intracellular lipid droplets also contain gangliosides in a pattern very similar to that of milk lipid globule membrane (Dylewski *et al.* 1984). Glycosphingolipids of the type found in the globule membrane and intracellular lipid droplets are synthesized by stepwise addition of carbohydrates to ceramide (Basu *et al.* 1980). Based on similarities in fatty acid composition, it appears that sphingomyelin, cerebrosides, and gangliosides of milk

Table 10.4. Structures of Glycosphingolipids of Bovine Milk Lipid Globule Membrane.

Glycosphingolipid[a]	Structure
Glucosylceramide	β-Glucosyl-(1→1)-ceramide
Lactosylceramide	β-Galactosyl-(1→4)-β-glucosyl-(1→1)-ceramide
GM$_3$ (hematoside)	Neuraminosyl-(2→3)-galactosyl-glucosyl-ceramide
GM$_2$	N-Acetylgalactosaminyl-(neuraminosyl)-galactosyl-glucosyl-ceramide
GM$_1$	Galactosyl-N-acetylgalactosaminyl-(neuraminosyl)-galactosyl-glucosyl-ceramide
GD$_3$	Neuraminosyl-(2→8)-neuraminosyl-(2→3)-galactosyl-glucosyl-ceramide
GD$_2$	N-Acetylgalactosaminyl-(neuraminosyl-neuraminosyl)-galactosyl-glucosyl-ceramide
GD$_{1b}$	Galactosyl-N-acetylgalactosaminyl-(neuraminosyl-neuraminosyl)-galactosyl-glucosyl-ceramide

[a] Abbreviations for gangliosides are those of Svennerholm (1963).
SOURCE: Huang (1973), Keenan (1974A) and Bushway and Keenan (1978).

are synthesized from a common pool of ceramides (Morrison *et al.* 1965; Hladik and Michalec 1966; Kayser and Patton 1970; Morrison 1970; Keenan 1974A,B; Bushway and Keenan 1978).

Protein Composition

Our knowledge of the protein composition of membranes has advanced markedly since the second edition of this book was published. This is largely due to the introduction of sodium dodecyl sulfate as an agent for the disaggregation of membrane samples. Under appropriate conditions, this detergent effects nearly complete solubilization of membranes, leading to the formation of polypeptide–dodecyl sulfate complexes. When membrane proteins treated with sodium dodecyl sulfate are separated by electrophoresis in polyacrylamide gels containing dodecyl sulfate, the relative mobility of the proteins correlates with their M_r (Weber and Osborn 1969). This occurs because the native charges of the proteins are masked by the large amount of negatively charged detergent bound; separation of the dodecyl sulfate–protein complexes is therefore effected by molecular sieving through the pores of the polyacrylamide gel.

A number of investigators have used this method for the characterization of milk lipid globule membrane proteins. Prior to this development, various soluble and insoluble fractions of globule membrane were characterized, but results were usually average values for an ag-

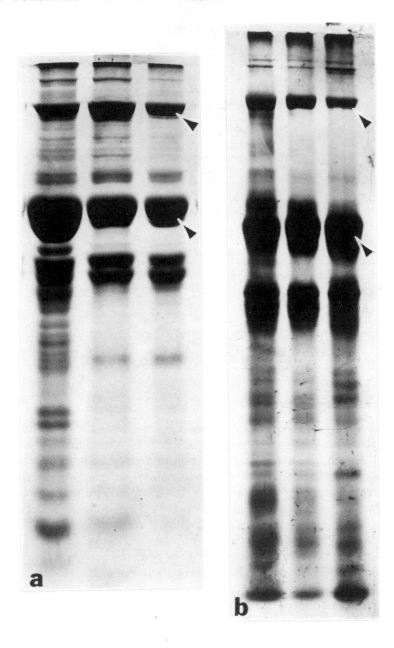

Figure 10.7 Polypeptide patterns of material associated with lipid globules of cow's milk. The sodium dodecyl sulfate-polyacrylamide gel in (a) was stained with coomassie blue, and polypeptides in the gel in (b) were visualized with the more sensitive silver stain (Merril *et al.* 1981). In both (a) and (b) the left lane contains proteins extracted directly from washed milk

gregate of different proteins and were not characteristic of individual proteins. These earlier studies were important to more recent developments in this area and in fact, in many cases, early fractionation methods have been used as starting points for the purification of individual proteins. Since this early work is discussed in detail in the second edition of this book (Brunner 1974), attention here will be focused on recent developments.

Several groups of investigators have now reported separation of the polypeptides of milk lipid globules or the isolated globule membrane in sodium dodecyl sulfate–polyacrylamide gels (Kobylka and Carraway 1972, 1973; Anderson 1974; Anderson *et al.* 1974B; Kitchen 1974, 1977; Mather and Keenan 1975; Mangino and Brunner 1975; Mather *et al.* 1977; Freudenstein *et al.* 1979; Eigel *et al.* 1984). In these reports, there are variations in the number of polypeptides detected and in the molecular weights calculated for these polypeptides. In spite of these variations, which may be due in part to methodological differences, patterns obtained by most groups show striking similarities in the number and distribution of the major polypeptides and glycoproteins of the membrane. All groups have detected major size classes of polypeptides of M_r about 155,000 and about 67,000. Although calculated molecular weights varied, relative mobilities show that all groups, in fact, detected the same classes of polypeptide. A third major polypeptide size class, consisting of at least two polypeptides of M_r about 44,000 and 48,000, was observed in some but not all studies. These polypeptides can be selectively extracted from lipid globules with solutions of high ionic strength (Mather and Keenan 1975). Moreover, variable amounts of these polypeptides remain associated with the core lipid or remain in the supernatant when the membrane is sedimented. When lipid globules are directly extracted with solutions containing sodium dodecyl sulfate and the extract is analyzed by electrophoresis, these polypeptides are invariably observed as major components (Kobylka and Carraway 1972; Mather and Keenan 1975).

Typical electrophoretic patterns for lipid globules and milk lipid globule membrane are illustrated in Figure 10.7. In addition to these

lipid globules, the center lane contains proteins of milk lipid globule membrane which was released by churning and collected by ultracentrifugation, and the right lane contains milk lipid globule membrane material insoluble in 1.5 M KC1 and 1% Triton X-100. Arrowheads denote xanthine oxidase (upper, M_r 155,000) and butyrophilin (lower, M_r 67,000). Note that many distinct polypetides associated with washed lipid globules are depleted during isolation of the membrane and that xanthine oxidase and butyrophilin become concentrated in the salt- and detergent-insoluble material.

major polypeptide size classes, several other polypeptides have been detected in electrophoretic gels stained with coomassie blue, the dye most commonly used for staining proteins. Mather and Keenan (1975) enumerated at least 21 bands ranging in molecular weight from about 250,000 to 11,000. With application of larger amounts of sample on gels, we have detected additional polypeptides in the 10,000- to 40,000-dalton range. Under these conditions, regions of gels containing proteins of higher molecular weight were so overloaded that detection of individual constituents was impossible. The number of different polypeptides which occur in native milk lipid globule membrane is not known, but it is certainly many more than the 21 to 30 detected to date. By the use of isoelectric focusing in polyacrylamide gels, in which proteins are separated according to charge, Mather (1978) detected over 40 polypeptides in milk lipid globule membrane. However, many of these proteins were undoubtedly isoelectric variants of the major proteins such as xanthine oxidase and butyrophilin. To gain a more accurate picture of the protein composition, one could separate the membrane proteins by either of the two-dimensional electrophoretic systems described (O'Farrell 1975; O'Farrell *et al.* 1977). In comparison with one-dimensional separation, resolution is much superior in these systems because proteins are separated in one dimension by charge (electrofocusing) and in the second dimension by size. Several such two-dimensional maps of bovine and guinea pig milk lipid globule membrane have been published (Mather 1978; Franke *et al.* 1981; Jarasch *et al.* 1981; Bruder *et al.* 1982; Johnson and Mather 1985). However, in all of these studies, either the gels were loaded with insufficient protein or the methods used for detection were not sufficiently sensitive to enable detection of the less abundant proteins. Minor membrane proteins could be detected better with the recently developed sensitive silver stain (Merril *et al.* 1981) shown in Figure 10.7B or by the *in vitro* incorporation of radioactive label into the proteins and subsequent detection of separated constituents by autoradiography or fluorography.

Evaluation of the protein complexity of globule membranes is complicated by the association of a plasmin-like protease with the membrane (Hofmann *et al.* 1979). One must thus ascertain whether constituents detected are native to the membrane or are proteolytically derived fragments of other membrane proteins. In this regard, the major coomassie blue–stained proteins are distinctly different from each other, as judged by immunological characterization (Mather *et al.* 1980, 1982; Franke *et al.* 1981; Jarasch *et al.* 1981; Bruder *et al.* 1982; Greenwalt and Mather 1985) and by peptide mapping procedures (Heid 1983). In contrast, however, polypeptides of M_r 43,500 and 48,000 (Mather and

Keenan 1975) appear to be closely related (Mather *et al.* 1980; Johnson *et al.* 1985).

In addition to coomassie blue–stained polypeptides, five (Kitchen 1974; Mather and Keenan 1975), six (Kobylka and Carraway 1972; Anderson *et al.* 1974B, or seven (Mather *et al.* 1980; glycoproteins have been detected on one-dimensional gels of globule membrane proteins. These glycoproteins have been identified by using Schiff reagent to detect carbohydrates after periodate oxidation. It is a common observation that some glycoproteins stain poorly, if at all, with coomassie blue (Kobylka and Carraway 1972; Mather and Keenan 1975). Four of the membrane glycoproteins migrate with coomassie blue–stained polypeptides, whereas three others appear to migrate in gel regions where polypeptides are not detected with coomassie blue. Using radioactive lectins to "stain" gels, Murray *et al.* (1979) detected at least seven different glycoproteins in bovine milk lipid globule membrane, and at least eight different glycoproteins or variants have been detected by electrofocusing (Mather 1978). The true complexity of milk lipid globule membrane with respect to glycoprotein composition has not received detailed study.

Studies of globule membrane proteins of species other than the cow have been limited. Human and guinea pig milk lipid globule membranes have polypeptide patterns qualitatively similar to that of the cow (Martel *et al.* 1973; Freudenstein *et al.* 1979; Murray *et al.* 1979; Johnson *et al.* 1985). Constituents with mobilities similar to the M_r 155,000 and 67,000 polypeptides in bovine globule membrane have been detected in human, pig, goat, and sheep globule membrane preparations. Butyrophilin from rat and guinea pig has a slightly lower M_r and more basic isoelectric points (Heid *et al.* 1983; Johnson *et al.* 1985). However, butyrophilin from all the above species appears to be closely related, using peptide mapping and immunological properties as criteria (Heid *et al.* 1983; Johnson and Mather 1985). Xanthine oxidase from the cow, human, goat, and guinea pig also appears similar, but not identical, with respect to molecular weight, isoelectric points, and immunological characteristics (Jarasch *et al.* 1981; Heid 1983; Johnson *et al.* 1985).

In addition, all species examined contain at least one glycoprotein of high M_r which stains with the periodic acid-Schiff (PAS) reagent, but poorly, if at all, with coomassie blue e.g., component I in bovine membranes (Mather and Keenan 1975), PAS-O in human membranes (Shimizu and Yamauchi 1982), and PAS-I in guinea pig membranes (Johnson and Mather 1985; Johnson *et al.* 1985). In human MFGM this glycoprotein has an M_r probably in excess of 500,000, so that the glycoprotein remains unresolved at the top of the stacking or separating gels after electrophoresis in polyacrylamide (Burchell et al. 1983; Ceriani et al. 1983). Despite differences in molecular size, the high M_r glycoproteins of human, bovine, and guinea pig milk lipid globule membranes contain some similar immunological determinants, using cross-reactivity with monoclonal antibodies as a criterion (Greenwalt et al. 1985A; Johnson and Mather 1985).

Over the past several years, much progress has been made in the purification and characterization of milk lipid globule membrane proteins. While we know little regarding the functional significance of many of the membrane proteins, the rapid rate at which knowledge is accumulating suggests that the significance of individual proteins in the secretion and stabilization of milk lipid globules may soon become apparent.

The polypeptide size class of M_r 155,000 has been found to consist, at least in part, of xanthine oxidase (Waud et al. 1975; Mangino and Brunner 1977A; Mather et al. 1977, 1982; Jarasch et al. 1981; Bruder et al. 1982; Sullivan et al. 1982). Xanthine oxidase is a complex, iron-sulfur, molybdenum flavoprotein with multifunctional enzymatic activities (Bray 1975; Coughlan 1980). This enzyme oxidizes purines, pyrimidines, aldehydes, and pterins and has been used to generate superoxide radical. The biological function of this enzyme, which occurs in several different body tissues as well as in milk, is obscure. The electrophoretic band containing xanthine oxidase accounts for 10 to 20% of the protein associated with milk lipid globules. At least four isoelectric variants, focusing in the pH range of 7.0 to 7.5, are observed in xanthine oxidase purified to apparent homogeneity (Jarasch et al. 1981; Bruder et al. 1982; Sullivan et al. 1982). Part of this enzyme is associated stably with the membrane (Briley and Eisenthal 1974; Bruder et al. 1982), and when the membrane is extracted with nonionic detergents, xanthine oxidase is one of two polypeptides selectively enriched in the detergent-insoluble fraction (coat fraction) (Freudenstein et al. 1979). A considerable portion of xanthine oxidase is released in soluble form on lysis of lipid globules and recovery of the membrane (Briley and Eisenthal 1974; Mather et al. 1977; Jarasch et al. 1981). In mammary tissue the enzyme is in a predominantly soluble form (Bruder et al. 1982; Mather et al. 1982).

The significance of this dual manifestation of membrane-bound and soluble forms of xanthine oxidase is not understood. Several workers have suggested that xanthine oxidase generates free radicals and hydrogen peroxide in vivo which may act as antimicrobial agents (Björck and Claesson 1979). Alternatively, superoxide radicals may alter lipid fluidity by promoting peroxidation reactions during the envelopment of lipid globules by apical plasma membrane (Jarasch et al. 1981). However, while xanthine oxidase has been used to promote lipid peroxidation in *in vitro* systems, we could find no evidence supporting a role of the native form of xanthine oxidase, in milk lipid globule membranes or the associated soluble fraction, in lipid oxidation (Bruder *et al.* 1982). Furthermore, the enzyme is present in tissues predominantly as a dehydrogenase which is unable to utilize molecular oxygen as electron acceptor and generate superoxide radicals (Stirpe and Della Corte 1969; Della Corte and Stirpe 1972). The enzyme is converted to an oxidase only during or after secretion into milk (Battelli *et al.* 1973). Attempts to explain the biological function of xanthine oxidase in tissues, before secretion of the enzyme, should therefore take these considerations into account.

Antibodies to xanthine oxidase have been used to show that this enzyme is concentrated in apical regions of mammary epithelial cells, but that it is also found throughout the epithelial cell cytoplasm (Jarasch *et al.* 1981). This enzyme has an unusual distribution in that in tissues other than mammary gland, it is detected only in capillary endothelial cells (Jarasch *et al.* 1981; Bruder *et al.* 1982). Interestingly, many species, including man, cow, rabbit, and guinea pig, produce autoantibodies to xanthine oxidase, which leads to quite high serum antibody titers to the enzyme in individual animals (Bruder *et al.* 1984). These antibodies may be generated following tissue damage, when endothelial cells release soluble proteins into the systemic circulation, although it is not clear why individuals do not become tolerant to the antigen.

While bovine milk is a rich source of xanthine oxidase, milks from some species do not necessarily contain appreciable amounts of enzymatically active xanthine oxidase. For example, human milk contains only traces of xanthine oxidase activity as measured by oxidation of xanthine or hypoxanthine (Zikakis and Treece 1971; Zikakis *et al.* 1976), yet a band corresponding in electrophoretic mobility to xanthine oxidase is a major constituent of human milk lipid globule membrane (Freudenstein *et al.* 1979; Murray *et al.* 1979). Evidence that the membrane-bound form of xanthine oxidase in bovine lipid globule membrane contains small amounts of tightly bound fatty acid has been obtained (Keenan *et al.* 1982). Whether this property promotes attraction between the membrane or membrane-associated coat and the surface of the globule core remains to be determined.

Polypeptide size class 67,000 is the major electrophoretically resolved constituent of bovine milk lipid globule membrane. This band consists, in large part, of a hydrophobic, difficult to solubilize protein which has been named "butyrophilin" so as to reflect its association with and affinity for milk lipid (Franke *et al.* 1981). This protein is insoluble or sparingly soluble in aqueous solutions after treatment with many detergents, chaotropic agents, or solutions of high ionic strength (Mather et al. 1977; Freudenstein et al. 1979) Butyrophilin appears to have a propensity for tight interaction with other proteins, and for this reason has proven difficult to purify (Freudenstein et al. 1979) except by separation in and elution from dodecyl sulfate-polyacrylamide gels (Franke et al. 1981; Keenan et al. 1982). So far, no enzymatic activity has been ascribed to butyrophilin (Mather et al. 1980). Bovine butyrophilin has at least four isoelectric variants focusing in the pH range of 5.2 to 5.3 in the presence of urea (Franke et al. 1981; Heid et al. 1983). This protein, which has been shown to be a glycoprotein, tenaciously binds phospholipids (Freudenstein et al. 1979). Like xanthine oxidase, butyrophilin appears to contain tightly, perhaps covalently, associated fatty acids (Keenan et al. 1982). Antibodies to butyrophilin have been used to show that, within mammary epithelial cells, this protein is observed at the apical cell surface, including those portions of the apical plasma membrane over budding milk lipid globules (Franke et al. 1981; Johnson and Mather 1985) and in small quantities on intracellular lipid droplets (Deeney et al. 1985). Butyrophilin was not detected elsewhere in mammary tissue or in any of the several other tissues examined. Differential scanning calorimetry has been used to show that butyrophilin undergoes an irreversible endothermic transition at 58°C; on denaturation butyrophilin forms a disulfide-stabilized aggregate (Appell et al. 1982). While function cannot yet be ascribed to butyrophilin, its properties and location are such as to imply a role for this constituent in the envelopment of lipid globules with membrane during secretion.

Polypeptides in the M_r 44,000 to 48,000 size class are not tightly associated with milk lipid globule membrane, as evidenced by their depletion from the membrane during isolation and their extractability from lipid globules with salt solution (Mather and Keenan 1975). The fraction solubilized by extraction with salt solution contains at least two glycoproteins, one of which may be identical to the glycoprotein that Basch et al. (1976) purified and characterized. The proteins extracted from lipid globules with magnesium chloride were found to have M_rs of about 43,500 and 48,000 (Mather and Keenan 1975), while the salt soluble glycoprotein purified by Basch et al. (1976) was found to have an M_r of about 49,500 on dodecyl sulfate-polyacrylamide gels. The purified protein, termed "glycoprotein B," was found to have single N- and C-terminal amino acids (serine and leucine, respectively). This protein contains 14% carbohydrate, including sialic acide, mannose, galactose, glucose, galactosamine, and glucosamine.

A protein of M_r 55,000 in guinea pig milk lipid globule membranes appears to be similar but not identical to the bovine proteins using peptide mapping techniques and solubility in aqueous solutions as comparative criteria (Johnson et al. 1985). This protein is synthesized in a membrane-bound form and becomes progressively solubilized after incorporation into intracellular membranes (Mather et al. 1984), a property shared by several other peripheral membrane proteins, e.g., the glycoprotein GP2 in the pancreas (Scheffer et al. 1980).

In an attempt to determine if microfilaments have a role in lipid globule secretion, we searched for actin, a major constituent of cytoplasmic microfilaments, in milk lipid globule membrane (Keenan et al. 1977A). A fraction which contained two polypeptides with apparent molecular weights of about 48,000 and 50,000 was isolated. Proteins in this fraction resembled actin in that they interacted with myosin, aggregated at high ionic strength, and were tightly bound by deoxyribonuclease I. However, these proteins were distinguished from known actins by their molecular weights, amino acid composition, inability to stimulate myosin adenosine triphosphatase activity, and the ultrastructure of their aggregated forms. These proteins displayed a tenacious association with lipids.

Several glycoprotein fractions and purified glycoproteins of milk lipid globule membrane have been characterized in recent years. Early studies with glycoprotein fractions from the membrane have been reviewed, and this information will not be repeated here (Brunner 1974; Patton and Keenan 1975; Anderson and Cawston 1975). Kanno et al. (1975) obtained a soluble glycoprotein fraction from the globule membrane which they characterized by determining its carbohydrate composition and sedimentation velocity. This fraction appeared to be homogeneous by sedimentation velocity analysis but was found to have multiple N-terminal amino acids (Kanno et al. 1975) and to be heterogeneous by isoelectric focusing and immunoelectrophoresis (Kanno et al. 1977). Examination on dodecyl sulfate-polyacrylamide gels revealed this preparation to contain seven glycoproteins and at least one nonglycosylated protein (Shimizu et al. 1976). Kanno and Yamauchi (1978) found that antibodies to this soluble glycoprotein fraction from milk lipid globule membrane reacted with a protein(s) found in whey which was not identical to any of the major whey proteins. Keenan et al. (1977B) used lithium diiodosalicylate to obtain a soluble glycoprotein fraction from globule membranes. By dodecyl sulfate electrophoresis this fraction was found to contain three glycoproteins, with apparent M_rs of 215,000, 135,000 and 86,000 in 10% (w/v) polyacrylamide gels. The M_r 215,000 polypeptide was the major constituent of this fraction.

This fraction contained carbohydrates typical of membrane glycoproteins and was especially rich in galactose and sialic acid. Glycoproteins in this fraction avidly bound the lectin Concanavalin A.

Apart from glycoprotein B, discussed above, several other glycoproteins have been purified to apparent homogeneity and characterized from bovine, human, and guinea pig milk lipid globule membranes. Glycoproteins of apparent M_r 155,000, 70,000, and 39,000 have been purified from human membranes (Imam et al. 1981, 1982) and characterized with respect to amino acid and sugar composition. The glycoprotein of M_r 70,000 was localized in mammary tissue and various mammary carcinomas (Imam et al. 1984) and may be identical to human butyrophilin. However, the presence of sialic acid in this preparation (Imam et al. 1981) discredits this possibility, since both human and bovine butyrophilin do not contain this sugar (Heid et al. 1983). Also, the isolate of Imam et al. (1981) was obtained in soluble form by the extraction of human milk lipid globules with $MgCl_2$ solutions. These conditions were originally shown to extract peripheral proteins from bovine milk lipid globule membranes, leaving the bulk of butyrophilin (component 12) in the membrane residue (Mather and Keenan 1975).

Shimizu and Yamauchi (1982) purified the mucin-like glycoprotein of high M_r from human milk lipid globule membranes (PAS-O) by gel filtration of detergent/urea extracts. The purified glycoprotein was composed of about 50% by weight of carbohydrate and contained high levels of serine and threonine. The principal sugars detected were fucose, galactose, N-acetylglucosamine, N-acetylgalactosamine, and sialic acid, properties characteristic of mucin-like glycoproteins, with carbohydrate chains covalently linked to the peptide chain via O-glycosidic bonds. PAS-O contained the sugar sequence D-galactosyl-β-(1→3)-N-acetyl-D-galactosamine, the Thomsen-Friedenreich antigen (Springer and Desai 1974; Cartron et al. 1978; Shimizu and Yamauchi 1982), which is specifically recognized by peanut agglutinin (PNA), the lectin from Arachis hypogaea (Lotan et al. 1975). Binding of PNA occurs after removal of terminal sialic acid residues, and expression of lectin-binding sites on the surface of cells has been used to monitor clinically the course of various malignancies. Discussion of this topic is beyond the scope of this chapter, and the reader is referrred to a review by Springer (1984) for further information.

Various other investigators have isolated PAS-O or fragments of this glycoprotein from either milk lipid globule membranes (Fischer et al. 1984) or human skim milk (Ormerod et al. 1983). The preparation of Ormerod et al., called "epithelial membrane antigen," was heterodisperse and appears to consist of proteolytic cleaved peptides of the native protein. Fischer *et al.* (1984) isolated a desialylated preparation of PAS-O by neuraminidase digestion, solubilization in Triton X-100, and PNA affinity chromatography. The results of sugar analysis of this preparation were generally in good agreement with those of Shimizu and Yamauchi (1982), although Fischer *et al.* were apparently unaware of the earlier Japanese work.

Several groups have prepared monoclonal antibodies to PAS-O (Taylor-Papadimitriou *et al.* 1981; Foster *et al.* 1982B; Ceriani *et al.* 1983). Many of these antibodies appear to recognize carbohydrate epitopes (Burchell *et al.* 1983; McIlhinney *et al.* 1985) which are expressed on the apical surface of lactating mammary cells and also, in some cases, on human mammary carcinomas (Arklie *et al.* 1981; Foster *et al.* 1982A,B; Burchell *et al.* 1983). Monoclonal antibodies to a mucin-like glycoprotein, PAS-I, in guinea pig milk lipid globule membrane (Johnson and Mather 1985; Johnson *et al.* 1985) also cross-react with human PAS-O and can be used to detect infiltrating duct carcinomas of the human breast (Greenwalt *et al.* 1985A). Clearly, these mucin-like glycoproteins and their associated carbohydrate chains have clinical potential as tumor markers (Epenetos *et al.* 1982; Wilkinson *et al.* 1984).

The high M_r glycoprotein, PAS-I, of guinea pig milk lipid globule membranes has been purified and appears to be a mucin-like glycoprotein with some similarities to human PAS-O (Greenwalt *et al.*, unpublished). Guinea pig PAS-I contains serine and threonine as the principal amino acids (accounting for 30 mol % of the total) and mannose, galactose, *N*-acetylglucosamine, *N*-acetylgalactosamine, and sialic acid as the principal sugars. The presence of mannose and *N*-acetylglucosamine suggests that some N-linked carbohydrate chains covalently bound to asparagine residues are also present in addition to O-linked moieties.

As discussed above, bovine milk lipid globule membranes also contain a high M_r glycoprotein which stains with the PAS reagent but not with coomassie blue (Mather *et al.* 1980). Snow *et al.* (1977) released a glycoprotein from aqueous suspensions of bovine globule membranes or from lipid globules by exposure to chloroform-methanol and purified it to near homogeneity by gel filtration. This glycoprotein had an apparent M_r of 70,000 by dodecyl sulfate-polyacrylamide gel electrophoresis and was found to be heavily glycosylated, containing about 50% by weight carbohydrate. Some properties of this glycoprotein were similar to those of human PAS-O and guinea pig PAS-I. However, whether the isolate of Snow *et al.* was identical to bovine PAS-I is uncertain, since estimates of the apparent M_r of this preparation are only a third of those from bovine PAS-I in unfractionated membranes. One of the authors (I.H.M.) considered the possibility that this preparation consisted of the glycoprotein PAS-IV (Mather *et al.* 1980), which has an apparent M_r of 76,000. However, the recent isolation and characterization of PAS-IV (Greenwalt and Mather 1985; Greenwalt *et al.* 1985B) has clearly shown that the two preparations contain different proteins.

PAS-IV is a hydrophobic glycoprotein containing approximately 5% carbohydrate on a weight basis. Purified preparations contained mannose, galactose, and sialic acid as the principal sugars (Greenwalt and Mather 1985). The glycoprotein contained a high proportion of amino acids with nonpolar residues and displayed hydrophobic properties in aqueous solution. By several criteria, PAS-IV appeared to be an integral component of milk lipid globule membranes, including resistance to digestion by exogenous proteases when bound to membrane and separation in the detergent phase in Triton X–114 solutions at room temperature. By the use of specific polyclonal and monoclonal antibodies, the distribution of PAS-IV was determined in the mammary gland and other tissues. Interestingly, like xanthine oxidase, PAS-IV was detected in capillary endothelial cells in many tissues, including mammary gland, heart, liver, spleen, pancreas, salivary gland, and small intestine. Epithelial expression of this glycoprotein appeared to be restricted to the mammary gland and lung (Greenwalt and Mather 1985; Greenwalt *et al.* 1985B). As in the case of xanthine oxidase, the significance of this tissue distribution remains unknown, although it is curious that two components of the epithelially derived lipid globule membrane share a common distribution in cells of different ontogeny.

Harrison *et al.* (1975) released glycopeptides from lipid globules by treatment with pronase, fractionated these glycopeptides, and characterized their carbohydrate structures. They obtained evidence for both O- and N-linked oligosaccharide chains in lipid globule membrane glycoproteins. Further work by Newman *et al.* (1976) led to the establishment of the structure of a tetrasaccharide released from globule membrane as β-D-galactosyl(1→3)-N-acetyl-D-galactosamine (the Thomsen-Friedenreich antigen discussed earlier) substituted by sialic acid at position C 3 of galactose and C 6 of N-acetyl-D-galactosamine. Several workers, (Farrar and Harrison 1978; Glöckner *et al.* 1976; Newman and Uhlenbruck 1977; Farrar *et al.* 1980) confirmed this structure in bovine lipid globule membranes, demonstrated it in human globule membranes, presumably associated with PAS-O, (Shimizu and Yamauchi 1982), and elucidated structures of two trisaccharides released from bovine globule membrane glycoproteins by alkaline borohydride (O-linked oligosaccharides).

The amino acid composition of milk lipid globule membranes, as determined by several groups, has been summarized elsewhere (Patton and Keenan 1975). Some differences are evident in the data from different laboratories, which, as discussed by Mangino and Brunner (1977B), reflect differences in both preparative methods and analytical precision. In all studies, milk lipid globule membrane was characterized by high levels of glutamic and aspartic acids and leucine and low levels of sulfur amino acids. Using a statistical difference index for comparison, Mangino and Brunner found a high degree of compositional homology between globule membrane, various plasma membranes, and membrane-associated proteins. They speculated that this homology reflected the evolutionary convergence of proteins necessitated by the lipid bilayer environment of membranes.

Enzymes of Milk Lipid Globule Membrane

Numerous enzymatic activities have been measured in lipid globule membranes, as summarized in Table 10.5. Several of the enzymes with high specific activities in the globule membrane, such as 5'-nucleotidase, phosphodiesterase I, and adenosine triphosphatase, are characteristically found in plasma membranes and in at least some tissues serve as marker enzymes for plasma membrane. Marker enzymes specific for the plasma membrane of mammary gland have not been established, but 5'-nucleotidase is known to be enriched in plasma membrane–rich fractions from lactating mammary gland (Huang and Keenan 1972C; Huggins and Carraway 1976; Huggins et al. 1980). As discussed in a previous section, xanthine oxidase is an abundant protein of the globule membrane. It can catalyze the oxidation of NADH (Bray 1975), and evidence has been obtained that part of the NADH-cytochrome c reductase (as well as NADPH-cytochrome c reductase) activity of milk lipid globule membrane is due to xanthine oxidase (Jarasch et al. 1977; Bruder et al. 1982). However, the NADH-ferricyanide reductase activity of the membrane appears to be separate from that of xanthine oxidase (Bruder et al. 1982). Milk lipid globule membranes have been found to have a cytochrome-linked redox system, and the cytochrome in this system consists of two components, cytochrome b_5 and cytochrome P-420 (Jarasch et al. 1977; Bruder et al. 1978). No cytochrome P-450 was detected in globule membrane or in endoplasmic reticulum from mammary gland, suggesting that cytochrome P-420 is a native membrane constituent and not a degradation product of P-450 (Bruder et al. 1978).

Several enzymatic activities of the membrane are not specific to plasma membranes but instead are widely distributed throughout intracellular membranes. Activities of several enzymes normally associated with lysosomes have been reported in lipid globule membranes,

Table 10.5. Enzymatic Activities Detected in Bovine Milk Lipid Globule Membrane.

Enzyme	Reference[a]
Alkaline phosphatase EC 3.1.3.1	1, 2, 3
Acid phosphatase EC 3.1.3.2	1, 2, 3
5'-Nucleotidase EC 3.1.3.5	1, 2
Phosphodiesterase I EC 3.1.4.1	1, 2, 3
Inorganic pyrophosphatase EC 3.6.1.1	1, 2
Nucleotide pyrophosphatase EC 3.6.1.9	1, 2
Phosphatidic acid phosphatase EC 3.1.3.4	4
Adenosine triphosphatase EC 3.6.1.3	1, 2
Lipoamide dehydrogenase EC 1.6.4.3	1
Cholinesterase EC 3.1.1.8	1, 2
Aldolase EC 4.1.2.13	1, 2
Xanthine oxidase EC 1.2.3.2	1
Thiol oxidase EC 1.8.3.2	1, 2
γ-Glutamyl transpeptidase EC 2.3.2.1	1, 2, 5
UDP-glucose hydrolase EC 3.2.1.__	1, 2, 6
UDP-galactose hydrolase EC 3.2.1.__	1, 2, 6
NADH-cytochrome c reductase EC 1.6.99.3	2, 7
NADH-ferricyanide reductase EC 1.6.99.3	7
NADPH-cytochrome c reductase EC 1.6.99.1	7
Glucose-6-phosphatase EC 3.1.3.9	2, 4
Galactosyl transferase EC 2.4.1.__	8
Plasmin EC 3.4.21.7	9
β-Glucosidase EC 3.2.1.21	2
β-Galactosidase EC 3.2.1.23	2
Catalase EC 1.11.1.6	2
Ribonuclease I EC 3.1.4.22	2

[a]Where possible, reference is made to a review where the primary references have been cited. References: (1) Patton and Keenan (1975), (2) Anderson and Cawston (1975), (3) Diaz-Maurino and Nieto (1977), (4) Dowben et al. (1967), (5) Baumrucker (1979), (6) Keenan and Huang (1972A), (7) Jarasch et al. (1977), Bruder et al. (1978, 1982), (8) Powell et al. (1977), (9) Hofmann et al. (1979).

notably aldolase, acid phosphatase, and β-glycosidases. With all except acid phosphatase, specific activities are low and variable and may be due to the presence of leukocytes entrained in lipid globule preparations (Anderson and Cawston 1975; Anderson 1977). In contrast to other lysosomal enzymes, acid phosphatase is present in high specific activities, and there is reason to believe that this enzyme is a true constituent of the primary milk lipid globule membrane (Anderson and Cawston 1975). There is controversy regarding the presence of galactosyl transferase in the globule membrane. This enzyme is normally considered to be a marker for the Golgi apparatus (Roth and Berger 1982), although recent work indicates that the enzyme also may be located on the surface of some cells (Roth et al. 1985; Shaper et al. 1985). Galactosyl transferase has been detected in preparations of bovine (Powell *et al.* 1977) and human milk lipid globule membranes (Martel-Pradal and Got 1972; Martel and Got 1976), but others could not detect this enzyme in bovine membranes (Keenan and Huang 1972A). As discussed in a previous section, whether such enzymes occur in the primary surface-derived globule membrane or are present in material from intracellular lipovesicles, Golgi-derived secretory vesicles, or are entrained in cytoplasmic crescents must be determined. Globule membrane preparations liberate phosphorus from glucose-6-phosphate (Dowben *et al.* 1967; Kitchen 1974), but whether this is due to a specific glucose-6-phosphatase or a nonspecific phosphatase is unknown, especially since endoplasmic reticulum from mammary gland may not have a specific glucose-6-phosphatase (Keenan *et al.* 1972B, 1974B). Recently, evidence that plasmin, an alkaline protease found in milk, is associated with the globule membrane has been obtained (Hofmann *et al.* 1979). When incubated under appropriate conditions, globule membrane proteins undergo autoproteolysis, presumably catalyzed by plasmin.

While many different enzymes have been detected in lipid globule membrane, few of these activities have been extensively characterized and even fewer of the enzymes have been purified. Xanthine oxidase is the outstanding exception; numerous investigators have purified this enzyme, and its enzymatic activities have been extensively studied (Bray 1975; Coughlan 1980). Others have developed purification methods (Waud *et al.* 1975; Mangino and Brunner 1977A; Nathans and Hade 1978; Sullivan *et al.* 1982). 5'-Nucleotidase has been partially purified, and evidence of two distinct enzymes, both of which were tenaciously associated with phospholipid, was obtained (Huang and Keenan 1972A). Mather *et al.* (1980) found 5'-nucleotidase to be resolved into several distinct isozymes on electrofocusing. Both membrane-bound and detergent-solubilized forms of 5'-nucleotidase are inhibited by Concanavalin A, confirming the glycoprotein nature of this enzyme (Carraway and Carraway 1976; Snow *et al.* 1980). Both K^+ and Mg^{2+} were found to stimulate adenosine triphosphatase activity of globule membranes but, in contrast to an earlier report (Dowben *et al.* 1967), Na^+ did not activate the adenosine triphosphatase of the globule membrane and ouabain, an inhibitor of the Na^+-activated activity, was without effect (Patton and Trams 1971; Huang and Keenan 1972B). Other enzymes, such as acid and alkaline phosphatases, are found both in the globule membrane and in milk serum; these two enzymes have been extensively purified from milk (Patton and Keenan 1975; McPherson and Kitchen 1983). Some properties of the membrane-bound form of alkaline phosphatase have been described, including the interesting observation that sucrose inhibits this enzyme (Diaz-Mauriño and Nieto

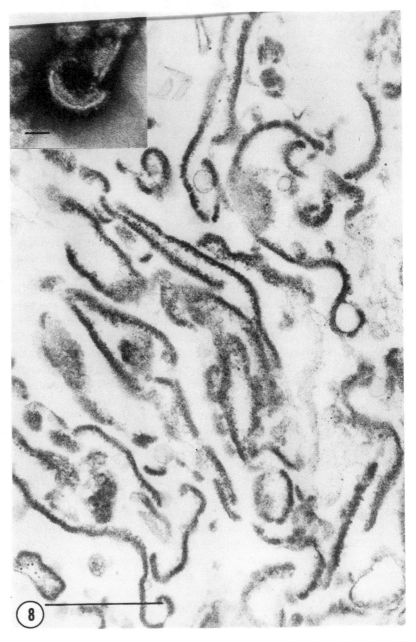

Figure 10.8 Milk lipid globule membranes released by churning of washed globules and collected by ultracentrifugation retain densely staining coat material along one face of the bilayer membrane. As seen in this electron micrograph of glutaraldehyde and osmium tetroxide–fixed material, the

1976). The remaining enzymes of the globule membrane have received little attention other than activity measurements. We know little or nothing about the functional significance of any of the globule membrane enzymes with respect to secretion of lipid globules or in effecting changes in milk after secretion.

MOLECULAR ORGANIZATION OF THE MILK LIPID GLOBULE MEMBRANE

Several early investigators of lipid globule membrane proposed structural models for the organization of the material on the surface of lipid globules. The more popular models pictured the membrane (or interfacial material) as consisting of various layers, each of which was believed to be composed of a particular class of constituents. Apparently these investigators did not consider the possibility that lipid globule membrane is a true biological membrane. This is surprising, since the concept of the lipid bilayer dates to a paper by Gorter and Grendel (1925) and was extensively developed by Danielli and Davson beginning in 1935 (Robertson 1981). These early concepts of globule membrane structure have been discussed by Brunner (1969, 1974). For discussions on current concepts of membrane structure, several publications can be consulted (Singer and Nicholson 1972; Bretscher 1973; Singer 1974; Rothman and Lenard 1977; Robertson 1981).

Several different lines of morphological evidence show the milk lipid globule membrane to have a typical bilayer membrane structure with some specializations, including the presence of a densely staining proteinaceous coat material along one face of the membrane (Figure 10.8). This structure is evident in thin sections of isolated membrane or of intact lipid globules fixed with glutaraldehyde and postfixed with osmium tetroxide (Figure 10.4, 10.8). Examination of preparations negatively stained with phosphotungstate or ammonium molybdate has revealed isolated globule membrane existing at least partially in the form of plate-like structures; in contrast, isolated plasma membrane is seen

---

isolated membranes exist primarily as open sheets, although a few vesicular profiles are evident. The insert is an electron micrograph of isolated lipid globule membrane negatively stained with phosphotungstate. The cup-shaped membrane profile reveals the coat material along the membrane face to have a plaque-like structure. Bars = 0.5 μm and 0.1 μm (insert); magnification × 69,000. (Micrographs were generously provided by Prof. Dr. W. W. Franke, German Cancer Research Center, Heidelberg, F.R.G.)

to be largely vesiculated (Keenan *et al.* 1970; Freudenstein *et al.* 1979; Diaz-Maurîno and Nieto 1977). Examination of thin-sectioned material confirms that a large proportion of the membrane does not form vesicles (Keenan *et al.* 1971; Wooding and Kemp 1975A; Jarasch *et al.* 1977; Freudenstein *et al.* 1979; Franke *et al.* 1981), and that this inhibition of vesiculation may well be due to the presence of the coat material along the inner face of the membrane. Morphological observations of freeze-fractured replicas of lipid globule membrane have shown that the intramembranous particles are aggregated, leaving large areas of the membrane faces devoid of these particles (Zerban and Franke 1978; Peixoto de Menezes and Pinto da Silva 1978; Pinto da Silva *et al.* 1980). There is a stark contrast between the apparently reduced particle density on lipid globule membrane faces and the high particle densities on apical plasma membrane. These observations suggest that membrane constituents which form intramembraneous particles are cleared from, or rearranged in plasma membrane regions which envelop lipid globules during secretion.

A variety of techniques have been used to provide evidence for the asymmetric distribution of carbohydrates with respect to the plane of the lipid bilayer of lipid globule membrane. Monis *et al.* (1975) found that surfaces of lipid globules in human and rat milk bound ruthenium red, a dye which selectively binds to anionic groups. Since they found carbohydrates, including sialic acid, in cream fractions of rat milk, they believed that ruthenium red was selectively bound to complex carbohydrates present on globule surfaces. Using specific lectins bound to gold granules, Horisberger *et al.* (1977) obtained convincing evidence for the uniform distribution of carbohydrates over the outer face of the membrane on bovine and human milk lipid globules. Sasaki and Keenan (1979) used many different carbohydrate-selective staining procedures to show a similar distribution of carbohydrates over the outer face of rat milk lipid globule membrane.

Biochemical studies have also suggested an asymmetric orientation of constituents in lipid globule membrane. By comparison of specific activities of enzymes in washed lipid globules and released membrane, Patton and Trams (1971) suggested that the active site of Mg^{2+}-adenosine triphosphatase was accessible to substrates on both faces of the membrane and that of 5'-nucleotidase on the outer membrane face. Recent evidence from studies of Concanavalin A inhibition of globule membrane and plasma membrane 5'-nucleotidase support an outer surface localization for the active site of this enzyme (Carraway and Carraway 1976; Snow *et al.* 1980). Kobylka and Carraway (1973) observed that exposure of lipid globules to proteolytic enzymes resulted in cleavage of all major membrane-associated proteins. They concluded that milk lipid globule membrane does not act as a permeability barrier to proteases and suggested that the membrane did not exist on the globule surface in intact form, a view also expressed by Shimizu et al. (1979). In contrast, Mather and Keenan (1975) observed major differences in rates of hydrolysis of proteins when isolated globule membrane or intact globules were incubated with trypsin. Many membrane proteins of intact globules resisted trypsin hydrolysis in comparison with the same proteins in isolated membrane. The contrast between this study and that of Kobylka and Carraway (1973) appears to be due to inadequate inactivation of proteases before dissolution of membrane from lipid globules in the latter study. Mather and Keenan (1975) also found that more membrane proteins were accessible to lactoperoxidase-catalyzed iodination in isolated globule membranes than in intact globules. Based on their observations, Mather and Keenan concluded that, of the major constituents, polypeptides of M_r about 67,000 (butyrophilin), 48,000, and 44,000 were exposed on the lipid globule surface and that a polypeptide of M_r 155,000 (subsequently identified as xanthine oxidase) was accessible only when membrane was released from globules. Patton and Hubert (1983), using similar techniques with goat milk lipid globules, obtained qualitatively similar results.

By the use of histochemical techniques and immunoelectrophoresis, Nielsen and Bjerrum (1977) identified four major protein complexes in milk lipid globule membrane. They found that xanthine oxidase was located on the internal face of the membrane and that the other three complexes, as well as Mg^{2+}-adenosine triphosphatase and 5'-nucleotidase, were accessible on the outer surface of the membrane. Based on available information, it cannot be determined if any proteins of the globule membrane span the lipid bilayer, as do certain proteins of other membranes (Singer 1974; Rothman and Lenard 1977).

There is an apparent anomaly in the situation with butyrophilin. As discussed in a previous section, there is evidence that this protein is a constituent of the coat material along the inner face of the membrane (Freudenstein et al. 1979; Franke et al. 1981; Deeney et al. 1985) and that it is present on the cytoplasmic face of precursor lipovesicles prior to secretion (Deeney et al. 1985), yet is available to trypsin attack in intact milk lipid globules. It is possible that some butyrophilin molecules span the lipid bilayer so as to be accessible on the globule surface (Heid et al. 1983). Mather and Keenan (1975) did observe more extensive labeling of this constituent with ^{125}I in isolated membranes than in intact globules. Butyrophilin is markedly trypsin sensitive, and this may be partly responsible for the seemingly anomalous observations (Appell et al. 1982).

Biochemical results support morphological observations indicating that most of the oligosaccharide chains of glycoproteins are exposed on the outer face of the lipid globule membrane. Nearly the same amounts of sialic acid were released, at about the same rates, when intact globules or isolated membrane were incubated with neuraminidase (Mather and Keenan 1975). Concanavalin A, a lectin specific for α-D-mannopyranoside or α-D-glucopyranoside residues, bound to intact globules and isolated globule membrane to nearly the same extent (Keenan et al. 1974A). These results imply that much, but not necessarily all, of the protein-bound carbohydrate of the membrane is on the outer membrane face. However, the latter result must be interpreted with caution, since some Concanavalin A binding sites on the outer surface of milk lipid globules may be cryptic and, at least in the case of goat lipid globules, some of the glycoproteins do not bind Concanavalin A either in intact globules or after electrophoretic separation (Patton and Hubert 1983). Certainly it remains possible that some glycoprotein carbohydrate is present in the coat material along the inner face of the membrane, as may be the case if butyrophilin is present in the coat. The possible occurrence of carbohydrates in coat material was suggested by morphological observations when avidin-ferritin was used to localize biotinylated carbohydrates (Sasaki and Keenan 1979). When lipid globules were treated to remove proteins which mask gangliosides, evidence that ganglioside carbohydrates are oriented along the outer face of the bilayer was obtained (Tomich et al. 1976).

From available results, it appears that milk lipid globule membrane is similar to plasma membrane in that there is asymmetric disposition of constituents with respect to the plane of the lipid bilayer. The milk lipid globule membrane is distinguished by the presence of coat material along the inner face of the membrane, and apparently by reduced numbers of intramembranous particles, which are believed to be the morphological equivalent of transmembrane proteins. While much progress has been made in this area, a great deal of further study will be necessary before a detailed picture of the molecular organization of milk lipid globule membrane can be formulated. Certainly detailed models, such as that proposed by McPherson and Kitchen (1983), appear premature.

MILK SERUM LIPIDS

As mentioned previously, a small amount of the lipid of cow's milk, about 1.5 to 4%, remains in the serum phase when milk is centrifugally separated (Huang and Kuksis 1967). In commercial milk processing, this observation was of economic concern. It was long believed that commercial separators were not efficient enough to recover all of the very small lipid globules of milk and that these globules remained in skim milk, thus reducing potential yields of cream. Much effort was expended unsuccessfully in attempts to develop separators which would increase the recovery of milk lipid in the cream fraction. The first evidence that the lipid of skim milk was present in membranes which sedimented, rather than floated, in a centrifugal field appears to be a paper by Patton *et al.* (1964). Since then, several groups have studied the nature and origin of this membrane material. Membranes of various types are present in milk serum, and the possibility that a portion of the skim milk lipid is present in very small lipid globules, which have a density sufficiently high so that they sediment on centrifugation, has not been conclusively ruled out. Much of the research in this area has been reviewed, extensively by Anderson and Cawston (1975) and more briefly by Patton and Keenan (1975).

Huang and Kuksis (1967) found milk serum to contain about equal proportions of phospholipids and triglycerides and lower amounts of diglycerides, free fatty acids, and cholesterol. They found that major phospholipids were distributed about equally in milk lipid globule membrane and in milk serum, and that corresponding phospholipids in these two fractions were very similar in fatty acid composition. With goat's milk, Patton and Keenan (1971) found 42% of the total milk phospholipid in the serum phase, and the same five major phospholipids were present, in the same relative proportions, in both lipid globules and milk serum. In contrast, glucosyl- and lactosylceramides of milk serum differ from those of the globule membrane in fatty acid composition (Kayser and Patton 1970). The amount of phospholipid and cholesterol in milk serum is reduced when goats are milked at hourly intervals, suggesting that the amount of membrane in milk serum may vary with the time that secreted milk is stored in the gland between milkings (Patton *et al.* 1973). Kitchen (1974) isolated membrane material from milk serum by precipitating caseins with rennet, removing the curd, and subsequently precipitating membrane material with ammonium sulfate and recovering it for analysis by centrifugation. A potential problem with this approach is that some membrane material may be entrained in the curd. In a comparison of this membrane fraction with lipid globule membrane, Kitchen (1974) found the former to have higher levels of phospholipid, cholesterol, and carbohydrate on a protein basis and higher specific activities of nucleotide pyrophosphatase, γ-glutamyl transpeptidase, and sulfhydryl oxidase. Higher activities of γ-glutamyl transpeptidase were also found in milk serum membranes by Baumrucker (1979). Qualitatively, the two membranes have similar complements of enzymes, with the possible exception that ATPase activity is absent from serum membranes (Plantz and Patton 1973). By the use of preparative electrofocusing, several enzymes including xanthine oxidase, γ-glutamyl transpeptidase, and alkaline phosphatase in serum and globule membranes were shown to have similar isoelectric points (Mather *et al.* 1980; Janolino and Swaisgood 1984). The absence of enzymes characteristic of Golgi apparatus or endoplasmic reticulum in serum membranes (Plantz *et al.* 1973) was taken as evidence for a plasma membrane origin of the serum membrane fraction.

Major proteins of globule and serum membranes are immunochemically identical (Nielsen and Bjerrum 1977), and electrophoretic profiles of the proteins from either membrane are similar (Kitchen 1974). The major quantitative difference is the presence of higher amounts of a protein of M_r 85,000 in serum membrane fractions. In summary, the information suggests that milk serum membranes are related, but not identical, to milk lipid globule and plasma membranes.

Morphological studies of milk serum membranes have led to seemingly conflicting results. Some conflict appears to have arisen because of differences in the methods used for collection of the membrane material. When milk is subjected to high-speed centrifugation, casein micelles sediment into a large pellet. On top of this pellet is a small, friable layer of membrane material which has been termed "fluff" (Stewart *et al.* 1972). Larger and more dense membrane fragments are found within the casein micelle pellet, while the fluff layer appears to contain smaller and less dense membrane fragments and vesicles. Stewart *et al.* observed membrane vesicles, open membrane sheets, and tubular sacs, which they believed to be microvilli, in the fluff fraction of cow's milk. They also observed membranes entrained in the casein pellet. Micrographs of Stewart *et al.* were not of sufficient resolution to reveal the fine detail of membrane structure. Plantz and Patton (1973) also observed membrane vesicles in a fluff fraction, and their micrograph shows some evidence of coat material on the inner face of the membrane vesicles. Wooding (1974) questioned the presence of microvilli in the fluff fraction; he observed structures similar to those identified as microvilli (Stewart *et al.* 1972) but concluded that these originated from elongated membrane fragments which he observed to bleb from the surface of milk lipid globules. Wooding (1971B, 1972, 1974) also identified membrane vesicles in skim milk as originating from blebbing globule membrane vesicles. In both the vesicles and elongated membrane fragments, he observed a coat material similar to that of milk lipid globule membrane. In addition to vesicles and elongated membrane profiles, Wooding and colleagues (1977; Christie and Wooding 1975) have identified two other structures in the membrane fraction of milk serum. They found cell fragments, which were surrounded by a membrane and contained morphologically recognizable endoplasmic reticulum, mitochondria, and lipid droplets, to be numerous in goat's milk. These structures were believed to be responsible for the lipid bio-synthetic activity observed in membrane fractions from skim milk (Christie and Wooding 1975). The cell fragments were not abundant in cow's milk. Instead, cow's milk was found to contain membrane-limited structures which had a dense content and microvillar-like projections (Wooding *et al.* 1977). It was suggested that these structures were residues of dead cells.

In none of the morphological studies reported above were intact cells observed. This may be because of the sampling methods used, since cells would be expected to sediment rapidly and be found primarily in the lower regions of, or under, the casein micelle pellet. Cells of various types are present in milk, and Anderson *et al.* (1974A, 1975) did identify leukocytes in the sedimented skim milk membranes. These investigators showed that experimental endotoxin infusion or infection of the mammary gland altered the appearance and composition of skim milk membranes. They concluded that skim milk membranes originate from multiple sources including leukocytes, cell debris from mammary gland, and fragments of milk lipid globule membrane. Milk serum membrane material is therefore probably of heterogeneous origin, arising from cells, cell fragments, and lipid globule membrane fragments. In addition, one may find fragments of secretory vesicle membranes, released from milk-secreting cells, in milk serum (Dylewski and Keenan, 1983). It is also possible that the nature of the milk serum membrane material varies between species and, within a species, with factors such as stage of lactation and health of the mammary gland.

CREAMING AND AGGLUTINATION

In undisturbed milk, lipid globules rise and form a cream layer. In the past, this was so commonly known that it could have been left unstated. Since the preponderance of milk today is marketed as homogenized milk, most of the people in urban populations will not have observed the creaming phenomenon. Since creaming is one of the most readily observable physical properties of the lipid globules of milk, this phenomenon has been widely studied. Brunner (1974) reviewed literature in this area through 1971, and Mulder and Walstra (1974) have discussed some of the more recent studies.

Being less dense than the serum phase of milk, lipid globules rise during quiescent storage. This rate of rise follows Stoke's law but is much faster than that predicted for the sedimentation rate of spherical particles. The reason for the faster than predicted rate is that lipid globules agglutinate into clusters which may exceed 800 μm in diameter (Brunner 1974). The rate of formation of the cream layer is dependent on many parameters, including the previous temperature and mechanical manipulations of the milk, size and composition of the lipid globules, age and stage of lactation of the cows, and season of the year during which the milk is produced (Brunner 1974; Mulder and Walstra 1974; Bottazzi et al. 1975; Walstra and Oortwijn 1975; Bottazzi and Premi 1977). The clustering of fat globules can be prevented by heating or homogenization, and from these observations at least two separate factors were recognized by earlier workers—one sensitive to temperature and one to homogenization (Euber and Brunner 1984). The heat-labile component was shown to be immunoglobulin M (IgM), and most of the globule-clustering activity appeared to reside in the F_{ab} domains, which are involved in the binding of antibody to antigen. In an elegant series of experiments, Euber and Brunner (1984) showed that, in part, IgM in milk binds and agglutinates fat globules via the specific recognition of antigens on the milk lipid globule membrane surface. These workers identified the homogenization-labile component as skim milk membrane and proposed a model to explain creaming in which lipid globules cluster in milk on the formation of a cross-linked complex of fat globules, skim milk membrane fragments, and IgM molecules. The skim milk membrane and lipid globule membrane antigens recognized by milk IgM were not identified, although inhibition studies indicated that carbohydrate moieties including glucosamine, galactosamine, and sialic acid were involved.

Fat globules in goat, pig, and buffalo milk appear to lack clustering ability (Brunner 1974; Mulder and Walstra 1974); in molecular terms, this could be due to species differences in IgM or proteins of the globule membrane or to the absence of IgM or membrane receptor proteins.

HOMOGENIZATION

In the United States, virtually all milk for fluid consumption and many fluid milk products packaged or prepared commercially are homogenized to prevent creaming. Since lipolytic degradation occurs rapidly in raw milk after homogenization, virtually all fluid milk is both pasteurized and homogenized. In the classic and most commonly used method for homogenization, milk is forced through very narrow slits called "homogenization valves" at pressures of up to 2500 psi at high flow rates. Other methods can be used to homogenize milk, but these are not common. By being forced through the slits, lipid globules are disrupted to form much smaller globules. Lipid globules in homogenized milk commonly have diameters of 1 μm or less, although the size distribution of these globules is influenced by the pressure of homogenization, the valve type and number (one- or two-stage) used, the flow rate, and other variables (Brunner 1974; Mulder and Walstra 1974; Kurzhals 1973; Walstra 1975). There are various theories about the processes responsible for the disruption of lipid globules during homogenization; of these, turbulence and cavitation have been accepted as the most plausible explanations (Precht 1973; Brunner 1974; Mulder and Walstra 1974), although Phipps (1974) has obtained evidence that cavitation may not be involved. With reduction in globule diameter there is an increase in surface area of lipid globules of four to six or more times (Brunner 1974; Mulder and Walstra 1974). Homogenized milk differs in several characteristics from raw or pasteurized milk, including increased viscosity, whiter appearance, increased foaming capacity, and increased surface tension. In addition, homogenized milk is more prone to the development of off-flavors from lipolytic and photochemical degradation, and curd obtained after addition of rennet is softer than that obtained from unhomogenized milk. Factors causing or contributing to these changes have been discussed (Brunner 1974).

Several factors appear to be responsible for the maintenance of lipid globules in dispersion in pasteurized, homogenized milk. The inactivation of fat globule clustering after heating or homogenization has been discussed. In addition, homogenization-induced alterations in the structure of globule membrane serve to maintain globules in the dispersed state. The increased surface area on globules is too large to be covered by the original membrane, and it was early believed that milk serum constituents must be adsorbed onto globule surfaces. Jackson and Brunner (1960) were among the first to show that caseins and whey proteins were associated stably with the lipid globules in homogenized milk, an observation which has been confirmed by others (Itoh and Nakanishi 1974; Darling and Butcher 1978; Keenan et al. 1983B; McPherson et al. 1984C). Electron micrographs have shown association of casein micelles and smaller structures, which may be casein submicelles, with lipid globules in homogenized milk (Buchheim 1970A; Henstra and Schmidt 1970; Keenan et al. 1983B). Stable associations of caseins and whey proteins with small globules would be expected to increase the density and slow the rate of rise. In fact, only about 75% of the total lipid in homogenized milk can be floated by prolonged centrifugation at forces in excess of 150,000 g; in comparison, 97% or more of the lipid in raw milk can be floated by centrifugation (Keenan *et al.* 1983B). This suggests that homogenization increases the density of a portion of the lipid globules to a point at least equal to the density of skim milk.

Biochemical and morphological evidence that a large proportion of the original lipid globule membrane remains stably associated with the surface of globules in homogenized milk has been obtained (Keenan *et al.* 1983B; McPherson *et al.* 1984C). This membrane is morphologically similar to the original milk lipid globule membrane in that the unit-like membrane and internal coat structures remain discernible, but the overall appearance of these structures is less distinct, suggesting that the membrane is spread out over a larger surface area than in unprocessed milk.

MELTING, SOLIDIFICATION, AND CRYSTAL FORMATION

Since the response of milk lipids to heating or cooling is important in many processing operations and in determining the properties of several dairy and dairy-based manufactured products, much research has been conducted in this area. Much of this research occurred prior to 1970 and was reviewed in the second edition of this book (Brunner 1974). In the past several years, there has been renewed interest in temperature-induced transitions in milk lipids; this has been due, at least partially, to emphasis on the manufacture of butter and butter-like products with decreased hardness. Since milk lipid consists of a complex mixture of triglycerides, it does not have clearly defined melting and solidification points. Moreover, depending on the rate and final temperature to which it is cooled or heated, the same milk lipid may solidify and melt at different temperatures. Many methods have been applied to study the physical state of milk lipid, including dilatometry, calorimetry, and differential thermal analysis. With all methods, it has been found that the composition of the lipids and the rate and the extent of temperature change influence melting and solidification behavior. These studies have been reviewed (Brunner 1974; Mulder and Walstra 1974; Sherbon 1974).

When milk is cooled, lipid crystals form and grow. These crystals may have different polymorphic forms, only one of which is stable under a particular set of conditions. Other polymorphic crystalline forms are unstable and are transformed into the stable form (Mulder and Walstra 1974). This phenomenon, known as "polymorphism," has been extensively studied (Mulder and Walstra 1974; Brunner 1974), and the reader who desires information is referred to these sources. Our purpose in introducing this topic is to suggest that cooling-induced crystallization of triglycerides may lead to the formation of individual shells or layers of triglycerides within the globule. Investigators (Buchheim 1970B; Buchheim and Precht 1979; Knoop 1972) who examined replicas of lipid globules by freeze-fracture electron microscopy suggested that laminar layers of crystallized lipid may form, beginning at the periphery of globules. Based on the solidification and crystallization characteristics of milk triglycerides, one would expect triglycerides to segregate, according to molecular structure, into different laminar layers. The work of Timms (1980) on the phase of behavior of milk lipid also suggests that segregation of triglycerides, according to solidification ranges, could occur within lipid globules.

Practical use of the solidification and crystallization characteristics of milk lipids has been made in the manufacture of butter which is more easily spread than butter made conventionally. Based on the knowledge that temperature and mechanical manipulation can influence crystallization behavior, various methods of working butter have been devised to produce a softer product (Taylor *et al.* 1971; Schaap *et al.* 1981). Another approach has been to separate triglyceride fractions according to solidification or melting ranges and reblend fractions to achieve a softer butter (McGillivray 1972; Black 1975; Frede *et al.* 1980).

REFERENCES

Amato, P. A. and Loizzi, R. F. 1981. The identification and localization of actin and actin-like filaments in lactating guinea pig mammary gland alveolar cells. *Cell Motility* *1,* 329–347.

Anderson, M. 1974. Milk fat globule membrane composition and dietary change: Supplements of coconut oil fed in two physical forms. *J. Dairy Sci. 57,* 399–404.

Anderson, M. 1977. Source and significance of lysosomal enzymes in bovine milk fat globule membrane. *J. Dairy Sci. 60,* 1217–1222.

Anderson, M. and Brooker, B. E. 1975. Loss of material during the isolation of milk fat globule membrane. *J. Dairy Sci. 58,* 1442–1448.

Anderson, M., Brooker, B. E., Andrews, A. T. and Alichanidis, E. 1974A. Membrane material isolated from milk of mastitic and normal cows. *J. Dairy Sci. 57,* 1448–1458.

Anderson, M., Brooker, B. E., Andrews, A. T. and Alichanidis, E. 1975. Membrane material in bovine skim-milk from udder quarters infused with endotoxin and pathogenic organisms. *J Dairy Res. 42,* 401–417.

Anderson, M. and Cawston, T. E. 1975. Reviews of the progress of dairy science. The milk-fat globule membrane. *J. Dairy Res. 42,* 459–483.

Anderson, M., Cawston, T. and Cheeseman, G. C. 1974B. Molecular-weight estimates of milk-fat-globule-membrane protein-sodium dodecyl sulphate complexes by electrophoresis in gradient acrylamide gels. *Biochem. J. 139*, 653–660.

Appell, K. C., Keenan, T. W. and Low, P. S. 1982. Differential scanning calorimetry of milk fat globule membranes. *Biochim. Biophys. Acta 690*, 243–250.

Arklie, J., Taylor-Papadimitriou, J., Bodmer, W., Egan, M. and Millis, R. 1981. Differentiation antigens expressed by epithelial cells in the lactating breast are also detectable in breast cancers. *Int. J. Cancer 28*, 23–29.

Bailie, M. J. and Morton, R. K. 1958. Comparative properties of microsomes from cow's milk and from mammary gland. 2. Chemical composition. *Biochem. J. 69*, 44–53.

Bargmann, W. and Knoop, A. 1959. On the morphology of milk secretion. Light and electron microscopic studies on the mammary gland of the rat. *Z. Zellforsch. 49*, 344–388. (German)

Bargmann, W., Fleischhauer, K. and Knoop, A. 1961. On the morphology of milk secretion. II. A review together with a model of the secretion mechanism. *Z. Zellforsch. 53*, 545–568. (German)

Bargmann, W. and Welsch, U. 1969. On the ultrastructure of the mammary gland. *In: Lactogenesis*. M. Reynolds and S.J. Folley (Editors). University of Pennsylvania Press, Philadelphia, pp. 43–52.

Basch, J. J., Farrell, H. M. and Greenberg, R. 1976. Identification of the milk fat globule membrane proteins. I. Isolation and partial characterization of glycoprotein B. *Biochim. Biophys. Acta 448*, 589–598.

Basu, S., Basu, M., Chien, J. L. and Presper, K. A. 1980. Biosynthesis of gangliosides in tissues. *In: Structure and Function of Gangliosides*. L. Svennerholm, H. Dreyfus, and P. F. Urban (Editors). Plenum Press, New York, pp. 213–226.

Battelli, M. G., Lorenzoni, E. and Stirpe, F. 1973. Milk xanthine oxidase type D (dehydrogenase) and type O (oxidase). Purification, interconversion and some properties. *Biochem. J. 131*, 191–198.

Bauer, H. 1972. Ultrastructural observations on the milk fat globule envelope of cow's milk. *J. Dairy Sci. 55*, 1375–1387.

Baumrucker, C. R. 1979. Gamma-glutamyl transpeptidase of bovine milk membranes: distribution and characterization. *J. Dairy Sci. 62*, 253–258.

Baumrucker, C. R. and Keenan, T. W. 1973. Membranes of mammary gland. VII. Stability of milk fat globule membrane in secreted milk. *J. Dairy Sci. 56*, 1092–1094.

Bhavadasan, M. K. and Ganguli, N. C. 1976. Dependence of enzyme activities associated with milk fat globule membrane on the procedure used for membrane isolation. *Ind. J. Biochem. Biophys. 13*, 252–254.

Björck, L. and Claesson, O. 1979. Xanthine oxidase as a source of hydrogen peroxide for the lactoperoxidase system in milk. *J. Dairy Sci. 62*, 1211–1215.

Black, R. G. 1975. Partial crystallization of milkfat and separation of fractions by vacuum filtration. *Aust. J. Dairy Technol. 30*, 153–156.

Blanc, B. 1981. Biochemical aspects of human milk—comparison with bovine milk. *World Rev. Nutr. Diet. 36*, 1–89.

Bottazzi, V., Battistotti, B., Dellaglio, F. and Corradini, C. 1975. Use of cold-stored milk for grana cheesemaking II. Estimation of the creaming capacity of fat in milk by a new method as a function of its origin and the season. *Sci. Tech. Lattiero-Casearia 26*, 249–259. (Italian)

Bottazzi, V. and Premi, L. 1977. Relation between 5'-nucleotidase and agglutination of fat globules. *Sci. Tech. Lattiero-Casearia 28*, 7–15. (Italian)

Bouhours, J.-F. and Bouhours, D. 1979. Galactosylceramide in the major cerebroside of human milk fat globule membrane. *Biochem. Biophys. Res. Commun. 88*, 1217–1222.

Bray, R. C. 1975. Molybdenum iron-sulfur flavin hydroxylases and related enzymes. *In: The Enzymes,* Vol XII. 3rd ed. P. D. Boyer (Editor). Academic Press, New York, pp. 299–419.

Bretscher, M. S. 1973. Membrane structure: Some general principles. *Science 181,* 622–629.

Briley, M. S. and Eisenthal, R. 1974. Association of xanthine oxidase with the bovine milk-fat-globule membrane. Catalytic properties of the free and membrane-bound enzyme. *Biochem. J. 143,* 149–157.

Bruder, G., Fink, A. and Jarasch, E.-D. 1978. The b-type cytochrome in endoplasmic reticulum of mammary gland epithelium and milk fat globule membranes consists of two components, cytochrome b_5 and cytochrome P-420. *Exp. Cell Res. 117,* 207–217.

Bruder, G., Heid, H., Jarasch, E.-D., Keenan, T. W. and Mather, I. H. 1982. Characteristics of membrane-bound and soluble forms of xanthine oxidase from milk and endothelial cells of capillaries. *Biochem. Biophys. Acta 701,* 357–369.

Bruder, G., Jarasch, E.-D. and Heid, H. W. 1984. High concentrations of antibodies to xanthine oxidase in human and animal sera: Molecular characterization. *J. Clin. Invest. 74,* 783–794.

Brunner, J. R. 1969. Milk lipoproteins. *In: Structural and Functional Aspects of Lipoproteins in Living Systems.* E. Tria and A.M. Scanu (Editors). Academic Press, New York, pp. 545–578.

Brunner, J. R. 1974. Physical equilibria in milk: The lipid phase. *In: Fundamentals of Dairy Chemistry,* 2nd ed. B.H. Webb, A. H. Johnson, and J.A. Alford (Editors). AVI Publishing Co., Westport, Conn, pp. 474–602.

Brunner, J. R., Duncan, C. W. and Trout, G. M. 1953. The fat-globule membrane of nonhomogenized and homogenized milk. I. The isolation and amino acid composition of the fat-membrane proteins. *Food Res. 18,* 454–462.

Buchheim, W. 1970A. Distribution of butterfat and casein in completely and partly homogenized milk. *Kieler Milch. Forsch. 22.* 323–327. (German)

Buchheim, W. 1970B. The submicroscopical structure of milk fat and its importance for buttermaking. *18th Int. Dairy Congr. 1E,* 73.

Buchheim, W. 1982. Paracrystalline arrays of milk fat globule membrane-associated proteins as revealed by freeze-fracture. *Naturwissenschaft 69.,* 505–507.

Buchheim, W. and Precht, D. 1979. Electron microscopic study of the crystallization process in fat globules during the ripening of cream. *Milchwissenschaft 34,* 657–662. (German)

Burchell, J., Durbin, H. and Taylor-Papadimitriou, J. 1983. Complexity of expression of antigenic determinants, recognized by monoclonal antibodies HMFG-1 and HMFG-2, in normal and malignant human mammary epithelial cells. *J. Immunol. 131,* 508–513.

Bushway, A. A. and Keenan, T. W. 1978. Composition and synthesis of three higher ganglioside homologs in bovine mammary tissue. *Lipids 13,* 59–65.

Calberg-Bacq, C.-M. Francois, C., Gosselin, L., Osterrieth, P. M. and Rentier-Delrue, F. 1976. Comparative study of the milk fat globule membrane and the mouse mammary tumour virus prepared from the milk of an infected strain of Swiss albino mice. *Biochim. Biophys. Acta 419,* 458–478.

Carraway, C. A. and Carraway, K. L. 1976. Concanavalin A perturbation of membrane enzymes of mammary gland. *J. Supramol. Struct. 4,* 121–126.

Cartron, J.-P., Andreu, G., Cartron, J., Bird, G. W. G., Salmon, C. and Gerbal, A. 1978. Demonstration of T-transferase deficiency in Tn-polyagglutinable blood samples. *Eur. J. Biochem. 92,* 111–119.

Ceriani, R. L., Peterson, J. A., Lee, J. Y., Moncada, R. and Blank, E. W. 1983. Characterization of cell surface antigens of human mammary epithelial cells with monoclonal antibodies prepared against human milk fat globule. *Somat. Cell Genet. 9*, 415–427.

Christie, W. W. and Wooding, F. B. P. 1975. The site of triglyceride biosynthesis in milk. *Experientia 31*, 1445–1447.

Coughlan, M. P. 1980. Aldehyde oxidase, xanthine oxidase and xanthine dehydrogenase. Hydroxylases containing molybdenum, iron-sulphur and flavin. In: *Molybdenum and Molybdenum-Containing Enzymes*. M.P. Coughlan (Editor). Pergamon Press, Oxford, pp. 119–185.

Dalton, A. J. and Haguenau, F. (Editors). 1973. *Ultrastructure of Animal Viruses and Bacteriophages: An Atlas*. Academic Press, New York.

Darling, D. F. and Butcher, D. W. 1978. Milk-fat globule membrane in homogenized cream. *J. Dairy Res. 45*. 197–208.

Deeney, J. T., Valivullah, H. M., Dapper, C. H., Dylewski, D. P. and Keenan, T. W. 1985. Microlipid droplets in milk secreting mammary epithelial cells: Evidence that they originate from endoplasmic reticulum and are precursors of milk lipid globules. *Eur. J. Cell Biol.* 38, 16–26.

Della Corte, E. and Stirpe, F. 1972. The regulation of rat liver xanthine oxidase. Involvement of thiol groups in the conversion of the enzyme activity from dehydrogenase (type D) into oxidase (type O) and purification of the enzyme. *Biochem. J. 126*, 739–745.

Diaz-Mauriño, T. and Nieto, M. 1976. Milk fat globule membranes. Inhibition by sucrose of the alkaline phosphomonoesterase. *Biochim. Biophys. Acta 448*, 234–244.

Diaz-Mauriño, T. and Nieto, M. 1977. Milk fat globule membranes: Chemical composition and phosphoesterase activities during lactation. *J. Dairy Res. 44*, 483–493.

Dowben, R. M., Brunner, J. R. and Philpott, D. E. 1967. Studies on milk fat globule membranes. *Biochim. Biophys. Acta 135*, 1–10.

Dylewski, D. P. and Keenan, T. W. 1983. Compound exocytosis of casein micelles in mammary epithelial cells. *Eur. J. Cell Biol. 31*, 114–124.

Dylewski, D. P., Dapper, C. H., Valivullah, H. M., Deeney, J. T. and Keenan, T. W. 1984. Morphological and biochemical characterization of possible intracellular precursors of milk lipid globules. *Eur. J. Cell Biol. 35*, 99–111.

Eigel, W. N., Butler, J. E., Ernstrom, C. A., Farrell, H. M., Harwalkar, V. R., Jenness, R. and Whitney, R. McL. 1984. Nomenclature of proteins of cow's milk: Fifth revision. *J. Dairy Sci. 67*, 1599–1631.

Elias, P. M., Friend, D. S. and Goerke, J. 1979. Membrane sterol heterogeneity. Freeze-fracture detection with saponins and filipin. *J. Histochem. Cytochem. 27*, 1247–1260.

Epenetos, A. A., Mather, S., Granowska, M., Nimmon, C. C., Hawkins, L. R., Britton, K. E., Shepherd, J., Taylor-Papadimitriou, J., Durbin, H., Malpas, J. S. and Bodmer, W. F. 1982. Targeting of iodine-123-labelled tumour-associated monoclonal antibodies to ovarian, breast, and gastrointestinal tumours. *Lancet 2*, 999–1004.

Euber, J. R. and Brunner, J. R. 1984. Reexamination of fat globule clustering and creaming in cow milk. *J. Dairy Sci. 67*, 2821–2832.

Fahmi, A. H., Sirry, W. G. and Safwat, A. 1956. The size of fat globules and the creaming power of cow, buffalo, sheep and goat milk. *Ind. J. Dairy Sci. 9*, 124–130.

Farrar, G. H. and Harrison, R. 1978. Isolation and structural characterization of alkali-labile oligosaccharides from bovine milk-fat-globule membrane. *Biochem. J. 171*, 549–557.

Farrar, G. H., Harrison, R. and Mohanna, N. A. 1980. Comparison of lectin receptors on the surface of human and bovine milk fat globule membranes. *Comp. Biochem. Physiol. 67B*, 265–270.

Feldman, J. D. 1961. Fine structure of the cow's udder during gestation and lactation. *Lab. Invest. 10*, 238-255.

Fischer, J., Klein P.-J., Farrar, G. H., Hanisch, F.-G. and Uhlenbruck, G. 1984. Isolation and chemical and immunological characterization of the peanut-lectin-binding glycoprotein from human milk-fat-globule membranes. *Biochem. J. 224*, 581-589.

Foster C.S., Dinsdale, E. A., Edwards, P. A. W. and Neville, A. M. 1982A. Monoclonal antibodies to the human mammary gland. Distribution of determinants in breast carcinomas. *Virchows Arch. [Pathol. Anat.] 394*, 295-305.

Foster, C. S., Edwards, P. A. W., Dinsdale, E. A. and Neville, A. M. 1982B. Monoclonal antibodies to the human mammary gland. Distribution of determinants in non-neoplastic mammary and extra mammary tissues. *Virchows Arch [Pathol. Anat.] 394*, 279-293.

Franke, W. W. and Keenan, T. W. 1979. Interaction of secretory vesicle membrane coat structures with membrane free areas of forming milk lipid globules. *J. Dairy Sci. 62*, 1322-1325.

Franke, W. W., Heid, H. W., Grund, C., Winter, S., Freudenstein, C., Schmid, E., Jarasch, E.-D. and Keenan, T. W. 1981. Antibodies to the major insoluble milk fat globule membrane-associated protein: Specific location in apical regions of lactating epithelial cells. *J. Cell Biol. 89*, 485-494.

Franke, W. W., Lüder, M. R., Kartenbeck, J., Zerban, H. and Keenan, T. W. 1976. Involvement of vesicle coat material in casein secretion and surface regeneration. *J. Cell Biol. 69*, 173-195.

Frede, E., Peters, K.-H. and Precht, D. 1980. Improvement of the consistency of butter by means of fat fractionation and of a special tempering treatment of the cream. *Milchwissenschaft 35*, 287-292. (German)

Freudenstein, C., Keenan, T. W., Eigel, W. N., Sasaki, M., Stadler J., and Franke, W. W. 1979. Preparation and characterization of the inner coat material associated with fat globule membranes from bovine and human milk. *Exp. Cell Res. 118*, 277-294.

Fujino, Y. and Fujishima, T. 1972. Nature of ceramide in bovine milk. *J. Dairy Res. 39*, 11-14.

Fujino, Y., Nakano, M. and Saeki, T. 1970. The chemical structure of glycolipids of bovine milk. *Agr. Biol. Chem. (Japan) 34*, 442-447.

Glöckner, W. M., Newman, R. A., Dahr, W. and Uhlenbruck, G. 1976. Alkali-labile oligosaccharides from glycoproteins of different erythrocyte and milk fat globule membranes. *Biochim. Biophys. Acta 443*, 402-413.

Gorter, E. and Grendel, F. 1925. On bimolecular layers of lipoids on the chromocytes of the blood. *J. Exp. Med. 41*, 439-443.

Greenwalt, D. E., Johnson, V. G., Kuhajda, F. P., Eggleston, J. C. and Mather, I. H. 1985A. Localization of a membrane glycoprotein in benign fibrocystic disease and infiltrating duct carcinomas of the human breast with the use of a monoclonal antibody to guinea pig milk fat globule membrane. *Am. J. Pathol. 118*, 351-359.

Greenwalt, D. E., Johnson, V. G. and Mather, I. H. 1985B. Specific antibodies to PAS-IV, a glycoprotein of bovine milk-fat-globule membrane, bind to a similar protein in cardiac endothelial cells and epithelial cells of lung bronchioles. *Biochem J. 228*, 233-240.

Greenwalt, D. E. and Mather, I. H. 1985. Characterization of an apically derived epithelial membrane glycoprotein from bovine milk, which is expressed in capillary endothelia in diverse tissues. *J. Cell Biol. 100*, 397-408.

Harrison, R., Higginbotham, J. D. and Newman, R. 1975. Sialoglycopeptides from bovine milk fat globule membrane. *Biochim. Biophys. Acta 389*, 449-463.

Hay, J. D. and Morrison, W. R. 1971. Polar lipids in bovine milk. III. Isomeric *cis* and *trans* monoenoic and dienoic fatty acids, and alkyl and alkenyl ethers in phosphatidyl choline and phosphatidyl ethanolamine. *Biochim. Biophys. Acta 248,* 71–79.

Hayashi, S. and Smith, L. M. 1965. Membranous material of bovine milk fat globules. I. Comparison of membranous fractions released by deoxycholate and by churning. *Biochemistry 4,* 2550–2556.

Heid, H. W. 1983. Biochemical and immunological characterization of the proteins of the milk fat globule membrane. Ph. D. dissertation, University of Heidelberg, F. R. G. (German)

Heid, H. W., Winter, S., Bruder, G., Keenan, T. W. and Jarasch, E.-D. 1983. Butyrophilin, an apical plasma membrane–associated glycoprotein characteristic of lactating mammary glands of diverse species. *Biochim. Biophys. Acta 728,* 228–238.

Helminen, H. J. and Ericsson, J. L. E. 1968. Studies on mammary gland involution. I. On the ultrastructure of the lactating mammary gland. *J. Ultrastruct. Res. 25,* 193–213.

Henstra, S. and Schmidt, D. G. 1970. On the structure of the fat–protein complex in homogenized cow's milk. *Neth. Milk Dairy J. 24,* 45–51.

Herald, C. T. and Brunner, J. R. 1957. The fat globule membrane of normal cow's milk. I. The isolation and characteristics of two membrane-protein fractions. *J. Dairy Sci. 40,* 948–956.

Hladik, J. and Michalec, C. 1966. Ceramide-monohexosides and ceramide-dihexosides in lipoproteins of the membrane of fat globules in bovine milk. *Acta Biol. Med. Germania 16,* 696–699.

Hofmann, C. J., Keenan, T. W. and Eigel, W. N. 1979. Association of plasminogen with bovine milk fat globule membrane. *Int. J. Biochem. 10,* 909–917.

Hohe, K. A., Dimick, P. S. and Kilara, A. 1985. Milk lipoprotein lipase distribution in the major fractions of bovine milk. *J. Dairy Sci. 68,* 1067–1073.

Hollmann, K. H. 1974. Cytology and fine structure of the mammary gland. *In: Lactation,* Vol. I. B.L. Larson and V.R. Smith (Editors). Academic Press, New York, pp. 3–95.

Hood, L. F. and Patton, S. 1973. Isolation and characterization of intracellular lipid droplets from bovine mammary tissue. *J. Dairy Sci. 56,* 858–863.

Horisberger, M., Rosset, J. and Vonlanthen, M. 1977. Location of glycoproteins on milk fat globule membrane by scanning and transmission electron microscopy, using lectin-labelled gold granules. *Exp. Cell Res. 109,* 361–369.

Huang, C. M., and Keenan, T. W. 1972A. Preparation and properties of 5′-nucleotidases from bovine milk fat globule membranes. *Biochim. Biophys. Acta 274,* 246–257.

Huang, C. M., and Keenan, T. W. 1972B. Adenosine triphosphatase activity of bovine milk fat globule membranes. *Comp. Biochem. Physiol. 43B,* 277–282.

Huang, C. M., and Keenan, T. W. 1972C. Membranes of mammary gland. II. 5′-Nucleotidase activity of bovine mammary plasma membranes. *J. Dairy Sci. 55,* 862–864.

Huang, R. T. C. 1973. Isolation and characterization of the gangliosides of butter milk. *Biochim. Biophys. Acta 306,* 82–84.

Huang, T. C. and Kuksis, A. 1967. A comparative study of the lipids of globule membrane and fat core and of the milk serum of cows. *Lipids 2,* 453–470.

Huggins, J. W. and Carraway, K. L. 1976. Purification of plasma membranes from rat mammary gland by a density perturbation procedure. *J. Supramol. Struct. 5,* 59–63.

Huggins, J. W., Trenbeath, T. P., Chesnut, R. W., Carraway, C. A. C. and Carraway, K. L. 1980. Purification of plasma membranes of rat mammary gland. Comparisons of subfractions with rat milk fat globule membrane. *Exp. Cell Res. 126,* 279–288.

Imam, A., Laurence, D. J. R. and Neville, A. M. 1981. Isolation and characterization of a major glycoprotein from milk-fat-globule membrane of human breast milk. *Biochem J. 193,* 47–54.

Imam, A., Laurence, D. J. R. and Neville, A. M. 1982. Isolation and characterization of two individual glycoprotein components from human milk-fat-globule membranes. *Biochem. J. 207.* 37–41.

Imam, A., Taylor, C. R. and Tökés, Z. A. 1984. Immunohistochemical study of the expression of human milk fat globule membrane glycoprotein 70. *Cancer Res. 44,* 2016–2022.

Itoh, T. and Nakanishi, T. 1974. Milk protein fractions contributing to the globule membrane of fat emulsions. *J. Agr. Chem. Soc. Japan 48,* 239–244.

Jackson, R. H. and Brunner, J. R. 1960. Characteristics of protein fractions isolated from the fat/plasma interface of homogenized milk. *J. Dairy Sci. 43,* 912–919.

Janolino, V. G. and Swaisgood, H. E. 1984. Isolation, solubilization, fractionation by electrofocusing, and immobilization of skim milk membranes. *J. Dairy Sci. 67,* 1161–1168.

Janssen, M. M. T. and Walstra, P. 1982. Cytoplasmic remnants in milk of certain species. *Neth. Milk Dairy J. 36,* 365–368.

Jarasch, E.-D., Bruder, G., Keenan, T. W. and Franke, W. W. 1977. Redox constituents in milk fat globule membranes and rough endoplasmic reticulum from lactating mammary gland. *J. Cell Biol. 73,* 223–241.

Jarasch, E.-D., Grund, C., Bruder, G., Heid, H. W., Keenan, T. W. and Franke, W. W. 1981. Localization of xanthine oxidase in mammary-gland epithelium and capillary endothelium. *Cell 25,* 67–82.

Jeffers, K. R. 1935. Cytology of the mammary gland of the albino rat. I. Pregnancy lactation and involution. *Am. J. Anat. 56,* 257–277.

Jenness, R. G. 1974. The composition of milk. *In: Lactation,* Vol. III, B. L. Larson and V.R. Smith (Editors). Academic Press, New York, pp. 3–107.

Jenness, R. and Koops, J. 1962. Preparation and properties of a salt solution which simulates milk ultrafiltrate. *Neth. Milk Dairy J. 16,* 153–164.

Jenness, R. and Patton, S. 1959. *Principles of Dairy Chemistry.* John Wiley and Sons, New York.

Jensen, R. G., Clark, R. M. and Ferris, A. M. 1980. Composition of the lipids in human milk: A review. *Lipids 15,* 345–355.

Johnson, V. G., Greenwalt, D. E., Heid, H. W., Mather, I. H. and Madara, P. J. 1985. Identification and characterization of the principal proteins of the fat-globule membrane from guinea-pig milk. *Eur. J. Bichem. 151,* 237–244.

Johnson, V. G., and Mather, I. H. 1985. Monoclonal antibodies prepared against PAS-I, butyrophilin and GP-55 from guinea-pig milk-fat-globule membrane bind specifically to the apical pole of secretory-epithelial cells in lactating mammary tissue. *Exp. Cell Res. 156,* 144–158.

Kanno, C., Shimizu, M. and Yamauchi, K. 1975. Isolation and physicochemical properties of a soluble glycoprotein fraction of milk fat globule membrane. *Agr. Biol. Chem. (Japan) 39,* 1835–1842.

Kanno, C., Shimizu, M. and Yamauchi, K. 1977. Polydispersity and heterogeneity of the soluble glycoprotein isolated from bovine milk fat globule membrane. *Agr. Biol. Chem. (Japan) 41,* 83–87.

Kanno, C. and Yamauchi, K. 1978. Antigenic identity between the soluble glycoprotein of milk fat globule membrane and a heat-stable protein fraction of whey. *Agr. Biol. Chem. (Japan) 42,* 1697–1705.

Katz, F. N., Rothman, J. E., Knipe, D. M. and Lodish, H. F. 1977. Membrane assembly: Synthesis and intracellular processing of the vesicular stomatitis viral glycoprotein. *J. Supramol. Struct. 7,* 353–370.

Kayser, S. G. and Patton, S. 1970. The function of very long chain fatty acids in membrane structure: Evidence from milk cerebrosides. *Biochem. Biophys. Res. Commun. 41,* 1572–1578.

Keenan, T. W. 1974A. Composition and synthesis of gangliosides in mammary gland and milk of the bovine. *Biochim. Biophys. Acta 337*, 255–270.

Keenan, T. W. 1974B. Membranes of mammary gland. IX. Concentration of glycosphingolipid galactosyl and sialyltransferases in Golgi apparatus from bovine mammary gland. *J. Dairy Sci. 57*, 187–192.

Keenan, T. W. and Dylewski, D. P. 1985. Aspects of intracellular transit of serum and lipid phases of milk. *J. Dairy Sci. 68*, 1025–1040.

Keenan, T. W., Dylewski, D. P., Woodford, T. A. and Ford, R. H. 1983A. Origin of milk fat globules and the nature of the milk fat globule membrane. *In: Developments in Dairy Chemistry*, Vol. 2: *Lipids*, P.F. Fox (Editor). Applied Science Publishers, London, pp. 83–118.

Keenan, T. W., Franke, W. W. and Kartenbeck, J. 1974A. Concanavalin A binding by isolated plasma membranes and endomembranes from liver and mammary gland. *Febs. Lett. 44*, 274–278.

Keenan, T. W., Franke, W. W., Mather, I. H. and Morré, D. J. 1978. Endomembrane composition and function in milk formation. *In: Lactation*, Vol. IV. B.L. Larson (Editor) Academic Press, New York, pp. 405–436.

Keenan, T. W., Freudenstein, C. and Franke, W. W. 1977A. Membranes of mammary gland. XIII. A lipoprotein complex derived from bovine milk fat globule membrane with some preparative characteristics resembling those of actin. *Cytobiologie 14*, 259–278.

Keenan, T. W., Heid, H. W., Stadler, J., Jarasch, E.-D. and Franke, W. W. 1982. Tight attachment of fatty acids to proteins associated with milk lipid globule membrane. *Eur. J. Cell Biol. 26*, 270–276.

Keenan, T. W. and Huang, C. M. 1972A. Membranes of mammary gland. IV. Glycosidase activity of milk fat globule membranes. *J. Dairy Sci. 55*, 1013–1015.

Keenan, T. W. and Huang, C. M. 1972B. Membranes of mammary gland. VI. Lipid and protein composition of Golgi apparatus and rough endoplasmic reticulum from bovine mammary gland. *J. Dairy Sci. 55*, 1586–1596.

Keenan, T. W., Huang, C. M. and Morré, D. J. 1972A. Gangliosides: Nonspecific localization in the surface membranes of bovine mammary gland and rat liver. *Biochem. Biophys. Res. Commun. 47*, 1277–1283.

Keenan, T. W., Huang, C. M. and Morré, D. J. 1972B. Membranes of mammary gland. V. Isolation of Golgi apparatus and rough endoplasmic reticulum from bovine mammary gland. *J. Dairy Sci. 55*, 1577–1585.

Keenan, T. W., Huang, C. M. and Morré, D. J. 1972C. Membranes of mammary gland. III. Lipid composition of Golgi apparatus from rat mammary gland. *J. Dairy Sci. 55*, 51–57.

Keenan, T. W., Moon, T.-W. and Dylewski, D. P. 1983B. Lipid globules retain globule membrane material after homogenization. *J. Dairy Sci. 66*, 196–203.

Keenan, T. W. and Morré, D. J. 1970. Phospholipid class and fatty acid composition of Golgi apparatus isolated from rat liver and comparison with other cell fractions. *Biochemistry 9*, 19–25.

Keenan, T. W., Morré, D. J. and Huang, C. M. 1972D. Distribution of gangliosides among subcellular fractions from rat liver and bovine mammary gland. *Febs. Lett. 24*, 204–208.

Keenan, T. W., Morré, D. J. and Huang, C. M. 1974B. Membranes of the mammary gland. *In: Lactation*, Vol. II. B. L. Larson and V. R. Smith (Editors). Academic Press. New York, pp. 191–233.

Keenan, T. W., Morré, D. J., Olson, D. E., Yunghans, W. N. and Patton, S. 1970. Biochemical and morphological comparison of plasma membrane and milk fat globule membrane from bovine mammary gland. *J. Cell Biol. 44*, 80–93.

Keenan, T. W., Olson, D. E. and Mollenhauer, H. H. 1971. Origin of the milk fat globule membrane. *J. Dairy Sci. 54,* 295–299.

Keenan, T. W., Powell, K. M., Sasaki, M., Eigel, W. N., and Franke, W. W. 1977B. Membranes of mammary gland. XIV. Isolation and partial characterization of a high molecular weight glycoprotein fraction from bovine milk fat globule membrane. *Cytobiologie 15,* 96–115.

Keenan, T. W., Sasaki, M., Eigel, W. N., Morré, D. J., Franke, W. W., Zulak, I. M. and Bushway, A. A. 1979. Characterization of a secretory vesicle–rich fraction from lactating bovine mammary gland. *Exp. Cell Res. 124,* 47–61.

King, N. 1955. *The Milk Fat Globule Membrane.* Commonwealth Agricultural Bureaux, Farnham Royal, Bucks, England.

Kitchen, B. J. 1974. A comparison of the properties of membranes isolated from bovine skim milk and cream. *Biochim. Biophys. Acta 356,* 257–269.

Kitchen, B. J. 1977. Fractionation and characterization of the membranes from bovine milk fat globules. *J. Dairy Res. 44,* 469–482.

Knoop, E. 1972. Electron microscopical studies on the structure of milk fat and protein. *Milchwissenschaft 27,* 364–373.

Knudson, C. M., Stemberger, B. H. and Patton, S. 1978. Effects of colchicine on ultrastructure of the lactating mammary cell: Membrane involvement and stress on the Golgi apparatus. *Cell Tiss. Res. 195,* 169–181.

Kobylka, D. and Carraway, K. L. 1972. Proteins and glycoproteins of the milk fat globule membrane. *Biochim. Biophys. Acta 288,* 282–295.

Kobylka, D. and Carraway, K. L. 1973. Proteolytic digestion of proteins of the milk fat globule membrane. *Biochim. Biophys. Acta 307,* 133–140.

Kurosumi, K., Kobayashi, Y. and Baba, N. 1968. The fine structure of mammary glands of lactating rats, with special reference to the apocrine secretion. *Exp. Cell Res. 50,* 177–192.

Kurzhals, H. A. 1973. Evaluation of the effect of homogenization on milk. *Milchwissenschaft 28,* 637–645. (German)

Kuzdzal-Savoie, S. 1979. Comparative studies of the lipids of milk. *Cah. Nutr. Diet. 14,* 185–196. (French)

Linzell, J. L. and Peaker, M. 1971. Mechanism of milk secretion. *Physiol. Rev. 51,* 564–597.

Lis, D. and Monis, B. 1978. Glycosaminoglycans of the fat globule membrane of cow milk. *J. Supramol. Struct. 8,* 173–176.

Lotan, R., Skutelsky, E., Danon, D. and Sharon, N. 1975. The purification, composition, and specificity of the anti-T lectin from peanut *(Arachis hypogaea). J. Biol. Chem. 250,* 8518–8523.

Mangino, M. E. and Brunner, J. R. 1975. Molecular weight profile of fat globule membrane proteins. *J. Dairy Sci. 58,* 313–318.

Mangino, M. E. and Brunner, J. R. 1977A. Isolation and partial characterization of xanthine oxidase associated with the milk fat globule membrane of cow's milk. *J. Dairy Sci. 60,*841–850.

Mangino, M. E. and Brunner, J. R. 1977B. Compositional homology of membrane-protein systems and membrane-associated proteins: Comparison with milk fat globule membrane and "membrane"-derived xanthine oxidase. *J. Dairy Sci. 60,* 1208–1216.

Martel, M. B., Dubois, P. and Got, R. 1973. Membranes of human milk lipid globules. Preparation, morphology and chemical composition. *Biochim. Biophys. Acta 311,* 565–575. (French)

Martel, M. B. and Got, R. 1976. Transfer of galactose by human milk lipid globule membranes. *Biochim. Biophys. Acta 436,* 789–799. (French)

Martel-Pradal, M. B. and Got, R. 1972. Presence of marker enzymes for plasma membrane, Golgi apparatus and endoplasmic reticulum in lipid globules of human milk. *Febs. Lett. 21*, 220–222. (French)

Mather, I. H. 1978. Separation of the proteins of bovine milk-fat globule membrane by electrofocusing. *Biochim. Biophys. Acta 514*, 25–36.

Mather, I. H., Bruder, G., Jarasch, E.-D., Heid, H. W. and Johnson, V. G. 1984. Protein synthesis in lactating guinea-pig mammary tissue perfused in vitro. II. Biogenesis of milk-fat-globule membrane proteins. *Exp. Cell Res. 151*, 277–282.

Mather, I. H. and Keenan, T. W. 1975. Studies on the structure of milk fat globule membrane. *J. Membrane Biol. 21*, 65–85.

Mather, I. H. and Keenan, T. W. 1983. Function of endomembranes and the cell surface in the secretion of organic milk constituents. *In: Biochemistry of Lactation*. T. B. Mepham (Editor). Elsevier/North-Holland, Amsterdam, pp. 231–283.

Mather, I. H., Sullivan, C. H. and Madara, P. J. 1982. Detection of xanthine oxidase and immunologically related proteins in fractions from bovine mammary tissue and milk after electrophoresis in polyacrylamide gels containing sodium dodecyl sulphate. *Biochem. J. 202*, 317–323.

Mather, I. H., Tamplin, C. B. and Irving, M. G. 1980. Separation of the proteins of bovine milk-fat-globule membrane by electrofocusing with retention of enzymatic and immunological activity. *Eur. J. Biochem. 110*, 327–336.

Mather, I. H., Weber, K. and Keenan, T. W. 1977. Membranes of mammary gland. XII. Loosely associated proteins and compositional heterogeneity of bovine milk fat globule membrane. *J. Dairy Sci. 60*, 394–402.

McCarthy, M. and Headon, D. R. 1979. Lipid and protein composition of a membrane-rich fraction of butter oil. *J. Dairy Res. 46*, 511–521.

McGillivray, W. A. 1972. Softer butter from fractionated fat or by modified processing. *N.Z. J. Diary Sci. Technol. 7*, 111–112.

McIlhinney, R. A. J., Patel, S. and Gore, M. E. 1985. Monoclonal antibodies recognizing epitopes carried on both glycolipids and glycoproteins of the human milk fat globule membrane. *Biochem. J. 227*, 155–162.

McPherson, A. V., Dash, M. C. and Kitchen, B. J. 1984A. Isolation of bovine milk fat globule membrane material from cream without prior removal of caseins and whey proteins. *J. Dairy Res. 51*, 113–121.

McPherson, A. V., Dash, M. C. and Kitchen, B. J. 1984B. Isolation and composition of milk fat globule membrane material. I. From pasteurized milks and creams. *J. Dairy Res. 51*, 279–287.

McPherson, A. V., Dash, M. C. and Kitchen, B. J. 1984C. Isolation and composition of milk fat globule membrane material. II. From homogenized and ultra heat treated milks. *J. Dairy Res. 51*, 289–297.

McPherson, A. V. and Kitchen, B. J. 1983. Reviews of the progress of dairy science: The bovine milk fat globule membrane—its formation, composition, structure and behaviour in milk and dairy products. *J. Dairy Res. 50*, 107–133.

Merril, C. R., Dunau, M. L. and Goldman, D. 1981. A rapid sensitive silver stain for polypeptides in polyacrylamide gels. *Anal. Biochem. 110*, 201–207.

Monis, B., Rovasio, R. A. and Valentich, M. A. 1975. Ultrastructural characterization by ruthenium red of the surface of the fat globule membrane of human and rat milk with data on carbohydrates of fractions of rat milk. *Cell Tiss. Res. 157*, 17–24.

Montesano, R., Ravazzola, M. and Orci, L. 1983. Filipin labelling of lipid droplets in lactating rat mammary gland. *Cell Biol. Intern Rep. 7*, 194.

Morré, D. J. 1977. The Golgi apparatus and membrane biogenesis. *In: Cell Surface Reviews*, Vol. 4. G. Poste and G. L. Nicolson (Editors). North-Holland, Amsterdam, pp. 1–83.

Morré, D. J., Kartenbeck, J. and Franke, W. W. 1979. Membrane flow and interconversions among endomembranes. *Biochim. Biophys. Acta 559*, 71–152.

Morrison, W. R. 1968. The distribution of phospholipids in some mammalian milks. *Lipids 3*, 101–103.

Morrison, W. R. 1970. Milk lipids. *In: Topics in Lipid Chemistry.* F. D. Gunstone (Editor). Wiley-Interscience, New York, pp. 51–106.

Morrison, W. R., Jack, E. L. and Smith, L. M. 1965. Fatty acids of bovine milk glycolipids and phospholipids and their specific distribution in the diacylglycerophospholipids. *J. Am. Oil Chem. Soc. 42*, 1142–1147.

Morrison, W. R. and Smith, L. M. 1964. Identification of ceramide monohexoside and ceramide dihexoside in bovine milk. *Biochim. Biophys. Acta 84*, 759–761.

Morton, R. K. 1954. The lipoprotein particles in cow's milk. *Biochem. J. 57*, 231–237.

Mulder, H. and Walstra, P. 1974. *The Milk Fat Globule.* Commonwealth Agricultural Bureaux, Farnham Royal, Bucks, England.

Murray, L. R., Powell, K. M., Sasaki, M., Eigel, W. N. and Keenan, T. W. 1979. Comparison of lectin receptor and membrane coat-associated glycoproteins of milk lipid globule membranes. *Comp. Biochem. Physiol. 63B*, 137–145.

Nathans, G. R. and Hade, E. P. K. 1978. Bovine milk xanthine oxidase. Purification by ultrafiltration and conventional methods which omit addition of proteases. Some criteria for homogeneity of native xanthine oxidase. *Biochim. Biophys. Acta 526*, 328–344.

Nemanic, M. K. and Pitelka, D. R. 1971. A scanning electron microscope study of the lactating mammary gland. *J. Cell Biol. 48*, 410–415.

Newman, R. A. and Harrison, R. 1973. Characterisation of the surface of bovine milk fat globule membrane using microelectrophoresis. *Biochim. Biophys. Acta 298*, 798–809.

Newman, R. A., Harrison, R. and Uhlenbruck, G. 1976. Alkali-labile oligosaccharides from bovine milk fat globule membrane glycoprotein. *Biochim. Biophys. Acta 433*, 344–356.

Newman, R. A. and Uhlenbruck, G. G. 1977. Investigation into the occurrence and structure of lectin receptors on human and bovine erythrocyte, milk-fat globule and lymphocyte plasma-membrane glycoproteins. *Eur. J. Biochem. 76*, 149–155.

Nickerson, S. C., Akers, R. M. and Weinland, B. T. 1982. Cytoplasmic organization and quantitation of microtubules in bovine mammary epithelial cells during lactation and involution. *Cell Tiss. Res. 223*, 421–430.

Nickerson, S. C. and Keenan, T. W. 1979. Distribution and orientation of microtubules in milk secreting epithelial cells of rat mammary gland. *Cell Tiss. Res. 202*, 303–312.

Nickerson, S. C., Smith, J. J. and Keenan, T. W. 1980A. Role of microtubules in milk secretion—action of colchicine on microtubules and exocytosis of secretory vesicles in rat mammary epithelial cells. *Cell Tiss. Res. 207*, 361–376.

Nickerson, S. C., Smith, J. J. and Keenan, T. W. 1980B. Ultrastructural and biochemical response of rat mammary epithelial cells to vinblastine sulphate. *Eur. J. Cell Biol. 23*, 115–121.

Nielsen, C. S. and Bjerrum, O. J. 1977. Crossed immunoelectrophoresis of bovine milk fat globule membrane protein solubilized with non-ionic detergent. *Biochim. Biophys. Acta 466*, 496–509.

O'Farrell, P. H. 1975. High resolution two-dimensional electrophoresis of proteins. *J. Biol. Chem. 250*, 4007–4021.

O'Farrell, P. Z., Goodman, H. M. and O'Farrell, P. H. 1977. High resolution two-dimensional electrophoresis of basic as well as acidic proteins. *Cell 12*, 1133–1142.

Ormerod, M. G., Steele, K., Westwood, J. H. and Mazzini, M. N. 1983. Epithelial membrane antigen: Partial purification, assay and properties. *Br. J. Cancer 48*, 533–541.

Palmer, L. S. and Samuelson, E.-G. 1924. The nature of the substances adsorbed on the surface of the fat globules in cow's milk. *Proc. Soc. Exp. Biol. Med. 21*, 537–539.

Patton, S. 1973. Origin of the milk fat globule. *J. Am. Oil Chem. Soc. 50*, 178–185.

Patton, S. 1982. Release of remnant plasma membrane from milk fat globules by Triton X-100. *Biochim. Biophys. Acta 688*, 727–734.

Patton, S., Durdan, A. and McCarthy, R. D. 1964. Structure and synthesis of milk fat. VI. Unity of the phospholipids in milk. *J. Dairy Sci. 47*, 489–495.

Patton, S. and Fowkes, F. M. 1967. The role of the plasma membrane in the secretion of milk fat. *J. Theoret. Biol. 15*, 274–281.

Patton, S., Hood, L. F. and Patton, J. S. 1969. Negligible release of cardiolipin during milk secretion by the ruminant. *J. Lipid Res. 10*, 260–266.

Patton, S. and Hubert, J. 1983. Binding of Concanavalin A to milk fat globules and release of the lectin–membrane complex by Triton X-100. *J. Dairy Sci. 66*, 2312–2319.

Patton, S. and Jensen, R. G. 1976. *Biomedical Aspects of Lactation.* Pergamon Press, New York.

Patton, S. and Keenan, T. W. 1971. The relationship of milk phospholipids to membranes of the secretory cell. *Lipids 6*, 58–61.

Patton, S. and Keenan, T. W. 1975. The milk fat globule membrane. *Biochim. Biophys. Acta 415*, 273–309.

Patton, S., Long, C. and Sokka, T. 1980. Effect of storing milk on cholesterol and phospholipid of skim milk. *J. Dairy Sci. 63*, 697–700.

Patton, S. and McCarthy, R. D. 1963. Structure and synthesis of milk fat. V. A postulated sequence of events from analyses of mammary tissue lipids. *J. Dairy Sci. 46*, 916–921.

Patton, S., Plantz, P. E. and Thoele, C. A. 1973. Factors influencing phospholipids and cholesterol in skim milk: Effect of short interval milkings. *J. Dairy Sci. 56*, 1473–1476.

Patton, S. and Trams, E. G. 1971. The presence of plasma membrane enzymes on the surface of bovine milk fat globules. *Febs. Lett. 14*, 230–232.

Peaker, M. 1978. Ion and water transport in the mammary gland. *In: Lactation,* Vol. IV. B.L. Larson (Editor). Academic Press, New York, pp. 437–462.

Peaker, M. 1983. Secretion of ions and water. *In: Biochemistry of Lactation.* T. B. Mepham (Editor). Elsevier/North-Holland, Amsterdam, pp. 285–305.

Pease, D. C. and Porter, K. R. 1981. Electron microscopy and ultramicrotomy. *J. Cell Biol. 91*, 287s–292s.

Peixoto de Menezes, A. and Pinto da Silva, P. 1978. Freeze-fracture observations of the lactating rat mammary gland. *J. Cell Biol. 76*, 767–778.

Pfleger, R. C., Anderson, N. G. and Snyder, F. 1968. Lipid class and fatty acid composition of rat liver plasma membranes isolated by zonal centrifugation. *Biochemistry 7*, 2826–2833.

Phipps, L. W. 1974. Cavitation and separated flow in a simple homogenizing valve and their influence on the break-up of fat globules in milk. *J Dairy Res. 41*, 1–8.

Pinto da Silva, P., Peixoto de Menezes, A. and Mather, I. H. 1980. Structure and dynamics of the bovine milk fat globule membrane viewed by freeze fracture. *Exp. Cell Res. 125*, 127–139.

Pitelka, D. R. and Hamamoto, S. T. 1977. Form and function in mammary epithelium: The interpretation of ultrastructure. *J. Dairy Sci. 60*, 643–654.

Plantz, P. E. and Patton, S. 1973. Plasma membrane fragments in bovine and caprine skim milks. *Biochim. Biophys. Acta 291*, 51–60.

Plantz, P. E., Patton, S. and Keenan, T. W. 1973. Further evidence of plasma membrane material in skim milk. *J. Dairy Sci. 56*, 978–983.

Powell, J. T., Jälfors, U. and Brew, K. 1977. Enzymic characteristics of fat globule membranes from bovine colostrum and bovine milk. *J. Cell Biol. 72*, 617–627.

Precht, D. 1973. Theories on physical phenomena of homogenization. *Kieler Milch. Forsch. 25,* 29–47. (German)

Puri, B. R., Parkash, S. and Chandan, R. C. 1961. Studies in physico-chemical properties of milk. Part IX. Variation in fat globule size–distribution curves of cow and buffalo milk, on the removal of fat and addition of goat milk. *Ind. J. Dairy Sci. 14,* 31–35.

Ray, T. K., Skipski, V. P., Barclay, M., Essner, E. and Archibald, F. M. 1969. Lipid composition of rat liver plasma membranes. *J. Biol. Chem. 244,* 5528–5536.

Richardson, C. D. and Vance, D. E. 1976. Biochemical evidence that Semliki forest virus obtains its envelope from the plasma membrane of the host cell. *J. Biol. Chem. 251,* 5544–5550.

Robertson, J. D. 1981. Membrane structure. *J. Cell Biol. 91,* 189s–204s.

Rodriguez Boulan, E. and Sabatini, D. D. 1978. Asymmetric budding of viruses in epithelial monolayers: a model system for study of epithelial polarity. *Proc. Natl. Acad. Sci, USA 75,* 5071–5075.

Roth, J. and Berger, E. G. 1982. Immunocytochemical localization of galactosyltransferase in HeLa cells: Codistribution with thiamine pyrophosphatase in *trans*-Golgi cisternae. *J. Cell Biol. 93,* 223–229.

Roth, J., Lentze, M. J. and Berger, E. G. 1985. Immunocytochemical demonstration of ecto-galactosyltransferase in absorptive intestinal cells. *J. Cell Biol. 100,* 118–125.

Rothman, J. E. and Lenard, J. 1977. Membrane asymmetry. *Science 195,* 743–753.

Rüegg, M. and Blanc, B. 1981. The fat globule size distribution in human milk. *Biochim. Biophys. Acta 666,* 7–14.

Saacke, R. G. and Heald, C. W. 1974. Cytological aspects of milk formation and secretion. *In: Lactation,* Vol. II. B. L. Larson and V. R. Smith (Editors). Academic Press, New York, pp. 147–189.

Sasaki, M., Eigel, W. N. and Keenan, T. W. 1978. Lactose and major milk proteins are present in secretory vesicle–rich fractions from lactating mammary gland. *Proc. Natl. Acad. Sci. USA 75,* 5020–5024.

Sasaki, M. and Keenan, T. W. 1978. Membranes of mammary gland XV. 5-thio-D-glucose decreases lactose content and inhibits secretory vesicle maturation in lactating rat mammary gland. *Exp. Cell Res. 111,* 413–425.

Sasaki, M. and Keenan, T. W. 1979. Ultrastructural characterization of carbohydrate distribution on milk lipid globule membrane. *Cell Biol. Intern. Rep. 3,* 67–74.

Schaap, J. E., Hagedoorn, H. G. and Rutten, G. A. M. 1981. Effect of storage time, temperature and working on the firmness and spreadability of butter. *Zuivelzicht 73,* 38–40. (Dutch)

Scheffer, R. C. T., Poort, C. and Slot, J. W. 1980. Fate of the major zymogen granule membrane–associated glycoproteins from rat pancreas. A biochemical and immunocytochemical study. *Eur. J. Cell Biol. 23,* 122–128.

Schlesinger, M. J. 1981. Proteolipids. *Ann. Rev. Biochem. 50,* 193–206.

Schwartz, D. P., Burgwald, L. H., Shamey, J. and Brewington, C. R. 1968. Quantitative determination of lanosterol and dihydrolanosterol in milk fat. *J. Dairy Sci. 51,* 929.

Shaper, N. L., Mann, P. L. and Shaper, J. H. 1985. Cell surface galactosyltransferase: Immunochemical localization. *J. Cell Biochem. 28,* 229–239.

Sherbon, J. W. 1974. Crystallization and fractionation of milk fat. *J. Am. Oil Chem. Soc. 51,* 22–25.

Shimizu, M., Kanno, C. and Yamauchi, K. 1976. Dissociation of the soluble glycoprotein of bovine milk fat globule membrane by sodium dodecyl sulfate. *Agr. Biol. Chem. (Japan) 40,* 1711–1716.

Shimizu, M., Uryu, N. and Yamauchi, K. 1981. Presence of heparan sulfate in the fat globule membrane of bovine and human milk. Agr. Biol. Chem. (Japan) *45,* 741–745.

Shimizu, M. and Yamauchi, K. 1982. Isolation and characterization of mucin-like glyco-protein in human milk fat globule membrane. *J. Biochem. 91*, 515-524.

Shimizu, M., Yamauchi, K. and Kanno, C. 1979. Proteolytic digestion of milk fat globule membrane proteins. *Milchwissenschaft 34*, 666-668.

Singer, S. J. 1974. The molecular organization of membranes. *Ann. Rev. Biochem. 43*, 805-833.

Singer, S. J. and Nicolson, G. L. 1972. The fluid mosaic model of the structure of cell membranes. *Science 175*, 720-731.

Snow, L. D., Colton, D. G. and Carraway, K. L. 1977. Purification and properties of the major sialoglycoprotein of the milk fat globule membrane. *Arch. Biochem. Biophys. 179*, 690-697.

Snow, L. D., Doss, R. C. and Carraway, K. L. 1980. Cooperativity of the Concanavalin A inhibition of bovine milk fat globule membrane 5'-nucleotidase. Response to ex-traction of nucleotidase and of putative cytoplasmic surface coat components. *Biochim. Biophys. Acta 611*, 333-341.

Springer, G. F. 1984. T and Tn, General carcinoma autoantigens. *Science 224*, 1198-1206.

Springer, G. F. and Desai, P. R. 1974. Common precursors of human blood group MN specificities. *Biochem. Biophys. Res. Commun. 61*, 470-475.

Stein, O. and Stein, Y. 1967. Lipid synthesis, intracellular transport, and secretion. II. Electron microscopic radioautographic study of the mouse lactating mammary gland. *J. Cell Biol. 34*, 251-263.

Stemberger, B. H. and Patton, S. 1981. Relationships of size, intracellular location, and time required for secretion of milk fat droplets. *J. Dairy Sci. 64*, 422-426.

Stemberger, B. H., Walsh, R. M. and Patton, S. 1984. Morphometric evaluation of lipid droplet associations with secretory vesicles, mitochondria and other components in the lactating cell. *Cell Tiss. Res. 236*, 471-475.

Stewart, P. S., Puppione, D. L. and Patton, S. 1972. The presence of microvilli and other membrane fragments in the non-fat phase of bovine milk. *Z. Zellforsch, 123*, 161-167.

Stirpe, F. and Della Corte, E. 1969. The regulation of rat liver xanthine oxidase. Conver-sion in vitro of the enzyme activity from dehydrogenase (type D) to oxidase (type O). *J. Biol. Chem. 244*, 3855-3863.

Sullivan, C. H., Mather, I. H., Greenwalt, D. E. and Madara, P. J. 1982. Purification of xanthine oxidase from the fat-globule membrane of bovine milk by electrofocusing. Determination of isoelectric points and preparation of specific antibodies to the enzyme. *Mol. Cell. Biochem. 44*, 13-22.

Svennerholm, L. 1963. Chromatographic separation of human brain gangliosides. *J. Neu-rochem. 10*, 613-623.

Swope, F. C. and Brunner, J. R. 1965. Identification of ribonucleic acid in the fat-globule membrane. *J. Dairy Sci. 48*, 1705-1707.

Swope, F. C. and Brunner, J. R. 1968. The fat globule membrane of cow's milk: a reas-sessment of isolation procedures and mineral composition. *Milchwissenschaft 23*, 470-473.

Taylor, M. W., Dolby, R. M. and Russell, R. W. 1971. The reworking of butter. *N.Z. J. Dairy Sci. Technol. 6*, 172-176.

Taylor-Papadimitriou, J., Peterson, J. A., Arklie, J., Burchell, J., Ceriani, R. L. and Bodmer, W. F. 1981. Monoclonal antibodies to epithelium-specific components of the human milk fat globule membrane: Production and reaction with cells in cul-ture. *Int. J. Cancer 28*, 17-21.

Tedman, R. A. 1983. Ultrastructural morphology of the mammary gland with observa-tions on the size distribution of fat droplets in milk of the Weddell seal *Leptony-chotes weddelli* (Pinnipedia) *J. Zool. Lond. 200*, 131-141.

Thompson, M. P., Brunner, J. R., Stine, C. M. and Lindquist, K. 1961. Lipid components of the fat-globule membrane. *J. Dairy Sci. 44*, 1589–1596.

Timms, R. E. 1980. The phase behaviour and polymorphism of milk fat, milk fat fractions and fully hardened milk fat. *Aust. J. Dairy Technol. 35*, 47–53.

Tomich, J. M., Mather, I. H. and Keenan, T. W. 1976. Proteins mask gangliosides in milk fat globule and erythrocyte membranes. *Biochim. Biophys. Acta 433*, 357–364.

Vasic, J. and DeMan, J. M. 1966. High melting glycerides and the milk fat globule membrane. *Proc. 17th Int. Dairy Congr. C*, 167–172.

Walstra, P. 1968. Estimating globule-size distribution of oil-in-water emulsions by spectroturbidimetry. *J. Coll. Interf. Sci. 27, 493–500.*

Walstra, P. 1969A. Studies on milk fat dispersion, II. The globule-size distribution of cow's milk. *Neth. Milk Dairy J. 23*, 99–110.

Walstra, P. 1969B. Studies on milk fat dispersion. III. The distribution function of globule size in cow's milk and the process of milk fat formation. *Neth. Milk Dairy J. 23*, 111–123.

Walstra, P. 1974. High-melting triglycerides in the fat globule membrane; an artifact? *Neth. Milk Dairy J. 28*, 3–9.

Walstra, P. 1975. Effect of homogenization on the fat globule size distribution in milk. *Neth. Milk Dairy J. 29*, 279–294.

Walstra, P. and Oortwijn, H. 1969. Estimating globule-size distribution of oil-in-water emulsions by Coulter counter. *J. Coll. Interf. Sci. 29*, 424–431.

Walstra, P. and Oortwijn, H. 1975. Effect of globule size and concentration on creaming in pasteurized milk. *Neth. Milk Dairy J. 29*, 263–278.

Walstra, P., Oortwijn, H. and Degraaf, J. J. 1969. Studies on milk fat dispersion. I. Methods for determining globule-size distribution. *Neth. Milk Dairy J. 23*, 12–36.

Waud, W. R., Brady, F. O., Wiley, R. D. and Rajagopalan, K. V. 1975. A new purification procedure for bovine milk xanthine oxidase: Effect of proteolysis on the subunit structure. *Arch. Biochem. Biophys. 169*, 695–701.

Weber, K. and Osborn, M. 1969. The reliability of molecular weight determinations by dodecyl sulfate-polyacrylamide gel electrophoresis. *J. Biol. Chem. 244*, 4406–4412.

Wellings, S. R., Deome, K. B. and Pitelka, D. R. 1960A. Electron microscopy of milk secretion in the mammary gland of the C3H/Crgl mouse. I. Cytomorphology of the prelactating and the lactating gland. *J. Natl. Cancer Inst. 25*, 393–421.

Wellings, S. R., Grunbaum, B. W. and Deome, K. B. 1960B. Electron microscopy of milk secretion in the mammary gland of the C3H/Crgl mouse. II. Identification of fat and protein particles in milk and in tissue. *J. Natl. Cancer Inst. 25*, 423–437.

Whittlestone, W. G. 1952. The distribution of fat-globule size in sow's milk. II. The influence of stage of lactation. *J Dairy Res. 19*, 335–338.

Wilkinson, M. J. S., Howell, A., Harris, M., Taylor-Papadimitriou, J., Swindell, R. and Sellwood, R. A. 1984. The prognostic significance of two epithelial membrane antigens expressed by human mammary carcinomas. *Int. J. Cancer 33*, 299–304.

Wooding, F. B. P. 1971A. The mechanism of secretion of the milk fat globule. *J. Cell Sci. 9*, 805–821.

Wooding, F. B. P. 1971B. The structure of the milk fat globule membrane. *J. Ultrastruct. Res. 37*, 388–400.

Wooding, F. B. P. 1972. Milk microsomes, viruses and the milk fat globule membrane. *Experientia 28*, 1077–1079.

Wooding, F. B. P. 1973. Formation of the milk fat globule membrane without participation of the plasmalemma. *J. Cell Sci. 13*, 221–235.

Wooding, F. B. P. 1974. Milk fat globule membrane material in skim milk. *J. Dairy Res. 41*, 331–337.

Wooding, F. B. P. 1977. Comparative mammary fine structure. *In: Comparative Aspects of Lactation.* M. Peaker (Editor). Academic Press, London, pp. 1–41.

Wooding, F. B. P. and Kemp. P. 1975A. Ultrastructure of the milk fat globule membrane with and without triglyceride. *Cell Tiss. Res. 165,* 113–127.

Wooding, F. B. P. and Kemp. P. 1975B. High-melting-point triglycerides and the milk-fat globule membrane. *J Dairy Res. 42,* 419–426.

Wooding, F. B. P., Morgan, G. and Craig, H. 1977. "Sunbursts" and "christiesomes": Cellular fragments in normal cow and goat milk. *Cell Tiss. Res. 185,* 535–545.

Wooding, F. B. P., Peaker, M. and Linzell, J. L. 1970. Theories of milk secretion: Evidence from the electron microscopic examination of milk. *Nature 226,* 762–764.

Zerban, H. and Franke, W. W. 1978. Milk fat globule membranes devoid of intramembranous particles. *Cell Biol. Intern. Rep. 2,* 87–98.

Zikakis, J. P., Dougherty, T. M. and Biasotto, N. O. 1976. The presence and some properties of xanthine oxidase in human milk and colostrum. *J. Food Sci. 41,* 1408–1412.

Zikakis, J. P. and Treece, J. M. 1971. Xanthine oxidase polymorphism in bovine milk. *J. Dairy Sci. 54,* 648–654.

11

Milk Coagulation
and Protein Denaturation

Rodney J. Brown

INTRODUCTION

Research reports on milk stability have been published frequently since about 1919. Early studies were prompted by the need to ensure sufficient heat stability for evaporated milk to withstand heat sterilization (Sommer and Hart 1919, 1922). Between 1919 and 1960, most attention was directed to the influence of milk salts on heat stability (Miller and Sommer 1940; Pyne 1958; Pyne and McHenry 1955). It was not until the early 1960s that the importance of heating time and pH on coagulation of milk was appreciated (Rose 1961A,B). More recent work has been concerned with factors which affect the stability of milk proteins.

PROTEIN DENATURATION

Several reviews covering either general protein denaturation or specific aspects of denaturation have been published (Brandts 1969; Edsal and Wyman 1958; Flory 1969; Fox and Morrissey 1977; Kim and Baldwin 1982; Lapanje 1978; Pace 1975; Pyne 1962; Rose 1963, 1965; Schmidt 1980; Tanford 1961, 1968, 1970). These reviews are still pertinent even though our understanding of protein structure has moved away from the rigid x-ray crystalography-based models upon which these reviews were founded (Gurd and Rothgeb 1979; Karplus and McCammon 1983; Tanford 1980). Our understanding of denaturation will gradually become more complete as the intricacies of protein structure are unraveled.

A protein molecule is a group of atoms in an orderly arrangement with respect to each other. This orderliness is stipulated by chemical bonds between specific atoms and by other attractive and repulsive forces between atoms (Flory 1969). A protein molecule can be regarded as the total of amino acid residues, each interacting independently with the surrounding medium (Franks and Eagland 1975). This is a good starting point in discussing denaturation of proteins.

A long chain of amino acids attached end-to-end has many possible ways to fold. The final shape, or conformation, of a folded protein molecule is determined by its unique sequence of amino acids and by the effects of environmental conditions on amino acid side chains. The conformation selected is the one that is most stable because it has the lowest free energy (Bloomfield 1979). This conformation is designated the "native state" of the protein.

A protein molecule has the same conformation whenever it exists under the same conditions, and protein molecules with the same sequence of amino acids have identical conformations under identical conditions (Flory 1969; Mangino 1984). Some structures in protein chains are seen frequently in a variety of proteins and have been given names such as "α-helix" and "β-sheet." Others, referred to as "unordered structure" (Swaisgood 1982), are regions of protein folding which may be found only once, but are structurally stable. Much space in globular proteins is filled with such unordered structure (Flory 1969).

Denaturation is a drastic change from the native conformation which does not alter the amino acid sequence. This change must be steep; it must take place over a narrow range of temperature or concentration of the denaturing agent (Tanford 1968). Under conditions which completely denature protein, the chain is completely unfolded to a random coil conformation. It no longer has a fixed conformation. During denaturation the protein chain seeks the lowest free energy conformation compatible with its new conditions. Many hydrophobic amino acid side chains buried in native globular proteins are exposed to solvent in denatured proteins. Hydrophobic effects are important in maintaining protein stability (Brandts 1969).

Heat or change of solvent, including pH change, is used to denature proteins experimentally. Guanidine hydrochloride and urea are commonly used denaturing solvents. Urea is rarely able to denture proteins completely to a random coil conformation, as guanidine hydrochloride does (Tanford 1968). An exception is β-lactoglobulin, which can be completely denatured in urea (Kauzmann and Simpson 1953). Proteins denatured by pH change or heat are not completely random coils, like guanidine hydrochloride–denatured proteins. They have areas of random coil and areas which retain their ordered structure. Thermal denaturation is promoted at lower pH. Denaturation at lower pH is also less likely to be reversible (Tanford 1968). Denaturation by pH change alone is sometimes possible. For example, α-lactalbumin is partially denatured at pH 4 (Kronman et al. 1968).

NATIVE STATES OF MILK PROTEINS

Caseins

Casein makes up about 80% of protein in milk and is found primarily as a colloidal dispersion of large protein–mineral complexes called "casein micelles." Because of its high concentration relative to the other proteins, casein dominates in determining the characteristics of milk during processing (Schmidt 1980). Proteins in micelles have little secondary or tertiary structure but have a complex quaternary structure. Quaternary structure in casein micelles provides stability which is derived from tertiary structure in most globular proteins. Various casein proteins serve as "solvents" for each other, providing an environment protected from solvent or other outside influences (Brown 1984; McMahon and Brown 1984A).

Separate submicelles are combined to form each casein micelle, with the outside surface containing a high concentration of κ-casein. α_{s1}-, α_{s2}-, and β-caseins also contain hydrophobic regions which are represented on micelle surfaces, but κ-casein is predominant (Schmidt 1980). The inorganic portion of casein micelles helps to stabilize them by neutralizing negative charges of phosphorylated caseins with the micelle (calcium ions) and by providing a framework for the proteins (colloidal calcium phosphate and citrate) (McMahon and Brown 1984A; Kinsella 1984; Rose 1965).

Casein micelle proteins are primarily α_{s1}-, α_{s2}-, β-, and κ-caseins in approximate proportions 3:.8:3:1. α_{s1}-Casein has eight or nine phosphate groups, depending on the genetic variant. α_{s2}-Casein is the most hydrophilic of the caseins. It has two disulfide bonds which, by severe heat treatment, can be caused to interact with those of β-lactoglobulin. It also has 10 to 13 phosphate groups and is very sensitive to the calcium ion concentration (Kinsella 1984; Swaisgood 1982).

β-Casein is an extremely dipolar and amphiphilic molecule. It is mostly random coil, being 16% proline, and has two separate hydrophilic and hydrophobic domains. It has four or five phosphate groups, depending on the genetic variant (Swaisgood 1982). β-Casein is often associated with serum proteins of milk, as well as with casein micelles. Both heating and cooling of milk have been reported to move β-casein from serum into the micelles (Dzurec and Zall 1985). Cooling is the method most often used experimentally to release β-casein from micelles, and milk stored at 4°C can have as much as 40% of the β-casein dissociated from the micelles (Schmutz and Puhan 1981). Addition of calcium to milk causes β-casein to remain in micelles regardless of temperature treatment (Carpenter and Brown 1985). β-Casein is cleaved by proteinases in milk to yield α-caseins and components 5, 8-fast, and 8-slow of the proteose-peptone fraction of milk proteins (Pearce 1980; Swaisgood 1982).

Like α_{s2}-casein, κ-casein has two disulfide bonds which can form cross-links with β-lactoglobulin. The N-terminal two-thirds of the molecule is hydrophobic and contains the two disulfide bonds. The C-terminal end is hydrophilic, polar, and charged. It varies in the number of attached carbohydrate moieties and has only one phosphate group. These characteristics make κ-casein ideal for the surface of casein micelles, where it is most often found. It is not susceptible to calcium ion binding, as the other caseins are, and when present on the surface of micelles, it protects the other caseins from calcium (McMahon and Brown 1984A; Swaisgood 1982).

Serum Proteins

About 20% of milk protein is soluble in the aqueous phase of milk. These serum proteins are primarily a mixture of β-lactoglobulin, α-lactalbumin, bovine serum albumin, and immunoglobulins. Each of these globular proteins has a unique set of characteristics as a result of its amino acid sequence (Swaisgood 1982). As a group, they are more heat sensitive and less calcium sensitive than caseins (Kinsella 1984). Some of these characteristics (Table 11.1) cause large differences in susceptibility to denaturation (de Wit and Klarenbeek 1984).

β-Lactoglobulin is a globular protein which under normal milk storage conditions (less than 4°C and between pH 5 and 7) is a dimer of two identical monomers (de Wit and Karlenbeek 1984). About 47% of the molecule is unordered structure at the pH of fresh milk (Kinsella 1984). Each 18,400-dalton monomer has two disulfide bridges and one free thiol group. The thiol groups, especially the free ones, are impor-

Table 11.1.   Some Characteristics of Serum Proteins.

Protein	Weight contribution (g/liter)	Molecular weight ($\times 10^8$/liter)	Relative number	Cysteine residues per molecule	Lysine residues per molecule
β-Lactoglobulin	3.3	18,400	100	5	15
α-Lactalbumin	1.2	14,200	50	8	12
Immunoglobulin G	0.5	160,000	1.9	64	180
Bovine serum albumin	0.3	66,000	2.6	35	59

SOURCE: de Wit and Klarenbeek (1984).

tant to this discussion of denaturation because of their ability to interact with κ-casein and other proteins during heating.

Eight cysteine residues of α-lactalbumin are linked together in four disulfide bridges (de Wit and Karenbeek 1984). Based on its homology with hen's egg white lysozyme, we can safely assume that α-lactalbumin is a globular protein with a cleft to match that containing the active site of lysozyme. The involvement of α-lactalbumin in synthesis of lactose makes a further assumption tempting, but α-lactalbumin does not function as an enzyme. It acts only as a coenzyme with galactosyl transferase (Swaisgood 1982).

Serum albumin has 35 cystein residues which are found as 17 intrachain disulfide linkages and one free sulfhydryl group. Except for the immunoglobulins, serum albumin is the largest milk protein (Walstra and Jenness 1984).

Immunoglobulins of classes IgG1, IgG2, IgA, and IgM are measurable in milk. IgG has the familiar immunoglobulin structure, with two heavy and two light chains. IgA is found in milk as a dimer of two IgA complexes, linked by one J and one SC chain. Negative IgM is a pentamer of IgM complexes attached to one J chain (Walstra and Jenness 1984).

DENATURATION OF INDIVIDUAL MILK PROTEINS

General Considerations

Almost all dairy products are subjected to heat treatments and a variety of changes in pH, concentration, etc. between the collection of milk from cows and the sale of retail products. Heat is commonly used to control bacterial growth, but some products are heated to remove moisture or to change the texture or flavor. Severity of heating varies according to which product is being heated, and milk proteins are affected accordingly.

Mild heat treatments (up to 60°C) mainly affect hydrophobic bonding within and between proteins. Such effects are important in those milk proteins which have large hydrophobicities, such as β-casein and β-lactoglobulin (de Wit and Klarenbeek 1984; Payens and Vreeman 1982).

de Wit (1981) and de Wit and Klarenbeek (1981) analyzed the thermal behavior of major whey proteins up to 150°C by differential scanning calorimetry. They observed two distinct heat effects. The first, near 70°C, was attributed to denaturation and the second, near 130°C, to unfolding of the remaining protein structure.

Unfolding of protein molecules is an endothermic process that can be measured quantitatively, and independent of aggregation, as enthalpy of denaturation. Apparent transition temperature and denaturation temperature (transition temperature with heating rate effects removed) are also useful parameters of unfolding (de Wit and Klarenbeek 1984). Aggregation is a separate and usually irreversible process which follows unfolding. Unfolding is usually reversible if heating is stopped before aggregation begins. Protein unfolding and aggregation behave differently with respect to heating, pH, protein concentration, and concentrations of salts or other denaturing substances. Susceptibility to denaturation is largely determined by pH, and extent of aggregation is more dependent on the presence of calcium ions (de Wit 1981). Many reports include unfolding and aggregation as one parameter called "denaturation."

Caseins

Enzymic casein denaturation and coagulation are not covered in this chapter. Coagulation of the casein complex in milk initiated by enzymic cleavage of κ-casein has been recently reviewed (McMahon and Brown 1984B) and is covered in Chapter 12.

The caseinate system in milk is unique among major protein systems in its ability to withstand high temperatures. Studies using ultracentrifugation, viscosity measurements, and gel permeation chromatography have shown that micelles aggregate initially when heated and then dissociate until the onset of coagulation, when rapid and extensive aggregation occurs (Fox 1981A). Strands of protein form between casein micelles after heating for 30 min at pH 6.8 and 100°C. β-Lactoglobulin molecules are cross-linked and attached at κ-casein on micelle surfaces by disulfide bonds (Creamer et al. 1978). Even limited proteolysis of κ-casein destabilizes micelles (Fox and Hearn 1978C). Addition of α_{s2}-casein also reduces heat stability of casein micelles (Kudo 1980B).

β-Casein is very hydrophobic and, therefore, temperature sensitive. Low temperature or removal of calcium causes dissociation of β-casein from the micelle and destabilizes the remaining micelle (Carpenter and Brown 1985; Dalgleish 1982). Soluble β-casein can form aggregates of up to 40 monomers when heated. The C-terminal (hydrophobic) portions of β-casein monomers clump together, and the N-terminal (hydrophilic) portions extend outward into the surrounding aqueous medium (Kinsella 1984).

Glycopeptides have been found in milk at temperatures above 50°C (Hindle and Wheelock 1970), and peptides similar to macropeptides from chymosin (EC 3.4.23.4) hydrolysis are produced in milk heated to 120°C for 20 min (Alais et al. 1967). Under severe ultra-high-temperature pasteurization conditions (up to 154.4°C for 9 sec), casein is solubilized (Morgan and Mangino 1979). Less intense treatments (from 137.8°C for 1 sec up) cause serum proteins to precipitate with casein during centrifugation. Lorient (1979) found casein molecules crosslinked with each other through amino groups when milk was heated to 120°C.

Inorganic phosphate is released when casein is heated. Dephosphorylated casein is less able to bind calcium and is more heat labile (Howat and Wright 1934). α_s-Caseins are especially sensitive to the calcium concentration because of their high phosphorylation levels and small amounts of secondary and tertiary structure (Kinsella 1984).

Lowering the pH of milk to 4.6 solubilizes colloidal calcium phosphate. This removes its neutralizing effect, allowing electrostatic interactions between micelles. Under these conditions, micelles coagulate and precipitate from solution. Kudo (1980C) showed that release of whey proteins and κ-casein from casein micelle surfaces as the pH is increased from 6.2 to 7.2 allows micelles to stick together and precipitate from solution.

Serum Proteins

β-**Lactoglobulin**   With a denaturation temperature of 78°C, β-lactoglobulin is the least denaturable of the serum proteins (Table 11.2). It exhibits a second thermal change near 140°C caused by a breakdown of disulfide bonds and additional unfolding of the molecule (de Wit 1981; Watanabe and Klostermeyer 1976). A change in pH between 6 and 7.5 shifts denaturation between 78° and 140°C, the total denaturation at the two temperatures being nearly constant. pH 6 favors dena-

Table 11.2. Denaturation Characteristics of Some Milk Proteins (Heated at 21.4°K/min in 0.7 M Phosphate Buffer at pH 6).

Protein	T_{tr}^a 21.4°K/min (°C)	T_d^b 0°K/min (°C)	ΔH^c (J/g)	ΔH^c (KJ/mol)	ΔH^c SE
α-Lactalbumin	68	62	17.8	253	17
β-Lactoglobulin	83	78	16.9	311	15
Immunoglobulin (IgG)	89	72	13.9	500	15
Bovine serum albumin	70	64	12.2	803	14

[a]Transition temperature at 21.4°K/min.
[b]Denaturation temperature extrapolated to 0°K/min.
[c]Defatted, according to Chen (1967).
SOURCE: de Wit and Klarenbeek (1984).

turation at 78°C and pH 7.5 favors it at 140°C (de Wit and Klarenbeek 1984). Unfolding of β-lactoglobulin below 78°C is reversible (de Wit 1981).

β-Lactoglobulin is very pH sensitive. Denaturation is slower at pH 4 than at pH 6 or 9 (Hillier et al. 1979). Differential scanning calorimetry indicates that heat stability of β-lactoglobulin decreases as pH is increased from 3 to 7.5. Below pH 3 β-lactoglobulin is an 18,300-dalton monomer (McKenzie 1971). Each 36,000-dalton dimer of β-lactoglobulin contains two thiol groups and four disulfide linkages. Decreased stability above pH 6 parallels increased thiol activity of β-lactoglobulin at high pH. The thiol groups are unreactive when β-lactoglobulin is in the native state, but a marked increase in activity with heat induces reversible dimer-to-monomer dissociation (de Wit and Klarenbeek 1984).

β-Lactoglobulin warmed to 40°C between pH 5 and 7 dissociates from dimers to monomers (Sawyer 1969; McKenzie 1971), which unfold and then polymerize by sulfhydryl interchange. These polymers then aggregate further (Harwalker 1980A). Creamer et al. (1978) found that β-lactoglobulin complexes form in milk heated at 100°C for 30 min at pH 6.5. Heating milk at pH 6.8 resulted in less compact, thread-like strands of β-lactoglobulin because of net negative charges on individual protein molecules at higher pH. As pH is increased above 6.8, the ability of β-lactoglobulin's free thiol groups to interact with other thiols increases because of a conformational change in the molecule (Dunnill and Green 1966). An increase in pH above 7.5 causes irreversible denaturation in β-lactoglobulin (Kinsella 1984).

β-Lactoglobulin is the most prominent sulfhydryl-containing milk protein. Heat treatment of milk causes a deterioration in flavor related to free sulfhydryl groups of β-lactoglobulin which appears before the protein is completely denatured (Hutton and Patton 1952). Prolonged exposure to heat causes more extensive unfolding of individual protein chains, leading to cleavage of disulfide linkages and exchange reactions with other proteins. Heating of β-lactoglobulin to denaturation allows its sulfhydryl groups to become very active. Addition of p-chloromercuribenzoate (0.28 mM) to skim milk before heating does not affect the rate of denaturation of β-lactoglobulin below 78°C, but at higher temperatures it is reduced by as much as 100-fold (Lyster 1970).

Harwalker (1980A,B) observed the effects of ionic environment and pH on heated β-lactoglobulin by adjusting the ionic strength ($\Gamma/2$) of protein solutions from 0.01 to 1 with sodium chloride. β-Lactoglobulin at pH 2.5 stayed in solution after heating for 30 min at 90°C at all $\Gamma/2$ levels and remained in solution when adjusted to pH 4.5 or when trichloroacetic acid was added to a concentration of 2.4%. All samples heated at pH 4.5 precipitated. Protein solutions heated at pH 6.5 were opalescent at 0.01 $\Gamma/2$ and at 0.05 and 0.01 $\Gamma/2$, 33 and 60% of the protein precipitated. Starch-urea gel electrophoresis at pH 9.2 and polyacrylamide gel electrophoresis at acidic and basic pH showed that β-lactoglobulin with an $\Gamma/2$ between 0.01 and 0.1 at pH 4.5 and 6.5 followed the usual heat denaturation mechanism of reversible dissociation and denaturation followed by irreversible aggregation and precipitation. At pH 2.5 a different mechanism was used, resulting in predominantly unfolded molecules with no apparent aggregation (Harwalker 1980B).

In the absence of calcium, β-lactoglobulin solubility increases as pH is increased from 6.4 to 7. Addition of calcium at any pH causes a decrease in solubility. At any ratio of pH and calcium ion concentration, the increased charge on the protein induced by pH change is balanced by added calcium ions to hold the level of denaturation constant (de Wit 1981). This suggests that calcium-induced precipitation of β-lactoglobulin occurs by an isoelectric mechanism. However, Hillier *et al.* (1979) found that an increase in the calcium concentration up to 0.4 mg/ml slowed heat denaturation of β-lactoglobulin, but additional calcium had little effect.

Lyster (1970) found denaturation of β-lactoglobulin to be second order with respect to time. The kinetic constant K_2 in log^{-1} sec^{-1} is described by the equation

$$\text{Log } K_2 = 37.95 - 14.51 \, (10^3/t)$$

between 68° and 90°C and by

$$\text{Log } K_2 = 5.98 - 2.86 \, (10^3/t)$$

between 90° and 135°C, where t is the temperature in degrees Kelvin. These equations represent mixed herd milk. Samples of milk containing genetic variants A and B of β-lactoglobulin have constants, K_2, which are 50% lower and higher, respectively. Milk of the AB variant has a constant near that of mixed herd milk. Hillier and Lyster (1979) repeated the high-temperature portion of this work in skim milk. They found

$$\text{Log } K_2 = 4.25 - 1.91 \, (10^3/t)$$

for β-lactoglobulin A between 100° and 150°C and

$$\text{Log } K_2 = 3.48 - 1.67 \, (10^3/t)$$

between 95° and 150°C.

Harwalker (1980B) determined enthalpies of denaturation for β-lactoglobulin at pH 4.5 and pH 6.5 of 3.57 and 2.56 cal/g. The denaturation temperature was shifted from 83.2° to 77.3°C over this pH range. Increasing the pH from 6.5 to 7.3 decreases the 80°C denaturation temperature to 74°C (Ruegg *et al.* 1977). Denaturation of β-lactoglobulin at pH 2.5 follows a different, first-order mechanism or two consecutive first-order reactions with an activation energy of ca. 43 kcal/mol.

α-Lactalbumin.   de Wit and Klarenbeek (1984) used differential scanning calorimetry to follow the unfolding of whey proteins during heating (Table 11.2). With a denaturation temperature of 62°C, β-lactalbumin is the least stable of whey proteins, but it requires the largest amount of heat per gram for unfolding. The long-held notion that β-lactalbumin is the most stable serum protein (de Wit 1981; Larson and Rolleri 1955) is explained by noting that it is the only protein in Table 11.2 whose heat denaturation at pH 6 is reversible. It is stable against heat-induced aggregation because it renatures easily when cooled. Ruegg *et al.* (1977) reported that α-lactalbumin denaturation was 80 to 90% reversible. Heating of α-lactalbumin causes a reversible conformational change related to four pairs of disulfide bonds within the molecule (Lyster 1979). Addition of *p*-chloromercuribenzoate (0.28 mM) to skim milk before heating reduces the rate of denaturation of α-lactalbumin from 25-fold at 85°C to about 3-fold at 155°C.

Removal of calcium ions makes unfolding of α-lactalbumin irreversible. The denaturation temperature of α-lactalbumin decreased 20°C when calcium ions were removed by a chelator (Bernal and Jelen 1984). Hillier *et al.* (1979) found that an increase in the calcium concentration up to 0.4 mg/ml slowed the heat denaturation of α-lactalbumin, but additional calcium had little effect. There is a slow conformational change at pH 4 as calcium is released from carboxyl groups on the protein surface (Kronman *et al.* 1964). Failure to measure the heat of denaturation for α-lactalbumin at pH 3 shows the protein chain is already unfolded at low pH (de Wit and Klarenbeek 1984).

Denaturation of α-lactalbumin is slower at pH 4 than at pH 6 or 9 (Hillier *et al.* 1979), but β-lactalbumin is partially denatured at pH 4 without heating (Kronman *et al.* 1966). Addition of NaOH or HCl to skim milk before heating has no effect on the denaturation of α-lactalbumin at either 78° or 100°C within the pH range 6.2 to 6.9 (Lyster 1970). The rate of denaturation increases above and below this range.

Lyster (1970) determined by immunodiffusion of heated skim milk that denaturation of α-lactalbumin is first order, with a kinetic constant K_1 in sec^{-1} between 90° and 155°C, given by

$$\text{Log } K_1 = 7.15 - 3.60 \ (10^3/t)$$

where t is the temperature in degrees Kelvin. Hillier and Lyster (1979) followed the kinetics of α-lactalbumin denaturation in skim milk and cheese whey. They found a bend in the Arrhenius plot at about 95°C. Between 70° and 95°C they found

$$\text{Log } K_1 = 20.60 - 8.75 \ (10^3/t)$$

and between 70° and 95°C

$$\text{Log } K_1 = 6.02 - 3.24 \ (10^3/t)$$

Serum Albumin. With a denaturation temperature of 64°C (Table 11.2), bovine serum albumin is denatured almost as easily as α-lactalbumin. Since its denaturation is not as reversible as that of α-lactalbumin, it appears to be the most easily denatured serum protein (de Wit and Klarenbeek 1984). It precipitates between 40° and 50°C as a result of hydrophobicity-directed unfolding (Lin and Koenig 1976; Macritchie 1973). Some serum albumin remains undenatured even after prolonged heating at 65°C. This may be because already denatured albumin is able to protect native proteins from denaturation (Terada *et al.* 1980).

Bovine serum albumin is denatured at pH 4 because of repulsion of acidic amino acids (Haurowitz 1963). As with α-lactalbumin, failure to measure the heat of denaturation for bovine serum albumin at pH 3 indicates that it is already unfolded by acid (de Wit and Klarenbeek 1984). It is more stable at pH 7.5 than at pH 6 because of increased activity of thiol groups at high pH. Denaturation is enhanced more by calcium ions than by other anions (Shimada and Matsushita 1981). Fatty acids appear to stabilize bovine serum albumin against heat denaturation (Gumpen *et al.* 1979).

Immunoglobulins. Immunoglobulins in milk are very heat labile, especially below pH 6 (de Wit and Klarenbeek 1984). But they have other characteristics which make them interesting. IgM, unless heat denatured, acts as a specific agglutinin against some streptococci strains (Mulder and Walstra 1974). IgM is a component of a cryoglobulin in milk which causes cold agglutination of milk fat and attachment of bacteria to milk fat globules (Walstra and Jenness 1984).

Casein and Serum Protein Interactions

Heat denaturation of β-lactoglobulin is accompanied by alterations in the properties of κ-casein (Zittle *et al.* 1962). κ-Casein and β-lactoglobulin interact through disulfide linkages when heated together or when κ-

casein is added to β-lactoglobulin (Morr 1965; Morr *et al.* 1962; Sawyer 1969). This interaction occurs over a narrow pH range of 6.7 to 7.0, with the optimum at pH 6.8 (de Wit 1981) and at 85° to 90°C (Smits and Brouwershaven 1980).

Formation of κ-casein and β-lactoglobulin complexes decreases as the ionic strength is decreased and as the pH is increased from 6.8 to 7.3 (Smits and Brouwershaven 1980). Complex formation is favored by calcium salts. More severe heat treatments increase the sensitivity of serum proteins to calcium ions. These variables implicate ionic interactions, along with disulfide interchange and hydrophobic interactions, in the formation of κ-casein and β-lactoglobulin complexes in heated milk. Dziuba (1979) reported that neither thiol groups of casein nor amino groups play a role in the interaction between β-lactoglobulin and micellar casein. His conclusion was that most of the interaction was hydrophobic.

It has been reported that α_{s2}-casein forms disulfide bridges when heated with β-lactoglobulin and can interfere with the ability of κ-casein to bind to β-lactoglobulin (Kinsella 1984; Kudo 1980B). Farah (1979) noted that the total amount of whey protein attached to casein increases as heat treatment is intensified, but that the ratio of whey proteins attached remains constant.

Direct interaction between α-lactalbumin and κ-casein when heated is limited, if it occurs at all (Hartman and Swanson 1965), but the complex formed between α-lactalbumin and β-lactoglobulin is able to interact with κ-casein (Elfagm and Wheelock 1977; Hunziker and Tarassuk 1965). The degree of denaturation of α-lactalbumin is greater when heated with β-lactalbumin than when heated alone. This effect increases as pH increases from 6.4 to 7.2 and is more pronounced at temperatures between 70° and 85°C. The degree of denaturation of β-lactoglobulin is not affected by the presence of α-lactalbumin, but the presence of casein facilitates the formation of α-lactalbumin and β-lactoglobulin complexes (Elfagm and Wheelock 1978A,B). Direct interaction between β-lactoglobulin and κ-casein is reduced in the presence of α-lactalbumin (Baer *et al.* 1976; Elfagm and Wheelcock 1977, 1978B). κ-Casein also complexes with α_{s1}-casein and β-casein (Doi *et al.* 1979), which may interfere with other κ-casein complexes.

Most current models put κ-casein on the outer casein micelle surface (Heth and Swaisgood 1982; McMahon and Brown 1984A; Shahani 1974). This allows the possibility that heat-induced coagulation of milk is the result of serum proteins interacting with κ-casein on the micelle surface and with each other to interconnect micelles. The observation that chymosin cannot release macropeptides from κ-casein in heated milk (Morrissey 1969; Shalabi and Wheelock 1976, 1977) suggests that

κ-casein is physically inaccessible to enzyme. This theory is also supported by Creamer *et al.* (1978), who, with an electron microscope, observed protein complexes formed by heating skim milk at 100°C for 30 min. These complexes are large, containing hundreds of individual protein molecules attached to casein micelles. At higher pH levels the complexes change, becoming more filamentous and associating less with micelles.

COAGULATION IN MILK OR DAIRY PRODUCTS

General Comments

Denaturation of a combination of individual proteins in milk or dairy products results in coagulation. Such coagulation must be considered in the context of interactions among the different proteins and in the presence of additional milk components. For example, α-lactalbumin is more susceptible to denaturation in milk than in whey, but the opposite is true of β-lactoglobulin (Elfagam and Wheelock 1977). Many factors influence milk stability in these complex systems (Holt *et al.* 1978; Pyne and McHenry 1955; Rose 1961A,B, 1963; Sweetsur and White 1974; Tessier and Rose 1964) during heat treatments, ranging from 72°C for 15 sec for pasteurization to 120°C for 20 min or 142°C for several seconds for sterilization (Creamer and Matheson 1980; Douglas *et al.* 1981). Casein is most important in determining the properties of milk products because of its high concentration (Payens 1978), but β-lactoglobulin has a larger effect on a molar basis (de Wit 1981; Fox and Morrissey 1977; Rose 1961A,B, 1963). Many aspects of milk stability have been summarized previously (Fox 1981A, 1982; Fox and Morrissey 1977; Kinsella 1984; Parry 1974; Pyne 1962; Rose 1963, 1965; Tumerman and Webb 1965).

Salt Balance and pH

Most protein denaturation reactions are very pH dependent (Brandts 1969). Proteins normally expand as pH moves away from their isoelectric point (Tanford 1968). An increase in net charge causes repulsion between like groups. Most samples of milk require a maximum time for heat coagulation when adjusted to pH 6.7 and a minimum time at pH 6.9, as shown in Figure 11.1 (Fox 1982; Rose 1961A). Such curves of coagulation time at a fixed temperature versus pH are typical of milk designated as "type A" (Rose 1961B; Tessier and Rose 1964A). Some milk samples from individual cows fail to show minimum and

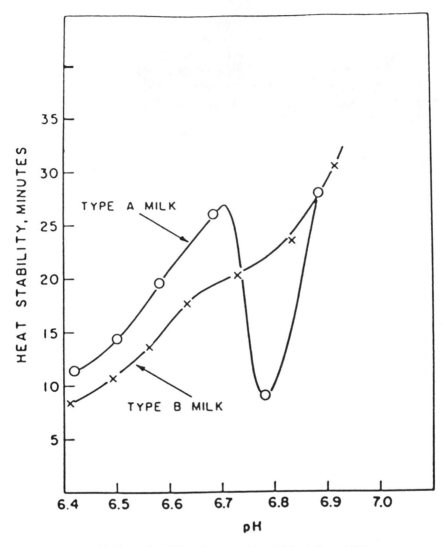

Figure 11.1.   pH Heat Stability Curves of Individual Cow Milks o_____o = Type A mild, x_____x = Type B milk. (*From Tessier and Rose 1964.*)

maximum points on the curve, but instead increase in coagulation time as pH is increased from 6.2 to 7.4. Such milk is referred to as "type B."

Type A accounts for a different proportion of milk samples, depending on geographic location, time, degree of agitation (Hyslop and Fox 1981), headspace atmosphere above the sample (Sweetsur and White

1975), species (Fox 1982), stage of lactation (Rose 1961A), feed composition (Feagan et al. 1972), and other less definable factors (Fox 1982). The temperature of the assay has been shown to affect the shape of these curves to greater or lesser degrees in separate studies (Fox and Hearn 1978B; Hyslop and Fox 1981; Sweetsur and White 1974). Sweetsur and White (1974) converted type B to type A behavior by using higher temperatures. In several studies, 20% of the milk was determined to be type B in Canada (Tessier and Rose 1964), ca. 1% in Australia (Feagan et al. 1972) and Ireland (Fox 1982), and ca. 70% in Japan (Fox 1982).

Sweetsur and White (1974) showed that type B milk coagulates by a one-step mechanism and that type A milk coagulates by a two-step mechanism. Parker et al. (1979) proposed a separate mechanism for coagulation of each type of milk. Although their mechanism is mathematically sound and accounts for the physical observations, there is some difficulty in accounting chemically for what happens during coagulation.

Fox (1981B) argues that pH change caused by heating is primarily responsible for heat coagulation of milk. Acidity of milk increases by ca. 0.1 pH unit for each $10°C$ temperature rise (Kruk 1979; Miller and Sommer 1940). The decrease in pH upon heating is partially due to changes in the buffer capacity of milk salts and the release of carbon dioxide. According to Fox (1981A,B), when milk is heated at elevated temperatures for prolonged periods of time, additional acidity develops as a result of (1) the production of organic acids, principally formic, from lactic acid, (2) the release of hydrogen ions by precipitation of primary and secondary calcium phosphate, and (3) the release of hydrogen ions by hydrolysis of casein phosphate and its subsequent precipitation as $Ca_3(PO_4)_2$. These reactions contribute 50, 20 and 30%, respectively, to pH decline (Pyne and McHenry 1955).

The pH effect in coagulation of milk is a function of κ-casein concentration on the micelle surfaces and the β-lactoglobulin concentration in milk serum. Tessier and Rose (1964) eliminated the minimum in coagulation time versus pH curves of type A milk samples by adding κ-casein, thus converting it to type B. They also converted type B milk to Type A by salting out some κ-casein or by adding β-lactoglobulin. Binding of calcium phosphate to the surface of β-lactoglobulin–coated casein micelles during heating has been implicated in the coagulation of milk (Fox and Hoynes 1975; Morrissey 1969).

When milk at a pH of less than 6.5 is heated for 20 to 30 min at $100°C$, it coagulates to form a gel. Casein micelles isolated from such milk have denatured whey protein attached to micelle surfaces (Creamer et al. 1978). Such micelle surfaces aggregates of whey proteins and κ-casein may serve to join micelles to each other. When milk pH is greater than 6.7, more heat treatment (20 to 30 min at 130°C) is required for precipitation and a gel does not form. This led Creamer and Matheson (1980) to conclude that coagulation occurs by two different mechanisms, one below ca. pH 6.7 and another above. At low pH, whey proteins denature onto micelles and link them together. At high pH, coagulation occurs only when the caseins in the micelles have changed enough to cross-link with each other. Denatured serum proteins do not precipitate separately in milk, but coprecipitate with caseins on acidification, salting out, or ultracentrifugation (Edmundson and Tarassuk 1956A,B; Fox et al. 1967; Rowland 1933; Sullivan et al. 1957).

Heating milk reduces soluble and ionic calcium and phosphate concentrations by converting the soluble calcium phosphate to the colloidal state (Rose and Tessier 1959). Mattick and Hallett (1929) reported that heating for 30 min at 57° to 60°C causes 0.6% of the total calcium in milk to become insoluble; between 63° and 65°C, 2% is insolubilized; and at higher temperatures this can be increased to 3.6%. Hilgeman and Jenness (1951) reported a much higher figure: a 25% soluble calcium loss in 30 min at 78°C. Different methods of calcium measurement and variations in time between heating and calcium measurement could account for the discrepancies between the reported values. When heated milk is cooled, it becomes unsaturated with respect to calcium and phosphate (Rose and Tessier 1959). Because of its association with casein micelles, heat-precipitated calcium phosphate in milk does not sediment (Evenhuis and de Vries 1956). Colloidal calcium phosphate slowly dissolves to restore equilibrium (Fox et al. 1967; Kannan and Jenness 1961; Pyne 1958).

Colloidal and serum salt levels affect casein integrity. Raising or lowering the temperature or pH of milk affects this equilibrium. Removal of calcium from milk by electrodialysis causes the amount of casein in serum to increase exponentially. Removal of 70% of the calcium results in all of the casein being released from micelles into the serum as submicelles (Lonergan 1978). Fox and Hearn (1978A) mimicked this situation by dialysis against distilled water and found a leveling of the denaturation time versus pH curve (type A shifted toward type B).

β-Lactoglobulin sensitizes casein to calcium phosphate binding. It has been suggested (Fox and Hearn 1978A; Sweetsur and White 1974) that calcium phosphate on their surfaces at pH 6.9 makes casein micelles or β-lactoglobulin–coated micelles susceptible to coagulation by calcium ions. The reduced amount of protein that can be precipitated from milk which has been heated to 110°C and the higher concentration of κ-casein than α_s-casein in the nonprecipitating protein support this theory (Fox *et al.* 1967). Kudo (1980C) showed that as milk is moved from pH 6.2 to pH 7.2, casein micelles lose whey proteins and κ-casein from their surfaces and become more susceptible to heat coagulation.

In the absence of casein, as in whey or whey powder, gelation is possible by heating at 80°C for 2.5 to 21 min. The time to form gels decreases as the concentration of thiol groups increases and is favored by low pH. Gels from almost instantaneously below pH 6 but are more of a coagulum of denatured protein than a gel. Those gels formed above pH 6 are made of polypeptide chains cross-linked by disulfide bonds and can be dissolved by addition of sulfhydryl reagents (Hillier *et al.* 1980).

Insolubilization of whey proteins by heat in the absence of casein is controlled by their ionic environment (de Rham and Chanton 1984). Acid precipitation produces a dense precipitate. Irreversible aggregation of whey proteins has been observed during storage of milk acidified to pH 3.4 to 4.6 for 10 days at 35° to 45° (Argyle *et al.* 1976). Between pH 5.5 and 6, whey solutions are stable up to boiling temperature for 5 min. Harwalker (1978, 1979) heated acid and whey protein solutions in distilled water adjusted to pH 2.5, 4.5, and 6.5 for 20 to 30 min at 90°C. Proteins at pH 2.5 remained soluble after heating even though they were denatured. Sulfhydryl reactions are inactive and less net positive charge for attraction between proteins exists at pH 2.5. Looking at the proteins individually, β-lactoglobulin and serum albumin were altered at pH 2.5, but α-lactalbumin was not.

Calcium precipitation of whey proteins produces a hydrated coagulum. Schmidt *et al.* (1979) found hardness of whey protein gels to be greatest when 11.1 mM calcium chloride was added. de Rham and Chanton (1984) observed that calcium concentrations critical for precipitation of whey proteins were independant of the calcium-to-protein ratios.

Addition of appropriate phosphate or citrate salts as stabilizers in skim or concentrated skim milk can increase stability during heating. The difficulty is in determining what is appropriate. Added salts must move the milk along the coagulation time versus pH curve to a stable point (Sweetsur and Muir 1980A). Indiscriminate addition of buffers to milk can easily move past the proper point. Mattick and Hallett (1929) reported that about 3.5% of the total phosphorus in milk is insolubilized in 30 min at 79° to 81°C. Hilgeman and Jenness (1951) also indicated that soluble phosphorus is insolubilized at higher temperatures.

Forewarming

Forewarming or preheating of milk to increase stability before further processing has long been used in the industrial concentration and sterilization of milk (Darling 1980; Fox 1982; Sweetsur and Muir 1980A; Sweetsur and White 1974). Heat treatments to 140°C for 10 sec are ineffective in protecting milk against coagulation during subsequent sterilization. The maximum heat stability may be obtained by heating at 110° to 120°C for 120 to 240 sec. Newstead and Baucke (1983) concluded that forewarming is a first-order reaction, and they found an activation energy of 50 ± 8 KJ/mole. Darling (1980) found an activation energy of 144 KJ/mole. The lower energy level suggests sulfhydryl interchange reactions in the whey proteins. Forewarming may be related to the ability of heat-denatured bovine serum albumin to protect native serum albumin from denaturation (Terada *et al.* 1980). Payens (1978) suggested that precipitation of whey proteins on casein micelle surfaces during forewarming prevents later coagulation by severely diminishing the number of κ-casein sites available for clotting.

Effects of Concentration

Heat denaturation of protein solutions is normally retarded by concentration. Concentration of milk to total solids levels of 9, 28 and 44% decreases apparent denaturation by 40, 60, and 80% (Whitney 1977). Individual proteins are affected differently by concentration, α-lactalbumin being denatured more easily as solids are increased and both A and B genetic variants of β-lactoglobulin being denatured less easily (Hillier *et al.* 1979).

Stability of the complex protein system of milk or whey is decreased by concentration (Fox 1982; Muir and Sweetsur 1978; Sweetsur and Muir 1980B). In addition to closer packing of casein micelles and other proteins in concentrated milk, calcium phosphate is precipitated so that the pH decreases (Fox 1982). The pH effect causes protein which would be soluble at a normal solids concentration to precipitate. Casein in milk concentrated to three times its original solids level forms a flocculent after 1 to 3 weeks at −8°C (Lonergan 1978).

Effects of Other Substances

The variation in lactose concentration in milk is small, but experiments with the adjustment of lactose concentration have shown that lactose plays a part in the stabilization of milk. Addition of lactose increases both denaturation temperatures of β-lactoglobulin (de Wit 1981). Milk stability can be increased by hydrolyzing lactose or by adding other sugars. Replacement of lactose with glucose increases stability, but replacement with sucrose does not (Kudo 1980A; Lonergan 1978). Denaturation of α-lactalbumin and β-lactoglobulin is inhibited by lactose and other sugars (Hillier *et al.* 1979).

The naturally occurring and variable concentration of urea in milk affects coagulation (Kudo 1980A). Pyne (1958) and Muir *et al.* (1978) stabilized milk by adding urea to it before heating. β-Lactoglobulin can be completely denatured in urea (Kauzmann and Simpson 1953). Kudo (1980A) found a greatly enhanced stabilizing effect when lactose and urea were added at the same time.

REFERENCES

Alias, C., Kiger, N. and Jolles, P. 1967. Action of heat on cow κ-casein. Heat caseino-glycopeptide. *J. Dairy Sci. 50,* 1738–1743.

Argyle, P. J., Jones, N., Chandan, R. C. and Gordon, J. F. 1976. Aggregation of whey proteins during storage of acidified milk. *J. Dairy Res. 43,* 45–51.

Baer, A., Orz, M. and Blanc, B. 1976. Serological studies on heat-induced interactions of α-lactalbumin and milk proteins. *J. Dairy Res. 43,* 419–432.

Bernal, V. and Jelen, P. 1984. Effect of calcium binding on thermal denaturation of bovine α-lactalbumin. *J. Dairy Sci. 67,* 2452–2454.

Bloomfield, V. A. 1979. Association of proteins. *J. Dairy Res. 46,* 241–252.

Brandts, J. F. 1969. Conformational transitions of proteins in water and in aqueous mixtures. *In: Structure and Stability of Biological Macromolecules.* S. N. Timascheff and G.D. Fasman (Editors). Marcel Dekker, New York, pp. 213–290.

Brown, R. J. 1984. Casein micelle structure. Symposium at the 79th American Dairy Science Association Meeting. College Station, Texas, June 24–27.

Carpenter, R. N. and Brown, R. J. 1985. Separation of casein micelles from milk for rapid determination of casein content. *J. Dairy Sci. 68,* 307–311.

Chen, R. F. 1967. Removal of fatty acids from serum albumin by charcoal treatment. *J. Biol. Chem. 242,* 173–181.

Creamer, L. K., Berry, G. P. and Matheson, A. R. 1978. The effect of pH on protein aggregation in heated skim milk. *N.Z. J. Dairy Sci. Technol. 13,* 9–15.

Creamer, L. K. and Matheson, A. R. 1980. Effect of heat treatment on the proteins of pasteurized skim milk. *N.Z. J. Dairy Sci. Technol 15,* 37–49.

Dalgleish, D. G. 1982. The enzymatic coagulation of milk. *In: Developments in Dairy Chemistry, Vol. 1: Proteins.* P.F. Fox (Editor). Applied Science Publishers, London, pp. 157–183.

Darling, D. F. 1980. Heat stability of milk. *J. Dairy Res. 47,* 199–210.

de Rham, O. and Chanton, S. 1984. Role of ionic environment in insolubilization of whey protein during heat treatment of whey products. *J. Dairy Sci. 67,* 939–949.

de Wit, J. N. 1981. Structure and functional behavior of whey proteins. *Neth. Milk Dairy J. 35,* 47–64.

de Wit, J. N. and Klarenbeek, G. 1981. A differential scanning calorimetric study of the thermal behavior of bovine β-lactoglobulin at temperatures up to 160°C. *J. Dairy Res. 48,* 293–302.

de Wit, J. N. and Klarenbeek, G. 1984. Effects of various heat treatments on structure and solubility of whey proteins. *J. Dairy Sci. 67*, 2701–2710.

Doi, H., Ibuki, F. and Kanamori, M. 1979. Interactions of κ-casein components with α_{s1}- and α_{s2}-caseins. *Agri. Biol. Chem. 43*, 1301–1308.

Douglas, F. W., Greenberg, R., Farrell, H. M. and Edmonson, L. F. 1981. Effects of ultra-high-temperature pasteurization on milk proteins. *J. Agri. Food Chem. 29*, 11–15.

Dunnill, P. and Green, D. W. 1966. Sulphydryl groups and the N$\rightleftharpoons$R conformational change in β-lactoglobulin. *J. Mol. Biol. 15.*, 147–151.

Dziuba, J. 1979. The share of functional casein groups in the formation of a complex with β-lactoglobulin. *Acta Alimentaria Polonica 5*, 97–115.

Dzurec, D. J. and Zall, R. R. 1985. Effect of heating, cooling, and storing milk on casein and whey proteins. *J. Dairy Sci. 68*. 273–280.

Edmundson, L. F. and Tarassuk, N. P. 1956A. Studies on the colloidal proteins of skim-milk: I. The effect of heat and certain salts on the centrifugal sedimentation of milk proteins. *J. Dairy Sci. 39*, 36–45.

Edmundson, L. F. and Tarassuk, N. P. 1956B. Studies on the colloidal proteins of skim-milk: II. The effect of heat and disodium phosphate on the composition of the casein complex. *J. Dairy Sci. 39*, 123–128.

Edsal. J. T. and Wyman, J. 1958. *Biophysical Chemistry*, Vol. 1: *Thermodynamics, Electrostatics, and the Biological Significance of the Properties of Matter.* Academic Press, New York.

Elfagm, A. A. and Wheelock, J. V. 1977. Effect of heat on α-lactalbumin and β-lactoglobulin in bovine milk. *J. Dairy Res. 44*, 367–371.

Elfagm, A. A. and Wheelock, J. V. 1978A. Interaction of bovine α-lactalbumin and β-lactoglobulin during heating. *J. Dairy Sci. 61*, 28–32.

Elfagm, A. A. and Wheelock, J. V. 1978A. Heat interaction between α-lactalbumin, β-lactoglobulin and casein in bovine milk. *J. Dairy Res. 61*, 159–163.

Evenhuis, N. and de Vries, T. R. 1956. The condition of calcium phosphate in milk III. *Neth. Milk Dairy J. 10*, 101–113.

Farah, Z. 1979. Changes in proteins in UHT unheated milk. *Milchwissenschaft 34*, 484–487.

Feagan, J. T., Bailey, L. F., Hehir, A. F., McLean, D. M. and Ellis, N. J. S. 1972. Coagulation of milk proteins. I. Effect of genetic variants of milk proteins on rennet coagulation and heat stability of normal milk. *Aust. J. Dairy Technol. 27*, 129–134.

Flory, P. J. 1969. *Statistical Mechanics of Chain Molecules*. John Wiley & Sons, New York.

Fox, K. K., Harper, M. K., Holsinger, V. H. and Pallansch, M. J. 1967. Effects of high heat treatment on stability of calcium casein aggregates in milk. *J. Dairy Sci. 50*, 443–450.

Fox, P. F. 1981A. Heat-induced changes in milk preceding coagulation. *J. Dairy Sci. 64*, 2127–2137.

Fox, P. F. 1981B. Heat stability of milk: Significance of heat-induced acid formation in coagulation. *Irish J. Food Sci. Technol. 5*, 1–11.

Fox, P. F. 1982. Heat-induced coagulation of milk. *In: Developments in Dairy Chemistry*, Vol. 1: *Proteins*. P.F. Fox (Editor). Applied Science Publishers, London, pp. 189–223.

Fox, P. F. and Hearn, C. M. 1978A. Heat stability of milk: Influence of dilution and dialysis against water. *J. Dairy Res. 45*, 149–157.

Fox, P. F. and Hearn, C. M. 1978B. Heat stability of milk: Influence of denaturable proteins and detergents on pH sensitivity. *J. Dairy Res. 45*, 159–172.

Fox, P. F. and Hearn, C. M. 1978C. Heat stability of milk: Influence of κ-casein hydrolysis. *J. Dairy Res. 45*, 173–181.

Fox, P. F. and Hoynes, M. C. T. 1975. Heat stability of milk: Influence of colloidal calcium phosphate and β-lactoglobulin, J. Dairy Res. 42, 427–435.

Fox, P. F. and Morrissey, P. A. 1977. The heat stability of milk. J. Dairy Res. 44, 627–646.

Franks, F. and Eagland, D. 1975. The role of solvent interactions in protein conformation. CRC Crit. Rev. Biochem. 3, 165–219.

Gumpen, S., Hegg, P. O. and Martens, M. 1979. Thermal stability of fatty acid–serum albumin complexes studied by differential scanning calorimetry. Biochim. Biophys. Acta 574, 189–196.

Gurd, F. R. N. and Rothgeb, T. M. 1979. Motions in proteins. In: Advances in Protein Chemistry, Vol. 33. C.B. Anfinsen, J. T. Edsall, and F.M. Richards (Editors). Academic Press, New York, pp. 74–165.

Hartman, G. H. and Swanson, A. M. 1965. Changes in mixtures of whey protein and κ-casein due to heat treatments. J. Dairy Sci. 48, 1161–1167.

Harwalker, V. R. 1978. Application of differential scanning calorimetry to the study of thermal denaturation of β-lactoglobulin in solution. J. Dairy Sci. 61 (suppl. 1), 107.

Harwalker, V. R. 1979. Comparison of physico-chemical properties of different thermally denatured whey proteins. Milchwissenschaft 34, 419–422.

Harwalker, V. R. 1980A. Measurement of thermal denaturation of β-lactoglobulin at pH 2.5. J. Dairy Sci. 63, 1043–1051.

Harwalker, V. R. 1980B. Kinetics of thermal denaturation of β-lactoglobulin at pH 2.5. J. Dairy Sci. 63, 1052–1057.

Haurowitz, F. 1963. Albumins, globulins and other soluble proteins. In: The Chemistry and Function of Proteins. F. Haurowitz (Editor). Academic Press, New York, pp. 1–455.

Heth, A. A. and Swaisgood, H. E. 1982. Examination of casein micelle structure by a method for reversible covalent immobilization. J. Dairy Sci. 65, 2047–2054.

Hilgeman, M. and Jenness, R. 1951. Observations on the effect of heat treatment upon the dissolved calcium and phosphorus in skimmilk. J. Dairy Sci. 34, 483–484.

Hillier, R. M. and Lyster, R. L. J. 1979. Whey protein denaturation in heated milk and cheese whey. J. Dairy Res. 46, 95–102.

Hillier, R. M., Lyster, R. L. J. and Cheeseman, G. C. 1979. Thermal denaturation of α-lactalbumin and β-lactoglobulin in cheese whey: Effect of total solids concentration and pH. J. Dairy Res. 46, 103–111.

Hillier, R. M., Lyster, R. L. J. and Cheeseman, G. C. 1980. Gelation of reconstituted whey powders by heat. J. Sci. Food Agr. 31, 1152–1157.

Hindle, E. J. and Wheelock, J. V. 1970. The release of peptides and glycopeptides by action of heat on cows' milk. J. Dairy Res. 37, 397–405.

Holt, C., Muir, D. D. and Sweetsur, A. W. M. 1978. Seasonal changes in the heat stability of milk from creamery silos in south-west Scotland. J. Dairy Res. 45, 183–190.

Howat, G. R. and Wright, N. C. 1934. The heat coagulation of caseinogen: I. The role of phosphorus cleavage. Biochem. J. 28, 1336–1345.

Hunziker, H. G. and Tarassuk, N. P. 1965. Chromatographic evidence for heat-induced interaction of α-lactalbumin and β-lactoglobulin. J. Dairy Sci. 48, 733–734.

Hutton, J. T. and Patton, S. 1952. The origin of sulfhydryl groups in milk proteins and their contributions to "cooked" flavor. J. Dairy Sci. 35, 699–705.

Hyslop, D. B. and Fox, P. F. 1981. Heat stability of milk: Interrelationship between assay temperature, pH and agitation. J. Dairy Sci. 48, 123–129.

Kannan, A. and Jenness, R. 1961. Relation of milk serum proteins and milk salts to the effects of heat treatment on rennet clotting. J. Dairy Sci. 44, 808–822.

Karplus, M. and McCammon, J. A. 1983. Dynamics of proteins: Elements and function. In: Annual Review of Biochemistry, Vol. 52. E.S. Snell, P.D.

Boyer, A., Meister and C.C. Richardson (Editors). Annual Reviews, Palo Alto, Calif., pp. 263–300.

Kauzmann, W. and Simpson, R. B. 1953. The kinetics of protein denaturation. III. The optical rotations of serum albumin, β-lactoglobulin and pepsin in urea solutions. *J. Am. Chem. Soc. 75,* 5154–5157.

Kim, P. S. and Baldwin, R. L. 1982. Specific intermediates in the folding reactions of small proteins and the mechanism of protein folding. *In: Annual Review of Biochemistry,* Vol. 51. E.S. Snell, P. D. Boyer, A. Meister and C.C. Richardson (Editors). Annual Reviews, Palo Alto, Calif., pp. 459–489.

Kinsella, J. E. 1984. Milk proteins: Physicochemical and functional properties. *CRC Crit. Rev. Food Sci. Nutr. 21,* 197–262.

Kronman, M. J., Andreotti, R. E. and Vitols, R. 1964. Inter- and intramolecular interactions of α-lactalbumin. II. Aggregation reactions at acid pH. *Biochemistry 3,* 1152–1160.

Kronman, M. J., Blum, R. and Holmes, L. G. 1966. Inter- and intramolecular interactions of α-lactalbumin. VI. Optical rotation dispersion properties. *Biochemistry 5,* 1970–1978.

Kruk, A. 1979. Relationship between casein hydration degree and thermal stability of milk. *Acta Alimentaria Polonica 5,* 147–156.

Kudo, S. 1980A. Influence of lactose and urea on the heat stability of artificial milk systems. *N.Z. J. Dairy Sci. Technol. 15,* 197–200.

Kudo, S. 1980B. The influence of α_{s2}-casein on the heat stability of artificial milks. *N.Z. J. Dairy Sci. Technol. 15,* 245–254.

Kudo, S. 1980C. The heat stability of milk: Formation of soluble proteins and protein-depleted micelles at elevated temperatures. *N.Z. J. Dairy Sci. Technol. 15,* 255–263.

Lapanje, S. 1978. *Physiochemical Aspects of Protein Denaturation,* John Wiley & Sons, New York.

Larson, B. L. and Rolleri, G. D. 1955. Heat denaturation of the specific serum proteins in milk. *J. Dairy Sci. 38,* 351–360.

Lin, V. J. C. and Koenig, J. L. 1976. Raman studies of bovine serum albumin. *Biopolymers 15,* 203–218.

Lonergan, D. A. 1978. Use of electrodialysis and ultrafiltration procedures to improve protein stability of frozen concentrated milk. Ph.D. dissertation, University of Wisconsin, Madison.

Lorient, D. 1979. Covalent bonds formed in proteins during milk sterilization: Studies on caseins and casein peptides. *J. Dairy Res. 46,* 393–396.

Lyster, R. L. J. 1970. The denaturation of α-lactalbumin and β-lactoglobulin in heated milk. *J. Dairy Res. 37,* 233–243.

Lyster, R. L. J. 1979. Milk and dairy products. *In: Effects of Heating on Food Stuffs.* R.J. Priestly (Editor). Applied Science Publishers, London, pp. 353–372.

Macritchie, F. 1973. Effects of temperature on dissolution and precipitation of proteins and polyamino acids. *J. Colloid Interface Sci. 45,* 235–241.

Mangino, M. E. 1984. Physicochemical aspects of whey protein functionality. *J. Dairy Sci. 67,* 2711–2722.

Mattick, E. C. V. and Hallett, H. S. 1929. The effect of heat on milk. *J. Agr. Sci. 19,* 452–462.

McKenzie, H. A. 1971. Whey proteins and minor proteins: β-Lactoglobulins. *In: Milk Proteins: Chemistry and Molecular Biology,* Vol. 2. H.A. McKenzie (Editor). Academic Press, New York, pp. 257–330.

McMahon, D. J. and Brown, R. J. 1984A. Composition, structure and integrity of casein micelles: A review. *J. Dairy Sci. 67,* 499–512.

McMahon, D. J. and Brown, R. J. 1984B. Enzymic coagulation of casein micelles: A review. *J. Dairy Sci. 67*, 919–929.

Miller, P. G. and Sommer, H. H. 1940. The coagulation temperature of milk as affected by pH, salts, evaporation and previous heat treatment. *J. Dairy Sci. 23*, 405–421.

Morgan, J. N. and Mangino, M. E. 1979. The effect of ultra high temperature processing on the proteins of whole milk. *J. Dairy Sci. 62*, (suppl. 1), 229.

Morr, C. V. 1965. Effect of heat upon electrophoresis and ultracentrifugal sedimentation properties of skimmilk protein fractions. *J. Dairy Sci. 48*, 8–13.

Morr, C. V., Van Winkle, Q. and Gould, I. A. 1962. Application of polarization of fluorescence technique to protein studies. III. The interaction of κ-casein and β-lactoglobulin. *J. Dairy Sci. 45*, 823–826.

Morrissey, P. A. 1969. The rennet hysteresis of heated milk. *J. Dairy Res. 36*, 333–341.

Muir, D. D., Abbot, J. and Sweetsur, A. W. M. 1978. Changes in the heat stability of milk protein during the manufacture of dried skim-milk. *J. Food Technol. 13*, 45–53.

Muir, D. D. and Sweetsur, A. W. M. 1978. The effect of concentration on the heat stability of skim-milk. *J. Dairy Res. 45*, 37–45.

Mulder, H. and Walstra, P. 1974. Creaming and separation. *In: The Milkfat Globule.* Commonwealth Agriculture Bureau, Bucks., England, pp. 168–173.

Newstead, D. F. and Baucke, A. G. 1983. Heat stability of recombined evaporated milk and reconstituted concentrated skim milk: Effects of temperature and time of preheating. *N.Z. J. Dairy Sci. Technol. 18*, 1–11.

Pace, C. N. 1975. The stability of globular proteins. *CRC Crit. Rev. Biochem. 3*, 1–43.

Parker, T. G., Horne, D. S. and Dalgleish, D. G. 1979. Theory for the heat-induced coagulation of a type A milk. *J. Dairy Res. 46*, 377–380.

Parry, R. M., Jr. 1974. Milk coagulation and protein denaturation. *In: Fundamentals of Dairy Chemistry*, 2nd ed. B. H. Webb, A. H. Johnson and J. A. Alford (Editors). AVI Publishing Co., Westport, Conn., pp. 603–655.

Payens, T. A. J. 1978. On different modes of casein clotting; the kinetics of enzymatic and non-enzymatic coagulation compared. *Neth. Milk Dairy J. 32*, 170–183.

Payens, T. A. J. and Vreeman, H. J. 1982. Casein micelles and micelles of κ- and β-casein. *In: Solution Behavior of Surfactants*, Vol. 1. K. L. Mital and E. J. Fendler (Editors). Plenum Press, New York, pp. 543–571.

Pearce, R. J. 1980. Heat-stable components in the Aschaffenburg and Drewry total albumin fraction from bovine milk. *N.Z. J. Dairy Sci. Technol. 15*, 13–22.

Pyne, G. T. 1958. The heat coagulation of milk: II. Variations in sensitivity of casein to calcium ions. *J. Dairy Res. 25*, 467–474.

Pyne, G. T. 1962. Some aspects of the physical chemistry of the salts of milk. *J. Dairy Res. 29*, 101–130.

Pyne, G. T. and McHenry, K. A. 1955. The heat coagulation of milk. *J. Dairy Res. 22*, 60–68.

Rose, D. 1961A. Variations in the heat stability and composition of milk from individual cows during lactation. *J. Dairy Sci. 44*, 430–441.

Rose, D. 1962B. Factors affecting the pH-sensitivity of the heat stability of milk from individual cows. *J. Dairy Sci. 44*, 1405–1413.

Rose, D. 1963. Heat stability of bovine milk: A review. *Dairy Sci. Abstr. 25*, 45–52.

Rose, D. 1965. Protein stability problems. *J. Dairy Sci. 48*, 139–146.

Rose, D. and Tessier, H. 1959. Composition of ultrafiltrates from milk heated at 80 to 230°F in relation to heat stability. *J. Dairy Sci. 42*, 969–980.

Rowland, S. J. 1933. The heat denaturation of albumin and globulin in milk. *J. Dairy Res. 5*, 46–53.

Ruegg, M., Moor, U. and Blanc, B. 1977. A calorimetric study of the thermal denaturation of whey proteins in simulated milk ultrafiltrate. *J. Dairy Res. 44*, 509–520.

Sawyer, W. H. 1969. Complex between β-lactoglobulin and κ-caseins: A review. *J. Dairy Sci. 52*, 1347–1355.

Schmidt, D. G. 1980. Colloidal aspects of casein. *Neth. Milk Dairy J. 34*, 42–64.

Schmidt, R. H., Illingworth, B. L., Deng, J. C. and Cornell, J. A. 1979. Multiple regression and response surface analysis of the effects of calcium chloride and crysteine on heat-induced whey protein gelation. *J. Agr. Food Chem. 27*, 529–532.

Schmutz, M. and Puhan, Z. 1981. Chemischphysikalische veranderungen wahrend der tiefkuhllagerung von milch. *Dtsch. Molkerei Z. 17*, 552–564.

Shahani, K. M. 1974. Recent advances in the chemistry and physics of milk products for standardisation of processing and manufacturing techniques. *XIX Int. Dairy Congr. 2*, 306–322.

Shalabi, S. I. and Wheelock, J. V. 1976. The role of α-lactalbumin in the primary phase of chymosin action on heated casein micelles. *J. Dairy Res. 43*, 331–335.

Shalabi, S. I. and Wheelock, J. V. 1977. Effect of sulphydryl blocking agents on the primary phase of chymosin action on heated casein micelles and heated milk. *J. Dairy Res. 44*, 351–355.

Shimada, K. and Matsushita, S. 1981. Efforts of salts and denaturants on thermocoagulation of proteins. *J. Agr. Food Chem. 29*, 15–20.

Smits, P. and Brouwershaven, J. 1980. Heat-induced association of β-lactoglobulin and casein micelles. *J. Dairy Res. 47*, 313–325.

Sommer, H. H. and Hart, E. B. 1919. The heat coagulation of milk. *J. Biol. Chem. 40*, 137–151.

Sommer, H. H. and Hart, E. B. 1922. The heat coagulation of milk. *J. Dairy Sci. 6*, 525–543.

Sullivan, R. A., Hollis, R. A. and Stanton, E. K. 1957. Sedimentation of milk proteins from heated milk. *J. Dairy Sci. 40*, 330–833.

Swaisgood, H. E. 1982. Chemistry of milk proteins. *In: Developments in Dairy Chemistry*, Vol. 1: *Proteins*. P. F. Fox (Editor). Applied Science Publishers, London, pp. 1–52.

Sweetsur, A. W. M. and Muir, D. D. 1980A. The use of permitted additives and heat-treatment to optimize the heat-stability of skim milk and concentrated skim milk. *J. Soc. Dairy Technol. 33*, 101–105.

Sweetsur, A. W. M. and Muir, D. D. 1980B. Effect of concentration by ultrafiltration on the heat stability of skim-milk. *J. Dairy Res. 47*, 27–335.

Sweetsur, A. W. M. and Muir, D. D. 1980B. Effect of concentration by ultrafiltration on the heat stability of skim-milk. *J. Dairy Res. 47*, 327–335.

Sweetsur, A. W. M. and White, J. C. D. 1974. Studies on the heat stability of milk protein. I. Interconversion of type A and type B milk heat-stability curves. *J. Dairy Res. 41*, 349–358.

Sweetsur, A. W. M. and White, J. C. D. 1975. Studies on the heat stability of milk protein. III. Effect of heat-indicated acidity on milk. *J. Dairy Res. 42*, 73–88.

Tanford, C. 1961. *Physical Chemistry of Macromolecules.* John Wiley and Sons, New York.

Tanford, C. 1968. Protein denaturation. *In: Advances in Protein Chemistry*, Vol. 23. C. B. Anfinsen, M. L. Anson, J. T. Edsall and F.M. Richards (Editors). Academic Press, New York, pp. 122–275.

Tanford, C. 1970. Protein denaturation. *In: Advances in Protein Chemistry*, Vol. 24. C. B. Anfinsen, J. T. Edsall and F. M. Richards (Editors). Academic Press, New York, pp. 1–93.

Tanford, C. 1980. *The Hydrophobic Effect*, 2nd ed. John Wiley and Sons, New York.

Terada, H., Watanabe, K. and Kametani, F. 1980. Possible role of denatured albumin in formation of "heat-resistant" serum albumin. *Bull. Chem. Soc. Japan 53*, 3138–3142.

Tessier, H. and Rose, D. 1964. Influence of κ-casein and β-lactoglobulin on the heat stability of skimmilk. *J. Diary Sci. 47*, 1047–1051.

Tumerman, L. and Webb, B. H. 1965. Coagulation of milk and protein denaturation. *In: Fundamentals of Dairy Chemistry*. B.H. Webb and A.H. Johnson (Editors). AVI Publishing Co., Westport, Conn., pp. 506–582.

Walstra, P. and Jenness, R. 1984. *Dairy Chemistry and Physics*. John Wiley and Sons, New York.

Watanabe, K. and Klostermeyer, H. 1976. Heat-induced changes in sulphydryl and disulphide levels of β-lactoglobulin A and the formation of polymers. *J. Dairy Res. 43*, 411, 418.

Whitney, R. M. 1977. *Food Emulsions*. H. Graham (Editor). AVI Publishing Co., Westport, Conn.

Zittle, C. A., Thompson, M. P., Custer, J. H. and Cerbulis, H. 1962. κ-Casein-β-lactoglobulin interaction in solution when heated. *J. Dairy Sci. 45*, 807–810.

12

Milk-Clotting Enzymes and Cheese Chemistry Part I—Milk-Clotting Enzymes

Rodney J. Brown and C. A. Ernstrom

INTRODUCTION

Milk-clotting enzymes, obtained from animal, plant, and microbial sources have been used since antiquity for the manufacture of cheese and other foods. Until recently the calf gastric enzyme chymosin (rennin) in the form of a crude extract, paste or powder was used almost exclusively in commercial cheesemaking. Since 1961 there has developed a substantial shortage of calf stomaches (vells), and cheese makers have resorted to other enzyme preparations to meet the needs of an expanding cheese industry (Food and Agriculture Organization. 1968). It is questionable whether calf vells available in the United States could fill more than one-third of the country's need for milk-clotting enzymes. However, chymosin still remains the enzyme of choice and the standard against which all others are evaluated.

Recent reviews (Green 1977; Phelan 1977; Visser 1981) have dealt with the role of milk-clotting enzymes in cheese manufacture, while Foltmann (1981) has provided an excellent discussion of the structure of chymosin and its enzymic properties.

Many proteolytic enzymes will clot milk (Berridge 1954), therefore, it is not surprising that milk-clotting enzymes have been obtained from virtually every class of living organism. The Food and Nutrition Board of the United States National Research Council (1981) has adopted a nomenclature system in which the term "rennet" may be applied to all milk-clotting enzyme preparations (except porcine pepsin) used for cheesemaking. Rennet is defined as "aqueous extracts made from the fourth stomaches of calves, kids or lambs." Bovine rennet is an "aqueous extract made from the fourth stomach of bovine animals, sheep or goats." Microbial rennet, "followed by the name of the organism" identifies milk-clotting preparations derived from microorganisms. Should a suitable milk-clotting preparation be derived from plants the nomenclature system could include a provision for "plant rennet."

United States Standards of Identity for Cheddar Cheese (Food and Drug Administration 1985) allow the use of "rennet and/or other clotting enzymes of animal, plant or microbial origin." Milk-clotting enzymes that are currently on the "generally recognized as safe" (GRAS) list include rennet and bovine rennet (FDA 1984B). Those permitted as secondary direct food additives include microbial rennets derived from *Endothia parasitica, Bacillus cereus, Mucor pusillus* var. Lindt and *Mucor miehei* var. Cooney *et* Emerson. (FDA 1984B). Petitions have been filed to affirm porcine pepsin as GRAS, but official action has not been taken. Until a ruling is made, porcine pepsin may be used (National Research Council 1981).

CLOTTING ENZYMES FROM ANIMALS

Several proteases from animal organs have been investigated for their milk-clotting potential, but only chymosin, porcine pepsin, and bovine pepsin are of interest to the cheese industry.

Chymosin (EC 3.4.23.4)

Pang and Ernstrom (1986) reported that milk-clotting activity is present in the abomasum of bovine fetuses as early as the sixth month of development, and increases in potency as the fetus approaches full term. In their experiments, the total recoverable milk-clotting activities from fetal abomasa during the sixth, seventh, eighth, and ninth months of gestation were 2, 7, 12, and 31% of that found in high-quality vells from milk-fed calves slaughtered at three to four days of age. It was not determined whether the large increase in activity between a nine month old fetus and three to four day old milk-fed calf was related to the birth process or was stimulated by milk consumption. At birth, chymosin is present in gastric mucosa at 2 to 3 mg/g but its production declines after one week (Foltmann 1981).

Procedures for extraction of chymosin from vells were described by Ernstrom and Wong (1974). Crude rennet extract contains active chymosin and an inactive precursor (prochymosin). Addition of acid to the extract facilitates conversion of prochymosin to chymosin and allows the extract to reach maximum activity. Even though activation at lower pH is faster, poor stability of chymosin below pH 5.0 in the presence of sodium chloride causes reduced yields (Mickelsen and Ernstrom 1967; Rand and Ernstrom 1964).

Activation of prochymosin involves the splitting of peptides from the N-terminal end of prochymosin with simultaneous reduction in molecular weight from about 36,000 to 31,000. The rate of conversion increases markedly with decreasing pH below 5.0 (Rand and Ernstrom 1964). At pH 5.0, NaCl concentrations up to 2M increase the rate of activation. Milk-clotting activity plotted against activation time at pH 5.0 shows the course of activation (Fig. 12.1) to be autocatalytic. If activation is carried out in the presence of preformed chymosin, the S-shape disappears and the initial rate of the activation process increases with increasing concentration of preformed chymosin. Folt-

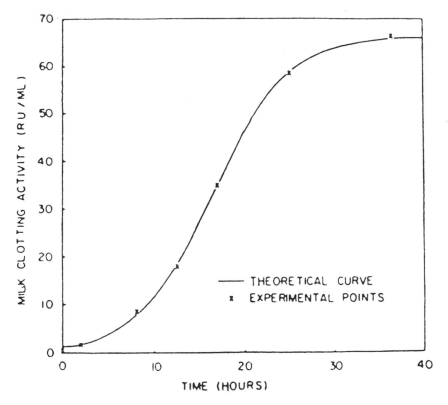

Figure 12.1 Activation of prorennin at pH 5.0 in 1.7M Sodium chloride compared to a theoretical curve calculated from the autocatalytic equation given by Herriott. (From Rand and Ernstrom 1964)

mann (1966) demonstrated that at pH 4.7 the course of activation is not purely autocatalytic, and suggested that the early part of activation may be second-order.

Foltmann (1959B) found that the optimum pH for chymosin stability is between 5.3 and 6.3, and that the enzyme is moderately stable at pH 2.0. As the pH is raised from 6.3 to neutrality the activity of the enzyme is destroyed at an increasing rate. A region of instability near pH 3.5 was also noted. This was confirmed by Mickelsen and Ernstrom (1967), who also found that at that pH, chymosin was even less stable in the presence of sodium chloride (Fig. 12.2). Foltmann (1959B) suggested that the region of instability near pH 3.5 may be due to

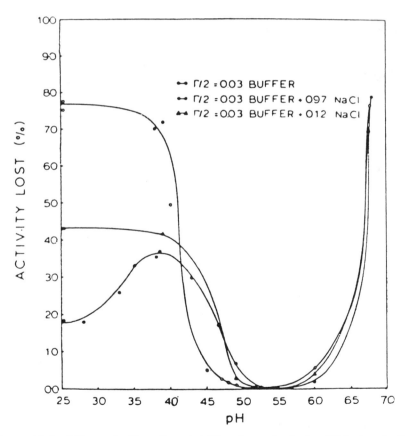

Figure 12.2 Effect of pH on the stability of rennin during 96 hr incubation at 30°C in buffers of $\Gamma/2 =$ 1.0 (0.03 due to buffer 0.97 due to Nacl). $\Gamma/2 =$ 0.15 (0.03 due to buffer, 0.12 due to NaCe), and $\Gamma/2 =$ 0.03 (no NaCe).

self-digestion, since this was near the pH optimum for the proteolytic activity of chymosin. This proved correct when Mickelsen and Ernstrom (1967) noted that activity losses at pH 3.8 paralleled an increase in ninhydrin color development.

Chymosin has now been produced in *Escherichia coli* (Chen *et al.* 1984; Hayenga *et al.* 1984) and *Saccharomyces cerevisiae* (Moir *et al.* 1985) by recombinant DNA techniques. Cheese-making trials comparing recombinant chymosin with calf rennet have found no significant differences between the two (Green *et al.* 1985). The impact of this development on the cheese industry will be felt when the various regulatory issues have been resolved.

Porcine Pepsin (EC 3.4.23.1)

A number of workers have recommended porcine pepsin as a satisfactory substitute for at least part of the rennet in making many varieties of cheese. Above pH 6.3 the milk-clotting activity of porcine pepsin decreases much more rapidly than that of chymosin. In fact, at pH 6.8, usual levels of pepsin may not clot milk. The pH of cheese milk is such that both the activity and the stability of porcine pepsin are far from optimum (pH 2.0), and the enzyme is actually being inactivated while in the milk (Emmons 1970). Slow coagulation and a weak set can be encountered if insufficient pepsin is used, which results in excessive fat losses and reduced yield. This difficulty can be overcome by using enough pepsin to provide firm coagulation in the appropriate time. Porcine pepsin is inexpensive relative to chymosin and greater usage is not cost prohibitive.

O'Keeffe *et al.* (1977) reported that some porcine pepsin survives Cheddar cheesemaking and contributes to casein breakdown during cheese curing. However, it has been shown that the breakdown they attributed to pepsin occurs in curd containing neither coagulant nor starter bacteria (Majeed 1984). More recent studies have shown that porcine pepsin does not survive in Cheddar cheese when the milk is set at pH 6.6 (Yiadom-Farkye 1986). This supports earlier reports of Green (1972) and Wang (1969).

Emmons (1970) experienced significant inactivation when commercial pepsin and pepsin-calf rennet mixtures were diluted with high-pH, hard water 10 min before adding them to the cheese vat. Mickelsen and Ernstrom (1972) reported that mixtures of porcine pepsin and calf rennet were stable between pH 5.0 and 6.0, but that pepsin activity was lost from the mixture above pH 6.0. This loss was shown to be entirely due to pepsin instability. Below pH 6.0 chymosin activity was destroyed by pepsin.

Porcine pepsin is secreted by hog stomach mucosa as catalytically inactive pepsinogen with a molecular weight of 40,400. Pepsinogen is stable in neutral and slightly alkaline solutions, but undergoes reversible denaturation above 55°C at pH 7.0 and at room temperature at pH 11.0. The conversion of pepsinogen to pepsin is catalyzed by pepsin below pH 5.0. Seven to nine peptide bonds are hydrolyzed during formation of pepsin, which splits off about 20% of the molecule. However, it is likely that cleavage of only one of these bonds in necessary to release the active enzyme. One of the polypeptides released during activation is a pepsin inhibitor (5000 MW), which is bound to pepsin between pH 5.0 and 6.0 and inhibits both the milk-clotting and protein-digesting activity of the enzyme. At lower pH the inhibitor dissociates from the enzyme and activity is restored. The inhibitor is destroyed by pepsin between pH 2.0 and 5.0, with a rate maximum near pH 4.0 (Ernstrom and Wong 1974).

Bovine Pepsin

Linklater (1961) reported that bovine pepsin accounted for only 0 to 6% of the milk-clotting activity of commercial rennet extracts. He used porcine pepsin as a reference standard. Bovine pepsin has increased in use as a coagulant because of the practice of extracting the stomach from older calves and adult cattle. More recently, Sellers (1982) reported that 85 to 95% of the proteolytic activity of calf rennet is due to chymosin and the remainder is from bovine pepsin. Adult bovine rennets preparations may contain 55 to 60% bovine pepsin. Mixtures of calf rennet and porcine pepsin may contain 40 to 45% chymosin, 5 to 10% bovine pepsin, and 50% porcine pepsin. Mixtures of adult bovine rennet and porcine pepsin typically contain 20 to 25% chymosin, 40 to 45% bovine pepsin, and 30 to 40% porcine pepsin activity (McMahon and Brown 1985).

In approving pepsin as a "safe and suitable" substitute for rennet, only porcine pepsin was considered by the Food and Drug Administration (1984B). This raises questions concerning the definition of rennet and the legal acceptability of bovine pepsin as a milk coagulant in the United States. In spite of this, substantial amounts of bovine pepsin are present in a high percentage of rennet extracts (Shovers et al. 1972).

Green (1972) reported that Cheddar cheese made entirely with bovine pepsin was only slightly inferior to that made with calf rennet and Fox and Walley (1971) found no significant difference between Cheddar cheese made with bovine pepsin and rennet. Fox (1969) found that the milk-clotting activity of bovine pepsin is less pH-dependent than that of porcine pepsin, and can coagulate milk up to pH 6.9. He suggested that bovine pepsin has proteolytic properties more like those of chymosin, and is less subject to pH denaturation than porcine pepsin.

Chicken Pepsin

A milk-clotting enzyme from chicken, which complies with Jewish law, has been manufactured in Israel since 1970. It is claimed that cheese made with this enzyme is superior to that made with *M. miehei, M. pusillus* or *E. parasitica* rennets (Gutfeld and Rosenfeld 1975). Chicken pepsin is more proteolytic and less heat stable than chymosin (Gordin and Rosenthal 1978). Stanley *et al.* (1980) reported that texture of Cheddar cheese made with chicken pepsin lacked firmness and was very bitter at three months of age. They also predicted, based on nitrogen in the whey, that yields using this enzyme would be significantly lower than those using chymosin or chymosin/porcine pepsin mixture. They concluded that chicken pepsin is unsuitable for Cheddar cheese manufacture. Green *et al.* (1984) evaluated chicken pepsin and came to the same conclusions.

CLOTTING ENZYMES FROM FUNGI

Fungal proteases have been investigated extensively in search of suitable milk clotting enzymes. Patents have been issued for production of rennets from *E. parasitica, M. Pusillus* var. Lindt and *M. miehei* var. Cooney *et* Emerson. These have been approved in the United States as secondary direct food additives (FDA. 1984B) and have experienced considerable commercial success in the United States as milk-clotting enzymes for cheese manufacture. Many other fungal sources have also been tried in the effort to find an inexpensive replacement for chymosin.

Endothia parasitica Rennet

Rennet from *E. parasitica* has been used with varying success for cheesemaking, and was reported to accelerate the ripening of Cheddar cheese (Shovers and Bavisotto 1967). Among the commercially used fungal rennets, *E. parasitica* rennet is the most proteolytic on α_s- and β-caseins and least proteolytic on κ-casein (Vannderpoorten and Weckx 1972). According to Emmons *et al.* (1978) the high proteolytic activity of *E. parasitica* rennet can cause a 1.2% loss of Cheddar cheese yield compared to cheese made from calf rennet. Substantial proteolytic activity is most probably the cause of bitterness in Cheddar, Edam, Til-sit, Taleggio, and some other varieties made with *E. parasitica* rennet (Stavlund and Kiermeier 1973; Thomasow *et al.* 1970). This rennet produced 18 peptides from the oxidized B chain of insulin compared to 12 for chymosin (Whitaker 1970). Even though bitterness is characteristic of several cheese varieties made with *E. parasitica* rennet, this enzyme works well in those varieties requiring high cooking temperatures (Ramet *et al.* 1969). The protease activity is probably destroyed during cooking and is unavailable to affect the cheese during curing. Changes in the pH of milk does not affect the milk-clotting activity of *E. parasitica* rennet as much as that of chymosin (Alais and Novak 1968; Larson and Whitaker 1970; Reps *et al.* 1970).

Whitaker (1970) reported that *E. parasitica* protease has its maximum stability between pH 3.8 and 4.8. Below pH 2.5, activity losses were associated with increase in ninhydrin reaction groups, which suggests autolysis of the molecule. Above pH 6.5 activity was rapidly lost, accompanied by decreased solubility and no increase in ninhydrin reaction groups. Thunell *et al.* (1979) found that *E. parasitica* protease is easily destroyed in whey at 68°C at all pH values from 5.2 to 7.0.

Mucor pusillus var. Lindt Rennet

M. pusillus var. Lindt protease has given satisfactory results as a chymosin substitute in the manufacture of a number of cheese varieties, but not all varieties of *M. pusillus* var. Lindt are capable of producing acceptable cheese (Babel and Somkuti 1968). The clotting activity of *M. pusillus* var. Lindt protease is more sensitive to pH changes between 6.4 and 6.8 than chymosin, but is much less sensitive than that of porcine pepsin (Richardson *et al.* 1967). The same authors reported that CaCl$_2$ added to milk affected the clotting activity of *M. pusillus* var. Lindt rennet more than it did that of chymosin rennet. They also reported that this rennet was more stable than chymosin between pH 4.75 and 6.25. *M. pusillus* var. Lindt rennet is not destroyed during the manufacture of Cheddar cheese, although less than 2% of the enzyme added to the milk remains in the curd. Nearly all of it is found in the whey (Holmes *et al.* 1977). Mickelsen and Fish (1970) found *M. pusillus* var. Lindt rennet to be much less proteolytic than *E. parasitica* rennet but more proteolytic than chymosin rennet on whole casein, α_s-casein and β-casein at pH 6.65.

Mucor miehei Rennet

It was reported by Prins and Nielsen (1970) that a proteolytic enzyme preparation from *M. miehei* resulted in Cheddar cheese of excellent quality, even after extended ripening times. The enzyme has a broad stability maximum between pH 4.0 and 6.0 and loses no activity during 11-hr incubation at pH 6.0 (Sternberg 1971).

M. miehei rennet is the most heat stable of all the commonly used milk-clotting enzymes (Thunell *et al.* 1979). None is destroyed during Cheddar cheese manufacture although, like *M. pusillus* var. Lindt rennet, less than 2% remains active in the cheese (Harper and Lee 1975; Holmes *et al.* 1977). It remains active in the whey and is concentrated in condensed whey products.

Because of problems encountered in blending whey products containing residual *M. miehei* rennet with materials containing casein, this rennet preparation has been modified to decrease its heat stability (Branner-Jorgensen *et al.* 1980; Cornelius 1982). This process involves treatment of the rennet with hydrogen peroxide under controlled conditions. Some enzymic activity is lost but the modified enzyme has about the same stability as calf rennet. Nearly all *M. miehei* rennet used by the cheese industry is now modified (Ramet and Weber 1981).

Rennet from other Fungal Sources

Additional fungal sources that have been investigated recently include *Absidia* (Abdel-Fattah *et al.* 1984; Sannabhadti and Srinivasan 1976), *Aspergillus* (Rotaru 1980; Foda 1982, 1983; Foda *et al.* 1975A,B), *Mucor racemosus* (Higashio and Yoshiaka 1981A, B, C, 1982), *Mucor bacilliformis* (Fraille *et al.* 1981), phycomycetous fungi (Diokno-Palo *et al.* 1979), *Basidiomycetes* (Kawai 1973), *Mucor mucedo* (Mashaly *et al.* 1981), *Acetomycetes* (Laxer *et al.* 1981) *Rhizopus* (Rao *et al.* 1979; Nand *et al.* 1980), *Penicillium* (Mabrouk *et al.* 1976, Abdel-Fattah and El-Hawwary 1972, Abdel-Fattah *et al.* 1972), *Streptomyces* (Abdel-Fattah *et al.* 1974), *Mucor lamprosporous* (Wilken and Bakker 1974), *Physarum* (Farr 1974), *Byssochlamys* (Reps *et al.* 1973) and *Mucor renninus* (Zvyagintsev *et al.* 1972).

CLOTTING ENZYMES FROM BACTERIA

The search for suitable chymosin substitutes has led to the investigation of a number of proteases produced by bacteria. Milk-clotting enzymes from *Bacillus cereus* (Choudhery and Mikolajcik 1969; Melachouris and Tuckey 1968), *Bacillus polymyxa* (Denkov and Vasileva 1979; Philippos and Christ 1977), *Bacillus mesentericus* (Antonova *et al.* 1978; Antonova *et al.* 1981; Dimitroff and Prodanski 1973; Nachev *et al.* 1973; Velcheva and Ghbova 1978; Velcheva *et al.* 1975A; Velcheva *et al.* 1975B; Goranova and Stefanova-Kondratenko 1975) and

Bacillus subtilis (Antonova *et al.* 1978; Cabrini *et al.* 1983; Irvine *et al.* 1969; Rao and Mathur 1979; Puhan 1969; Puhan and Irvine 1973) have received recent attention. *Pseudomonads* (Jackman and Patel 1983; Juffs 1974) *Bacillus licheniformis* (D'Souza *et al.* 1982) *Bacillus megaterium* (Hylmar *et al.* 1982) also have been investigated as possible rennet enzyme sources.

Acceptable cheese made with bacterial coagulants has been reported (Antonova *et al.* 1975; Cabrini *et al.* 1983; Dimitroff and Prodanski 1973; Hylmar *et al.* 1982; Kondratenko *et al.* 1977; Rao and Mathur 1979; Nachev *et al.* 1974). However, results have not been consistently favorable and no bacterial rennet is produced commercially even though milk-clotting enzymes from *B. cereus* have been approved (FDA. 1984A). Bacterial protease preparations are complex (Nachev *et al.* 1973; Nachev *et al.* 1974; Philippos and Christ 1977; Puhan 1969; Velcheva *et al.* 1975B; Velcheva and Gнbova 1978) and occasional reports of success in cheesemaking reflect the fact that suitable coagulating enzymes may exist as part of crude mixtures containing other highly proteolytic enzymes that are detrimental to cheesemaking (Shaker and Brown 1985A,B; Shehata *et al.* 1978; Velcheva and Gнbova 1978).

Melachouris and Tuckey (1968) reported that enzymes from *B. cereus* rapidly degraded whole casein, α-casein and particularly β-casein. Patents have been issued for the production and use of milk clotting enzymes from *B. subtilis* (Kondratenko *et al.* 1977). Puhan (1969) and Puhan and Irvine (1973) found that a protease from *B. subtilis* possessed high nonspecific proteolytic activity. However, it produced first-grade Canadian Cheddar cheese even though it caused extensive proteolysis during clotting which resulted in substantial fat and protein losses in the whey.

CLOTTING ENZYMES FROM HIGHER PLANTS

Many enzymes extracted from higher plants have been tried for clotting cheese milk (Burnett 1976), however, attempts to use them have been unsuccessful. Most plant proteases are strongly proteolytic and cause extensive digestion of the curd, which has resulted in reduced yields, bitter flavors, and pasty-bodied cheese.

Enzymes that have been investigated include papain (Arnon 1970) and chymopapain (Kunimitsu and Yasunobu 1970), ficin (Liener and Friedenson 1970), bromelain (Fuke and Matsuoke 1984; Murachi 1970), *Cynara cardunculus* (Barbosa *et al.* 1976), the 'litsusu' tree (*Wrightians calysina*) (Hosono *et al.* 1983), *Solanum torvum* (ElKoussy *et al.* 1976) ash gourd (*Benincasa cerifera*) (Eskin and Landman 1975; Gupta and Eskin 1977), *Cirsium arvense* (Poznanski *et al.* 1975), and Chinese gooseberries (Kiwi fruit) (Creamer 1972).

ENZYMIC COAGULATION OF MILK

Understanding milk-clotting is made more difficult by our rudimentary, and therefore often conflicting, views of casein micelles structure (Bloomfield and Morr 1973; Farrell 1973; Garnier 1973; McMahon and Brown 1984; Schmidt 1980; Slattery 1976; Swaisgood 1973). A complete explanation of milk-clotting will not be possible until more information, including the complete and correct structure of casein micelles, becomes available (Ekstrand *et al.* 1980).

Milk-clotting is a complex process, involving a primary enzymic phase in which κ-casein is altered and loses its ability to stabilize the remainder of the caseinate complex, a secondary non-enzymic phase in which aggregation of the altered caseinate takes place, a third step where the aggregate of casein micelles forms a firm gel structure and a possibly separate fourth step where the curd structure tightens and syneresis occurs (McMahon and Brown 1984B).

Proteolytic enzymes initiate milk-clotting by cleaving κ-casein molecules on the surfaces of casein micelles to form para-κ-casein and a macropeptide.

$$\kappa\text{-casein} \rightarrow \text{para-}\kappa\text{-casein} + \text{macropeptide}$$

Chymosin splits κ-casein specifically at the Phe_{105}-Met_{106} bond, with little other cleavage. Other milk-clotting enzymes are less specific, but have the same general effect (Fox 1981; Fisser 1981). Splitting of κ-casein destroys the stability of the milk system and casein begins to aggregate, forming a curd. Although some research (mostly done with very dilute milk) has suggested that the enzymic and nonenzymic phases of milk-clotting are not overlapping (Green and Morant 1981), aggregation of casein micelles starts long before proteolysis of κ-casein is completed (Payens and Wiersma 1980). Evidence now suggests that under normal cheesemaking conditions aggregation of micelles begins at the same time as enzymic cleavage of κ-casein begins (Reddy *et al.* 1986).

The enzymic phase is affected by the same factors that affect all enzyme reactions. Variation in the amount of enzyme has the largest effect (Castle and Wheelock 1972). Temperature and pH each affect the enzyme reaction in two ways. As temperature is increased the rate of κ-casein cleavage increases until the temperature becomes high enough to begin denaturing the enzyme. At this point the rate of reaction decreases rapidly and activity is not recoverable.

Each milk-clotting enzyme has an optimum pH at which it is most active. Moving the pH in either direction from that point decreases activity. Extremes of acid or base also denature the enzymes, but not as irreversibly as high temperature. In cheesemaking, the starter culture moves the pH down to the optimum of the milk-clotting enzyme. The different optimums for different enzymes must be considered in relation to pH at setting and culture activity (Brown 1981; Shalabi and Fox 1982).

The temperature coefficient (Q_{10}) for aggregation is much lower than that of the enzymic step (McMahon and Brown 1984B). The Q_{10} for the enzymic step of milk-clotting at pH 6.7 is between 1.8 and 2.0, while that for aggregation is about 11 to 12 (Cheryan et al. 1975A,B; McMahon and Brown 1984B). This fact has been used by many to study the secondary phase of milk-clotting after allowing the enzymic phase to go to completion at a lower temperature.

A recent addition to the variables to be considered in milk-clotting is concentration level of the milk (Leeuwen 1984). Work has been done and is in progress to better understand the effects of concentration, primarily by ultrafiltration (Dalgleish 1980; Payens 1984; Reuter et al. 1981).

MEASURING MILK-CLOTTING ACTIVITY

An ideal test for measuring milk-clotting activity has never been devised, but numerous methods have been tried. In practice, activity is determined by the speed with which the enzyme clots milk under a set of specified conditions. This differs from the usual procedure in enzyme chemistry where one measures the rate at which the products of an enzyme-catalyzed reaction appear, or conversely, the rate at which the substrate disappears.

Milk-Clotting Assays

It is customary to observe visually the formation of a clot, or rather the sudden fracture of a film of milk on the wall of a bottle or test tube. Apparatus for measuring clotting time in this way and standard substrates have been described by Sommer and Matsen (1935), Berridge (1952) and Bakker et al. (1968), and have been used for many years in rennet control laboratories.

The problem of standardizing the chymosin assay was discussed at length by Berridge (1952). Inasmuch as milk samples differ greatly in their susceptibility to chymosin action, analytical consistency may be approached, at least within one laboratory, by using a standard supply of milk solids nonfat (MSNF). The material should be prepared and stored in sealed containers at low temperature so the original properties of the powder are retained as nearly as possible. Berridge (1952) suggested reconstituting 60 g low-heat, MSNF in 500 ml 0.01 M CaCl$_2$. Clotting time is shorter in this substrate than in MSNF reconstituted in water. The Berridge substrate also gives more reproducible results since it is less subject to the many factors that cause variation in the clotting time of normal milk. Following reconstitution, the clotting time of the substrate continues to increase during storage at 2°C. Therefore, to achieve a constant clotting time, it is helpful to allow the reconstituted substrate to age for about 20 hr before use. Different samples of reconstituted MSNF introduced into the laboratory must be checked against the old sample with an enzyme solution of known activity.

A blood clot timer adapted for measuring milk-clotting time by de-Man and Batra has been used in industry control laboratories for standardizing rennet solutions. It is faster than visual methods, requires less substrate, and has an automatic end-point detector. The ratio of enzyme solution to substrate is higher than is used with most other clotting tests, therefore, care must be exercised to prevent differences in pH or salt concentration in the enzyme solution from affecting the clotting time (Ernstrom and Wong 1974).

Thomasow (1968) used a thrombelastograph for measuring the clotting time of milk. This device also measured the elasticity of the coagulum as it is formed, and gives information about the time of solidification of the curd. Clotting times measured with this device are longer than those determined visually.

Scott-Blair and Oosthuizen suggested measurement of viscosity change in milk as an index of clotting time. They showed that viscosity during the course of chymosin action first dropped to a minimum, then rose as incipient clotting started. The initial or decreasing viscosity phase followed zero kinetics for a useful period, during which time chymosin assays could be made. They found that plots of changes in specific viscosity against time during the action of chymosin on caseinates from a variety of sources produced the same slope. However, the slope varied with different rennet extracts, which they attributed to the presence of varying percentages of proteolytic enzymes other than chymosin (Ernstrom and Wong 1974). Gervais and Vermeire (1983) automated the Scott-Blair torsiometer and reduced the absolute error of measurement by three times.

McMahon and Brown (1982, 1983) evaluated a unique instrument (Formagraph) designed for following the progress of milk coagulation by recording the movement of small stainless steel loop pendulums as the milk samples in which they are immersed are continually moved from side to side. In comparison with the visual method of Sommer and Matsen (1935) the Formagraph had no difference in precision. The Formagraph has the advantage of unattended operation and can be used over a broader range of chymosin concentrations (.001 to .16 chymosin units/ml) than most methods. Observable coagulation times are slightly longer with the Formagraph than with either the Sommer Matsen method (1935) or with the viscosity method of Kopelman and Cogan (1976).

McMahon et al. (1984A, B, C) monitored coagulation of undiluted milk substrate directly with a spectrophotometer connected to a computer. Increase in turbidity was monitored at 600 nm after addition of enzyme to Berridge substrate (Berridge 1952). More information about the course of milk coagulation is available from this method than from any other devised thus far. Figure 12.3 shows the initial dip observed as κ-casein is cleaved followed by a rapid rise as aggregation begins. The maximum on the lower curve (point of inflection on the upper curve) matches the point of coagulation (τ) described by Payens (1978).

One of the most interesting new methods for following the progress of milk-clotting was reported by Hori (1985). A .1 mm $\times$ 106 mm platinum wire immersed in milk is heated by applying .7 amp of dc electricity and the temperature of the wire is monitored. As the milk surrounding the wire coagulates, the dissipation of heat away from the wire decreases and the temperature of the wire increases. This procedure has the advantages that it follows the progress of coagulation without disrupting the curd.

Many other methods have been tried to monitor the course of coagulation rather than just its beginning point. These include light scattering (Claesson and Claesson 1970), optical density (Butkus and Butene 1974), reflection photometry (Hardy and Fanni 1981), turbidimitry (Surkov et al. 1982), and other rheological measurements (Bachman et al. 1980; Bohlin et al. 1984; Garnot and Olson 1982; Marshall et al. 1982; Olson and Bottazzi 1977; Ramet et al. 1982; Richardson et al. 1971; Steinsholt 1973). Richardson et al. (1983) and Kowalchyk and Olson (1978) developed instruments for use in cheese plants with the aim of predicting cutting times during cheese manufacture.

Storch and Segelcke (1874) proposed that the product of coagulation time (t_c) and enzyme concentration (E) should be defined as a constant (k) (McMahon and Brown 1983, 1984B).

$$t_c E = k$$

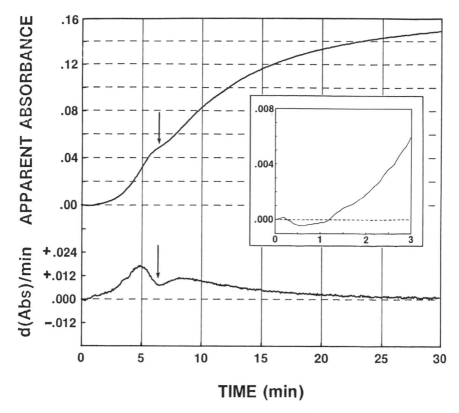

Figure 12.3 Plot of apparent absorbance (600nm) versus time after addition of chymosin to reconstitute nonfat dry milk (12g + 100 ml .01M CaCe₂). Arrows represent Formagraph coagulation time. Inset shows an expanded view of the first 3 min of coagulation. (From McMahon *et al.* 1984)

A linear plot of t_c vs $1/E$ is easily made from this relationship. But many have observed that in practice this relationship is useful only over a narrow range of enzyme concentration. This deficiency was corrected by Holter in 1932 by adding a factor *(x)* to correct for the time lag which he believed existed between enzymic cleavage of κ-casein and aggregation (Brown and Collinge 1986). Foltmann (1959A) rearranged Holter's equation, keeping k and x separate to show that k is a constant but that x varies according to measurement conditions.

$$t_c = (k/E) + x$$

Different instruments for measuring milk-clotting fit this equation differently, even with the same substrate sample and conditions of measurement.

McMahon *et al.* (1984C) fitted curves measured with the Formagraph instrument (McMahon and Brown 1982) to the equation:

$$G = 0 \qquad \text{for } t \leq \tau$$
$$G = G_\infty \exp[-t/k(t-\tau)] \qquad \text{for } t > \tau$$

and solved for τ. G is the width between branches of the Formagraph curve, G_∞ is the maximum width, t is the time after addition of enzyme and k is a constant. The value of τ is very close to the "actual coagulation time" defined by Payens (1978). By using G as a general measure of curd firmness, this method can be used with any instrument that gives a series of firmness readings rather than just an end point (Brown and Collinge 1986). If coagulation is monitored by the spectrophotometer method (McMahon *et al.* 1984A, B, C), then τ can be read directly.

If the t_c term in Foltmann's equation is replaced by τ, the intercept of the plot, x, approaches zero. This indicates that the difference between τ and t_c is not a lag time between enzymic action and aggregation, but is an artifact of the measurement method. The x term added to Storch and Segelcke's equation by Holter is not necessary if τ is used instead of t_c (Collinge and Brown 1986).

Micro Tests for Milk-Clotting Enzymes

Reyes (1971) described a very sensitive assay for measuring the milk-clotting activity of low concentrations of residual proteolytic enzymes in curd and whey. The substrate, buffered at pH 5.7, consisted of 1 g MSNF dissolved in a mixture of 70 ml 6.6×10^{-2} M cacodylic acid, 30 ml 6.6×10^{-2} M triethanolamine, and 1 ml 3 M $CaCl_2$.

Elliott and Emmons (1971) described a passive indirect hemagglutination test, and a corresponding inhibition test for measuring residual chymosin in cheese. They also produced high titer antisera for *E. parasitica* protease and *M. pusillus* protease, and suggested that these enzymes could also by quantitatively detected in cheese.

Lawrence and Sanderson proposed another micro-method for measuring chymosin and other proteolytic enzymes. Measurement of concentration was based on the rate of radial diffusion of the enzyme through a thin layer of caseinate-agar gel. The limit of diffusion was marked by a zone of precipitated casein (Ernstrom and Wong 1974). Holmes *et al.* (1977) developed a microdiffusion assay for residual proteolytic enzymes in curd and whey that is more sensitive than the method of Lawrence and Sanderson or the clotting-time assay of Reyes (1971).

NONSPECIFIC PROTEOLYSIS
OF MILK-CLOTTING ENZYMES

The purpose of adding milk-clotting enzymes to milk is to cleave κ-casein and begin coagulation of the milk. In addition to this action, all milk-clotting enzymes have general proteolysis capability. The level of such nonspecific proteolysis varies (Green 1972; Shaker and Brown 1985A, B; Visser 1981), but most of the substitutes are more proteolytic than chymosin (McMahon and Brown 1985; Green 1977).

The influence of milk-clotting enzymes on cheese curing has always been difficult to measure in the presence of the many changes brought about by microorganisms (Dulley 1974; Lawrence *et al.* 1972). Porcine pepsin does not survive cheesemaking, suggesting that the proteolytic activity of milk-clotting enzymes assist, but are not essential to cheese curing. Yiadom-Farkye (1986) found that α_{s1}-casein is degraded more extensively in cheese by chymosin than by porcine pepsin but that β-casein is degraded more by porcine pepsin. Reduction of chymosin or use of porcine pepsin resulted in better quality medium and aged Cheddar cheese. Excessive general proteolysis leads to excessive loss of fat and cheese yield and adversely affects flavor and texture (Green 1977; Sellers 1982).

REFERENCES

Abdel-Fattah, A. F. and EL-Hawwary, N. M. 1972. Purification and proteolytic action of milk-clotting enzyme produced by *Penicillium citrinum*. J. Gen. Appl. Microbiol. *18*, 341–348.

Abdel-Fattah, A. F., EL-Hawwary, N. M., and Amr, A. S. 1974. Milk-clotting enzymes of some streptomyces species. *Acta Microbiol. Pol. 6*, 27–32.

Abdel-Fattah, A. F., and Ismail, A. M. S. 1984. Production of rennin-like enzyme by *Absidia cylindrospora. Agric. Wastes 11*, 125–131.

Abdel-Fattah, A. F., Mabrouk, S. S. and EL-Hawwary, N. M. 1972. Production and some properties of rennin-like milk-clotting enzyme from *Penicillium citrinum. J. Gen. Microbiol. 70*, 151–155.

Alais, C. and Novak, G. 1968. Study of a microbial coagulating enzyme produced by *Endothia parasitica.* I. Biochemical properties of Pfizer coagulating enzyme (1) and rheological properties of curds formed in the milk. *Lait 48*, 393–418.

Antonova, T., Daov, T. and Dedova, P. 1975. Study of bacterial strains producing milk-coagulating enzymes. IV. Preparation and characterization of hard cheeses. *Prilozh. Mikrobiol. 6*, 5–10.

Antonova, T., Nachev, L., Kolev, D., Bodurska, I. and Manafova, N. 1981. Bacterial strains producing enzymes with milk-clotting activity. IX. Effect of certain factors on milk clotting. *Acta Microbiol. Bulg. 9*, 48–53.

Antonova, T., Nachev, L., Kosturkova, P., Daov, T. and Dedova, P. 1978. Study of bacterial strains producing milk-coagulating enzymes. VII. Characteristics of enzyme complex produced after different fermentation times. *Acta Microbiol. Bulg. 1*, 21–29.

Arnon, R. 1970. Papain. *In: Methods in Enzymology*, Vol. 19. G. E. Perlman and L. Lorand (Editors). Academic Press, New York, pp. 226–224.

Babel, F. J. and Somkuti, G. A. 1968. *Mucor pusillus* protease as a milk coagulant for cheese manufacture. *J. Dairy Sci. 51*, 937–937.

Bachman, S., Klimczak, B. and Gasyna, Z. 1980. Non-destructive viscometric studies of enzymic milk coagulation. III. The effect of pH, temperature and Ca-ions concentration on the secondary phase of milk coagulation. *Acta Aliment. Pol. 6*, 135–143.

Bakker, G., Scheffers, W. A. and Wiken, T. O. 1968. A new method for the determination of clotting times in milk. *Neth. Milk Dairy J. 22*, 16–21.

Barbosa, M., Valles, E., Vassal, L., and Mocquot, G. 1976. Use of *cynara cardunculus* L. extract as a coagulant in manufacture of soft and cooked cheeses. *Lait 56*, 1–17.

Berridge, N. J. 1952. Some observations on the determination of the activity of rennet. *Analyst 77*, 57–62.

Berridge, N. J. 1954. Rennin and the clotting of milk. *In: Advances in Enzymology*, Vol. 15. E. F. Nord (Editor). Interscience Publishers, New York, pp. 423–449.

Bloomfield, V. A. and Morr, C. V. 1973. Structure of casein micelles: Physical methods. *Neth. Milk Dairy J. 27*, 103–120.

Bohlin, L., Hegg, P. O. and Ljusberg-Wahren, H. 1984. Viscoelastic properties of coagulating milk. *J. Dairy Sci. 67*, 729–734.

Branner-Jorgensen, S., Schneider, P. and Eigtved, P. 1980. A method of modifying the thermal destabilization of microbial rennet and a method of cheese making using rennet so modified. U.K. Pat. Appl. 2,045,772A.

Brown, R. J. 1981. The mechanism of milk clotting. Proc. 2nd Bienn. Marschall Int. Cheese Conf. pp. 107–112.

Brown, R. J. and Collinge, S. K. 1986. Actual milk coagulation time and inverse of chymosin activity. *J. Dairy Sci. 69*, 956–958.

Burnett, J. 1976. A brief survey of plant coagulants. *Dairy Ind. Int. 41*, 162–164.

Butkas, K. and Butene, V. 1974. Instrument for recording the course of milk coagulation process. *XIX Int. Dairy Congr. Proc. 1E*, 507.

Cabrini, A., Capua, E. di, Mucchetti, G. and Neviani, E. 1983. Use of enzymes in cheesemaking. II. Partial characterization in vitro of a commercial proteinase and its use in Crescenza, Caciotta, Italico and Grana cheese production. *Latte 8*, 247–258.

Castle, A. V. and Wheelock, J. V. 1972. Effect of varying enzyme concentration on the action of rennin on whole milk. *J. Dairy Res. 39*, 15–22.

Chen, M. C. Y., Hayenga, K. J., Lawlis, V. B. and Snedecor, B. R. 1984. Microbially produced rennet, methods for its production and plasmid used for its production. Eur. Pat. Appl. EP,0,116,778,A1.

Cheryan, M., Van Wyk, P. J., Olson, N. F. and Richardson, T. 1975A. Continuous coagulation of milk using immobilized enzymes in a fluidized-bed reactor. *Biotechnol. Bioeng. 17*, 585–598.

Cheryan, M., Van Wyk, P. J., Olson, N. F. and Richardson, T. 1975B. Secondary phase and mechanism of enzymic milk coagulation. *J. Dairy Sci. 58*, 477–481.

Choudhery, A. K. and Mikolajcik, E. M. 1969. Rennin-like activity in milk of *Bacillus cereus*. *J. Dairy Sci. 52*, 896–896.

Claesson, O. and Claesson, E. 1970. Optical measurement of the rennin coagulation of milk. XVIII Int. Dairy Congr. Proc. 1E, 42.

Cornelius, D. A. 1982. Process for decreasing the thermal stability of microbial rennet. U.S. Pat. 4,348,482.

Creamer, L. K. 1972. Chinese gooseberry protease unsuitable as a rennet substitute. *New Zealand J. Dairy Sci. and Technol. 7*, 23–23.

Dalgleish, D. G. 1980. Effect of milk concentration on the rennet coagulation time. *J. Dairy Res. 47*, 231–235.

Denkov, T. and Vasileva, S. 1979. Study of milkozim bacterial rennet in the manufacture of white pickled cheese. *Nauchn. Tr. Inst. Mlech. Promish. 9*, 158–165.

Dimitroff, D. and Prodanski, P. 1973. Use of enzyme preparations of microbial origin in kachkaval cheese manufacture. Production of ewes' and cows' milk kachkaval cheese using an enzyme preparation from *Bacillus mesentericus. Milchwissenschaft 28*, 568–571.

Diokno-Palo, N., Palo, M. A., Cunanan, L. F. and Santos, P. S. 1979. Skim milk-coagulating activities of enzymes produced by phycomycetous fungi. *Philippine J. Sci. 108*, 137–151.

D'Souza, T. M. and Pereira, L. 1982. Production and immobilization of a bacterial milk-clotting enzyme. *J. Dairy Sci. 65*, 2074–2081.

Dulley, J. R. 1974. The contribution of rennet and starter enzymes to proteolysis in cheese. *Aust. J. Dairy Technol. 29*, 65–69.

Ekstrand, B., Larsson-Raznikiewicz, M. and Perlman, C. 1980. Casein micelle size and composition related to the enzymatic coagulation process. *Biochim. Biophys. Acta 630*, 361–366.

El-Koussy, L. A., Cheded, M. A., Foda, E. A., and Hamdy, A. M. 1976. Preparation of milk clotting enzymes from plant sources. III. Domiati cheesemaking using the extracted enzyme from *Solanum torvum. Agric. Res. Rev. 54*, 153–157

Elliott, J. A. and Emmons, D. B. 1971. Rennin detection in cheese with the passive indirect hemagglutination test. *Can. Inst. Food Tech. J. 4*, 16–18.

Emmons, D. B. 1970. Inactivation of pepsin in hard water. *J. Dairy Sci. 53*, 1177–1182.

Emmons, D. B., Beckett, D. C. and Binns, M. 1978. Proteolysis by milk-coagulating enzymes during cheesemaking. *XX Int. Dairy Congr. E*, 491–492.

Ernstrom, C. A. and Wong, N. P. 1974. Milk-clotting enzymes and cheese chemistry. *In: Fundamentals of Dairy Chemistry* (Second Edition). B. H. Webb, A. H. Johnson and J. A. Alford (Editors). AVI Publishing Co., Westport, pp. 662–771.

Eskin, N. A. M. and Landman, A. D. 1975. Study of milk clotting by an enzyme from ash gourd (*Benincasa cerifera*). *J. Food Sci. 40*, 413–414.

Farr, D. R. 1974. Milk clotting enzyme. U.S. Pat. 3,852,478.

Farrell, H. M. 1973. Models for casein micelle formation. *J. Dairy Sci. 56*, 1195–1206.

Foda, M. S. 1982. Characterization of rennin-like enzyme produced in submerged culture of *Aspergillus niger. Egyptian J. Microbiol. 17*, 105–114.

Foda, M. S. 1983. New microbial potential for production of fungal enzymes and proteins from whey. *Egyptian J. Microbiol. 18*, 151–160.

Foda, M. S., Ismail, A. A. and Khorshid, M. A. 1975A. Production of a new rennin-like enzyme by *Aspergillus Ochraceus. Milchwissenschaft 30*, 598–601.

Foda, M. S., Ismail, A. A., Khorshid, M. A. and El-Naggar, M. R. 1975B. Physiology and characterization of a fungal milk-clotting enzyme from *Aspergillus flavus. Acta Microbiol. Pol. 8*, 337–343.

Foltmann, B. 1959A. On the enzymatic and coagulation stages of the renneting process. XV Int. Dairy Congr. Proc. 2, 655.

Foltmann, B. 1959B. Studies on rennin. II. On the crystallisation, stability and proteolytic activity of rennin. Acta Chem. Scand. 13, 1927–1935.

Foltmann, B. 1966. A review on prorennin and rennin. Compt. Rend. Trav. *Lab. Carlsberg 35*, 143–299.

Foltmann, B. 1981. Mammalian milk-clotting proteases: Structure, function, evolution and development. *Neth. Milk Dairy J. 35*, 223–366.

Food and Agriculture Organization of the United Nations 1968. Report of the FAO Ad Hoc Consultation on World Shortage of Rennet in Cheese Making. Rome, Italy.

Food and Drug Administration, Dept. of Health and Human Services 1984A. Code of Federal Regulations 21 CFR 173.150. Washington, D.C.

Food and Drug Administration, Dept. of Health and Human Services 1984B. Code of Federal Regulations 21 CFR 184.1685. Washington, D.C.

Food and Drug Administration, Dept. of Health and Human Services 1985. Code of Federal Regulations.

Fox, P. F. 1969. Milk-clotting and proteolytic activities of rennet, and of bovine pepsin and porcine pepsin. *J. Dairy Res. 36*, 427–433.

Fox, P. F. 1981. Proteinases in dairy technology. *Neth. Milk Dairy J. 35*, 233–253.

Fox, P. F. and Walley, B. F. 1971. Bovine pepsin: Preliminary cheese making experiment. *Irish J. Agr. Res. 10*, 358–360.

Fraille, E. R., Muse, J. O. and Bernardinelli, S. E. 1981. Milk-clotting enzyme from *Mucor bacilliformis. European J. Applied Microbiol. Biotechnol, 13*, 191–193.

Fuke, I. and Matsuoka, H. 1984. Preparation of fermented soybean curd using stem bromelain. *J. Food Sci. 49*, 312–313.

Garnier, J. 1973. Models of casein micelle structure. *Neth. Milk Dairy J. 27*, 240–248.

Garnot, P. and Olson, N. F. 1982. Use of oscillatory deformation technique to determine clotting times and rigidities of milk clotted with different concentrations of rennet. *J. Food Sci. 47*, 1912–1915.

Gervais, A. and Vermeire, D. 1983. A critical study and improvement of the cheese curd torsiometer. *J. Texture Stud. 14*, 31–45.

Goranova, L. and Stefanova-Kondratenko, M. 1975. Effect of *Bacillus mesentericus* strain 76 clotting enzyme on casein fraction, relative to other enzymes of microbial or animal origin. *Lait 55*, 58–67.

Gordin, S. and Rosenthal, I. 1978. Efficacy of chicken pepsin as a milk clotting enzyme. *J. Food Prot. 41*, 684–688.

Green, M. L. 1972. Assessment of swine, bovine and chicken pepsins as rennet substitutes for Cheddar cheesemaking. *J. Dairy Res. 39*, 261–273.

Green, M. L. 1977. Review of the progress of dairy science: Milk coagulants. *J. Dairy Res. 44*, 159–188.

Green, M. L., Angal, S., Lowe, P. A. and Marston, F. A. O. 1985. Cheddar cheesemaking with recombinant calf chymosin synthesized in *Escherichia coli. J. Dairy Res. 52*, 281–286.

Green, M. L. and Morant, S. V. 1981. Mechanism of aggregation of casein micelles in rennet-treated milk. *J. Dairy Res. 48*, 57–63.

Green, M. L., Valler, M. J. and Kay, J. 1984. Assessment of the suitability for Cheddar cheesemaking of purified and commercial chicken pepsin preparations. *J. Dairy Res. 51*, 331–340.

Gupta, C. B. and Eskin, N. A. M. 1977. Potential use of vegetable rennet in the production of cheese. *Food Technol. 31*, 62–64.

Gutfeld, M. and Rosenfeld, P. P. 1975. The solution to Israel's rennet shortage. *Dairy Ind. 40*, 52–55.

Hardy, J. and Fanni, J. 1981. Application of reflection photometry to the measurement of milk coagulation. *J. Food Sci. 46*, 1956–1957.

Harper, W. J. and Lee, C. R. 1975. Residual coagulants in whey. *J. Food Sci. 40*, 282–284.

Hayenga, K. J., Lawlis, V. B. and Snedecor, B. R. 1984. Microbially produced rennet, methods for its production and reactivation, plasmids used for its production, and its use in cheesemaking. Eur. Pat. Appl. EP,0,114,507,A1.

Higashio, K. and Yoshioka, Y. 1981A. Studies on milk clotting enzyme from microorganisms. I. Screening test and identification of a potent fungus for producing milk clotting enzyme and improvement of its enzymic properties by using mutants. *J. Agric. Chem. Soc. Japan 55*, 561–571.

Higashio, K. and Yoshioka, Y. 1981B. Studies on milk clotting enzyme from microorganisms. II. Preparation and some properties of crude enzyme from *Mucor racemosus* No. 50 and its mutants. *J. Agric. Chem. Soc. Japan 55*, 573–581.

Higashio, K. and Yoshioka, Y. 1981C. Studies on milk clotting enzyme from microorganisms. III. Breakdown of casein fractions by milk clotting enzyme preparations of *Mucor recemosus* No. 50 and its mutants. *J. Agric. Chem. Soc. Japan 55*, 951–958.

Higashio, K. and Yoshioka, Y. 1982. Studies on milk clotting enzyme from microorganisms. VI. Cheesemaking with milk-clotting enzyme preparations from mutants of *Mucor racemosus* No. 50. *J. Agric. Chem. Soc. Japan, 55*, 951–958.

Holmes, D. G., Duersch, J. W. and Ernstrom, C. A. 1977. Distribution of milkclotting enzymes between curd and whey and their survival during Cheddar cheesemaking. *J. Dairy Sci. 60*, 862–869.

Hori, T. 1985. Objective measurement of the process of curd formation during rennet treatment of milks by the hot wire method. *J. Food Sci. 50*, 911–917.

Hylmar, B., Pokorna, L. and Peterkova, L. 1982. Utilization of *Bacillus megaterium* strains producing proteases with milk-clotting activity. *Prumysl Potravin 33*, 208–211.

Irvine, D. M., Puhan, Z. and Gruetzner, V. 1969. Protease complex from a mutated strain of *Bacillus subtilis* as a milk coagulant for cheese manufacture. *J. Dairy Sci. 52*, 889–889.

Jackman, D. and Patel, T. R. 1983. Heat-stable proteases of psychrotrophic pseudomonads: immunological and physico-chemical studies. *Proc. 6th Int. Congr. Food Sci. Technol. 2*, 72–73.

Juffs, H. S. 1974. Influence of proteinases produced by *Pseudomonas aeroginosa* and *Pseudomonas fluorescens* on manufacture and quality of Cheddar cheese. *Aust. J. Dairy Technol. 29*, 74–78.

Kawai, M. 1973. Productivity of proteolytic enzymes and distribution of its milk clotting activity among the basidiomycetes. *J. Agric. Chem. Soc. Japan 47*, 467–472.

Kondratenko, M. S., Nachev, L. T., Dedova, P. A. and Antonova, T. N. 1977. Preparation of cheese with a microbial coagulating enzyme. U.S. Pat. 4,048,339.

Kopelman, I. J. and Cogan, U. 1976. Determination of clotting power of milk clotting enzymes. *J. Dairy Sci. 59*, 196–199.

Kowalchyk, A. W. and Olson, N. F. 1978. Firmness of enzymatically-formed milk gels measured by resistance to oscillatory deformation. *J. Dairy Sci. 10*, 1375–1379.

Kunimitsu, D. K. and Yasunobu, K. T. 1970. Chymopapain B. In: *Methods in Enzymology*, Vol. 19. G. E. Perlman and L. Lorand (Editors). Academic Press, New York, pp. 244–252.

Larson, M. K. and Whitaker, J. R. 1970. *Endothia parasitica* protease, parameters affecting activity of the rennin-like enzyme. *J. Dairy Sci. 53*, 253–269.

Lawrence, R. C., Creamer, L. K., Gilles, J. and Martley, F. G. 1972. Cheddar cheese flavour. I. The role of starters and rennets. New Zealand *J. Dairy Sci. Technol. 7*, 32–37.

Laxer, S., Pinsky, A. and Bartoov, B. 1981. Further purification and characterization of a thermophilic rennet. *Biotechnol. Bioeng. 23*, 2483–2492.

Leeuwen, H. J. V., Freeman, N. H., Sutherland, B. J. and Jameson, G. W. 1984. Hard cheese from milk concentrate. PCT Int. Pat. Appl. WO,84,01,268,A1.

Liener, I. E. and Friedenson, B. 1970. Ficin. *In: Methods in Enzymology*, Vol. 19. G. E. Perlman and L. Lorand (Editors). Academic Press, New York, pp. 261–273.

Linklater, P. M. 1961. The significance of rennin and pepsin in rennet. Ph.D. Thesis. Univ. of Wisconsin, Madison.

Mabrouk, S. S., Amr, A. S. and Abdel-Fattah, A. F. 1976. A rennin-like enzyme from *Penicillium expansum. Agric. Biological Chem. 40,* 419–420.

Majeed, G. H. 1984. Survival of porcine pepsin during Cheddar cheesemaking and its effect on casein during cheese ripening. Ph.D. Thesis. Utah State University, Logan.

Marshall, R. J., Hatfield, D. S. and Green, M. L. 1982. Assessment of two instruments for continuous measurement of the curd-firming of renneted milk. *J. Dairy Res. 49,* 127–135.

Mashaly, R. I., Ramadan, B. I., Tahnoun, M. K., El-Soda, M., and Ismail, A. A. 1981. Milk clotting protease from *Mucor mucedo.* I. Factors affecting enzyme production. *Milchwissenschaft 36,* 677–679.

McMahon, D. J. and Brown, R. J. 1982. Evaluation of Formagraph for comparing rennet solutions. *J. Dairy Sci. 65,* 1639–1642.

McMahon, D. J. and Brown, R. J. 1983. Milk coagulation time: Linear relationship with inverse of rennet activity. *J. Dairy Sci. 66,* 341–344.

McMahon, D. J., and Brown, R. J. 1984A. Composition, structure and integrity of casein micelles: A review. *J. Dairy Sci. 67,* 499–512.

McMahon, D. J. and Brown, R. J. 1984B. Enzymic coagulation of casein micelles: A review. *J. Dairy Sci. 67,* 919–929.

McMahon, D. J. and Brown, R. J. 1985. Effects of enzyme type on milk coagulation. *J. Dairy Sci. 68,* 628–632.

McMahon, D. J., Brown, R. J. and Ernstrom, C. A. 1984A. Enzymic coagulation of milk casein micelles. *J. Dairy Sci. 67,* 745–748.

McMahon, D. J., Brown, R. J., Richardson, G. H. and Ernstrom, C. A. 1984B. Effects of calcium, phosphate, and bulk culture media on milk coagulation properties. *J. Dairy Sci. 67,* 930–938.

McMahon, D. J. Richardson, G. H. and Brown, R. J. 1984C. Enzymic milk coagulation: Role of equations involving coagulation time and curd firmness in describing coagulation. *J. Dairy Sci. 67,* 1185–1193.

Melachouris, N. P. and Tuckey, S. L. 1968. Properties of a milk-clotting microbial enzyme. *J. Dairy Sci. 51,* 650–655.

Mickelsen, R. and Ernstrom, C. A. 1967. Factors affecting stability of milk-clotting enzymes on caseins and cheese. *J. Dairy Sci. 50,* 645–710.

Mickelsen, R. and Ernstrom, C. A. 1972. Effect of pH on the stability of rennin-porcine pepsin blends. *J. Dairy Sci. 55,* 294–297.

Mickelsen, R. and Fish, N. L. 1970. Comparing proteolytic action of milk-clotting enzymes on caseins and cheese. *J. Dairy Sci. 53,* 704–710.

Moir, D. T., Mao, J. E., Duncan, M. J., Smith, R. A. and Kohno, T. 1985. Production of calf chymosin by the yeast *S. cerevisiae. Dev. Ind. Microbiol, 26,* 75–85.

Murachi, T. 1970. Bromelain enzymes *In: Methods in Enzymology,* Vol. 19. G. E. Perlman and L. Lorand (Editors). Academic Press, New York, pp. 273–284.

Nachev, L., Dobreva, E., Emanuilova, E., Antonova, T., Daov, T. and Dedova, P. 1974. Study of bacterial strains producing milk-coagulating enzymes. III. Characteristics of enzyme complexes. *Prilozh. Mikrobiol. 4,* 15–21.

Nachev, L., Velcheva, P. and Kolev, D. A. 1973. Bacterial enzyme complex with milk-coagulating activity. I. Preparation and some properties. *Prilozh. Mikrobiol. 1,* 31–37.

Nand, K., Srikanta, S., Rao, K. S. N., and Murthy, V. S. 1980. Comparison of the yield and quality of cheese made with rennet and treated enzyme preparations of *Rhizopus oligosporus*. *Nahrung 24*, 859–868.

National Research Council, Food and Nutrition Board 1981. Food Chemicals Codex, National Academy Press, Washington, D.C.

O'Keeffe, A. M., Fox, P. F. and Daly, C. 1977. Denaturation of porcine pepsin during Cheddar cheese manufacture. *J. Dairy Res. 44*, 335–343.

Olson, N. F. and Bottazzi, V. 1977. Rheology of milk gels formed by milk-clotting enzymes. *J. Food Sci. 42*, 669–673.

Osono, A., Otani, H. and Tokita, F. 1983. Studies on milk-clotting enzyme from the 'litsusu' tree (*Wrightiana calysina*): evidence for milk coagulation. *Japan J. Zootec. Sci. 54*, 720–728.

Pang, S. H. and Ernstrom, C. A. 1986. Milk clotting activity in bovine fetal abomasa. *J. Dairy Sci. 69*, 3005–3007.

Payens, T. A. J. 1978. On different models of casein clotting: The kinetics of enzymatic and non-enzymatic clotting compared. *Neth. Milk Dairy J. 32*, 170–183.

Payens, T. A. 1984. The relationship between milk concentration and rennet coagulation time. *J. Appl. Biochem. 6*, 232–239.

Payens, T. A. J. and Wiersma, A. K. 1980. On enzymatic clotting processes V. Rate equations for the case of arbitrary rate of production of the clotting species. *Biophys. Chem. 11*, 137–146.

Phelan, J. A. 1977. Milk coagulants -a critical review. *Dairy Ind. Int. 42*, 50–54.

Philippos, S. G. and Christ, W. 1977. Studies on some microbial milk-clotting enzymes for cheesemaking and their effect on cow's milk casein. II. Differentiation of the milk-clotting enzyme. *Milchwissenschaft 32*, 67–71.

Poznanski, S., Reps, A., and Dowlaszewicz, E. 1975. Coagulating and proteolytic properties of a protease extracted from *Cirsium arvense*. *Lait 55*, 669–682.

Prins, J. and Nielsen, T. K. 1970. Microbial rennet. *Mucor miehei*. *Process Biochem. 5*, 34–35.

Puhan, Z. 1969. Composition and properties of a rennet substitute from *Bacillus subtilis*. *J. Dairy Sci. 52*, 889–889.

Puhan, Z. and Irvine, D. M. 1973. Proteolysis by proteases of *Bacillus subtilis* used to make Canadian Cheddar cheese. *J. Dairy Sci. 56*, 317–322.

Ramet, J. P., El-Mayda, E., and Weber, F. 1982. A new continuous method for measuring the rigidity of milk gels. *Lait 62*, 511–520.

Ramet, J. P., Alais, C. and Weber, F. 1969. Study of a microbial coagulating enzyme produced by *Endothia parasitica*. II. Experimental production of soft and cooked cheeses with Pfizer coagulating enzyme. *Lait 49*, 40–52.

Ramet, J. P. and Weber, F. 1981. Cheesemaking properties of a thermolabile milk-clotting enzyme form *Mucor miehei*. *Lait 61*, 458–464.

Rand, A. G. and Ernstrom, C. A. 1964. Effect of pH and sodium chloride on activation of prorennin. *J. Dairy Sci. 47*, 1181–1187.

Rao, K. S. N., Krishna, N., Nand, K., Srikanta, S., Krishna-Swamy, M. A., and Murthy, V. S. 1979. Changes during manufacture and ripening of Cheddar cheese prepared with fungal rennet substitute of *Rhizopus oligosporus*. *Nahrung 23*, 621–626.

Rao, L. K. and Mathur, D. K. 1979. Assessment of purified bacterial milk clotting enzyme from *Bacillus subtilus* k-26 for Cheddar cheesemaking. *J. Dairy Sci. 62*, 378–383.

Reddy, D., Payens, T. A. and Brown, R. J. 1986. Effect of pepstatin on the chymosin-triggered coagulation of casein micelles. *J. Dairy Sci. 69 (Suppl. 1)*, 72.

Reps, A., Poznanski, S. and Kowalska, W. 1970. Characteristics of milk-coagulating proteases obtained from *Byssochlamys fulva* and *Endothia parasitica*. *Milchwissenschaft 25*, 146–150.

Reps, A., Poznanski, S., Rymaszewski, J., Jakubowski, J. and Jarmul, I. 1973. Production of milk-clotting enzymes by *Byssochlamys fluva* and *Endothia parasitica* moulds. *Roczniki Instytutu Przemyslu Mleczarskiego 15*, 73–85.

Reuter, H. Hisserich, D. and Prokopek, D. 1981. Study on the formal kinetics of rennet coagulation of milk concentrated by ultrafiltration. *Milchwissenschaft 36*, 13–18.

Reyes, J. 1971. A procedure for measuring residual rennin activity in whey and curd from freshly coagulated milk. M.S. Thesis. Utah State University, Logan.

Richardson, G. H., Nelson, J. H., Lubnow, R. E. and Schwarberg, R. L. 1967. Renninlike enzyme from *Mucor pusillus* for cheese manufacture. *J. Dairy Sci. 50*, 1066–1072

Richardson, G. H., Gandhi, N. R., Diratia, M. A. and Ernstrom, C. A. 1971. Continuous curd tension measurement during milk coagulation. *J. Dairy Sci. 51*, 182–186.

Richardson, G. H., Okigbo, L. M. and Thorpe, J. D. 1983. Continuous measurement of curd tension during cheese manufacture. *Proc. 6th Int. Congr. Food Sci. Technol. 2*, 148.

Rotaru, G. 1980. The milk clotting activity characterization of an enzymatic preparation from *Aspergillus niger. Bull. Univ. Galati Technol. Chimia Produselor Aliment. 3*, 43–48.

Sannabhadti, S. S. and Srinivasan, R. A. 1976. Use of milk clotting enzyme of *Absidia ramosa* in Cheddar cheese preparation. *J. Food Sci. Technol. India 13*, 305–309.

Schmidt, D. G. 1980. Colloidal aspects of casein. *Neth. Milk Dairy J. 34*, 42–64.

Sellers, R. L. 1982. Effect of milk-clotting enzymes on cheese yield. 5th Bienn. Cheese Ind. Conf., Utah State University, Logan.

Shaker, K. A. and Brown, R. J. 1985A. Effects of enzyme choice and fractionation of commercial enzyme preparations on protein recovery in curd. *J. Dairy Sci. 68*, 1074–1076.

Shaker, K. A. and Brown, R. J. 1985B. Proteolytic and milk clotting fractions in milk clotting preparations. *J. Dairy Sci. 68*, 1939–1942.

Shalabi, S. I. and Fox, P. F. 1982. Influence of pH on the rennet coagulation of milk. *J. Dairy Res. 49*, 153–157.

Shehata, A. E., Ismail, A. A., Hegazi, A. and Hamdy, A. M. 1978. Fractionation of commercial rennet enzymes on Sephadex G-100. *Milchwissenschaft 33*, 693–695.

Shovers, J. and Bavisotto, V. S. 1967. Fermentation derived enzyme substitute for animal rennet. *J. Dairy Sci. 50*, 942–942.

Shovers, J., Fossum, G. and Neal, A. 1972. Procedure for electrophoretic separation and visualization of milk-clotting enzymes in milk coagulants. *J. Dairy Sci. 55*, 1532–1534.

Slattery, C. W. 1976. Review: Casein micelle structure; an examination of models. *J. Dairy Sci. 59*, 1547–1556.

Sommer, H. H. and Matsen, H. 1935. The relation of mastitis to rennet coagulability and curd strength of milk. *J. Dairy Sci. 18*, 741–749.

Stanley, D. W., Emmons, D. B., Modler, H. W. and Irvine, D. M. 1980. Cheddar cheese made with chicken pepsin. *Can. Inst. Food Sci. Technol. J. 13*, 97–102.

Stavlund, K. and Kiermeier, F. 1973. Detection of rennet substitutes. *Z. Lebensm. Unter.mForsch. 152*, 138–144.

Steinsholt, K. 1973. The use of an Instron universal testing instrument in studying the rigidity of milk during coagulation by rennin. *Milchwissenschaft 28*, 94–97.

Sternberg, M. Z. 1971. Crystalline milk clotting protease from *Mucor miehei* and some of its properties. *J. Dairy Sci. 54*, 159–167.

Storch, V. and Segelcke, T. 1874. Milchforsch. Milchprax. 3:997. *Cited by* B. Foltmann. 1959. On the enzymatic and coagulation stages of the renneting process. *XV Int. Dairy Congr. Proc. 2*, 655.

Surkov, B. A., Klimovskii, I. I. and Krayushkin, V. A. 1982. Tirbidometric study of kinetics and mechanism of milk clotting by rennet. *Milchwissenschaft 37*, 393–395.

Swaisgood, H. E. 1982. The caseins. *CRC Crit. Rev. Food Technol. 3*, 375–414.

Thomasow, J. 1968. The Hellige thrombo-elastograph in studies of rennet coagulum. *Milchwissenschaft 23*, 725–731.

Thomasow, J., Mrowetz, G. and Schmanke, E. 1970. Experimental cheesemaking with rennet from *Endothia parasitica*. *Milchwissenschaft 25*, 211–217.

Thunell, R. K., Duersch, J. W. and Ernstrom, C. A. 1979. Thermal inactivation of residual milk clotting enzymes in whey. *J. Dairy Sci. 62*, 373–377.

Vanderpoorten, R. and Weckx, M. 1972. Breakdown of casein by rennet and microbial milk-clotting enzymes. Neth. Milk Dairy J. 26, 47–59.

Velcheva, P. and GHbova, D. 1978. Proteolytic activity of a milk-coagulating bacterial enzyme isolated from *Bacillus mesentericus* strain 90. *Acta Microbiol. Bulg. 1*, 12–20.

Velcheva, P., Kolev, D. A. and Chipileva, R. 1975B. Bacterial enzyme complex with milk-coagulating activity. IV. Action of its components on whole casein. *Prilozh. Mikrobiol. 5*, 44–51.

Velcheva, P., Kolev, D. A. and GHbova, D. 1975B. Bacterial enzyme complex with milk-coagulating activity. V. Effect of its components on β-and α_s-casein. *Prilozh. Mikrobiol. 6*, 19–30.

Visser, S. 1981. Proteolytic enzymes and their action on milk proteins. A review. *Neth. Milk Dairy J. 35*, 65–88.

Wang, J. T. 1969. Survival and distribution of rennin during Cheddar cheese manufacture. M.S. Thesis. Utah State University, Logan.

Whitaker, J. R. 1970. Protease of *Endothia parasitica*. In: *Methods in Enzymology*, Vol. 19. G. E. Perlman and L. Lorand (Editors). Academic Press, New York, pp. 436–445.

Wilken, T. O. and Bakker, G. 1974. Process of making a milk coagulating enzyme preparation. U.S. Pat. 3,857,969.

Yiadom-Farkye, N. 1986. Role of chymosin and porcine pepsin in Cheddar cheese ripening. Ph.D. Thesis. Utah State University, Logan.

Zvyagintsev, V. I., Krasheninin, P. F., Sergeeva, E. G., Buzov, I. P., Mosichev, M. S. and Rubtsova, N. A. 1972. Characteristics of cheeses made with *Mucor renninus 367* enzyme preparation. *Prikl. Biokhim. Mikrobiol. 8*, 913–917.

Part II—Cheese Chemistry

Mark E. Johnson

The origin of cheese is unknown, but it is likely to have originated by natural spoilage (souring) of milk and to have evolved into a process of preserving milk, a less stable source of nutrients. Cheese manufacturing and ripening processes involve a complex series of reactions in which biological, chemical, and physical factors affect and are affected by each other. These interacting factors, plus the composition and type of milk, create the various types of cheeses. Several hundred varieties of cheeses have been described, but there is a great deal of duplication because of the close similarity between many varieties of different national origins. There are many schemes by which cheeses are categorized. These usually involve factors such as the composition, manufacture, and ripening of the cheese.

Lawrence *et al.* (1984) suggested that all types of cheese can be best classified by their calcium content and pH. According to this classification scheme, the extent of acid production at various stages of cheese manufacture ultimately influences the body and texture of cheese. Cheeses can, therefore, be classified by manufacturing procedure rather than by flavor.

The chemistry of cheesemaking can be divided into several phases: the characteristics of milk, the cheesemaking process, and the ripening of cheese. The types of cheeses differ in specific aspects of the three phases, but there are substantial similarities. General principles will be emphasized in this chapter, with references to unique aspects of important types. The reader is referred to the previous volume of *Fundamentals of Dairy Chemistry* for early research reports (Ernstrom and Wong 1974). The details of the technical procedures for the manufacture of different cheeses are beyond the scope of this book. Selected references on the manufacture of a variety of cheeses include Van Slyke and Price (1952), Wilster (1980), Kosikowski (1977), Morris (1981), Reinbold (1972), Olson (1969), Emmons and Tuckey (1967), and Reinbold (1963).

MILK COMPOSITION

Many milk constituents affect the manufacturing and various characteristics of cheese, but milk fat and casein are of primary importance since they constitute most of the solids in cheese (e.g., 91% of the solids in Cheddar cheese). These two constituents, plus water, influence the yield of cheese from milk and the gross composition of cheese (Van Slyke and Price 1952). Formulas used to predict the cheese yield from milk include the concentration factors of casein and fat in milk, a minor correction factor for other milk constituents, and the added salt and moisture content of cheese (Van Slyke and Price 1952; Lelievre et al. 1983; Banks et al. 1984).

Standards of identity for cheese varieties of greater commercial value have been established in the United States (FDA 1984). These regulations set the limits on the moisture content and minimum fat in the dry matter of cheese (FDM; or fat on the dry basis, FDB). To conform to the legal standards and for economic reasons, the cheesemaker will sometimes have to adjust the proportions of constituents of the cheese milk (Barbano 1984). The composition of milk (standardization) is regulated in basically two ways: removal or addition of fat as cream and addition of casein as nonfat dry milk, skim milk, or condensed skim milk (Johnson 1984). Traditionally, cheesemakers have only adjusted the fat concentrations of milk, but because the composition of milk varies (particularly the casein), they are now adjusting the casein-to-fat ratio of milk. This ratio in milk influences the FDB of cheese, although the relationship is not exact and may vary among manufacturing plants. The amount of fat retained in cheese is a function of the amount of fat in the milk and, most importantly, of the cheesemaking practices. This value differs among plants and reflects different efficiencies and processing conditions. The moisture content of cheese usually increases as the FDB decreases, and the moisture level can be controlled indirectly by altering the FDB level (Lawrence and Gilles 1980). In Cheddar cheese, an increase of 0.05 in the casein-to-fat ratio generally results in a decrease of about 1.4% in the FDB and an increase of about 0.8% in moisture (Lawrence and Gilles 1980).

The casein concentration in milk is affected by heredity, feed, season, state of lactation, and milk storage. The first four factors are dealt with in previous chapters. Milk storage has been given considerable attention in the last few years, and there is evidence that proteinases can substantially alter caseins in milk.

Cold storage of milk causes solubilization of colloidal calcium phosphate and a shift in caseins from the micellar to the soluble state. Soluble caseins are lost in the whey during cheesemaking. The soluble caseins constitute less than 15% of the total casein in normal milk directly from the udder, but during storage at 4°C the concentration of soluble casein has been shown to increase up to 42% of the total casein (Ali et al. 1980A). Most of the increase resulted from solubilization of β-casein, with 30 to 60% of this fraction being found in the soluble phase. Solubilization of the caseins and of colloidal calcium phosphate reached a maximum after approximately 48-hr storage but reversed slightly during further storage. The solubilization during cold storage could be reversed by heating at 60°C for 30 min or 72°C for 30 to 60 sec. However, the milk equilibrium system never completely attained its initial state (Ali et al. 1980A). Dissociation of the caseins, especially β-casein, from micelles is enhanced by cold solubilization of colloidal calcium phosphate, with cleavage of bridges between the salt and β-casein, as well as breakage of hydrophobic bonds (Ali et al. 1980A–C; Pierre and Brule 1981).

The changes in protein and salt equilibrium during storage at 4°C causes an increase in rennet clotting time, reduction of firmness of the rennet clot, and loss of cheese yield (Ali et al. 1980A). The extent of change is proportional to the degree of casein solubilization.

The growth of psychrotrophic bacteria in refrigerated milk is of concern to cheesemakers. These bacteria can produce extracellular heat-resistant lipases and proteinases that can act directly on micellar casein and the fat globules in milk. Cousin (1982) and Law (1979) have reviewed the effects of psychrotrophic bacteria and their enzymes. Substantial growth of psychrotrophic bacteria is necessary to cause significant losses in cheese yields, since 10^7 colony-forming units per milliliter cause only a low degree of β- and α_{s1}-casein breakdown (Law et al. 1979). The heat-resistant enzymes of psychrotrophic bacteria may be responsible for off-flavors (rancidity and bitterness) during ripening of cheese (Cousin 1982; Cousin and Marth 1977).

Proteolysis of casein may be substantial under certain conditions, such as late lactation and mastitic infections. Under these conditions, the number of somatic cells increases. The most noticeable effect of high somatic cell counts is loss of cheese yield. Everson (1984) identified a loss of 0.045 kg of cheese per 45.36 kg of milk for every 10^5/ml increase of somatic cell count. Somatic cell counts above 4×10^5/ml were also correlated with enhanced lipolysis and with an increased tendency for rancid dairy products (Everson 1984). The effects on casein and milk fat are the result of an increase in the activity of lipoprotein lipase and alkaline milk proteinase (plasmin) concomitant with an increase in somatic cells (Andrews 1983A; Ali *et al.* 1980C; Jurczak and Sciubisz 1981.) However, Jellema (1975) found no significant relationship between mastitis and lipolysis.

Plasmin hydrolyzes casein and is thermostabile during pasteurization (Humbert and Alais 1979). The most significant effects of plasmin relate to cheese yield and hydrolysis of β-casein during ripening of certain cheese varieties. According to Noomen (1978) and Trieu-Cuot and Gripon (1982), plasmin activity is especially important at the surface of Camembert cheese, where the high pH is favorable to the activity of plasmin. Plasmin has been implicated in the ripening of Swiss (Richardson and Pearce 1981) and Romano cheese (Guinee and Fox 1984). Both β-and α_{s1}-caseins are hydrolyzed by plasmin, with β-casein being slightly more susceptible to hydrolysis (Andrews 1983B). High levels of γ-caseins in cheese are indicative of plasmin activity.

Cheeses made from milks with high fat contents tend to have higher moisture levels in relation to protein content. This ratio, also called "moisture in the nonfat substance (MNFS)," is an important compositional factor influencing the quality of cheese (Pearce and Gilles 1979). It is the relative wetness of the casein in the cheese, rather than the percentage of moisture in the cheese as a whole, that influences the course of the ripening process (Lawrence and Gilles 1980).

Milk fat plays a very important role in the development of texture in cheese. Reduced-fat cheeses tend to be firmer and more elastic than cheeses with a higher fat content. Undoubtedly the presence of a more dense protein matrix results in a firmer cheese. The precise role of fat in cheese texture is not well understood, since problems of increased firmness can be partially overcome by increasing the MNFS. Studies by Green *et al.* (1981) on the texture of cheeses made from concentrated milk suggest a possible role of fat in cheese firmness. Reduced fat in the curd would result in a smaller fat–protein interfacial area and an increased separation between fat globules. The capacity of the fat and protein phases of cheese to move in relation to each other would be reduced and would consequently result in a firmer cheese.

THE CHEESEMAKING PROCESS

Prior to cheesemaking, milk is generally clarified and may or may not be homogenized or pasteurized, depending on the type of cheese. The first step in the actual cheesemaking process is the formation of the coagulum. This has been reviewed in the previous chapter. The second step is the separation of curd from the whey. The last stage of cheese manufacture is the ripening process.

Clarification

Milk is clarified by high-speed centrifugation to remove extraneous matter held in suspension. Clarification occurs prior to heat treatment of the milk to prevent dissolution of the extraneous matter. Although clarification removes somatic cells, the elevated levels of lipoprotein lipase activators and plasmin that may be associated with increased numbers of white blood cells in the milk are not eliminated. Therefore, increased lipolysis of milk fat by lipoprotein lipase and proteolysis of casein by plasmin may not be deterred.

Milk for Swiss cheese is invariably clarified to remove sediment and thereby reduce the number of eyes (Reinbold 1972). Extraneous matter acts as loci for gas accumulation.

Bactofugation, a process based on centrifugal separation of bacteria and their spores, is practiced in the Netherlands. Since the spores of lactate-fermenting *Clostridia* (butyric acid bacteria) are removed, there is less risk that Gouda cheese will develop the "late blowing" defect caused by the metabolism of these bacteria (Van den Berg *et al.* 1980).

Pasteurization

The primary purpose of pasteurization is to destroy pathogens that may be present in milk. Pasteurization also destroys the majority of all bacteria present, including coliforms, lactic acid streptococci, yeasts, and molds. However, many bacteria can survive pasteurization, including spores of *Bacillus* and *Clostridium* species, as well as the vegetative cells of some species of *Propionibacterium, Lactobacillus,* and *Micrococcus.* Normally, the interior of cheese is anaerobic, a necessary condition for the growth of obligate anaerobes, clostridia, and propionibacteria. If present in sufficient numbers, clostridia and propionibacteria can cause gassiness. As facultative anaerobes, *Lactobacillus* and *Micrococcus* may be found on the surface and throughout the cheese, and may be important for the development of flavor in ripened cheeses.

Pasteurization inactivates many enzymes, including alkaline phosphatase and lipoprotein lipase. The absence of active alkaline phosphatase in cheese is often used to determine if the milk has been properly pasteurized prior to cheesemaking. Since pasteurization kills most of the lactic acid bacteria in milk, the lactic acid developed during cheesemaking is due almost entirely to the added starter culture. Hence, it is much easier to control the rate of acid development during cheesemaking and produce a uniform product. Pasteurization, however, may not destroy phage active against streptococci in starter cultures (Chopin 1980).

An increase in cheese yield can occur with pasteurized milk (Walstra and Jenness 1984). This is due to casein–whey protein interaction and a greater retention of moisture. The exact mechanism of the heat-induced association between whey proteins and casein micelles is not known. When purified β-lactoglobulin and casein micelles are heated together, they complex with each other primarily through intermolecular S–S bonds between β-lactoglobulin and κ-casein (Smits and Van Brouwershaven 1980). When purified α-lactalbumin is heated with casein in the absence of β-lactoglobulin, little association with casein micelles occurs (Baer et al. 1976). Smits and Van Brouwershaven (1980) have proposed that β-lactoglobulin reacts with κ-casein because of the presence of SH groups, whereas α-lactalbumin cannot because it contains only S–S bonds. The SH group functions as a catalyst in the formation of heat-induced intermolecular S–S bonds through S–S interchange reactions. When whole milk is heated, it is thought that α-lactalbumin and β-lactoglobulin form a complex, which in turn reacts with κ-casein (Elfagm and Wheelock 1977). The association between denatured whey proteins and κ-casein prevents rennet from clotting the milk.

The casein micelles become surrounded by whey proteins and cannot interact with one another, thus reducing whey syneresis. This results in a soft curd that retains more moisture. The yield of cheese is increased due to the incorporation of whey proteins and the higher moisture content. Overheated milk requires longer rennet coagulation times. If milk is heated for 30 min at 75°C, it will not clot at all (Ustunol and Brown 1985).

Federal definitions and standards of identity for the various kinds and groups of cheeses, issued by the Food and Drug Administration (1984), require that if the milk used is not pasteurized, the cheese must be cured for not less than 60 days at a temperature not lower than 1.67°C. These conditions allow any pathogens that might be present to die or become inactive during storage.

The major objection to using pasteurization is that aged cheeses develop flavor more slowly and to a lesser extent than does raw milk cheese (Kristoffersen 1985). This has led many cheesemakers to use heat-treated milk (60° to 68.5°C for 15 sec or less) instead of pasteurized milk. It is believed that such heat treatment is sufficient to control undesirable bacteria but not to completely inactivate or destroy native milk enzymes and certain bacteria necessary for proper ripening of the cheese. Attempts to make Swiss cheese from fully pasteurized milk have not been successful, so raw milk or milk heated to 67.7° to 70°C is used (Reinbold 1972).

Pasteurization will not cure the problems associated with the use of milk of poor bacteriological quality. Lipases and proteinases associated with psychrotrophs are sometimes heat resistant even though the bacteria themselves are destroyed. These enzymes can be responsible for rancid, bitter, and unclean flavors.

Homogenization

The use of homogenized milk for cheesemaking has been reviewed by Peters (1964). The advantages of homogenized milk in the manufacture and ripening of cheese are (1) lower fat losses in whey and therefore a higher yield, (2) reduced fat leakage of cheese at room temperatures, and (3) increased rate of fat hydrolysis and, therefore, desired flavor production in blue cheese.

Homogenized milk is generally not used for cheesemaking because of the cost and potential increase in hydrolytic rancidity in cheese. There are a few major exceptions; cheese spreads, cream, Neufchâtel, and blue cheese (Kosikowski 1977).

The effects of homogenization on milk components have been summarized by Walstra and Jenness (1984) and Harper (1976). Homogenization disrupts fat globules and results in an increase in fat surface area (about 4–10 times). Casein micelles adsorb on the fat surface and constitute part of the fat globule membrane. The curd tension of milk is thus lowered. Walstra and Jenness (1984) have described the effect of homogenization on rennet coagulation.

Partial coverage of fat globules with casein makes them behave, to some extent, like large casein micelles. Renneting causes the homogenized fat globules to aggregate, and because homogenization has increased effectively the content of micellular casein, aggregation occurs more rapidly. This has serious consequences in the formation of cheese curd. Green (1984) has suggested that when the initial rate of casein micelle aggregation is increased, the primary aggregating particle is larger. However, because of their size, these particles aggregate with other particles much more slowly. The overall effect is that the formation of a continuous gel occurs much more slowly, i.e., the rennet clotting time is increased. Scanning electron micrographs show that curds made from homogenized milk have a finer protein network than curds from unhomogenized milk (Green et al. 1983). Since only one surface of the casein micelles associated with the fat globules is free to react with adjacent casein micelles, the strength and continued shrinkage of the curd decrease. Whey loss from the curd (syneresis) is slower and the curd retains more moisture. These effects can be overcome by concentrating the milk or by adding casein in the form of low-heat nonfat dry milk or concentrated skim milk (Maxcy et al. 1955). Adding calcium chloride even to levels twice as high as those allowed does not overcome the problem of a soft coagulum (Maxcy et al. 1955).

A higher yield of cheese is obtained when homogenized milk is used for cheesemaking. This is due to increased fat and moisture retention of the curd. However, the fat in the whey cannot be recovered by centrifugal processes because of the failure of homogenized fat globules to cluster.

Homogenization is beneficial for cream, Neufchâtel cheese, and cheese spreads in producing a smoother-bodied cheese that does not leak fat at room temperature. Cheddar cheese is softer, smoother, and more elastic when made from homogenized milk (Emmons et al. 1980). Homogenization of milk for cheesemaking is done during or following pasteurization. The pasteurization time and temperature may be increased because homogenization may activate lipase not destroyed by traditional pasteurization conditions. Cream, rather than whole milk, is generally homogenized and mixed with skim milk prior to cheesemaking. In the hot pack method of manufacturing Neufchâtel cheese, the curd is homogenized.

Almost all of the blue cheese made today is manufactured from homogenized milk or from a blend of homogenized thin cream (14–20% fat) and skim milk. The benefits of homogenization of the milk for blue cheese have been reviewed by Morris (1981). Homogenization of the milk for blue cheese causes considerable improvement in its ripening and flavor development. The body of blue cheese made from homogenized milk is more porous, allowing accelerated growth of the essential mold *Penicillium roqueforti*. Lipase activity of the mold results in a marked increase in free fatty acids and subsequent formation of methyl ketones essential for typical blue cheese flavor. The cheese is lighter in color and the body is softer.

Milk Coagulation or Clotting

The physical and chemical characteristics of cheese curd depend on the method used to form the curd matrix. The curd is formed in basically one of two ways: acid or enzymatic coagulation. In acid curd cheeses (cottage, baker's, cream), the curd is formed by direct addition of acid to the milk or by lactic acid produced by the fermentation of lactose. As the pH of the milk approaches the isoelectric point of casein (pH

4.6), casein micelles begin to aggregate. Scanning electron microscopy shows that casein micelles aggregate into chains, then into strands and clusters, and eventually into amorphous masses during the manufacture of both acid curd (Glaser *et al.* 1980) and rennet curd (Kimber *et al.* 1974). The final degree of chaining is always higher in rennet curd, which binds water more strongly than acid curd (Kalab 1979). Acid curd is extremely fragile (due to loss of calcium), and tends to shatter more and to contract less than that formed by rennet. These are desirable characteristics in the manufacture of cream cheese, where the curd is broken by vigorous stirring. This disperses the curd as fine particles, giving the cheese a smooth texture. Cottage cheese curd is cut into cubes to allow some curd firmness to develop before stirring. Little syneresis occurs in cottage cheese curd, and the curd retains more moisture. Small amounts of rennet are used in manufacturing acid curd cheeses to increase the rate of whey expulsion. If the pH drops below the isoelectric point (pH 4.6) of casein, cottage cheese curd will retain more moisture and the cheese may have a soft, pasty body (Olson 1979). A high pH may produce a curd that is too firm and rubbery. These effects are due to the amount of calcium complexed with casein. Calcium cross-linkages may be involved in firming the casein matrix of the curd, or the charge-neutralizing effect of calcium may allow proteins to interact through hydrophobic bonding (Lawrence *et al.* 1983). Cutting the curd too small increases the loss of fat in the whey, as the fat exposed at the surface is not held in the curd. Curd sizes can vary from the size of rice grains (Swiss) to 1.27 cm or larger (Brie, Camembert). Cream and Neufchâtel cheeses are not cut but stirred, forming micrograins of curd. These small particles are separated from the whey by centrifugation or filter cloth.

The rate of syneresis is accelerated by increased rennet levels, increased temperature, stirring, and the development of acid by the starter bacteria (Patel *et al.* 1972; Lawrence 1959). Subsequent to cutting and prior to stirring, the curd is usually allowed to "heal," a process whereby a thin skin forms on the surface of the cut curd. The healing process firms the curd, making it more resistant to physical damage during stirring and cooking. If the curd is heated too rapidly, the outer layers on the curd particles dehydrate and shrink rapidly, retarding the escape of whey. Hence, the rate of heating may have a significant effect on syneresis.

After cutting and healing, the curd and whey mixture is heated. The cooking process is always accompanied by stirring to allow even heating and encourage whey syneresis. Freshly cut curd is soft and sticky and must be carefully stirred to prevent matting. The curd will become firmer and less sticky as syneresis proceeds. Concentrating the curd particles by physically removing a portion of the whey will also increase whey syneresis (Lawrence 1959). The effect of stirring on syneresis is probably due to physical stresses on the curd. Stirring produces small pressure gradients, and these become larger when part of the whey is removed.

⌐Combined or single effects of heating and acid production by the starter bacteria increase whey syneresis and establish moisture levels for a given variety of cheese⌐ Almost 96% of the moisture lost in Cheddar cheese during cooking occurs in the first 30 min (Lawrence 1959). A comprehensive review of syneresis has been written by Walstra *et al.* (1985).

Rennet is inactivated at the high cooking temperatures used in Swiss and Mozzarella but is still active in Cheddar curd cooked to 39°C (Matheson 1981). Residual rennet activity has implications for the subsequent ripening of the cheese.

Most varieties of cheese are cooked by applying heat to the outside of the vessel containing the curd and whey slurry. Gouda cheese curd is heated by first draining a portion of the whey and then adding hot water. The proportion of whey removed and water added is varied to control the amount of residual lactose in the curd. "Washing" of the curd is also used in cottage and brick cheese manufacture to remove lactic acid and lactose, but in these cases the cheese curds have first been heated.

The starter culture used in cheesemaking depends on the type of cheese and the temperature to which the curd is heated. *Streptococcus lactis* or *S. cremoris* are used in cheese varieties heated to 40°C or less, since no acid development occurs with these cultures above that temperature (Sellars and Babel 1970). High-temperature homolactic bacteria such as *S. thermophilus*, *Lactobacillus bulgaricus*, or *L. helveticus* are used in the manufacture of cheese varieties heated to higher temperatures.

The relationship between pH, mineral retention, and basic cheese structure has been illustrated by Lawrence *et al.* (1984). Hill *et al.* (1985) have developed mathematical models of the association between pH at draining and mineral content of whey. The calcium, phosphorus, magnesium, and nonprotein nitrogen content of whey increased with decreasing pH, while sodium and potassium levels were not affected. Mineral and nonprotein nitrogen concentrations in the whey were not associated with cooking temperature.

High mineral content of cheese curd at draining promotes the development of elastic texture. Minimum mineral loss from the curd occurs after draining. Cheese varieties with "eyes" (Swiss, Gouda) require elastic curds to permit round eye formation. These cheeses are drained at relatively high pH (6.4–6.5). Parmesan and Romano-type cheeses are drained at low pH (6.0–6.1), forming a granular, inelastic curd structure.

Residual lactose in the curd after draining affects the body, texture, flavor, and final pH of the cheese. The amount of lactose remaining in the curd depends on the amount of acid developed during cooking. Lactose can be removed from curd by washing (Colby, brick, Gouda). It has not been conclusively established that prolonged contact of the curd with the whey will allow lactose to diffuse into cheese curd and result in higher residual lactose (Lawrence and Gilles 1982). Fermentation of lactose by the lactic acid bacteria continues until their metabolism is stopped or the curd is depleted of lactose. Residual lactose fermented by nonstarter bacteria may result in undesirable fermentations and off-flavors. The browning of Mozzarella and processed cheese has been shown to be the result of residual lactose and galactose (Bley *et al.* 1985; Johnson and Olson 1985).

Separation of Curd from Whey

There are basically two ways in which whey is separated from curd: (1) continuous filtering of whey through screens, retaining the curd either as a solid block (Cheddar, Mozzarella, brick) or as granules (Colby, stirred curd Cheddar, cottage), and (2) forming a mass of curd under the whey, which is subsequently drained. There are many variations of the former method, allowing varying degrees of openness in the cheese.

Close contact between curd grains can be achieved by pressing them together, fusing curd grains into a more or less homogeneous and coherent mass. For actual fusing, new bonds between para-casein micelles must be formed. This is possible only if the pH continues to decrease while the grains are being pressed together. If pressing occurs after the final pH value has been obtained, a coherent mass is not produced (Walstra and Jenness 1984).

In Swiss and Gouda-type cheeses, the curd is first formed into a mass and pressure is applied. The whey is drained, but pressure on the curd is maintained. As the pH drops from 6.4–6.5 at draining to 5.2–5.3 in the finished cheese, the curd fuses into a very tight, smooth structure.

After the curd and whey are physically separated and the optimum pH level is reached, the curd is salted. Salt improves the flavor of cheese, retards microbial metabolism, and helps expel moisture from the curd. Salt is either added directly to the curd (Cheddar, Colby) or the preformed block of cheese is placed in a brine solution (almost all other cheese types).

Salt equilibrium throughout a block of cheese is a slow process even in Cheddar cheese, which is salted after the curd is milled into relatively small pieces (Morris *et al.* 1985). Cheeses which are salted by immersion in brine or by having salt rubbed on the exterior of the cheese require much more time for the salt to reach equilibrium throughout the cheese. Variations in moisture content within a block of cheese may be responsible for the uneven distribution of salt. Transport of salt into cheese is a process of impeded mutual diffusion, consisting of sodium chloride penetration into cheese and outward migration of water (Geurts *et al.* 1980). Concentration gradients are established with greater salt levels at the surface and higher moisture levels at the center. The concentration of sodium chloride in brine is the driving force of diffusion. Higher diffusion rates occur in lower-fat cheeses and cheeses with higher moisture content, and are increased by higher temperatures (Geurts *et al.* 1980). In addition, the larger the relative surface area of the cheese, the greater the rate of diffusion (Geurts *et. al.* 1980). Salt content, or more importantly, salt-in-moisture (S/M) of the cheese, influences fermentation of residual lactose and hydrolysis of protein during ripening. The final pH of the cheese and the cheese flavor is dependent on the S/M level. Cheeses with S/M values of <4% are acidic and tend to develop bitterness, while cheeses with S/M values of >6% exhibit less acidity after salting and arrested flavor development (Lawrence and Gilles 1982; Thomas and Pearce 1981). Salt sensitivities of the starter bacteria and amount of residual lactose will ultimately determine the final pH of the curd.

Cheese Ripening

It is the paracasein matrix that determines the body and texture of cheese. The products derived from paracasein breakdown are acted upon chemically and enzymatically to give cheese much of its characteristic flavor. Hydrolysis of paracasein causes the body of the cheese to lose its firm, tough, curdy properties and become soft and smooth. This process, along with the development of flavor, is called "ripening." Recently, many detailed reviews of various phases of cheese ripening have been published (Grappin *et al.* 1985; Rank *et al.* 1985; Adda *et al.* 1982; Aston and Dulley 1982; Green and Manning 1982; Lawrence and Gilles 1982; Law 1981, 1984).

Proteolysis of casein begins with the addition of rennet to the milk and the formation of a coagulum. Calf rennet is actually 80% chymosin and 20% bovine pepsin A (Grappin *et al.* 1985). Rennet can remain active in Cheddar and Camembert cheeses for up to three months, but not in Mozzarella or Swiss cheeses due to the high cooking temperature (Matheson 1981). The major contribution of rennet activity to the softening of the cheese is the hydrolysis of α_{s1}-casein. Creamer and Olson (1982) have suggested a model of Cheddar cheese microstructure in which an extensive network involving α_{s1}-casein molecules traverses the cheese and have stated that cleavage of α_{s1}-casein weakens the protein network. It is generally accepted that chymosin plays the major role in the initial breakdown of α_{s1}-casein. The peptide α_{s1}-1 is the first and principal degradation product of α_{s1}-casein by rennet (Creamer and Richardson 1974).

The other major casein in cheese is β-casein, but it is generally not hydrolyzed by rennet in low-pH cheeses. Alkaline milk protease (plasmin) plays the major role in the hydrolysis of β-casein (Richardson and Pearce 1981). The plasmin level in cheese is related to the pH of the curd at whey drainage, since plasmin dissociates from casein micelles as the pH is decreased. Richardson and Pearce (1981) found two or three times more plasmin activity in Swiss cheese than in Cheddar cheese. Swiss cheese curds are drained at pH 6.4 or higher, while Cheddar cheese curds are drained at pH 6.3 or lower. Proteolysis of β-casein is significantly inhibited by 5% sodium chloride. The inhibitory influence of sodium chloride is most likely due to alteration of β-casein or a reduction in the attractive forces between enzyme and substrate (Fox and Walley 1971).

The gross proteolysis of casein is probably due solely to rennet and plasmin activity (O'Keeffe et al. 1978). Bacterial proteases and peptides are responsible for subsequent breakdown of the large peptides produced by rennet and plasmin into successively smaller peptides and finally amino acids (O'Keeffe et al. 1978). If the relative rate of proteinase activity by rennet, plasmin, and bacterial proteases exceeds that of the bacterial peptidase system, bitterness in the cheese could result. Bitter peptides can be produced from α_{s1}- or β-casein by the action of rennet or the activity of bacterial proteinase on β-casein (Visser et al. 1983). The proteolytic breakdown of β-casein and the subsequent development of bitterness are strongly retarded by the presence of salt (Fox and Walley 1971; Stadhouders et al. 1983). The principal source of bitter peptides in Gouda cheese is β-casein, and more particularly the C-terminal region, i.e., $\beta(193-209)$ and $\beta(193-207)$ (Visser et al. 1983). In model systems, bitter peptides are completely debittered by a peptidases system of S. cremoris (Visser et al. 1983).

Mills and Thomas (1980) have provided direct evidence that the level of starter proteinase has a role in the development of bitterness in Cheddar cheese. Using cultures containing different proportions of proteinase-positive and proteinase-negative variants of S. cremoris and S. lactis, they showed that cheeses containing 45–75% proteinase-negative cells developed significantly less bitterness than cheeses containing only proteinase-positive cells. They also provided indirect evidence that bacterial peptidase activity could remove bitter peptides.

The activity of rennet in some maturing cheese is essential for normal cheese ripening. The practice of using less rennet in making cheese with concentrated milk has shown that the cheese does not develop characteristic sharp flavors (Chapman *et al.* 1974). The role of rennet in flavor development of cheese is to produce peptides that are degraded subsequently by the bacterial flora of the cheese.

The softening of surface-ripened cheeses (Brie, Camembert, Limburger) has been generally attributed to proteolytic enzymes produced by the surface flora migrating into the cheese and causing protein breakdown. Recent reports by Noomen (1983) and LeGraet *et al.* (1983) have modified this assumption. Softening of these cheeses is due to the combined effects of deacidifying activity of the surface flora and casein hydrolysis by rennet and plasmin. Deacidifying occurs as the surface flora hydrolyze casein to alkaline breakdown products such as ammonia, which diffuse from the surface to the center. As the pH is raised, caseins become more negatively charged, resulting in electrostatic repulsion between casein molecules. This weakens the protein network, and the cheese becomes more gel-like. Lactic acid catabolism at the surface causes migration of lactic acid from the center of the cheese and a concomitant translocation of calcium and phosphorus to the surface (LeGraet *et al.* 1983.)

Amino acids are generally not considered to be important flavor components of several varieties of cheese, although they are important precursors of a variety of flavor components: volatile sulfur compounds, amines, aldehydes, and ammonia (Adda *et al.* 1982; Aston and Dulley 1982; Forss 1979; Langsrud and Reinbold 1973). Free proline levels in Swiss cheese are important in producing the typical sweet cheese flavor. Cheeses with a proline content of <100 mg/100 g cheese lacked the sweet flavor, while levels of >300 mg/100 g produced a cheese of excessive sweetness (Mitchell 1981).

The fermentation of residual lactose in cheese curd after whey drainage is an integral part of the ripening process. This metabolism produces lactic acid that inhibits the growth of many undesirable microorganisms (Babel 1977) and lowers the redox potential of the cheese. The formation of active sulfhydryl groups (H_2S, methanthiol) which are essential for cheese flavor development requires low redox potential (Kristoffersen 1985; Green and Manning 1982). H_2S is probably produced by bacterial metabolism (Sharpe and Franklin 1962), but the exact mechanism of methanthiol formation is unclear. Green and Manning (1982) have suggested that methanthiol is produced by purely chemical means. Kristoffersen (1985) argues that, in addition to reduced conditions, active lactic acid bacteria are essential to full flavor development.

Excessive or insufficient acid development during manufacture can produce variability in the moisture content of cheese and defects in flavor, body, texture, color, and finish (Van Slyke and Price 1952). The rate of lactose fermentation varies with the type of cheese, but the conversion to lactic acid is virtually complete during the first weeks of aging (Van Slyke and Price 1952; Turner and Thomas 1980). Very small amounts of lactose and galactose may be found in cheese months after manufacture. (Huffman and Kristoffersen 1984; Turner and Thomas 1980; Harvey *et al.* 1981; Thomas and Pearce 1981). Turner and Thomas (1980) showed that the fermentation of residual lactose in Cheddar cheese is affected by the storage temperature, the salt level in the cheese and the salt tolerance of the starter used.

The lactic streptococci used in cheese manufacture produce only the L(+) isomer of lactic acid (Lawrence *et al.* 1976). However, ripened cheeses contain both D(−) and L(+) lactate isomers (Turner and Thomas 1980). Nonstarter bacteria (pediococci and lactobacilli) form D(−) lactate from residual lactose or by conversion of L(+) lactate (Thomas and Crow 1983).

S. thermophilus metabolizes lactose to L(+) lactic acid but utilizes only the glucose moiety of lactose, leaving the galactose moiety in the cheese (Tinson *et al.* 1982). In Swiss cheese manufacture, *S. thermophilus* metabolizes the lactose and *L. helveticus* metabolizes the galactose to D(−) and L(+) lactic acid (Turner *et al.* 1983). The L(+) lactate isomer is preferentially utilized by propionibacteria to form acetic and propionic acids, which are essential for the development of the characteristic flavor in Swiss cheese (Langsrud and Reinbold 1973).

Lipolysis is one of the major biochemical reactions occurring during the ripening of blue-veined cheeses (Coghill 1979), Camembert (Schwartz and Parks 1963), and several Italian-type cheeses (Woo and Lindsay 1984). The flavor of mold-ripened cheeses is largely due to the accumulation of fatty acids and the subsequent formation of methyl ketones from fatty acids by β-oxidation and decarboxylation (Kinsella and Hwang 1976; Karahadian *et al.* 1985). Lipases of *Penicillium caseicolum* (Camembert) and *P. roqueforti* (blue, Roquefort) are mainly responsible for the release of free fatty acids, while in Italian cheese varieties, added lipases from animal sources (pregastric esterases) hydrolyze the triglycerides. Rennet pastes often used in the manufacture of Italian-type cheeses contain chymosin and other proteases, as well as lipases.

Breakdown of milk fat probably occurs in all cheeses, but the rate and extent of hydrolysis varies considerably between cheese varieties

(Shahani 1971). There is still considerable debate over the contribution of fat and its breakdown products to flavor in Cheddar cheese (Law 1984; Aston and Dulley 1982).

Hydrolytic rancidity flavor defects in Swiss, brick, and Cheddar cheeses have been linked to high concentrations of individual short chain free fatty acids (Woo *et al.* 1984). Lipases from psychrotrophic bacteria have been implicated in causing rancidity in cheese (Cousin 1982; Kuzdzal-Savoie 1980), although most starter streptococci and lactobacilli isolated from cheese are also capable of hydrolyzing milk fat (Paulsen *et al.* 1980; Umemoto and Sato 1975). Growth of *Clostridium tyrobutyricum* in Swiss cheese causes the release of butyric acid and subsequent rancid-off flavors (Langsrud and Reinbold 1974). The endogenous lipoprotein lipase is also responsible for hydrolytic rancidity in nonpasteurized milk.

REFERENCES

Adda, J., Gripon, J. C. and Vassal, L. 1982. The chemistry of flavour and texture generation in cheese. *Food Chem. 9,* 115–129.

Ali, A. E., Andrews, A. T. and Cheeseman, G. C. 1980A. Influence of storage of milk on casein distribution between the micellar and soluble phase and its relationship to cheesemaking parameters. *J. Dairy Res. 47,* 371–382.

Ali, A. E., Andrews, A. T. and Cheeseman, G. C. 1980B. Factors influencing casein distribution in cold-stored milk and their effects on cheesemaking parameters. *J. Dairy Res. 47,* 383–391.

Ali, A. E., Andrews, A. T. and Cheeseman, G. C. 1980C. Influence of elevated somatic cell count on casein distribution and cheesemaking. *J. Dairy Res. 47,* 393–400.

Andrews, A. T. 1983A. Proteinases in normal bovine milk and their action on caseins. *J. Dairy Res. 50,* 45–55.

Andrews, A. T. 1983B. Breakdown of caseins by proteinases in bovine milks with high somatic cell counts arising from mastitis or infusion with bacterial endotoxin. *J. Dairy Res. 50,* 57–66.

Aston, J. W. and Dulley, J. R. 1982. Cheddar cheese flavor. *Aust. J. Dairy Technol. 37,* 59–64.

Babel, F. J. 1977. Antibiosis by lactic culture bacteria. *J. Dairy Sci. 60,* 815–820.

Baer, A., Oroz, M. and Blanc, B. 1976. Serological studies on heat-induced interactions of α-lactalbumin and milk proteins. *J. Dairy Res. 43,* 419–432.

Banks, J. M., Muir, D. D. and Tamime, A. Y. 1984. Equations for estimation of the efficiency of Cheddar cheese production. *Dairy Industries Int. 49*(4), 14–17.

Barbano, D. M. 1984. Mozzarella cheese composition, yield, and how composition control influences profitability. Paper No. 1984-1. 21st Annual Marschall Invitational Italian Cheese Seminar. Marschall Products, Madison, Wisc.

Bley, M. E., Johnson, M. E. and Olson, N. F. 1985. Factors affecting nonenzymatic browning of process cheese. *J. Dairy Sci. 68,* 555–561.

Chapman, H. R., Bines, V. E., Glover, F. A. and Skudder, P. J. 1974. Use of milk concentrated by ultrafiltration for making hard cheese, soft cheese and yoghurt. *J. Soc. Dairy Technol. 27,* 151–155.

Chopin, M. C. 1980. Resistance of 17 mesophilic lactic *Streptococcus* bacteriophages to pasteurization and spray-drying. *J. Dairy Res. 47,* 131–139.

Coghill, D. 1979. The ripening of blue-vein cheese: A review. *Aust. J. Dairy Technol. 34,* 72–75.

Cousin, M. A. 1982. Presence and activity of psychrotrophic microorganisms in milk and dairy products: A review. *J. Food Protection 45,* 172–207.

Cousin, M. A. and Marth, E. H. 1977. Cheddar cheese made from milk that was precultured with psychrotrophic bacteria. *J. Dairy Sci. 60,* 1048–1056.

Creamer, L. K. and Olson, N. F. 1982. Rheological evaluation of maturing Cheddar cheese. *J. Food Sci. 47,* 631–636, 646.

Creamer, L. K. and Richardson, B. C. 1974. Identification of the primary degradation product of α_{s1}-casein in Cheddar cheese. *N.Z. J. Dairy Sci. Technol. 9,* 9–13.

Elfagm, A. A. and Wheelock, J. V. 1977. Effect of heat on α-lactalbumin and β-lactoglobulin. *J. Dairy Res. 44,* 367–371.

Emmons, D. B., Kalab, M., Larmond, E. and Lowrie, R. J. 1980. Milk gel structure. X. Texture and microstructure in Cheddar cheese made from whole milk and from homogenized low-fat milk. *J. Texture Studies 11,* 15–34.

Emmons, D. B. and Tuckey, S. L. 1967. *Cottage Cheese and Other Cultured Milk Products.* Pfizer Cheese Monographs, Vol. III. Pfizer, Inc., New York.

Ernstrom, C. A. and Wong, N. P. 1974. Milk clotting enzymes and cheese chemistry. *In: Fundamentals of Dairy Chemistry.* B. H. Webb, A. H. Johnson and J. A. Alford (Editors). AVI Publishing Co., Westport, Conn., pp. 662–753.

Everson, T. 1984. Concerns and problems of processing and manufacturing in super plants. *J. Dairy Sci. 67,* 2095–2099.

Food and Drug Administration, Department of Health, Education, and Welfare. 1984. Code of Federal Regulations Title 21. U.S. Government Printing Office, Washington, D.C.

Forss, D. A. 1979. Review of the progress of dairy science: Mechanisms of formation of aroma compounds in milk and milk products. *J. Dairy Res. 46,* 691–706.

Fox, P. F., and Walley, B. F. 1971. Influence of sodium chloride on the proteolysis of casein by rennet and by pepsin. *J. Dairy Res. 38,* 165–170.

Geurts, T. J., Walstra, P. and Mulder, H. 1980. Transport of salt and water during salting of cheese. 2. Quantities of salt taken up and moisture lost. *Neth. Milk Dairy J. 34,* 229–254.

Glaser, J., Carroad, P. A. and Dunkley, W. L. 1980. Electron microscopic studies of casein micelles and curd microstructure in cottage cheese. *J. Dairy Sci. 63,* 37–48.

Grappin, R., Rank, T. C. and Olson, N. F. 1985. Primary proteolysis of cheese proteins during ripening. A review. *J. Dairy Sci. 68,* 531–540.

Green, M. L. 1984. Milk coagulation and the development of cheese texture. *In: Advances in the Microbiology and Biochemistry of Cheese and Fermented Milk.* F. L. Davies and B. A. Law (Editors). Elsevier Applied Science Publishers, LD., London, pp. 1–33.

Green, M. L., Glover, F. A., Scurlock, E. M. W., Marshall, R. J. and Hatfield, D. S. 1981. Effect of use of milk concentrated by ultrafiltration on the manufacture and ripening of Cheddar cheese. *J. Dairy Res. 48,* 333–341.

Green, M. L. and Manning, D. J. 1982. Development of texture and flavor in cheese and other fermented products. *J. Dairy Res. 49,* 737–748.

Green, M. L., Marshall, R. J. and Glover, F. A. 1983. Influence of homogenization of concentrated milks on the structure and properties of rennet curds. *J. Dairy Res. 50,* 341–348.

Green, M. L., Scott, K. J., Anderson, M., Griffen, M. C. A. and Glover, F. A. 1984. Chemical characterization of milk concentrated by ultrafiltration. *J. Dairy Res. 51,* 267–278.

Guinee, T. P. and Fox, P. F. 1984. Studies on Romano-type cheese: General proteolysis. *Irish J. Food Sci. Technol. 8*, 105–114.

Harper, W. J. 1976. Processing induced changes. *In: Dairy Technology and Engineering.* W. J. Harper and C. W. Hall (Editors). AVI Publishing Co., Westport, Conn., pp. 539–596.

Harvey, C. D., Jenness, R. and Morris, H. A. 1981. Gas chromatographic quantitation of sugars and nonvolatile water-soluble organic acids in commercial Cheddar cheese. *J. Dairy Sci. 64*, 1648–1654.

Hill, A. R., Bullock, D. H. and Irvine, D. M. 1985. Composition of cheese whey. Effect of pH and temperature at dipping. *Can. Inst. Food Sci. Technol. J. 18*, 53–57.

Huffman, L. M. and Kristoffersen, T. 1984. Role of lactose in Cheddar cheese manufacturing and ripening. *N.Z. J. Dairy Sci. Technol. 19*, 151–162.

Humbert, G. and Alais, C. 1979. Review of the progress of dairy science: The milk proteinase system. *J. Dairy Res. 46*, 559–571.

Jellema, A. 1975. Note on susceptibility of bovine milk to lipolysis. *Neth. Milk Dairy J. 29*, 145–152.

Johnson, M. E. 1984. Methods of standardizing milk for cheesemaking. Paper No. 1984-2. 21st Annual Marschall Invitational Italian Cheese Seminar. Marschall Products, Madison, Wisc.

Johnson, M. E. and Olson, N. F. 1985. Nonenzymatic browning of Mozzarella cheese. *J. Dairy Sci. 68*, 3143–3147.

Jurczak, M. E. and Sciubisz, A. 1981. Studies on the lipolytic changes in milk from cows with mastitis. *Milchwissenschaft 36*, 217–219.

Kalab, M. 1979. Microstructure of dairy foods. 1. Milk products based on protein. *J. Dairy Sci. 62*, 1352–1364.

Karahadian, C., Josephson, D. B. and Lindsay, R. C. 1985. Contribution of *Penicillium* sp. to the flavors of Brie and Camembert cheese. *J. Dairy Sci. 68*, 1865–1877.

Kimber, A. M., Broker, B. E., Hobbs, D. G. and Prentice, J. H. 1974. Electron microscope studies of the development of structure in Cheddar cheese. *J. Dairy Res. 41*, 389–396.

Kinsella, J. E. and Hwang, D. 1976. Biosynthesis of flavors by *Penicillium roqueforti. Biotechnol. Bioeng. 18*, 927–938.

Kosikowski, F. 1977. *Cheese and Fermented Milk Foods.* F. V. Kosikowski and Associates, Brooktondale, N.Y.

Kristoffersen, T. 1985. Development of flavor in cheese. *Milchwissensch 40*, 197–199.

Kuzdzal-Savoie, S. 1980. Determination of free fatty acids in milk and milk products. *In: Flavor Impairment of Milk and Milk Products due to Lipolysis.* J. H. Moore (Editor). Int. Dairy Fed. Annu. Bull. Doc. No. 118.

Langsrud, T. and Reinbold, G. W. 1973. Flavor development and microbiology of Swiss cheese—a review. III. Ripening and flavor production. *J. Milk Food Technol. 36*, 593–609.

Langsrud, T. and Reinbold, G. W. 1974. Flavor development and microbiology of Swiss cheese—a review. IV. Defects. *J. Milk Food Technol. 37*, 26–41.

Law, B. A. 1979. Reviews of the progress of dairy science: Enzymes of psychrotrophic bacteria and their effects on milk and milk products. *J. Dairy Res. 46*, 573–588.

Law, B. A. 1981. The formation of aroma and flavor compounds in fermented dairy products. *Dairy Sci. Abstr. 43*, 143–154.

Law, B. A. 1984. Flavour development in cheeses. *In: Advances in the Microbiology and Biochemistry of Cheese and Fermented Milk.* F. L. Davies and B. A. Law (Editors). Elsevier Applied Science Publishers LD., London, pp. 187–208.

Law, B. A., Andrews, A. T., Cliffe, A. J. Sharpe, M. E. and Chapman, H. R. 1979. Effect of proteolytic raw milk psychrotrophs on Cheddar cheesemaking with stored milk. *J. Dairy Res. 46*, 497–509.

Lawrence, A. J. 1959. Syneresis of rennet curd. Part II. Effect of stirring and of the volume of whey. *Aust. J. Dairy Technol. 14*, 169–172.

Lawrence, R. C. and Gilles, J. 1980. The assessment of the potential quality of young Cheddar cheese. *N.Z. J. Dairy Sci. Technol. 15*, 1–12.

Lawrence, R. C. and Gilles, J. 1982. Factors that determine the pH of young Cheddar cheese. *N.Z. J. Dairy Sci. Technol. 17*, 1–14.

Lawrence, R. C., Gilles, J. and Creamer, L. K. 1983. The relationship between cheese texture and flavor. *N.Z. J. Dairy Sci. Technol. 18*, 175–190.

Lawrence, R. C., Heap, H. A. and Gilles, J. 1984. A controlled approach to cheese technology. *J. Dairy Sci. 67*, 1632–1645.

Lawrence, R. C., Thomas, T. D. and Terzaghi, B. E. 1976. Reviews of the progress of dairy science: Cheese starters. *J. Dairy Res. 43*, 141–193.

LeGraet, Y., Lepienne, A., Brûlé, G. and Ducruet, P. 1983. Migration du calcium et des phosphates inorganiques dans les fromages a pâte molle de type Camembert au cours de l'affinage. *Le Lait 63*, 317–332.

Lelievre, J., Freese, O. J. and Gilles, J. 1983. Prediction of Cheddar cheese yield. *N.Z. J. Dairy Sci. Technol. 18*, 169–172.

Matheson, A. R. 1981. The immunological determination of chymosin activity in cheese. *N.Z. J. Dairy Sci. Technol. 16*, 33–41.

Maxcy, R. B., Price, W. V. and Irvine, O. M. 1955. Improving curd-forming properties of homogenized milk. *J. Dairy Sci. 38*, 80–86.

Mills, O. E. and Thomas, T. D. 1980. Bitterness development in Cheddar cheese: Effect of level of starter proteinase. *N.Z. J. Dairy Sci. Technol. 15*, 131–141.

Mitchell, G. E. 1981. The production of selected compounds in a Swiss-type cheese and their contribution to cheese and flavor. *Aust. J. Dairy Technol. 36*, 21–25.

Morris, H. A. 1981. *Blue-Veined Cheeses*. Pfizer Cheese Monographs, Vol. VII. Pfizer, Inc., New York.

Morris, H. A., Guinee, T. P. and Fox, P. F. 1985. Salt diffusion in Cheddar cheese. *J. Dairy Sci. 68*, 1851–1858.

Noomen, A. 1978. Activity of proteolytic enzymes in simulated soft cheeses (Meschanger type). 1. Activity of milk protease. *Neth. Milk Dairy J. 32*, 26–48.

Noomen, A. 1983. The role of surface flora in the softening of cheeses with a low initial pH. *Neth. Milk Dairy J. 37*, 229–232.

O'Keefe, A. M., Fox, P. F. and Daly, C. 1978. Proteolysis in Cheddar cheese: Role of coagulant and starter bacteria. *J. Dairy Res. 45*, 465–477.

Olson, N. F. 1969. *Ripened Semisoft Cheeses*. Pfizer Cheese Monographs, Vol. IV. Pfizer, Inc., New York.

Olson, N. F. 1979. Cheese. *In: Microbial Technology*. H. J. Peppler and D. Perlman (Editors). Academic press, New York, pp. 39–77.

Patel, M. C., Lund, D. B. and Olson, N. F. 1972. Factors affecting syneresis of renneted milk gels. *J. Dairy Sci. 55*, 913–918.

Paulsen, P. V., Kowalewska, J., Hammond, E. G. and Glatz, B. A. 1980. Role of microflora in production of free fatty acids and flavor in Swiss cheese. *J. Dairy Sci. 63*, 912–918.

Pearce, K. N. and Gilles, J. 1979. Composition and grade of Cheddar cheese manufactured over three seasons. *N.Z. J. Dairy Sci. Technol. 14*, 63–71.

Peters, I. I. 1964. Homogenized milk in cheesemaking. Review Article 125. *Dairy Sci. Abst. 26*, 457–461.

Pierre, A. and Brule, G. 1981. Mineral and protein equilibria between the colloidal and soluble phases of milk at low temperature. *J. Dairy Res. 48*, 417–428.

Rank, T. C., Grappin, R. and Olson, N. F. 1985. Secondary proteolysis of cheese during ripening: A review. *J. Dairy Sci. 68*, 801–805.

Reinbold, G. W. 1963. *Italian Cheese Varieties*, Vol. I. Pfizer, Inc., New York.

Reinbold, G. W. 1972. *Swiss Cheese Varieties*. Pfizer Cheese Monographs, Vol. V. Pfizer, Inc., New York.

Richardson, B. C. and Pearce, K. N. 1981. The determination of plasmin in dairy products. *N.Z. J. Dairy Sci. Technol. 16*, 209–220.

Schwartz, D. P. and Parks, O. N. 1963. Methyl ketones in Camembert cheese. *J. Dairy Sci. 46*, 1136.

Sellars, R. L. and Babel, F. J. 1970. *Cultures for the Manufacture of Dairy Products*. Chr. Hansen's Laboratory, Milwaukee, Wisc.

Shahani, K. M. 1971. Lipases and flavor development. Paper 1971-8 in Proceedings of the 8th Annual Marschall Invitational Italian Cheese Seminar, Madison, Wisc. May 10–11. Marschall Products, Miles Laboratories, Madison, Wisc.

Sharpe, M. E. and Franklin, J. G. 1962. Production of hydrogen sulphide by lactobacilli with special reference to strains isolated from Cheddar cheese. *VIII Int. Cong. Microbiol.* B11.3.

Smits, P. and Van Brouwershaven, J. H. 1980. Heat-induced association of β-lactoglobulin and casein micelles. *J. Dairy Res. 47*, 313–325.

Stadhouders, J., Hup, G., Exterkate, F. A. and Visser, S. 1983. Bitter flavor defect in cheese. *Neth. Milk Dairy J. 37*, 157–167.

Thomas, T. D. and Crow, V. L. 1983. Mechanism of D(-)-lactic acid formation in Cheddar cheese. *N.Z. J. Dairy Sci. Technol. 18*, 131–141.

Thomas, T. D. and Pearce, K. N. 1981. Influence of salt on lactose fermentation and proteolysis in Cheddar cheese. *N.Z. J. Dairy Sci. Technol. 16*, 253–259.

Tinson, W., Hillier, A. J. and Jago, G. R. 1982. Metabolism of *Streptococcus thermophilus* 1. Utilization of lactose, glucose and galactose. *Aust. J. Dairy Technol. 37*, 8–13.

Trieu-Cuot, P. and Gripon, J. C. 1982. A study of proteolysis during Camembert cheese ripening using isoelectric focusing and two-dimensional electrophoresis. *J. Dairy Res. 49*, 501–510.

Turner, K. W., Morris, H. A. and Martley, F. G. 1983. Swiss-type Cheese II. The role of thermoduric lactobacilli in sugar fermentation. *N.Z. J. Dairy Sci. Technol. 18*, 117–123.

Turner, K. W. and Thomas, T. D. 1980. Lactose fermentation in Cheddar cheese and the effect of salt. *N.Z. J. Dairy Sci. Technol. 15*, 265–276.

Umemoto, Y. and Sato, Y. 1975. Relation of Cheddar cheese ripening to bacterial lipolysis. *Agr. Biol. Chem. 39*, 2115–2122.

Ustunol, Z. and Brown, R. J. 1985. Effects of heat treatment and posttreatment holding time on rennet clotting of milk. *J. Dairy Sci. 68*, 526–530.

Van den Berg, G., Hup, G., Stadhouders, J. and de Vries, E. 1980. Rapport R112. Application of the "Bactotherm" process (self-desludging bactofuge, Type MRPX 314 SGV, in combination with bactofugate sterilizer) in the manufacture of Gouda cheese. Technological effects on cheese manufacture and methods of controlling butyric acid fermentation. Bedrijven van het Nederlands Institute voor Zvivelonderzoek.

Van Slyke, L. L. and Price, W. V. 1952. Changes during the ripening of Cheddar cheese. *In: Cheese*, 2nd ed. Orange Judd Pub. Co., New York, pp. 318–333.

Visser, S., Hup, G., Exterkate, F. A. and Stadhouders, J. 1983. Bitter flavor in cheese. 2. Model studies on the formation and degradation of bitter peptides by proteolytic enzymes from calf rennet, starter cells and starter cell fractions. *Neth. Milk Dairy J. 37*, 169–180.

Visser, S., Slangen, K. J., Hup, G. and Stadhouders, J. 1983. Bitter flavor in cheese. 3. Comparative gel-chromatographic analysis of hydrophobic peptide fractions from twelve Gouda type cheeses and identification of bitter peptides isolated from a cheese made with *Streptococcus cremoris* strain H.P. *Neth. Milk Dairy J. 37*, 181–192.

Walstra, P. and Jenness, R. 1984. *Dairy Chemistry and Physics.* John Wiley and Sons, New York.

Walstra, P., van Dijk, J. M. and Geurts, T. J. 1985. The syneresis of curd. 1. General considerations and literature review. *Neth. Milk Dairy J. 39*, 209–246.

Wilster, G. H. 1980. *Practical Cheesemaking.* Oregon State University Book Stores, Inc., Corvalis, Ore.

Woo, A. H., Kollodge, S. and Lindsay, R. C. 1984. Quantification of major free fatty acids in several cheese varieties. *J. Dairy Sci. 67*, 874–878.

Woo, A. H. and Lindsay, R. C. 1984. Concentrations of major free fatty acids and flavor development in Italian cheese varieties. *J. Dairy Sci. 67*, 960–968.

13

Fermentations

Joseph F. Frank and Elmer H. Marth

INTRODUCTION

"Fermentation" is a term that has been used to refer to various processes involving limited biochemical changes brought about by microorganisms or their enzymes. Prescott and Dunn (1957) reviewed the changes in meaning which the term has undergone since its derivation from the Latin word for "boil," which was used to describe the fermentation of wine. Milk fermentation can be defined as any modification of the chemical or physical properties of milk or dairy products resulting from the activity of microorganisms or their enzymes. This activity can involve metabolizing cells, extracellular enzymes, or intracellular enzymes released after cell lysis. Milk fermentations contribute to desirable flavors and textures in products such as cheese and yogurt or result in spoiled and degraded products. To ensure development of desired fermentations, microbial cultures with known properties are added to milk or dairy product substrates. Fermentations initiated by natural milk contaminants are often inconsistent and consequently are undesirable for industrial purposes. The unique organoleptic properties of fermented dairy products result from the highly specific metabolic activity of starter culture bacteria in converting lactose to lactic acid and from the curd-forming properties of the casein–micelle complex.

MILK AS A FERMENTATION MEDIUM

Nutritional Properties of Milk

Milk is a suitable growth medium for many microorganisms because of the variety of substrates available for fermentation (lactose, fat, and various proteins), as well as the presence of growth stimulants such as vitamins and minerals. Growth of some microorganisms in milk is limited by their inability to use lactose or milk proteins or by a high iron requirement. Addition of glucose and yeast extract to milk stimulates the growth of many microorganisms.

Fermentable Substrates. Lactose is the only carbohydrate existing in sufficient quantity in milk to support microbial growth. The microorganisms which grow most rapidly in milk are those which can ferment this sugar. Some microorganisms can also satisfy their energy requirements through fermentation of amino acids derived from milk proteins or fermentation of fatty acids produced by hydrolysis of milk fat. Citric acid is also an important fermentable compound in milk, although lactic acid bacteria obtain no energy from its use. However, citric acid fermentation does result in formation of various flavor compounds.

Nitrogen Availability. Sources of nitrogen in milk include proteins, proteoses, peptones, peptides, amino acids, urea, ammonia, and various other nonprotein compounds. A significant amount of the total nitrogen in milk occurs in forms which are readily available for microbial metabolism. These include amino acid N (0.4%), urea N (1.2%), and ammonia N (0.6%), which together represent 2.0% of the total nitrogen in milk (Miller and Kandler 1967). Although these compounds, together with proteoses and peptones, are present in sufficient quantity for initiation of microbial growth, this quantity is insufficient for sustained growth. When rapid fermentation is desired, as is true when lactic starter cultures are used, cultures with proteolytic activity are usually used or the milk is supplemented with a protein hydrolysate (Speck 1962). Nonproteolytic microorganisms may be useful in milk fermentations if they are inoculated in sufficiently high numbers so that growth is not necessary to obtain sufficient activity, or if they are combined with a proteolytic culture. Adding yeast extract to milk stimulates growth of lactic acid acid bacteria, mainly because of the free amino acids but also because of the presence of purine ribosides, nucleosides, nucleotides, and mineral components (Lawrence *et al.* 1976; Selby Smith *et al.* 1975).

Vitamins and Minerals. Milk is a rich source of vitamins and other organic substances that stimulate microbial growth. Niacin, biotin, and pantothenic acid are required for growth by lactic streptococci (Reiter and Oram 1962). Thus the presence of an ample quantity of B-complex vitamins makes milk an excellent growth medium for these and other lactic acid bacteria. Milk is also a good source of orotic acid, a metabolic precursor of the pyrimidines required for nucleic acid synthesis. Fermentation can either increase or decrease the vitamin content of milk products (Deeth and Tamime 1981; Reddy *et al.* 1976). The folic acid and vitamin B_{12} content of cultured milk depends on the species and strain of culture used and the incubation conditions (Rao *et al.* 1984). When mixed cultures are used, excretion of B-complex vitamins by some species can stimulate growth of others (Lawrence *et al.* 1976).

The role of minerals in milk as microbial nutrients is often overlooked. Minerals such as magnesium, calcium, manganese, zinc, iron, silicon, potassium, cobalt, copper, and molybdenum are important in microbial fermentations (Weinberg 1977). These minerals, as well as several others, are present in milk. However, supplementation of milk with various trace elements may be necessary to obtain maximum microbial growth or maximum production of secondary metabolites. Trace minerals in milk are not necessarily available for microbial growth. Both cow's milk and human milk contain small amounts of iron (<1.5 μg/ml). However, human milk contains a much larger quantity of the iron-binding protein lactoferrin, making the unavailability of iron an important growth inhibitor (Masson and Heremans 1971). Since the major function of iron in microbial metabolism involves oxygen use and electron transport, bacterial species which rely on aerobic metabolism for maximum growth rates are likely to be affected by the lack of available iron in milk (Byers and Arveneaux 1971). Garibaldi (1971) reported that the iron requirement of a fluorescent pseudomonad decreases as the incubation temperature is lowered. Whether iron availability is a limiting factor in the rate of low-temperature milk spoilage has not been determined. Addition of trace amounts of iron, as well as of magnesium, molybdenum, and selenium, to milk stimulates acid production by lactic streptococci (Olson and Qutub 1970).

Other minerals in milk which may be present in growth-limiting quantities are manganese, cobalt, and zinc. Addition of manganese to milk stimulates growth of *Leuconostoc cremoris* (Anderson and Leesment 1970). Supplementation of milk with trace amounts of cobalt is necessary to achieve maximum microbial synthesis of vitamin B_{12}. Supplementation of cheese whey with copper, manganese, zinc, iron, and molybdenum increases yeast mass production (Bayer 1983). Increased concentrations of certain minerals in milk can also inhibit microbial growth. Several strains of microorganisms used in starter cultures for Swiss cheese are sensitive to increased levels of copper, which can result from the use of copper vats (Maurer *et al.* 1975), and to increased levels of cadmium, which may be present from environmental contamination (Korkeala *et al.* 1984). Olson and Qutub (1970) observed that addition of 4 ppm or more of copper, iodine, or mercury to milk inhibited acid production by lactic streptococci.

Microbial Inhibitors in Milk

Microbial inhibitors in raw milk include lactoferrin, lysozyme, the lactoperoxidase/thiocyanate/hydrogen peroxide system, specific immunoglobulins, folate- and vitamin B_{12}-binding systems, and others. A detailed discussion of various antimicrobial factors in milk is presented by Reiter (1978) who noted that concentrations of microbial inhibitors in milk from different species vary considerably. The bacteriostatic nature of raw cow's milk decreases with storage and does not appear to be significant after heat treatment.

Effect of Heat Treatment

Heat treatment of milk changes its characteristics as a culture medium. Heating milk at 62° to 72°C for 30 to 40 min stimulates growth of starter cultures, as does heating milk at 90°C for 60 to 180 min and autoclaving at 120°C for 15 to 30 min. Heating milk at 72°C for 45 min, at 82°C for 10 to 45 min, or at 90°C for 1 to 45 min inhibits growth (Greene and Jezeski 1957A). The stimulatory effect of mild heat treatments results from expulsion of oxygen, with subsequent lowering of the oxidation-reduction potential, destruction of inhibitors, partial protein hydrolysis, and serum protein denaturation. Inhibitory heat treatments are associated with formation of cysteine and toxic volatile sulfides. The stimulatory effect of additional heating results from lowered volatile sulfide concentrations as well as additional protein hydrolysis (Greene and Jezeski 1957B).

FERMENTATION OF LACTOSE

Lactose, the major substrate for microbial fermentation in milk, is a disaccharide constituting about 40% of the solids of whole milk. Milk is the only significant source of lactose in the environment, though lactose can appear in small amounts in blood and urine. The fact that a great variety of microorganisms is able to utilize lactose may be of evolutionary significance, since there are few naturally occurring ecological niches where lactose exists at fermentable concentrations. The gastrointestinal tract of warm-blooded animals and the streak canal of the mammary gland are the two natural environments in which the ability to ferment lactose is advantageous.

Most microorganisms which ferment lactose use one of two strategies to initiate fermentation: either hydrolysis catalyzed by β-D-galactosidase (β-D-galactoside galactohydrolase, EC 3.2.1.23) also known as "lactase," or hydrolysis of phosphorylated lactose by β-D-phosphogalactosidase (β-D-phosphogalactoside galactohydrolase). A few microorganisms, such as some *Pseudomonas* spp., *Bacterium anitratum,* and

Penicillium chrysogenum, oxidize lactose to lactobionic acid (Wallenfels and Mulhotra 1961).

Distribution and Properties of β-D-Galactosidase

β-D-Galactosidase catalyzes hydrolysis of the 1,4-β-galactosidic bond of lactose to produce glucose and galactose. β-Galactosidase is found in various plants, animals, and microorganisms. The general properties and distribution of this enzyme have been reviewed by Pomeranz (1964) and by Wallenfels and Mulhotra (1961). Examples of microorganisms which produce a cell-bound β-D-galactosidase are *Escherichia coli* and other enteric bacilli (Lederberg 1950), *Streptococcus thermophilus* (Somkuti and Steinberg 1979B), *S. lactis* (McFeters *et al.* 1967), *Lactobacillus bulgaricus* (Itoh *et al.* 1980), *L. plantarum* (Hasan and Durr 1974), *Aeromonas formicans* (Rohlfing and Crawford 1966), *Xanthomonas campestris* (Frank and Somkuti 1979), *Bacillus subtilis* (Anema 1964), a marine *Pseudomonas* sp. (Hidalgo *et al.* 1977), *Neurospora crassa* (Johnson and DeBusk 1970), and various *Aspergillus, Mucor, Kluyveromyces,* and *Candida* species (Wierzbicki and Kosikowski 1973A; Goncalves and Castillo 1982). Extracellular β-D-galactosidase is produced by *Aspergillus oryzae* (Park *et al.* 1979), *A. foetidus* (Borglum and Sternberg 1972), *A. niger,* and the thermophilic fungi, *Spicaria* and *Scopulariopsis* (Pastore and Park 1979). Preparations of various microbial β-galactosidases are commercially available for use in producing low-lactose dairy products or products of increased sweetness.

β-Galactosidase acts not only as a hydrolase but also as a transgalactosidylase; that is, it can transfer galactosidyl residues to various acceptors that contain a hydroxyl group. Possible acceptors include water, alcohols, glucose, galactose, lactose, and oligosaccharides. Theoretically, seven different aldodisaccharides can be formed from hydrolyzed lactose. Toba and Adachi (1978) isolated five of these disaccharide transfer products of *Saccharomyces (Kluyveromyces) fragilis* and *A. niger* β-galactosidase. As many as 12 oligosaccharides can result from the action of β-galactosidase on lactose (Toba and Adachi 1978). The continued activity of β-galactosidase results in hydrolysis of these oligosaccharides (Burvall *et al.* 1979). Since the transgalactosylation reaction is competitive, transfer products formed vary according to the concentration of water, glucose, galactose, lactose, other disaccharides, and oligosaccharides (Roberts and Pettinati 1957; Wierzbicki and Kosikowski 1973B). Since water is by far the predominant hydroxyl compound in most reaction systems, formation of a significant amount of oligosaccharides indicates that the enzyme has a relatively high affinity for mono- and disaccharides. Attempts to hydrolyze lactose in concentrated milk or whey products could result in conversion of 5 to 13% of the total sugar to oligosaccharides (Burvall et al. 1979). It is possible that consumption of oligosaccharides can cause digestive problems.

The Phosphoenolpyruvate: Phosphotransferase System

Many microorganisms which ferment lactose transport it into the cell in the form of lactose-P (glucosyl-β-(1-4)-galactoside-6-phosphate), which is then hydrolyzed by the enzyme β-D-phosphogalactosidase (β-Pgal) to glucose and galactose-6-P. This transport mechanism requires phosphoenolpyruvate (PEP) as a source of phosphate, and thus it is called the "PEP-dependent phosphotransferase system (PEP:PTS)." The general characteristics of this system have been described in reviews by Roseman (1972, 1975) and Postma and Roseman (1976). Transport of lactose into a microbial cell via the PEP:PTS was first found in *Staphylococcus aureus* (Egan and Morse 1966). A similar system for transport of glucose was first reported for *Escherichia coli* (Kundig et al. 1964). A generalized representation of the PEP:PTS is presented in Figure 13.1. At least four proteins are required in this system. Enzyme I, which accepts the phosphate from PEP, and the HPr protein, which is phosphorylated by enzyme I, are both nonspecific (used in phosphorylation of all sugars). Proteins designated as II-A, II-B, and III are sugar-specific and inducible. The II-B proteins are membrane bound and found in all PEP:PTS systems. Either the membrane-bound II-A proteins or the soluble III proteins can complete the system. The proteins (II-A and III) are phosphorylated by phospho-HPr and in turn transfer this phosphate to the II-B protein, which binds to and phosphorylates the sugar. The phosphorylated sugar is then released into the cytoplasm.

Roseman (1969) and Dills et al. (1980) discussed the benefits to the bacterial cell of using the PEP:PTS for sugar transport. At least two major physiological advantages can be theorized: ease of regulation through control of the internal pool of PEP if the sugar is fermented by anaerobic glycolysis, and conservation of metabolic energy. Since high-energy phosphate in the form of adenosine triphosphate (ATP) or PEP is required for active transport of sugar across the bacterial membrane, conserving this phosphate by releasing the phosphorylated sugar into the cytoplasm saves the equivalent of one ATP (the ATP which would be used to initiate fermentation). Since anaerobic fermentation of lactose yields only 4 moles of ATP per mole of lactose, use of

Cytoplasm | Membrane | Growth Medium

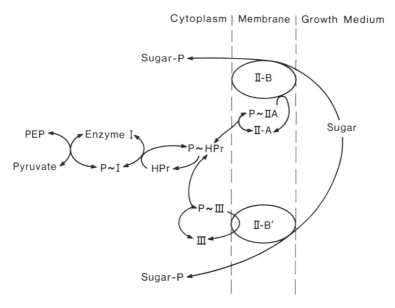

Figure 13.1 A generalized representation of the (PEP:PTS). (Adapted from Roseman 1975.)

one ATP equivalent for both transport and initiation of fermentation is extremely advantageous. Romano *et al.* (1970, 1979) and Kundig (1976) observed that the PEP:PTS for a specific sugar is found only in bacteria that use the Embden-Meyerhof-Parnas (EMP) pathway for fermentation of the specific sugar. The EMP pathway provides the PEP necessary for energizing this system, whereas, as noted by Dills *et al.* (1980), strict aerobes which use hexoses via the Entner-Duodoroff pathway have forms of energy other than PEP more readily available for use in sugar transport. Thus it is not advantageous for strict aerobes to use the PEP:PTS.

A PEP:PTS specific for lactose was first found in *S. aureus* (Johnson and McDonald 1974) and has since been found in *S. lactis* and *S. cremoris* (McKay *et al.* 1970), *Streptococcus mutans* (Calmes 1978), *S. salivarius* (Hamilton and Lo 1978), *S. thermophilus* (Hemme *et al.* 1980A), homofermentative lactobacilli (Premi *et al.* 1972) and a strain of *Klebsiella* (Hall 1979). Many lactic acid bacteria produce both β-galactosidase and β-Pgal enzymes, indicating the presence of both permease and phosphotransferase active transport systems. Of 13 *Lactobacillus* spp. studied by Premi *et al.* (1972), both lactose-hydrolyzing enzymes were found in 11. Only *Lactobacillus casei* exhibits just β-Pgal activity.

Chassy and Thompson (1983) demonstrated the presence of a lactose-specific PEP:PTS in *L. casei.* Most *Lactobacillus* spp. exhibit much greater β-galactosidase activity than β-Pgal activity. Of 31 strains of *S. thermophilus* surveyed, Somkuti and Steinberg (1979B) found that 29 produced β-galactosidase only, one produced β-Pgal only, and one produced both enzymes. Hemme *et al.* (1980B) observed both β-galactosidase and β-Pgal activity in all ten strains of *S. thermophilus* studied. The two strains of *S. thermophilus* studied by Farrow (1980) exhibited much greater β-galactosidase than β-Pgal activity.

The lactose-hydrolyzing enzymes found in some bacterial strains correspond to the source of isolation. Strains of Group N streptococci isolated from nature (nondairy strains) generally produce both β-galactosidase and β-Pgal, with some atypical isolates, such as *S. lactis* ATCC 7962, producing very small amounts of β-Pgal (Farrow 1980; Okamoto and Morichi 1979; Farrow and Garvie 1979). However, strains isolated from dairy starter cultures produce only β-Pgal (Molskness *et al.* 1973; Farrow 1980). Since Group N streptococci use the EMP pathway, one would expect strains using the PEP:PTS with lactose-P hydrolyzed by β-Pgal to be the most rapid lactose fermenters. Studies by Farrow (1980), Cords and McKay (1970), and Crow and Thomas (1984) support this generalization. Rapid lactose fermentation is a highly desirable trait in dairy starter cultures. It is possible that past selection for this characteristic has produced lactic streptococci which are physiologically different from wild-type strains in that only the dairy cultures depend exclusively on the PEP:PTS for lactose transport (Farrow 1980). The ability of lactic streptococci to produce PEP:PTS proteins is plasmid-associated and therefore readily manipulated by environmental selection (McKay *et al.* 1976; Efstathiou and McKay 1976; McKay 1982, 1983).

Lactose Metabolism in Homofermentative Lactic Streptococci

The most important fermentative reaction used in dairy processing is the homofermentative conversion of lactose to lactic acid. The efficient manufacture of high-quality cultured products, including most cheese varieties, yogurt, and cultured buttermilk, requires a rapid and consistent rate of lactic acid production. Lactic acid helps to preserve, contributes to the flavor, and modifies the texture of these products. Nearly all starter cultures used to produce acidified dairy products contain one or more strains of lactic streptococci, because these organisms can produce the desired acidity without causing detrimental changes in flavor or texture. Strains of lactic streptococci can be classified as "fast" or "slow" acid producers when grown in milk. Research on lactose metabolism of "fast" or "dairy industry" strains of *S. lactis* and *S. cremoris* has been reviewed by Law and Sharpe (1978), Cogan (1980), and Lawrence *et al.* (1976). Lactose metabolism as it is thought to occur in starter culture strains of group N streptococci is summarized in Figure 13.2. This fermentation scheme also applies to *S. mutans*.

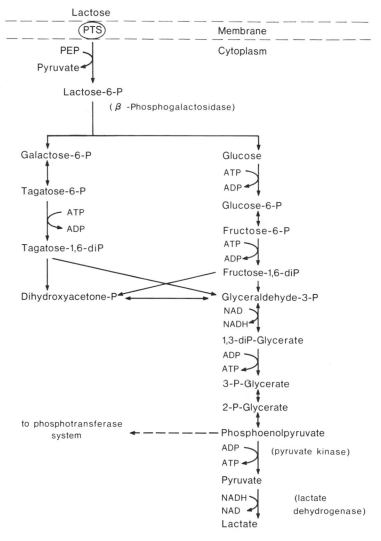

Figure 13.2 Metabolism of lactose as it is believed to occur in starter culture strains of Group N streptococci. (Adapted from Lawrence *et al.* 1976 and Thompson 1979.)

Two pathways are involved in fermentation of lactose to pyruvate: the EMP pathway for metabolism of glucose (anaerobic glycolysis) and the tagatose-6-P pathway for metabolism of galactose-6-P. The tagatose-6-P pathway was first found in *S. aureus* by Bissett and Anderson (1973, 1974A), who later also found it to be active in group N streptococci (1974B). Hamilton and Lebtag (1979) reported finding this pathway in *S. mutans*. Other microorganisms which metabolize lactose via the PEP:PTS probably also use the tagatose-6-P pathway to degrade galactose-6-P originating either from lactose-6-P hydrolysis or from the PEP:PTS transport of galactose into the cell (Park and McKay 1982). Metabolism of lactose using the PEP:PTS–EMP-tagatose-6-P system, as outlined in Figure 13.2, nets the organism four ATP molecules per molecule of lactose. Reduction of pyruvate to lactate is catalyzed by lactate dehydrogenase (LDH) and produces the nicotinamide adenine dinucleotide (NAD) necessary to achieve an NAD–reduced NAD (NAD–NADH) balance. Products other than lactate are produced when LDH activity is decreased.

Control of Lactose Metabolism

Discussion of the genetic regulation of lactose metabolism is beyond the scope of this chapter, but this subject has been reviewed by McKay (1982, 1983). The key enzymes involved in metabolic control of lactose metabolism in Group N streptococci are pyruvate kinase and LDH. The activity of pyruvate kinase controls the intracellular concentration of PEP, as well as ATP production. PEP availability, in turn, controls lactose transport. The pyruvate kinase of lactic streptococci is activated by glycolytic and tagatose-6-P pathway intermediate compounds up to 1,3-diP-glycerate (Thomas 1976A) and is inhibited by inorganic phosphate (Thompson and Torchia (1984). Dependence of pyruvate kinase activity on glycolytic intermediate compounds allows the cell to maintain PEP potential (PEP plus 3-P-glycerate and 2-P-glycerate) under starvation conditions (Thompson and Thomas 1977). Consequently, the energy necessary for sugar transport is immediately available to the cell. Low pyruvate kinase activity in starved cells also provides a continuous supply of ATP for cell maintenance (Thompson and Thomas 1977). When excess sugar is available, pyruvate kinase activity increases and intracellular PEP concentration decreases (Thomas 1976B; Yamada and Carlsson 1975B). Pyruvate kinase thus has the ability to synchronize the rate of ATP production with the rate of energy source uptake (Thompson 1978).

NAD-dependent lactate dehydrogenase (nLDH) activity controls formation of end products from pyruvate in lactic streptococci. Garvie (1980) reviewed the types of LDH, their properties, and their function. NAD-independent LDHs convert lactate to pyruvate by using an unknown hydrogen acceptor. Although many lactic acid bacteria, including group N streptococci, produce this enzyme, its physiological function is unknown (Anders *et al.* 1970B). nLDHs, which are found in many microorganisms, convert pyruvate to lactate and are highly active in lactic acid bacteria. The end products of fermentation in these bacteria are controlled by nLDH activity. Fructose 1,6-diP is a required activator for the nLDH of *L. casei, L. curvatus* (Hensel *et al.* 1977), *S. mutans* (Brown and Wittenberger 1972), *S. faecalis, S. lactis* (Crow and Pritchard 1977), and *S. cremoris* (Jonas *et al.* 1972). Tagatose 1,6-diP can also activate the nLDH of *S. lactis, S. cremoris, S. faecalis,* and some lactobacilli (Thomas 1975). When lactic streptococci are grown in continuous culture with limiting glucose (or lactose), they have low intracellular fructose diP and tagatose diP levels, and the nLDH is deactivated (Thomas *et al.* 1979; Yamada and Carlsson 1975A). Under these conditions, a shift from production of lactate to production of formate, acetate, and ethanol occurs. Such a shift does not occur in static culture. LDH is also inactive when streptococci capable of growth on ribose are grown on this substrate, since neither fructose diP nor tagatose diP is an intermediate compound in ribose degradation. Consequently, lactic acid bacteria, which are normally homofermentative when growing on glucose or lactose, can be heterfermentative when using other energy sources or when carbohydrate is limited.

Under anaerobic conditions, with excess glucose or lactose available, lactic streptococci produce over 95% lactic acid, because the nLDH is sufficiently active so that there is little excess pyruvate available for conversion to other end products (Thomas *et al.* 1979). When growth-limiting amounts of lactose or glucose are present, excess pyruvate is converted to formate through the action of pyruvate formate-lyase, resulting in production of ATP and ultimately ethanol and acetate. (Thomas *et al.* 1979). Pyruvate formate-lyase is inhibited by the trios phosphate intermediate compounds of glycolysis (Fordyce *et al.* 1984). A mutant of *S. lactis* deficient in LDH produces acetoin as a major end product (McKay and Baldwin 1974). These alternative pathways of pyruvate metabolism in homofermentative streptococci are outlined in Figure 13.3. In the presence of oxygen, formation of acetyl-CoA may result from pyruvate dehydrogenase rather than from pyruvate formate-lyase activity (Fordyce *et al.* 1984). Pyruvate dehydrogenase may be involved in generating the NADH required for oxygen use by

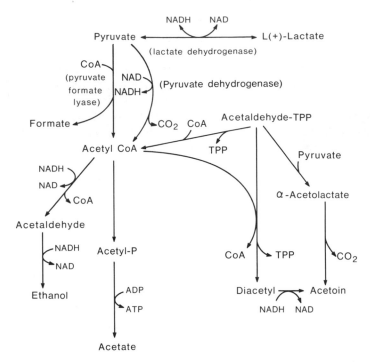

Figure 13.3 Alternative pathways of pyruvate metabolism by homofermentative lactic streptococci. CoA = coenzyme A; TPP = thiamine pyrophosphate. (Adapted from Thomas *et al.* 1979.)

NADH oxidase (Grufferty and Condon 1983). Not all lactic streptococci express these alternative pathways. However, these deficient strains grow poorly under conditions producing low nLDH activity.

Effect of pH on Fermentation End-Products

When grown in alkaline media, certain species of lactic acid bacteria decrease production of LDH, resulting in increased formation of formate, acetate, and ethanol as end products. This phenomenon has been observed in *S. faecalis* subsp. *liquefaciens* (Gunsalus and Niven 1942), *Streptococcus durans, S. thermophilus,* (Platt and Foster 1958), and *Lactobacillus bulgaricus* (Rhee and Pack 1980). Data of Rhee and Pack (1980) indicate that a phosphoroclastic split of pyruvate occurs under alkaline conditions to yield ATP. The enzymes involved in this reaction (pyruvate formate-lyase and acetate kinase) require alkaline conditions for optimum activity. A shift from homo- to heterofermentation because of increased pH has not been observed for Group N streptococci.

Lactose Metabolism in *Streptococcus thermophilus*

Strains of *S. thermophilus* generally use only the glucose portion of lactose. Galactose, which results from lactose hydrolysis via β-galacto-sidase, is excreted from the cell (Tinson *et al.* 1982). Consequently products cultured with *S. thermophilus* may contain residual free galactose, which is available for fermentation by other microorganisms to products other than lactic acid (Tinson *et al.* 1982). Galactose-fermenting variants of *S. thermophilus* have been isolated and may be useful in some fermentations (Thomas and Crow 1984).

Effect of Oxygen on Metabolism of Lactic Acid Bacteria

Lactose fermentation in lactic acid bacteria occurs anaerobically. However, presence of oxygen can significantly affect growth of these microorganisms. Observations of Keen (1972) indicate that in continuous culture, oxygen can either inhibit or stimulate growth of lactic streptococci. Small amounts of oxygen stimulate growth between pH 5.05 and 5.45 and have no effect at lower pH values. Between pH 5.45 and 6.35, small amounts of oxygen result in accumulation of toxic amounts of hydrogen peroxide. Oxygen also possesses some toxicity independent of pH. When exposed to oxygen, lactic streptococci (Anders *et al.* 1970A), *S. faecalis* (Dolin 1955), *L. plantarum* (Gotz *et al.* 1980), and *L. casei* (Walker and Kilgour 1965) all produce hydrogen peroxide through the action of NADH oxidase. This enzyme catalyzes the reaction

$$NADH + H^+ + O_2 \rightarrow NAD + H_2O_2$$

Another enzyme, NADH peroxidase, is usually present to remove at least some of the hydrogen peroxide from the growth medium through the reaction

$$NADH + H^+ + H_2O_2 \rightarrow NAD + 2H_2O$$

These reactions not only reduce the toxic effect of oxygen but may also supply the cell with NAD, so that the NAD–NADH balance of the cell can be maintained with low LDH activity (Gotz *et al.* 1980). Whether or not hydrogen peroxide accumulates in the growth medium under aerobic conditions depends on the relative activity of NADH oxidase and NADH peroxidase. Lactic streptococci with relatively low NADH peroxidase activity can produce enough hydrogen peroxide to inhibit their own growth (Anders *et al.* 1970A). *S lactis,* when grown aerobically on galactose, lactose, or maltose, produced autoinhibitory levels of hydrogen peroxide (Grufferty and Condon 1983). Such levels were not reached when glucose was the substrate.

Small amounts of hydrogen peroxide in raw milk can activate the lactoperoxidase-catalyzed oxidation of thiocyanate to produce a bacterial inhibitor (Hogg and Jago 1970). Inhibitory compounds resulting from oxygen metabolism can produce initially slow starter culture growth in industrial dairy fermentations if the milk has been excessively agitated.

The Leloir Pathway for Galactose Metabolism

Microorganisms which initiate lactose metabolism by hydrolysis with β-galactosidase usually metabolize the resulting glucose via the EMP pathway and galactose via the Leloir pathway. This includes galactose-fermenting variants of *S. thermophilus* (Thomas and Crow 1984). The Leloir pathway results in formation of glucose 6-P, which then enters the EMP pathway (Cardini and Leloir 1952; Maxwell *et al.* 1962). This pathway can be summarized as follows:*

(galactokinase)
galactose + ATP → galactose 1-P + ADP

(galactose 1-P uridyl transferase)
galactose 1-P + UDP-glucose ←→ glucose 1-P + UDP-galactose

(UDP-galactose-4-epimerase)
UDP-galactose ←→ UDP-glucose

(phosphoglucomutase)
glucose 1-P → glucose 6-P

Group N streptococci produce enzymes for both the Leloir and tagatose 6-P pathways (Bissett and Anderson 1974B) In these organisms, the Leloir pathway is functional during galactose use if galactose is transported via a permease system rather than a phosphotransferase system (Thomas *et al.* 1980). Since the permease system in *S. lactis* has a ten-fold greater affinity for galactose than the phosphotransferase system, when low concentrations of galactose are present, metabolism occurs primarily through the Leloir pathway (Thompson 1980). *S. cremoris* grows relatively slowly at low galactose concentrations and apparently transports galactose only via a phosphotransferase system. Various streptococci, including *S. lactis, S. cremoris,* and *S. pyogenes,* exhibit a heterolactic fermentation when growing on free galactose, re-

---

*UDP = uridine diphosphate.

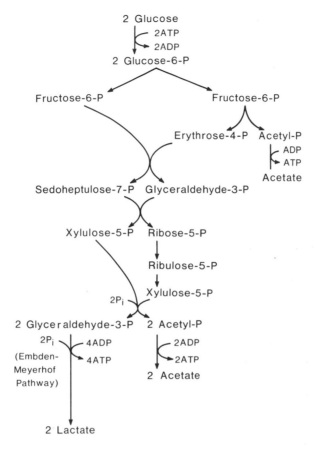

Figure 13.5 Fermentation of glucose by *Bifidobacterium* spp. (Adapted from Stanier 1970 and Gottschalk 1979.)

ure 13.5) involves two phosphoketolase enzymes: one specific for fructose 6-P, which produces acetyl phosphate and erythrose 4-phosphate, and the other specific for xylulose 5-P, which produces acetyl-P and glyceraldehyde 3-P. Since this pathway yields 5 moles of ATP for every two moles of glucose metabolized, it is more efficient than homolactic fermentation (Stanier *et al.* 1970).

Optical Configuration of Lactic Acid

Stereoisomers of lactic acid produced by lactic acid bacteria are useful for species identification. The optical configuration of lactic acid (Table 13.1) depends on the stereospecificity of the LDH. Some microorga-

Table 13.1. Stereoisomers of Lactic Acid Produced by Various Microorganisms.

Organism	Configuration of Lactic Acid
Streptococcus spp.	L(+)
Leuconostoc spp.	D(−)
Pediococcus spp.	DL
Lactobacillus	
L. bulgaricus	D(−)
L. helveticus	DL
L. acidophilus	L(+)
L. casei	L(+)
L. plantarum	DL
L. brevis	DL
Bifidobacterium spp.	L(+)

SOURCE: Buchanan and Gibbons (1974), Wood (1961), and Alm (1982).

nisms produce to LDHs of differing stereospecificity, with the result that DL lactic acid is produced (Stamer 1979). Some lactobacilli with an L(+) LDH can also produce a lactate racemase to establish an L(+)-D(−) equilibrium (Hiyama *et al.* 1968). Lactobacilli with an L(+) LDH will produce less than 5% D(−) lactic acid (Stetter and Kandler 1973). L(+) lactic acid is the predominant stereoisomer in cultured dairy products, although yogurt can contain substantial amounts of the D(−) form (Alm 1982).

End Products of Pyruvate Metabolism

Many microorganisms other than the homofermentative lactic acid bacteria previously discussed use sugars, including lactose, via the EMP (glycolytic) pathway. Diversity among these microorganisms arises from the various strategies used to metabolize pyruvate. These metabolic options can be classified according to the major end products summarized in Table 13.2. The mechanisms which produce these end products are outlined in Figure 13.6. These fermentations are all significant to the dairy industry. Alcoholic fermentation by yeasts, via the intermediate acetaldehyde, is important in the manufacture of certain fermented milks such as kefir and kumiss. Industrial production of alcohol from whey is also possible. Mixed acid and butanediol fermentations associated with growth of coliform bacteria can cause spoilage of milk, cheese, and other dairy products. Presence of acetoin, an intermediate compound in butanediol fermentation, is an important diagnostic characteristic for identification of enteric bacteria. Butyric acid and acetone–butanol fermentations can cause late gas defect in

Table 13.2. Types of Sugar Fermentations in Which Pyruvate is Produced via the Embden-Meyerhof Pathway.

Type of Fermentation	Principal Products	Associated Microorganisms
Homolactic	Lactic acid	*Streptococcus, Lactobacillus*
Alcoholic	Ethanol, CO_2	Yeast
Mixed acid	Lactic acid, acetic acid, formic acid, ethanol, CO_2, H_2	Enteric bacteria (*Escherichia, Salmonella*)
Butanediol	Butanediol, ethanol, lactic acid, acetic acid, CO_2, H_2	Enteric bacteria (*Enterobacter, Serratia*)
Butyric acid	Butyric acid, acetic acid, CO_2, H_2	Various anaerobes (*Clostridium, Butyribacterium*)
Acetone-butanol	Acetone, butanol, ethanol, isopropanol, butyric acid, acetic acid, CO_2, H_2	Some *Clostridium* spp. (*C. acetobutyicum*)
Propionic acid	Propionic acid, acetic acid, succinic acid, CO_2	*Propionibacterium*

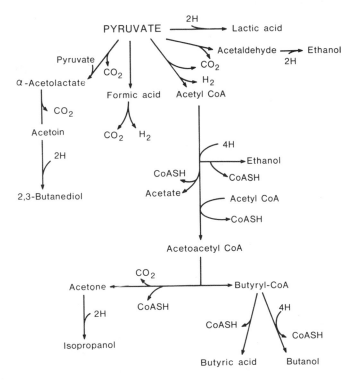

Figure 13.6 Mechanism for production of a diversity of end products from fermentation of pyruvate. (Adapted from Stanier 1970.)

cheese, which results from the growth of *Clostridium* spp. (Kosikowski 1977).

Propionic Acid Fermentation

Fermentation of lactic acid to yield propionic acid, carbon dioxide, acetic acid, and succinic acid is important for proper eye formation and flavor development in Emmental, Gruyere, and Swiss-type cheese varieties. This fermentation is associated with *Propionibacterium* spp.; subspecies of *Propionibacterium freudenreichii* are of greatest significance. These organisms can also be used for industrial production of vitamin B_{12} and propionic acid.

Although propionibacteria do not grow well under normal aerobic conditions, they can use small amounts of oxygen and are thus considered microaerophilic. Their growth characteristics and nutritional requirements have been reviewed by Hettinga and Reinbold (1972A). Various amino acids are beneficial but not necessary for growth, whereas pantothenic acid, biotin, iron, magnesium, cobalt, and perhaps other constituents of yeast extract are required. Growth of propionibacteria can be inhibited by certain concentrations of calcium or sodium propionate, calcium or sodium lactate, acetate, or sodium chloride, and by glucose that has been improperly heated. Copper from cheese-manufacturing equipment can also be inhibitory (Maurer *et al.* 1975). Growth and production of propionic acid is inhibited at pH 5.0 and is optimum from pH 6.65 to 7.0.

The metabolic pathway for production of propionic acid as it occurs in propionibacteria is summarized in Figure 13.7. The metabolism of these microorganisms has been reviewed by Hettinga and Reinbold (1972B). Various carbon sources, including glucose and lactate, can serve as fermentation substrates, with the organism generating four moles of ATP per mole of glucose and one mole of ATP per mole of lactate (Bauchop and Elsdon 1960). The fermentation reactions can be summarized by the equations

$$1.5 \text{ glucose} \rightarrow 2 \text{ propionate} + \text{acetate} + CO_2 + H_2O$$
$$3 \text{ lactate} \rightarrow 2 \text{ propionate} + \text{acetate} + CO_2 + H_2O$$

Of the six moles of ATP generated from 1.5 moles of glucose, three are derived from formation of pyruvate, one from formation of acetate from acetyl phosphate, and two from reduction of fumerate to succinate.

Propionibacteria can produce pyruvate either from glucose, primarily via the EMP pathway, or from oxidation of lactate by using a flavoprotein as a hydrogen acceptor (Gottschalk 1979). As shown in Figure

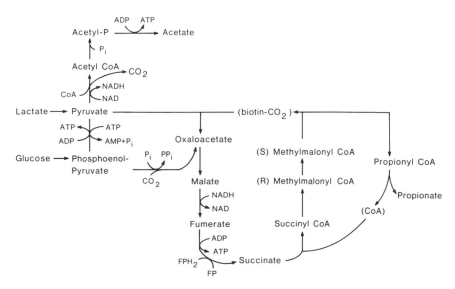

Figure 13.7   Metabolic pathway for production of propionic acid by propionibacteria.

13.7, pyruvate can then either be converted to acetyl-CoA and carbon dioxide through the action of pyruvate dehydrogenase, or it can be used to form oxaloacetate and ultimately propionate. The mechanism for formation of propionate from pyruvate involves two connected cycles. In the first cycle, a biotin-bound carboxyl group is transferred from methylmalonyl-CoA to pyruvate to form oxaloacetate and propionyl-CoA. The transcarboxylase which catalyses this reaction is unique because neither ATP nor a divalent metal is required. This enzyme acts only on the S enantiomer of methylmalonyl-CoA, and since the R enantiomer is formed from succinyl-CoA, the enzyme methylmalonyl-CoA racemase is necessary for the cycle to function. Another interesting aspect of this cycle is reduction of fumerate to succinate. The fumerate reductase system is the only known mechanism by which strict anaerobes can produce ATP by electron transport phosphorylation. This energy-producing process involves fumerate reductase, menaquinone, and cytochrome b (de Vries *et al.* 1974). The second cycle in the propionic acid fermentation is completed by transfer of CoA from propionyl-CoA to succinate. This produces the end product propionate and the intermediate compound succinyl-CoA.

If the propionate fermentation were maintained only by incoming pyruvate, the only end product would be propionic acid; otherwise the cycle would be broken. However, many strains of propionibacteria produce significant quantities of succinate as an end product. This is possible because excess oxaloacetate can be added to the metabolic cycle by carbon fixation (Figure 13.7). Carbon fixation is catalyzed by phosphoenolpyruvate carboxytransphosphorylase and requires inorganic phosphate (Hettinga and Reinbold 1972B).

PROTEIN DEGRADATION

Microbial degradation of milk proteins is important both in the manufacture of cultured products and because it contributes to loss of quality in raw milk and other manufactured dairy products.

Although milk does contain some nonprotein nitrogenous compounds, which are readily available for microbial metabolism, these are not prevalent enough to sustain prolonged or maximum growth. Thus all microorganisms important in either milk fermentation or spoilage must produce enzymes that hydrolyze milk proteins to compounds that can be assimilated. Proteinases can be extracellular, intracellular, or surface-bound. Extracellular enzymes are released from the microbial cell into the growth medium, intracellular enzymes are contained inside the cell, and surface-bound enzymes are either attached to the cell wall or are trapped in the periplasmic space. Surface-bound proteinases are able to act on substrates outside the cell and release amino acids and peptides close to the cell membrane for efficient absorption. Glenn (1976) has proposed that all enzymes which are transported across the cell membrane be considered extracellular even though they may be surface-bound. Details of the transport and use of proteins by bacteria have been presented by Law (1980).

Microbial proteinases can be classified by mechanism of action. Hartley (1960) divided them into four groups: serine proteinases, thio proteinases, metalloproteinases, and acid proteinases. Morihara (1974) classified enzymes within these groups according to substrate specificity. Enzymes which split peptide substrates at the carboxyl side of specific amino acids are called "carboxyendopeptidases," and those which split peptide substrates at the amino side of specific amino acids are called "aminoendopeptidases." Acid proteinases, such as rennin and pepsin, split either side of specific aromatic or hydrophobic amino acid residues. The action of proteolytic enzymes on milk proteins has been reviewed by Visser (1981).

Proteolysis by Lactic Streptococci

Thomas and Mills (1981) have reviewed the literature on the proteolytic enzymes of lactic streptococci. Lactic streptococci present in starter cultures require various preformed amino acids for growth. Freshly drawn milk contains 5–20% of the concentration of free amino acids necessary for maximum starter growth, as well as readily available peptides (Lawrence et al. 1976). Consequently, proteolysis is necessary for starter cultures to attain maximum growth and rapid coagulation of milk. When grown in milk, protease-deficient variants of lactic streptococci will only reach 10–25% of the maximum cell density reached by parent strains. Addition of hydrolyzed casein reverses the slow acid production of protease-deficient cultures (Pearce et al. 1974). Proteinase activity in starter cultures is associated with plasmid deoxyribonucleic acid, which accounts for the instability of this characteristic (Efstathiou and McKay 1976).

During cheese ripening, proteases associated with starter culture organisms are released into cheese after cell lysis (Law et al. 1974). The proteolytic activity associated with lysed lactic streptococci is necessary for proper flavor development in Cheddar and other cheese varieties. The role of streptococcal proteases and peptidases appears to be production of flavor compound precursors such as methionine and other amino acids, rather than direct production of flavor compounds (Law et al. 1976A). Additional discussion of cheese ripening is presented in Chapter 12.

Lactic streptococci initiate casein degradation through the action of cell wall–associated and cell membrane–associated proteinases and peptidases. Small peptides are taken into the cell and hydrolyzed to their constituent amino acids by intracellular peptidases (Law and Sharpe 1978). Peptides containing four to seven residues can be transported into the cell by S. cremoris (Law et al. 1976B). S. lactis and S. cremoris have surface-bound peptidases and thus are not totally dependent on peptide uptake for protein use (Law 1979B). Some surface peptidases of S. cremoris are located in the cell membrane, whereas others are located at the cell wall–cell membrane interface (Exterkate 1984). Lactic streptococci have at least six different aminopeptidase activities, and can be divided into three groups based on their aminopeptidase profiles (Kaminogawa et al. 1984).

Proteolysis by Yogurt Cultures

The proteolysis of casein by starter culture organisms is important for proper flavor and texture development in yogurt. This topic has been reviewed by Tamime and Deeth (1980) and Rasic and Kurman (1978). In a yogurt culture, Lactobacillus bulgaricus is better able to hydrolyze casein, whereas S. thermophilus has significant peptidase activity for hydrolyzing the products of initial casein breakdown. Consequently, the proteolytic activities of the two starter culture bacteria complement each other and contribute to cooperative growth (Moon and Reinbold 1976). Most of the proteolysis in yogurt takes place during incubation and cooling, with a 1:1 ratio of *L. bulgaricus* to *S. thermophilus* producing the greatest amount of free amino acids (Tamime and Deeth 1980). As the proportion of *S. thermophilus* increases in the starter culture, the amount of free proline in the yogurt increases so that at a 1:3 ratio of *L. bulgaricus* to *S. thermophilus,* proline accounts for 71% of the free amino acids after 24 hr (Kapac-Parkaceva *et al.* 1975). Under similar conditions and with a 1:1 starter culture ratio, tyrosine, phenylalanine, and leucine make up 56% of the free amino acids. *L. bulgaricus* possesses the enzyme threonine aldolase, which may contribute to flavor development in yogurt through production of acetaldehyde from threonine (Hickey *et al.* 1983). Lactobacilli, such as *L. acidophilus,* which produce alcohol dehydrogenase, will not produce cultured milk with a typical yogurt flavor since alcohol dehydrogenase reduces acetaldehyde to ethanol (Marshall and Cole 1983).

Proteolysis by Lactobacilli

Proteolysis associated with lactobacilli other than *L. bulgaricus* is of interest because of its importance in cheese ripening. *L. casei* causes increased proteolysis in Cheddar cheese, which can result in improved flavor (Yates *et al.* 1955). The effect of proteolysis on cheese ripening has been reviewed by Castberg and Morris (1976), Reiter and Sharpe (1971), and Marth (1963). The maximum proteolytic activity of *L. casei* grown in milk occurs between pH 5.5 and 6.5 (Brandsaeter and Nelson 1956A). Peptidase activity of this organism is stimulated by cobalt ions at pH 5.5 (Brandseater and Nelson 1956B). El Soda *et al.* (1978B) determined the presence of an endopeptidase and three cytoplasmic exopeptidases in *L. casei.* The exopeptidases, a dipeptidase, an aminopeptidase, and specific carboxypeptidases were activated by divalent cobalt and manganese ions (El Soda *et al.* 1978A).

Proteases of *L. bulgaricus* and *L. helveticus* contribute to the ripening of Swiss cheese (Langsrud and Reinbold 1973). Strains of thermoduric lactobacilli are generally more proteolytic than *S. thermophilus* (Dyachenko *et al.* 1970). The proteinase activity of *L. bulgaricus* is optimal at pH 5.2–5.8 and is associated with the cell envelope (Argyle *et al.* 1976). Some strains of *L. brevis* (Dacre 1953) and *L. lactis* (Bottazzi 1962) are also proteolytic. Surface-bound aminopeptidase from *L. lactis,* characterized by Eggiman and Bachman (1980), is activated by cobalt and zinc ions and has optimum activity at pH 6.2–7.2. A surface-bound proteinase and carboxypeptidase are also present in *L. lactis.*

Proteolysis by Micrococci

Micrococci comprise approximately 78% of the nonlactic bacteria in raw milk Cheddar cheese (Alford and Frazier 1950). The proteolytic system of *Micrococcus freudenreichii* functions optimally at 30°C and at a pH near neutrality (Baribo and Foster 1952). An analysis of proteinases present in 1-year-old Cheddar cheese indicates that micrococci may contribute to proteolytic activity (Marth 1963). Proteolytic micrococci also contribute to the ripening of surface-ripened cheeses such as brick and Camembert (Lenoir 1963; Langhus *et al.* 1945). Micrococcal proteases probably contribute to development of ripened cheese flavor when ripening temperatures are above 10°C (Moreno and Kosikowski 1973). This effect results from degradation of β-casein.

Proteolysis by Brevibacterium linens

B. linens is an aerobic organism found in large numbers on the surface of Limburger, Trappist, and other surface-ripened cheeses. Its proteolytic activity at the cheese surface results in release of free amino acids, which contributes to an amino acid gradient of decreasing concentration from the surface of the cheese (Schmidt *et al.* 1976). *B. linens* produces both intracellular and extracellular proteinases (Foissy 1974). Cell extracts of *B. linens* contain at least six different peptide hydrolases (Torgerson and Sorhaug 1978). Foissy (1978A) purified the extracellular aminopeptidase of *B. linens;* maximum activity of this enzyme occurred at pH 9.6 and 26°–30°C (Foissy 1978B). The aminopeptidase was activated by cobalt and inhibited by low concentrations of heavy metals (Foissy 1978C).

Proteolysis of Molds

In blue-vein cheeses, the proteolytic activity of *Penicillium roqueforti* is a major contributor to the ripening process. Blue cheese lacking in proteolytic activity exhibits a tough and crumbly texture instead of being soft and smooth (Kinsella and Hwang 1976B). The different proteolytic activities of various strains of *P. roqueforti* are reflected in different flavor and texture characteristics of ripened cheese (Coghill 1979). *P. roqueforti* has both extracellular and intracellular protease systems. The extracellular system consists of three exopeptidases, which are characterized as an acid and an alkaline carboxypeptidase, an alkaline aminopeptidase (Gripon and Debest 1976; Gripon 1977A,B), and two endopeptidases, which are characterized as an acid (aspartyl) protease (Zevaco *et al.* 1973; Modler *et al.* 1974) and a metalloprotease (Gripon and Hermier 1974). The aspartyl proteinase releases mainly short peptides from α_{s1}-casein and mainly high molecular weight peptides from β-casein (Le Bars and Gripon 1981). This protease has little effect on release of free amino acids in the ripening cheese (Gripon et al. 1977). Paquet and Gripon (1980) characterized the intracellular peptide hydrolases of five strains of P. roqueforti. No aminopeptidase activity was observed. Other activities including those of acid, neutral, and alkaline endopeptidases were found in various degrees in all strains.

Penicillium caseicolum produces an extracellular aspartyl proteinase and a metalloproteinase with properties very similar to those of the extracellular enzymes produced by P. roqueforti (Trieu-Cout and Gripon 1981; Trieu-Cout et al. 1982). Breakdown of casein in mold-ripened cheese results from the synergistic action of rennet and the proteases of lactic streptococci and penicillia (Desmazeaud and Gripon 1977). Peptidases of both lactic acid bacteria and penicillia contribute to formation of free amino acid and nonprotein nitrogen (Gripon et al. 1977).

Proteolysis by Psychrotrophic Bacteria

Psychrotrophic bacteria grow relatively rapidly in milk kept at 7°C or less. Growth of these organisms, which include species of Pseudomonas, Aeromonas, Flavobacterium, Acinetobacter, Bacillus, Micrococcus, and other genera, limits the shelf life of milk and milk products (Witter 1961). Many of these microorganisms, especially Pseudomonas spp., are actively proteolytic. Proteolytic pseudomonads most often isolated from dairy products include Pseudomonas fluorescens, P. fragi, and P. putrefaciens (Law 1979A). (P. putrefaciens is no longer recognized as a Pseudomonas species in the eighth edition of Bergey's Manual [Buchanon and Gibbons 1974].) P. fluorescens produces both extracellular and endocellular proteases (Peterson and Gunderson 1960). An extracellular protease of P. fluorescens is described as a heat-stable thiol proteinase inhibited by zinc and various other metal ions (Alichanidis and Andrews 1977). In other studies, the extracellular protease has been reported to be a metalloenzyme containing zinc and calcium atoms (Juan and Cazzulo 1976; Richardson 1981). This enzyme has a half-life of 37.5 sec at 150°C when heated in milk. Heat stability is a common property of proteases produced by Pseudomonas spp. (Richardson and Whaiti 1978; Adams et al. 1975; Griffiths et al. 1981). Some, but not all, of these heat-stable proteases are inactivated by heating at 50° to 60°C (Stepaniak and Fox 1983). P. fragi produces an extracellular neutral endopeptidase similar to the protease of P. fluores-

cens in that it is stabilized by ionic calcium (Porzio and Pearson 1975; Noreau and Drapeau 1979). A surface-bound aminoendopeptidase has been isolated from various *Pseudomonas* spp. (Murgier *et al.* 1976). Other gram-negative psychrotrophic organisms, including *Aeromonas* spp., can be highly proteolytic (Denis and Veillet-Poncet 1980). Characteristics of proteases produced by psychrotrophic bacteria have been reviewed by Cousin (1982) and Fox (1981).

The proteolytic systems of psychrotrophic bacteria selectively attack β- and α_s-caseins (Cousin and Marth 1977A), whereas whey proteins are relatively unaffected. Growth of psychrotrophic bacteria in milk results in decreased stability of casein, as measured by rennet coagulation time and heat stability (Cousin and Marth 1977B). Growth of psychrotrophs in milk also causes an increased rate of acid production by starter cultures as a result of increased quantities of readily available nitrogen compounds (Cousin and Marth 1977C,D).

Psychrotrophic *Bacillus cereus* can cause spoilage of pasteurized milk that is free of the more rapidly growing gram-negative psychrotrophs. Proteolytic *B. cereus* produces a defect of pasteurized milk known as "sweet curdling" (Overcast and Atmaran 1974). The action of *B. cereus* proteases on milk proteins has been described by Choudhery and Mikolajcik (1971).

Bitter Flavor Resulting from Proteolysis

Bitterness is a common defect that is frequently associated with proteolysis. This off-flavor results from formation of bitter-tasting peptides usually derived from β- or α_s-caseins (Visser 1981). These peptides have been isolated from bitter cheese (Hamilton *et al.* 1974) and casein digests (Minamiura *et al.* 1972). Bitter flavor in raw and pasteurized milk can result from growth of psychrotrophic bacteria (Patel and Blankenagel 1972; Garm *et al.* 1963). Law (1979A) concluded that there was little evidence to support the conjecture that psychrotrophic proteases are active in stored cultured products, since these enzymes are usually neutral proteases with temperature optima greater than 30°C. Bitter flavor in cheese can result from the action of rennet or starter culture proteinases (Visser 1981; Thomas and Mills 1981). Data of Mills and Thomas (1980) indicate that development of bitterness in Cheddar cheese corresponds to high levels of starter culture proteinase. Visser (1981) has postulated that bitterness occurs when high densities of intact starter culture cells persist in the cheese. Proteinases that produce bitter peptides are active on the surface of cells, whereas peptide-degrading peptidases are primarily active inside the cell, where they are isolated from the cheese until the cells lyse. "Bitter" strains of lactic streptococci have lower peptidase activity at pH 5 than "nonbitter" strains (Sullivan *et al.* 1973).

FAT DEGRADATION

Hydrolysis of triglycerides to produce free fatty acids and glycerol is catalyzed by native milk and microbial lipases. Fatty acids are an important flavor component of cultured and high-fat dairy products. However, when they are present at abnormally high levels, they cause rancid off-flavors. In some cultured products such as Cheddar cheese and yogurt, fatty acids produced from amino acid degradation may be more prevalent than fatty acids derived from milk fat (Lawrence *et al.* 1976; Tamime and Deeth 1980). In other products, such as mold-ripened cheese, microbial lipases play a crucial role in flavor development. A survey of microorganisms from different sources, including milk and butter, indicated that lipolytic activity is common; only 27 of 650 cultures did not exhibit any lipolysis (Ruban *et al.* 1978).

Lipolysis by Lactic Acid Bacteria

The relative importance of lipase from lactic streptococci in producing free fatty acids during cheese ripening is still uncertain. Reiter *et al.* (1967) concluded that milk lipase is more important in the initial hydrolysis of milk fat, even though lactic acid bacteria have lipolytic activity (Fryer *et al.* 1967). Starter culture bacteria can produce free fatty acids from mono- and diglycerides in partially hydrolyzed milk fat (Stadhouders and Veringa 1973). The study of aseptically manufactured cheese indicates that weak hydrolysis of fat by starter culture organisms does contribute to increased concentrations of free fatty acids in aged cheese (Reiter *et al.* 1967).

The lipolytic activity of yogurt cultures is discussed in reviews by Tamime and Deeth (1980) and Rasic and Kurman (1978). Although these organisms have only a weakly active cytoplasmic lipase, its activity is thought to contribute to the increase in free fatty acids observed during storage of yogurt. Tamime and Deeth (1980) conclude that most of the volatile acid content of yogurt is derived from nonfat milk components.

Lipolysis by Molds

Lipolysis by *P. roqueforti* is necessary for flavor development in blue-vein cheese. *P. roqueforti* produces intracellular and extracellular lipases, the extracellular activity being the more important (Kinsella and Hwang 1976B). Both acid and alkaline extracellular lipases are produced (Coghill 1979; Lamberet and Menassa 1983). Strain differences and ripening times may contribute to different amounts of these lipases being present in the cheese, resulting in flavor variations (Lamberet and Menassa 1983).

The flavor of blue cheese is produced by a combination of free fatty acids and methyl ketones derived from fatty acids. The partial oxidation of fatty acids to methyl ketones occurs via the β-oxidation pathway (Kinsella and Hwang 1976A).

Lipolysis by Psychrotrophs

Gram-negative psychrotrophic bacteria that are actively lipolytic include *Pseudomonas fragi, P. fluorescens, Achromobacter lipolyticum, Flavobacterium* spp., *Alcaligenes* spp., and *Acinetobacter* spp. Research on lipase production by these organisms has been reviewed by Cousin (1982) and Severina and Bashkatova (1981). Optimum lipase production is achieved by *Pseudomonas* spp. when they are grown in organically complex media at a near neutral pH with aeration. The optimum temperature for lipase production by psychrotrophs is usually $20°-30°C$, although the enzyme is often active at and below refrigeration temperatures (Anderson 1980; Cousin 1982). Lipase of *P. fragi* has been characterized by Nashif and Nelson (1953A–C). This enzyme preferentially hydrolyzes triglycerides at the one and three positions (Mencher and Alford 1967). Severina and Bashkatova (1981) conclude that *P. fragi* lipase exhibits a broad range of strain differences in terms of optimum pH and temperature, inhibitors, activators, and thermostability. Extracellular lipases produced by *P. fluorescens* have been isolated and the major enzyme characterized by Fox and Stepaniak (1983). The enzyme was heat stable, with optimum activity at alkaline pH. Other lipases produced by common raw milk psychrotrophic bacteria exhibit similar properties (Fitz-Gerald and Deeth 1983). Lipase activity of *P. fluorescens* is higher in cultures incubated below the optimum growth temperature (Anderson 1980). This increased activity at low temperatures may result from decreased proteolytic inactivation of the enzyme.

Many lipases produced by psychrotrophic bacteria retain activity after pasteurization and ultra-high-temperature (UHT) heat treatments (Cousin 1982; Adams and Brawly 1981). Butter made from cream which supported growth of lipase-producing psychrotrophs became rancid within two days (Kishonti and Sjostrom 1970). UHT milk processed from raw milk contaminated with lipase from a *Pseudomo-*

nas sp. developed rancidity in one to seven months (Adams and Brawly 1981). Pasteurized milk cheese made from raw milk which supported growth of a *Pseudomonas* sp. became rancid after four months (Law *et al.* 1976C).

Psychrotrophic *Pseudomonas* spp. are capable of producing extracellular proteases, glycosidases, and phospholipases which can degrade various components of the milk fat globule membrane, making it more susceptible to lipolysis (Alkanhal *et al.* 1985; Chrisope and Marshall 1976; Marin *et al.* 1984).

Development of Fruity Flavors

Fruity flavor in dairy products is the result of ethyl ester formation, usually catalyzed by esterases from psychrotrophic or lactic acid bacteria. Ester formation by *P. fragi* involves liberation of butyric and caproic acids from the one and three positions of milk triglycerides and the subsequent enzymatic esterification of these fatty acids with ethanol (Hosono *et al.* 1974; Hosono and Elliott 1974). Consequently, among the esters formed, ethyl butyrate and ethyl hexanoate predominate. *Pseudomonas*-produced fruity flavor can occur in fluid milk, cottage cheese, and butter.

Fruity flavor in Cheddar cheese is also associated with high levels of ethyl butyrate and ethyl hexanoate (Bills *et al.* 1965). However, this defect is usually caused by esterase activity from lactic acid bacteria, especially *S. lactis* and *S. lactis* subsp. *diacetylactis* (Vedamuthu *et al.* 1966). Fruity-flavored cheeses tend to have abnormally high levels of ethanol, which is available for esterification (Bills *et al.* 1965). Streptococcal esterase activity in cheese is affected by the level of glutathione, which suggests a dependence on free sulfhydral groups for activity (Harper *et al.* 1980).

METABOLISM OF CITRIC ACID

Although citric acid is present in milk in small amounts (0.07–0.4%), it is a required substrate for production of desirable butter-like flavor and aroma compounds in cultured products. Because seasonal variation in the citrate content of milk is sufficient to affect the flavor of cultured products (Mitchell, 1979), milk may need to be supplemented with citrate to produce cultured products with consistent flavor. Citric acid is metabolized by many organisms found in milk, including *S. lactis* subsp. *diacetylactis, Leuconostoc* spp., *Bacillus subtilis*, various lactobacilli, various yeasts, coliforms, and other enteric bacteria.

Citrate Use

Studies of citrate metabolism have been reviewed by Collins (1972) and Rodulpo *et al.* (1976). The following discussion applies to citrate use in lactic acid bacteria, especially as studied in *Leuconostoc* spp. and *S. lactis* subsp. *diacetylactis*. The pathway for citrate metabolism used by these organisms is presented in Figure 13.8. The major end products are carbon dioxide, acetic acid, diacetyl, acetoin, and 2,3 butanediol. Diacetyl is the major flavor and aroma compound produced. Not only does this pathway not yield energy for the organism, data of Cogan *et al.* (1981) indicate that energy may be required for its functioning.

Cellular uptake of citric acid is accomplished by the enzyme citrate permease. There is evidence that this enzyme is plasmid-associated in *S. lactis* subsp. *diacetylactis* (Cogan 1981; Kempler and McKay 1979).

Once citrate enters the cell, it is degraded to pyruvate, acetate, and carbon dioxide (Figure 13.8). Citrate lyase, one of the enzymes required for citrate metabolism, is induced by citrate in *Leuconostoc* spp. and heterofermentative lactobacilli (Mellerick and Cogan 1981). Since pyruvate is formed from citrate without the simultaneous production of reduced NAD, it does not have to be diverted to reoxidizing NAD, as is true for pyruvate formed from sugar fermentation. This "surplus"

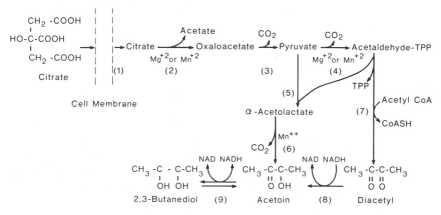

Figure 13.8 Pathway for metabolism of citrate by *Leuconostoc* spp. and *S. lactis* subsp. *diacetylactis*. (1) Citrate permease, (2) citrate lyase, (3) oxaloacetic acid decarboxylase, (4) pyruvate decarboxylase, (5) α-acetolactate synthetase, (6) α-acetolactate carboxylase, (7) diacetyl synthetase, (8) diacetyl reductase, and (9) acetoin reductase.

pyruvate can be used either in synthesis of cellular consituents or in formation of products such as acetoin and diacetyl. Oxidized NAD can still be produced from acetoin and diacetyl through the action of reductase enzymes. Synthesis of both acetoin and diacetyl requires formation of an acetaldhyde–thiamine pyrophosphate (TPP) complex (hydroxyethylthiamine pyrophosphate). This reaction requires a divalent metal in addition to TPP and results in release of carbon dioxide. The acetaldehyde–TPP complex either combines with pyruvate to form α-acetolactate, which is then decarboxylated to produce acetoin, or it combines with acetyl-CoA to produce diacetyl (Figure 13.8). Microorganisms able to produce diacetyl form only small amounts but normally produce large amounts of acetoin. This results from either a lack of available acetyl-CoA or a relatively high activity of diacetyl reductase. Studies by Cogan (1981) and Cogan et al. (1981) present data which support the active diacetyl reductase hypothesis.

Control of Diacetyl and Acetoin Production

Presence of citrate in growth media partially represses both diacetyl reductase and acetoin reductase, allowing these compounds to accumulate (Cogan 1981; Mellerick and Cogan 1981). Once citrate is depleted, reductase activities increase, resulting in loss of flavor. Addition of citrate to milk in excess of the amount needed for fermentation will delay loss of diacetyl flavor. Other conditions conducive to production and maintenance of diacetyl in cultured products include low pH (4.3–5.5), aeration, and storage at low temperatures (Sandine et al. 1972). To obtain the necessary low pH, acid-producing cultures of the S. lactis/ S. cremoris type are combined with citrate fermentors such as Leuconostoc spp. and S. lactis subsp. diacetylactis. Citrate use by Leuconostoc lactis is optimal at pH 5.3 and optimum acetoin production occurs at pH 4.5 (Cogan et al. 1981). Increased acetoin production at low pH may be related to a decrease in the apparent Michaelis constant (K_m) of LDH, which occurs as the pH is lowered. At a lower K_m, the LDH is more readily saturated by pyruvate, thus making more pyruvate available for acetoin production. Cogan et al. (1981) also investigated the effect of sugars on citrate metabolism. Glucose and lactose stimulate citrate uptake in L. lactis, probably by supplying required energy. Metabolic intermediate compounds of these sugars inhibit α-acetolactate synthetase and decarboxylase activities, and thus decrease acetoin production. However, at low pH the inhibition of α-acetolactate synthetase decreases, explaining why acetoin is produced by this organism only at low pH (Cogan et al. 1984).

Acetoin Production in Nonlactic Microorganisms

Enterobacter aerogenes, B. subtilis, P. fluorescens, and *Serratia marcescens* produce acetoin by decarboxylation of α-acetolactate. However, yeasts and *E. coli* form acetoin from the acetaldehyde–TPP complex and free acetaldehyde (Rodopulo *et al.* 1976). These organisms do not decarboxylate α-acetolactate, but use it to produce valine and pantothenic acid. In lactic acid bacteria, α-acetolactate is not used for valine or pantothenic acid synthesis, since these substances are required for growth (Law *et al.* 1976B; Reiter and Oram 1962). In those microorganisms which can synthesize valine, this amino acid inhibits α-acetolactate synthesis (Rodopulo *et al.* 1976).

Diacetyl Reductase

Various cultured products such as cottage cheese and cultured buttermilk require a minimum diacetyl concentration for acceptable flavor. Active diacetyl reductase in these products reduces flavor quality by converting diacetyl to acetoin (Figure 13.8). Potential sources of diacetyl reductase in cultured products include starter culture organisms, as well as contaminants such as gram-negative psychrotrophs, coliforms, yeasts, and lactobacilli (Seitz *et al.* 1963; Wang and Frank 1981; Keenan and Lindsay 1968). In products not heat treated after culturing, lactic acid bacteria including starter culture organisms and contaminants are likely to be a major source of diacetyl reductase (Sadovski *et al.* 1980; Hogarty and Frank 1982). *Leuconostoc* spp. generally have lower diacetyl reductase activity than *S. lactis* subsp. *diacetylactis* (Seitz *et al.* 1963).

BACTERIAL EXOPOLYSACCHARIDES

Growth of polysaccharide-producing microorganisms can result in a defect known as "ropy" milk and appearance of surface slime on products such as cottage cheese. Causative organisms include various *Enterobacter, Pseudomonas, Alcaligenes, Klebsiella,* and *Leuconostoc* organisms, as well as some strains of Group N streptococci. The major slime-producing microorganisms isolated from raw milk held at 10°C include *Klebsiella oxytoca* and *Pseudomonas aeruginosa* (Cheung and Westhoff 1983). Bacterial slime production is desirable in some Scandinavian cultured milks such as "taettamelk," "langmjolk," and "filli." Ropy variants of *S. cremoris* and *S. lactis* produce the slime found in these products (Forsen *et al.* 1973; Sundman 1953). This slime may consist primarily of glycoprotein rather than polysaccharide (Macura and Townsley 1984). Limited polysaccharide production by yogurt cultures can improve the consistency and viscosity of yogurt, thereby eliminating the need for additional stabilizers (Rasic and Kurmann 1978).

Composition of Exopolysaccharides

Bacterial exopolysaccharides are produced either as capsules closely associated with the cell or as slime unattached to the cell. Monomers found in bacterial polysaccharides include neutral hexoses such as D-glucose, D-galactose, and D-mannose, methyl pentoses such as L-fucose and L-rhamnose, and polyols such as ribotol and glycerol (Sutherland 1977A). Polysaccharides isolated from yogurt cultures are composed of arabinose, mannose, glucose, and galactose (Rasic and Kurman 1978). Lactobacilli can produce polysaccharides containing only glucose (glucan) or glucose and mannose (Dunican and Seeley 1965). *Enterobacter aerogenes* produces capsules composed of galactose, mannose, and glucuronic acid (Troy *et al.* 1971). *Leuconostoc* spp. produce glucan (dextran) (Smith 1970).

Exopolysaccharide Biosynthesis

Glucan and fructan (levan) are synthesized outside the bacterial cell. Synthesis of glucan from sucrose is catalyzed by extracellular dextransucrase. This enzyme transfers the glucosyl portion of sucrose to the 6 position of the glucose at the end of a glucan chain (Sutherland 1977A). Fructose is a product of this reaction. Fructan is synthesized from sucrose through a similar process catalyzed by levansucrase. Glucose is an end product of this reaction.

Heteropolysaccharides such as those produced by enteric organisms and pseudomonads are synthesized at the cell membrane. This synthesis involves nucleotide diphosphate sugars and requires an isoprenoid lipid carrier, as diagrammed in Figure 13.9 (Sutherland 1977A,B, 1979). Data of Forsen and Haiva (1981) indicate that polysaccharide production by Group N streptococci may follow a similar biosynthetic pathway. The level of nucleotide diphosphate sugar such as uridine diphosphate-glucose (UDP-glucose) provides a possible control mechanism for heteropolysaccharide synthesis (Figure 13.9). UDP-glucose can be epimerized to UDP-galactose or oxidzed to UDP-glucuronic acid; both molecules are found in exopolysaccharides. UDP-mannose can function in a manner similar to that of UDP-glucose. The availabil-

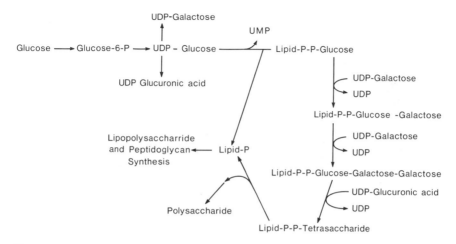

Figure 13.9 Mechanism for biosynthesis of microbial heteropolysaccharides. UDP = uridine diphosphate; UMP = uridine monophosphate; P-P = pyrophosphate; P = phosphate. (Adopted from Sutherland 1979.)

ity of isoprenoid lipid may further control polysaccharide production (Sutherland 1977A). The composition of a heteropolysaccharide is independent of available substrate and energy. Greater amounts of polymer are produced when excess carbohydrate is present or if the growth medium is deficient in nitrogen, phosphorus, or sulfur (Sutherland 1977A,B). Cultural conditions such as high or low incubation temperatures and high pH promote exopolysaccharide production by lactic starter cultures (Rasic and Kurman 1978). The ability of Group N streptococci to produce slime is easily lost through repeated transfer, providing evidence that this characteristic is plasmid-linked (Macura and Townsley 1984).

PRODUCTION OF MALTY FLAVOR

Malty flavor and odor have been reported as defects in milk, cream (Hammer and Cordes 1921), and butter (Vertanen and Nikkla 1947). These defects result from growth of *S. lactis* var. *maltigenes,* an organism which differs from *S. lactis* only in its ability to produce aldehydes and alcohols from amino acids. Morgan *et al.* (1966) identified 2-methylpropanal and 3-methylbutanal as the principal components of malty flavor. Milk cultured with *S. lactis* var. *maltigenes* also contains relatively large amounts of ethanol, 2-methylpropanol, and 3-methyl-butanol. Additional research of Sheldon *et al.* (1971) indicates that numerous aldehydes and alcohols, including phenylacetaldehyde and phenylethyl alcohol, contribute to the flavor and aroma of malty milk. *Lactobacillus maltaromicus* also can produce malty-flavored milk. This microorganism produces aldehydes and alcohols which are similar to those produced by the *S. lactis* variant (Miller *et al.* 1974).

The mechanism for synthesis of alcohols and aldehydes from amino acids has been discussed in a review by Morgan (1976). Both *S. lactis* and its malty variant can reversibly form keto acids from the amino acids valine, leucine, isoleucine, methionine, and phenylalanine. However, unlike *S. lactis, S. lactis* var. *maltigenes* can decarboxylate these keto acids to form aldehydes and reduce the aldehydes to their corresponding alcohols through the action of alcohol dehydrogenase in the presence of NADH.

PRODUCTION OF MUSTY OR POTATO-LIKE FLAVOR

Musty or potato-like flavor and aroma have been observed as a defect in milk (Hammer and Babel 1957) and Gruyere de Comte cheese (Dumont *et al.* 1975). This off-flavor results from the production of nitrogenous cyclic compounds by *Pseudomonas taetrolens* and *P. perolens* (Morgan 1976). Musty-flavored compounds produced by these organisms include 2,5-dimethylpyrazine and 2-methoxy-3-isopropylpyrazine. The Gruyere de Comte with potato off-flavor contained 3-methoxy-2-propyl pyridine, as well as alkyl pyrazine compounds (Dumont *et al.* 1975). Murray and Whitfield (1975) postulated that alkyl pyrazines are formed in vegetables by condensation of amino acids such as valine, isoleucine, and leucine with a 2-carbon compound. Details of the synthetic mechanism in pseudomonads are unknown.

FERMENTATIONS IN MILK

Cultured Milks

The ability of bacteria to affect the flavor and consistency of milk was used to make special milk drinks many years before microorganisms were seen by van Leeuwenhoek. Most of these fermented drinks originated in southern Russia and in countries at the eastern end of the Mediterranean Sea. They were made by producing conditions in milk favorable for the desired fermentation or by inoculating milk with small amounts of product from an earlier fermentation. Usually the unwashed containers provided the starter flora, especially when skin bottles were used. Under these conditions, a mixture of bacteria was present, and flavors resulted which were difficult to reproduce with pure cultures, although the essential bacteria of these fermented drinks were eventually known. In all instances, the lactic acid fermentation was the basic fermentation. Sometimes it occurred in combination with production of gas, a milk alcohol fermentation, and some proteolysis (Corminboef 1933A,B).

One of the more popular fermented milks worldwide is yogurt, which is also known in Bulgaria and Turkey as "yaourt," in Armenia as "matzoon," and in Egypt as "leben." In this product, fermentation is brought about entirely by lactic acid bacteria, including *Thermombacterium yoghurt* of Orla-Jensen (probably *L. bulgaricus*), *L. bulgaricus*, and *S. thermophilus*. *S. lactis* may be present but is not essential; *L. acidophilus* is sometimes added and then the product is called "acidophilus yogurt." Fresh milk is pasteurized, cooled to 40° to 45°C, inoculated, and held at that temperature until the fermentation is complete; it is then refrigerated. Incubation at 29° to 32°C for 12 to 14 hr produces a yogurt with more distinctive flavor than when the short incubation period of 3–4 hr is used. The body of yogurt may be improved by heating fresh milk at 90°C for 5 min or more, homogenizing the milk after the culture is added, or adding nonfat dry milk (3%). Yogurt is made from cow's milk, which is pasteurized and may be partially evaporated, or nonfat dry milk may be added. In some places, yogurt is made from sheep or buffalo milk. The Egyptian leben is said to contain a lactose-fermenting yeast which produces a mild alcohol fermentation. Two fermented milks called "kajobst" and "kajovit" are produced by a modified yogurt fermentation.

Modern yogurt is almost invariably made from a mixed culture of *S. thermophilus* and *L. bulgaricus*. The bacteria, although in diminished number, survive throughout the normal shelf life of yogurt (Hamann and Marth 1984). The two cultures are often maintained separately. (See the earlier discussion in this chapter on the relationship of these two organisms when they grow together in yogurt.) Schulz and Hingst (1954) and Schulz et al. (1954) identified acetaldehyde as the compound responsible for the characteristic flavor and aroma of yogurt. Addition of 0.001 to 0.005% acetaldehyde to milk soured with *S. thermophilus* imparts a yogurt-like aroma and taste to the product. Sour milks possessing a well-developed yogurt flavor and aroma contain a relatively high concentration of acetaldehyde, and those lacking these qualities contain less. For a more detailed discussion of yogurt, the interested reader should consult the reviews by Tamime and Deeth (1980) and Deeth and Tamime (1981).

A popular American cultured milk is buttermilk. If cultured buttermilk is made with *L. bulgaricus,* it is desirable to temper the sharp acid flavor from the bulgaricus culture by growing a streptococcus in association with it. However, maintenance of the two in mixed culture is difficult; the lactic streptococci cease growing at a pH of 4.0 to 4.2, whereas *L. bulgaricus* decreases the pH considerably below these values. The desired result may be accomplished by growing the microorganisms in separate containers, followed by proportioning and mixing to give a product with the desired flavor and texture. The product made with *L. bulgaricus* is known as "Bulgarian buttermilk." More commonly, buttermilk making employs either *S. lactis* or *S. cremoris* grown together with *L. cremoris* or possibly *S. lactis* subsp. *diacetylactis.* One function of the homofermentative streptococcus is to provide a sufficiently acid environment for flavor and aroma compounds to be produced by the heterofermentative streptococcus (Michaelian *et al.* 1938; Mather and Babel 1959). Production of the flavor and aroma compounds is described earlier in this chapter. If the starter culture is permitted to develop 0.80 to 0.85% acidity and if incubation is at 21°C, associative growth of the homo- and heterofermentative streptococci will produce a buttermilk with excellent flavor and body characteristics.

Kefir, usually made from cow's milk, may contain about 1.2% alcohol and is peculiar in that the fermentation is brought about by kefir "grains," which resemble miniature cauliflowers but are the size of kernels of wheat. These grains consist of casein, yeasts, and bacteria. The microorganisms include lactose-fermenting *Torula* yeasts, *S. lactis, Betabacterium caucasicum* (probably a variant of *Lactobacillus brevis*), and glycogen-containing, rod-shaped kefir bacilli. The grains increase in size in the fermenting milk and may be strained out, dried, kept for long periods, and used as inocula. The fermentation may be completed in closed bottles so that gas is retained and the milk becomes effervescent, although this is more likely to be done with kumiss.

In Russia, a popular milk drink traditionally made from unpasteurized mare's milk is known as "kumiss." The fermentation is caused principally by *L. bulgaricus,* lactose-fermenting *Torula* yeasts, and *Lactobacillus leichmannii.*

Kuban fermented milk, a product of southern Russia, is made from pasteurized milk by a combined lactic and alcohol fermentation. The microflora includes a lactic streptococcus resembling *S. lactis* var. *hollandicus,* a lactic rod of the *L. bulgaricus* type, and three yeast types.

Taette milk is used in the Scandinavian peninsula. A slime-producing fermentation is induced by a variant of *S. lactis* designated as *S. lactis* var. *taette.* This is possibly identical to *S. lactis* var. *hollandicus,* which has been used to make Edam cheese, and with other streptococci associated with ropy milk.

A milk drink known as "saya" is prepared from fresh unheated milk, ripened first by *S. lactis* and later by a lactobacillus. In saya milk, considerable carbon dioxide and vigorous proteolysis are produced. Characteristic of the fermentation is a six-day ripening period at 11°C. Corminboeuf (1933A,B) has described numerous milk beverages, including mazun, groddu, skorup, and tättemjolk. Fermented milks also are described by Emmons and Tuckey (1967) and Kosikowski (1977).

Undesirable Fermentations in Milk

The flavor and body of cultured milks are distinguished by a delicate balance between components of the cultured product. Unless conditions of culture are carefully controlled, this balance may not be achieved even when pure cultures are employed. Empirical formulations relating to proper cultural and environmental conditions constitute the art of fermentations. Apart from problems arising from the use of improper conditions of culture, other defects may occur in the milk products because of contamination of a milk supply by unwanted organisms and because some organisms required in fermentations can produce antibiotics (bacteriocius).

Ropy and Slimy Milk. Ropy or slimy milk of bacterial origin becomes apparent only after milk has been stored and is therefore distinguishable from the stringy milk associated with mastitis (Harding and Prucha 1920). Ropiness may be evident only as a slightly abnormal viscosity, or it may be so pronounced that the affected milk can be drawn out in fine threads a yard long and in some instances may assume a gel-like consistency. Thickening may be confined to the top layer of the milk.

Among the bacteria causing ropiness are active gelatin liquefiers, including some of the hay bacillus type. More frequently, however, ropy milk is caused by some members of the coliform group, some lactic streptococci, or some gram-negative psychrotrophs. The common occurrence of the defect in the presence of certain streptococci has led to the assignment of distinguishing names to these organisms. *S. lactis* var. *taette, S. lactis* var. *hollandicus,* and strains of the common *S. lactis* are the essential organisms in Swedish ropy milk and in certain Edam cheese starter cultures. Among the organisms associated with development of ropiness are these: *Alcaligenes viscolactis,* (Hammer and Hussong 1931), *S. lactis* var. *hollandicus* (Hammer 1930), certain corynebacteria, and some organisms of the *Escherichia-Enterobacter* group (Marth *et al.* 1964; Sarles and Hammer 1933). Ordinary milk streptococci, such as *S. lactis, S. cremoris,* or *S. thermophilus,* may at times cause ropiness. Certain strains of streptococci easily acquire and lose their ability to produce ropiness. Rope-producing strains are more oxygen exacting than non-rope-producing ones and develop less volatile acid. Induction of rope-producing properties in bacteria by means of bacteriophage has been observed. The flavor of ropy milk, unless the defect is associated with lactic fermentation, is indistinguishable from that of normal milk; nor is the milk unwholesome.

The immediate cause of the ropy or slimy condition is the bacterial formation of gums or mucins by bacteria. Gums are the more common cause. These are probably galactans produced by fermentation of lactose. Some of the active peptonizing bacteria produce sliminess by formation of mucins, which are combinations of proteins with a carbohydrate radical. Development of sliminess is closely associated with capsule formation (Hammer 1930).

Emmerling in 1900 and Schardinger in 1902 (Marth, 1974) observed that *E. aerogenes* produced slime in milk. The slime dissolved readily in water, yielding a gelatinous solution; it was optically inactive and did not reduce Fehling's solution. Hydrolysis with dilute acids yielded a reducing sugar; oxidation with nitric acid yielded both mucic and oxalic acids. The gummy substance was called "arabogalactan." Metabolic processes involved in polysaccharide formation by bacteria are described earlier in this chapter.

Antibiotic (Bacteriocin) Production in Milk

Production of antibiotic-like substances in cultured dairy products has been associated with homo- and heterolactic streptococci, some of the lactobacilli, and *Brevibacterium linens.* Production may be unwanted and fortuitous, as in commercial starter cultures, or it may be desired and encouraged.

Nisin. Elaboration by *S. lactis* of a substance inhibitory to *L. bulgaricus* was reported by Rogers (1928). The substance, later named "nisin" and produced by some strains of *S. lactis,* is a large polypeptide with a molecular weight of approximately 10,000 (Mattick and Hirsch 1944). Lanthionine and a structural isomer of cystathionine, two sulfur-containing amino acids, were recovered from hydrolysates by Newton *et al.* (1953), who concluded that nisin resembles the antibiotic subtilin. A partially purified preparation (MW 7000) of Cheeseman and Berridge (1959) lacked amino or carboxyl end groups but contained side chains with the epsilon amino group of lysine and the imino group of histidine. Baribo and Foster (1951) observed that an endocellular inhibitory substance (probably nisin) was liberated when cultures of S. lactis were acidified. Cultures boiled or autoclaved for 10 min at pH 4.8 retained their activity, whereas those heated at pH 7.4 rapidly lost about 50% of their activity.

Nisin dissolves in an aqueous solution at pH 7, 5.6, and 4.2 to the extent of 75, 1000, and 12,000 μg/ml, respectively (Hawley 1957A,B). Solubility is substrate dependent.

Assays of nisin are based on its inhibitory action toward *Streptococcus agalactiae* in a tube dilution test (Hirsch 1950). One Reading unit is defined as that amount of antibiotic preparation dissolved in N/20 HCl which causes the same inhibition as a standard preparation. A 0.1% solution of the standard contains 10,000 Reading units, i.e., it is inhibitory at 1:10,000 dilution. Modified tests using litmus milk and *S. cremoris* (Galesloot and Pette 1956) and the 1-hr resazurin test and *S. cremoris* (Friedmann and Epstein 1951) have been described. Modified microbiological assays have been reported by several workers (Beach 1952; Macquot and Lefebure 1959). Nisin is distinguished from chemical preservatives by means of its antibacterial spectrum, particularly with respect to its activity toward some yeasts and lactobacilli (Czeszar and Pulay 1956). The subject of nisin is discussed in greater detail in reviews by Marth (1966) and Hurst (1972).

Antibacterial Spectrum.   Hawley reported that various species and strains of the genera *Staphylococcus, Streptococcus, Neisseria, Bacillus, Clostridium,* and *Corynebacterium* are inhibited by nisin (Hawley 1957A, B). Mattick and Hirsch (1947) added actinomycetes, pneumococci, mycobacteria, and *Erysipelothrix* to this list. The nisin concentration required for complete inhibition is organism specific and ranges from 0.25 to 500 units per milliliter. Inhibition of *L. casei* by antibiotics from *S. lactis* and *S. cremoris* was observed by Baribo and Foster (1951). Inhibition of *Propionibacterium* by nisin but not of coliform bacteria was reported by Galesloot (1957).

Nisin Inactivation.   *L. plantarum* isolates from milk and cheese reduced nisin activity in these substrates (Kooy 1952) and *S. faecalis* and *S. lactis* isolates from raw milk destroyed nisin (Galesloot 1956). Galesloot (1956) observed that use of nisin-producing starters in cheese manufacture is unattractive if large numbers of Group N streptococci are present. Galesloot (1956) also observed that some cultures of *Leuconostoc* were antagonistic. Nisin- (but not subtilin-)destroying nisinase, an enzyme, has been recovered from some strains of *S. thermophilus* (Alifax and Chevalier 1962).

Relationship between Starter Cultures and Streptococcal Antibiotics. When antibiotic-producing and non-antibiotic-producing strains of *S. lactis* and *S. cremoris* from commercial starter cultures were mixed in equal proportions, the antibiotic-producing strains soon predominated (Hoyle and Nichols 1948). Domination occurred after 24 to 48 hr. (Collins 1961). Emergence of a predominant strain may be accompanied by a loss of starter activity and renders the starter more susceptible to complete inactivation by bacteriophage.

Heterolactic Streptococci. Ritter in 1945 noted that two strains of *S. lactis* were inhibited by five strains of betacocci (*Leuconostoc* sp.) when grown at 20°C (Marth 1974). Later, Mather and Babel (1959) found that a creaming mix made up in part of skim milk cultured with *L. cremoris,* when added to cottage cheese, inhibited such spoilage organisms as *P. fragi, Pseudomonas putrefaciens,* or coliforms but not the yeasts *G. candidum* and *Candida pseudotropicalis.* Marth and Hussong (1963) showed that filtrates from cultured skim milk, in which four strains of *L. cremoris* were allowed to ferment citrate, inhibited to different degrees *Staphylococcus aureus, E. aerogenes, A. viscolactis, E. coli, P. fragi,* and *P. fluorescens,* but in no instance did they inhibit the yeasts *Torula glutinis, S. cerevisiae, K. fragilis,* or *Mycotorula lipolytica.* Dilution of the filtrates to the level at which they might be present in cottage cheese eliminated the inhibition of all bacteria except one of two strains of *P. fragi* and one of five strains of *P. fluorescens* (Marth and Hussong 1962). Collins (1961) noted that three of six strains of *S. lactis* subsp. *diacetylactis* formed an antibacterial substance similar to that produced by *S. cremoris.* Some strains of *S. lactis, S. cremoris,* and *S. lactis* subsp. *diacetilactis* were inhibited by the substance. These observations suggest that care must be exercised to combine only suitable strains in compounding a mixed-strain starter culture.

Lactobacilli. Kodama in 1951 isolated an antimicrobial substance designated "lactolin" from *L. plantarum* (Marth 1974), and Wheater *et al.* (1951) obtained a substance designated as "lactobacillin" from organisms resembling *L. helveticus.* Lactobacillin inhibited *C. butyricum* in Gruyere cheese (Hirsch *et al.* 1952). Cultures of *L. helveticus* and a fitrate from the cultures reduced gas formation by *E. coli* (Meewes and Milosevic 1962) and inhibited *Propionibacterium shermanii* and other propionibacteria (Winkler 1953). Pasteurized cultures and culture filtrates of *L. acidophilus* inhibited growth of *E. coli* (Orla-Jensen *et al.* 1926). Concentrated sterile culture filtrates prevented and growing *L. acidophilus* cultures halted development of *E. coli, P. fluorescens, Shigella* sp., *Salmonella* sp., and aerobic spore-forming bacilli (Marth 1974). Lactocidin, the antibiotic produced by *L. acidophilus*, was isolated by Vincent *et al.* (1959). It has a wide spectrum of antibiotic activity. Winkler (1953) reported on the inhibition of propionibacteria by milk cultures of *L. acidophilus*. Some lactobacilli, according to DeKlerk and Coetzer (1961), produce substances inhibitory to other bacteria of the same genus.

Antibiotics in Cultured Milks.   Some cultured milks exhibit antibiotic activity, the causative organisms for which are obscure. Thus acidophilus milk is antagonistic to *E. coli* and bactericidal to *Mycobacterium tuberculosis* (Marth 1974); yogurt inhibits *Erysipelothrix rhusiopathiae* (Marth 1974), *E. coli* (Marth 1974) and human, bovine and bacille Calmette-Guerin (BCG) strains of *M. tuberculosis* (Tacquet *et al.* (1961); kumiss is bacteriostatic or bactericidal to *E. coli, S. aureus, B. subtilis, B. cereus, E. aerogenes* and other organisms; kefir is inhibitory to *E. coli, S. aureus* and *B. subtilis;* and kuräng inhibits mycobacteria and organisms in the genus *Bacillus* (Marth 1974).

Antibiotics from Surface-Ripened Cheese.   An antimicrobial agent attributed to *B. linens* reportedly appears in surface-ripened cheese stored at 2° to 4°C for 8 weeks and is inhibitory to *S. aureus, B. cereus,* and *Clostridium botulinum.* Strains of *B. linens* yielded culture fluids inhibitory to the germination and outgrowth of *C. botulinum* type A spores (Grecz *et al.* 1959). Organisms other than *B. linens* on the surface of cheese contribute minor antimicrobial activity (Grecz *et al.* 1959). The inhibitory substance from *B. linens* withstood heating at 121°C for 25 min. Its properties differed from those of nisin. Surface-ripened cheeses tend to be more resistant than other cheeses to spoilage by growth of molds. Beattie and Torrey (1984) attributed this resistance to production of volatile sulfur compounds by *B. linens.*

Frozen Starter Cultures

Although lactic starter cultures undoubtedly were frozen earlier, research interest in the use of this technique to produce cultures for commercial purposes began during the mid-1950s. Initial attempts involved two approaches: (1) inoculation of milk with the desired culture, incubation in the normal manner so that approximately 0.85% lactic acid was produced, cooling the culture, packaging, and then freezing it at approximately −29°C; and (2) inoculation of milk with the desired culture, followed immediately by packaging and freezing. In the first procedure, neutralization of acid with sodium hydroxide was claimed to improve both survival of bacteria and activity of survivors.

Several investigators claimed success when frozen ripened (incubated to produce approximately 0.85% titratable acid before freezing) or unripened (incubated briefly or not at all before freezing) cultures were used directly to produce bulk cultures or cheese (Martin and Cardwell 1960; Simmons and Graham 1959). Some workers acknowledged that lactic cultures, when frozen as described, lost activity during frozen storage (Richardson and Calbert 1959). Loss of activity and consequent variability in performance of cultures prepared by these techniques prompted development of alternative procedures to provide frozen cultures which are more uniformly dependable. Even though the procedures just described were not entirely successful for preparing conventional lactic starter cultures, it has been claimed that kefir grains can be preserved by freezing and storage at −18°C. According to Toma and Meleghi (Marth 1974), grains held at that temperature for 9 months were easily reactivated and provided normal alcohol and lactic acid fermentation.

Commercial producers of frozen starter cultures now (1) grow the desired culture in a suitable medium; (2) harvest and concentrate the cells by centrifugation; (3) resuspend cells at a desired concentration, based on activity (e.g., ability to produce acid), in a suitable medium; (4) package the culture; and (5) freeze it at −196°C (liquid nitrogen). The culture is then shipped to the user in containers which hold liquid nitrogen or dry ice so that a low temperature can be maintained during storage. Quantities of frozen material are packaged so that the user can prepare his own bulk culture or can directly inoculate milk that is to be made into a fermented food.

Use of this procedure (1) eliminates the need for carrying starters in a factory and thus reduces the hazards of culture failure through infection by bacteriophage or for other reasons; (2) enables the manufacturer of cultures to exert suitable control of quality so that the user is assured of a pure and active culture; (3) avoids repeated transfer of the culture, thereby minimizing problems of strain dominance, and allows the user to mix pure cultures in the exact proportions desired; and (4) enables the user to store cultures which are almost instantly active.

Among the first to describe a procedure for preparing frozen concentrated cultures were Foster (1962) and Lamprech and Foster (1963). In their process, they grew *S. lactis* or *S. lactis* subsp. *diacetylactis* separately at 25°C in a tryptone-yeast extract-glucose-magnesium phosphate medium. Cells early in the maximum stationary phase (10 to 15 hr of incubation) were recovered by centrifugation, resuspended in sterile skim milk to a concentration of 25 to 55 × 10^9 cells per milliliter, and the pH was adjusted to 7.0. The suspensions were then frozen and stored at −20°C. Under these conditions, cultures retained sufficient viability and activity during 10 months of storage to serve satisfactorily as direct inoculum for making buttermilk. Shortly after the report by Lamprech and Foster (1963), Cowman and Speck (1965) noted less loss of viability and greater retention of proteolytic activity and of the ability to produce acid when concentrates of S. *lactis* cells suspended in skim milk were frozen and stored at −196°C (liquid nitrogen) rather than at −20°C. Other investigators also have reported excellent retention of viability and activity by different lactic cultures when they were frozen and stored at −196°C. Included are single and mixed-strain cultures of S. *lactis* (Baumann and Reinbold 1964, 1966), mixtures of L. *cremoris* and S. *lactis* subsp. *diacetylactis* (Waes 1970), and single strains of S. *lactis, S. cremoris* and S. *lactis* subsp. *diacetylactis* (Gibson *et al.* 1966; Keogh 1970).

Vallea and Mocquot (1968) reported on a process used to prepare frozen concentrated cultures of L. *helveticus* and S. *thermophilus*. In their procedure, cultures are grown in cheese whey fortified with papain, yeast extract, manganese sulfate, dried milk, and corn steep liquor; the pH is controlled during growth at 5.0 to 6.5, depending on the organism; cells are recovered by centrifugation and resuspended in skim milk, and suspensions are frozen and stored at −30°C. Continuous pH control during growth of S. *cremoris* to be used for frozen concentrated cultures was advocated by Peebles *et al.* (1969) to obtain maximum cell yields. Higher prefreezing populations (approximately 10^{10}/ml) were obtained when ammonium hydroxide rather than sodium hydroxide was added during growth to neutralize the acid. Concentrates of some S. *cremoris* cultures retained activity for 231 days when frozen and stored at −196°C. Frozen (−196°C) concentrated cultures of L. *cremoris* grown in a tryptone-yeast extract-glucose-citrate broth medium also retained viability and the ability to produce diacetyl during 30 days of storage in liquid nitrogen. Facilities needed and procedures used to produce frozen (−196°C) concentrated starter cultures commercially have been described by Ziemba (1970).

Commercial production and distribution of starter cultures by these methods require (1) liquid nitrogen for freezing and temporary storage; (2) special containers to hold cultures in liquid nitrogen or dry ice during distribution; and (3) a distribution system to deliver cultures promptly to the user. Although facilities to handle cultures this way are available in the United States, they are not found in other parts of the world. Consequently, work has been done on procedures to preserve lactic cultures by storing them at temperatures in the range of −20°

to $-40°C$. Kawashima *et al.* and Kawashima and Maeno completed extensive studies in which they stored lactic-acid bacteria at $-15°$ to $-20°C$ (Marth 1974). Their results showed that (1) *S. lactis* retained greater activity than *L. bulgaricus* after 6 months of storage, although differences disappeared on subculture; (2) satisfactory preservation of yogurt cultures (*L. bugaricus* and *S. thermophilus*) was obtained when they were grown in sterile skim milk fortified with 0.5% calcium carbonate and then diluted with sterile skim milk before freezing; (3) cultures of *S. lactis*, *S. cremoris*, or *S. thermophilus* in skim milk containing 10% total solids exhibited increased resistance to frozen storage and more rapid growth after thawing when L-glutamic acid was added, but corn-steep liquor, yeast extract, and glucose were not beneficial; (4) addition of L-glutamic acid provided no protection to *L. bulgaricus* or *L. acidophilus* during frozen storage but stimulated their growth when the cultures were thawed.

Stadhouders and van der Waals prepared *S. lactis* subsp. *diacetylactis*, *L. cremoris*, and mixtures of the two for freezing by subculturing them five times in reconstituted skim milk and then freezing and storing ripened cultures at $-20°$ or $-40°C$ (Marth 1974). After storage for 20 weeks, approximately 50% of the initial activity was lost. In contrast, when the ripened culture was transferred to skim milk before freezing, 17 to 31% of the activity was lost, depending on the heat treatment received by the milk before freezing (loss of activity was less in steamed than in pasteurized milk).

When mixed cultures are frozen, there may be changes in the proportion of each organism that will remain viable under different conditions. Jabrait (1969) froze a mixed culture composed of *S. lactis* and *L. acidophilus* used to manufacture bioghurt (a German yogurt-like food). When the culture was stored at $0°C$, 53% of the viable cells were *S. lactis* and the remainder were *L. acidophilus*. At $-30°C$, 57% were *S. lactis*, but at $-35°$ and $-40°C$ the proportions of the two organisms were approximately equal. The total viable population at $0°$, $-20°$, and $-35°C$ were 10^{10}, 10^{10}, and 10^9 per milliliter, respectively.

Production of concentrated cultures and their storage at $-20°$ to $-40°C$ also has been advocated by several investigators. Usually the procedures differ somewhat from those originally proposed by Lamprech and Foster (1963). Two examples of differences are as follows: (1) growth of lactic streptococci on tryptone-lactose agar, followed by preparation of cell suspensions, recovery of cells by centrifugation, resuspension of cells in glycerol:water (1:1), and storage at $-30°C$; such cultures, when held for up to eight months, do not lose viability or activity and are suitable for use as direct inoculum in cheese making;

(2) growth of cultures in papain-digested milk enriched with yeast extract and lactose, recovery of cells by centrifugation, suspension of the concentrate in a glycerol-skim milk mixture, freezing, and storage at −30°C until the concentrate is used for commercial purposes. Similar processes with slight additional modifications also have been described (Marth 1974).

Interaction of Starter Cultures and Foodborne Pathogens

Cultured dairy foods seldom cause foodborne illness in the consumer. If an active starter culture is used, common foodborne pathogens, even if present in the milk, do not grow well and often are inactivated during the fermentation or early during the storage life of the product. Even if some cultured products are recontaminated after manufacture, pathogens generally do not survive well. Several examples will illustrate these points.

Goel et al. (1971) added E. coli and E. aerogenes to commercially prepared yogurt, buttermilk, sour cream, and cottage cheese and then stored the products at 7.2°C. Viable coliforms were nearly always absent from yogurt after one or two days of storage. They persisted somewhat longer in buttermilk and sour cream, whereas cottage cheese seldom had a deleterious effect on the bacteria. Although the strains of E. coli used were not known to be pathogenic, it is likely that such strains would behave as did the test strains. Minor and Marth (1972) did similar tests with S. aureus and found that when fewer than 1000 cells were added per gram of yogurt, sour cream, or buttermilk, viable S. aureus seldom could be recovered from the products after one or two days of storage at 7°C. Use of a higher initial inoculum generally resulted in survival of S. aureus for four to five days, regardless of the product.

Park and Marth (1972B) prepared a series of cultured milks which contained Salmonella typhimurium. Survival of salmonellae in the products stored at 11°C ranged from less than three days to more than nine days, depending on species of starter culture, strain of a given species, level of inoculum used to prepare the cultured product, temperature at which the product was cultured, and amount and speed of acid production. In other studies, Park et al. (1970) noted that S. typhimurium survived for up to seven to ten months in Cheddar cheese made with a slow acid-producing starter culture and stored at 13° or 7°C, respectively. In contrast, Goepfert et al. (1968) and Hargrove et al. (1969) found that S. typhimurium survived for three to seven months

in Cheddar cheese made with normal starter cultures. Additional information on behavior of salmonellae in fermented dairy foods can be found in a review by Marth (1969).

Results of tests on survival of foodborne pathogens in cultured products suggest that the starter culture is an important factor in determining inhibition of the pathogen in the food. The starter culture also is important in governing growth of the pathogens, if present in milk, during fermentation.

Reiter *et al.* (1964) showed that growth of *S. aureus* in raw, steamed, and pasteurized milk was inhibited by a lactic starter culture. When they neutralized the lactic acid as it was produced, inhibition of the staphylococcus was still evident. Jezeski *et al.* (1967) also observed that growth of *S. aureus* in steamed or sterile reconstituted nonfat dry milk was inhibited by an actively growing *S. lactis* culture. Enterotoxin was detected in *S. aureus-S. lactis* mixed cultures when *S. lactis* was inactivated by bacteriophage but not when the lactic streptococcus grew normally. Further information on *S. aureus* has been summarized by Minor and Marth (1976).

Park and Marth (1972A) inoculated skim milk with *S. typhimurium* and with different lactic-acid bacteria. They noted that *S. cremoris, S. lactis,* and mixtures of the two repressed growth but did not inactivate *S. typhimurium* during 18 hr of incubation at 21° or 30°C when the lactic inoculum was 0.25%. An increase in inoculum to 1% resulted in inactivation of *S. typhimurium* by some of the mixed cultures during incubation at 30°C. Both *S. lactis* subsp. *diacetylactis* and *L. cremoris* were less inhibitory to *S. typhimurium* than were *S. cremoris* or *S. lactis.* When added at the 1% level, *S. thermophilus* was more detrimental to *S. typhimurium* at 42°C than was *L. bulgaricus.* Mixtures of these two lactic acid bacteria, when added at levels of 1.0 and 5.0%, caused virtually complete inactivation of *S. typhimurium* between the 8th and 18th hr of incubation at 42°C. Daly *et al.* (1972) observed that *S. lactis* subsp. *diacetylactis* inhibited (in most instances, more than 99%) growth of the following spoilage or pathogenic bacteria in milk or broth: *P. fluorescens, P. fragi, P. viscosa, P. aeruginosa, Alcaligenes metalcaligenes, A. viscolactis, E. coli, S. marcescens, Salmonella senftenberg, Salmonella tennessee, S. aureus, Clostridium perfringens, Vibrio parahaemolyticus,* and *S. liquefaciens.* Growth of *S. aureus* in a variety of foods and of *P. putrefaciens* in cottage cheese also was inhibited successfully by *S. lactis* subsp. *diacetylactis.*

In 1971 and again in 1983, illness was caused by enteropathogenic strains of *E. coli* present in imported (from France) Camembert cheese (Kornacki and Marth 1982B). The first of these outbreaks prompted a

series of studies to determine the behavior of this pathogen in some cheese.

Park *et al.* (1973) studied the fate of six strains of enteropathogenic *E. coli* in the manufacture and ripening of Camembert cheese. *E. coli* was inoculated at a rate of about 100/ml of pasteurized milk. Growth was slow until after the curd was cut and hooped. Populations in excess of 10^4/g occurred in some cheeses 5 hr after cheesemaking began. *E. coli* populations began to decrease after overnight growth. The pH also decreased at this time to about 5.0 or less. (The pH of the cheese averaged 4.65 at this point.) Salting of the cheese and 1 day of ripening at 15.6°C resulted in a further decrease in viable *E. coli*. The decrease persisted during the rest of the week (15.6°C) and during storage at 10°C. Between zero and nine weeks were required at 10°C to reduce *E. coli* populations to a nondetectable level, depending on the strain used. The strain which persisted longest was *E. coli* 0128:B12, which survived for more than seven weeks but less than nine weeks at 10°C. When *S. cremoris* C_1 was substituted for a commercial lactic starter culture, *E. coli* populations exceeded 10^4/g at nine weeks of storage at 10°C. When cheese was made from milk with penicillin, which prevented adequate acid production by the starter culture, *E. coli* populations reached a level of 10^9/g in 24 hr and decreased to 10^7/g by nine weeks of storage at 10°C.

Frank *et al.* (1977) also studied survival of different strains of enteropathogenic *E. coli* during Camembert cheese manufacture. In these studies, one pathogenic strain (B_2C) and two nonpathogenic strains (H-52 and B) survived in the cheese past four weeks at levels ranging from 1/g to 1000/g. Hence, these organisms could have been present in the cheese at the time of consumption. Growth and survival curves were nearly parallel when inocula of 10^2, 10^3, and 10^4 *E. coli* B_2C per milliliter were used. When less starter culture was used (0.25% v/v instead of 2.0% v/v), *E. coli* strain B_2C increased four log cycles in number compared to two log cycles during normal manufacture. During ripening of cheese, numbers of *E. coli* B_2C decreased less when less starter was used. Under these conditions, the pH was lowered to only 6.4 instead of to 5.0 after 6 hr of manufacture. However, the cheese had a normal pH thereafter. Because ripening proceeds from the outside to the inside of the cheese wheel, the authors determined survival of *E. coli* at different places in the cheese wheel during ripening. From the bactericidal unripened core outward, conditions became more favorable for survival, and growth of *E. coli* was observed at the outer surface. Survival of several enteropathogenic *E. coli* strains was related also to pH. The higher the pH, the longer the strains survived in the cheese.

Rash and Kosikowski (1982) examined the behavior of enteropathogenic *E. coli* in Camembert cheese made from milk after it was concentrated by ultrafiltration. This work also demonstrated the importance of pH in controlling growth of the pathogen during cheesemaking and in determining survival of *E. coli* during ripening of the cheese.

Frank *et al.* (1978) also studied survival of three enteropathogenic *E. coli* strains in the manufacture and ripening of brick cheese. Growth of two of these strains was 10 times greater during the initial hours of manufacture than in Camembert cheese manufacture, probably as a result of a higher temperature and less rapid decrease in pH during brick cheese manufacture. Inhibition of enteropathogenic *E. coli* occurred at pH values of 5.2–5.5, as opposed to 5.8–5.2, for the same strains in Camembert cheese manufacture. The strains studied were inactivated more quickly in Camembert than in brick cheese. However, the pH of unripened Camembert cheese was lower. Little difference in survival was noted when the pH of the brick cheese ranged from 5.15 to 5.3. The pH of most of the brick cheese samples was 5.3 after two weeks of ripening. After seven weeks (two weeks of ripening at 15.5°C; five weeks at 7°C), *E. coli* populations ranged between 20,000 and 700/g. Growth of *E. coli* on the surface of brick cheese was much less than on the surface of Camembert cheese. A major difference between the surface environment of brick cheese and that of Camembert cheese is the absence of *Penicillium camemberti* on brick cheese. Yeasts and micrococci predominate on the surface of brick cheese during the early ripening stages and cause the pH to rise to 5.4–5.5. On Camembert cheese surfaces the pH is much higher. The microflora on brick cheese surfaces may inhibit growth of *E. coli* by competing for nutrients or by producing metabolites toxic to them.

A study was completed by Kornacki and Marth (1982A) on survival of enteropathogenic and nonpathogenic *E. coli* in Colby-like cheese made from pasteurized whole milk artificially contaminated with 100 to 1000 *E. coli* per milliliter. Numbers of *E. coli* increased 100- to 1000-fold, depending on the strain, to about 1×10^6 per gram of curd, in most instances, by the end of the cook (3.5 to 3.9 hr). After this point, numbers of *E. coli* decreased over a period of 2–13 weeks. Survival of *E. coli* in Colby-like cheese differed among the strains tested with enterotoxigenic strain B_2C surviving better than enterotoxigenic strain H 10407, which survived better than enteroinvasive strain 4608. When results obtained with Colby-like cheese are compared to those of Frank *et al.* (1977), it appears that Camembert cheese was more inhibitory to *E. coli* than was Colby or brick. For instance, when *E. coli* B_2C

was added to milk at a rate of about 100/ml, it survived better in Colby than in Camembert cheese. Furthermore, numbers of *E. coli* B$_2$C in Colby-like cheese after seven weeks were comparable to those in brick cheese at about the same time (20,000/g). Frank *et al.* (1978) suggested that the lower pH of unripened Camembert cheese accounted for the poorer survival of *E. coli* in Camembert than in brick cheese. This is also likely to be true for Colby-like cheese.

Storage of Colby-like cheese at 3 ± 1°C and 10 ± 1°C had little effect on survival of *E. coli.* The pH of the cheese appeared to have the most important role in determining survival of *E. coli* in Colby-like cheese. Washing curds for more than 20 min resulted in cheese with high pH values and high coliform counts (1 × 10^8/g), which persisted at high levels for many weeks. Results of this study emphasized the importance of pH regulation, proper sanitation, and proper manufacturing procedures in the manufacture of cheese and suggest that intermediate moisture cheeses like Colby could be potential vehicles for transmission of *E. coli* under certain conditions. Additional information on enteropathogenic *E. coli* in milk and milk products can be found in a review by Kornacki and Marth (1982B).

Milk- and cheese-associated outbreaks of listeriosis occurred in the United States in 1983 and 1985, respectively. The causative agent, *Listeria monocytogenes,* is widespread in the environment and can be recovered from soil, improperly fermented silage, dairy cows, and other animals (Mitscherlich and Marth 1984). Experiments by Ryser *et al.* (1985) indicated that *L. monocytogenes* did not grow during the cottage cheese-making process; some viable cells remained in the curd after the cooking procedure, and they survived in creamed and uncreamed cottage cheese throughout 28 days of refrigerated storage. Additional information on foodborne illness in general and related to dairy products in particular can be found in reviews by Marth (1981, 1985).

The mechanism(s) by which lactic acid bacteria inhibit or inactivate other bacteria is not totally clear. Daly *et al.* (1972), Speck (1972), and Gilliland and Speck (1972) have cited evidence which suggests that the following may be involved: (1) production of antibiotics such as nisin, diplococcin, acidophilin, lactocidin, lactolin, and perhaps others; (2) production of hydrogen peroxide by some lactic acid bacteria; (3) depletion of nutrients by lactic acid bacteria, which makes growth of pathogens difficult or impossible; (4) production of volatile acids; (5) production of acid and reduction in pH; (6) production of D-leucine; and (7) lowering the oxidation-reduction potential of the substrate.

INDUSTRIAL FERMENTATIONS WITH MEDIA DERIVED FROM MILK

Among purely industrial fermentations, milk and its products, for historic and economic reasons, have received only limited attention. Decentralization of casein and cheese manufacture in the early days weakened the competitive position of the low-solids by-product, whey, relative to that of grains and molasses. With changing economic and market trends, by-products of milk which are suited for many industrial fermentations may become more competitive. In times of unusual demand, such as wars produce, these by-products are of considerable industrial interest.

Whey can be used to manufacture lactic acid. Lactose was once the carbohydrate of choice for antibiotic production in the United States but now is of limited use in a few countries outside of the United States. Whey has been used on a large scale for microbiological synthesis of riboflavin, butanol, and acetone. Currently there is renewed interest in this fermentation. Use of whey to manufacture alcohol and yeast is of commercial interest, and its use to produce fat has been studied. Whey is a suitable substrate for the microbiological synthesis of vitamin B_{12} in several fermentations. Enzymatic digests of casein continue to be used on a limited basis in production of antibiotics and can be used in any fermentation which requires a source of amino acids. Casein and the nitrogenous components of whey give rise to large yields of riboflavin in flavinogenesis by means of the fungus *Eremothecium ashbyii*. Skim milk is recommended as a medium for microbiological synthesis of the antibiotic, nisin.

Bacterial oxidations may yield useful products. Vinegar may be obtained from whey in the acetic acid fermentation. Lactobionic acid may be obtained in high yields by the action of *Pseudomonas graveolens* on the lactose in whey. Fermented whey can be used as a food or beverage. The reader who is interested in a more detailed discussion of these and some other fermentations employing whey as a substrate should consult the discussion by Marth (1970).

Production of Lactic Acid

Production of lactic acid using whey sometimes is industrially important. In this fermentation the culture of choice is *L. bulgaricus* because (1) it is homofermentative, producing almost theoretical yields of lactic acid; (2) it is thermophilic and, having an optimum growth temperature between 45° and 50°C, it can be grown in a pasteurized rather than a sterile medium with little danger that the medium will become contaminated; (3) it is acid tolerant and, in a batch process, infrequent neutralization of the medium is satisfactory; and (4) it grows under aerobic or anaerobic conditions.

Fermentation with *L. bulgaricus* is likely to be sluggish in whey, and hence the bacterium is sometimes grown in association with a yeast (*Mycoderma*). The function of the yeast is not clearly defined; perhaps it produces growth factors needed by *L. bulgaricus*, an organism which is highly fastidious in its nutritional requirements. In this connection, it should be remembered that certain strains of *L. bulgaricus* are unusual in that they cannot use the pyrimidine derivative uracil, but instead require orotic acid; nor can the species in general use pantothenic acid, but instead requires pantetheine; and finally, the species cannot use biotin, but instead requires unsaturated fatty acids such as oleic or linoleic. For maximum growth, *L. bulgaricus*, like other species in this genus, probably requires more manganese than is usually present in milk. The fermentation has been described in some detail (Burton 1937A,B; Olive 1936).

Butanol-Acetone-Riboflavin Fermentation

Besides volatile solvents, fermentation with *Clostridium acetobutylicum* yields an appreciable amount of riboflavin. The medium should contain 1.5 and 2.0 ppm iron (Meade *et al.* 1945) and certain salts of organic acids or calcium carbonate (Yamasaki 1939, 1941; Yamasaki and Yositome 1938) for maximal synthesis of riboflavin and greatest fermentation rates.

The butanol-acetone fermentation, using whey supplemented with yeast extract, is substantially a butanol fermentation. Volatile solvents from the fermentation consist of 80% butanol, 13% acetone, and 5% ethanol. Approximately 30% of the lactose that is fermented is converted to butanol, 5% to acetone, 2% to ethanol, and the balance largely to CO_2 and small quantities of butyric and acetic acids, acetylmethylcarbinol, and hydrogen.

C. acetobutylicum is not too exacting in its growth requirements. Asparagine is needed to effect normal production of solvents in what otherwise would be an acid fermentation. Both biotin and p-aminobenzoic acid are required in trace quantities, 0.001 and 0.05 $\mu g/ml$, respectively. Iron is essential in small but variable quantities for attainment of maximal fermentation rates (Leviton 1946, 1949; Meade *et al.* 1945). The requirement for iron varies with the composition of the medium; potassium is also required (Davies 1942A,B, 1943; Davies and Stephenson 1941). Trace quantities of manganese sulfate, lithium chloride, strontium chloride, tin chloride, and zinc chloride aid fermentation in whey (Meade *et al.* 1945).

Yeast extract, liver extract, or cornmeal can be added to whey, thus achieving normal fermentation and avoiding addition of iron. The presence of these added solids (1%) in whey ensures normal fermentation and high yields of riboflavin.

The physiological state of the organism influences the yield of riboflavin, as well as that of volatile solvents, and consequently requires control (Leviton 1946). Transfer of inocula when approximately 25% of the gaseous products of fermentation have envolved is conducive to high yields. Butanol-acetone fermentation was of considerable importance during and shortly after World War II. For years it was of little or no importance, but recently it has again become of commercial interest.

Production of Alcohol

Lactose-fermenting yeasts have been known for some time, but their use to produce ethyl alcohol and yeast from whey received serious attention only much later. Certain *Torula* species yield more alcohol than might have been expected from statements in the literature. Four kefir yeasts, two *Torula* species, one of *Torulopsis* and one additional yeast species, produced alcohol yielding 68 to 80% of the theoretical quantity. A maximal yield, 80.3%, based on a theoretical yield of 4 moles of alcohol per mole of lactose fermented, was obtained with a strain of *Torula cremoris* in a 21.7-hr fermentation at 30 to 32°C (Marth 1974).

Rogosa *et al.* (1947) extended the scale of operation and employed, in addition to *Torula* species, *K. fragilis, Saccharomyces lactis, S. anamensis, Zygosaccharomyces lactis, Mycotorula lactis* and *Candida pseudotropicalis.* Again *T. cremoris* gave the highest yields. The yield of alcohol averaged 90.73% under laboratory conditions and 84% under pilot plant conditions. Additional details of this fermentation are provided by Marth (1970). The advent of ultrafiltration of whey has resulted in permeate (free of whey proteins) which can be fermented. A process claimed to be 5 to 30 times more productive than batch fermentations and employing a continuous fermentor has been described by Mehala *et al.* (1985) as a means of producing alcohol from permeate.

Alcoholic Beverages

Wort of whey supplemented with malt, has been used as a raw material to prepare beer. According to Marth (1974), Dietrich added 5.4% malt wort to dilute whey (2.5% solids), precipitated the albumin at 90°C, and filtered the mixture. He then inoculated the filtrate with a strain of the yeast *S. lactis,* and after 5 to 7 days obtained a product with true beer taste and character.

Whey can be fortified with sucrose and fermented with yeast to yield an alcoholic whey. Upon freeze concentration, a whey liquor with 10 to 69% alcohol can be obtained. Alcohol fermentation carried out in whey supplemented with brown sugar yields a whey cordial (Baldwin 1868). Wine-like beverages have been made from whey, but they do not appear to be competitive with wines made from grape juice.

Microbiological Fat Synthesis

Some microorganisms can synthesize appreciable amounts of fat during growth on various substrates, including whey. However, because of an abundant supply of plant and animal fats, fermentations to produce fat have not become of commercial importance. The reader who is interested in this aspect of fermentation should consult the discussion by Marth (1974) for further details.

Production of Yeast

Conversion of lactose into edible protein for animal or human consumption has appeal because of trends in nutrition which emphasize the importance of protein in diets. The high content of purines and pyrimidines in yeast cells is a limitation in consumption of yeasts by humans. These materials in the diet can lead to high levels of uric acid in blood, which may then lead to gout. Principles underlying microbiological conversion of sugars to protein have been available for many years.

Demmler (1950) used whey and a mixed culture containing predominantly *Candida utilis* in a continuous process to produce yeast in high yields. The fermentation was conducted in a Waldhof-type fermentation tank equipped with a rotating sparger. Under normal operating conditions, an average yield ranging from 13 to 15 g of yeast per liter of whey was obtained. In addition, 1.24 g of heat-coagulable whey proteins were obtained in association with the yeast. The drum-dried yeast product contained 59.4% protein, 4.7% fat, 26.6% invert sugar, 9.2% ash, 3.17-3.4% P_2O_5, 8.6% moisture, and 0.2% sulfur. The purine content was lower than the average for other yeasts. The drum-dried product was more digestible than the spray-dried product, presumably because cell walls of the yeast were destroyed in the drum-drying operation. However, preliminary heat treatment before spray processing eliminated this difference.

Lactose-fermenting yeasts contain the following vitamins in milligrams percent on a dry basis: vitamin A, traces; B_1, 12.8; B_2, 4.4; nicotinic acid, 8.3; ascorbic acid, 7.8; and provitamin A, 40.5 (Springer 1950).

The yeast fermentation methodology has been perfected and is now reasonably economical (Wasserman 1960A,B; Wasserman and Hampson 1960; Wasserman et al. 1958, 1961). Peak oxygen requirements of 100 to 120 ml of O_2 per liter of whey per minute, corresponding to a solution rate of 1 lb/min, were realized in both laboratory and plant investigations in which specially designed sprayer-agitation combinations were employed. In laboratory experiments, supplementation of whey with 0.5 to 1% ammonium sulfate, 0.5% dipotassium phosphate, and 0.1% yeast extract, together with the use of a heavy inoculum constituting 25 to 30% of the weight of sugar present, resulted in both maximal assimilation of available carbon and nitrogen and maximal assimilation rates. Thus time was reduced from the usual 12 to 24 hr to 3 to 4 hr without diminishing yeast yield or quality (high protein content). Calculation based on the quantity of lactose and lactic acid carbon converted to yeast carbon showed that a theoretical yield of 27 g of yeast (containing 45% carbon) per liter of whey was possible. Actual yields of 85% of the theoretical yield were obtained. Stated otherwise, about 0.55 lb of dry yeast could be obtained per pound of lactose.

Exceedingly important in yeast fermentation are the propagators with their aerator-agitator combinations. These govern the oxygen absorption rate of the medium, which must correspond to the peak oxygen demand of the growing culture. Wasserman and Hampson (1960) observed a dependency of the oxygen absorption rate on agitator design and speed and aeration rate. With the Waldhof fermentor, good growth was obtained even when the desired oxygen absorption rate (five mmoles of O_2 per liter per minute) was not realized.

Of the nitrogenous components of whey, yeast uses ammonia nitrogen and about two-thirds of the heat-noncoagulable organic nitrogenous compounds, to the exclusion of the heat-coagulable nitrogenous substances (Wasserman 1960A).

In reproducing its own substance, the yeast cell produces an abundance of nucleic acids. Thus, not all the nitrogen in yeast is protein nitrogen, although calculation of protein concentration is based on this assumption. It is estimated that nucleoproteins make up 20 to 40% of bacterial nitrogen.

Lactose in the Production of Penicillin

The reason for the startling increase in the demand for lactose during World War II was the discovery that this carbohydrate was uniquely suitable to produce penicillin in high yields. The demand for lactose continued to increase largely in connection with an expanding penicillin industry. However, at present in the United States, lactose is rarely used as a media ingredient for fermentations in the antibiotics industry. Producers of antibiotics in some European countries continue to use lactose in some of their fermentation media. Readers interested in a more extensive discussion of this topic can consult an earlier reference by Marth (1974).

Microbiological Synthesis of Nisin

Nisin is distinguished from most antibiotics because it is an assimilable polypeptide that can be tolerated in large dosage by humans and appears to have no effect on the intestinal microflora. The antibiotic occurs in some cultured milks made with *S. lactis,* as well as in raw milk and in some milk products such as cheese. Interest has centered on its ability, when present in cheese, to minimize (although not in all instances) spoilage (gas production and flavor defects) caused by butyric organisms (Eastoe and Long 1959; Hirsch and Grindsted 1954; Hirsh *et al.* 1951; Ramseier 1960).

Nisin has been applied successfully to prepare sterile beverage-quality chocolate milk. The antibiotic serves as a sterilization aid because it inhibits outgrowth of heat-damaged spores and so permits use of less drastic heat treatments for sterilization (Heinemann *et al.* 1964).

Skim milk is a suitable medium to produce nisin (Hawley and Hall 1960). It is inoculated with a suitable active strain of *S. lactis,* and after 40 to 48 hr, during which pH values between 4.5 and 5.5 are established, coagulated proteins containing nisin are separated by centrifugation. This preparation is useful commercially. It may be dried and the nisin extracted with acidified acetone. Methods for further purification are given by Cheeseman and Berridge (1957). In a patent, Hawley and Hall (1960) describe a process in which sterilized skim milk is cultured with *S. lactis* until the titer of nisin at pH 6.0 and 6.3 reaches 1000 Reading units (also designated as "International Units") per milliliter. Paracasein is precipitated with $CaCl_2$ and chymosin (rennin), and the resulting whey is adjusted to pH 4.0 to 4.5 with HCl and drained. The combined whey and curd washings adjusted to pH 5.0 are transferred to a circulating system of vertical foam tubes, and 0.1% Tween is added. The collected foam contains ca. 40,000 Reading Units/ml. Solid nisin is prepared by saturating 500 ml of foam with 27 ml of acetone. The resulting precipitate is extracted with 500 ml of methanol, and nisin in the extract is precipitated with 1000 ml of acetone. The dried precipitate has an activity of 1.4×10^6 Reading Units/g.

Production of Vitamins

Microbiological Synthesis of Riboflavin.   Three types of microorganisms can synthesize riboflavin in significant quantities. *C. acetobutylicum* produces quantities of up to 50 mg/liter. *Candida guilliermondi* and related species synthesize it under suitable conditions in quantities exceeding 100 mg/liter. The yeast-like fungi *Ashbya gossypii* and *Eremothecium ashbyii* are the most productive and under proper conditions will synthesize riboflavin in quantities of up to 2.4 g/liter. Whey supplemented with other nutrients can serve as a substrate for the fermentations that yield riboflavin. Because riboflavin is available more economically from other sources, fermentation is not commonly used to produce the vitamin. A more extensive discussion of these fermentations can be found in an earlier article by Marth (1974).

Microbiological Synthesis of Vitamin B_{12}.   Microbiological synthesis affords the only known means for bulk production of pure vitamin B_{12} and concentrates of the vitamin. Several reports concerned chiefly with vitamin B_{12} yields in actinomycete cultures are available (Garey and Downing 1951; Garibaldi *et al.* 1951; Hall and Tsuchiya 1951; Hall *et al.* 1951; Saunders *et al.* 1951).

A strain of *Bacillus megaterium*, when grown in suitable substrates including whey, can synthesize the vitamin. Garibaldi *et al.* (1951) obtained yields of 0.8 mg/liter, corresponding to a glucose consumption of 10 g.

A low cobalt concentration was shown by Hendlin and Ruger (1950) to limit synthesis of vitamin B_{12}. Cobalt comprises about 4% of the molecule. Working with 13 cultures, including a strain of *Streptomyces griseus*, unidentified rumen and soil isolates, a strain of *Mycobacterium smegmatis*, and *Pseudomonas* species, Hendlin and Ruger (1950) found that addition of 1 to 2 ppm of cobalt increased yield by threefold.

Hargrove and Leviton (1955) and Leviton (1956A,B) found that bacteria in the genus *Propionibacterium* elaborated vitamin B_{12}-active substances in concentrations equal to or greater than those reportedly obtained with other organisms. The active compound produced was identified as hydroxocobalamine.

Leviton and Hargrove (1952) compared lactose and glucose as sources of energy in several vitamin B_{12} fermentations. Employing different strains of *B. megaterium* and several unidentified rumen isolates, they found that lactose brought about higher yields and faster fermentation than did glucose. With *Streptomyces olivaceus* as the organism and clarified whey as the lactose source, lactose and glucose were compared in enzymatically hydrolyzed casein-yeast extract media, in distillers' soluble media, and in ammonium caseinate media. All media were fortified with Co^{2+}. Highest yields were obtained with the lactose-containing media.

Laboratory-scale experiments which used *L. casei* symbiotically with *Propionibacterium freudenreichii* in the fermentation of whey gave an average yield of 2.2 mg of vitamin per liter; the maximum was 4.3 mg/liter. Production of vitamin B_{12} is not species-specific. All species of *Propionibacterium*, when cultivated under the same conditions, produce active substances, but in different quantities. *P. freudenreichii* and *P. zeae* synthesized sufficient quantities to warrant their consideration for commercial exploitation. Because propionic acid bacteria are active during Swiss cheese ripening, it was anticipated, and actually demonstrated, that production of vitamin B_{12} in Swiss cheese is influenced by the same factors that influence its production in pure culture, particularly by the cobalt content of milk (Hargrove and Leviton 1955).

Propionic acid bacteria require, for maximal growth rates, a highly degraded source of amino acids. In caseinate media and even in peptone media, rates are likely to be relatively slow. For maximal yields of vitamin B_{12}, a high degree of anaerobiosis is not required. Because assimilation is largely anaerobic, a high ratio of vitamin concentration to total cell mass is obtained. Thus this fermentation is particularly suitable for preparation of the pure vitamin, since the cell mass contains all of the vitamin and furnishes a highly concentrated initial source for further treatment. As a first step in further treatment, harvested cells may be coagulated and then lysed in a 50% (by volume) acetone solution or in mixtures of butyl and ethyl alcohols (Leviton 1956B).

Sewage wastes contain as much as 4 ppm of vitamin B_{12} (Hoover *et al.* 1952B; Miner and Wolnak 1953). Although frowned on for aesthetic reasons as a source of vitamin B_{12} for human nutrition, wastes from activated sludge processes may well provide the cheapest source for preparation of vitamin B_{12} concentrates used in cattle feed. Symbiotic growth of lactic and acetic acid bacteria has been recommended for producing sour milk products biologically enriched with vitamin B_{12} (Rykshina 1961). Acetic acid bacteria cultured in whey fortified with cobalt salts led to an 80-fold increase in vitamin B_{12}. Propionic acid bacteria in skim milk supplemented with dimethylbenzimidazole increased the vitamin content by 300-fold.

In view of work by Barker *et al.* (1960A,B) and Weissbach *et al.* (1961), it appears that the natural cobamide produced in bacterial cultures is not vitamin B_{12} but rather coenzyme B_{12}. Berry and Bullerman (1966) and Bullerman and Berry (1966A,B) described a two-stage process for production of vitamin B_{12} by *Propionibacterium shermanii.* Maintenance of anaerobiosis (first step) during the first half of the fermentation is accompanied by formation of the macro-ring portion of the B_{12} molecule. During the aerobic phase (second step) in the second half of the fermentation, the organism attains its greatest population and also attaches the nucleotide portion and thus completes synthesis of the B_{12} molecule. Use of aerobiosis during the second phase of fermentation precludes addition of the B_{12} precursor (5,6-dimethylbenzimidazole) to the medium.

The process of Bullerman and Berry involves (1) preparing a medium containing 6 to 8% whey solids, 0.5 to 1% yeast extract, and 15 ppm cobalt; (2) adding a 10% inoculum of *P. shermanii* and holding the temperature at 29°C; (3) adjusting the pH daily so that it is returned to 6.5 to 7.0; (4) sparging with CO_2 for 84 hr and then with air for 84 hr; and (5) drying the fermented material. The dried product thus obtained contained 365 μg of B_{12} per gram, whereas the maximum yield in the unconcentrated liquid approximated 15 μg/ml.

"Oxidative" Fermentations

Whey does not lend itself to direct production of acetic acid by species of the genus *Acetobacter.* Furthermore, use of combined inocula of yeasts and *Acetobacter* species has not proved fruitful. However, Haeseler has described an operable procedure, in which an alcoholic followed by an acetic acid fermentation yielded a vinegar with satisfactory qualities (Marth 1974).

Although production from whey of a 5 to 7% acid vinegar may prove feasible, vinegar with as much as 10% acid seems unlikely because of adverse effects from the high salt concentration in concentrated whey. Use of newer methods for concentration (ultrafiltration and reverse osmosis) may overcome this difficulty and enable production of vinegar with 10 to 12% acetic acid. The process described by Haeseler yielded a whey vinegar containing only 4% acid. This product, yellow-brown in color, had a malt-vinegar character with only a weak whey taste and slight saltiness, which were not detrimental. The possibility of slime formation and overoxidation with whey as a substrate were considered detrimental to the use of quick vinegar processes. Whey vinegar is produced commercially in Europe.

Production of lactobionic acid from lactose through bacterial oxidation is of some interest because of the properties of this substance. Lockwood and Stodola (1950), using *P. graveolens,* recovered lactobionic acid in 77% yield from a fermentation mixture containing the following per liter: 96 g of anhydrous lactose, 0.62 g of KH_2PO_4, 0.25 g of $MgSO_4 \cdot 7H_2O$, 2.1 g of urea, 28 g of $CaCO_3$, 5 ml of cornsteep liquor, and 0.3 ml of soybean oil.

The sequestering and emulsifying properties of lactobionic acid suggest a commercial potential for this product. In addition, it is a solubilizing agent for calcium salts. Solutions of calcium lactobionate containing up to 70% of the salt have been prepared and may be useful as a source of calcium (Kastens and Baldauski 1952).

Other Fermentations Using Whey

Making whey cheese is perhaps one of the earliest fermentations which used whey (or its components) as a substrate. Examples of such cheese include Schottengsied, Primost (Mysost), Ricotta, and Gjetost (made from goat's milk whey). Some kinds of whey cheese, however, do not involve a fermentation step.

Whey has been suggested as a culture medium for growth of lactic acid bacteria. Czulak (1960) reported whey could be used to grow *P. roqueforti*, and Lundstedt and Fogg (1962) found it suitable for growth of *S. lactis* subsp. *diacetylactis*. They noted further that when citrated whey was cultured with *S. lactis* subsp. *diacetylactis* and added to creamed cottage cheese, a pleasing diacetyl flavor and aroma developed in two to six days when the cheese was held refrigerated.

More recently, Richardson *et al.* (1977) have used phosphate-supplemented fresh whey as a medium to propagate lactic starter cultures. The pH of the medium is maintained at 6.0 to 6.3 through addition of ammonia. Use of this medium results in a starter culture with superior activity.

Use of fermented whey as a food has been suggested. Jagielski (1871) combined whey and lactose with an appropriate culture and produced a whey kumiss. Later, Krul'kevich mixed equal volumes of whey and buttermilk with kumiss yeasts, *L. bulgaricus*, and *L. acidophilus*. The finished product is claimed to resemble kumiss (Marth 1974). A condensed whey food composed, in part, of whey fermented by *L. bulgaricus* and *P. shermanii* has been described in a patent issued to Meade *et al.* (1945). Additional information on fermenting whey is given by Friend and Shahani (1979).

Other uses for whey based on fermentation include production of (1) lactase enzyme from *K. fragilis* (or other organisms able to use whey), as described by Myers and Stimpson (1956) and Wendorff *et al.* (1970); (2) a high-vitamin, high-protein product containing little or no lactose and prepared by fermenting whey with an organism able to use lactose (e.g., *K. fragilis*), followed by drying the fermented material; and (3) an animal feed suitable for ruminants by fermenting whey with *L. bulgari-*

cus at a pH of 5.8 to 6.0, concentrating the fermented whey to 30 to 80% solids, and neutralizing the concentrate to pH 7 to 8 (Marth 1974).

Attempts to improve the quality of whey include those of Johnstone and Pfeffer (1959), who increased its nitrogen content with a nitrogen-fixing strain of *E. aerogenes,* and Davidov and Rykshina (1961), who used whey fortified with $CoCl_2$, fermented it with acetic acid bacteria, and observed an 80-fold increase in vitamin B_{12}.

Addition of a whey paste plus a nisin-producing strain of *S. lactis* to silage has been suggested as a means of preventing development of butyric acid bacteria in the fodder. Further information on conversion of dairy and other food processing wastes to useful products appears in a review by Cousin (1980).

DAIRY WASTE DISPOSAL

Wet oxidation of dairy waste is one of the most difficult tasks that microorganisms are required to do. The microbiological system must oxidize the carbon and hydrogen of organic compounds to carbon dioxide and water, respectively, and must at the same time conserve its own mass. In other words, the cellular mass must neither increase nor decrease over long periods. That this ultimate objective is closely approached in practice testifies to the remarkable power of the metabolic capacity of microorganisms.

Dairy wastes fall into two categories, one of which may be described as an intrinsic waste, and the other as a conditional waste. All dairy factories experience losses that are intrinsically a part of factory operation. For example, a dairy factory that receives 10,000 lb of milk daily may produce each working day about 1250 gal of waste with a milk solids concentration of 0.1%. Cheese plants, on the other hand, produce whey as a by-product of cheesemaking; although whey contains half the nutrients of the milk from which it was derived, it must be treated as a conditional waste—conditional upon the absence of a suitable market for its use. A more detailed discussion on disposal of dairy wastes can be found in a review by Arbuckle (1970).

Treatment of Dairy Waste by Aeration

The magnitude of the chemical or biological oxygen demand of solutions of organic matter determines whether or not these solutions may be safely added to bodies of water. Chemical oxygen demand (COD) is the amount of oxygen, determined chemically, necessary for complete oxidation of an organic substance, and is usually reported in parts per million (ppm) (Porges *et al.* 1950). For milk wastes, biochemical oxidation demand (BOD) and COD are practically equal.

As oxidants, either permanganate or dichromate may be employed under standard conditions of concentration, temperature, and time. These reagents have been studied critically; only the results with dichromate were found to reflect accurately the BOD of dairy wastes (Fritz 1960A,B).

Aeration techniques are successful only if oxygen can be supplied at a sufficiently high rate to lower the COD to an acceptable value. Extensive investigations on the biochemical and chemical oxidation of dairy wastes have shown that each pound of dry organic matter in dairy waste requires about 1.2 lb of oxygen for complete oxidation (Hoover and Porges 1952; Hoover *et al.* 1952A; Porges 1956). During the period of rapid assimilation, bacteria need about 37.5% of their complete oxygen requirement, or 0.45 lb; and in the process, 0.52 lb of new cell material is formed per pound of waste solids. To oxidize this newly formed sludge, 0.75 lb of oxygen is required, the difference between the oxygen required for complete oxidation of 1 lb of waste solids and that required for assimilation. During endogenous respiration at 32.2°C, sludge is consumed at an hourly rate of approximately 1%. Thus, if an amount of sludge equal to 0.52 lb of newly formed cells is to be oxidized in time t_1, no less an amount of sludge than that given below would be required to maintain this condition: equilibrium weight of sludge per pound of organic matter $= 52/t_1$. If the parts of oxygen required to oxidize the organic matter in 1 million parts of waste volume—the ppm COD—is known, the total oxygen requirement in pounds for any given waste volume, V, in gallons is easily calculated. The weight of organic solids is equal to 83.3% of the total oxygen requirement (COD), and hence the equilibrium sludge weight is given by the following equation: sludge $= (52 \times V \times \text{ppm COD} \times 8.34 \times 0.833 \times 10^{-6})/t_1$.

If, for example, a waste volume, V, of 10,000 gal with a ppm COD of 1500 is processed in $t = 20$ hr, the equilibrium sludge weight is 270 lb. The calculation is oversimplified and is about 10% too low, assuming, as it does, that endogenous respiration and assimilation occur simultaneously during the entire operation. Actually, there is always a retention time during which cellular substance is consumed without replenishment.

The hourly oxygen requirement for sludge respiration is equal to the sludge dissipation rate multiplied by the pounds of oxygen (1.44) required for oxidation of each pound of ash-free sludge. The hourly oxygen requirement for assimilation is given by the quotient of total oxygen required for assimilation and the time required to introduce the waste. The hourly oxygen requirement during assimilation is equal to the sum of the two aforementioned requirements, and may be expressed in terms of the volume, V, of influent, the ppm COD, the feed time, t_2, and the endogenous respiration time, t_1, thus: O_2 (lb/hr) $=$ $(5.2V \times \text{ppm COD} \times 10^{-6})t_1 + (3.13V \times \text{ppm COD} \times 10^{-6})/t_2$.

This equation summarizes some of the arguments and data contained in the literature (Porges *et al.* 1960). The aeration device must be designed to furnish the solution with oxygen at the required rate. The tank must be designed to accommodate milk waste and sludge. Allowances must be made for a certain proportion of free space (freeboard), and settling space. The design, construction, and operation of dairy waste disposal units have been described (Porges 1958; Porges *et al.* 1960).

Processing of Whey Wastes

Whey solids compared with milk solids contain a greater proportion of lactose and a much smaller proportion of nitrogen. Consequently, in the processing of whey wastes even under conditions of adequate aeration, the rate of assimilation may be limited by the COD–nitrogen imbalance. Jasewicz and Porges (1958) observed that when sludge (2000 ppm COD) was used to treat dilute whey waste (1000 ppm COD) under highly aerobic conditions, no additional nitrogen was necessary for complete whey removal, since the essential nitrogen was supplied during endogenous respiration. Addition of ammonium sulfate to aerators was recommended to compensate for the additional load imposed on them when whey is wasted along with the normal load. In studies using whey, it was found that under the laboratory schedule of daily feedings, both supplemented and unsupplemented sludges gradually deteriorated and presented serious bulking problems after three months. This suggested that supplementation with nitrogen alone was not enough. In a 61-day study of the COD balance in a system to which whey was added 48 times to aerated sludge, it was observed that whey wastes may be readily treated under certain conditions without nitrogen addition. An average of 75% of the influent whey COD was relieved when no provisions were made for removal of sludge from the effluent. The sludge accounted for all but 2 to 3% of the effluent COD. Calculations based on a sludge oxidation rate of 6.3% per day showed that dynamic equilibrium would be possible if 100 units of sludge were used to treat 10 units of whey.

REFERENCES

Adams, D. M., Barach, J. T. and Speck, M. L. 1975. Heat resistant proteases produced in milk by psychrotrophic bacteria of dairy origin. *J. Dairy Sci. 58*, 828-834.

Adams, D. M. and Brawley, T. G. 1981. Heat resistant bacterial lipases and ultra-high temperature sterilization of dairy products. *J. Dairy Sci. 64*, 1951-1957.

Alford, J. A. and Frazier, W. C. 1950. Occurrence of micrococci in Cheddar cheese made from raw and from pasteurized milk. *J. Dairy Sci. 33*, 107-114.

Alichanidis, E. and Andrews, A. T. 1977. Some properties of the extracellular protease produced by the psychrotrophic bacterium *Pseudomonas fluorescens* strain AR-11. *Biochim. Biophys. Acta 485*, 424-433.

Alifax, R. and Chevalier, R. 1962. Studies on nisinase produced by *Streptococcus thermophilus. J. Dairy Res. 29*, 233-240.

Alkanhal, H. A., Frank, J. F. and Christen, G. L. 1985. Microbial protease and phospholipase C stimulate lipolysis of washed cream. *J. Dairy Sci. 68*, 3162-317.

Alm, L. 1982. Effect of fermentation on L(+) and D(−) lactic acid in milk. *J. Dairy Sci. 65*, 515-520.

Anders, R. F., Hogg, D. M. and Jago, G. R. 1970A. Formation of hydrogen peroxide by Group N streptococci and its effect on their growth and metabolism. *Appl. Microbiol. 19*, 608-612.

Anders, R. F., Jonas, H. A. and Jago, G. R. 1970B. A survey of the lactate dehydrogenase activities in Group N streptococci. *Aust. J. Dairy Technol. 5*, 73-76.

Anderson, R. 1980. Microbial lipolysis at low temperatures. *Appl. Environ. Microbiol. 39*, 36-40.

Anderson, I. and Leesment, H. 1970. The influence of manganese on the activity of aroma bacteria in starters. *XVIII Int. Dairy Congr. 1E*, 114.

Anema, P. J. 1964. Purification and properties of β-galactosidase of *Bacillus subtilis. Biochim. Biophys. Acta 89*, 495-502.

Arbuckle, W. S. 1970. Disposal of dairy wastes. *In: Byproducts from Milk*, B. H. Webb, and E. O. Whittier (Editors). AVI Publishing Co., Westport, Conn., pp. 405-421.

Argyls, P. J., Mathison, G. E. and Chandan, R. C. 1976. Production of cell-bound proteinase by *Lactobacillus bulgaricus* and its location in the bacterial cell. *J. Appl. Bacteriol. 41*, 175-184.

Baldwin, A. E. 1868. Improved process of treating milk to obtain useful products. U.S. Patent 78,640.

Baribo, L. E. and Foster, E. M. 1951. The production of a growth inhibitor by lactic streptococci. *J. Dairy Sci. 34*, 1136-1144.

Baribo, L. E. and Foster, E. M. 1952. The intracellular proteinases of certain organisms from cheese and their relationship to the proteinases in cheese. *J. Dairy Sci. 35*, 149-160.

Barker, H. A., Smyth, R. D., Weissbach, H., Munch-Peterson, A., Toohey, J. I., Ladd, J. N., Volcani, B. E. and Marilyn Wilson, R. 1960A. Assay, purification, and properties of adenylcobamide coenzyme. *J. Biol. Chem. 235*, 181-190.

Barker, H. A., Smyth, R. D., Weissbach, H., Toohey, J. I., Ladd, J. N. and Volcani, B. E. 1960B. Isolation and properties of crystalline cobamide coenzymes containing benzimidazole or 5,6-dimethylbenzimidazole. *J. Biol. Chem. 235*, 480-488.

Bauchop, T. and Elsdon, S. R. 1960. The growth of microorganisms in relation to their energy supply. *J. Gen. Microbiol. 23*, 457-469.

Baumann, D. P. and Reinbold, G. W. 1964. Preservation of lactic cultures. *J. Dairy Sci.* (abstract). *47*, 674.

Baumann, D. P. and Reinbold, G. W. 1966. Freezing of lactic cultures. *J. Dairy Sci. 49*, 259–264.

Bayer, K. 1983. Trace element supplementation of cheese whey for the production of feed yeast. *J. Dairy Sci. 66*, 214–220.

Beach, A. S. 1952. An agar diffusion method for the assay of nisin. *J. Gen. Microbiol. 6*, 60–63.

Beattie, S. E. and Torrey, G. S. 1984. Volatile compounds produced by *Brevibacterium linens* inhibit mold spore germination (abstract). *J. Dairy Sci. 67* (suppl. 1), 84.

Berry, E. C. and Bullerman, L. B. 1966. Use of cheese whey for vitamin B_{12} production. II. Cobalt, precursor, and aeration limits. *Appl Microbiol. 14*, 356–357.

Bills, D. D., Morgan, M. E., Reddy, L. M. and Day, E. A. 1965. Identification of compounds responsible for fruit flavor defect of experimental Cheddar cheeses. *J. Dairy Sci. 48*, 1168–1170.

Bissett, D. L. and Anderson, R. L. 1973. Lactose and D-galactose metabolism in *Staphylococcus aureus:* Pathway of D-galactose 6-phosphate degradation. *Biochem. Biophys. Res. Commun. 52*, 641–645.

Bissett, D. L. and Anderson, R. L. 1974A. Genetic evidence for the physiological significance of the D-tagatose 6-phosphate pathway of lactose and D-galactose degradation in *Staphylococcus aureus. J. Bacteriol. 119*, 698–704.

Bissett, D. L. and Anderson, R. L. 1974B. Lactose and D-galactose metabolism in Group N streptococci: Presence of enzymes for both the D-glucose 1-phosphate and D-tagatose 6-phosphate pathways. *J. Bacteriol. 117*, 318–320.

Borglum, G. B. and Sternberg, M. Z. 1972. Properties of a fungal lactase. *J. Food Sci. 37*, 619–624.

Bottazzi, V. 1962. Proteolytic activity of some strains of thermophilic lactobacilli. *Proc. 16th Int. Dairy Congr. B*, 522.

Brandsaeter, E. and Nelson, F. E. 1956A. Proteolysis by *Lactobacillus casei.* I. Proteinase activity. *J. Bacteriol. 72*, 68–72.

Brandsaeter, E. and Nelson, F. E. 1956B. Proteolysis by *Lactobacillus casei.* II. Peptidase activity. *J. Bacteriol. 72*, 73–78.

Brown, A. T. and Wittenberger, C. L. 1972. Fructose 1,6 diphosphate–dependent lactate dehydrogenase from a cariogenic streptococcus: Purification and regulatory properties. *J. Bacteriol. 110*, 604–615.

Buchanan, R. E. and Gibbons, N. E. 1974. *Bergey's Manual of Determinative Bacteriology*, 8th ed. Williams and Wilkins Co., Baltimore.

Bullerman, L. B. and Berry, E. C. 1966A. Use of cheese whey for vitamin B_{12} production. I. Whey solids and yeast extract levels. *Appl. Microbiol. 14*, 353–355.

Bullerman, L. B. and Berry, E. C. 1966B. Use of cheese whey for vitamin B_{12} production. III. Growth studies and dry-weight activity. *Appl. Microbiol. 14*, 358–360.

Burton, L. V. 1937A. By products of milk: Methods of conversion which will help solve the burdensome surplus-milk problem. Part I. *Food Ind. 9*, 571–575, 617.

Burton, L. V. 1937B. Part II. Conversion of calcium lactate to lactic acid and production of whey powders. *Food Ind. 9*, 634–636.

Burvall, A., Asp, N. G. and Dahlqvist, A. 1979. Oligosaccharide formation during hydrolysis of lactose with *Saccharomyces lactis* lactase (Maxilact): Part 1. Quantitative aspects. *Food Chem. 4*, 243–250.

Byers, B. R. and Arveneaux, J. E. L. 1971. Microbial transport and utilization of iron. *In: Microorganisms and Minerals.* E. D. Weinberg (Editor). Marcel Dekker, New York, pp. 215–249.

Calmes, R. 1978. Involvement of phosphoenolpyruvate in the catabolism of caries-conducive disaccharides by *Streptococcus mutans:* Lactose transport. *Infect. Immun. 19*, 934–942.

Cardini, C. E. and Leloir, L. F. 1952. Enzymic phosphorylation of galactosamine and galactose. *Arch. Biochem. Biophys. 45*, 55-64.

Castberg, H. B. and Morris, H. A. 1976. Degradation of milk proteins by enzymes from lactic acid bacteria used in cheesemaking: A review. *Milchwissenschaft 31*, 85-90.

Chassy, B. M. and Thompson, J. 1983. Regulation of lactose-phosphoenolpyruvate-dependent phosphotransferase system and β-D-phosphogalactoside galactohydrolase activities in *Lactobacillus casei. J. Bacteriol. 154*, 1195-1203.

Cheeseman, G. C. and Berridge, N. J. 1957. An improved method of preparing nisin. *Biochem. J. 65*, 603-608.

Cheeseman, G. C. and Berridge, N. J. 1959. Observations on the molecular weight and chemical composition of nisin A. *Biochem. J. 71*, 185-194.

Cheung, B. A. and Westhoff, D. C. 1983. Isolation and identification of ropy bacteria in raw milk. *J. Dairy Sci. 66*, 1825-1834.

Choudhery, A. K. and Mikolajcik, E. M. 1971. Activity of *Bacillus cereus* proteinases in milk. *J. Dairy Sci. 53*, 363-366.

Chrisope, G. L. and Marshall, R. T. 1976. Combined action of lipase and phospholipase C on a model fat globule emulsion and raw milk. *J. Dairy Sci. 59*, 2024-2030.

Cogan, T. M. 1980. Mesophilic lactic streptococci: A review. *Lait 60*, 397-425. (French)

Cogan, T. M. 1981. Constitutive nature of the enzymes of citrate metabolism in *Streptococcus lactis* subsp. *diacetylactis. J. Dairy Res. 48*, 489-495.

Cogan, T. M., Fitzgerald, R. J. and Doonan, S. 1984. Acetolactate synthase of *Leuconostoc lactis* and its regulation of acetoin production. *J. Dairy Res. 51*, 597-604.

Cogan, T. M., O'Dowd, M. and Mellerick, D. 1981. Effect of pH and sugar on acetoin production from citrate by *Leuconostoc lactis. Appl. Environ. Microbiol. 41*, 1-8.

Coghill, D. 1979. The ripening of blue vein cheese: A review. *Aust. J. Dairy Technol. 34*, 72-75.

Collins, E. B. 1961. Domination among strains of lactic streptococci with attention to antibiotic production. Appl. Microbiol. *9*, 200-205.

Collins, E. B. 1972. Biosynthesis of flavor compounds by microorganisms. *J. Dairy Sci. 55*, 1022-1028.

Cords, B. R. and McKay, L. L. 1974. Characterization of lactose-fermenting revertants from lactose-negative *Streptococcus lactis* C2 mutants. *J. Bacteriol. 119*, 830-839.

Corminboeuf, F. G. 1933A. Historical considerations of the acidic fermentation in milk and of its microflora. Part I. *Sci. Agr. 13*, 466-470. (French)

Corminboeuf, F. G. 1933B. Historical considerations of the acidic fermentation in milk and on its microflora. Part II. *Sci. Agr. 13*, 596-607. (French)

Cousin, M. A. 1980. Converting food processing wastes into food or feed through microbial fermentation. *Ann. Rep. Ferment. Proc. 4*, 31-65.

Cousin, M. A. 1982. Presence and activity of phychrotrophic microorganisms in milk and dairy products: A review. *J. Food Prot. 45*, 172-207.

Cousin, M. A. and Marth, E. H. 1977A. Changes in milk protein caused by psychrotrophic bacteria. *Milchwissenschaft 32*, 337-341.

Cousin, M. A. and Marth, E. H. 1977B. Psychrotrophic bacteria cause changes in stability of milk to coagulation by rennet or heat. *J. Dairy Sci. 60*, 1042-1047.

Cousin, M. A. and Marth, E. H. 1977C. Lactic acid production by *Streptococcus lactis* and *Streptococcus cremoris* in milk precultured with psychrotrophic bacteria. *J. Food Prot. 40*, 406-410.

Cousin, M. A. and Marth, E. H. 1977D. Lactic acid production by *Streptococcus thermophilus* and *Lactobacillus bulgaricus* in milk precultured with psychrotrophic bacteria. *J. Food Prot. 40*, 475-479.

Cowman, R. A. and Speck, M. L. 1965. Ultra-low temperature storage of lactic streptococci. *J. Dairy Sci. 48*, 1531-1532.

Crow, F. L. and Pritchard, G. C. 1977. Fructose 1,6 diphosphate activated lactate dehydrogenase from *Streptococcus lactis;* Kinetic properties and factors affecting activation. *J. Bacteriol. 131,* 82–91.

Crow, V. L. and Thomas, T. D. 1984. Properties of a *Streptococcus lactis* strain that ferments lactose slowly. *J. Bacteriol. 157,* 28–34.

Czeszar, J. and Pulay, G. 1956. Standardization of methods to analyze milk in France. *14th Int. Dairy Congr., Proc. 3*(2), 423–427. (French)

Czulak, J. 1960. Growth of *Penicillium roqueforti* on a whey medium. *Aust. J. Dairy Technol. 15,* 118–120.

Dacre, J. C. 1953. Cheddar cheese flavor and its relation to tyramine production by lactic acid bacteria. *J. Dairy Res. 20,* 217–223.

Daly, C., Sandine, W. E. and Elliker, P. R. 1972. Interactions of food starter cultures and food-borne pathogens: *Streptococcus diacetilactis* versus food pathogens. *J. Milk Food Technol. 35,* 349–357.

Davidov, R. B. and Rykshina, Z. P. 1961. An inexpensive source of vitamin B_{12} for use as animal feed (abstract). *Milchwissenschaft 16,* 434. (German)

Davies, R. 1942A. Studies on the acetone-butyl alcohol fermentation. II. Intermediates in the fermentation of glucose by *Clostridium acetobutylicum. Biochem. J. 36,* 582–596.

Davies, R. 1942B. Studies on the acetone-butyl alcohol fermentation. III. Potassium as an essential factor in the fermentation of maize meal by *Clostridum acetobutylicum* (BY). *Biochem. J. 36,* 596–599.

Davies, R. 1943. Studies on the acetone-butanol fermentation. IV. Acetonacetic acid decarboxylase of *Clostridium acetobutylicum* (BY). *Biochem. J. 37,* 230–238.

Davies, R. and Stephenson, M. 1941. Studies on the acetone-butyl alcohol fermentation. I. Nutritional and other factors involved in the preparation of active suspensions of *Clostridium acetobutylicum* (Weizmann). *Biochem. J. 35,* 1320–1331.

Deeth, H. C. and Tamime, A. Y. 1981. Yogurt: Nutritive and therapeutic aspects. *J. Food Prot. 44,* 78–86.

DeKlerk, H. C. and Coetzer, J. N. 1961. Antibiosis among lactobacilli. *Nature 192,* 340–341.

Demmler, G. 1950. Growth of yeast in whey using the Waldhof procedure. *Milchwissenschaft 5,* 11–17. (German)

Denis, F. and Veillet-Poncet, L. 1980. Characteristics of the proteolytic enzymatic system of *Aeromonas hydrophilia* LP 50. *Lait 60,* 238–253. (French)

Desmazeau, M. J. and Gripon, J. C. 1977. General mechanism of protein breakdown during cheese ripening. *Milchwissenschaft 32,* 731–734.

De Vries, W., Aleem, M. T. H. and Hemri-Wagner, A. 1974. The functioning of cytachrome b in the election transport to fumerate in *Propionbacterium freudenreichii* and *Propionbacterium pentosaceum. Arch. Microbiol. 112,* 271–276.

Dills, S. S., Apperson, A., Schmidt, M. R. and Sater, M. H., Jr. 1980. Carbohydrate transport in bacteria. *Microbiol. Rev. 44,* 385–418.

Doelle, H. W. 1975. *Bacterial Metabolism.* Academic Press, New York.

Dolin, M. I. 1955. The DPNH-oxidizing enzymes of *Streptococcus faecalis.* II. The enzymes utilizing oxygen, cytochrome c, peroxide and 2,6-ichlorophenol or ferricyanide as oxidants. *Arch. Biochem. Biophys. 55,* 415–435.

Dumont, J. P., Roger, S. and Adda, J. 1975. Identification of a nitrogenous heterocyclic compound responsible for a potato-like off-flavor in Gruyere de Comte. *Lait 55,* 479–487.

Dunican, L. K. and Seeley, H. W., Jr. 1965. Extracellular polysaccharide synthesis by members of the genus *Lactobacillus:* Conditions for formation and accumulation. *J. Gen. Microbiol. 40,* 297–308.

Dyachenko, P. F., Shchedushnov, E. V. and Nassib, T. G. 1970. Characteristics of proteolytic activity of thermophilic lactic acid bacteria used for cheesemaking. *XVIII Int. Dairy Congr. 1E*, 274.

Eastoe, J. E. and Long, J. E. 1959. The effect of nisin on the growth of cells and spores of *Clostridium welchii* in gelatine. *J. Appl. Bacteriol. 22*, 1–7.

Efstathiou, J. P. and McKay, L. L. 1976. Plasmids in *Streptococcus lactis:* Evidence that lactose metabolism and proteinase activity are plasmid linked. *Appl. Environ. Microbiol. 32*, 38–44.

Egan, J. B. and Morse, M. L. 1966. Carbohydrate transport in *Staphylococcus aureus.* III. Studies in the transport process. *Biochim. Biophys. Acta 112*, 63–73.

Eggimann, B. and Bachmann, M. 1980. Purification and partial characterization of an aminopeptidase from *Lactobacillus lactis. Appl. Environ. Microbiol. 40*, 876–882.

El Soda, M., Bergere, J. L. and Desmazeaud, M. J. 1978A. Detection and localization of peptide hydrolases in *Lactobacillus casei. J. Dairy Res. 5*, 519–524.

El Soda, M., Desmazeaud, M. J. and Bergere, J. L. 1978B. Peptide hydrolases of *Lactobacillus casei:* Isolation and general properties of various peptidase activities. *J. Dairy Res. 45*, 445–455.

Emmons, D. B., and Tuckey, S. L. 1967. *Cottage Cheese and Other Cultured Milk Products.* Pfizer, Inc., New York.

Exterkate, F. A. 1975. An introductory study of the proteolytic system of *Streptococcus cremoris* strain HP. *Neth. Milk Dairy J. 29*, 303–318.

Exterkate, F. A. 1979. Accumulation of proteinase in the cell wall of *Streptococcus cremoris* strain AM, and regulation of its production. *Arch. Microbiol. 120*, 247–254.

Exterkate, F. A. 1984. Location of peptidases outside and inside the membrane of *Streptococcus cremoris. Appl. Environ. Microbiol. 47*, 177–183.

Farrow, J. A. E. 1980. Lactose hydrolysing enzymes in *Streptococcus lactis* and *Streptococcus cremoris* and also in some other species of streptococci. *J. Appl. Bacteriol. 49*, 493–503.

Farrow, J. A. E., and Garvie, E. 1979. Strains of *Streptococcus lactis* which contain β-galactosidase. *J. Dairy Res. 46*, 121–125.

Fitz-Gerald, C. H. and Deeth, H. C. 1983. Factors influencing lipolysis by skim milk cultures of some psychrotrophic microorganisms. *Aust. J. Dairy Technol. 38*, 97–101.

Foissy, H. 1974. Examination of *Brevibacterium linens* by an electrophoretic zymogram technique. *J. Gen. Microbiol. 80*, 197–207.

Foissy, H. 1978A. Aminopeptidase from *Brevibacterium linens:* Production and purification. *Milchwissenschaft 33*, 221–223.

Foissy, H. 1978B. Some properties of aminopeptidase from *Brevibacterium linens. FEMS Microbiol. Lett. 3*, 207–210.

Foissy, H. 1978C. Aminopeptidase from *Brevibacterium linens:* Activation and inhibition. *Z. Lebensm. Unters.-Forsch. 166*, 164–166.

Fordyce, A. M., Crow, V. L. and Thomas, T. D. 1984. Regulation of product formation during glucose or lactose limitation in nongrowing cells of *Streptococcus lactis. Appl. Environ. Microbiol. 48*, 332–337.

Forsen, R. and Haiva, V. 1981. Induction of stable slime-forming and mucoid states by p-fluorophenylalanine in lactic streptococci. *FEMS Microbiol. Lett. 12*, 409–413.

Forsen, R., Raunio, V. and Myllymaa, R. 1973. Studies on slime-forming Group N streptococcus strains. *Acta U. Ouluensis*, Series A12 (Biochemica No. 3), 4–19.

Foster, E. M. 1962. Symposium on lactic starter cultures. VI. Culture preservation. *J. Dairy Sci. 45*, 1290–1294.

Fox, P. F. 1981. Proteinases in dairy technology. *Neth. Milk Dairy J. 35*, 233–253.

Fox, P. F. and Stephaniak, L. 1983. Isolation and some properties of extracellular heat-stable lipases from *Pseudomonas fluorescens* strain AFT 36. *J. Dairy Res. 50*, 77–89.

Frank, J. F., Marth, E. H. and Olson, N. F. 1977. Survival of enteropathogenic and nonpathogenic *Escherichia coli* during the manufacture of Camembert cheese. *J. Food Prot. 40*, 835–842.

Frank, J. F., Marth, E. H. and Olson, N. F. 1978. Behavior of enteropathogenic *Escherichia* during manufacture and ripening of brick cheese. *J. Food Prot. 41*, 111–115.

Frank, J. F. and Somkuti, G. A. 1979. General properties of beta-galactosidase of *Xanthomonas campestris*. *Appl. Environ. Microbiol. 38*, 554–556.

Friedmann, R. and Epstein, C. 1951. The assay of the antibiotic nisin by means of a reductase (resazurin) test. *J. Gen. Microbiol. 5*, 830–839.

Friend, B. A. and Shahani, K. M. 1979. Whey fermentation. *N.Z. J. Dairy Sci. Technol. 14*, 143–152.

Fritz, A. 1960A. Determination of the strength of dairy wastes. Part I. *Milchwissenschaft 15*, 237–242. (German)

Fritz, A. 1960B. Determination of the strength of dairy wastes. Part II. *Milchwissenschaft 15*, 609–612. (German)

Fryer, T. F., Reiter, B. and Lawrence, R. C. 1967. Lipolytic activity of lactic acid bacteria. *J. Dairy Sci. 50*, 388–389.

Galesloot, T. E. 1956. Lactic acid bacteria which destroy the antibioticum (nisin) of *S. lactis*. *Ned. Melk. Zuiveltijdschr. 10*, 143–154. (Dutch)

Galesloot, T. E. 1957. The effect of nisin upon the growth of bacteria which are concerned or possibly concerned in bacterial processes in cheese and processed cheese. *Ned. Melk. Zuiveltijdschr. 11*, 58–73. (Dutch)

Galesloot, T. E. and Pette, J. W. 1956. The estimation of the nisin content of antibiotic starters and cultures and of cheese made by means of antibiotic starters. *Ned. Melk. Zuiveltijdschr. 10*, 137–142. (Dutch)

Garey, J. C. and Downing, J. F. 1951. Microbiological synthesis of vitamin B_{12} by a species of *Streptomyces*. (abstract). *119th Meeting Am. Chem. Soc.*, p. 22A.

Garibaldi, J. A. 1971. Influence of temperature on the iron metabolism of a fluorescent pseudomonad. *J. Bacteriol. 105*, 1036–1038.

Garibaldi, J. A., Ijichi, K., Lewis, J. C. and McGinnis, J. 1951. Fermentation process for production of vitamin B_{12}. U.S. Patent 2,576,932.

Garm, O., Lunaas, T. and Velle, W. 1963. The causes of bitter flavour in milk. *Meieriposten 52*, 253–258.

Garvie, E. I. 1978. Lactate dehydrogenase of *Streptococcus thermophilus*. *J. Dairy Res. 45*, 515–518.

Garvie, E. I. 1980. Bacterial lactate dehydrogenases. *Microbiol. Rev. 44*, 106–139.

Gibson, C. A., Landerkin, G. B. and Morse, P. M. 1966. Effects of additives on the survival of lactic streptococci in frozen storage. *Appl. Microbiol. 14*, 665–669.

Gilliland, S. E. and Speck, M. L. 1972. Interactions of food starter cultures and foodborne pathogens: Lactic streptococci versus staphylococci and salmonellae. *J. Milk Food Technol. 35*, 307–310.

Glenn, A. R. 1976. Production of extracellular proteins by bacteria. *Ann. Rev. Microbiol. 30*, 41–62.

Goel, M. C., Kulshrestha, D. C., Marth, E. H., Francis, D. W., Bradshaw, J. G. and Read, R. B., Jr. 1971. Fate of coliforms in yogurt, buttermilk, sour cream, and cottage cheese during refrigerated storage. *J. Milk Food Technol. 34*, 54–58.

Goepfert, J. M., Olson, N. F. and Marth, E. H. 1968. Behavior of *Salmonella typhimurium* during manufacture and curing of Cheddar cheese. *Appl. Microbiol, 16*, 862–866.

Goncalves, J. A. and Castillo, F. J. 1982. Partial purification and characterization of β-D-galactosidase from *Kluyveromyces marxianus*. *J. Dairy Sci.* 65, 2088-2094.

Gottschalk, G. 1979. *Bacterial Metabolism*. Springer-Verlag, New York.

Gotz, F., Sedewitz, B. and Elstner, E. F. 1980. Oxygen utilization by *Lactobacillus plantarum*. I. Oxygen consuming reactions. *Arch. Microbiol.* 125, 209-214.

Grecz, N., Wagenaar, R. O. and Dack, G. M. 1959. Inhibition of *Clostridium botulinum* by culture filtrates of *Brevibacterium linens*. *J. Bacteriol.* 78, 506-510.

Greene, V. W. and Jezeski, J. J. 1957A. Studies on starter metabolism. II. The influence of heating milk on the subsequent response of starter cultures. *J. Dairy Sci.* 40, 1053-1061.

Greene, V. A. and Jezeski, J. J. 1957B. Studies on starter metabolism. III. Studies on cysteine-induced stimulation and inhibition of starter cultures in milk. *J. Dairy Sci.* 40, 1062-1070.

Griffiths, M. W., Phillips, J. D. and Muir, D. D. 1981. Thermostability of proteases and lipases from a number of species of psychrotrophic bacteria of dairy origin. *J. Appl. Bacteriol.* 50, 289-303.

Gripon, J. C. 1977A. Proteolytic system of *P. roqueforti*. IV. Properties of an acid carboxypeptidase. *Ann. Biol. Biochim. Biophys.* 17, 283-298.

Gripon, J. C. 1977B. The proteolytic system of *Penicillium roqueforti*. V. Purification and properties of an alkaline aminopeptidase. *Biochemie* 59, 679-686.

Gripon, J. C. and Debest, B. 1976. Electrophoretic studies of the exocellular proteolytic system of *Penicillium roqueforti*. *Lait* 56, 423-438. (French)

Gripon, J. C. Desmazeaud, M. J., Le Bars, D. and Bergere, J. L. 1977. Role of proteolytic enzymes of *Streptococcus lactis*, *Penicillium roqueforti*, and *Penicillium caseicolum* during cheese ripening. *J. Dairy Sci.* 60, 1532-1538.

Gripon, J. D. and Hermier, J. 1974. The proteolytic system of *Penicillium roqueforti*. III. Purification, properties and specificity of the protease inhibited by EDTA. *Biochemie* 56, 1323-1332.

Grufferty, R. C. and Condon, S. 1983. Effect of fermentation sugar on hydrogen peroxide accumulation by *Streptococcus lactis* C10. *J. Dairy Res.* 50, 481-489.

Gunsalus, I. C. and Niven, C. F., Jr. 1942. The effect of pH on lactic acid fermentation. *J. Biol. Chem.* 145, 131-136.

Hall, B. G. 1979. Lactose metabolism involving phospho-β-galactosidase in *Klebsiella*. *J. Bacteriol.* 138, 691-698.

Hall, H. H., Benjamin, J. C., Wiesen, C. F. and Tsuchiya, H. M. 1951. Production of vitamin B_{12} by microorganisms, especially *Streptomyces olivaceus* (abstract). *119th Meeting Am Chem. Soc.* p. 22A.

Hall, H. H. and Tsuchiya, H. M. 1951. Method for producing vitamin B_{12}. U.S. Patent 2,561,364.

Hamann, W. T. and Marth, E. H. 1984. Survival of *Streptococcus thermophilus* and *Lactobacillus bulgaricus* in commercial and experimental yogurts. *J. Food Prot.* 47, 781-786.

Hamilton, I. R. and Lebtag, H. 1979. Lactose metabolism by *Streptococcus mutans*: Evidence for induction of the tagatose 6-phosphate pathway. *J. Bacteriol.* 140, 1102-1104.

Hamilton, I. R. and Lo, G. C. 1978. Co-induction of β-galactosidase and the lactose-phosphoenolpyruvate phosphotransferase system in *Streptococcus salivarius* and *Streptococcus mutans*. *J. Bacteriol.* 136, 900-908.

Hamilton, J. S., Hill, R. D. and Van Leeuwen, H. 1974. A bitter peptide from Cheddar cheese. *Agr. Biol. Chem.* 38, 375-379.

Hammer, B. W. 1930. Observations on ropiness in butter cultures. *J. Dairy Sci.* 13, 69-77.

Hammer, B. W. and Babel, F. J. 1957. *Dairy Bacteriology.* John Wiley and Sons, New York.

Hammer, B. W. and Cordes, W. A. 1921. Burnt or caramel flavor of dairy products. Iowa Agr. Exp. Sta. Bull. 68, 146–156.

Hammer, B. W. and Hussong, R. V. 1931. Observations on the heat resistance of some ropy milk organisms. *J. Dairy Sci. 14,* 27–39.

Harding, H. A. and Prucha, M. J. 1920. An epidemic of ropy milk. Ill. Agr. Expt. Sta. Bull. 228.

Hargrove, R. E. and Leviton, A. 1955. Process for the manufacture of vitamin B_{12}. U.S. Patent 2,715,602.

Hargrove, R. E., McDonough, F. E. and Mattingly, W. A. 1969. Factors affecting survival of *Salmonella* in Cheddar and Colby cheese. *J. Milk Food Technol. 32,* 480–484.

Harper, W. J., Carmona de Catril, A. and Chen, J. L. 1980. Esterases of lactic streptococci and their stability in cheese slurry system. *Milchwissenschaft 35,* 129–132.

Hartley, B. S. 1960. Proteolytic enzymes. *Ann. Rev. Biochem. 29,* 45–72.

Hasan, N. and Durr, I. F. 1974. Induction of β-galactosidase in *Lactobacillus plantarum.* *J. Bacteriol. 120,* 66–73.

Hawley, H. B. 1957A. Nisin in food technology—1. *Food Manuf. 32,* 370–376.

Hawley, H. B. 1957B. Nisin in food technology—2. *Food Manuf. 32,* 430–434.

Hawley, H. B. and Hall, R. H. 1960. Production of nisin. U.S. Patent 2,935,503.

Heinemann, B., Stumbo, C. R. and Scurlock, A. 1964. Use of nisin in preparing beverage-quality sterile chocolate-flavored milk. *J. Dairy Sci. 47,* 8–12.

Hemme, D., Nardi, M. and Jette, D. 1980A. β-Glactosidase and phospho-β-galactosidases of *Streptococcus thermophilus. Lait 60,* 595–618. (French).

Hemme, D., Wahl, D. and Nardi, M. 1980B. Variations of enzyme systems by *Streptococcus thermophilus. Lait 60,* 111–129. (French)

Hendlin, D. and Ruger, M. L. 1950. The effect of cobalt on the microbial systhesis of LLD-active substances. *Science 111,* 541–542.

Hengstenberg, W., Egan, J. B. and Morse, M. L. 1967. Carbohydrate transport in *Staphylococcus aureus.* V. The accumulation of phosphorylated carbohydrate derivatives and evidence for a new enzyme-splitting lactose phosphate. *Proc. Natl. Acad. Sci. USA 58,* 274–279.

Hensel, R., Mayr, R., Stetter, K. O. and Kandler, O. 1977. Comparative studies of lactic acid dehydrogenase in lactic acid bacteria. I. Purification and kinetics of the allosteric L-lactic acid dehydrogenase from *Lactobacillus casei* spp. *casei* and *Lactobacillus curvatus. Arch. Microbiol. 112,* 81–93.

Hettinga, D. H. and Reinbold, G. W. 1972A. The propionic-acid bacteria—A review. I. Growth. *J. Milk Food Technol. 35,* 295–301.

Hettinga, D. H. and Reinbold, G. W. 1972B. The propionic-acid bacteria—A review. II. Metabolism. *J. Milk Food Technol. 35,* 358–372.

Hickey, M. W., Hillier, A. J. and Jago, G. R. 1983. Enzymatic activities associated with lactobacilli in dairy products. *Aust. J. Dairy Technol. 38,* 154–158.

Hidalgo, C., Reyes, J. and Goldschmidt, R. 1977. Induction and general properties of β-galatosidase and β-galactoside permease in *Pseudomonas* BAL-31. *J. Bacteriol. 129,* 821–829.

Hirsch, A. 1950. The assay of the antibiotic nisin. *J. Gen. Microbiol. 4,* 70–83.

Hirsch, A., Grinsted, E., Chapman, H. R. and Mattick, A. T. R. 1951. A note on the inhibition of an anaerobic sporeformer in Swiss-type cheese by a nisin-producing streptococcus. *J. Dairy Res. 18,* 205–206.

Hirsch, A., McClintock, M. and Mocquot, G. 1952. Observations on the influence of inhibitory substances produced by the lactobacilli of Gruyere cheese on the development of anaerobic spore-formers. *J. Dairy Res. 19,* 179–186.

Hirsch, A. and Grinsted, E. 1954. Methods for the growth and enumeration of anaerobic spore-formers from cheese, with observations on the effect of nisin. *J. Dairy Res.* *21*, 101–110.

Hiyama, T., Fukui, S. and Kitahara, K. 1968. Purification and properties of lactate racemase from *Lactobacillus* sake. *J. Biochem.* *64*, 99–107.

Hogarty, S. L. and Frank, J. F. 1982. Low temperature activity of lactic streptococci isolated from cultured buttermilk. *J. Food Prot.* *45*, 1208–1211.

Hogg, D. M. and Jago, G. R. 1970. The influence of aerobic conditions on some aspects of the growth and metabolism of Group N streptococci. *Aust. J. Dairy Technol.* *25*, 17–18.

Hoover, S. R., Jasewicz, L., Pepinsky, J. B. and Porges, N. 1952B. Activated sludge as a source of vitamin B_{12} for animal feed. *Sewage Ind. Wastes 24*, 38–44.

Hoover, S. R., Jasewicz, L. and Porges, N. 1952A. Biochemical oxidation of dairy wastes. IV. Endogenous respiration and stability of aerated dairy waste sludge. *Sewage Ind. Wastes 24*, 1144–1149.

Hoover, S. R. and Porges, N. 1952. Assimilation of dairy wastes by activated sludge. II. The equation of synthesis and rate of oxygen utilization. *Sewage Ind. Wastes 24*, 306–312.

Hosono, A. and Elliott, J. A. 1974. Properties of crude ethylester-forming enzyme preparations from some lactic acid and psychrotrophic bacteria. *J. Dairy Sci. 57*, 1432–1437.

Hosono, A., Elliott, J. A. and Morgan, W. A. 1974. Production of ethylesters by some lactic acid and psychrotrophic bacteria. *J. Dairy Sci. 57*, 535–539.

Hoyle, M. and Nichols. A. A. 1948. Inhibitory strains of lactic streptococci and their significance in the selection of cultures for starter. *J. Dairy Res. 15*, 398–408.

Hurst, A. 1972. Interactions of food starter cultures and food-borne pathogens: The antagonism between *Streptococcus lactis* and spore-forming microbes. *J. Milk Food Technol. 35*, 418–423.

Itoh, T., Ohashi, M., Toba, T. and Adachi, S. 1980. Purification and properties of β-galactosidase from *Lactobacillus bulgaricus*. *Milchwissenschaft 35*, 593–597.

Jabrait, A. 1969. Influence of coagulation and freeze-drying on survival of lactic acid bacteria in bioghurt. *Lait 49*, 520–532. (French)

Jagielski, V. 1871. Improvement in dietetic compounds from milk. U.S. Patent 117,889.

Jasewicz, L. and Porges, N. 1958. Aeration of whey wastes. I. Nitrogen supplementation and sludge oxidation. *Sewage Ind. Wastes 30*, 555–561.

Jezeski, J. J., Tatini, S. R., DeGarcia, P. C. and Olson, J. C., Jr. 1967. Influence of *Streptococcus lactis* on growth and enterotoxin A production by *Staphylococcus aureus* in milk (abstract). *Bacteriol. Proc. 12*, A66.

Johnson, H. N. and DeBusk, A. G. 1970. The β-galactosidase system of *Neurospora crassa*. I. Purification and properties of the pH 4.2 enzyme. *Arch. Biophys. 138*, 408–411.

Johnson, K. G. and McDonald, I. J. 1974. β-D-Phosphogalactosidase galactohydrolase from *Streptococcus cremoris* HP: Purification and properties. *J. Bacteriol. 117*, 667–674.

Johnstone, D. B. and Pfeffer, M. 1959. Aerobic fermentation of whey by a nitrogen-fixing strain of *Aerobacter aerogenes*. *Nature 183*, 992–993.

Jonas, H. A., Anders, R. F. and Jago, G. R. 1972. Factors affecting the activity of lactate dehydrogenase of *Streptococcus cremoris*. *J. Bacteriol. 111*, 397–403.

Juan, S. M. and Cazzulo, J. J. 1976. The extracellular protease from *Pseudomonas fluorescens*. *Experientia 32*, 1120–1122.

Kaminogawa, S., Ninomiya, T. and Yamauchi, K. 1984. Aminopeptidase profiles of lactic streptococci. *J. Dairy Sci. 67*, 2483–2492.

Kapac-Parkaceva, N., Bauer, O. and Cizbanovski, T. 1975. Effects of different ratios of starter bacteria on amino acids spectrum in yoghurt made from cows milk. *Mljekar-stro 25*, 33–42, cited in *Dairy Sci. Abstr. 37*, 722.

Kastens, M. L. and Baldauski, F. A. 1952. Chemicals from milk. *Ind. Eng. Chem. 44*, 1257–1268.

Keen, A. R. 1972. Growth studies on lactic streptococci. III. Observations on continuous growth behaviour in reconstituted skim-milk. *J. Dairy Res. 39*, 151–159.

Keenan, T. M. and Lindsay, R. C. 1968. Diacetyl production and utilization by *Lactobacillus* species. *J. Dairy Sci. 51*, 188–191.

Kempler, G. M. and McKay, L. L. 1979. Characterization of plasmid deoxyribonucleic acid in *Streptococcus lactis* subsp. *diacetylactis:* Evidence for plamid-linked citrate utilization. *Appl. Environ. Microbiol. 37*, 316–323.

Kempler, G. M. and McKay, L. L. 1981. Biochemistry and genetics of citrate utilization in *Streptococcus lactis* ssp. *diacetylactis*. *J. Dairy Sci. 64*, 1527–1539.

Kennedy, E. P. and Scarborough, G. A. 1967. Mechanism of hydrolysis of O-nitrophenyl-β-galactoside in *Staphylococcus aureus* and its significance for theories of sugar transport. *Proc. Natl. Acad. Sci. USA 58*, 225–228.

Keogh, B. P. 1970. Survival and activity of frozen starter cultures for cheese manufacture. *Appl. Microbiol. 19*, 928–931.

Kinsella, J. E. and Hwang, D. 1976A. Biosynthesis of flavors by *Penicillium roqueforti*. *Biotechnol. Bioeng. 18*, 927–938.

Kinsella, J. E. and Hwang, D. H. 1976B. Enzymes of *Penicillium roqueforti* involved in the biosynthesis of cheese flavor. *Crit. Rev. Food Sci. Nutr. 8*, 191–228.

Kishonti, E. and Sjostrom, G. 1970. Influence of heat resistant lipases and proteases in psychrotrophic bacteria on product quality. *18th Int. Dairy Congr. 1E*, 501.

Kooy, J. S. 1952. Strains of *Lactobacillus plantarum* which will inhibit the activity of the antibiotics produced by *Streptococcus lactis*. *Ned. Melk. Zuiveltijdschr. 6*, 323–330. (Dutch)

Korkeala, H., Soback, S. and Hirn, J. 1984. Effect of cadmium on the growth of *Lactobacillus lactis*, *L. helviticus* and *Streptococcus thermophilus* in milk. *J. Dairy Res. 51*, 591–596.

Kornacki, J. L. and Marth, E. H. 1982A. Fate of nonpathogenic and enteropathogenic *Escherichia coli* during the manufacture of Colby-like cheese. *J. Food Prot. 45*, 310–316.

Kornacki, J. L. and Marth, E. H. 1982B. Foodborne illness caused by *Escherichia coli:* A review. *J. Food Prot. 45*, 1051–1067.

Kosikowski, F. 1977. *Cheese and Fermented Milk Foods*. F. V. Kosikowski and Associates, Brookton, NY.

Kundig, W. 1976. The bacterial phosphoenolpyruvate phosphotransferase system. *In: The Enzymes of Microbiological Membranes*. A. Martonasi (Editor). Plenum Press, New York, pp. 31–53.

Kundig, W., Ghosh, S. and Roseman, S. 1964. Phosphate bound to histidine in protein as an intermediate in a novel phosphotransferase system. *Proc. Natl. Acad. Sci. USA 52*, 1067–1074.

Lamberet, G. and Menassa, A. 1983. Purification and properties of an acid lipase from *Penicillin roqueforti*. *J. Dairy Res. 50*, 459–468.

Lamprech, E. D. and Foster, E. M. 1963. The survival of starter organisms in concentrated suspensions. *J. Appl. Bacteriol. 26*, 359–369.

Langhus, W. L., Price, W. V., Sommer, H. H. and Frazier, W. C. 1945. The "smear; of brick cheese and its relation to flavor development. *J. Dairy Sci. 28*, 827–838.

Langsrud, T. and Reinbold, G. W. 1973. Flavor development and microbiology of Swiss cheese—A review. III. Ripening and flavor production. *J. Milk Food Technol. 36*, 593–609.

Larsen, L. D. and McKay, L. L. 1978. Isolation and characterization of plasmid DNA in *Streptococcus cremoris. Appl. Environ. Microbiol. 36,* 944-952.

Law, B. A. 1979A. Enzymes of psychrotrophic bacteria and their effects on milk and milk products. *J. Dairy Res. 46,* 573-588.

Law, B. A. 1979B. Extracellular peptidases in Group N streptococci used as cheese starters. *J. Appl. Bacteriol. 46,* 455-463.

Law, B. A. 1980. Transport and utilization of proteins by bacteria. *In: Micro-organisms and Nitrogen Sources.* J. W. Payne (Editor). John Wiley and Sons, New York, pp. 381-409.

Law, B. A., Castanon, M. J. and Sharpe, M. E. 1976A. The contribution of starter streptococci to flavour development in Cheddar cheese. *J. Dairy Res. 43,* 301-311.

Law, B. A., Sezgin, E. and Sharpe, M. E. 1976B. Amino acid nutrition of some commercial cheese starters in relation to their growth in peptone supplemented whey media. *J. Dairy Res. 43,* 291-300.

Law, B. A., and Sharpe, M. E. 1978. Streptococci in the dairy industry. *In: Streptococci.* F. A. Skinner and L. B. Quesnal (Editors). Academic Press, New York, pp. 263-278.

Law, B. A., Sharpe, M. E. and Chapman, H. R. 1976C. The effect of lipolytic gram-negative psychrotrophs in stored milk on the development of rancidity in Cheddar cheese. *J. Dairy Res. 43,* 459-468.

Law, B. A., Sharpe, M. E. and Reiter, B. 1974. The release of intracellular dipeptidase from starter streptococci during Cheddar cheese ripening. *J. Dairy Res. 41,* 137-146.

Lawrence, R. C., Thomas, T. D. and Terzaghi, B. E. 1976. Reviews of the progress of dairy science: Cheese starters. *J. Dairy Res. 43,* 141-143.

Le Bars, D. and Gripon, J. E. 1981. Role of *Penicillium roqueforti* proteinases during blue cheese ripening. *J. Dairy Res. 48,* 479-487.

Lederberg, J. 1950. The β-D-galactosidase of *Escherichia coli* strain K-12. *J. Bacteriol. 60,* 381-392.

Lenoir, J. 1963. The development of microflora during the ripening of Camembert cheese. *Lait 43,* 262-270. (French)

Leviton, A. 1946. The microbiological synthesis of riboflavin—A theory concerning its inhibition. *J. Am. Chem. Soc. 68,* 835-840.

Leviton, A. 1949. Microbiological production of riboflavin. U.S. Patent 2,477,812.

Leviton, A. 1956A. Process for the microbiological synthesis of vitamin B_{12} active substances. U.S. Patent 2,753,289.

Leviton, A. 1956B. Process for the preparation and concentration of vitamin B_{12} active substances. U.S. Patent 2,764,521.

Leviton, A. and Hargrove, R. E. 1952. Microbiological synthesis of vitamin B_{12} by propionic acid bacteria. *Ind. Eng. Chem. 44,* 2651-2655.

Lockwood, L. B. and Stodola, F. H. 1950. Process of culturing bacteria. U.S. Patent 2,496,297.

Lundstedt, E. and Fogg, W. B. 1962. Citrated whey starters. II. Gradual formation of flavor and aroma in creamed cottage cheese after the addition of small quantities of citrated cottage cheese whey cultures of *Streptococcus diacetilactis. J. Dairy Sci. 45,* 1327-1331.

Macquot, G. and Lefebvre, E. 1959. A simple procedure to detect nisin in cheese. *J. Appl. Bacteriol. 19,* 322-323.

Macura, D. and Townsley, P. M. 1984. Scandinavian ropy milk—Identification and characterization of endogenous ropy lactic streptococci and their extracellular excretion. *J. Dairy Sci. 67,* 735-744.

Marin, A., Mawhinney, T. P. and Marshall, R. T. 1984. Glycosidic activities of *Pseudomo-*

nas fluorescens on fat extracted skim milk, buttermilk, and milk fat globule membranes. *J. Dairy Sci. 67.* 52–59.

Marshall, V. M. and Cole, W. M. 1983. Threonine aldolase and alcohol dehydrogenase activities in *Lactobacillus bulgaricus* and *Lactobacillus acidophilus* and their contribution to flavour production in fermented foods. *J. Dairy Res. 50,* 375–379.

Marth, E. H. 1963. Microbiological and chemical aspects of Cheddar cheese ripening. *J. Dairy Sci. 46,* 869–890.

Marth, E. H. 1966. Antibiotics in foods—naturally occurring, developed, and added. *In: Residue Reviews,* Vol. 12. F. A. Gunther (Editor). Springer-Verlag, New York, pp. 65–161.

Marth, E. H. 1969. Salmonella and salmonellosis associated with milk and milk products. A review. *J. Dairy Sci. 52,* 283–315.

Marth, E. H. 1970. Fermentation products from whey. *In: Byproducts from Milk.* B. H. Webb and E. O. Whittier (Editors). AVI Publishing Co., Westport, Conn., pp. 43–74.

Marth, E. H. 1974. Fermentations. *In: Fundamentals of Dairy Chemistry,* 2nd ed. B. W. Webb, A. H. Johnson and J. A. Alford (Editors). AVI Publishing Co., Westport, Conn., pp. 772–858.

Marth, E. H. 1981. Foodborne hazards of microbial origin. *In: Food Safety.* H. R. Roberts (Editor). John Wiley and Sons, New York, pp. 15–65.

Marth, E. H. 1985. Pathogens in milk and milk products. *In: Standard Methods for the Examination of Dairy Products,* 15th ed. G. H. Richardson (Editor). American Public Health Association, Washington, D.C., pp. 43–87.

Marth, E. H. and Hussong, R. V. 1962. Effect of skim milks cultured with different strains of *Leuconostoc citrovorum* on growth of some microorganisms associated with cottage cheese spoilage (abstract). *J. Dairy Sci. 45,* 652–653.

Marth, E. H. and Hussong, R. V. 1963. Effect of skim milks cultured with different strains of *Leuconostoc citrovorum* on growth of some bacteria and yeasts. *J. Dairy Sci. 46,* 1033–1037.

Marth, E. H., Ingold, D. L. and Hussong, R. V. 1964. Ropiness in milk caused by a strain of *Escherichia intermedia. J. Dairy Sci. 47,* 1265–1266.

Martin, J. H. and Cardwell, J. T. 1960. Use of frozen ripened lactic cultures in the manufacture of cottage cheese. *J. Dairy Sci. 43,* 438–439.

Masson, P. L. and Heremans, J. R. 1971. Lactoferrin in milk from different species. *Comp. Biochem. Physiol. 39B,* 119–129.

Mather, D. W. and Babel, F. J. 1959. Inhibition of certain types of bacterial spoilage in creamed cottage cheese by the use of a creaming mixture prepared with *Streptococcus citrovorus. J. Dairy Sci. 42,* 1917–1926.

Mattick, A. T. R. and Hirsch, A. 1944. A powerful inhibitory substance produced by group N streptococci. *Nature 154,* 551.

Mattick, A. T. R. and Hirsch, A. 1947. Further observations on an inhibitory substance (nisin) from lactic streptococci. *Lancet 253,* 5–7.

Maurer, L., Reinbold, G. W. and Hammond, E. G. 1975. Effect of copper on microorganisms in the manufacture of Swiss cheese. *J. Dairy Sci. 58,* 1630–1635.

Maxwell, E. S., Kurahashi, K. and Kalckar, H. M. 1962. Enzymes of the Leloir pathway. *Meth. Enzymol. 5,* 174–189.

McFeters, G. A., Sandine, W. E. and Elliker, P. R. 1967. Purification and properties of *Streptococcus lactis* β-galactosidase. *J. Bacteriol. 93,* 914–919.

McKay, L. L. 1982. Regulation of lactose metabolism. *In: Developments in Food Microbiology,* Vol. 1. R. Davies (Editor). Applied Science Publisher, Essex, England, pp. 153–182.

McKay, L. L. 1983. Functional properties of plasmids in lactic streptococci. *Antonie van Leuwenhoek 49,* 259–274.

McKay, L. L. and Baldwin, K. A. 1974. Altered metabolism in a *Streptococcus lactis* C2 mutant deficient in lactic dehydrogenase. *J. Dairy Sci. 57*, 181–185.

McKay, L. L. and Baldwin, K. A. 1975. Plasmid distribution and evidence for a proteinase plasmid in *Streptococcus lactis* C2. *Appl. Microbiol. 29*, 546–548.

McKay, L. L. and Baldwin, K. A. 1978. Stabilization of lactose metabolism in *Streptococcus lactis* C2. *Appl. Environ. Microbiol. 36*, 360–367.

McKay, L. L., Baldwin, K. A. and Efstathiou, J. D. 1976. Transductional evidence for plasmid linkage of lactose metabolism in *Streptococcus lactis* C2. *Appl. Environ. Microbiol. 32*, 45–52.

McKay, L. L., Miller, A., Sandine, W. E. and Elliker, P. R. 1970. Mechanism of lactose utilization by lactic acid streptococci: Enzymatic and genetic analyses. *J. Bacteriol. 120*, 804–809.

Meade, R. E., Pollard, H. L. and Rodgers, N. E. 1945. Process for manufacturing a vitamin concentrate. U.S. Patent 2,369,680.

Meewes, K. H. and Milosevic, S. 1962. On the effect of a culture filtrate of *Lactobacillus helveticus* on the ability of *Escherichia coli* to produce gas. *Milchwissenschaft 17*, 678–679. (German)

Mehala, M. A., Cheryan, M. and Argondelis, A. 1985. Conversion of whey permeate to ethanol. Improvement of fermentor productivity using membrane reactors. *Cult. Dairy Prod. J. 20*, 9–12.

Mellerick, D. and Cogan, T. M. 1981. Induction of some enzymes of citrate metabolism in *Leuconostoc lactis* and other heterofermentative lactic acid bacteria. *J. Dairy Res. 48*, 497–502.

Mencher, J. R. and Alford, J. A. 1967. Purification and characterization of the lipase of *Pseudomonas fragi. J. Gen. Microbiol. 48*, 317–328.

Michaelian, M. B., Hoecker, W. H. and Hammer, B. W. 1938. Effect of pH on the production of acetylmethylcarbinol plus diacetyl in milk by the citric acid fermenting streptococci. *J. Dairy Sci. 21*, 213–218.

Miller, A., III, Morgan, M. E. and Libbey, L. M. 1974. *Lactobacillus maltaromicus*, a new species producing a malty arome. *Int. J. Systematic Bacteriol. 24*, 346–354.

Miller, I. and Kandler, O. 1967. Proteolysis and liberation of free amino acids by lactic acid bacteria in milk. I. Changes of the N fractions. *Milchwissenschaft. 22*, 150–159.

Mills, O. E. and Thomas, T. D. 1980. Bitterness development in Cheddar cheese: Effect of the level of starter proteinase. *N.Z. J. Dairy Sci. Technol. 15*, 131–141.

Minamiura, N., Matsumera, Y., Fukumoro, J. and Yamamoto, T. 1972. Bitter peptides in cow milk casein digests with amino acid sequence of a bitter peptide. *Agr. Biol. Chem. 36*, 588–595.

Miner, C. S., Jr. and Wolnak, B. 1953. Process of preparing vitamin B_{12}-active product from sewage sludge. U.S. Patent 2,646,386.

Minor, T. E. and Marth, E. H. 1972. Fate of *Staphyloccus aureus* in cultured butter-milk, sour cream, and yogurt during storage. *J. Milk food Technol. 35*, 302–306.

Minor, T. E. and Marth, E. H. 1976. *Staphylococci and Their Significance in Foods*. Elsevier Scientific Publishing Co., Amsterdam.

Mitchell, G. E. 1979. Seasonal variation in citrate content of milk. *Aust. J. Dairy Technol. 34*, 158–160.

Mitscherlich, E. and Marth, E. H. 1984. *Microbial Survival in the Environment: Bacteria and Rickettsiae Important in Human and Animal Health*. Springer-Verlag, Heidelberg.

Modler, H., Brunner, J. R. and Stine, C. M. 1974. Extracellular protease of Pencillium roqueforti. II. Characterization of a purified enzyme preparation. *J. Dairy Sci. 57*, 528–534.

Molskness, T. A., Lee, D. R., Sandine, W. E. and Elliker, P. R. 1973. β-D-Phosphogalactoside galactohydrolase of lactic stretococci. *Appl. Microbiol.* 25, 373-380.

Moon, N. J. and Reinbold, G. W. 1976. Commensalism and competition in mixed cultures of *Lactobacillus bulgaricus* and *Streptococcus thermophilus. J. Milk Food Technol.* 39, 337-341.

Moreno, V. and Kosikowski, F. V. 1973. Peptides, amino acids and amines liberated from β-casein by micrococcal cell-free preparations. *J. Dairy Sci.* 56, 39-44.

Morgan, M. E. 1976. The chemistry of some microbially induced flavor defects in milk and dairy foods. *Biotechnol. Bioeng.* 18, 953-965.

Morgan, M. E., Lindsay, R. C., Libbey, L. M. and Pereira, R. L. 1966. Identity of additional aroma constituents in milk cultures of *Streptococcus lactis* var. *maltigenes. J. Dairy Sci.* 49, 15-18.

Morihara, K. 1974. Comparative specificity of microbial proteinases. *Adv. Enzymol.* 41, 179-244.

Mou, L., Sullivan, J. J. and Jago, G. R. 1975. Peptidase activities in Group N streptococci. *J. Dairy Res.* 42, 147-155.

Murgier, M., Pelissier, C. and Lazdunski, A. 1976. Existence, localization, and regulation of the biosynthesis of aminoendopeptidase in gram-negative bacteria. *Eur. J. Biochem.* 65, 517-520.

Murray, K. E. and Whitfield, F. B. 1975. The occurrence of 3-alky-2-methoxy-pyrazines in raw vegetables. *J. Sci. Food Agr.* 26, 973-986.

Myers, R. P. and Stimpson, E. G. 1956. Production of lactase. U.S. Patent 2,762,749.

Nashif, S. A. and Nelson, F. E. 1953A. The lipase of *Pseudomonas fragi.* I. Characterization of the enzyme. *J. Dairy Sci.* 39, 459-470.

Nashif, S. A. and Nelson, F. E. 1953B. The lipase of *Pseudomonas fragi.* II. Factors affecting lipase production. *J. Dairy Sci.* 36, 471-480.

Nashif, S. A. and Nelson, F. E. 1953C. The lipase of *Pseudomonas fragi.* III. Enzyme action in cream and butter. *J. Dairy Sci.* 36, 481-488.

Newton, G. G. F., Abraham, E. P. and Berridge, N. J. 1953. Sulfur-containing amino-acids of nisin. *Nature 171,* 606.

Noreau, J. and Drapeau, G. R. 1979. Isolation and properties of the protease from the wild-type and mutant strains of *Pseudomonas fragi. J. Bacteriol.* 140, 911-916.

Okamoto, T. and Morichi, T. 1979. Distribution of β-galactosidase and β-phosphogalactosidase activity among lactic streptococci. *Agr. Biol. Chem.* 43, 2389-2390.

O'Leary, V. S. and Woychik, J. H. 1976. Utilization of lactose, glucose, and galactose by a mixed culture of *Streptococcus thermophilus* and *Lactobacillus bulgaricus* in milk treated with lactase enzyme. *Appl. Environ. Microbiol.* 32, 89-94.

Olive, T. R. 1936. Waste lactose is raw material for a new lactic acid process. *Chem. Met. Eng.* 43, 481-483.

Olson, H. C. and Qutub, A. H. 1970. Influence of trace minerals on the acid production by lactic cultures. *Cult. Dairy Prod. J.* 5, 12-17.

Orla-Jensen, S., Orla-Jensen, A. D. and Spur, B. 1926. The butter aroma bacteria. *J. Bacteriol.* 12, 333-342.

Otto, R., Devos, W. M. and Gavrieli, J. 1981. Plasmid DNA in *Streptococcus cremoris* Wg2: Influence of pH on selection in chemostats of a variant lacking a protease plasmid. *Appl. Environ. Microbiol.* 43, 1272-1277.

Overcast, W. W. and Atmaran, K. 1974. The role of *Bacillus cereus* in sweet curdling of fluid milk. *J. Milk Food Technol.* 37, 233-236.

Pack, M. Y., Vedamuthu, E. R., Sandine, W. E. and Elliker, P. R. 1968. Hydrogen peroxide-catalase milk treatment for enhancement and stabilization of diacetyl in lactic starter cultures. *J. Dairy Sci.* 51, 511-516.

Pacquet, J. and Gripon, J. C. 1980. Intracellular peptide hydrolases of *Penicillium roqueforti. Milchwissenschaft 35,* 72-74.

Park, H. S. and Marth, E. H. 1972A. Behavior of *Salmonella typhimurium* in skimmilk during fermentation by lactic acid bacteria. *J. Milk Food Technol.* 35, 482–488.

Park, H. S. and Marth, E. H. 1972B. Survival of *Salmonella typhimurium* in refrigerated cultured milk. *J. Milk Food Technol.* 35, 489–495.

Park, H. S., Marth, E. H., Goepfert, J. M. and Olson, N. F. 1970. The fate of *Salmonella typhimurium* in the manufacture and ripening of low-acid Cheddar cheese. *J. Milk Food Technol.* 33, 280–284.

Park, H. S., Marth, E. H. and Olson, N. F. 1973. Fate of enteropathogenic strains of *Escherichia coli* during the manufacture and ripening of Camembert cheese. *J. Milk Food Technol.* 36, 532–546.

Park, Y. H. and McKay, L. L. 1982. Distinct galactose phosphoenolpyruvate-dependent phosphotransferase system in *Streptococcus lactis. J. Bacteriol.* 149, 420–425.

Park, Y. K., Desanti, M. S. S. and Pastore, G. M. 1979. Production and characterization of β-galactosidase from *Aspergillus oryzae. J. Food Sci.* 4, 100–103.

Pastore, G. M. and Park, Y. K. 1979. Screening of high β-galactosidase producing fungi and characterizing the hydrolysis properties of a selected strain. *J. Food Sci.* 44, 1577–1579.

Patel, G. B. and Blankenagel, G. 1972. Bacterial counts of raw milk and flavor of the milk after pasteurization and storage. *J. Milk Food Technol.* 35, 203–206.

Pearce, L. E., Skipper, N. A. and Jarvis, B. D. W. 1974. Proteinase activity in slow lactic acid producing variants of *Streptococcus lactis. Appl. Microbiol.* 27, 933–937.

Peebles, M. M., Gilliland, S. E. and Speck, M. L. 1969. Preparation of concentrated lactic streptococcus starters. *Appl. Microbiol.* 17, 805–810.

Peterson, A. C. and Gunderson, M. F. 1960. Some characteristics of proteolytic enzymes from *Pseudomonas fluorescens. Appl. Microbiol.* 8, 98–103.

Platt, T. B., and Foster, E. M. 1958. Products of glucose metabolism by homo-fermentative streptococci under anaerobic conditions. *J. Bacteriol.* 75, 453–459.

Pomeranz, Y. 1964. Lactase (beta-D-galactosidase). I. Occurrence and properties. *Food Technol.* 18, 682–687.

Porges, N. 1956. Waste treatment by optimal aeration—Theory and practice in dairy waste disposal. *J. Milk Food Technol.* 19, 34–38.

Porges, N. 1958. Practical application of laboratory data to dairy waste treatment. *Food Technol.* 12, 78–80.

Porges, N., Michener, T. S., Jr., Jasewicz, J. and Hoover, S. R. 1960. Dairy waste treatment by aeration. *Agriculture Handbook.* Agriculture Research Service, Washington, D.C.

Porges, N., Pepinsky, J. B., Hendler, N. C. and Hoover, S. R. 1950. Biochemical oxidation of dairy wastes. I. Methods of study. *Sewage Ind. Wastes* 22, 318–325.

Porzio, M. A. and Pearson, A. M. 1975. Isolation of an extracellular neutral proteinase from *Pseudomonas fragi. Biochim. Biophys. Acta* 384, 235–241.

Postma, P. W. and Roseman, S. 1976. The bacterial phosphoenolpyruvate: Sugar phosphotransferase system. *Biochim. Biophys. Acta* 457, 213–257.

Premi, L., Sandine, W. E. and Elliker, P. R. 1972. Lactose-hydrolyzing enzymes of *Lactobacillus* species. *Appl. Microbiol.* 24, 51–57.

Prescott, S. C. and Dunn, C. G. 1957. *Industrial Microbiology.* McGraw-Hill, New York.

Ramseier, H. R. 1960. The action of nisin on *Clostridium butyricum* Prazm. *Arch. Mikrobiol.* 37, 57–94. (German)

Rao, D. R., Reddy, A. V., Pulusani, S. R. and Cornwell, P. E. 1984. Biosynthesis and utilization of folic acid and vitamin B_{12} by lactic culture in skim milk. *J. Dairy Sci.* 67, 1169–1174.

Rash, K. E. and Kosikowski, F. V. 1982. Influence of lactic acid starter bacteria on enteropathogenic *Escherichia coli* in ultrafiltration prepared Camembert cheese. *J. Dairy Sci.* 65, 537–543.

Rasic, J. and Kurman, J. A. 1978. *Yoghurt—Scientific Grounds, Technology, Manufacture and Preparation*, Vol. 1. Technical Dairy Publishing House, Copenhagen.

Reddy, K. P., Shahani, K. M. and Kulkarni, S. M. 1976. B-complex vitamins in cultured and acidified yogurt. *J. Dairy Sci. 59*, 191–195.

Reiter, B. 1978. Review of the progress of dairy science: Antimicrobial systems in milk. *J. Dairy Res. 45*, 131–147.

Reiter, B., Fewins, B. G., Fryer, T. F. and Sharpe, M. E. 1964. Factors affecting the multiplication and survival of coagulase positive staphylococci in Cheddar cheese. *J. Dairy Res. 31*, 261–272.

Reiter, B. and Oram, J. D. 1962. Nutritional studies on cheese starters. I. Vitamins and amino acid requirements of single strain starters. *J. Dairy Res. 29*, 63–77.

Reiter, B. and Sharpe, M. E. 1971. Relationship of the microflora to the flavour of Cheddar cheese. *J. Appl. Bacteriol. 34*, 63–80.

Reiter, B., Fryer, T. F., Pickering, A., Chapman, H. R., Lawrence, R. C. and Sharpe, M. E. 1967. The effect of the microbiol flora on the flavour and free fatty acid composition of Cheddar cheese. *J. Dairy Res. 34*, 257–272.

Rhee, S. K. and Pack, M. Y. 1980. Effect of environmental pH on fermentation balance of *Lactobacillus bulgaricus. J. Bacteriol. 144*, 217–221.

Richardson, B. C. 1981. The purification and characterization of a heat-stable protease from *Pseudomonas fluorescens* B52. *N.Z. J. Dairy Sci. Technol. 16*, 195–207.

Richardson, B. C. and Te Whaiti, I. E. 1978. Partial Characterization of heat-stable extracellular proteases of some psychrotrophic bacteria from raw milk. *N.Z. J. Dairy Sci. Technol. 13*, 172–176.

Richardson, G. H. and Calbert, H. E. 1959. A storage study of a lyophilized and a frozen lactic culture (abstract). *J. Dairy Sci. 42*, 907.

Richardson, G. H., Cheng, C. T. and Young, R. 1977. Lactic bulk culture system using whey-based bacteriophage-inhibitory medium and pH control. I. Applicability to American style cheese. *J. Dairy Sci. 60*, 378–386.

Richter, R. L., Brank, W. S., Dill, C. W. and Watts, C. A. 1979. Ascorbic acid stimulation of diacetyl production in mixed-strain lactic acid cultures. *J. Food Prot. 42*, 294–296.

Roberts, H. R. and Pettinati, J. D. 1957. Concentration effects in the enzymatic conversion of lactose to oligosaccharides. *J. Agr. Food Chem. 5*, 130–134.

Rodopulo, A. K., Kavadze, A. V. and Pisarnitskii, A. F. 1976. Biosynthesis and metabolism of acetoin and diacetyl. *Appl. Biochem. Microbiol. 12*, 249–255.

Rogers, L. A. 1928. The inhibiting effect of *Streptococcus lactis* on *Lactobacillus bulgaricus. J. Bacteriol. 16*, 321–325.

Rogosa, M., Browne, H. H. and Whittier, E. O. 1947. Ethyl alcohol from whey. *J. Dairy Sci. 30*, 263–270.

Rohlfing, S. R. and Crawford, I. P. 1966. Purification and characterization of the β-galactosidase of *Aeromonas formicans. J. Bacteriol. 91*, 1085–1097.

Romano, A. H., Eberhand, S. J., Dingle, S. L. and McDowell, T. D. 1970. Distribution of the phosphoenolpyruvate:glucose phosphotransferase system in bacteria. *J. Bacteriol. 104*, 808–813.

Romano, A. H., Trifone, J. D. and Brustolon, M. 1979. Distribution of the phosphoenolpyruvate:glucose phosphotransferase system in fermentative bacteria. *J. Bacteriol. 139*, 93–97.

Roseman, S. 1969. The transport of carbohydrates by a bacterial phosphotransferase system. *J. Gen. Physiol. 54*, 138s–179s.

Roseman, S. 1972. A bacterial phosphotransferase system and its role in sugar transport. *In: The Molecular Basis of Biological Transport*. J. F. Woissner, Jr. and J. Huijing (Editors). Academic Press, New York, pp. 181–218.

Roseman, S. 1975. The bacterial phosphoenolpyruvate:sugar phosphotransferase system. *In: Energy Transformation in Biological Systems.* CIBA Foundation Symposium 31. Associated Scientific Publishers, New York, pp. 225-241.

Ruban, E. L., Lobyreva, L. B., Sviridenko, Y. Y., Marchenkova, A. I. and Umanskii, M. S. 1978. Lipolytic activity of microorganisms isolated from different sources. *Appl. Biochem. Microbiol. 14,* 393-396.

Rykshina, Z. P. 1961. Biological means to enrich fermented milk products with vitamin B_{12} (abstract). *Milchwissenschaft 16,* 434. (German)

Ryser, E. T., Marth, E. H. and Doyle, M. P. 1985. Survival of *Listeria monocytogenes* during manufacture and storage of cottage cheese. *J. Food Prot. 48,* 746-750.

Sadovski, A. Y., Gordin, S. and Foreman, I. 1980. Psychrotrophic growth of microorganisms in a cultured milk product. *J. Food Prot. 43,* 765-768.

Sandine, W. E., Daly, C., Elliker, P. R. and Vedamuthu, E. R. 1972. Causes and control of culture-related flavor defects in cultured dairy products. *J. Dairy Sci. 55,* 1031-1039.

Sarles, W. B. and Hammer, B. W. 1933. Species of *Escherichia-Aerobacter* organisms responsible for some defects in dairy products. *J. Bacteriol. 25,* 461-467.

Saunders, A. P., Otto, R. H. and Sylvester, J. C. 1951. The production of B_{12} by various strains of actinomycetes (abstract). *119th Meeting Am. Chem. Soc.* p. 21A.

Schmidt, R. H., Morris, H. A., Castberg, H. B. and McKay, L. L. 1976. Hydrolysis of milk proteins by bacteria used in cheesemaking. *J. Agr. Food Chem. 24,* 1106-1113.

Schulz, M. E. and Hingst, G. 1954. Contributions to the chemistry of yogurt. Part I. Acetaldehyde—color reactions in the examination of yogurt. *Milchwissenschaft 9,* 330-336. (German)

Schulz, M. E., Vosz, E. and Kley, W. 1954. Contributions to the chemistry of yogurt. Part II. Studies on the application of the acetaldehyde—color reactions to evaluate yogurt. *Milchwissenschaft 9,* 361-365.

Seitz, E. W., Sandine, W. E., Elliker, P. R. and Day, E. A. 1963. Distribution of diacetyl reductase among bacteria. *J. Dairy Sci. 43,* 346-350.

Selby Smith, J., Hillier, A. J., Lees, G. J. and Jago, G. R. 1975. The nature of the stimulation of the growth of *Streptococcus lactis* by yeast extract. *J. Dairy Res. 42,* 123-138.

Severina, L. O. and Bashkatova, N. A. 1981. Lipases of gram-negative bacteria (review). *Appl. Biochem. Microbiol. 17,* 131-143.

Sheldon, R. M., Lindsay, R. C., Libbey, L. M. and Morgan, M. E. 1971. Chemical nature of malty flavor and aroma produced by *Streptococcus lactis* var. *maltigenes. Appl. Microbiol. 22,* 263-266.

Simmons, J. C. and Graham, D. M. 1959. Maintenance of active lactic cultures by freezing as an alternative to daily transfer. *J. Dairy Sci. 42,* 363-364.

Smith, E. E. 1970. Biosynthetic relation between the soluble and insoluble dextrans produced by *Leuconostoc mesenteroides* NRRL B-1229. *FEBS Lett. 12,* 33-37.

Somkuti, G. A. and Steinberg, D. H. 1979A. Adaptability of *Streptococcus thermophilus* to lactose, glucose, and galactose. *J. Food Prot. 42,* 885-887.

Somkuti, G. A. and Steinberg, D. H. 1979B. β-D-Galactoside galactohydrolase of *Streptococcus thermophilus:* Induction, purification, and properties. *J. Appl. Biochem. 1,* 357-368.

Speck, M. L. 1962. Symposium on lactic starter cultures. IV. Starter culture growth and action in milk. *J. Dairy Sci. 54,* 1253-1258.

Speck, M. L. 1972. Control of food-borne pathogens by starter cultures. *J. Dairy Sci. 55,* 1019-1022.

Springer, R. 1950. The components of whey-yeast and their importance in pharmacy. *Pharmazie 5*, 113–115. (German)

Stadhouders, J. and Veringa, H. A. 1973. Fat hydrolysis by lactic acid bacteria in cheese. *Neth. Milk Dairy J. 27*, 77–91.

Stamer, J. R. 1979. The lactic acid bacteria: Microbes of diversity. *Food Technol. 33*, 60–65.

Stanier, R. Y., Doudoroff, M. and Adelberg, E. A. 1970. *The Microbial World*, 3rd ed. Prentice-Hall, Englewood Cliffs, N.J.

Stepaniak, L. and Fox, P. F. 1983. Thermal stability of an extracellular proteinase from *Pseudomonas fluorescens* AFT 36. *J. Dairy Res. 50*, 171–184.

Stetter, K. O. and Kandler, O. 1973. Formation of DL-lactic acid by lactobacilli and characterization of a lactic acid racemase from several streptobacteria. *Arch. Mikrobiol. 94*, 221–247.

Sullivan, J. J., Mou, L., Rood, J. I. and Jago, G. R. 1973. The enzymic degradation of bitter peptides by starter streptococci. *Aust. J. Dairy Technol. 28*, 20–26.

Sundman, V. 1953. On the microbiology of Finnish ropy sour milk. *13th Int. Dairy Congr. 3*, 1420–1427.

Sutherland, I. W. 1977A. Bacterial exopolysaccharides—their nature and production. *In: Surface Carbohydrates of the Prokaryotic Cell.* I. Sutherland (Editor). Academic Press, New York, pp. 27–96.

Sutherland, I. W. 1977B. Microbial exopolysaccharide synthesis. *In: Extracellular Microbial Polysaccharides.* P. A. Sanford and A. Laskin (Editors) ACS, Symposium Series 45, American Chemical Society, Washington, D.C. pp. 40–57.

Sutherland, I. W. 1979. Microbial exopolysaccharides: Control of synthesis and acylation. *In: Microbial Polysaccharides and Polysaccharases.* R. C. W. Berkeley, G. W. Gooday and D. C. Ellwood, (Editors). Academic Press, New York, pp. 1–34.

Tacquet, A., Tison, F. and Devulder, B. 1961. Bactericidal action of yogurt on mycobacteria. *Ann. Inst. Pasteur. 100*, 581–587. (French)

Tamime, A. Y., and Deeth, H. C. 1980. Yogurt: Technology and biochemistry. *J. Food Prot. 43*, 939–977.

Thomas, T. D. 1975. Tagatose-1,6-diphosphate activation of lactate dehydrogenase from *Streptococcus cremoris. Biochem. Biophys. Res. Commun. 63*, 1035–1042.

Thomas, T. D. 1976A. Activator specificity of pyruvate kinase from lactic streptococci. *J. Bacteriol. 125*, 1240–1242.

Thomas, T. D. 1976B. Regulation of lactose fermentation in Group N streptococci. *Appl. Environ. Microbiol. 32*, 474–478.

Thomas, T. D. and Crow, V. L. 1984. Selection of galactose-fermenting *Streptococcus thermophilus* in lactose-limited chemostat cultures. *Appl. Environ. Microbiol. 48*, 186–191.

Thomas, T. D., Ellwood, D. C. and Longyear, V. M. C. 1979. Change from homo-to heterolactic fermentation by *Streptococcus lactis* resulting from glucose limitation in anaerobic chemostat cultures. J. Bacteriol. 138, 109–117.

Thomas, T. D. and Mills, O. E. 1981. Proteolytic enzymes of starter bacteria. *Neth. Milk Dairy J. 35*, 255–273.

Thomas, T. D., Turner, K. W. and Crow, V. L. 1980. Galactose fermentation by *Streptococcus lactis* and *Streptococcus cremoris:* Pathways, products, and regulation. *J. Bacteriol. 144*, 672–682.

Thompson, J. 1978. *In vivo* regulation of glycolysis and characterization of suger:phosphotransferase systems in *Streptococcus lactis. J. Bacteriol. 136*, 465–476.

Thompson, J. 1980. Galactose transport systems in *Streptococcus lactis. J. Bacteriol. 144*, 683–691.

Thompson, J. and Thomas, T. D. 1977. Phosphoenolpyruvate and 2-phosphoglycerate:

Endogenous energy sources for sugar accumulation by starved cells of *Streptococcus lactis. J. Bacteriol. 130*, 583–595.

Thompson, J. D., Turner, K. W. and Thomas, T. D. 1978. Catabolite inhibition and sequential metabolism of sugars by *Streptococcus lactis. J. Bacteriol. 133*, 1163–1174.

Thomson, J. and Torchia, D. A. 1984. Use of ^{31}P nuclear magnetic resonance spectroscopy of ^{14}C fluorography in studies of glycolysis and regulation of pyruvate kinase in *Streptococcus lactis. J. Bacteriol. 158*, 791–800.

Tinson, W., Hillier, A. J. and Jago, G. R. 1982. Metabolism of *Streptococcus thermophilus*. 1. Utilization of lactose, glucose and galactose. *Aust. J. Dairy Technol. 37*, 8–13.

Toba, T. and Adachi, S. 1978. Hydrolysis of lactose by microbial β-galactosidases. Formation of oligosaccharides with special reference to 2-o-β-D-galactopyranosyl-D-glucose. *J. Dairy Sci. 61*, 33–38.

Torgersen, H. and Sorhaug, T. 1978. Peptide hydrolases of *Brevibacterium linens. FEMS Microbiol. Lett. 4*, 151–153.

Trieu-Cuot, P. and Gripon, J. C. 1981. Casein hydrolysis by *Penicillium caseicolum* and *Penicillium roqueforti* proteinases: A study with isoelectric focusing and two-dimensional electrophoresis. *Neth. Milk Dairy J. 35*, 353–357.

Trieu-Cuot, P., Archiere-Haze, M. and Gripon, J. C. 1982. Effect of aspartyl proteinases of *Penicillium caseicolum* and *Penicillium roqueforti* on caseins. *J. Dairy Res. 49*, 487–500.

Troy, F. A., Freeman, F. E. and Heath, E. C. 1971. The biosynthesis of capsular polysaccharide in *Aerobacter aerogenes. J. Biol. Chem. 246*, 118–133.

Vallea, E. and Mocquot, G. 1968. Preparation of a concentrated suspension of thermophilic lactic acid bacteria for use in cheesemaking. *Lait 48*, 631–643. (French)

Vedamuthu, E. R., Sandine, W. E. and Elliker, P. R. 1966. Flavor and texture in Cheddar cheese. I. Role of mixed strain lactic starter cultures. *J. Dairy Sci. 49*, 144–150.

Vincent, J. G., Veomett, R. C. and Riley, R. F. 1959. Antibacterial activity associated with *Lactobacillus acidophilus. J. Bacteriol. 78*, 477–484.

Virtanen, A. I. and Nikkla, O. E. 1947. "Malty" flavor in starter and butter. *J. Dairy Res. 15*, 89–93.

Visser, S. 1981. Proteolytic enzymes and their action on milk proteins. A review. *Neth. Milk Dairy J. 35*, 65–88.

Waes, G. 1970. Preservation of lactic acid bacteria at temperatures of $-20°$, $-30°$ and $-196°$C. *Rev. Agr. 23*, 1097–1109. (French)

Walker, G. A. and Kilgour, G. L. 1965. Pyridine nucleotide oxidizing enzymes of *Lactobacillus casei*. II. Oxidase and peroxidase. *Arch. Biochem. Biophys. 111*, 534–539.

Wallenfels, K. and Mulhotra, O. P. 1961. Galactosidases. *Adv. Carbohydrate Chem. 16*, 239–298.

Wang, J. J. and Frank, J. F. 1981. Characterization of psychrotrophic bacterial contamination in commercial buttermilk. *J. Dairy Sci. 64*, 2154–2160.

Wasserman, A. E. 1960A. Whey utilization. II. Oxygen requirements of *Saccharomyces fragilis* growing in whey medium. *Appl. Microbiol. 8*, 291–293.

Wasserman, A. E. 1960B. Whey utilization. IV. Availability of whey nitrogen for the growth of *Saccharomyces fragilis. J. Dairy Sci. 43*, 1231–1234.

Wasserman, A. E. and Hampson, J. W. 1960. Whey utilization. III. Oxygen absorption rates and the growth of *Saccharomyces fragilis* in several propagators. *Appl. Microbiol. 8*, 293–297.

Wasserman, A. E., Hampson, J. W., Alvare, N. F. and Alvare, N. J. 1961. Whey utilization. V. Growth of *Saccaromyces fragilis* in whey in a pilot plant. *J. Dairy Sci. 44*, 387–392.

Wasserman, A. E., Hopkins, W. J. and Porges, N. 1958. Whey utilization—growth conditions for *Saccharomyces fragilis. Sewage Ind. Wastes 30*, 913–920.

Weinberg, E. D. 1977. Introduction. *In: Microorganisms and Minerals.* E. D. Weinberg (Editor). Marcel Dekker, New York.

Weissbach, H., Redfield, B. and Peterkofsky, A. 1961. Conversion of vitamin B_{12} to coenzyme B_{12} in cell-free extracts of *Clostridium tetanomorphum. J. Biol. Chem. 236*, PC40–PC42.

Wendorff, W. L., Amundson, C. H. and Olson, N. F. 1970. Nutrient requirement and growth conditions for production of lactase enzyme by *Saccharomyces fragilis. J. Milk Food Technol. 33*, 451–455.

Wheater, D. M., Hirsch, A. and Mattick, A. T. R. 1951. "Lactobacillin," an antibiotic from lactobacilli. *Nature 168*, 659.

Wierzbicki, L. E. and Kosikowski, F. V. 1973A. Lactase potential of various microorganisms in whey. *J. Dairy Sci. 56*, 26–31.

Wierzbicki, L. E. and Kosikowski, F. V. 1973B. Formation of oligosaccharides during β-galactosidase action on lactose. *J. Dairy Sci. 56*, 1400–1404.

Winkler, S. 1953. Antibiotic activity of lactobacilli against propionic acid bacteria. *13th Int. Dairy Congr. Proc. 3*, 1164–1167. (German)

Witter, L. D. 1961. Psychrophilic bacteria—A review. *J. Dairy Sci. 44*, 983–1015.

Yamada, T. and Carlsson, J. 1975A. Regulation of lactate dehydrogenase and change of fermentation products in streptococci. *J. Bacteriol. 124*, 55–61.

Yamada, T. and Carlsson, J. 1975B. Glucose-6-phosphate dependent pyruvate kinase in *Streptococcus mutans. J. Bacteriol. 124*, 562–563.

Yamasaki, I. 1939. Flavins that are formed during the acetone-butyl alcohol fermentation. Part I. Flavins from rice. *Biochem. Z. 300*, 160–166. (German)

Yamasaki, I. 1941. Flavin formation by the acetone-butyl alcohol bacteria. IV. *Biochem. Z. 307*, 431–441. (German)

Yamasaki, I., and Yositome, W. 1938. Formation of the vitamin B_{12}-complex from cereals by the acetone-butyl alcohol bacteria. *Biochem. Z. 297*, 398–411. (German)

Yates, A. R., Irvine, O. R. and Cunningham, J. D. 1955. Chromatographic studies on proteolytic bacteria in their relationship to flavor development in Cheddar cheese. *Can. J. Agr. Sci. 35*, 337–343.

Zevaco, C. and Desmazeaud, J. 1980. Hydrolysis of β-casein and peptides by intracellular neutral protease of *Streptococcus diacetylactis. J. Dairy Sci. 63*, 15–24.

Zevaco, C., Hermeir, J. and Gripon, J. C. 1973. Proteolytic system of *Penicillium roqueforti.* II. Purification and properties of the acid protease. *Biochemie 55*, 1353–1360.

Ziemba, J. V. 1970. Top-quality cultures made in unique plants. *Food Eng. 42* (1), 68–71.

14

Chemistry of Processing

Charles V. Morr and Ronald L. Richter

INTRODUCTION

The vast knowledge of milk chemistry has been extensively used by the dairy manufacturing industry to develop and perfect the modern technology required to produce the high-quality milk products to which we are accustomed. A thorough understanding of the chemistry of milk and milk components is essential for designing processing equipment and treatments needed for the manufacture and distribution of high-quality dairy products. Knowledge and application of milk chemistry is also indispensible for fractionating milk into its principal components, as in the manufacture of milk proteins, lactose, and milk fat products, for use as functional and nutritional ingredients by the food industry (Fox 1970; Harper 1981).

This chapter summarizes the chemistry of the major milk processing treatments and the chemical properties of major milk components that determine the functional and sensory characteristics of milk products. Most of these chemical phenomena are treated in greater detail in other chapters.

FLUID MILK PRODUCTS

All fluid milk and further processed dairy products are subjected to a series of processing treatments beginning with milking, pumping, cooling, mixing, and storage on the farm; transportation to the processing plant; and clarification, separation, standardization, pasteurization, vacuum off-flavor removal, homogenization, and packaging at the processing plant. Other processing treatments, e.g., acid and rennet coagulation, fermentation, vacuum evaporation, drying, churning, freezing, and sterilization, are used for manufacturing a variety of further processed dairy products.

Cooling and Agitation

Modern pipeline milking systems convey the milk into a refrigerated tank equipped with mechanical agitation to cool it rapidly and maintain it at $\leq 5°C$. Milk is normally held on the farm for 24–72 hr and is then transported to the processing plant. Upon arrival at the processing plant, milk is examined for temperature, flavor, and acidity. It is then sampled and tested for fat, total solids, and antibiotics content. If acceptable, the milk is received and stored in refrigerated tanks until it is processed further.

Cooling, agitation, and pumping of cold milk cause a number of chemical and physicochemical changes in the milk fat system (Brunner 1974; Harper 1976; Reimerdes 1982). For example, up to 65–70% of the milk fat crystallizes within 30 min, and crystallization is complete within 2–3 hr at 0–5°C. Upon cooling, milk fat globules adsorb whey proteins, mainly immunoglobulin, that promote their clustering and creaming, according to factors included in Stokes' law. Excessive agitation of cold milk and incorporation of air during pumping and agitation cause partial removal of the protective milk fat globule membrane, resulting in partially denuded globules that are more susceptible to lipase-catalyzed rancidity, churning, and development of oxidative off-flavors.

Cooling milk also causes important changes in the chemical and physicochemical properties of casein micelles (Morr 1975; Farrell and Thompson 1974; Brunner 1974; Harper 1976; Reimerdes 1982). These changes include release of proteolytic enzymes from micelles, which attack milk proteins and render them susceptible to slow coagulation and incomplete curd formation during cheese manufacture and may also result in flavor and texture defects in cheese and cultured milk products. Partial hydrolysis of milk proteins by residual proteolytic enzymes has also been implicated in the age thickening of ultra-high-temperature (UHT) sterile milk products (Harwalker 1982). Casein micelles in cooled milk undergo partial disaggregation to release β-casein and other casein components that may function as lipolytic enzymes to promote hydrolytic rancidity of the milk fat. This release of casein subunits causes several important changes in the physicochemical properties of the casein micelles. For example, casein micelles undergo increased solvation at 0°–5°C compared to 35°–40°C (Morr 1973A), and they also release inorganic phosphorus upon cooling. Similarly, the calcium content of cooled milk micelles is substantially lowered from the values observed at 35°–40°C. The ratio of micelle to total casein content in milk is lowered from about 85–95% at 35°–40°C to 75–80%

at $0°$–$5°C$. As a result of these changes, casein micelles reversibly disaggregate from 2–3 μm aggregates at $35°$–$40°C$ to sizes ranging from 100 to 250 ηm at $0°$–$5°C$, become more translucent, and are less electrondense upon cooling (Morr 1973B).

Because of these chemical and physicochemical changes in the casein micelles upon cooling of milk, the milk becomes more viscous and displays an increased tendency to foam. Also, casein micelles in cold milk commonly exhibit incomplete coagulation upon acidification and treatment with rennet (Harper 1976; Muller 1982A; Marshall 1982; Morr 1982).

Prolonged storage of raw milk and use of high-speed pumps, agitators, and blenders has resulted in an increase in the prevalence of off-flavors due to chemical, biochemical, and microbiochemical deterioration (Harper 1976; Reimerdes 1982; Tobias 1976). Milk is subject to spoilage, when stored at $0°$–$5°C$, by psychrotrophic bacterial enzymes that hydrolyze lactose, milk fat, and proteins (Cousin 1982; Law 1981, 1982). Storage and agitation of cold raw milk promote development of rancid flavors due to milk fat hydrolysis. In addition, fermentation of lactose yields lactic acid and other products that contribute to spoilage of raw milk. Exposure of milk to oxygen during blending, mixing, and pumping promotes oxidation of milk fat and development of off-flavors from the oxidation products. Contact of milk with trace metals also promotes lipid oxidation and development of oxidized flavors. Heat processing of milk activates whey protein sulfhydryl groups that function as antioxidants to inhibit lipid oxidation. Exposure of milk to sunlight and fluorescent light promotes milk fat deterioration, with development of associated off-flavors. However, before the reaction proceeds far enough to cause lipid oxidation, the riboflavin and whey proteins undergo chemical reactions that result in their degradation (Thomas 1981; Harper 1976).

Clarification

Clarification is normally one of the initial steps in processing fluid milk and is important for removing somatic cells, bacteria, and other foreign particles. It is accomplished by passing milk through a rapidly rotating clarifier bowl to sediment suspended particles.

Separation and Standardization

Milk is normally warmed to $35°$–$40°C$ or slightly higher to melt the milk fat before separation. Cold milk separators are generally less efficient than warm milk separators for removing milk fat due to unfavorable viscosity and density conditions. Cold milk separators are therefore operated at lower flow rates than warm milk separators to accomplish satisfactory milk fat separation (Harper et al. 1976; Brunner 1974). The cream fraction is centrifugally removed from milk by mechanical separators that operate at 5000 to 10,000 × g (Brunner 1974). Factors in Stokes' law that relate to fat separation include centrifugal force, milk fat globule size and density, density of the suspending medium, and viscosity. Factors that promote milk fat globule clustering or lower buoyant density of milk fat globules tend to improve separation efficiency. Efficiently operating separators normally produce skim milk with ≤0.10% milk fat.

Standardization of the milk fat and total solids contents of milk is accomplished by blending cream or skim milk with separated milk. Modern technology has developed continuous standardization processes that use turbidity or infrared absorption measuring devices to monitor and adjust the composition of the product as it leaves the separator. It is important that milk be accurately standardized to meet governmental legal requirements and to manufacture dairy products with optimal functional and quality attributes.

Pasteurization

Fluid milk is pasteurized by heating under various time and temperature conditions that meet U.S. Public Health Service requirements. These requirements include 63°C for 30 min, 72°C for 15 sec to 100°C for 0.01 sec (HTST), or 138°C for 2 sec (UHT) (Jones and Harper 1976). Higher temperatures than those just mentioned are required to pasteurize adequately milk products that contain >10% milk fat such as ice cream mix and cream products. Heat inactivation of alkaline phosphatase is used to monitor pasteurization and determine that the product has been given adequate heat treatment (Johnson 1974). However, phosphatase may be reactivated, especially in UHT-pasteurized cream and other products having a high milk fat content. Pasteurization so effectively destroys pathogenic microorganisms in milk that it almost never is a source of foodborne disease. In addition, pasteurization increases the shelf life of products by destroying microorganisms that cause spoilage. It is not uncommon to obtain a 10- to 14-day shelf life for pasteurized milk, provided it has been manufactured and handled under proper conditions.

Although batch and HTST pasteurization produces a cooked flavor in milk by activation of whey protein sulfhydryl groups, it is not severe enough to denature whey proteins or cause other heat-induced chemical reactions.

Vacuum Removal of Off-flavors

Fluid milk is commonly subjected to a combination steam injection/infusion and vacuum flash evaporation process to remove volatile off-flavor compounds. The process is designed to remove the same amount of water by the flash treatment as is added during steam injection/infusion, so that the composition of the milk remains unchanged. This treatment is most effective for removing volatile, water-soluble flavor compounds, such as those from weeds and feed consumed by the cow. The additional heat from this process usually provides further improvement in product shelf life.

Homogenization

Homogenization is accomplished by subjecting milk at $\geq 60°C$ to a two-stage process at combined pressures of 140 to 175 kg/cm^2 (2000 to 2500 lb/in^2) (Brunner 1974; Harper 1976), which provides sufficient shear and turbulence to subdivide milk fat globules into those with diameters ranging from 0.1 to 3 μm. This process results in several changes in the product, including a sixfold or more increase in milk fat globule surface area; adsorption of major amounts of casein and casein micelles onto the newly created milk fat globule surface; enhanced foaming properties; lowered heat stability of high-fat milk products; decreased curd tension; increased viscosity; increased susceptibility to lipolytic enzyme action; increased susceptibility to the formation of light-activated flavor; and decreased susceptibility to development of oxidized flavor.

Packaging and Distribution

Processed fluid milk is promptly cooled to 0°–5°C and packaged in glass, paper, or plastic containers by modern filling machines. Selection of packaging material is important, since each type of packaging material provides a different degree of protection to the milk against light exposure during storage and distribution (Harper 1976). Absorbed light radiation causes chemical deterioration of riboflavin and whey proteins, with production of compounds that catalyze oxidation of milk fat and formation of light-activated flavor (Henderson 1971; Harper 1976; Thomas 1981). When handled under proper light and temperature conditions, processed fluid milk products should have a ≥ 10–14-day shelf life. During this storage period, however, psychrotrophic microorganisms multiply and produce chemical compounds that cause gradual loss of product quality (Law 1981, 1982; Cousin 1982).

In addition to flavor deterioration, light radiation causes loss of certain water-soluble vitamins in milk during storage. For example, up to 50% of the riboflavin is destroyed by exposing milk to sunlight for 2 hr. Although fluorescent light destroys only a slight amount of riboflavin in milk, its decomposition products promote destruction of vitamin C through oxidation (Bender 1978). Vitamin losses are more prevalent in milk products stored in clear plastic containers which provide minimum protection against light-induced chemical reactions. For these reasons, there is a generally recognized need either to use packaging materials that provide the milk with greater protection against light radiation than is obtained from clear plastic or glass containers, or to avoid exposing the product to light radiation.

ICE CREAM

Ice cream is manufactured by rapidly freezing and simultaneously whipping an approximately equal volume of air into the formulated mix (Berger, 1976; Keeney and Kroger, 1974). Ice cream mix contains a minimum of 10% milk fat and 20% total milk solids, except when chocolate, fruit or nuts, are added. In addition to milk solids, ice cream mix normally contains 10–15% sucrose, 5–7% corn sweetener, 0.2–0.3% stabilizer gum, ≤0.1% emulsifier, and small amounts of natural or artificial color and flavor ingredients.

Ingredients and Their Functionality

Milk fat and milk solids-not-fat (MSNF) are most commonly obtained from cream and condensed skim milk, but may also be obtained from a combination of fluid milk, condensed whole milk, frozen cream, frozen condensed milk, nonfat dry milk, dry whole milk, and butter. Sweeteners used in the mix normally include a combination of liquid or dry sucrose, corn sweetener, high-fructose corn sweetener, and corn syrup solids. Ice cream stabilizers are formulated to contain one or more polysaccharide hydrocolloids, e.g., carboxymethyl cellulose, locust bean gum, carageenin, alginate, and other gums. Ice cream emulsifiers normally contain monoglycerides and diglycerides of palmitic and stearic acids or polyoxyethylene derivatives of sorbitan tristearate and sorbitan monooleate (Berger 1976).

Milk fat, which is one of the most important ingredients in the ice cream formulation, imparts richness, flavor, and body to the product and also contributes a smooth texture by physically limiting the size of ice crystals (Arbuckle 1977). During the freezing process, a portion of the milk fat is released from milk fat globules by partial churning to form a thin film surrounding the air cells (Berger 1976), thereby affecting the physical characteristics of the final product. Fresh cream is the most desirable source of milk fat, since it contributes optimum flavor and is normally free of oxidized, rancid, or stale flavors common to stored, frozen cream products.

MSNF are essential for developing small air cells and ice crystals to provide desirable body and texture (Berger 1976; Keeney and Kroger 1974). It is likely that soluble casein and casein micelles provide this important function by lowering surface tension to facilitate incorporation of air into the product. Milk proteins probably also function to limit growth of ice crystals during the freezing and hardening steps of ice cream manufacture by their strong affinity for free water and their ability to impart viscosity to the unfrozen phase of the product. Concentrated skim milk and whole milk are the most common sources of MSNF for ice cream manufacture.

Sucrose and corn sweeteners are added primarily to contribute sweetness to the product. In addition, these added carbohydrates function to control the freezing temperature of the mix and to assist proteins and stabilizers in providing viscosity to limit ice and lactose crystal growth during freezing of the product. These added sweeteners also stabilize the frozen product against lactose and ice crystal growth during storage of the product, when it may be exposed to extreme temperature fluctuations. Low dextrose-equivalent (D.E.) corn sweeteners increase the solids content of the mix without providing excessive sweetness or adversely lowering the freezing point of the mix. These latter ingredients contribute a chewy body and a high thermal shock stability to the frozen product. Higher D.E. corn sweeteners are used when body and texture similar to those of an all-sucrose ice cream are desired (Berger 1976).

The primary function of stabilizers in ice cream is to bind water and provide added viscosity to limit ice and lactose crystal growth, especially during storage under temperature fluctuation conditions. Stabilizers also assist in aerating the mix during freezing and improve body, texture, and melting properties in the frozen product.

Emulsifiers facilitate air incorporation into the mix during freezing and thus contribute to a smooth texture in the frozen product. They also control dryness of the frozen product by affecting the release of free milk fat from milk fat globules during the freezing process (Govin and Leeder 1971).

Processing the Mix

The required amounts of dry and liquid ingredients are introduced into the mix formulation vat, where they are blended and heated to obtain a homogeneous mix. Stabilizers, emulsifiers, and other ingredients that are difficult to disperse are dry-blended with a small amount of the sucrose and slurried in a small amount of water to facilitate their uniform dispersion into the mix. The homogeneous mix is pasteurized at $\geq 79°C$ for 25 sec or $\geq 68°C$ for 30 min. These pasteurization temperatures, which are higher than those of fluid milk products, facilitate dissolution and solvation of protein and carbohydrate hydrocolloids and destroy pathogenic and spoilage microorganisms.

The mix is then homogenized at 105 to 210 kg/cm^2 (1500 to 3000 lb/in^2) to subdivide milk fat globules to sizes ranging from 0.5 to 2 μm in diameter. This process is essential to produce a mix with adequate aeration properties so that the final product will contain ≤ 175-μm-diameter air cells to contribute a smooth texture. The homogenized mix is cooled and "aged" to fully hydrate the hydrocolloids, e.g., milk proteins, stabilizers and corn sweetners, and to provide adequate viscosity to the mix.

Freezing the Mix

Most commercial ice cream is manufactured in continuous freezers that provide rapid freezing, control of air incorporation (overrun), and drawing temperature. The process involves rapid and continuous whipping and freezing of roughly equivalent volumes of ice cream mix and air. The mix and air are metered into the freezer in the proper proportions to obtain ice cream of desired overrun, which is computed as the percentage increase in volume or the percentage decrease in the density of the mix. Upon entering the freezer, the mix is rapidly whipped by rotating mutator and ice scraper blades. The mix is simultaneously chilled to $-4°$ to $-7°C$ to freeze roughly half of its water content. The water is frozen on the inside wall of the freezer barrel, which is chilled to $-30°$ to $-50°C$ by contact with liquid ammonia or Freon. The ice crystals that form on the inside wall of the freezer barrel are continuously removed by the rapidly rotating scraper blades. The ice crystals must be kept at ≤ 50 μm in diameter to produce ice cream with a smooth texture. The rapidly rotating mutator and scraper blades whip air into the partially frozen mix to form air cells about 175 μm in diameter that provide optimum body and texture to the ice cream. Both the amount and the distribution of air in the frozen product are important in providing body and texture in ice cream. The partially frozen product is continuously extruded from the freezer into bulk containers or smaller-sized packages, which are held in the hardening tunnel or room at −20° to −35°C until the freezing process has been largely completed. The hardening process must be completed as rapidly as possible to provide small ice crystals that result in a smooth-textured ice cream. Hardening may require up to several days to complete, depending upon air temperature and velocity, package dimensions, and product composition. As more and more of the water freezes, proteins, lactose, and other solutes become highly concentrated in the unfrozen phase of the product. At this point, the proteins and stabilizers function to inhibit crystallization of lactose, which is present in a supersaturated state. Retention of lactose and other solutes in the highly concentrated, unfrozen solution is necessary to prevent coagulation of casein micelles that would result in product with a "curdy meltdown" defect. Lactose crystallization in ice cream is responsible for a defect known as "sandiness."

Prolonged storage of ice cream and exposure to severe temperature fluctuation commonly causes "shrinkage," which is a defect due to partial thawing and loss of moisture and air. An additional defect common to ice cream after prolonged storage is oxidized flavor, which is caused by autoxidation of milk fat. This defect is especially important in ice cream products that contain frozen or dried milk ingredients.

BUTTER

Historically, butter has been produced by churning chilled cream until the oil-in-water (O/W) milk fat emulsion is broken and the milk fat forms butter granules that separate from the aqueous buttermilk phase. Several continuous buttermaking processes are now available to manufacture butter (Brunner 1974; Harper and Seiberling 1976).

Processing the Cream

Cream is pasteurized at 71° to 77°C for 30 min, cooled to 5°–10°C, and held for several hours to provide the required distribution of liquid and crystalline milk fat. For continuous butter making, pasteurized cream, containing about 40% milk fat, is processed by one of the following methods: (1) cream is reseparated to produce 80% milk fat cream, which is then homogenized to destabilize the milk fat emulsion; (2) cream is heated and agitated to partially destabilize its emulsion and reseparated to recover the milk fat phase as butter oil.

Conventional Churning

Pasteurized, tempered 40% cream is churned for 30–45 min to destabilize its emulsion and form butter granules. Although several theories have been proposed to explain the churning mechanism, it is likely that the excessive abrasion that results from the violent mixing of the cream during the churning process removes a sufficient portion of the milk fat globule membrane to render the milk fat globules "sticky." Further churning simply causes the sticky milk fat globules to agglomerate and form butter granules. The churning process is continued until the butter granules reach the size of a pea. The buttermilk is then drained and cold water is added to wash and temper the butter granules. After a short churning treatment, the wash water is drained and salt is added and thoroughly worked throughout the butter. The final product contains ≥80% milk fat as a legal requirement and is composed of 2–50% intact milk fat globules and small water droplets dispersed throughout the free milk fat continuous phase.

Continuous Buttermaking

There are several different continuous buttermaking processes in operation today. For example, one process involves passing 40% cream through a churning cylinder that produces sufficient agitation and foaming to destabilize the emulsion and form butter granules within only several minutes. A second process churns 80% fat cream on a continuous basis to form butter granules rapidly. Butter churned by these two processes contains 6–30% intact milk fat globules dispersed in free milk fat (Brunner 1974). A third process emulsifies a dilute salt solution into butter oil to form a water-in-oil (W/O) emulsion, by a process similar to that used for manufacturing margarine. Butter produced by this process contains only large milk fat crystals dispersed in free milk fat, but no intact milk fat globules.

EVAPORATED MILK

Evaporated milk is manufactured by forewarming and concentrating milk under vacuum and standardizing, homogenizing, canning, and sterilizing the concentrate.

Standardization

Milk is clarified and standardized to contain a 1:2.28 milk fat:total solids ratio (Hall and Hedrick 1966). This ratio is necessary to produce a final product that meets federal government requirements, e.g., $\geq 7.5\%$ milk fat and 25% total milk solids contents (CFR, 1982).

Forewarming

The standardized milk is forewarmed by heating to 71° to 88°C for 10 to 30 min or to 149°C for 1 to 2 sec (Edmondson 1970). These heat treatments are essential to stabilize the milk protein and mineral components against heat-induced precipitation during subsequent sterilization of the concentrate (Morr 1975; Fox 1981; Hall and Hedrick 1966; Edmondson 1970; Darling 1980).

These drastic heat treatments cause important chemical and physicochemical changes in the major milk protein and mineral components, some of which are essential to the processing of evaporated milk, whereas others are detrimental to the color, flavor, and nutritional quality of the product. These reactions include increase in titratable acidity and reduction of pH caused by lactose destruction and formation of organic acids; formation of colloidal phosphate; denaturation of whey proteins; interaction of denatured β-lactoglobulin and other whey proteins with the κ-casein component of casein micelles by disulfide interchange; Maillard reaction between lactose and milk proteins to reduce available lysine and produce brown pigments, furfural, maltol, and other organic compounds; aggregation of casein micelles; casein dephosphorylation; and destruction of water-soluble vitamins. Activated whey protein sulfhydryl groups result in development of heated flavor and function as antioxidants that inhibit oxidation of milk fat during storage of sterile and dried milk products. The rates of most of the chemical and physicochemical reactions just mentioned are accelerated at the higher milk solids content of concentrated and evaporated milk (Fox 1981, 1982).

Whey proteins are sensitive to heating at temperatures above 60°C and are converted from their native globular conformational state to a random conformation that renders them susceptible to protein–protein and protein–ion interaction (Morr 1975; Parry 1974). Denatured β-lactoglobulin and other whey proteins preferentially complex with casein micelles by disulfide interchange with their κ-casein components and by Ca-mediated bonding. This interaction stabilizes whey proteins against aggregation and precipitation in heated milk systems. Heating milk to temperatures of ≥ 90°C causes formation of casein micelle–whey protein aggregates with a molecular weight of $\geq 100,000$ and is more effective in this respect than heating milk by UHT conditions, e.g., 149°C for 1–2 sec (Morr 1975).

Heating milk to $\geq 100°C$ for 15 to 20 min, as in the forewarming of milk for evaporated milk manufacture, effectively denatures and complexes whey proteins with casein micelles and causes simultaneous aggregation and disaggregation of these complexed milk protein aggregates (Morr 1975).

The β-lactoglobulin content of milk appears to be the most important compositional factor controlling its heat stability. It probably functions by interacting with κ-casein on the casein micelles, thus altering their susceptibility to heat-induced coagulation. In addition, β-lactoglobulin may interact simultaneously with other denatured whey protein components, thereby attaching them to the casein micelles to stabilize them against heat-induced precipitation.

It is necessary to forewarm milk to impart adequate heat stability to the concentrate to permit it to withstand subsequent sterilization treatments. The heat-induced casein micelle-whey protein complexes in forewarmed milk are less sensitive to heat than native whey proteins and thus provide the required stability to the concentrate. The forewarming treatment also stabilizes the milk mineral system by complexing Ca and Mg ions with casein micelles and by converting ionic forms to the less reactive form of colloidal phosphate (Morr 1975).

As indicated in Table 14.1, the water-soluble vitamins, e.g., thiamine, vitamin B_{12}, and vitamin C, are susceptible to loss by heat processing of milk (Bender 1978; Harper 1976). UHT sterilization and

Table 14.1. Effect of Processing Treatments Upon Vitamin Destruction in Milk Products.

Process Treatment	Vitamin Destruction (%)			
	Thiamine	Biotin	Vitamin B_{12}	Vitamin C
HTST pasteurized	<10	0	<10	10
UHT sterilized	<10	0	20	10
Conventional sterilized	35	0	>90	50
Condensed	10	10	30	15
Evaporated[a]	40	10	90	60
Drum dried	15	10	30	30
Spray dried	10	10	30	20

[a]Condensed 2.25:1 prior to sterilization.
SOURCE: Adapted from Graham (1974).

HTST pasteurization cause similar losses of these vitamins, which are less than those caused by conventional sterilization and drum drying. Deaeration before heat processing stabilizes the water-soluble vitamins against heat-induced destruction and storage loss.

Although milk proteins are widely recognized for their excellent nutritional quality, drastic heat processing causes substantial loss of available lysine by the browning reaction (Bender 1978). For example, drum drying causes a 40% loss of available lysine; sterilization destroys 15–20% of the available lysine in evaporated milk, but spray drying has little effect on the available lysine content of milk. Severe heating of milk, as at 120°C, in the presence of lactose or other reducing sugars, lowers protein digestibility (Graham 1974). This effect is more pronounced with casein, which contains a high lysine content, than with whey proteins.

Concentration

Forewarmed milk is concentrated 2.25:1 by vacuum evaporation at a product temperature of $\leq 45°-50°C$, which is used to minimize chemical and physicochemical changes in the product. A variety of evaporators are available for concentrating milk, including standard evaporators that rely upon convection to mix the product; falling-film evaporators; mechanically aided, thin-film evaporators; and centrifugal evaporators. Thin-film evaporators generally operate at higher vacuum than standard evaporators and therefore require only 10–15 sec to concentrate the milk. Centrifugal evaporators require product residence times ranging from 0.5 to 2 sec, and thus are capable of producing a concentrate with minimal heat damage. Most of the chemical and physicochemical reactions just discussed occur at accelerated rates upon concentration of milk (Morr, 1975; Parry, 1974).

Homogenization

The concentrated milk is homogenized at 140 to 210 kg/cm^2 (2000 to 3000 lb/in^2) at about 48°C (Hall and Hedrick 1966). This process is essential to provide adequate physical stability to the milk fat emulsion system to withstand prolonged storage at room temperature (Brunner 1974). However, homogenization lowers the heat stability of concentrated milk products (Parry 1974), which may be due to increased adsorption of casein micelles onto the newly created milk fat globule surfaces, thus making them more sensitive to heat-induced aggregation.

Mineral Adjustment

It is necessary to adjust the Ca, Mg, phosphate, and citrate content of the concentrate to control aggregation and precipitation of the proteins and minerals during sterilization. By controlling protein aggregation, this adjustment provides optimum viscosity to stabilize the protein, mineral, and milk fat emulsion systems during prolonged storage of the sterile product. Some milk concentrates are stabilized by addition of Ca and Mg salts, whereas others are stabilized by addition of phosphate or citrate salts (Parry, 1974). Chemical compounds approved for addition to evaporated milk include calcium chloride, sodium citrate, and disodium phosphate (CFR 1982).

Sterilization

The concentrated milk described above is canned and sterilized in a retort heater at 115° to 118°C for 15 to 20 min. The cans are rotated as they proceed through the preheater, sterilizer, and cooler sections of the retort to ensure uniform heating of their contents.

The drastic time–temperature treatment used to sterilize the concentrate accelerates the aforementioned chemical and physicochemical reactions. Of special importance is the aggregation of the proteins that is responsible for increasing the viscosity of the product. This process must be properly controlled by adjusting the ionic composition and the temperature of the heat treatment to provide adequate product viscosity without forming protein-mineral grains that would impart a gritty, grainy texture. Additionally, the heat treatment causes substantial Maillard browning, resulting in the formation of brown color and a scorched flavor that reduces the acceptability of the product.

UHT STERILE MILK

Most of the developmental research and production of UHT sterile milk, which does not require refrigerated storage, has been in Europe, where home refrigeration is minimal. Until now, the U.S. dairy industry has been slow to promote this new product, since home refrigeration is universally available and the consumer is not accustomed to its highly heated flavor.

Typical time and temperature combinations for UHT milk products are 132°C for ≥ 1 sec for milk and 132°C for ≥ 2 sec for cream (Burton,

1979). In practice, higher temperatures, in the range of 135° to 150°C, are used to produce commercially sterile milk products, which must then be aseptically packaged to retain sterility. The product is heated either directly, by steam injection or steam infusion, or indirectly by a plate, tubular, or scraped-surface heat exchanger. Raw milk is commonly preheated to 80°–85°C by an indirect heater and then heated to the final temperature by steam injection or steam infusion. Direct heating provides instantaneous distribution of heat throughout the product so that the final temperature is achieved in a fraction of a second. Indirect heating requires a longer time period to reach the final temperature and suffers from the major problem of product "burn-on," which further slows the heat exchange rate. The small amount of product dilution from added steam by the direct heating process is compensated for by its removal during the evaporative cooling stage of the process.

Although all of the heat treatments described cause a severely heated flavor defect by activation of whey protein sulfhydryl groups and formation of Maillard browning reaction products, the intensity of this defect is much less than that of conventionally sterilized milk products, such as evaporated milk. Also, the other heat-induced chemical and physicochemical reactions considered earlier occur to a much more limited extent than in conventionally sterilized milk products. UHT sterilization causes whey protein denaturation and aggregation and casein micelle aggregation (Freeman and Mangino 1981), as well as some Maillard browning and vitamin destruction. In this latter regard, the indirect heating process causes more product deterioration than occurs by direct UHT processing. The heated flavor defect, which is especially evident following UHT processing, gradually disappears and is replaced by a typical UHT "stale" flavor.

Sterile milk is aseptically homogenized to stabilize the milk fat emulsion system without adversely affecting its heat stability, which would occur if the order of these two process treatments was reversed. Indirectly heated UHT milk contains about 7–8 μg oxygen per gram compared to only about 1 μg oxygen per gram for directly heated UHT milk. This difference in oxygen content is largely due to removal of oxygen by the vacuum treatment during direct UHT processing. Ascorbic and folic acids are more stable in directly heated UHT milk, with its lower oxygen content, than in indirectly heated UHT milk, especially when the product is packaged in oxygen-impermeable containers. Heat-induced, activated whey protein sulfhydryl groups mentioned earlier function as antioxidants in these products to stabilize these latter nutrients further and inhibit autoxidation of milk fat.

UHT STERILE MILK CONCENTRATE

Although UHT sterile milk concentrate has been proposed as a replacement for conventionally sterilized milk concentrate, such as evaporated milk, this product has not been commercially developed. One reason for the lack of interest by the dairy industry is the tendency of this product to age thicken during storage. Although this problem has been largely circumvented, the previously described flavor defect of sterile milk products also applies to UHT sterile milk concentrate. In addition, sterile milk concentrate must be diluted and mixed with water before use, which is a major disadvantage. The product would compete with UHT sterile milk rather than with evaporated milk, since the latter product, with its serious flavor and color limitations, is not suitable for beverage use. Application of modern heating, homogenizing, and packaging technology developed for UHT milk processing may improve opportunities for UHT sterile milk concentrates, especially in regions of low milk production. Since the product would compete with UHT pasteurized and UHT sterile milk, the severity of heat processing must be minimal to maintain adequate quality.

Forewarming and Concentration

Although no uniform processing scheme has been reported for this product, it seems logical to forewarm milk at 115°C for 2 min and concentrate it 3:1 (v/v) by vacuum evaporation (Parry 1974). While the severity of the heat treatment must be held to a minimum, the forewarming treatment must be adequate to stabilize proteins to withstand subsequent sterilization and to prevent age thickening during storage (Parry 1974; Morr 1975). Treatments that lower heat stability of milk proteins, e.g., high forewarming temperature and added Ca ions, tend to stabilize the concentrate against age thickening. However, increasing the total milk solids concentration lowers both heat and age thickening stability of the concentrate. The most successful approach for stabilizing UHT sterile milk concentrate against age thickening is addition of sodium or potassium polyphosphate before sterilization of the product (Parry 1974).

Sterilization

The forewarmed, concentrated milk is sterilized by direct or indirect UHT heat exchangers at time and temperature combinations approximating 135°C for 30 sec (Parry 1974). The sterilized product would then be aseptically homogenized at 280 kg/cm$_2$ to stabilize the milk fat emulsion against physical separation during subsequent storage. As indicated before, it is advantageous to homogenize the product after sterilization, since this order provides added heat stability to the protein system.

Packaging

The sterile milk concentrate is aseptically placed in suitable containers, probably metal cans or modern laminated, foil-lined, plastic containers. Modern aseptic packaging equipment used for manufacturing UHT sterile milk should be satisfactory for packaging the sterile milk concentrate.

Storage

Age thickening is a common defect of UHT sterile milk concentrate, which has received less drastic heat treatment than conventionally sterilized evaporated milk (Harwalker 1982). Age thickening is promoted by high milk solids content, addition of alkali to raise the pH, and addition of citrate, phosphate, and other anions that lower the Ca ion activity. Conversely, addition of Ca ions improves stability of the product against age thickening.

Electron microscopic data indicate that UHT sterile milk concentrate contains casein micelle-denatured whey protein complexes that are about double the size of native casein micelles (Parry 1974; Morr 1975). These complexes undergo extensive aggregation during storage that eventually causes age thickening. Added polyphosphate compounds do not prevent interaction of casein micelles and denatured whey proteins during sterilization but function mainly by preventing their aggregation during storage that is responsible for age thickening.

FROZEN MILK CONCENTRATE

Surplus milk is commonly stored as frozen skim milk and whole milk concentrate and used as ingredients in ice cream and other formulated food products. There has also been some interest in producing frozen milk concentrates to substitute for pasteurized and sterile fluid milk products (Webb 1970).

The major defect, which limits exploitation of frozen milk concentrates as consumer products, is the instability of the casein micelle system (Keeney and Kroger 1974; Morr 1975). The casein micelles gradually destabilize during storage of the frozen milk concentrate.

Several compositional and processing variables affect the physical stability of the casein micelles in frozen milk concentrates. These factors include pH, mineral composition, total solids content, forewarming treatment, homogenization and fat content, freezing rate, storage temperature, and fluctuation of storage temperature (Keeney and Kroger 1974; Webb 1970).

The physical stability of the casein micelle system is closely related to the degree of lactose crystallization from the unfrozen phase of the frozen concentrate. Crystallization of lactose from the unfrozen solution temporarily raises its freezing point, causing additional water to freeze, thus increasing the concentration and promoting destabilization of casein micelles.

The stability of casein micelles in frozen milk concentrate is enhanced by removing Ca ions or by adding Ca ion-complexing chemicals, such as hexametaphosphate. Both of these treatments dissipate the casein micelles and form soluble casein consisting largely of nondescript aggregates and submicelles (Morr 1975). Addition of hydrocolloids, such as ice cream stabilizers, also inhibits lactose crystallization and thereby stabilizes casein micelles in frozen milk concentrates. Enzymatic hydrolysis of lactose prevents its crystallization during frozen storage and increases the stability of casein micelles in frozen milk concentrate (Webb 1970).

CHEESE

Since Cheddar cheese is the major cheese produced in the United States, this discussion pertains to the chemistry of the processing treatments used in its manufacture. Additional details of cheese manufacture are presented in Chapter 12.

Standardization and Pasteurization of Milk

Raw milk is standardized to the proper fat and total milk solids content to produce a final product with a minimum of 50% fat on a solids basis and ≤39% moisture (CFR 1982; Packard 1975). Cheese is made from pasteurized or raw milk, but raw milk cheese must be aged a minimum of 60 days at ≥1.7°C (CFR, 1982). Minimum temperature and time combinations are normally used for pasteurization of milk for cheese manufacture in order not to interfere with casein micelle coagulation and curd formation. Milk is sometimes heated only to subpasteurization temperatures to dispel dissolved gases, reduce bacterial populations, and kill certain pathogens, thus resulting in a cheese product with improved flavor (Babel 1976).

Acid Development

The milk processed as just described is pumped into the cheese vat, tempered to 30°–31°C, and inoculated with an active homofermentative lactic streptococcus starter culture. The lactic starter culture produces acid to facilitate rennet coagulation of casein micelles, and its enzymes hydrolyze lactose and milk proteins during ripening of the cheese. The developed acidity in the milk partially dissolves the colloidal phosphate from the casein micelles, altering their charge and making them more susceptible to rennet-induced coagulation (Morr 1973A).

After sufficient acid has been developed by the lactic starter culture, rennet or clotting agents of microbial origin are added to the milk to coagulate the casein micelles by a two-step process (Morr 1975). The first step of the reaction involves hydrolysis of the κ-casein 105–106 peptide bond to release the highly acidic and hydrophilic glycomacropeptide (GMP) fragment. Release of GMP from κ-casein lowers the zeta potential of the micelles, facilitating their coagulation. The resulting coagulum is weakly structured and is further processed, as discussed below, to expel whey and convert it into the final cheese product (Ernstrom and Wong 1974).

Cheese Curd Processing

The rennet-coagulated milk gel is cut into small cubes and cooked by gradually warming to about 40°C. The combination of elevated temperature and lowered pH during the cooking process causes casein micelles and submicelles to associate strongly, resulting in curd particle shrinkage and expulsion of whey by syneresis. After drainage of the whey from the cheese vat, the curd is matted and cheddared by conventional or modern automated technology (Olson 1975, 1981).

Conventional Matting and Cheddaring

The cheddared cheese curd is milled into thin strips, salted, placed in cheese hoops, and pressed overnight to expel additional whey and fuse and curd strips together. The pressed cheese is then removed from the hoops and coated with wax or wrapped in a plastic film.

Automated Matting and Cheddaring

Curd and whey are pumped from the cheese vat onto a continuous draining and matting conveyor system to allow the curd to mat and cheddar. The matted curd is milled, salted, and drawn by vacuum onto the top of a "block-forming" tower maintained at 30°–32°C. The milled curd pieces fuse together as they move downward through the "column" of curd for about 30 min. As they move down through the tower, the combined effect of vacuum and pressure serves to remove excess whey, which further facilitates fusion of the curd. The fused curd is extruded from the bottom of the tower and cut into 40-lb (18 kg) rectangular blocks, which are immediately packaged in oxygen- and water-impermeable plastic bags under a vacuum of 736 mm Hg. This vacuum treatment further compacts the cheese block, facilitating continued curd fusion and production of a close-textured cheese. The vacuum treatment also contributes to a smooth-surfaced cheese block, which aids in preventing mold growth and spoilage (Domnitz 1984).

Ripening

Cheese is ripened for 6 months to 1 year or longer at 5° to 15°C and 70–75% relative humidity. Cheese ripening is a complex process involving a combination of chemical, biochemical, and physical reactions. Proteolytic enzymes, e.g., rennet and lactic starter culture enzymes, hydrolyze caseins to produce flavor compounds and proper body. Lipase and lactase enzymes also hydrolyze their respective substrates to produce a large number of characteristic flavor compounds (Reiter and Sharpe 1971; Harper 1959; Law 1981; Schmidt *et al.* 1976), including free fatty acids, methanethiol, methanol, dimethyl sulfide, diacetyl, acetone, and others (Moskowitz 1980).

CULTURED MILK PRODUCTS

Cultured milk products are manufactured by fermentation of milk or cream by lactic culture microorganisms that produce desirable flavor and rheological properties which are influenced by the composition of the milk or cream, and by the processing conditions used (Richter 1977; Foster *et al.* 1957; Marth 1974). Cultured buttermilk may be made from skim milk but is sometimes made from milk containing 1.0 to 3.5% milk fat. Some cultured milk products often contain added MSNF and plant gum or modified starch stabilizers to increase viscosity and control whey syneresis. Dextran-producing culture microorganisms are sometimes used to provide needed viscosity to the cultured milk product without the need to add MSNF or stabilizers. Up to 0.1% citric acid or sodium citrate is commonly added as a substrate for diacetyl-producing culture microorganisms when that flavor agent is desired and salt is frequently added as a flavorant. Sour cream is similar to cultured buttermilk in processing requirements and composition, except that it contains $\geq 18\%$ milk fat.

The milk or cream formulation is heated at minimal time and temperature combinations, e.g., $74\,^\circ$C for 30 min to $91\,^\circ$C for 3 min, to pasteurize the product. These heat treatments, which denature substantial amounts of whey proteins (Harland et al 1955), provide increased viscosity and prevent whey syneresis in the cultured product (Foster et al. 1957). In addition, the heat treatment destroys inhibitory substances in the milk, thereby stimulating subsequent growth of the culture microorganisms. Milk and cream formulations are then homogenized at 140 to 175 kg/cm^2 to stabilize their milk fat emulsion system. Double homogenization at sequential temperatures of $74\,^\circ$ and $43\,^\circ$C promotes milk fat globule clustering and subsequent development of viscosity in sour cream (Emmons and Tuckey 1967).

Pasteurized, homogenized milk or cream formulations for buttermilk or sour cream are cooled to about $21\,^\circ$C, inoculated with a culture containing one or more strains of acid- and flavor-producing bacteria, and incubated at this temperature. Lactic acid, the major product formed during initial stages of the fermentation process, lowers the pH to ≤ 6.0. At this point, the culture microorganisms ferment citric acid to form the desirable flavor compound diacetyl, which increases in concentration until the pH falls to ≤ 4.6 and the titratable acidity reaches $\geq 0.85\%$ (Pack et al 1968). The cultured product is immediately cooled to $\leq 7\,^\circ$C to inhibit culture microorganisms and to maintain optimum diacetyl concentration. Agitation and pumping of the product at this point should be minimized to avoid air incorporation, loss of viscosity, and syneresis.

Yogurt is manufactured from milk and contains 0.5 to 3.5% milk fat, 10–14% MSNF, and low concentrations of added stabilizer gums. Fruit-containing yogurts often have up to 15–22% of added carbohydrate which functions mainly as a sweetener. the MSNF content of the formulation may be increased to the values just mentioned by adding nonfat dry milk (NFDM) or concentrating it by vacuum evaporation. The standardized milk formulation is pasteurized at $74\,^\circ$C for 30 min or $90\,^\circ$C for ≥ 5 min, cooled to incubation temperature, inoculated with *Lactobacillus bulgaricus* and *Streptococcus thermophilus* culture microorganisms, and incubated for 12–16 hr at $30\,^\circ$C or for 4–6 hr at $41\,^\circ$–$45\,^\circ$C. Major flavor compounds formed in yogurt include lactic and acetic acids, acetaldehyde, and diacetyl (Vedamuthu 1974). The final pH of 4.1 to 4.2 is achieved by controlled cooling of the product to retard growth and further fermentation by the culture microorganism.

DEHYDRATED MILK PRODUCTS

Nonfat dry milk (NFDM), lowfat dry milk (LFDM), and dry whole milk (DWM) are produced and used as functional ingredients in dairy, bakery, confectionary, and other food applications. Instantized NFDM is made primarily for home and institutional beverage and miscellaneous product applications (Hall 1976; Hall and Hedrick 1966).

Standardization

Milk or skim milk is standardized to meet milk fat and total milk solids compositional requirements in the final dry milk product; e.g., NFDM must contain $\leq 1.5\%$ milk fat; DWM must contain $\geq 26\%$ and $< 40\%$ milk fat, and both must contain $\leq 5\%$ moisture on an MSNF basis (CFR, 1982). Dry milk products must also meet U.S. Department of Agriculture and American Dry Milk Institute standards for titratable acidity, scorched particle content, solubility, and bacteria content. Product users may also have specialized requirements for color, flavor, and dispersibility.

Forewarming

Standardized milk or skim milk is preheated to meet product specifications on whey protein denaturation. For example, high-heat milk powder must contain $\leq 1.5\%$ mg of undenatured whey proteins per gram and low-heat milk powder must contain ≥ 6.0 mg of undenatured whey proteins per gram (Pallansch 1970). Low-heat milk powder is manufactured from milk that has been forewarmed at a minimal temperature and time, such as 72° to 74°C for 15 sec, to denature less than 10% of the whey proteins (Hall and Hedrick 1966; Pallansch 1970). High-heat milk powder is made from milk that has been forewarmed under more drastic conditions, such as 90°C for 10 to 30 min, to denature a higher percentage of whey proteins.

Vacuum Evaporation

Forewarmed milk or skim milk is concentrated by vacuum evaporation to contain 40–60% total milk solids. The degree of concentration is determined to meet dryer design requirements and is used to control the particle size, density, dispersibility, and solubility of the powder.

Drying

Skim milk or whole milk concentrate is normally preheated, homogenized at 140 to 280 kg/cm^2, and atomized into the dryer as minute liquid droplets. Drying is accomplished by passing large volumes of heated air at inlet temperatures in the range of 163° to 204°C through the dryer. The liquid droplets rapidly absorb energy from the air and are instantly dried, lowering the outlet air temperature to about 38° to 49°C (Hall 1976). The temperature of the liquid droplets approaches that of the outlet air as the drying process nears completion (Parry 1974). For this reason, outlet air temperature is a critical parameter in controlling heat damage to the dry milk product. The dried particles acquire sufficient density to fall to the bottom of the dryer, where they are removed and cooled to inhibit the Maillard reaction, whey protein denaturation, and other chemical and physicochemical reactions that lower product quality. DWM must be promptly cooled, degassed, and stored under vacuum or inert atmosphere to inhibit development of stale and oxidized off-flavors. Both NFDM and DWM are packaged in containers that prevent absorption of moisture from the atmosphere during subsequent storage.

Instantizing

Most NFDM used in home and institutional applications is instantized to agglomerate the powder particles and thereby improve their dispersibility (Pallansch 1970; Hall and Hedrick 1966). The instantizing process involves partial rehydration of spray-dried milk powder with steam to 11 to 16% moisture to convert its lactose to the α-monohydrate form, which imparts stickiness and facilitates particle agglomeration (Gillies 1974). The agglomerated milk powder particles are then redried on a fluidized bed dryer. Instantized milk powder may be produced by specially designed spray driers that provide simultaneous drying and agglomeration of the particles. The agglomerated milk powder particles are removed from the spray dryer, and drying is completed on a belt dryer.

Instantized milk powder normally exhibits low bulk density but higher water dispersibility than conventionally spray-dried powder. However, the extra heat exposure from the agglomeration and redrying treatments causes additional Maillard reaction, whey protein denaturation, and related chemical and physicochemical reactions that tend to lower product quality.

Solubility Factors

Spray drying per se has little effect on the chemical and physicochemical properties of milk powder when the severity of heat treatments has been minimized (Parry 1974). The solubility of the milk powder product is controlled by altering the temperature and time of heat treatments during forewarming, vacuum evaporation, spray drying, and storage. Chemical and physicochemical reactions involving whey proteins and casein micelles are important in controlling the solubility of dry milk. High-temperature storage of milk powder causes further loss of solubility by a mechanism that mainly involves changes in casein micelles. High-temperature treatments of the skim milk or milk prior to drying, e.g., during forewarming and evaporation, also contribute to loss of powder solubility, but the mechanism responsible for this reaction involves whey protein denaturation and its interaction with casein micelles.

Formation of free milk fat in DWM powder particles as a result of atomization and drying may also contribute significantly to poor solubility and dispersibility. Free milk fat, which is extractable from the powder by a 50:50 mixture (v/v) of ethyl and petroleum ether (Brunner, 1974), probably coats the powder particles and prevents their rehydration. The physical state of the milk fat, as controlled by the liquid-to-solid ratio and the presence of free milk fat on the particle surface, strongly influences DWM particle dispersibility. Spray coating of DWM particles with lecithin or other surfactants and dispersion in warm water improve their dispersibility.

Browning Reaction

The Maillard browning reaction between available milk protein α- and ϵ-amino groups and lactose, is promoted during the forewarming, vacuum evaporation, drying, and storage of milk powder. This reaction produces a number of chemical compounds including carbon dioxide, formic acid, maltol, and furfural that cause discoloration and off-flavor development in the product (Parry 1974; Gordon and Kalan 1974).

MILK PROTEIN FRACTIONATION

Several casein and whey protein products are commercially manufactured from skim milk and whey (Fox 1970; Richert 1975; Morr 1975, 1979; Muller 1982A,B; Marshall 1982). These protein products are used as nutritional and functional ingredients throughout the food industry (Hugunin and Ewing 1977).

Casein and caseinates are generally made from skim milk by adding hydrochloric or sulfuric acid or by lactic acid fermentation. The isoelectrically precipitated casein is washed and dried or neutralized with sodium hydroxide, potassium hydroxide, or calcium hydroxide to produce the corresponding caseinate and spray dried. Rennet casein is produced by treating skim milk with rennet to coagulate the casein micelles, and the coagulum is cut, cooked, drained, and washed, as in cheese manufacture, and dried (Fox 1970; Muller 1982A,B).

Proteins that remain in whey after removing casein from milk are recovered as whey protein concentrates by precipitation with added polyphosphate or other polyvalent anionic compounds, ultrafiltration, ion exchange adsorption, gel filtration, or a combined acid and heat precipitation process. Whey protein concentrates are also manufactured by a combined process involving electrodialysis, concentration, lactose crystallization, and drying (Richert 1975; Morr 1979; Marshall 1982; Anon. 1982; Muller 1982B).

Lactose is recovered from skim milk or whey concentrates or from whey ultrafiltration retentate by crystallization technology (Nickerson 1970). Lactose is also hydrolyzed by chemical and enzymatic processes to form syrups with increased sweetness and improved functionality (Hobman 1984; Zadow 1984).

REFERENCES

Anon. 1982. Corning/Kroger combine technology to exploit lactose-hydrolyzed whey. *Food Prod. Dev. 16*, 34–35.

Arbuckle, W. S. 1977. *Ice Cream*, 3rd ed. AVI Publishing Co., Westport, Conn.

Babel, F. J. 1976. Technology of dairy products manufactured with selected microorganisms. *In: Dairy Technology and Engineering*. W. J. Harper and C. W. Hall (Editors). AVI Publishing Co., Westport, Conn., pp. 213–271.

Bender, A. E. 1978. *Food Processing and Nutrition*. Academic Press, New York.

Berger, K. G. 1976. Ice cream. *In: Food Emulsions*. S. Friberg (Editor). Marcel-Dekker, New York, pp. 141–213.

Brunner, J. R. 1974. Physical equilibria in milk: The lipid phase. *In: Fundamentals of Dairy Chemistry* 2nd ed. B. H. Webb, A. H. Johnson and J. A. Alford (Editors). AVI Publishing Co., Westport, Conn., pp. 474–592.

Burton, H. 1979. An introduction to the ultra-high-temperature processing of milk and milk products. *In: International Conference on UHT Processing and Aseptic Packaging*. North Carolina State University, Raleigh, pp. 1–20.

CODE OF FEDERAL REGULATIONS, 21 CFR. 1982. U.S. Government Printing Office, Washington, D.C.

Cousin, M. A. 1982. Presence and activity of psychrotrophic microorganisms in milk and dairy products: A review. *J. Food Prot. 45*, 172–207.

Darling, D. F. 1980. Heat stability of milk. *J. Dairy Res. 47*, 199–210.

Domnitz, D. 1984. Personal communication.

Edmondson, L. F. 1970. Sterilized products. *In: By-products from Milk*, 2nd ed. B. H. Webb and E. O. Whittier (Editors). AVI Publishing Co., Westport, Conn., pp. 226–266.

Emmons, D. B. and Tuckey, S. L. 1967. *Cottage Cheese and other Cultured Milk Products*. Chas. Pfizer and Co. Inc., New York.

Ernstrom, C. A. and Wong, N. P. 1974. Milk-clotting enzymes and cheese chemistry. *In:*

Fundamentals of Dairy Chemistry, 2nd ed. B. H. Webb, A. H. Johnson and J. A. Alford (Editors). AVI Publishing Co., Westport, Conn., pp. 662–753.

Farrell, H. M. and Thompson, M. P. 1974. Physical equilibria: Proteins. *In: Fundamentals of Dairy Chemistry*, 2nd ed. B. H. Webb, A. H. Johnson and J. A. Alford (Editors). AVI Publishing Co., Westport, Conn., pp. 442–471.

Foster, E. M., Nelson, F. E., Speck, M. L., Doetsch, R. N. and Olson, J. C., Jr. 1957. *Dairy Microbiology*. Prentice Hall, Inc., Englewood Cliffs, N.J.

Fox, K. K. 1970. Casein and whey protein. *In: By-products from Milk*, 2nd ed. B. H. Webb and E. O. Whittier (Editors). AVI Publishing Co., Westport, Conn., pp. 331–353.

Fox, P. F. 1981. Heat-induced changes in milk preceding coagulation. *J. Dairy Sci. 64*, 2127–2137.

Fox, P. F. 1982. Heat-induced coagulation of milk. *In: Developments in Dairy Chemistry*, Vol. 1: *Proteins*. P. F. Fox (Editor). Applied Science Publishers, New York, pp. 189–223.

Freeman, N. W. and Mangino, M. E. 1981. Effects of ultra-high temperature processing on size and appearance of casein micelles in bovine milk. *J. Dairy Sci. 64*, 1772–1780.

Gillies, M. T. 1974. *Dehydration of Natural and Simulated Dairy Products*. Food Technology Review No. 15. Noyes Data Corp., Park Ridge, N.J.

Gordon, W. G. and Kalan, E. B. 1974. Proteins of milk. *In: Fundamentals of Dairy Chemistry*, 2nd ed. B. H. Webb, A. H. Johnson and J. A. Alford (Editors). AVI Publishing Co., Westport, Conn., pp. 87–118.

Govin, R. and Leeder, J. G. 1971. Action of emulsifiers in ice cream utilizing the HLB concept. *J. Food Sci. 36*, 718–722.

Graham, D. M. 1974. Alteration of nutritive value resulting from processing and fortification of milk products. *J. Dairy Sci. 57*, 738–745.

Hall, C. W. 1976. Heat and heat-transfer processes. *In: Dairy Technology and Engineering*. W. J. Harper and C. W. Hall (Editors). AVI Publishing Co., Westport, Conn., pp. 429–503.

Hall, C. W. and Hedrick, T. I. 1966. *Drying Milk and Milk Products*. AVI Publishing Co., Westport, Conn.

Harland, H. A., Coulter, S. T., Townley, U. Y. and Jenness, R. 1955. A quantitative evaluation of changes occurring during heat treatment of skimmilk at temperatures ranging from 170 to 300°F. *J. Dairy Sci. 38*, 1199–1207.

Harper, W. J. 1959. Chemistry of cheese flavors. *J. Dairy Sci. 42*, 207–213.

Harper, W. J. 1976. Processing-induced changes. *In: Dairy Technology and Engineering*. W. J. Harper and C. W. Hall (Editors). AVI Publishing Co., Westport, Conn., pp. 539–596.

Harper, W. J. 1981. Advances in chemistry of milk. *J. Dairy Sci. 64*, 1028–1037.

Harper, W. J. and Seiberling, D. A. 1976. General processes for manufactured products. *In: Dairy Technology and Engineering*. W. J. Harper and C. W. Hall (Editors). AVI Publishing Co., Westport, Conn., pp. 185–212.

Harper, W. J., Seiberling, D. A. and Blaisdell, J. L. 1976. Fluid flow and flow processes. *In: Dairy Technology and Engineering*. W. J. Harper and C. W. Hall (Editors). AVI Publishing Co., Westport, Conn., pp. 387–428.

Harwalker, V. R. 1982. Age gelation of sterilized milks. *In: Developments in Dairy Chemistry*, Vol. 1: *Proteins*. P. F. Fox (Editor). Applied Science Publishers, New York, pp. 229–265.

Henderson, J. L. 1971. *The Fluid-Milk Industry*, 3rd ed. AVI Publishing Co., Westport, Conn.

Hobman, P. G. 1984. Review of processes and products for utilization of lactose in deproteinated milk serum. *J. Dairy Sci. 67*, 2630–2653.

Hugunin, A. G. and Ewing, N. L. 1977. *Dairy Based Ingredients for Food Products.* Dairy Research, Rosemont, Ill.

Johnson, A. H. 1974. The composition of milk. *In: Fundamentals of Dairy Chemistry,* 2nd ed. B. H. Webb, A. H. Johnson and J. A. Alford (Editors). AVI Publishing Co., Westport, Conn., pp. 1–45.

Jones, V. A. and Harper, W. J. 1976. General processes for fluid milks. *In: Dairy Technology and Engineering.* W. J. Harper and C. W. Hall (Editors). AVI Publishing Co., Westport, Conn., pp. 141–184.

Keeney, P. G. and Kroger, M. 1974. Frozen dairy products. *In: Fundamentals of Dairy Chemistry,* 2nd ed. B. H. Webb, A. H. Johnson and J. A. Alford (Editors). AVI Publishing Co., Westport, Conn., pp. 873–908.

Law, B. A. 1981. The formation of aroma and flavor compounds in fermented dairy products. *Dairy Sci. Abstr. 43*, 143.

Law, B. A. 1982. Microbial proteolysis of milk proteins. *In: Food Proteins.* P. F. Fox and J. J. Condon (Editors). Applied Science Publishers, New York, pp. 307–328.

Marshall, K. R. 1982. Industrial isolation of milk proteins: Whey proteins. *In: Developments in Dairy Chemistry,* Vol. 1: *Protein.* P. F. Fox (Editor). Applied Science Publishers, New York, pp. 339–367.

Marth, E. H. 1974. Fermentations. *In: Fundamentals of Dairy Chemistry,* 2nd Ed. B. H. Webb, A. J. Johnson and J. A. Alford (Editors). AVI Publishing Co., Westport, Conn. pp. 772–858.

Morr, C. V. 1973A. Milk ultracentrifugal opalescent layer. 1. Composition as influenced by heat, temperature and pH. *J. Dairy Sci. 56*, 544–552.

Morr, C. V. 1973B. Milk ultracentrifugal opalescent layer. 2. Physico-chemical properties. *J. Dairy Sci. 56*, 1258–1266.

Morr, C. V. 1975. Milk proteins in dairy and food processing. *J. Dairy Sci. 58*, 977–984.

Morr, C. V. 1976. Whey protein concentrates: An update. *Food Tech. 30*, 18, 19, 22, 42.

Morr, C. V. 1979. Utilization of milk proteins as starting materials for other food stuffs. *J. Dairy Res. 46*, 369–376.

Morr, C. V. 1982. Functional properties of milk proteins and their use as food ingredients. *In: Developments in Dairy Chemistry,* Vol. 1: *Proteins.* P. F. Fox (Editor). Applied Science Publishers, New York, p. 375–397.

Moskowitz, G. J. 1980. Flavor development in cheese. *In: The Analysis and Control of Less Desireable Flavors in Foods and Beverages.* G. Charalambous (Editor). Academic Press, New York, pp. 53–70.

Muller, L. L. 1982A. Manufacture of casein, caseinates and coprecipitates. *In: Developments in Dairy Chemistry,* Vol. I: *Protein.* P. F. Fox (Editor). Applied Science Publishers. New York, pp. 315–335.

Muller, L. L. 1982B. Milk proteins-manufacture and utilization. *In: Food Proteins.* P. F. Fox and J. J. Condon (Editors). Applied Science Publishers, New York, pp. 179–189.

Nickerson, T. A. 1970. Lactose. *In: By-products from Milk,* 2nd ed. B. H. Webb and E. O. Whittier (Editors). AVI Publishing Co., Westport, Conn., pp. 273–319.

Olson, N. F. 1975. Mechanized and continuous cheese making processes for cheddar and other ripened cheese. *J. Dairy Sci. 58*, 1015–1021.

Olson, N. F. 1981. Trends in cheese manufacture. *J. Dairy Sci. 64*, 1063–1069.

Pack, M. Y., Vedamuthu, E. R., Sandine, W. E. Elliker, P. R. and Leesment, H. 1968. Effect of temperature on growth and diacetyl production by aroma bacteria in single-and mixed-strain lactic cultures. *J. Dairy Sci. 51*, 339–344.

Packard, V. S. 1975. *Processed Foods and the Consumer: Additives, Labeling, Standards and Nutrition.* University of Minnesota Press, Minneapolis.

Pallansch, M. J. 1970. Dried products. *In: By-Products from Milk,* 2nd ed. B. H. Webb and E. O. Whittier (Editors). AVI Publishing Co., Westport, Conn., pp. 124-175.

Parry, R. M. 1974. Milk coagulation and protein denaturation. *In: Fundamentals of Dairy Chemistry,* 2nd ed. B. H. Webb, A. H. Johnson and J. A. Alford (Editors). AVI Publishing Co., Westport, Conn., pp. 603-655.

Reimerdes, E. H. 1982. Changes in the proteins of raw milk during storage. *In: Developments in Dairy Chemistry,* Vol. 1: *Proteins.* P. F. Fox (Editor). Applied Science Publishers, New York, pp. 271-285.

Reiter, B. and Sharpe, M. E. 1971. Relationship of microflora to the flavor of cheddar cheese. *J. Appl. Bacteriol. 34,* 63-80.

Richert, S. H. 1975. Current milk protein manufacturing processes. *J. Dairy Sci. 58,* 985-993.

Richter, R. L. 1977. Manufacture of superior quality buttermilk. Cultured Dairy Prod. J. *12,* 22-28.

Schmidt, R. H., Morris, H. A., Castberg, H. B. and McKay, L. M. 1976. Hydrolysis of milk protein by bacteria used in cheese making. *J. Agr. Food Chem. 24,* 1106-1113.

Thomas E. L. 1981. Trends in milk flavors. *J. Dairy Sci. 64,* 1023-1027.

Tobias, J. 1976. Organoleptic properties of dairy products. *In: Dairy Technology and Engineering.* W. J. Harper and C. W. Hall (Editors). AVI Publishing Co., Westport, Conn., pp. 75-140.

Vedamuthu, E. R. 1974. Cultures for buttermilk, sour cream and yogurt with special comments on acidophilus yogurt. Cultured Dairy Prod. J. *9,* 16-21.

Webb, B. H. 1970. Condensed products. *In: By-products from Milk,* 2nd ed. B. H. Webb and E. O. Whittier (Editors). AVI Publishing Co., Westport, Conn., pp. 83-118.

Zadox, J. G. 1984. Lactose: Properties and uses. *J. Dairy Sci. 67,* 2654-2679.

Index

Acetaldehyde, 759
Acetoin production, 686
Acetone-butanol fermentation, 673, 707
Acetyl carnitine, 16
Acetyl choline esterase, 107
Acetyl-coenzyme A, 174, 177, 367
N-Acetyl-D-glucominidase, 107
N-Acetyl glucosamine, 16
CMP-N-Acetyl neuraminate galactosyl-
   glycoprotein transferase, 106
N-Acetylneuraminic acid, 106
Acid-base equilibria, 410–414
α_1-Acid glycoprotein, *see* Orosomucoid
Acid degree value (ADV), 221, 233, 235
Acid protease, 107
Acid phosphatase, 107
Acidified milks, 45
Acidophilus milk, 47
Acidophilus yogurt, 47
Actin, 525
Adenosine triphosphate (ATP), 107, 660,
   664, 665, 669, 675
FMN Adenyltransferase, 106
Age thickening, 754, 755
Agitation, 740
Alanine aminotransferase, 106
Alcoholic beverages, 708
Alcoholic fermentation, 673, 708
Aldehydes in milk, 14
Alkaline phosphatase, 107, 742
Aluminum in milk, 11
Amine oxidase, 106
Amines in milk, 8, 16
Amino acids, 348
  cheese ripening, 647
  essential, 348
  nonessential, 348
α-Amino-N in milk, 15
Ammonia in milk, 15
α-Amylase, 107
ß-Amylase, 107
Anaerobic glycolysis, 664

Antibiotics, 740
  cultured milks, 697
  lactic streptococci, 694, 696
  lactobacilli, 696
  production, 694, 711
Anti-tumor effect, 387
Arsenic in milk, 11
Arylesterase, 106
Ascorbic acid, 367, *see also* Vitamin C
  as antioxidant, 247–249
  effect on E_h, 417
Aspartate aminotransferase, 106
Autoxidation, 236–262
  acidity, 258
  ascorbic acid, 247–(25?)
  carbonyls, 260–262
  fluid milk, 244–245
  homogenization, 258–259
  hydroperoxides, 237–239
  light, 256–258
  measurement, 241–242
  mechanism, 237–241
  metals, 245–247
  off-flavors, 239–240
  oxygen levels, 253–254
  pasteurization, 254–256
  prevention by antioxidants, 242–244
    sulfhydryls, 254–256
    α-tocopherol, 250–252
    vacuum storage, 253–254
  products of, 238–239
  temperature, 252
  xanthine oxidase, 244–245

Bacillus cereus, 610, 617, 618
Bacillus subtilis, 618
Bactofugation, 638
Barium in milk, 11, 12
Beer's law, 444–445
Berridge substrate, 621–622
Biekost, 352, 381
Bifidobacterium fermentation, 670, 671
Biological value (BV), 349

Biotin, 368
Bitter flavors, 681
  cheese, 616
Bitter peptides, 646
Blood clot times, 621
Blue cheese, 65
  effect of homogenization, 641
  lipolysis, 648
Bone remodeling, 372
Boron in milk, 11
Bound aldehydes, 183
Bovine serum albumin, 82
  composition, 95–97
  conformation, 117–119
  heterogeneity, 117–119
  sequence, 95–97
Brick cheese, 65, 644, 649, 704
Brie, 647
  curd size, 642
Bromelain, 618
Bromine in milk, 11
Browning reaction, 326–328, 332
  carmelization, 326
  inhibition, 327
  Maillard, 326, 332
  nutritive value, 327–328
Buffer capacity, 410–414
Bulk density, 761
Butanediol fermentation, 673
Butanol-acetone-riboflavin fermentation, 707
Butter, 39, 40, 57, 747
Buttermilk, 46
  cultured, 662, 692, 701, 758
Butteroil, 57, 748
Butyric acid
  fermentation, 673
  Swiss cheese, 649
Butyrophilin, 523, 524, 544, 547, 559

Cadmium in milk, 11
Calcium in milk, 8, 372–378
  binding by milk proteins, 145–147
  bioavailability, 372
  hypertension, 376–377
  intake, 375
  osteoporosis, 375
  quantitation, 6–7
  utilization, 329–330
Calcium phosphorus ratio, 372, 378
Calf vells, 609

Camel milk, 21
Camembert cheese, 64, 637, 645, 647, 648, 702–704
  curd size, 642
Carbon dioxide in milk, 13–14
Carbonate in milk, 8
  quantitation, 7–8
Carbonic dehydratase, 107
Carbonyl compounds in oxidized products, 260–262
Carboxylesterase, 106
Cardiolipin, 186
Carnitine in milk, 16
Carotinoids, 369
Casein, 4, 72, 585
  α_{s1}-Casein, 463–465
    association, 109–111
    charge, 464
    composition, 463–464
    conformation, 109, 469
    genetic variants, 83–85
    hydrodynamic properties, 109
    hydrophobicity, 464–465
    isoionic points, 142
    nomenclature, 82, 83–85
    precipitation by calcium, 465
    sequence, 83–85
    structure, 108–111
  α_{s2}-Casein, 466–467
    association, 112–113
    charge, 466
    composition, 466
    genetic variants, 85–86
    hydrodynamic properties, 109
    hydrophobicity, 466
    isoionic points, 142
    nomenclature, 82, 85–86
    self association, 467
    sequence, 85–86
    structure, 111–113
  ß-Casein, 465–466
    association, 114–115
    charge, 465
    composition, 465
    genetic variants, 87–89
    hydrodynamic properties, 109
    hydrophobicity, 465
    isoionic points, 142
    nomenclature, 82, 87–89
    sequence, 87–89
    structure, 113–115

titratable groups, 140
k-Casein, 467–469
  association, 115–117
  charge, 468
  composition, 467–468
  conformation, 468–469
  genetic variants, 89–91
  heterogeneity, 115–117
  hydrodynamic properties, 109
  hydrophobicity, 468
  isoionic points, 142
  milk clotting action, 619
  nomenclature, 82, 89–91
  polymerization, 467
  sequence, 89–91
  solubility in calcium, 467
  structure, 115–117
acid, 762
denaturation, 588
electrostatic interactions, 471–474
fractionation, 128–135
hydrogen bonding, 474–475
hydrophobic interactions, 470–471
interactions with serum proteins, 593
isolation from milk, 128
micelles, 461–493, 740, 745, 749, 750–753,
    755, 756, 762
  coat-core models, 481–483
  colloidal calcium phosphate in, 476–479
  destabilization by chymosin, 491–493
  forces stabilizing, 469–479
  internal structure models, 483–485
  protein components, 463–496
  structure, 479–493
  submicellar models, 485–491
milk serum, 499
nutritive value, 347
rennet, 762
secondary structure, 474–475
standards, 73
tertiary structure, 474–475
Caseinates, 762
Catabolite inhibition, 669
Catalase, 106
Cerebrosides, 186
Ceruloplasmin, 83, 105
Cesium in milk, 11, 12
Cheddar cheese, 66, 635, 643–646, 648–649,
    677–679, 682, 684, 701
Cheddaring, 757
Cheese, 39, 40, 58

classification, 59, 634
composition, 61
curd, 641, 642, 757
  cutting, 642
  mineral content, 643
  salt distribution, 645
  separation from whey, 644
dental caries, 363
eye formation, 643
flavor, 639–641, 647, 649
foods, 69
late blowing, 638
manufacture, 634–635, 641–645, 740, 756
  bacterial growth, 636
  clarification, 638
  cooking, 643
  cutting the curd, 642
  draining, 644
  fat in dry matter (FDM), 635
  fat on dry basis (FDB), 635
  heated milk, 639
  homogenization, 642
  lactose, 644
  milk composition, 635–637
  pasteurization, 638
  raw milk, 639
  salting, 645
  somatic cells, 636
  storage, 636
  syneresis, 642
ripening, 645–649, 758
  acid development, 648
  amino acids, 647
  deacidifying, 647
  lactose fermentation, 647
  lipolysis, 648
  proteolysis, 646
  rancidity, 649
  rennet, 647
salt in moisture (S/M), 645
spreads, 690
standards, 63
yield, effect of
  homogenization, 641
  pasteurization, 639
  somatic cells, 636
Chemical score, 349
Chloride in milk, 7–8
Chocolate milk, 45
Cholesterol, 187
  heart disease, 356
  milk factor, 357

Cholesterol (*cont.*)
  nutritive value, 353–357
Cholesterol esters, 202
Choline esterase, 107
Choline in milk, 16
Chromium in milk, 11
Churning, 747, 748
Chymopapain, 618
Chymosin, 609–613
  activation, 611
  extraction, 610
  milk-clotting action, 619, 491–493
  stability, 612
Cirsium arvense, 619
Citrate in milk, 7–8
Citric acid, 656
  metabolism, 684
Clarification, 741
  in cheesemaking, 638
Clostridia, 638
Clotting agents, 757
Coagulation of milk, 595–601
  effect of concentration, 600
  effect of forewarming, 600
  effect of pH, 595–597
  effect of salt balance, 595
  time, 622, 624
Cobalt in milk, 11
Colby cheese, 644, 704
Colloidal calcium phosphate, 476–479
Colloidal phase of milk, 500–501
Campesterol, 187
Composition of milk
  age of cow, 27
  breed differences, 24–26
  change during lactation, 26–27
  effect of milking procedure, 29–30
  effect of nutrition, 27–28
  effect of temperature, 28–29
  effect of udder infection, 29
  factors influencing, 21–30
  gross, 19–30
  inherited differences, 23–26
  interspecies differences, 20–21
  seasonal variation, 28–29
  variations in bovine, 20–30
  various species, 21
Concanavalin A, 550, 555, 558, 560
Concentration of milk, 751, 754
  effect of coagulation, 600
Condensed buttermilk, 55
  composition, 53

Condensed skim milk, 54
Contaminants in milk, 54
Continuous buttermaking, 748
Cooling, 740
Copper in milk, 11, 245–247
  oxidized flavor, 246–247
Cottage cheese, 60, 642, 687, 701
  curd size, 642
Coulter counter, 513
Cream, 50, 747
  coffee, 51
  composition, 52
  sour 46, 701
  sterile, 752
  whipping, 51
Cream cheese, 63, 641
  curd, 642
  effect of homogenization, 641
Creatine in milk, 15
Creatinine in milk, 15
Cultured dairy foods
  nutritive value, 385–387
Cultured milks, 690, 701, 758
  Scandinavian, 687, 692
Curd size, 642
Curd tension, 743
Cynara cardunculus, 618
Cytoplasmic cresents, 526

Dairy products and obesity, 346
Dairy spreads, 58
Dehydrolanosterol, 187
Denaturation
  casein, 585
  immunoglobulin, 593
  α-lactalbumin, 592
  ß-lactoglobulin, 589–591
  serum albumin, 593
  serum proteins, 593
Density of dairy products, 422–423
Density of milk, 419–423
  measurement, 420
  relation to composition, 419–420
  relation to temperature, 420–422
Dextrose equivalent, 745
Diacetyl, 685, 759
  production, 686
Diacetyl reductase, 687
Digestibility (D), 349
Dimethylsulfone in milk, 18
Disulfide bonds in casein, 475–476

Domiati cheese, 64
Donkey milk, 21
Dried milk products, 760
  buttermilk, 56
  composition, 53
  cream, 56
  milk, 56
    instantizing, 311
    lactose crystallization, 294, 311
    lumpiness, 294, 311
    whey, 75, 309–310, 326–327
    lactose crystallization, 309–310
  whole milk, 56, 760
Drying, 760

E_h of milk, see Oxidation-reduction potential
Eggnog, 45
Electrical conductivity of milk, 437–440
  effect of bacterial action, 439
  effect of dilution and concentration, 439–440
  effect of fat globules, 439
  effect of temperature, 438–439
  ions contributing, 439
  measurement, 437–438
  to determine added water, 440
Electrical conductivity of milk fat, 203
Electrodialysis, 313
Electrolytes
  chloride, 385
  potassium, 385
  sodium, 384
Electrophoresis, 143–144
Embden-Meyerhof-Parnas pathway, 661, 662, 669, 672, 674
Emulsifiers, 744, 745
Endothia parasitica, 610, 615, 624
  bitterness due to, 616
Enzymes in milk, 105–108
Enzymes in milk fat globule membrane, 553–557
Epilactose, 281
Escherichia coli, 613, 702
Essential fatty acids, 193–194
Esters in milk, 14
Ethanol in milk, 14
Evaporated milk, 39, 54, 748, 754
  composition, 53
Evaporators, 751

Exopolysaccharide, 687
  biosynthesis, 688, 689

Fat on dry basis (FDB), 635
Fat in dry matter (FDM), 635
Fat globule membrane, 740, 743, 748, see also Milk fat globule membrane
  proteins, 82, 100–113
Fat globules in milk, 2–3, see also Milk fat globules
Fat synthesis, microbial, 709
Fatty acid-synthetase complex, 174–175
Fermentable substrates, 656
Fermentation, 655
  effect on E_h, 417–418
  lactose, 332–333
Fermentation medium
  milk, 655
  whey, 332–333
Fermented milks, 46
Feta cheese, 64
Ficin, 618
Fishy flavor, 258
Flavin adenine dinucleotide, 365
Flavor
  defects due to
    lipolysis, 215, 233
    oxidation, 239, 240
  methional, 257
  rancid, 233 234
  sunlight, 256–258
Fluff, 562
Fluid milk, 40, 41, 42
  flavored, 45
  processing, 739
Fluorescence, 445–446
Fluorine in milk, 11
Foaming, 741, 743
  depression by lipolysis, 234
Folate-binding protein, 83, 105
Folic acid (folacin), 367
Foltmann's equation, 623–624
Foodborne disease, 742
  pathogens, 701
Forewarming, 749, 750, 754, 756, 760, 762
  effect on coagulation, 600
Formagraph, 621, 624
Fractionation of casein by
  chromatography, 130–135
  electrophoresis, 130
  solubility, 128–130

Free radicals, 237
Freezing ice cream, 746
Freezing point of milk, 432–437
    constituents affecting, 434–437
    effect of environmental factors, 435–436
    effect of handling, 436–437
    measurement, 433–434
Frozen custard, 71
Frozen milk concentrate, 755
Frozen starter cultures, 697
Fructose-bisphosphate aldolase, 107
Fruity flavors, 684

Galactose, 280, 283, 328–329
    cataracts caused by, 328
    galactosemia, 328–329
    lactose hydrolysis, 323–326, 328
    metabolism, 668
    oligosaccharide formation, 326
ß-D-Galactosidase, 659, 667, see also Lactase
Galactosyl transferase, 282–283
Gangliosides, 186, 201, 516
Gases in milk, 13, 14
Ghee, 57
Glucose, 280–283
Glucose-6-phosphatase, 107
Glucose phosphate isomerase, 107
ß-Glucuronidase, 107
γ-Glutamyl transferase, 106
Glycerides
    biosynthesis, 173–178
    structure, 178–182
Glycerol, 178–182
Glycerol kinase, 106
M_1-glycoproteins, 83, 104–105
M_2-glycoproteins, 83, 105
Glycolysis, 665
Goat milk, 1
    composition, 21
Golgi apparatus, 552, 529, 537, 554, 555
Gorgonzola cheese, 65
Gouda cheese, 638, 643–644, 646
Group N streptococci, 663, 664, 666, 668, 687
Guinea pig milk, 21

Half and half, 51
Haptogenic membrane, 517
Heat capacity, 440–441
    definition, 440
        measurement, 440
    relation to fat, 441
    variation with temperature, 441
Heat treatment
    effect on E_h, 419
    starter growth, 658
4-cis Heptenal, 260
Heterofermentative lactic fermentations, 669
High density lipoprotein, 226
Hippuric acid in milk, 16
Holter equation, 623
Homofermentative lactic streptococci, 662, 666
Homogenization, 743, 751
    activation of lipases, 224–225
    autoxidation, 258–259
    Blue cheese, 641
    cheese making, 640
    cheese yield, 641
    milk fat globule membrane, 564–565
    Neufchatel, 641
    xanthine oxidase, 358–360
Homolactic fermentation, 673
Hormones in milk, 19
Horse milk, 21
Human milk, 1
    composition, 21
Hydrocarbons, 173, 188
Hydrocolloids, casein binding, 148–149
Hydrolytic rancidity, 215
Hydroperoxides, 238, 241
Hydroxy acids, 172, 195
ß-Hydroxy butyrate, 174
Hydroxymethyl glutarate (HMG), 357
Hypocholesterolemic effect, 357

Ice cream, 39, 70, 744
    composition, 52
    ingredients, 744
    sandiness, 284
    standards, 71
Ice crystals, 747
Ice milk, 39, 70
    composition, 52
    standards, 71
Imitation dairy products, 389
    nutritive value, 389–391
Immunoglobulin M, 564
Immunoglobulins, 657
    denaturation, 593
    heterogeneity, 97–100

kinds, 97–100
nomenclature, 82
structure, 127
Indoxylsulfate in milk, 16, 18
Infrared absorption to measure milk constituents, 446–448
Inhibitory mechanisms, lactic bacteria, 705
Inorganic pyrophosphatase, 107
Instantizing, 761
  dried milk, 311
  lactose for, 331–332
Interfacial tension, *see* Surface tension
Iodine, 11, 12, 381
Iodine number, 241
Ion complexes in milk, 10
Ion exchange, 313–314, 323
Ions in milk, 10
Iron, 11, 245–247, 381
  oxidized flavor, 245–247
Italian type cheeses, 648

Kefir, 49, 672, 692
Keto acids, 172, 195
Ketones in milk, 14
Kininogen, 83, 105
Kuban, 692
Kumiss, 49, 672, 692
Kwashiorkor, 349
Kynurenine, 16

α-Lactalbumin
  denaturation, 592
  genetic variants, 93–95
  glycosylated, 93–95
  isoionic points, 142
  lactose biosynthesis, 282–283
  nomenclature, 82
  sequence, 93–95
  structure, 125–127, 497–498
Lactase, 280, 325–326, 328–330, 386
  activity, 329
  deficiency, 328–330
  sources, 325
  uses, 325–326
Lactate dehydrogenase, 106
Lactic acid, 706
  optical configuration, 671
Lactic acid bacteria, 667
  sugar usage, 669
Lactic streptococci, 648
Lactitol, 317–318
  palmitate, 318–319

Lactobacillus acidophilus, 387
Lactobacillus bifidus, 282, 330
Lactobacillus bulgaricus, 386, 387, 643
Lactobacillus helveticus, 643
Lactobionic acid, 281, 321
  production, 714
Lactoferrin, 83, 103–104, 657
Lactofil, 50
ß-Lactoglobulin, 82
  association, 119–124
  binding, 495–496
  denaturation, 494–495, 589–591
  genetic variants, 92–93
  interactions, 497
  isoionic points, 142
  nomenclature, 92–93
  self association, 493–494
  sequence, 92–93
  structure, 119–124
  sulfhydryl groups, 496–497
  titration curves, 140–142
Lactones, 196
Lactoperoxidase, 106
Lactoperoxidase/thiocyanate/hydrogen peroxide system, 657
Lactose, 39, 74, 279–342
  absorption, 328–330
  alpha, 284–294
    anhydrous, 291–294
    hydrate, 284–288
    manufacture, 311–314
    solubility, 295
  amorphous, 294
    use, 331
  amount in milk, 280
  analysis, 3, 297, 315–317
  beta, 289–290, 294–296
    manufacture, 295–296, 315
    solubility, 295
    use, 315
  biosynthesis, 282–283
  calcium utilization, 329–330
  chemical reactions, 317–328
  cheese making, 644
  cheese ripening, 647
  commercial, 312
  crystalline forms, 283–296
  crystalline habits, 284–287
  crystallization, 283–315, 745
    acceleration, 288–290
    dried milk, 294, 311

Lactose (*cont.*)
    dried whey, 309–310
    ice cream, 284
    inhibition, 288–289
    kinetics, 305, 314
    pH, 289, 299
    rate, 284–291, 305
    riboflavin, 290–291
    sweetened condensed milk, 284, 310–311
    temperature, 284
    viscosity, 305–306, 313
    whey, 305, 309–310
  decomposition, 296
  density, 293, 306
  derivatives, 317–323
  equilibria, 296–306
  equilibrium constant, 297
  fermentation, 332–333, 656, 658, 710
  food use, 330–333
  glass, *see* amorphous
  heat, *see* browning reaction
    desorption, 306
    effect of lactulose, 281, 319, 320
  hydrolysis, 316, 323–326
    acid, 323–324
    enzymatic, 325–326
  instantizing, 311, 331–332
  intolerance, 328–329, 362–364
  manufacture, 311–315
  melting point, 284, 292–293, 296
  metabolism, 662
    control, 664
    pathway, 663
  molecular structure, 280–282
  mutarotation, 296–306
  nutritional value, 328–330, 361–364
  occurrence, 279–280
  optical rotation, 284, 296–297
  oxidation, 321–322
  permeate, 309–310, 313, 332
  physical properties, 283–309
  physiological effects, 328–330
  recovery, 763
  reduction, 317–318
  refractive index, 306
  solubility, 295, 300–306
  specific rotation, 294, 296–297
  standards, 74
  structure, 280–282
  sweetness, 307–309
  synthase, 106
  uses, 311, 330–333
    pharmaceutical, 331–332
Lactosuria, 279, 328
Lactosyl urea, 322
Lactulose, 281, 319–320
  uses, 320
Lead in milk, 11
Lectin, 560
Leloir pathway, 668
Light absorption, 444–448
Light scattering by milk, 445–448
  contribution of constituents, 445–446
  measurement of constituents, 446–448
Limburger cheese, 64
Lipase, 215, *see also* Lipolysis
  activition, 224–227
    chemical, 227
    homogenization, 224–225
    thermal, 226–227
  distribution, 221–224
    colostrum, 222–224
    cow's milk, 221–222
    goat's milk, 224
    human milk, 223–224
    microbes, 223
  inhibition, 227–231
    chemical, 229–231
    colorimetry, 235–236
    gas-liquid chromatography, 236
    light, 228–229
    measurement, 234–236
    miscellaneous, 236
    radioactive substrates, 236
    surface tension, 235
    temperature, 227–228
    titration, 234–235
  milk, 221–236
  pH optimum, 231–233
  properties, 231–233
Lipids, 2, *see* Milk fat
Lipoamide dehydrogenase, 106
Lipoic acid in milk, 18–19
Lipolysis, 215–236, *see also* Lipase
  agitation induced, 217
  bacteria, 216, 223
  cheese ripening, 648
  estrous, 220
  farm equipment, 220–221
  feed, 218–219
  flavor defect, 215, 233

hormonal disturbances, 220
  lactation, 219
  mastitis, 219–220
  spontaneous, 217–218
Lipoprotein esterase, 107
Listeria monocytogenes, 705
Listeriosis, 705
Lithium in milk, 11
Low sodium milk, 43
Lowfat cheese, 68
  Lowfat milk, 43
Lysophospholipids, 186, 200
Lysozyme, 657

Magnesium, 379
  concentration, 8
  quantitation, 7
Manganese in milk, 11
Maillard reaction, 326, 332, 749, 751–753,
    761–762, *see also* Browning reac-
    tion
Malate dehydrogenase, 106
Malonyl coenzyme A, 174
Malted milk powder, 57
Malty flavor, 689
α-D-Mannosidase, 107
Margarine, 748
Mastitis, 219–220
Mellorine, 72
  standards, 71
Mercury in milk, 11
Metals and autoxidation, 245–247
Methanthiol, 18, 647
Methional, 257
Microbial inhibitors, 657
Microbial proteinases, 676
β_2-Microglobulin, 83, 104
Micro-tests, 624
Minor proteins, 82–83, 103–105
Milk
  carbohydrates, 3–4
  constituents, 1–19, *see also* Composition
    of milk
  clotting, 641
    action, 619–620
      k-Casein, 619
      chymosin, 619
      optimum pH, 620
      proteolytic enzymes, 619
      temperature coefficient, 620
    assays, 620–624

Berridge substrate, 621, 622
  blood clot times, 621
  Foltmann's equation, 623–624
  Formograph, 621, 624
  torsiometer, 621
  trombelastograph, 621
composition and cheese making, 635–637
energy, 345
factor, 357
fat, 171–205
  composition, 178–189
    fatty acids, 172, 190–191
    glycerides, 178–183
    glycerol ethers, 183
    keto acids, 196
    phospholipids, 183–188
  consistency, 205
  degradation, 682
  dielectric constant, 203
  digestibility, 352
  diglycerides, 182–183
  monoglycerides, 173
  nutritive value, 350–360
  physical properties, 203–204
    crystallization, 203
    electrical conductivity, 203
    melting range, 204
    nitrogen solubility, 203
    specific heat, 203
    thermal conductivity, 203
fermentation, 655
hypersensitivity, 350
intolerance, 363
lipids, 2–3
microsomes, 518
miscellaneous compounds, 13–14
proteins, 4–6, *see also* Casein, lactal-
    bumin and lactoglobulin
  adsorption of water, 147
  association with macromolecules, 148–
    149
  association with small molecules, 145–
    149
  classification and nomenclature, 81–83
  concentrations, 82
  degradation, 676
  electrochemical properties, 138–144
  electrophoretic behavior, 143–144
  isoionic points, 142
  reaction with lactose, 326–328
  structure and conformation, 108–127

Milk (*cont.*)
  salts, 5–10
  serum
    bovine serum albumin, 498
    caseins, 499
    equilibria with colloidal phase, 500–501
    lipid composition, 560–563
    proteins, 586
      characteristics, 586
      denaturation, 589, 593
      interactions, 593
      salts, 499–500
    sugar, *see* Lactose
    type A and type B, 596–597
    water, 2
Milk fat globule, 512–519
  composition, 512
  core, 515–517
  creaming and agglutination, 563–564
  electron micrographs, 520, 521, 526
  homogenization, 564–566
  isolation, 530–534
  membrane, 184–185, 517–563
    enzymes, 553–557
    glycoproteins, 545, 549–550
    gross composition, 534–536
    lipid composition, 534–536
    minerals, 536
    molecular organization, 557–563
    origin, 519–529
    protein composition, 541–553
  physical states, 566–567
  size distribution, 512–515
    breeds, 514
    lactation, 515
  surface, 517–519
Minerals, 656, 752
  cheese curd, 643
  cheese making, 643
  fat globule membrane, 536
Mixed acid fermentation, 673
Moisture in the non-fat substance (MNFS),
    637
Mold-ripened cheese, 679, 682
Molybdenum in milk, 11
Morphine in milk, 16
Mozzarella cheese, 67, 643–645
Mucor miehei, 610, 615–617
  stability, 616
Mucor pusillus, 610, 616, 624
  cheddar cheese, 616
Musty flavor, 690

NADH dehydrogenase, 106
Net protein utilization (NPU), 349
Neufchatel cheese, 641–642
  curd, 642
  homogenization, 641
Niacin, 366
Nickel in milk, 11
Nicotinamide adenine dinucleotide (NAD),
    366
Nisin, 694–711
Nitrogen in milk, 13, 14, 656
Nitrogeneous compounds in milk, 16
6-*trans*-Nonenal, 260
Nonfat dry milk, 39, 40, 55, 760
Non-protein nitrogen in milk, 15
Nucleic acids in milk, 18
5′-Nucleotidase, 107
Nucleotide pyro-phosphate, 107
Nucleotides in milk, 18
Nutritional equivalence, 390

Oligosaccharides, 660
  from lactose, 281–282, 326
  milk, 4, 280
Organic acids, 8, 14
Orosomucoid, 83, 105
Orotic acid, 15
Osteoporosis, 375–376
Oxidation-reduction equilibria, 414–419
Oxidation-reduction potential
  relation to pH, 415–416
  systems affecting, 415–417
Oxidized flavor, 256, 743
Oxygen in milk, 13, 14
  effect on E_h, 417

Packaging, 743, 755
Pantothenic acid, 367
Papain, 618
Parmesan cheese, 67
Pasteurization, 456, 742, 746–747, 751
  cheese making, 638
  cheese yield, 639
  effect on enzymes, 639
Provolone cheese, 67
Penicillin production, 710
Penicillium roqueforti, 641
Pepsin
  bovine, 614
  chicken, 615
  porcine, 613–614

inactivation, 613
survival, 613
Pepsinogen, 614
Peptide nitrogen in milk, 15
Peroxide determination, 241
pH of milk, 410–414
effect on coagulation, 595–597
effect on processing, 414
Phenylacetyl glutamine in milk, 16
Phosphate esters, 8, 17
Phosphate in milk
colloidal, 9
content, 8
inorganic, 9
quantitation, 7
partition, 9
Phosphodiesterase, 107
Phosphoenolpyruvate pathway, 670
Phosphoenolpyruvate phosphotransferase
system, 660
Phospholipids, 9, 199
content of milk, 184–185
fatty acids, 199
Phosphoprotein phosphatase, 107
Phosphoroclastic split, 666
Phosphorus, 378
concentration, 8
inorganic, 8
Phylloquinone, 371, *see also* Vitamin K
Physical properties, 409
Pig milk, 21
Pipeline milkers, 740
Plasmin, 107, 636–637
Poising capacity, 418–419
Polymorphism, 566
Potassium in milk, 8
Potato-like flavor, 690
Processing, 739
Prochymosin, 611, *see also* Chymosin
Prolactin in milk, 19
Propionic acid
fermentation, 673, 674
production, 675
Protected milk, 197–198
Protective foods, 343
Protein
conformation, 584
denaturation, 583
factors affecting, 584
general considerations, 587
solvents, 584

efficiency ratio (PER), 349
nutritive value, 347–350
quality, 349
quantitation, 5–6
Proteolysis
bitter flavors, 681
Brevibacterium linens, 679
cheese ripening, 646
lactic streptococci, 676
lactobacilli, 678
micrococci, 679
molds, 679
psychrotrophic bacteria, 680
yogurt cultures, 677
Pseudomonas, 321
Psychrotrophic bacteria, 741
Pumping, 740
Pyruvate kinase, 664
Pyruvate metabolism
end products, 672, 673
pathways, 666

Rabbit milk, 21
Radiation, 444
Radionuclides in milk, 12–13
Rancidity, 215, 224, 226, 234, 741
cheese ripening, 649
Raoult's law, 432
Rat milk, 21
Recommended daily allowance (RDA), 390
Refractive constant, 444
Refractive index, 442–444
contribution of constituents, 443
definition and limits, 441
estimate casein, 443
lactose, 306
measurement, 442–443
Reindeer milk, 21
Rennet, 609–610, *see also* Chymosin
cheese ripening, 647
from animals, 610–615
from bacteria, 617–618
from fungi, 615–617
from plants, 618–619
Retinol, 369, *see also* Vitamin A
Retort, 752
Rhodopsin, 368
Riboflavin, 365, *see also* Vitamin B$_2$
effect of light, 365–366
effect on E$_h$, 417
production, 712

Riboflavin kinase, 106
Ribonuclease, 107
Ricotta cheese, 68
Ropy milk, 693
Roquefort cheese, 65

Saccharomyces cerevisiae, 613
Safflower oil, 198
Salicyluric acid in milk, 16
Salmonella, 701
Salt balance
  effect on coagulation, 595
Salts in milk, 6
Sandiness, 284, 747
Saya milk drink, 693
Selenium in milk, 11
Separation, 741
Serum transferrin, 82, 103
Sheep milk, 1, 21
Sherbet, 72
  composition, 52
  standards, 71
Signets, 526
Silicone in milk, 11
Silver in milk, 11
ß-Sitosterol, 187
Skim milk, 43
  condensed, 54
Skim milk cheese, 68
Slime, bacterial, 687
Slimey milk, 693
Sodium caseinate, 73
Sodium in milk, 8
Solanum torvum, 618
Solubility, milk powder, 761
Somatic cells, 741
  cheese yield, 636
Sour cream, 46, 701
Special Milk Program (SMP), 345
Specific gravity, see Density
Specific heat, see Heat capacity
  milk fat, 203
Specific refractive index, 444
Specific rotation, lactose, 294, 296-297
Sphingomeylin, 198, 200-201, see also Phospholipids
Spray drying, 761
Stabilizers, 289, 744-746, 756, 758, 759
Standardization, 742, 749, 756, 760
Standards, 42
Starter culture, 757, 759

Sterile milk, see Ultra high temperature
  (UHT) milk
Sterile milk concentrate, 754
Sterilization, 750, 752, 754
Steroid hormones, 119
Sterols, 187
Stigmasterol, 187
Stilton cheese, 65
Streptococcus cremoris, 386, 643, 646
Streptococcus lactis, 234, 386, 643, 646
Streptococcus thermophilus, 358, 386, 643, 648
Strontium in milk, 11, 12
Substitute dairy products, see Imitation
  dairy products
Sugar phosphate, 17
Sugar transport, 660, 664
Sulfate in milk, 7, 8
Sulfhydryl oxidase, 106
Sulfhydryls, autoxidation, 254-256
Sulfur in milk, 17, 18
Sunlight flavor, 257
Superoxide dismutase, 106
Surface-ripened cheese, 647, 679
Surface tension, 428-432
  definition, 428, 429
  depression, 430-431
  effect of homogenization, 432
  effect of lipolysis, 431-432
  effect of processing, 431-432
  measurement, 429-430
Sweet acidophilus milk, 47
Sweetened condensed milk, 54
  lactose crystallization, 284, 310-311
Sweeteners, 746
Swiss Cheese, 66, 637-638, 640, 643, 645-647, 649, 674, 678
  amino acids, 647
  curd size, 642

Taette milk, 692
Tagatose-6-P-pathway, 664
Taurine in milk, 16
Temperature, effect on
  conductivity, 438
  density, 421-422
  pH, 411
  surface tension, 431
  viscosity, 425-426
Thermal conductivity, 441-442
Thiamin, 367, see also Vitamin B$_1$

Thiobarbituric acid (TBA), 241, 259
Thioesterase, 175
Thiosulfate sulfur transferase, 106
Thyroid, 381
Tin in milk, 11
Titanum in milk, 11
Titratable acidity, 411–414
Titration curve of milk, 412–414
Titration curves, 411–414
α-Tocopherol, 250–251, 371, *see also* Vitamin E
Torsiometer, 621
Trace elements, 10–12, 380
Triacylglycerol lipase, 106
Triglycerides, 176–182
  acids, 180
  determination, 178–182
  diet, 173
  hydroxy acids, 172
  keto acids, 172
  structure, 178–182
  synthesis, 173–178
Trombelastograph, 621
Trypsin inhibitor, 83, 105
Tryptophan, 349

Ultra high temperature milk (UHT) 43, 740, 752
  nutritive value, 387–388
Urea in milk, 15
Uric acid in milk, 15

Vacuum evaporation, 743, 751, 754, 759, 760, 762
Vanadium in milk, 11
Vegans, 366, 371
Vinegar, 714
Viscosity of milk, 424–428
  definition, 424
  effect of caseinate, 427–428
  measurement, 425
  relation to composition, 426–428
Vitamins, 346, 364–371, 656, 712, 750, 753
  fat-soluble
    A, 368–369
    D, 370

E, 731
K, 371
water-soluble
  B₁, 367
  B₂, 365
  B₆, 367
  B₁₂, 366
  biotin, 368
  C, 367
  folic acid, 367
  niacin, 366
  pantothenic acid, 367

Waste treatment, 716
  aerobic, 716
Water buffalo milk, 1
  composition, 21
Whey, 74
  acid, 75
  fermentation, 706–709, 712, 714, 715, 718
    medium, 332–333
  products, 76
  proteins, 4, 82, 91–100, 741–743, 749, 750, 753, 759, 762
    concentrate, 763
    fractionation, 135, 138
      by chromatography, 137–138
      by solubility, 135–137
    isolation, 135
    nutritive value, 348
  sweet, 75

Xanthine oxidase, 106, 358–360
  autoxidation, 244–245
  fat globule membrane, 554, 546–547
  homogenization, 358–359
  oxidized flavor, 244
Xerophthalmia, 368

Yak milk, 21
Yeast, 709
  extract, 656
Ymer, 50
Yogurt, 48, 662, 677, 682, 691, 701, 759

Zebra milk, 21
Zinc, 11, 382–383